GERMAN REARMAMENT AND THE WEST,
1932-1933

EDWARD W. BENNETT

German Rearmament and the West, 1932-1933

PRINCETON UNIVERSITY PRESS

Copyright © 1979 by Princeton University Press
Published by Princeton University Press, Princeton, New Jersey
In the United Kingdom: Princeton University Press,
Guildford, Surrey

All Rights Reserved

Library of Congress Cataloging in Publication Data will be
found on the last printed page of this book

This book has been composed in Linotype Baskerville

Clothbound editions of Princeton University Press books
are printed on acid-free paper, and binding
materials are chosen for strength and durability

Printed in the United States of America by
Princeton University Press, Princeton, New Jersey

To Barb

CONTENTS

Abbreviations	xi
Preface	xiii
Introduction	3

PART I

Chapter One The Political Implications of Reichswehr Plans

1. The Protest against Disarmament	11
2. Means and Goals of Evasion	15
3. The Reichswehr's Conflict with the Socialists	22
4. The Deeper Levels of Mobilization Planning	35
5. Brüning's Economic and Political Ideas	47
6. The Question of Disarmament Conference Tactics	51
7. The Question of the Cost of Arms	59
8. The Question of Wehrsport and the Wehrverbände	62
9. Brüning's Overthrow	70

Chapter Two Perceptions and Preoccupations of France and Britain

1. Western Detection of German Rearmament	78
2. The Specter of General von Seeckt	85
3. Distortion through British Hopes and Fears	89
4. France: The Quest for a British Alliance	92
5. Britain: The Repudiation of the Balance of Power	101
6. Other Nations in British Eyes	106
7. England's Military Decline	113
8. England's Spiritual Disarmament	116
9. England's Economic Ordeal	120
10. Chamberlain and the Treasury as Military Planners	125

Chapter Three The United States and the Legends of Lost Opportunity

1. The New Peace Conference	131
2. The United States: Ideals and Interests	137
3. Conflict between the "Anglo-Saxons" and the French	142
4. The Episode of the Bessinge Meeting	147
5. American Efforts for Further Talks	156
6. Negotiations without Germany	162

Chapter Four General von Schleicher's New Approach

1. Military Dominance in the New Cabinet ... 169
2. The Disappointments of Lausanne ... 176
3. Schleicher Intervenes Personally ... 180
4. Should Germany Openly Threaten Rearmament? ... 188
5. Schleicher Outflanks the Foreign Ministry ... 191
6. Berlin Plans a Meeting ... 196
7. Germany Apparently Alone ... 202

Chapter Five Britain Intervenes to Prevent a Break

1. The Actual Significance of the Consultative Declaration ... 208
2. The Actual Background of the September 19 Declaration ... 218
3. The September 19 Statement Misfires ... 226
4. A New British Conference Proposal ... 230
5. Current Reichswehr Plans and British Assessments ... 235
6. France Bids for International Support ... 242
7. "Anglo-Saxon" Alternatives ... 250
8. Conference Plans and Apprehensions ... 255
9. The Five-Power Conference, December 6-11, 1932 ... 262

Chapter Six Schleicher Reaches a Dead End

1. Domestic Politics and the Ott War Game ... 273
2. Schleicher's Political and Diplomatic Problems ... 282
3. Schleicher's Financial Problems ... 288
4. The Role of Blomberg and Reichenau in Hitler's Appointment ... 293

PART II

Chapter Seven Hitler's Accommodation with the Military

1. Hitler's Ideas ... 307
2. Hitler Bids for Military Support ... 320
3. Caution on the Part of Diplomats, Soldiers, and Chancellor ... 325
4. Was Preventive War a Possibility? ... 334
5. The Reichswehr Gets Its Money ... 338
6. Emergency Preparations and the Role of the SA ... 346

Chapter Eight Britain Reconsiders Its Policy

1. New Proposals and a New Face ... 356
2. The Background of the MacDonald Plan ... 360
3. Mussolini Intervenes ... 367
4. The Four-Power Pact and the Upset of British Policy ... 373

Contents — ix

    5. British Opinion and Hitler   379
    6. Crisis at the Conference   383
    7. Hitler Plans His Reply   395
    8. Alarm outside Germany   398
    9. Hitler's "Peace Speech"   401

Chapter Nine   France Attempts to Form a New Alliance

    1. Daladier Seeks Provision for Sanctions   406
    2. American Policy athwart the French Path   413
    3. British Unease and Henderson's Mission   422
    4. Vansittart Challenges German Policy   429
    5. Paul-Boncour's Scheme for Expert Discussions   434
    6. Anglo-French Problems and an Italian Solution   439
    7. The British Reject Sanctions   443

Chapter Ten   Hitler's Hand Is Forced, and He Frees It

    1. German Hopes and Concerns   449
    2. Soundings with Britain and France   453
    3. Neurath's Talks in Geneva   458
    4. Hitler Supports Foreign Ministry Views   460
    5. An "Ems Dispatch" Arrives in Berlin   466
    6. The British Take Position   471
    7. Hitler under Compulsion   475
    8. Davis Leads Simon to Moderate His Position   480
    9. Hitler's Diplomatic Triumph   485
    10. Hitler's New Armament Proposals   491
    11. Hitler Proceeds with His Own Plan   496
    12. The Road toward War   502

Conclusion   506

Appendix   513

Bibliography   517

Index   553

ABBREVIATIONS

AA	Auswärtiges Amt (Foreign Ministry)
ADAP	*Akten zur deutschen Auswärtigen Politik, 1918-1945*
AHR	*American Historical Review*
AW	Ausbildungswesen ("training affairs"—SA premilitary training organization)
BA	Bundesarchiv, Koblenz
BA-MA	Bundesarchiv-Militärarchiv, Freiburg
BDC	Berlin Document Center
CEH	*Central European History*
DBFP	*Documents on British Foreign Policy, 1919-1939*
DDB	*Documents diplomatiques belges*
DDF	*Documents diplomatiques français (1932-1939)*
DDP	Deutsche Demokratische Partei (German Democratic party)
DNVP	Deutschnationale Volkspartei (German National People's party)
DR	*Deutsche Rundschau*
DVP	Deutsche Volkspartei (German People's party)
FO	Foreign Office
FRUS	*Foreign Relations of the United States*
GG	*Geschichte und Gesellschaft*
HZ	*Historische Zeitschrift*
JCH	*Journal of Contemporary History*
JMH	*Journal of Modern History*
MDP	Ramsay MacDonald Papers
MGM	*Militärgeschichtliche Mitteilungen*
ML	Marineleitung (Naval Command)
NA-VGM	National Archives, Völkerbundsabteilung Gruppe Marine (League of Nations Section [Group Navy]) documents (on microfilm)
NSDAP	Nationalsozialistische Deutsche Arbeiterpartei (National Socialist German Workers' party)
OKH	Oberkommando des Heeres (High Command of the Army)
OKW	Oberkommando der Wehrmacht (High Command of the Armed Forces)
PRO	Public Record Office, London
RK	Reichskanzlei (Reich Chancellery)
SA	Sturmabteilungen (storm troops)
SDP	State Department Papers

SIA	*Survey of International Affairs*
SPD	Sozialdemokratische Partei Deutschlands (Social Democratic party of Germany)
SS	Schutzstaffeln (élite storm troops)
VfZ	*Vierteljahrshefte für Zeitgeschichte*
WKK	Wehrkreiskommando (Army Area Command)
Z	Zentrum (Center party)

PREFACE

No HISTORIAN can write without support and assistance, and in my case the support and assistance have been considerable. I would like first to express my gratitude to that unparalleled institution, the John Simon Guggenheim Memorial Foundation, for conferring a grant in 1965-1966, enabling me to undertake this study. One of my deepest wishes has been that the results of my research might justify, if belatedly, the confidence shown in me by this grant.

Thanks are also due to my erstwhile superiors in the Central Intelligence Agency for permitting me to take a year's leave of absence without pay for independent historical research under my Guggenheim fellowship. I chose to resign from this agency in 1971, but I wish to state that I learned much from my experience there, and that one can find more respect for, and practice of, disinterested research and analysis in that organization than any of its critics imagine.

Assistance, for the historian, means particularly help in gaining access to sources, his raw material. This involves both the granting of permission to see, use, and quote material, and aid in actually finding and working with it.

The archivists of the West German Bundesarchiv in Koblenz and the Bundesarchiv-Militärarchiv in Koblenz and Freiburg proved most willing to help by providing access to their records, and the latter also assisted by obtaining permission for me to see the papers of Joachim von Stülpnagel. I first saw some documents now in the Bundesarchiv-Militärarchiv by courtesy of the staff of the Dokumentenzentrale of the Militärgeschichtliches Forschungsamt in Freiburg. And I found the custodians of the extraordinary collection of papers at the Institut für Zeitgeschichte to be especially helpful and friendly.

The change of the British fifty-year rule to a thirty-year rule unexpectedly led to several visits to the Public Record Office (PRO) in London, where the staff is always most cooperative and courteous. Transcripts from Crown-copyright materials in the Public Record Office appear here by permission of the Controller of Her Majesty's Stationery Office. Among the many valuable sources at the PRO are now the diaries of Ramsay MacDonald, who left word that they were "meant as notes to guide and revive memory as regards happenings and must on no account be published as they are." His son, the Right Honorable Malcolm MacDonald, has accordingly decided that there should be no publication of the diaries in extenso, but he has consented to such use as I have made of this source, as well as to quotations from his father's other writings, and I am grateful for this. I appreciate having been able to see and use the Samuel Papers at the House of Lords Record Office and the Lord Cecil of Chelwood Papers at the British Library. I also wish to thank the Head of Special Collections at the Library of the University of Birmingham for permission to cite from the papers of Austen and Neville Chamberlain.

In Washington, I have seen and used State Department and War Department records at the National Archives, the former originally with the permission of the Department of State. I have also, in particular, made use of National Archives microfilms of captured German and Italian documents, and I think I may say that no one can appreciate more than I the value of the support given by the American Historical Association for the filming of these papers. The archivists at the National Archives gave every assistance, as did those at the Franklin D. Roosevelt Library in Hyde Park, New York.

The Manuscript Division of the Library of Congress kindly produced for me the papers of Norman Davis, bequeathed by him to the public, and the resources of the library generally have provided essential help for my project. I was also fortunate in being able to use both the general and the microfilm holdings of the University of Michigan Graduate Library. Thanks are due to the Librarian of Yale University for permission to quote from the Henry L. Stimson Papers, and to the Curator of Manuscripts of the Houghton Library of Harvard University, and to Mrs. Albert Levitt for permission to quote from the diary of Jay Pierrepont Moffat.

I am very grateful to Martin Gilbert for enabling me to see the papers of Sir Horace Rumbold and for sending me authorization for a quotation. And I appreciate the patience and kindness of Lady Wheeler-Bennett in answering my letters about her late husband's writings.

Many other individuals have given aid and advice. I should like to mention in particular Donald Watt; John Barnes; Miss G. L. Beech, Mrs. Jane Cox, and D. Crook of the Public Record Office; Dr. Thilo Vogelsang, Dr. Anton Hoch, Frau Strasser, and Hermann Weiss of the Institute für Zeitgeschichte; Dr. Wolfgang Mommsen, Herr Wagner, and Herr Verlande of the Bundesarchiv; Dr. Friedrich-Christian Stahl of the Bundesarchiv-Militärarchiv; Oberst i.G. Karl-Heinz Völker; Key Kobayashi of the Library of Congress; and Robert Wolfe, John Mendelsohn, Harry Rilley, and Thomas Oglesby of the National Archives. I am grateful to Professors Hans W. Gatzke and Jon Jacobson for criticisms and suggestions respecting an earlier version of this study. And I will always remain indebted to Professor William L. Langer for his teaching and his example.

The greatest debt of gratitude is that owing to my wife, who has supported this long project in every sense of the word. Without her it could have been neither begun nor completed.

Great Falls, Virginia
April 1978

GERMAN REARMAMENT AND THE WEST,
1932-1933

INTRODUCTION

THIS VOLUME describes German military planning in the early 1930s, and relates this to German politics, but it is also very definitely a study in international relations. For me, a non-German, the most compelling reasons for studying German history are, first, to learn about German ideas, so original in so many fields, and so different from our own, and second, to judge the extent of German responsibility for the two world wars, which have shaped our present-day world. One must also, however, consider the ideas and responsibilities of others, which can also hold compelling interest.

My first acquaintance with Germany, and the beginning of my curiosity about that country, came when, in the fall and winter of 1932, I traveled there with my family. As a boy of nine, I was rather oblivious to signs of depression and political strife. I spent long fascinating hours by myself at the Deutsches Museum in Munich, and talked there with a kindly, English-speaking museum guard, but was badly shaken when a Stuttgart policeman gave me a barrack-square tongue-lashing for walking on top of a four-foot retaining wall. From a window of our pension in the Schellingstrasse in Munich, I watched a Reichswehr company march down the street toward the Ludwigstrasse. Twelve years later, as an infantry replacement in the Belgian Ardennes, I again saw German soldiers. Many GIs in my battalion had been slain while attacking by daylight across a snow-covered field, but the Germans were now prisoners, and had suffered like us from the winter elements. In the following summer, I visited Munich again; the pension was destroyed, and on the wall at the back of the Feldherrnhalle, someone (as I recall) had written large his conclusion: "Ich schäme mich, dass ich ein Deutscher bin."

Were there grounds for a German to feel ashamed? I hardly doubted it then; I also visited Dachau at the time. But I had already studied enough history to suspect that the Germans were not all equally responsible for the war, and that the rest of the world was not entirely innocent. Responsibility would be decided not at Nuremberg, but by historians, who could weigh over the years a larger body of evidence, and who could consider the role of all the participants, and the play of all the forces. It has seemed to me that the origins of World War II, like those of World War I, could not be simply a German question, or simply a political question.

There have been, however, major differences between the debates on the origins of the two wars. With the first war, blame was assigned to the leaders of various nations. Aside from those who insist on indicting a ruling clique of heavy industrialists, few historians venture to question Adolf Hitler's primary responsibility for the second war. With Hitler identified as mainly responsible for the war that occurred, the question becomes one of how far others, aside from known Nazi leaders, may have supported his goals. Controversy has arisen, mainly in Germany, as to whether other German leaders before Hitler shared his willingness to embark on wars of conquest, or, as the question is more usually put, as to how much continuity there is in German history. Perhaps at first out of a lingering distrust of recent history, or latterly because of their rediscovery of the work of Eckart Kehr,[1] German historians have paid more attention to possible pre-1918 antecedents than to those in the immediate 1920-1933 period. A subsection of the continuity debate has concerned the responsibility or innocence of German officers, but where the military have received blame, it has been more for their political outlook than for their professional activity.[2] A controversy has also flourished, mainly in England, as to the wrongness or rightness of attempts to appease Germany. These debates have treated German or English political history almost in isolation, and partly for this reason and partly because of an initial lack of documents, the debates have failed to explain how national policies developed together, or to settle questions of actual responsibility, as opposed to those of ideology.

Although a tremendous quantity of literature exists dealing with interwar history, and especially with the problems just described, I believe that for two reasons, there is a need to review and reassess its conclusions.

First, anything written on recent political or military history is inevitably subject to revision. Revision does not happen merely because some historians like to advance new theories to challenge wartime or cold-war orthodoxies. A more fundamental cause of revision is that,

---

[1] Supplementing his pioneering work on interest-group politics in Imperial Germany, *Schlachtflottenbau und Parteipolitik 1894-1901: Versuch eines Querschnitts durch die innenpolitischen, sozialen under ideologischen Voraussetzungen des deutschen Imperialismus* (Berlin: Emil Ebering, 1930), articles by Kehr on German bureaucracy, militarism, and society are now readily available in *Der Primat der Innenpolitik*, Hans-Ulrich Wehler, ed. (Berlin: Walter de Gruyter, 1965). Kehr's sociocritical stress on the primacy of internal economic and political forces, as opposed to the traditional conservative doctrine of the "primacy of foreign policy," found little acceptance in the late Weimar Republic, and none at all under the Nazi regime; he died young, in the United States, in 1933.

[2] I have discussed some of this literature in a paper delivered at the Georgetown History Forum in Washington, D.C., 11 October 1975. For a relatively recent review, see Michael Geyer, "Die Wehrmacht der deutschen Republik ist die Reichswehr: Bemerkungen zur neueren Literatur," *Militärgeshichtliche Mitteilungen* (MGM), 14 (1973), 152-199.

some time after an event, additional documents may become accessible, revealing the policy planning and the intelligence information which officials long sought to keep secret. A great deal of material, now available, was not at hand when the earlier histories of the 1930s were written. This of course does not imply any fault on the part of the earlier writers, who wrote badly needed studies with what sources they had. But not only do we now have more source material, we often have better source material. Contemporary military records are largely replacing the often-apologetic postwar memoirs of German officers, so that we have a better idea of actual German strength, and a much better idea of the full scope of German military plans. Where most of our information on British policy formerly came from the papers of outsiders and from published diplomatic correspondence (largely incoming), today we have direct evidence on British policy formation and policy deliberations. Regarding French diplomatic history, we now possess a very extensive documentary publication.[3] Of course, another reason for rewriting is that we discover the mistakes and biases of participants and of fellow historians. All historians make simple mistakes that need correction, as I have found all too often in the process of my own writing. Beyond this, in the course of our work, we historians constantly and necessarily make hypotheses, and these are often later falsified by evidence new or overlooked. I myself prefer to make provisional hypotheses as I work with the evidence, rather than to start with a highly developed theoretical construct. Such constructs may make assumptions more explicit, but commitment to the position tends to become rigid and closed to differentiation or reconsideration.

Second, events are not isolated by international boundaries, or by distinctions between departments or sectors of a society. Ever since I first read William L. Langer's *European Alliances and Alignments* and *The Diplomacy of Imperialism* many years ago, I have been convinced that at least some historians should trace the interaction between the policies of several nations, rather than describing the policy of just one. Of course, one country may set the pace, as Germany did in the 1930s. But it takes two, or usually more than two, to have diplomatic relations. Also, interconnections do not begin only at the water's edge. Within each country, foreign policy has a relation to domestic politics and to military, economic, and other social considerations. Most of the time, it is not professional diplomats who make foreign policy, but political leaders responding to domestic pressures from soldiers, organ-

---

[3] French unpublished archival sources for the years after 1930 were not available in time for use in this study. They will undoubtedly supply much further information, although, owing to wartime destruction, they can never be as complete as British, German, or American material.

izational lobbyists, businessmen, newspaper owners, and the public. Bureaucratic organizations have functions and responsibilities that make them ardent and powerful champions of particular policies. The mere inauguration of their plans may affect the rest of society long before the plans are carried out.[4]

Whether one is interested in the responsibility for World War II, the continuity or discontinuity of German history with Hitler, or the justification of attempts to appease Germany, it seems essential to establish the nature of German arms plans and preparations. To do this is the first purpose of this study. German military plans, if they can be discovered, provide more evidence of intentions than the actual preparations accomplished, which may have been distorted or delayed by extraneous factors. Military plans can tell us much about the extent of German revisionism, and unlike, say, plans for reparations revision, military plans can reveal the willingness (or unwillingness) of German policy-makers to use force for revision. Most historians have concluded that Hitler was prepared to use force for offensive ends. Was this true also of other German leaders in the years just before Hitler? So far, most historians have assumed that it was not,[5] but it is important to learn the answer, which bears directly on the questions of continuity and the responsibility of non-Nazi German leaders for war.

Policies like rearmament (or reparations revision) are not simply indices of dissatisfaction or militance. They have a functional place in

[4] The importance of material evidence and the connection of domestic with foreign policy are among the many points involved in a recent debate between Andreas Hillgruber, Klaus Hildebrand, Hans-Ulrich Wehler, and others; see esp. Andreas Hillgruber, "Politische Geschichte in moderner Sicht," *Historische Zeitschrift* (*HZ*), 216 (1973), 529-552; Hans-Ulrich Wehler, "Moderne Politikgeschichte oder 'Grosse Politik der Kabinette'?" *Geschichte und Gesellschaft* (*GG*), 1 (1975), 344-369; Klaus Hildebrand, "Geschichte oder 'Gesellschaftsgeschichte'? Die Notwendigkeit einer politischen Geschichtsschreibung von den Internationalen Beziehungen," *HZ*, 223 (1976), 328-357; Hans-Ulrich Wehler, "Kritik und kritische Antikritik," *HZ*, 225 (1977), 347-384. Ostensibly an argument over conventional versus social science methods in political and esp. diplomatic history, the debate is also part of a larger continuing struggle over the picture Germans should be given of their past, or indeed, over the kind of Weltanschauung historians should propagate. See also Jürgen Kocka, "Theoretical Approaches to Social and Economic History of Modern Germany: Some Recent Trends, Concepts, and Problems in Western and Eastern Germany," *Journal of Modern History* (*JMH*), 47 (1975), 101-119; Wolfgang J. Mommsen, "Domestic Factors in German Foreign Policy before 1914," *Central European History* (*CEH*), 6 (1973), 3-43; Karl-Georg Faber, [Review essay], *History and Theory*, 16 (1977), 51-66; Thomas Nipperdey, "Wehlers 'Kaiserreich': Eine kritische Auseinandersetzung," *GG*, 1 (1975), 539-560; Volker R. Berghahn, "Der Bericht der Preussischen Oberrechnungskammer: Wehlers 'Kaiserreich' und seine Kritiker," *GG*, 2 (1976), 125-136.

[5] An exception is Gaines Post, Jr., whose *The Civil-Military Fabric of Weimar Foreign Policy* (Princeton: Princeton University Press, 1973) uncovers new evidence.

the course of events. If military policy reflects political ideas, it is at least as true that political events can reflect military policy. Since Clausewitz, at the latest, many Germans have regarded political and military affairs as a continuum, and both the victories that created Bismarck's Reich and the internal consolidation of that Reich rested on a particular military system. The secrecy of military planning, particularly when forbidden by the Treaty of Versailles, has made it hard to follow. Even since 1945, there has been a tendency to conceal that planning. I believe, however, that an elucidation of German military policy in the Weimar period can illuminate the German domestic politics of the time, and provide new explanations for the dissolution of parliamentary government and the acceptance of Hitler. Military planning may also do much to explain German foreign policy, even though, unlike 1866 and 1870, or the late 1930s, military power was not so much a tool of German diplomacy in the early 1930s as an objective of it.

If it is essential for a knowledge of German intentions and responsibilities to establish the nature of German military plans, it is essential for a knowledge of British, French, and American responsibilities and resolve to establish their awareness of German arms plans, their views on the way to deal with those plans, and the nature of their relations with each other and with Germany. Did they recognize a need to maintain a balance of power? In particular, did any of them show more insight and realism than the others, or were they equally at fault? A clarification of the extent and timing of their disagreements on the question of German armaments should help us understand how Germany was allowed to rearm.

Perhaps the greatest puzzle in the international history of the 1930s is, why and how did the other nations, especially Britain and France but also the United States, allow Germany to regain her power? It is not self-evident that Germany had to regain it. In 1932, whatever her plans and potential, Germany was still militarily very weak. The declared purpose of the Disarmament Conference, which met precisely in 1932 and 1933, was to reduce unnecessary armament, and not to permit any government to rearm. German rearmament presumably ran counter to both the ideals and practical national interests of other nations. The German government did not intend, of course, that other governments should perceive it as rearming, at least to any significant extent; that might encourage the others to take some kind of preventive measures while Germany was still weak. Yet it seemed that only extensive rearmament, such as could hardly be concealed, would reliably deter action by other governments. This dilemma runs like a crimson thread through the story of German policy in this period. Germany could hardly have escaped it without unintended help from other governments, and we may well ask how this came about.

The first aim of this study, then, is to investigate German military policy, as an index to German intentions and responsibilities and as a factor in Germany's domestic politics and foreign policy. The second is to explore the way the Western powers perceived and responded to German military policy, and how they interacted with each other and Germany, so that we may ascertain their responsibility and resolve, and learn how Germany was permitted to rearm.

PART I

CHAPTER ONE

The Political Implications of Reichswehr Plans

1. THE PROTEST AGAINST DISARMAMENT

Most Germans felt that the 1919 peace settlement had perpetrated a colossal series of wrongs. They considered the reparations prescribed to be an unbearable burden, and blamed them for German economic distress. They regarded the occupation of the Rhineland by foreign forces as a shame and a scandal, especially when carried out by French African troops. They believed the boundaries laid down in the East to be unjust and irrational, particularly those that established a Polish Corridor between East Prussia and the rest of Germany; the Corridor problem seemed all the more urgent because the lost territories were becoming increasingly Polonized with each passing year. The military disarmament imposed on Germany was a more telling blow than the rest of the treaty, because it forced the country, in the last analysis, to abide by all the other provisions that the other powers wished to maintain in force. German disarmament guaranteed the rest of the status quo established in 1919.

Germany's disarmament might have seemed less onerous if middle- and upper-class Germans had not believed that their country had had the best army in the world, that the nation had been unified through the agency of military force, and that a military career was the most honorable of professions. As under the Prussian monarchy, the military leaders still believed that the army was the rock on which the state rested, and in truth, in a new and unstable republic, armed force was highly important to state power. In addition, with the election of Paul von Hindenburg as President in 1925, the highest office in the republic was assumed by the former wartime commander of the armies. Hindenburg had no dictatorial ambitions, but he tended to regard himself as the heir to the powers of the Prussian kings. The legal fact that the Weimar constitution granted him special powers in an emergency was less important than the tradition that the monarch could choose the advisers and policies that he believed would best serve state interests. Hindenburg naturally looked to soldiers for advice. As they were at once responsible for national security and unable with existing resources to ensure it, they naturally pressed for increased military strength, despite treaty limitations.

The soldiers favored the political forces that claimed to support the army's objectives. The German National People's party (DNVP), representing conservative traditions, especially those of Prussia, called loudly for a German military revival, as did the populist and racist National Socialists (NSDAP). Although German officers professed to stand "above party," they found the outlook of the DNVP to be close to their own. But in the center of the spectrum, the business-oriented German People's party (DVP), most of the more republican German Democratic party (DDP), and the Catholic Center party (Z) also gave their backing to defense programs, and the support of these parties was in practice more useful than that of the far right, since they were usually represented in the government. Even among the Socialists or Social Democrats (SPD), some leaders sympathized with the army's objectives. A desire for German rearmament, both before and after Hitler's accession, was the conventional German view, rooted in traditional institutions and respected ideas.

This is not to say that the German people as a whole were militaristic. Germany had a significant pacifist movement, at least in the early 1920s,[1] and much of the SPD, the largest German party, and especially its leadership in the *Land* (state) of Prussia, strongly opposed rearmament. Ironically, the force of the Socialists' opposition would hasten their own downfall. They failed to develop a coherent policy of their own, but they were in a position to obstruct military preparations by parliamentary action, hostile publicity, and the refusal of administrative cooperation, and they often did so; as a result, the military came to regard the elimination of Socialist power as a patriotic necessity.

Aside from parties as such, interest groups were important powers in German society, and with respect to armament, one naturally thinks of the heavy industrialists. Did they join forces with the military, or egg the military on? Certainly the industrialists were nationalists, like the soldiers, and at least as anti-Socialist. As individual, upper-middle-class Germans, most industrialists assuredly favored rearmament, and as businessmen they were ready to turn a profit, if that could be done, by selling guns or armor plate. But the experience of World War I controls had made German businessmen wary of government intervention, and government arms contracts were not necessarily profitable; the military wanted industry to maintain facilities for full wartime production, and these were often underemployed and uneconomic

---

[1] On the pacifist movement: Rolf R. Schlueter, *Probleme der deutschen Friedensbewegung in der Weimarer Republik* (Ph.D. dissertation, University of Bonn, 1974); by January 1933, the Deutsche Friedensgesellschaft had only 5,000 members, one sixth of its former size (p. 304).

in peacetime, especially under Versailles restrictions. Military and industrial interests overlapped, but they were far from identical.[2]

During the depression of the early 1930s, military leaders became concerned over the financial difficulties of some of their arms suppliers. The government aided the Borsig (Tegel) firm with an advance of RM 3 million, eventually merging it with the state-owned Rheinmetall, and we shall see the head of the army's Ordnance Office pleading generally that German defenses would be threatened if the arms manufacturers were not saved from bankruptcy. But the effort to plan and launch rearmament antedated the financial difficulties; it did not spring from them. Military documents indicate that defense policy decisions, including for supporting arms manufacture, were based on military, not economic, grounds. Politically, industrial influence could add little to the military pressures for secret arms appropriations; socially, industrialists were largely separate from, and inferior to, the military caste. From 35 to 49 percent of the officers were sons of officers, and from 37 to 42 percent were sons of high civil officials or professional men, yet only 6 to 10 percent were sons of businessmen or factory owners.[3]

There is another possible reason for a lack of industrial influence: in the case of the land army, which in Germany was the "senior service," and on which we will concentrate here,[4] German military experts

[2] On anti-Socialism and anti-*dirigiste* attitudes in German industry: Charles S. Maier, *Recasting Bourgeois Europe* (Princeton: Princeton University Press, 1975), pp. 22, 33-39, 53-70; also Gerald D. Feldman, "The Social and Economic Policies of German Big Business, 1918-1929," *American Historical Review* (*AHR*), 75 (1969-1970), 47-55. Uneconomic production and overlapping interests: Michael Geyer, "Das Zweite Rüstungsprogramm (1930-1934)," *MGM*, 17 (1975), 132-133, 135.

[3] On aid to industry: Geyer, "Zweite Rüstungsprogramm," p. 133 and Doc. 6; Karl-Heinz Ludwig, "Strukturmerkmale nationalsozialistischer Aufrüstung bis 1935," in Friedrich Forstmeier and Hans-Erich Volkmann, eds., *Wirtschaft und Rüstung am Vorabend des Zweiten Weltkrieges* (Düsseldorf: Droste, 1975), p. 54. On origins of officers: Karl Demeter, *Das deutsche Offizierskorps in Gesellschaft und Staat 1650-1945*, 4th ed. (Frankfurt: Bernard & Graefe, 1965), pp. 57-61. The ranges of percentages come from three estimates by two different scholars for the Weimar/early Hitler period. Alfred Sohn-Rethel, a perceptive Marxist observer with experience in interest-group activity, notes that certain overexpanded and overrationalized industrial firms, notably Fritz Thyssen's Vereinigte Stahlwerke, found themselves unable to adjust to the shrunken markets of the depression. They could only be saved, he argues, if the state intervened on their behalf, esp. by giving arms orders. See his *Ökonomie und Klassenstruktur des deutschen Faschismus*, Johannes Agnoli, Bernhard Blanke, Niels Kadritzke, eds. (Frankfurt: Suhrkamp, 1973), pp. 47-50. But Sohn-Rethel uses this argument to explain why some parts of industry supported a Nazi-DNVP coalition; he does not suggest that industry influenced Reichswehr armament plans before 1933. On the role of heavy industry in bringing Hitler to power, see Chapter Six (concluding section) below.

[4] On naval history in this period: Jost Dülffer, *Weimar, Hitler und die Marine* (Düsseldorf: Droste, 1973). On the air force, see Karl-Heinz Völker, *Die Entwicklung*

did not think of "rearmament" primarily in terms of hardware, of armor and explosives; this was more of an Anglo-American habit Most German officers gave priority, rather, to the organization, training, and handling of men. One officer who did concern himself for years with advocating and preparing for industrial mobilization, General Georg Thomas, commented later in a retrospective survey of this work: "The experiences of the [First] World War were soon forgotten, and the military leadership primarily concerned itself with the operational or else the organizational and personnel aspects of national defense." German generals did not, of course, suppose that industrial power was irrelevant, and they took more interest in material armament after Hitler and Hjalmar Schacht made massive arms production possible. Despite all bureaucratic rivalries and mistaken judgments at the top, even German industrial mobilization itself far outstripped that elsewhere up until 1939. But in both world wars, Germany was able to gain early victories and then delay for years defeat at the hands of numerically and materially superior opponents, and much of this accomplishment was due to just that stress on men, organization, and tactics which Thomas deplored.[5]

The officers' claim to defense leadership also gained strength from the need to avoid disclosures. Any serious defense planning would violate the Treaty of Versailles, and if such a violation of the treaty became known, it would arouse opposition abroad and perhaps trigger intervention; therefore, secrecy was essential. German staff officers devised Germany's military plans and thus became the guardians of their secrecy. The officers had to inform cabinet ministers to some extent—usually slight—to secure authorizations, and they invited German dip-

---

der militärischen Luftfahrt in Deutschland 1920-1933 (in the same volume as Wiegand Schmidt-Richberg, *Die Generalstäbe in Deutschland 1871-1945*) (Stuttgart: Deutsche Verlags-Anstalt, 1962); the same author's *Die deutsche Luftwaffe 1933-1939* (Stuttgart: Deutsche Verlags-Anstalt, 1967); David Irving, *The Rise and Fall of the Luftwaffe: The Life of Luftwaffe Marshal Erhard Milch* (London: Weidenfeld & Nicolson, 1973); and Edward L. Homze, *Arming the Luftwaffe* (Lincoln: University of Nebraska Press, 1976).

[5] Thomas's statement: Georg Thomas, *Geschichte der deutschen Wehr- und Rüstungswirtschaft (1918-1943/45)*, Wolfgang Birkenfeld, ed. (Boppard am Rhein: Harald Boldt, 1966), pp. 80, 88-89. On the primacy of personnel, see also Max Graf Montgelas, "Potentielle und aktuelle Rüstungen," *Wissen und Wehr*, 12 (1932), 24-32, and the comments in Ludwig, "Strukturmerkmale," p. 59. On Thomas's efforts to organize industrial mobilization, see Berenice A. Carroll, *Design for Total War* (The Hague: Mouton, 1968), and on the idea that German businessmen avidly supported rearmament, see esp. pp. 67-71. On the later mobilization of the economy, Burton H. Klein, *Germany's Economic Preparations for War* (Cambridge, Mass.: Harvard University Press, 1959), p. 19, was corrected by Carroll, *Design*, p. 184; now see also Willi A. Boelcke, "Probleme der Finanzierung von Militärausgaben," in Forstmeier and Volkmann, pp. 14-38.

lomats to participate in war games, so as to obtain foreign policy advice and enlist the diplomats' support, inside the German government as well as in foreign negotiations. Manufacturers of forbidden arms necessarily knew about the particular contracts they were engaged on. But the staff officers kept all the threads in their own hands, and, so far as possible, concealed the scope of their preparations from outside scrutiny. This concealment did not entirely succeed, but it was successful enough to mislead contemporaries and affect later historical writing.

2. MEANS AND GOALS OF EVASION

There was a model for carrying out concealed rearmament, following a military defeat. Germans, and especially German officers, could recall the defeat of Prussia by Napoleon and the treaties of Tilsit of 1807 and of Paris of 1808, which had imposed an indemnity and an army of occupation, reduced Prussian territory by about half, and set the maximum strength of the Prussian army at 42,000 men. This catastrophe had been followed by a movement of regeneration, led by Baron vom Stein, and also, in military affairs, by Gerhard von Scharnhorst. According to the exaggerated legend, Scharnhorst initiated a so-called *Krümper* system, giving short-term basic training to successive batches of recruits, and thereby circumventing the attempt to limit army strength; once trained, these men or Krümper had constituted a reserve, and supposedly 150,000 of them had leaped to arms in 1813. Actually, Prussia only had about 36,000 Krümper by the fall of 1812, most of them old soldiers, but they did provide more cadre for training raw recruits, enabling Prussia to put approximately 280,000 men in the field, and to become a great power again. Of more lasting importance were the introduction in 1814 of universal peacetime training in the regular army, and Albrecht von Roon's 1860 reform of integrating all the younger trained men in wartime into regular units with expanded cadres, leaving only older men in the Landwehr or militia. After the Prussian army proved the value of these measures by rapidly defeating the professional forces of Austria and France, other continental European powers imitated the Prussian pattern, at least in its broad outlines. The idea of building reserves of trained conscripts in peacetime and then mobilizing them in wartime may well have been Germany's greatest contribution to military technique. In any case, after 1918 it was natural for patriotic Germans to draw historical parallels with the nineteenth century, and to think of ways in which, once again, a dictated peace might be overturned, and preeminence as a power won. Even Gustav Stresemann, who led Germany back into European political society, believed that his country could not regain its freedom and independence as long as it was not a great power and had no significant army.[6]

The victors of 1919 also remembered the story of the Krümper, and they saw to it that Part V of the Treaty of Versailles, the section concerning armament restrictions, not only limited the size of the German army to 100,000 men, but also specified that officers had to serve for twenty-five years and enlisted men for twelve years; each year, no more than five percent of the effectives could be discharged before their full term had expired, e.g., for reasons of health. In fact, Part V attempted to erect a whole series of obstacles to another Scharnhorst; its prohibitions included officers' schools, preparations for mobilization including the organization of reserves, the maintenance of an overall general staff, and the provision of military training by educational establishments, veterans organizations, shooting clubs, and "associations of every description." Part V also banned tanks, aircraft, and heavy artillery, and the flexibility of German planning was further limited by the specification that the army should consist of seven infantry and three cavalry divisions. The treaty also drastically limited the German navy; submarines were forbidden, and no German capital ship was to have a displacement of more than 10,000 tons. Part V completely forbade the German possession of an air force. Another part of the treaty permanently excluded the German forces from the territory bordering on and west of the Rhine. Also, it was not without importance that the Treaty of Versailles, including Part V, was part of the law of Germany.

The German Reichswehr, or armed forces, could evade many of these restrictions. It continued general-staff work under conditions of secrecy, in a division of the Ministry of Defense known as the Trup-

[6] On the Krümper system: Curt Jany, *Geschichte der Königlich-Preussischen Armee bis zum Jahre 1807*, 3 vols. (Berlin: Verlag von Karl Siegismund, 1928-1929), III, 465-468; William O. Shanahan, *Prussian Military Reforms, 1786-1813* (New York: Columbia University Press, 1945), pp. 159-178, 197-208; Rainer Wohlfeil, *Vom stehenden Heer des Absolutismus zur allgemeinen Wehrpflicht* (in Hans Meier-Welcker and Wolfgang von Groote, eds., *Handbuch zur deutschen Militärgeschichte 1648-1939*, Vol. II) (Frankfurt: Bernard & Graefe, 1964), pp. 122-123. On Prussian military reforms more generally, including through Roon: Gordon A. Craig, *The Politics of the Prussian Army, 1640-1945* (New York: Oxford University Press, 1956), pp. 38-65, 69-70, 74, 139-179. The incorporation of a militia with the regulars had actually been proposed by Karl Friedrich von der Knesebeck in 1803, while Scharnhorst wanted a separate militia: Shanahan, pp. 75-76, 121-122; Jany, III, 451-458; Wohlfeil, *Vom stehenden Heer*, pp. 98-99. Michael Salewski, in *Entwaffnung und Militärkontrolle in Deutschland 1919-1927* (Munich: R. Oldenbourg, 1966), pp. 382-392, describes Stresemann's outlook and surveys the effect of the post-Jena parallel on French as well as German thinking after 1918. On importance of military power to Stresemann, see also Hans W. Gatzke, *Stresemann and the Rearmament of Germany* (Baltimore: Johns Hopkins University Press, 1954), p. 109, and note 16 below.

penamt, or Troops Office. It concealed stocks of arms, and made preparations for initiating war production when that should become necessary. A locally recruited border militia, the Grenzschutz, was organized in the Eastern provinces. Forbidden weapons could be and were tested outside of Germany, in Soviet Russia. And, most important, while Germany could be compelled to reduce gradually to 100,000 men on active duty, the war veterans who were discharged could have been recalled in case of war; Part V restrictions would hardly be honored then. What counted was not peacetime strength, but mobilizable wartime strength. German military planning was based on the idea of a threefold, or possibly a ninefold, expansion of the seven infantry divisions, to twenty-one or sixty-three divisions, and it was this that lay behind the doctrine that every man should be capable of doing the job of his superior: lieutenants should be potential captains, captains potential majors, and so on. In the plans of the mid-twenties, mobilized divisions were to be of varying quality, with varying proportions of regulars to reserves.[7] Even in 1914, regular German divisions had, after mobilization, contained an average of 46 percent reserves.[8] As the 1914 army had been expanded by reserves, the 100,000-man army could be, too—at least as long as there were reserves young enough to fight.

These clandestine plans for wartime expansion had implications extending far beyond the military sphere. In order to fill the ranks of the expanded force, the army planned an illicit organization to com-

[7] Military activity in Russia: Hans W. Gatzke, "Russo-German Military Collaboration During the Weimar Republic," *AHR*, 63 (1957-1958), 565-597. For Wilhelm Groener's report on this subject, recorded by Hans Schäffer, see Gerhard Schulz, *Aufstieg des Nationalsozialismus: Krise und Revolution in Deutschland* (Frankfurt: Propyläen Verlag, 1975), pp. 849-850 (n. 116). According to Friedrich von Rabenau, *Seeckt: Aus seinem Leben* (Leipzig: Von Hase & Koehler, 1940), pp. 474-475, the expansion plans dated back to January 1921. On the trebling: Harold J. Gordon, Jr., *The Reichswehr and the German Republic, 1919-1926* (Princeton: Princeton University Press, 1957), p. 174. In 1926-1928, near-future mobilization plans were cut back to fifteen to sixteen divisions: National Archives, Washington, D.C., Microfilm T-77, Oberkommando der Wehrmacht Records (hereafter OKW), Wi/IF 5.502, att. to 550/26, 9 Aug 26, Serial 111/Frame 838883; Wi/IF 5.126, TA 223/28, 2 Feb 28, 18/728882-83. (In this study, microfilmed material is generally cited by serial and frame numbers; if serial numbers are not available, the first number is a reel number and is so described. With military documents, where the frame numbers were not stamped on the documents themselves, I also give the file number in use at the time of filming—in the cases above, Wi/IF 5.502 and Wi/IF 5.126—in the hope that this may facilitate tracing in German files.) On the consistent intention to mobilize a larger army, see also Wilhelm Deist, "Internationale und nationale Aspekte der Abrüstungsfrage 1924-1932," in Hellmuth Rössler, ed., *Locarno und die Weltpolitik 1924-1932* (Göttingen: Musterschmidt, 1969), p. 83; Rainer Wohlfeil and Edgar Graf von Matuschka, *Reichswehr und Republik (1918-1933)* (in Meier-Welcker and von Groote, eds., *Handbuch*, Vol. VI) (Frankfurt: Bernard & Graefe, 1970), pp. 207-212.
[8] National Archives, Microfilm T-78, Oberkommando des Heeres Records (hereafter, OKH), H1/326, att. to Generalstab des Heeres 497/35, 12 Oct 35, 395/6398224.

pile lists of veterans for a recall to arms. It also maintained secret stores of arms and encouraged secret military training outside the army. Such mobilization preparations were known in the army as *Landesschutz*. Landesschutz work was supported by Reichswehr funds and guided largely by ex-officers. It included preparatory steps for mustering the Grenzschutz to defend the borders, and indeed the fairly well known Grenzschutz was deliberately used as a cover for Landesschutz activities generally.[9] But aside from Grenzschutz work, Landesschutz also, and more significantly, involved the finding, training, and registering of reserves for expanding the regular army. Private paramilitary organizations, the so-called *Wehrverbände*, constituted one potential source of reserves, and close relations existed at first between the army and these groups, which were for the most part rightist and antirepublican. Such ties were reduced after the groups proved uncontrollable.[10] Later, army leaders would attempt to establish a more official and less partisan system, but they came to believe that it was impossible to do this under an open, federal, and democratic regime —to the misfortune of the regime—and they turned again to the Wehrverbände.

A consequence of illicit military planning was that many respectable members of society—and no one in Germany was more respectable than an officer—involved themselves in conspiratorial activity. Social pressures and, in the early 1920s, terrorist action, worked to discourage disclosures. Considering what their reporters must have known, newspapers evidently exercised considerable self-restraint. Officially, many people were tried and convicted of the crime of *Landesverrat* or treason against the external security of the state, when their crime was the disclosure of illegal rearmament activities. A state-of-siege mentality developed, exemplified in an article by the legal adviser of the Defense Ministry, which sought to show that leading pacifists were guilty of treason. The compromise of legality and condemnation of protest help to explain the later legal breakdown and acceptance of repression in

[9] On listing: Franz von Gaertner, *Die Reichswehr in der Weimarer Republik* (Darmstadt: Fundus Verlag, 1969), pp. 131-132. On cover: OKW, Wi/IF 5.498, TA 750/27, 21 May 27, 110/837821.

[10] On Landesschutz in general, see Craig, *Prussian Army*, pp. 400-401; Post, pp. 176-177; Wohlfeil and Matuschka, pp. 212-216; Thilo Vogelsang, *Reichswehr, Staat und NSDAP* (Stuttgart: Deutsche Verlags-Anstalt, 1962), pp. 18-23, 34-45. The scope of Landesschutz work is shown in National Archives, Microfilm T-79, Army Area Command Records (hereafter WKK), WK VII/741, WKKdo VII Ib 230/32, 25 Feb 32, att., 28/152-155. In some cases, Landesschutz came to mean secret defense work *aside from* that involving the Grenzschutz; see Vogelsang, *Reichswehr*, Annex 4. The parallelism of the names encourages what Gilbert Ryle would call a category mistake; the Grenzschutz was meant to be, eventually, an operational organization, but Landesschutz was a preparatory activity.

the Nazi period. At the time and among those concerned, however, belief in patriotic and national ideals seemed to make rearmament a high, noble, moral goal.[11]

There was certainly some reason for thinking that defense preparations were necessary, if only to meet immediate dangers. Polish irregulars had come and gone across the border in the immediate postwar period, and the French had moved into the Ruhr area in 1923. Germans noted the failure of the League to act when the Poles seized Vilna in 1920, or when the Lithuanians seized Memel in 1923. The officers responsible for the defense of Germany were naturally among the most concerned. Even if the German army could expand ninefold to sixty-three infantry divisions, France (according to German estimates) could mobilize in 1929 up to eighty-nine divisions, and Poland up to forty-five, while the Czechs had a potential seventeen divisions.[12] If Reichswehr officers were to take their profession seriously, they had to attempt to prepare for their country's defense; they could not be content to regard themselves as members of a domestic police force. It is also understandable that they considered anyone, German or other, who opposed a strengthening of German defenses as one who would deny fundamental German rights.

The defense of Germany could include, however, a "restoration of freedom" to the country, and in the eyes of some officers, it might be impossible, as during the Napoleonic era, to restore national freedom without defeating France in a War of Liberation. Freedom, to them, was not a birthright of an individual, but a power position for a nation, an independent status the nation had to win from its rivals. One key German staff officer, Joachim von Stülpnagel, chief of the First or Operations Section of the Truppenamt until 1926, wrote in 1924:

[11] Generally, see Robert G. L. Waite, *Vanguard of Nazism* (Cambridge, Mass.: Harvard University Press, 1952), pp. 212-227; Wolfgang Sauer, "Die Mobilmachung der Gewalt," in Karl Dietrich Bracher, Wolfgang Sauer, Gerhard Schulz, *Die nationalsozialistische Machtergreifung*, 2nd ed. (Cologne: Westdeutsche Verlag, 1962), pp. 777-779. A report by Abwehr Abteilung IIIc (OKH, H24/3, 15 Feb 30, 247/ 6226570-73) shows that the Abwehr, the Reichswehr's intelligence and security department, investigated an average of 345 Landesverrat cases a year from 1920 to 1929. Article on pacifists: [Adolf] von Carlowitz, "Pazifismus und Landesverrat," *Wissen und Wehr*, 12 (1931), 194-215; Carlowitz was a close collaborator of Kurt von Schleicher's. In private discussion, Schleicher made the same charge against a leading pacifist, while claiming himself, after his war experiences, to be "pacifistically inclined": Institut für Zeitgeschichte, Munich, ED-93, Hans Schäffer Diary (transcribed from shorthand by Schäffer and Ernst Göhle), entry for 29 Jan 32.

[12] Karl Ludwig von Oertzen, *Rüstung und Abrüstung* (Berlin: Mittler, 1929), pp. 88, 142, 178-179. A fairer comparison of these totals might have included on the German side thirty-four divisions of the Grenzschutz. But the Reichswehr and the German public were convinced that they were greatly outnumbered.

20—Political Implications of Reichswehr Plans

> Versailles was an injustice built on lies; it must someday be set aside by might through a military combination of all national forces unless—France gives in beforehand, eliminates injustice from the world, and makes agreements with us which do not restrict our national development. But this is not to be expected and would be contrary to the trend of German-French history.[13]

An August 1925 memorandum from the Second (Organization) Section of the Truppenamt was prefaced by the following statement:

> So long as there is no guarantee that Germany's legal claims can be made to prevail in the long run by peaceful means, and that the phantom of general disarmament will become reality, Germany must try to build up its armed forces again.
>
> The goal must be the preparation of a defensive war for the preservation of German life-necessities [*die Vorbereitung eines Verteidigungskriegs zur Wahrung deutscher Lebensnotwendigkeiten*] and for warding off any enemy attack.[14]

Although the second sentence speaks of defensive war, which might be fought against enemy attack, it might also be fought to protect national "life-necessities," and this could be a fairly elastic concept. A 1926 paper by Reichswehr disarmament experts listed Germany's immediate policy objectives from the viewpoint of *Realpolitik*. These included the replacement of foreign by German (military) sovereignty in the Rhineland, the firm acquisition of areas separated from her (i.e., Austria), and the reacquisition of "areas essential to the German economy" (the Corridor, Polish Upper Silesia, and the Saar). France would be the primary opponent of this policy, and land forces would therefore be "almost exclusively decisive"; Germany should seek British, American, and also Russian cooperation for the disarmament of France. Naval contention with America and England would come later, after the Franco-German problem had been solved "through either peace or war."[15]

German needs had in fact tended to be elastic ever since 1870. The distinction between "defensive" and "offensive" had no clarity for a

---

[13] Bundesarchiv-Militärarchiv (BA-MA), Freiburg im Breisgau, N42/32, Kurt von Schleicher Papers, T1 176/24, 20 Feb 24. See also BA-MA, N5/20, Stülpnagel Papers, ltr., Stülpnagel to Hasse, 26 Jun 25, and N5/10, lecture, February 1924, "Gedanken über den Krieg der Zukunft."

[14] OKH, H1/663, T2 84/25, 14 Aug 25, 409/6415507.

[15] Hans Rothfels et al., eds., *Akten zur deutschen Auswärtigen Politik, 1918-1945* (Göttingen: Vandenhoeck & Ruprecht, 1966-) (hereafter, *ADAP*), Series B, Vol. I, Pt. 1, No. 144. Most of Series C and D of this publication have been published in English as *Documents on German Foreign Policy, 1918-1945* (Washington: G.P.O./London: H.M.S.O., 1949-).

power that was surrounded by actual or potential rivals, whose borders did not coincide with national divisions, and that owed its existence largely to military prowess. Americans and even Englishmen, having lived beyond water barriers, sometimes find it hard to grasp the importance of military might for such a country. For it, the achievement of national sovereignty, self-determination, and economic expansion would also mean new military strength—and perhaps new fears and escalation elsewhere. For it, an offensive might indeed eventually become the best defense.

The Polish Corridor always seemed the issue most likely to lead to conflict, the area most in need of liberation. German diplomats explicitly refused to rule out the possibility of a war against Poland, and German army officers, not believing that any peaceful solution was possible, expected someday to remove by force the strategic encumbrance of the Corridor. Many diplomats and officers hailed from either East Prussia, the lands just to the west of the Corridor, or the Corridor itself, and these men presumably had a personal interest in the problem. But resentment of the Corridor was doubtless widespread throughout Germany, and loudly trumpeted protests against the "illogical" delineation of the border were spurred on by an underlying fear that with the drain of German population to the West, East Prussia, and the border areas, as well as the Corridor itself, would become ethnically Polish. As the world agricultural depression grew worse, the Defense Ministry worried that Poles, drawing financial support (as the ministry believed) from Polish institutions, would replace the bankrupt German population in the German border districts and make it impossible to muster the local Grenzschutz.[16]

Young officers were especially attracted to the idea of a war of liberation, and this state of feeling in turn moved army leaders to reemphasize their own commitment to a larger army. In 1930, the Minister of Defense, Wilhelm Groener, turned over to the civil prosecutor of the Reich the case of three lieutenants of the Ulm garrison who had spread National Socialist propaganda within the army. At the trial, the immediate superiors of one defendant implicitly endorsed his

[16] Views of diplomats and officers: Post, pp. 28-31, 43-44, 50-57, 98-100, 122-124, 167, 231, 236-237, 295, 312, 328-329, 330n. Stresemann indicated publicly that Germany reserved the right to fight a war with Poland: *ADAP*, B/I/1, Annex II (pp. 739-740): for a Bismarckian statement at a closed meeting, see Stephen A. Schuker, *The End of French Predominance in Europe* (Chapel Hill: University of North Carolina Press, 1976), p. 267. On Reichswehr concern over Polonization, see National Archives, Microfilm T-120, Alte Reichskanzlei Records (hereafter referred to as RK), Reichswehrminister 197/31 and att., 8 May 31, and Reichswehrminister 126/32, 21 Apr 32, K953/K249738-49, K249779-80. This was an old German worry, with implications for the future of the nation: see Wolfgang J. Mommsen, *Max Weber und die deutsche Politik 1890-1920* (Tübingen: J.C.B. Mohr, 1959), pp. 23-76.

statement that the army should lead the German people in a war of liberation. (One of these superiors, Colonel Ludwig Beck, would in 1944 lead the conspiracy against Adolf Hitler.) One witness estimated that 75 percent of the younger Reichswehr officers agreed with such ideas; another officer privately placed the figure at 90 percent for the whole officer corps.[17] There was so much sympathy for the lieutenants within the army that Groener felt it necessary to assure army leaders of his own desire to build up the army for national goals. The draft of his remarks stated: "Since I have been Reichswehr Minister, all my thoughts and endeavors have been directed toward only one goal, the liberation of our country." One of his listeners, General Curt Liebmann, understood him to say that his aim was "to further the armed forces and war preparedness so that army and nation would one day be *ready*." Groener explained that he could not say such things publicly; responsible leaders could not act like romantic youths. He also said that only a firmly united Wehrmacht, standing above parties, could be used "for the high purpose intended for it."[18] Groener did not have a merely defensive purpose in mind.

3. The Reichswehr's Conflict with the Socialists

As already suggested, the idea of a war of liberation did not by any means appeal to the whole of the German people. Only along the eastern border, where a Polish invasion was feared, was there a strong public interest in defense. All the prominent political personalities did support at least some preparation for the defense of the border. In June 1923, after considerable friction, Carl Severing, as Social Democratic Minister of Interior in the Prussian Land government, had reached an understanding on guidelines with the Minister of Defense, Otto Gessler, whereby the Prussian government agreed to the storing of arms and to preparations for establishing centers (*Meldestellen*) under civil control to which volunteers could report in case of emergency, while in return the army was supposed to break off relations with private paramilitary groups. Social Democrats, however, remained hostile

---

[17] F[rancis] L. Carsten, *The Reichswehr and Politics, 1918-1933* (Oxford: Oxford University Press, 1966), pp. 319-323; Vogelsang, *Reichswehr*, pp. 82-83, 90-93. Annex 7; Otto-Ernst Schüddekopf, ed., *Das Heer und die Republik* (Hanover: Norddeutsche Verlagsanstalt O. Goedel, 1955), p. 290; Klaus-Jürgen Müller, *Das Heer und Hitler* (Stuttgart: Deutsche Verlags-Anstalt, 1969), pp. 32-34. On the testimony, see Peter Bucher, *Der Reichswehrprozess: Der Hochverrat der Ulmer Reichswehroffiziere 1929-1930* (Boppard: Harald Boldt, 1967), pp. 137, 216-223.

[18] National Archives, Microfilm M-137, Groener Papers, Notes for speech to Wehrkreis commanders, 25 Oct 30, reel 25/Stück 229; Thilo Vogelsang, "Neue Dokumente zur Geschichte der Reichswehr 1930-1933," *Vierteljahrshefte für Zeitgeschichte (VfZ)*, 2 (1954), 408 (emphasis in source).

to any proposals for reserve training or the listing of men for mobilization (both of which were barred by the agreement), and they generally opposed rearmament. They shared the antimilitarism of social democratic parties in other countries, and of the more idealistic of the German Democrats, and they also believed that reactionary monarchical forces were dominant in the army, which appeared to them to provide financial support to the paramilitary groups of the right, while the latter in turn attacked the republic and the republican parties. This Socialist distrust was of course reciprocated: the common view in the officers' casinos was that the "scoundrels without a country [*vaterlandslose Gesellen*]" had stabbed the armed forces in the back in the closing stages of World War I by organizing strikes and encouraging desertions and mutiny.[19]

For the Defense or Reichswehr Ministry, however, the Socialists represented not so much a historic opponent as a major present problem. Until the last year of the Weimar Republic, the SPD had the largest delegation in the Reichstag, a position strengthened in the December 1924 and May 1928 elections, and although they did not for the most part join in government coalitions, they were in a position to cause difficulty on budgetary questions. Perhaps more important, the SPD did hold until July 1932 a dominant position in the Prussian government, and the Minister President (premier) and Minister of Interior of that Land, which encompassed three fifths of the German population, and most of the eastern border areas, were invariably Social Democrats. Albert Grzesinski, Severing's successor as Prussian Minister of Interior, also succeeded in filling almost all the top provincial and police posts with members of the SDP, DDP, and the Center party.[20] His ministry, with control of the Land police forces, was able to follow many of the illegal activities of the Reichswehr, and it made a point of doing so. Lieutenant Colonel Kurt von Schleicher reportedly stated in the fall of 1926 that the Reichswehr was sick of the Prussian government's "snooping." Shortly after this comment was made, Grzesinski produced a long memorandum charging that the Reichswehr was flagrantly violating the 1923 agreement, and he urged that the time had come to make new arrangements.[21]

[19] RK, Gessler-Severing Agreement, 30 Jun 23, K951/K248617-19. The tenor of the agreement indicates that it was drafted, at least mainly, by the Prussian side. On this agreement and attitudes generally: Otto Braun, *Von Weimar zu Hitler*, 2nd ed. (New York: Europa Verlag, 1940), pp. 265-269; Hans-Peter Ehni, *Bollwerk Preussen?* (Bonn-Bad Godesberg: Verlag Neue Gesellschaft, 1975), pp. 25-26, 138, 154; Carsten, *Reichswehr and Politics*, pp. 58-66, 124-135, 203-205, 217-220, 269-272.

[20] See Anthony Glees, "Albert Grzesinski and the Politics of Prussia, 1926-1930," *English Historical Review*, 89 (1974), 823.

[21] National Archives, Microfilm T-120, Stresemann Papers, Note by Stresemann, 29 Oct 26, 7334/H162853 (cited by Erich Eyck, *A History of the Weimar Republic*,

24—Political Implications of Reichswehr Plans

Grzesinski's proposal probably stemmed less from any recent discoveries than from the fact that General Hans von Seeckt, the aloof and aristocratic Chief of the Army Command (supreme army commander) had recently had to resign, ostensibly because he had authorized the son of the German Crown Prince to attend maneuvers. Seeckt's downfall seemed to many Social Democrats to open the way for republicanizing the army, perhaps all the more because negotiations were afoot for securing a possible SPD participation in a Grand Coalition. The speaker of the Reichstag, a member of the SPD, made a proposal to prevent company commanders and rightist Verbände from controlling the selection of recruits. When this was rejected by the Reichswehr and the ministers in office, the SPD leaders threatened to make disclosures. The culmination was a dramatic speech by Philipp Scheidemann, an SPD leader, in the Reichstag on December 16, 1926; Scheidemann described and attacked secret Reichswehr preparations and, in particular, Reichswehr ties with the Soviet Union. This speech, publicly airing official secrets, provoked cries of "Landesverrat!" or "Treason!" from the floor, the other parties dissociated themselves from Scheidemann's remarks, and even some of the Socialists had regrets. President von Hindenburg reacted strongly against the Social Democrats, and began to seek a government leaning more to the right.[22]

Seeckt's resignation had also encouraged the hopes of a younger group of army leaders, led by Colonel von Schleicher, who had probably had a hand in bringing about Seeckt's downfall. Seeckt had tended to make the Reichswehr an isolated élite force, and had shown relatively little interest in mobilization preparations.[23] Yet, to other officers, these preparations were both urgently needed and increasingly feasible. The impending withdrawal (by international agreement) of the inspectors of the Inter-Allied Control Commission would soon make an extensive program of Landesschutz activity easier to conceal and less likely to lead to diplomatic complications. Instead of relying on clandestine Soviet supplies and on disorganized volunteer efforts at home, it should now be possible to make organized, regular preparations under direct Reichswehr control.[24]

Harlan P. Hanson and Robert G. L. Waite, trans., 2 vols. [Cambridge, Mass.: Harvard University Press, 1962], II, 92); RK, ltr., Grzesinski to Braun, 6 Nov 26, K951/K248620-44.

[22] Gatzke, *Stresemann*, pp. 68-69, 72-78; Eyck, II, 94-108; Josef Becker, "Zur Politik der Wehrmachtsabteilung in der Regierungskrise 1926/27: Zwei Dokumente aus dem Nachlass Schleicher," *VfZ*, 14 (1966), 73; Michael Stürmer, *Koalition und Opposition in der Weimarer Republik 1924-1928* (Düsseldorf: Droste, 1967), pp. 172-174, 177-183.

[23] See Hans Meier-Welcker, *Seeckt* (Frankfurt: Bernard & Graefe, 1967), pp. 518-521, 537; cf. Vogelsang, *Reichswehr*, p. 48n. See also Chapter Two below.

[24] Soviet-German military relations continued to be active and friendly until

Such plans would require, however, wider public and official support, and Schleicher and some of his fellow officers looked for ways to secure this. They have sometimes been portrayed as accepting the Versailles restrictions, as seeking Socialist support, or as temporarily attempting, as F. L. Carsten has put it, a "move to the left." The supposed acceptance of Versailles seems to stem largely from the suspicions of conservative officers, and particularly from a false scent laid by Seeckt's partisan first biographer, in an effort to vilify Seeckt's successors. It is true that Schleicher wrote of showing the republican colors, red, black and gold, and some of his associates seem to have thought of separating secret work from the army, or of reducing the scope of secret mobilization preparations. But to Schleicher, such measures were for the show-window. Their purpose was to overcome the distrust of political moderates, and thereby to enable the Reichswehr to win ministerial sanction for a definite program, committing the whole nation, including Prussia, to defense preparation. Schleicher was no doctrinaire rightist or revanchist; he was often devious and also impulsive and changeable. He was quite willing to work with Socialists, if they could be enticed into supporting Reichswehr goals. But Schleicher saw in Scheidemann's speech a chance to isolate Scheidemann and other critics, and to form a majority government of the center and right; failing that, his office staff and Hindenburg's entourage were already considering the possibility of securing a nonparliamentary cabinet responsible only to the President.[25]

1933, but though the initial emphasis had been on obtaining munitions and supplies for Germany, after 1925 the German interest was more in testing and training activity in Russia, while the Soviets wanted German financial and technical aid. See Gatzke, "Russo-German Military Collaboration," pp. 565-597, also Geyer, "Zweite Rüstungsprogramm," pp. 127, 159 (notes 18, 19).

[25] On the views ascribed to Schleicher and his circle: Rabenau, pp. 482-483; Vogelsang, *Reichswehr*, pp. 50-51; Carsten, *Reichswehr and Politics*, Chapter Six. Carsten concludes that the move to the left was slight and of short duration. Schleicher's views are documented in Vogelsang, *Reichswehr*, Annex 3; those of two other oft-cited officers are shown in Vogelsang, *Reichswehr*, Annex 4, and Meier-Welcker, *Seeckt*, p. 537 (Otto Hasse); Rabenau, pp. 482-483; and Karl Dietrich Bracher, *Die Auflösung der Weimarer Republik*, 4th ed. (Villingen: Ring-Verlag, 1964), Annex 4; and note 34 below (Erich von Bonin). Both Hasse (who was not one of Schleicher's circle) and Bonin (who was replaced as head of the Organization Department [T2] in February 1927 by the conservative Ritter von Mittelberger) seem to have wanted broader popular contacts primarily in the interests of later national mobilization. On plans for a center-right or presidial (presidential) cabinet, and on the supposed hopes for Socialist support, cf. Becker, "Zur Politik," pp. 70-78 (incl. 72n); Stürmer, pp. 179-180, 182-183 (and note). On Schleicher's personality, see also Thilo Vogelsang, *Kurt von Schleicher* (Göttingen: Musterschmidt, 1965). Heinrich Brüning, *Memoiren 1918-1934* (Stuttgart: Deutsche Verlags-Anstalt, 1970) and John W. Wheeler-Bennett, *The Nemesis of Power*, 2nd ed. (New York: St. Martin's, 1964), based largely on information from Brüning and personal obser-

Already in November, before Scheidemann's speech, the Reichswehr civil and military chiefs were bidding for cabinet approval and promising to present the ministers with a clear description of their plans. General Wilhelm Heye, the new Chief of the Army Command, addressed the (now more conservative) cabinet in February 1927, trying to reassure the ministers and at the same time to get their approval for Landesschutz. Heye gave a fairly frank account of Reichswehr wishes. He said that although the Grenzschutz in the East had at first been largely in the hands of the (rightist) Wehrverbände, it and the stores of arms had now for the most part been taken over by the army. He stated that experience had shown it was necessary to prepare lists of men liable to serve; this had already been fully carried out in East Prussia, but the execution varied greatly elsewhere. The number of trained ex-soldiers was in any case declining, and no training of reservists was currently going on. Great difficulties had arisen in relations with the Prussian authorities (Heye admitted that some of the blame here fell to the army), and Prussia had now more or less ceased cooperating. This state of affairs could not continue; the cabinet must approve the necessary measures, and stand behind them. It would be desirable to make preparations so that, in the event of a war between other countries, Germany could openly rearm within ten or eleven months and then (as Heye put it) "throw a decision onto the scales." Thus Heye hoped for a situation in which the Reichswehr would be able not merely to repel attacks, but also to intervene deliberately in a war in pursuit of political advantages. He said that the army would do nothing at all that the cabinet did not authorize, but he added: "If the cabinet did decide positively, then by all means care must also be taken to ensure through legal measures that Landesverrat, such as has been possible until now, can be suppressed."[26]

The Reich ministers declared their fundamental agreement with the Reichswehr's ideas, and raised no objections to the idea of intervention in a war between other countries, but no specific decisions were taken. Later in 1927, Schleicher and Colonel Werner von Fritsch briefed German Foreign Ministry officials on the Reichswehr's need

---

vation, are extremely hostile to Schleicher. Vincenz Müller, *Ich fand das wahre Vaterland*, Klaus Mammach, ed. ([E.] Berlin: Deutscher Militärverlag, 1963), gives a more sympathetic picture.

[26] National Archives, Microfilm T-120, Foreign Ministry Records (hereafter AA), Extract from record of ministerial meeting, 26 Feb 27, 4565/E164073-83. OKW, Wi/IF 5.509, Chef TA 30/26, 1 Apr 26, 113/841660-61, had confirmed the ban in the 1923 Gessler-Severing agreement on contacts with the Wehrverbände. But (leaving aside the "April Fool's" date) this may have been "for the record" in a period when the German government was trying to allay Allied suspicions; see Gatzke, *Stresemann*, p. 49.

for authorized guidelines; presumably other interested ministries were also given further explanations. Schleicher warned that a fait accompli might be created in the East which the League would be unable to reverse, like that in Vilna. He said that both military and civil authorities had to get ready to meet a sudden attack, not only by providing for civil defense, the evacuation of border areas, and the mustering of the Grenzschutz, but also by preparing for the transformation of the peacetime army into a field army ready for action. Fritsch argued that, on both moral and practical grounds, the border population could not be left to defend itself, so that lists of potential soldiers had to cover the whole population of the Reich. And as Schleicher said, since the number of trained reservists was declining each year, "a certain military preparation, and ultimately military training itself—to begin with at least in the Grenzschutz—became simply a compelling need." Fritsch admitted that these measures were contrary to the Treaty of Versailles (particularly Article 178, which banned preparations for mobilization) and that the Allies would eventually learn of them, but he thought that there would be little problem if the measures were given a harmless cover, and represented merely an exercise of the right of self defense. He believed that the purpose of the measures should be concealed as much as possible from the officials who would carry them out, as well as from foreign eyes.[27] As though to drive home the point about the Polish danger, the Truppenamt in November also invited the Foreign Ministry to send a representative to a war game involving a Polish effort to seize East Prussia.[28]

Foreign Ministry officials worried over the effect possible disclosures would have on foreign relations, but they approved the Reichswehr proposals, stipulating only that there should be no Grenzschutz along the western border. One high official of the ministry, Gerhard Koepke, explained his support to Gustav Stresemann, the Foreign Minister, by pointing out that no government could refuse to approve minimum defense measures and that in any case, these measures were already being carried out in the East. If the Foreign Ministry did not get involved in good time, the army might again associate with the rightist Wehrverbände, and such one-sided ties carried the risk of disclosures by the left. Since the diplomatically risky illegal defense activity could not be prevented, it was best to involve the Social Democrats and the Prussian government in it.[29]

[27] AA, Briefing [by Schleicher] on Landesschutz, [prior to 28 Oct 27], K6/K000301-04; Oral remarks by Col. Freiherr von Fritsch, [prior to 21 Nov. 27], K6/K000305-08.
[28] See Post, pp. 209-215; Gatzke, *Stresemann*, p. 101.
[29] AA, Memo by Koepke, 21 Nov 27, K6/K000309-15; Memo by Koepke, 23 Jan 28, 4565/E164092-95. Post (p. 162n) considers that the 21 Nov 27 memo was actually

The military were genuinely concerned over the danger of a sudden Polish attack, they wished to be able to expand the army in any near-term emergency, and they were anxious to train reserves for the future. If they were to have a field army as well as a Grenzschutz, they would need to extend mobilization preparations to the whole country, and they would also need to secure stiffer laws against Landesverrat. To do these things, however, they would somehow have to tame the Prussian Social Democrats, so that the latter would stop spying on Reichswehr activities, serve the army by preparing lists of mobilizable men (some of whom would be sought out for training), and even help in suppressing criticism and the publication of secrets. It made a great difference whether the civil authorities regarded Landesschutz as "playing soldier" or actively supported the work. If the civil authorities refused to cooperate, the army would be virtually forced to rely on the private organizations of the far right.[30]

The immediate question was whether the Reichswehr Ministry and the Prussian government could agree on a new and mutually satisfactory set of guidelines. In discussions beginning in April 1927, the Prussian government showed, as before, some readiness to support the Grenzschutz along the eastern border, but no sympathy at all toward peacetime preparations for mobilization in the rest of the country. A Reich cabinet meeting on Landesschutz was called off in September 1927 because of the reservations of Otto Braun, the Minister President of Prussia, about the Reichswehr's current plans. But the Reichswehr persevered with its hope of an officially authorized national program, which was symbolized by the appointment, in January 1928, of Wilhelm Groener as Minister of Defense. Groener—whose appointment was supported by Schleicher—was "above party," an ex-general, and yet also acceptable to prorepublican Germans. Reichswehr draft guidelines of early 1928 show the army's far-reaching objectives: the draft called for continuous civilian cooperation, as in the protection of arms stores, in the collection and regular reporting of "statistical information of a personal and material nature" (i.e., respecting manpower, providing names of persons eligible for service, drawing on such sources as tax rolls), and in preparing for the establishment of Melde-

---

drafted by Dirk Forster, military expert in the Wilhelmstrasse. A later Reichswehr draft for the guidelines is on K6/K000384-87; it was approved by the Foreign Ministry in May 1928 (Unsigned memo, 12 May 28, K6/000388-89).

[30] In one case, a Landesschutz officer made a cautious inquiry (through an intermediary) of an approachable SPD man, and was told that individual Socialists becoming involved in Landesschutz would face exposure, ridicule, and perhaps economic boycott, although many individuals might go along if there were no "party terror" to fear. See RK, Address by Hasse, 14 Dec 28, K953/K248637-41; Br. 910, 7 Jan 29, K953/K249643.

stellen. (According to internal army directives, the Meldestellen were to form only the bottom level of an elaborate structure for securing military and civil manpower in a war emergency; the structure could collect volunteers and veterans under existing laws, backed by moral suasion, but if a need for mobilization arose, the domestic and international obstacles to conscription would probably disappear, and the system could then also be used to carry out conscription.) The military would choose the civilians to be involved in Landesschutz work, subject to the approval of the civil *Oberpräsidenten* (the governors of the Prussian provinces), and the civil authorities would share in the responsibility for maintaining secrecy. On the other hand, the army agreed that private organizations should not participate in military training. The Prussian government replied to the Reichswehr draft for guidelines with a draft of its own, which seemed to the army to be completely unsatisfactory as a basis for negotiation. In particular, the Prussian proposal would have confined activity, except for the storage of arms, to the border areas, so that there would have been no preparations for mobilizing personnel and production in the rest of the country.[31]

The formation of a Grand Coalition government in June 1928, with a Social Democrat, Hermann Müller, as Chancellor and Severing as Reich Minister of Interior, seemed to offer the Reichswehr new fulcrums for leverage on the Prussian government. These leaders were relatively cooperative, and in contradiction to recent Socialist campaign slogans, they agreed to support a Defense Ministry request for funds to begin replacing the navy's over-aged ships with "pocket battleships." In September, Schleicher thought that Socialist interest in developing a defense program was making prospects favorable for securing cabinet support for Landesschutz, although he still anticipated violent arguments and possibly a cabinet crisis. Instead of opposing Landesschutz arrangements, Severing began to use his good offices to bring about an agreement between the Reichswehr and the Prussian government. These cooperative Socialist ministers were unable, however, to win the support of the Social Democratic Reichstag delegation for the ship appropriation, and Severing also seemed to make little progress with his mediation efforts. The Prussian police uncovered evidence in October and November that guerrilla warfare training with National Socialist involvement was continuing in the West, including in the demilitarized zone, and this led to Prussian expressions

---

[31] Vogelsang, *Reichswehr*, pp. 52-55; AA, Memo by Koepke, 21 Nov 27, K6/K000312; Memo by Forster, 13 Apr 28, K6/K000370-72; Memo on Landesschutz, [prior to 12 May 28], K6/K000384-87. On the manpower system, BA-MA, file II H 288, TA 77/31 and att., 4 Feb 31. See also the sources in note 27.

of distrust and anger. The atmosphere improved somewhat at the end of 1928, and the negotiations were reported to be ripening in December, but only on April 26, 1929 did Severing inform the cabinet that Prussia was ready to work with a modified set of guidelines. It emerged at the cabinet meeting that the Prussian government would not assume a specific responsibility for the arrangements, but the Reichswehr obtained on this day the approval of the Reich cabinet itself for a set of guidelines covering the listing of reserves.[32]

Thilo Vogelsang, an authority on Reichswehr history, has suggested that this Reich cabinet decision put an end to most of the difficulties; this was not, however, the case. Instead of implementing the decision of the Reich cabinet, a decision in which Socialist ministers had participated, the Prussian government continued its resistance, and Grzesinski held that no agreement had replaced the voided guidelines of 1923. Schleicher enjoyed Groener's trust, and now, with the rank of brigadier general, he occupied a post at the minister's right hand, as head of the newly created Ministerial Office; this put him in a good position to pursue the struggle. He complained at a meeting in October 1929 that Minister President Braun had still not sent the agreed guidelines to the Oberpräsidenten of the Prussian provinces. In November, the Reichswehr Ministry appealed informally for intervention by Chancellor Müller, and on February 3, 1930, Schleicher sent another letter to the Reich chancellery, arguing the urgency of Prussian implementation at an early date. This was followed by a letter from Groener to the Chancellor on February 14, recounting the history of the controversy and stating that it was not consonant with Reich authority that the largest German Land should be permitted to exercise more or less passive resistance. Müller was unable (according to his own note on the letter) to settle the matter due to the pressure of the budgetary crisis preceding his resignation on March 28. An agreement was not achieved until the end of 1930, and even then, a letter from Severing (now again Prussian Minister of Interior) hinted at the likelihood of further difficulties. In November 1931, an internal Reichswehr briefing indicated that smooth cooperation with Prussian officials had still not been secured.[33]

[32] Groener Papers, ltr., Schleicher to Groener, 3 Sep 28, reel 25/Stück 224; Vogelsang, *Reichswehr*, pp. 56-58, Annex 4; Carsten, *Reichswehr and Politics*, pp. 300-301; RK, Note, 4 Dec 28, K953/K248611; Martin Vogt, ed., *Das Kabinett Müller II, 28. Juni 1928 bis 27. März 1930*, 2 vols. (Boppard: Harald Boldt, 1970), I, Doc. No. 181. In early 1927, after Scheidemann's disclosures, Hermann Müller admitted at a Reichstag Foreign Affairs Committee meeting that "he, too, had helped rock the Russian cradle": Gatzke, *Stresemann*, p. 86.

[33] Vogt, *Kabinett Müller II*, I, No. 181n; RK, Record of chiefs' meeting, 15 Oct 29, K951/K248965; Note on guidelines, 23 Nov 29, K953/K249707-09; ltr., Schleicher to Pünder, 3 Feb 30, K953K249719-20; ltr., Groener to Müller, 14 Feb 30, K953/

There were many reasons for the Prussian resistance to the Reichswehr's wishes. Despite Reichswehr claims that rightist ties had been severed, the officer corps and the unofficial Landesschutz workers obviously had little sympathy for the republic. One staff officer who honestly tried to foster contact with the republican Wehrverband, the Reichsbanner, met with Socialist suspicion of the military, came under attack from Junker elements, and finally was recalled and dismissed; at the same time, many Landesschutz personnel were drawn from the leading monarchist Wehrverband, the Stahlhelm, and ties with the Nazi SA (Sturmabteilungen—storm troops) were reported from time to time.[34] Some of the military arrogance that had produced incidents like the prewar Zabern affair still existed, as when an officer told a Prussian official in Allenstein, "What such a . . . [expletive deleted by source] minister as Severing says is a matter of indifference for us in the Reichswehr."[35] More concretely, mobilization preparations, as Albert Grzesinski noted, were obviously modeled on prewar patterns. This could evidently pave the way for a reintroduction of conscription, something the broad German public had little desire for. No doubt the Prussian leaders did not like to assume the politically thankless task of supporting military preparations in the nonborder areas where there was no direct concern over the Polish threat and where, in working-class circles, there was no social pressure for supporting mobilization preparations. The Social Democrats were being asked to go against the decisions of their party congresses (such as for the barring of punishment for disclosures of illegal rearmament), and in Prussia they were also being asked to become part of an apparatus to support military preparations and curb antimilitary publications and manifestations. The attempt to bring Reich authority to bear doubtless added to the resentment on the side of Braun and his colleagues.[36]

K249721-25; ltr., Severing to Groener, 23 Dec 30, K953/K249733-34; Brüning, *Memoiren*, p. 247 (the dating appears to be confused on p. 246); Otto Braun, pp. 267-268; OKW, Wi/IF 5.498, Briefing paper on relations with civil authorities, 25 Nov 31, 110/837784-86.

[34] The staff officer was Col. Erich von Bonin (see note 25). He was replaced in Königsberg in 1931 by Walther von Reichenau. See Carsten, *Reichswehr and Politics*, pp. 258-260, 271; Bracher, *Auflösung*, p. 276n; Harold C. Deutsch, *Hitler and His Generals: The Hidden Crisis, January-June 1938* (Minneapolis: University of Minnesota Press, 1974), p. 9. Schleicher had advised Bonin against joining a Republican Club in Königsberg (Schleicher Papers, N42/39, ltrs., Bonin to Schleicher and Schleicher to Bonin, 19 and 22 Mar 29).

[35] Stresemann Papers, Note by Stresemann, 29 Oct 26, 7334/H162852.

[36] See Gustav Adolf Caspar, *Die sozialdemokratische Partei und das deutsche Wehrproblem in den Jahren der Weimarer Republik* (Berlin: Mittler, 1959), esp. pp. 35, 57; also Otto Braun, pp. 267-268; Albert Grzesinski, *Inside Germany* (New York: Dutton, 1939), esp. pp. 89-95, 140-141, 144; and Grzesinski's November 1926 memo to Braun; RK, K951/K248620-44.

There were also larger principles at stake. The German military's conception of preparedness included ideas about indoctrination and the suppression of dissent that were incompatible with democratic institutions and that were bound to entail coercion. The concepts of the "nation in arms" and the "army of the nation" encouraged efforts to control and direct public opinion to support national defense. Aside from the usual teaching of patriotism in schools and popular culture, German nationalists in and out of government had made formidable propaganda efforts before and during World War I, and particularly in the final years of the war. Many army officers believed that the guidance and control had simply not been thorough enough, and some of them drafted memoranda describing the measures they thought should be taken. Although the Social Democrats were not privy to this material, they were certainly well aware of the desire to direct the nation's thinking.[37] The Prussian leaders had reason to resist being drawn into these plans.

The Reichswehr leaders, however, felt exasperated. From a military point of view, there were many grievances or irritations, such as the right of the police to conduct investigations on military territory, Prussian opposition to the elimination of small garrisons (the Prussians wanted them retained in the economic interest of the garrison towns), or the participation of a Prussian high official in a pacifist conference in Breslau. Grzesinski, as Prussian Minister of the Interior, supported unwelcome Socialist demands for political control of Reichswehr administration, and he also tried to induce Severing to secure a dissolution of the paramilitary formations of the radical left and right, including the Nazi SA and even the Stahlhelm, which not only provided personnel for Grenzschutz and Landesschutz, but also had as its official sponsor the President of the Reich, Paul von Hindenburg. Although the Reich banned only the Communist Red Front Fighters League (in May 1929), Grzesinski on his own authority

[37] See OKW, OKW 2139, "Deutschlands Lage für einen künftigen Verteidigungskrieg," n.d. [1925-1926?], 870/5616291-344; OKW 2140, "Die organisatorische Lage für eine personelle Heeresverstärkung vom Jahre 1931 an," [April 1925], 870/5616415; Wi/IF 5.498, TA 1207/31, att., 24 Nov 31, 110/837781; OKH, H15/103, Leiter des Beschaffungswesens 18/32, 11 Apr 32, 146/6118665; also Jutta Sywottek, *Mobilmachung für den totalen Krieg* (Opladen: Westdeutscher Verlag, 1976), pp. 9, 13-19. (On the "organisatorische Lage" paper, see also note 50, below.) Groener, sometimes seen as a liberal, was a strong advocate of indoctrination and an enemy of anyone who criticized the military; see RK, ltr., Groener to Braun, 14 Oct 29, K951/K248959-61; ltr., Groener to Severing, 11 Nov 29, K951/K249002-03; ltr., Groener to Severing, 26 Nov 29, K951/249005; ltr., Groener to Wirth, 11 Apr 30, K951/K249091-92; ltr., Groener to Brüning, 19 May 30, K951/K249119-20; ltr., Groener to Severing, 8 Dec 30, K951/K249320-22.

banned the Stahlhelm in the Rhineland and Westphalia in October 1929, on the ground that the organization had conducted maneuvers in the demilitarized zone. Even personal irritants played a part: Groener became incensed over the Prussian habit of sending complaints to him and asking him to report back on the action he had taken. And Groener and Schleicher themselves felt exposed to criticism within their own military interest group, their professional constituency, because of their past efforts to work with the recalcitrant Socialists.[38]

There were bigger issues from the military perspective, too. If the Grenzschutz was acceptable, then Landesschutz should be, for the population of the East could not be left to shift for itself without help from the rest of the Reich. And if national defense was not to be the monopoly of the rightist organizations, the officials of the whole country would have to show that they supported it. As Groener wrote to Müller in February 1930, "The gradual elimination of the widespread distrust of this activity and the widest possible inclusion of the elements in the German people that support the constitution can only be gradually accomplished through the responsible cooperation of the Land governments." In the Reichswehr view, the Prussian Socialists were shirking their duty. The Social Democrats did not simply refuse any contribution at all—that would have been politically hazardous in post-Versailles Germany—but what they kept on doing became even more trying to the army: they made counter-proposals, raised new charges that the army was engaged in politically dangerous activities, and referred questions to each other, and after the guidelines had been approved on the Reich level, they still failed to implement them. By

[38] RK, memo, Braun to Marx, 5 Apr 27, K951/K248664-73 (on police rights, with follow-up material, inter alia, on K248695, K248713-19, K248722-41, K248744-45, K248750-52, K248926, K248930, K248932-41); ltr., Braun to Müller, 28 Jul 28, K951/K248764-65 (on garrisons; also see K248766-68, K248769-71); ltr., Groener to Braun, 14 Oct 29, K951/K248959-61 (protesting a speech by Braun; also K249086); ltr., Groener to Severing, 11 Nov 29, K951/K249002-03 (on pacifists in Breslau; also K249091-92, K249119-20); ltr., Groener to Braun, 7 Dec 29, K951/K249017 (objections to Prussian requests for a report on action taken). On Grzesinski's role, see his book, pp. 93-95, 140-141, 144. On the Stahlhelm ban in the Rhineland and Westphalia: Ehni, pp. 141-143. On the sensitivity of Groener and Schleicher to criticism within the services, see Werner Conze, "Die politischen Entscheidungen in Deutschland 1929-1933," in W. Conze and Hans Raupach, eds., *Die Staats- und Wirtschaftskrise des Deutschen Reichs 1929/33* (Stuttgart: Ernst Klett, 1967), p. 231. When Schleicher argued that the Reichsbanner was just as paramilitary as the Stahlhelm, and should (according to Grzesinski's principles) be banned as well, Grzesinski replied: "We must not forget that the Reichsbanner is on our side. The state cannot afford to push its sense of justice so far as to rob itself of its own defense" (Glees, p. 833).

the end of 1929, military annoyance, not only with the behavior of Braun and Grzesinski, but also with the inability of Müller and Severing to control their comrades, had plainly reached explosive levels.[39]

After the 1926 Scheidemann speech, Schleicher's staff had advanced the idea that a coalition should be formed extending from the Center party to the DNVP, and they had considered that if that failed, a so-called presidial government might be established without party ties by someone enjoying the President's confidence. Schleicher returned in 1929 to the first of these alternatives, while keeping the second in mind. As a leader for the conservative coalition, Schleicher's choice finally fell on Heinrich Brüning, the financial expert of the Center party, and the general began to prepare Hindenburg's mind for a Brüning cabinet. Like Schleicher, Brüning in December 1926 had been interested in forming a conservative coalition, as had a friend of his, Gottfried Treviranus of the DNVP.[40] Alfred Hugenberg, the chairman of the DNVP, preferred a course of intransigent opposition, but Schleicher, who was also close to Treviranus, seems to have hoped that the latter could bring a considerable number of DNVP dissidents into the coalition. If the calculation failed, a presidial government could exercise the President's decree powers under Article 48 of the constitution, and this could make up for a lack of votes.

Schleicher's particular interest in ending Socialist and Prussian opposition is apparent if we look at the full text of notes for Groener for a briefing of Hindenburg, prepared under Schleicher's direction in early 1930. The notes argue against strengthening the SPD, and in favor of a government independent of parties, probably to be led by Brüning. Aside from general financial and political arguments for a change, specific grievances of the Defense Ministry were to be raised with the President, including the Socialist attitude toward naval construction, Prussian sabotage of Landesschutz work, and Prussian harassment of the armed forces by circulating a memorandum on Reichswehr-Stahlhelm collaboration in the Rhineland and Westphalia, legalizing the Reichsbanner (the prorepublican Wehrverband), exaggerating minor incidents, and interfering with the command prerog-

[39] The most comprehensive and forceful summary of the Reichswehr position is Groener's letter to Müller, 14 Feb 30, K953/K249721-25. This letter ended with a pointed request that Müller confirm that the 23 (sic: 26) April decision had entered into force, and see to it that Prussia declare a readiness to cooperate.

[40] Becker, "Zur Politik," pp. 73-75, 76 (n. 5); Brüning, *Memoiren*, pp. 116, 145-148; Schäffer Diary, 7 Jun 32 (Brüning testimony), 17 Jun 32 (Groener testimony). Individuals favoring a presidial solution included Joachim von Stülpnagel, Otto Gessler, and Werner von Alvensleben: Joachim von Stülpnagel, *75 Jahre meines Lebens* (Oberandorf/Obb.: J. Stülpnagel, 1960; copy in BA-MA), p. 271; Schleicher Papers, N42/29, ltr. from Alvensleben, 12 Mar 30.

atives of the Reichswehr Ministry (probably a reference to police entry into military installations).[41]

A budget crisis arose in March 1930. Brüning had hesitations about becoming chancellor, and he claimed later that he had worked conscientiously to uphold the existing Grand Coalition under Hermann Müller, hoping that it would last until autumn. At the same time, he made his party's support of the Young Plan legislation, and thus its continued participation in the Coalition, contingent on a prior commitment to financial reform, which meant primarily a balancing of payments into and out of the unemployment insurance scheme. But for months, as Groener wrote to a friend, Schleicher had been doing "excellent work behind the scenes," and Groener himself did his best, in talking with Brüning in early March, to induce Brüning to accept the chancellorship. The presidential entourage evidently wanted a change. The Social Democrats finally declined to accept a compromise settlement of the unemployment insurance question, and Hindenburg refused to give the Müller cabinet the use of decree powers. When the old cabinet resigned, Hindenburg immediately summoned Brüning. After a night's consideration, Brüning agreed to form a new government; he states that he feared the Reichswehr would otherwise find someone else "to resolve matters in their own fashion by direct means," i.e., without regard for parties and parliament.[42]

4. The Deeper Levels of Mobilization Planning

The army's desire for a change of government was clearly fed by a sense of urgency about getting the new guidelines for Landesschutz into operation. Schleicher wrote in February 1930 of how important it was, in the domestic and foreign political interest, that disclosures be

---

[41] Schleicher Papers, N42/29, Notes by Noeldechen, n.d. [probably January-February 1930]; also Ehni, pp. 154-156. As Carsten points out (*Reichswehr and Politics*, p. 307n), the text in Vogelsang, *Reichswehr*, Annex 5, omits the Reichswehr's specific grievances against the Prussian government and the Socialists. As to the Reichsbanner, it was considered in this period to be tainted with pacifism: Wohlfeil and Matuschka, p. 224. On the desire to exclude the Social Democrats, see Conze, "Politischen Entscheidungen," pp. 202-203, 209-210, 230.

[42] Schäffer Diary, 7, 17 Jun 32; Brüning, *Memoiren*, pp. 150-161; Reginald H. Phelps, "Aus den Groener Dokumenten," *Deutsche Rundschau* (*DR*), 76 (1950), 1013, 1014. On Brüning's role, cf. Rudolf Morsey, "Die Deutsche Zentrumspartei," in Erich Matthias and R. Morsey, eds., *Das Ende der Parteien 1933* (Düsseldorf: Droste, 1960), p. 293 and notes 9 and 10. Groener flatly told Hans Schäffer in June 1932 that Schleicher had brought down Müller's cabinet, just as he had now helped to bring down Brüning's. Just before the fall of the Müller government, Erich Marcks, one of Schleicher's aides, hoped for and apparently expected a replacement of party rule by a government that would really "govern": Otto Jacobsen, *Erich Marcks* (Göttingen: Musterschmidt, 1971), p. 45.

barred by getting the Prussian government to recognize the guidelines "soonest." Later, after a recognition appeared to have been finally obtained at the end of 1930, Groener expressed the hope that loyal cooperation from Prussia would now make it possible "to make up for the already-so-large amount of lost time."[43]

Why had the guidelines become so urgent? For one thing, staff officers considered the listing of manpower as more than just a future objective; it was a current need. The existing regular forces might suffice for some purposes, particularly to put down internal disorder, but army planners had never intended to fight a war, either in the immediate or the distant future, with a mere seven infantry and three cavalry divisions. They had always hoped to mobilize a larger force in the event of a real conflict, and they believed that the occasion for such a mobilization, for which manpower lists were essential, might arise at almost any time. For them, particularly after Seeckt's departure, the demand for mobilization also rested on a broad belief that the First World War had proven the need for a total mobilization of the nation's materiel and manpower. Moreover, beyond the ostensible aims of the guidelines themselves, the army had other plans, which were kept more secret, which were subject to pressures of time, and which would be furthered by the cover the mobilization organization would offer, as well as by the improved protection the guidelines would provide for official secrets. One category of secret planning, involving materiel rearmament, was tied to a projected schedule, while another, involving training programs, was designed to meet a foreseeable crisis arising at a definite and not-far-distant date. And aside from all these objectives, two foreign-policy developments appeared to be very timely for the army's plans. First, French troops were due to evacuate the occupied Rhineland, a step that was completed on June 30, 1930; this would make it more difficult for French intelligence to monitor German armament, and it would make German national defense more feasible in the event of war. Second, there was the long-anticipated World Disarmament Conference, which was finally scheduled for February 1932; German officers expected this conference either to concede, or, by failing, to provide an excuse for, a measure of German rearmament.[44]

[43] RK, ltr., Schleicher to Pünder, 3 Feb 30, K953/K249719; ltr., Groener to Brüning, 9 Jan 31, K953/K249732.

[44] On early Reichswehr plans for mobilization, see Craig, *Prussian Army*, pp. 400-405; Gaertner, pp. 130-135. Heye hinted at both material rearmament and the need for training plans in his 26 Feb 27 remarks to the ministers. On liberating effects of the evacuation of the Rhineland: OKW, Wi/IF 5.3049, H Wa A 611/30, 5 Sep 30, 399/1651123-25. On expectations from Disarmament Conference: Gaertner, p. 131; AA, Report by Schlimpert, 3 Jun 30, K6/K000480; OKW, Wi/IF 5.509, TA 562/31, 12 Jun 31, 113/841862.

These same developments might, however, take an adverse turn if the Reichswehr's more far-reaching aims were not kept secret.[45]

The first category of the Reichswehr's secret planning, that of increased weapons production and preparations for industrial mobilization, was no new concern. Discussions on industrial mobilization planning had begun in late 1924, and a "Statistical Society," purportedly private but actually directed by the army's Ordnance Office, was formed in January 1925.[46] In April 1926, the chief of the Truppenamt, the army general staff, issued a directive for planning a long-term armament and procurement program beginning in 1927-1928. This launched what was to be the first armament program, covering the years from 1928 to 1932. The aim was to provide materiel for a twenty-one-division army, although it soon became apparent that the initial program would not by itself suffice for this. This first program was not finally approved by General Heye until September 1928. Perhaps to avoid a repetition of this delay, Heye simultaneously ordered that plans for the second armament program should be completed by the fall of 1931, to be ready for inclusion in the 1933 budget, taking effect on April 1 of that year. The second program was intended to cover the years from 1933 to 1938.[47] Even the first program provided for significant rearmament: for example, the number of machine guns increased from 12,000 in February 1927 to more than 20,000 by April 1931, an increase more than six times the production authorized by the Allies for such a period, supplying 1,000 fewer light machine guns and 1,500 more heavy machine guns than were needed for a twenty-one-division mobilization.[48] In the fall of 1929, following a budget cut in that year, Groener was anxious to establish the principle that the standard defense budget should be at least RM700 million, with full freedom for the Reichswehr Ministry to apportion that total. Chan-

[45] In 1927, the Foreign Ministry was already anxious that German hands appear clean in view of the danger of a call for an investigation by the League, and in anticipation of a failure of the Disarmament Conference (AA, Memo by Koepke, 21 Nov 27, K6/K000309-15). The Inter-Allied Control Commission had in fact reported numerous German violations; see, e.g., Gatzke, *Stresemann*, p. 71. Its successor, a Conference of Ambassadors, reached similar conclusions in July 1927 (Gatzke, *Stresemann*, pp. 93-94); for details, see University of Michigan Library, Ann Arbor, Mich., Microfilm 6342, Marineleitung Records, Project II (hereafter referred to as ML), Shelf 5993, Cover ltr., 26 Aug 27, and summary report, reel 5/720-828.

[46] OKW, Wi/IF 5.23, Thomas, *Rüstungswirtschaft*, pp. 55-56.

[47] OKW, Wi/IF 5.1983, Chef TA 43/26, 15 Apr 26, 327/1162220-22; Wi/IF 5.126, [TA] 747/27, 19 May 27, 18/728961-62; TA 223/28, 2 Feb 28, 18/728882-84; Chef HL, TA 1791/28, 29 Sep 28, 18/728727-28; Chef HL, TA 1779/28, 29 Sep 28, 18/728729.

[48] OKW, Wi/IF 5.2234, Chef HL, TA 761/30, 22 Jul 30, 347/1186186, 1186189. The 1927 machine gun figure is from Heye's briefing of the ministers on 26 Feb 27 (AA, 4565/E164073); the Allied limit from Georges Castellan, *Le Réarmement clandestin du Reich, 1930-1935* (Paris: Plon, 1954), p. 255.

cellor Müller was sympathetic, and his cabinet finally voted RM700 million for 1930, but the resistance of other Socialists had not augured well for future budgetary demands. A paper from the army Ordnance Office, circulated in November 1929, stated that other staff elements agreed on the necessity for comprehensive preparations for industrial mobilization and added, "There should therefore be an attempt during the next armament period (after 1932) to provide significantly greater means for this purpose."[49]

The second and even more pressing category of secret planning involved the training of young men who could be mobilized. General Heye, in his remarks to the ministers in February 1927, had pointed out that a serious problem was arising, since even though the veterans of World War I might be listed for an eventual recall to arms, the number of war veterans who were young enough to serve was declining by 200,000 each year, and by accepted military standards they would all become overage between 1930 and 1933. No younger men were being trained, at least in any numbers. This problem caused increasing concern at the Bendlerstrasse, the Reichswehr Ministry headquarters, where the year of crisis was usually placed at 1931. Joachim von Stülpnagel had referred to the problem in a lecture in February 1924, and it was exhaustively discussed in an army memorandum in April 1925; this paper judged that after 1931, it would be too late to train a mass army, owing to the foreseeable lack of leaders, specialists, and instructors. Albert Grzesinski (in his November 1926 memorandum on the army's transgressions of the 1923 agreement) noted that this problem accounted for the military interest in illegal training. French intelligence officers were also aware of the difficulty. When Schleicher, in 1927, had cited declining reserves to explain to Foreign Ministry officials why "a certain military preparation, and ultimately military training itself" were becoming necessary, he had added, "Ways for carrying them out must be sought and found."[50]

---

[49] Geyer, "Zweite Rüstungsprogramm," pp. 127, 131-132; Vogt, *Kabinett Müller II*, Nos. 348, 350, 353, 359, 412, 433, 448, 449, 454, 489; OKW, Wi/IF 5.1983, H Wa A 990/29, [26?] Nov 29, 327/1162113-19.

[50] AA, Extract from record of ministerial meeting, 26 Feb 27, 4565/E164074; OKH, H1/663, T2 84/25, 14 Aug 25, 409/6415510; RK, ltr., Grzesinski to Braun, 16 Nov 26, K951/K248624; Stülpnagel Papers, N5/10, "Gedanken über den Krieg der Zukunft," February 1924; Castellan, *Réarmement clandestin*, pp. 404, 406; AA, Briefing [by Schleicher] on Landesschutz, [prior to 28 Oct 27], K6/K000303-04; the British general staff also noted that the veterans were becoming overage: E. L. Woodward et al., eds., *Documents on British Foreign Policy, 1919-1939* (London: H.M.S.O., 1946–) (hereafter, *DBFP*), Series Ia, Vol. V, No. 268. The April 1925 memorandum was "Die organisatorische Lage für eine personelle Heeresverstärkung vom Jahre 1931 ab: Welche Folgerungen sind aus ihr auf personellen Gebiet vom organisatorischen Standpunkt zu ziehen?" Several copies are extant (see also note 37 above); I am indebted to Professor Donald E. Emerson of the University of Washington for a

Ways were indeed found. One of them was to prepare young men for military training by exercise in the Teutonic recreation of "Volksport" or "Wehrsport" (e.g., marching, calisthenics, map reading, and small-caliber target shooting). In fact, the army had begun to conduct widespread if secret Wehrsport training in 1924, and a plan was developed in 1925 whereby the Ministry of Interior would seemingly supervise the activity, while the army would retain actual control. Army direction supposedly ended in 1926, with various (largely nonmilitary) sport associations, notably the Deutsche Turnerschaft, continuing the activity with funds channeled through the Interior Ministry. This arrangement did not work out to army satisfaction.[51] But properly conducted, Wehrsport would reduce the time required for overt military training in the army. Then, in a war emergency, when treaty restrictions no longer counted, men who had had Wehrsport preparation could be inducted and made into soldiers, after a fashion, in a few weeks.

Another expedient was to give actual military training outside the army, detailing officers for this purpose. Such training could appropriately follow Wehrsport preparation, and at first it was intended only for the training of reserve officers, noncommissioned officers, and specialists. External training constituted a more clear-cut treaty violation than Wehrsport, and it would have to be kept more concealed than either Wehrsport or Grenzschutz activities; it was in a way an extension of Grenzschutz preparation to the more defense-minded elements in other, less defense-oriented areas of the country. For a long time it could only be carried out in closed rooms or in suitably out-of-the-way training areas. There was hardly any hope for a large-scale external

---

reference to a copy attached to [T2] 159/25 of 27 Apr 25, which indicates the approximate date: Wi/IF 5.422, 98/822602-64. The memo's contents were described in a lecture to a small group on 4 Apr 25: Prof. Emerson notes that the lecturer was a Capt. Beschnitt. The memo projected a need for 2,800,000 men, including a sixty-three-division field army, an air force, thirty-six divisions of Grenzschutz, and replacement troops. Of these men, only 180,000 would be Reichswehr or ex-Reichswehr, 160,000 of which could perhaps be used as instructors, if the armed forces were not required for other purposes during the training of the first successive wave —a consideration that made the army anxious to avoid domestic strife. The memo also brought out the need for political conditions favorable for military training and for the physical and moral "Ertüchtigung" of the people.

51 Vogelsang, *Reichswehr*, pp. 44-45; Stülpnagel Papers, N5/20, ltr. to Hasse, 26 Jun 25; RK, TA 68/27, 9 Feb 27, and TA 257/27, 9 Feb 27, K951/K248660-63; Vogelsang, "Neue Dokumente," pp. 405, 423. Note also, Col. M. von Wiktorin, "Wehrhafte Jugenderziehung," *Wissen und Wehr*, 11 (1930), 613-623. The 1923 Gessler-Severing agreement implicitly recognized the need for a physical-training program for youth. The later voluntary labor service or Arbeitsdienst also served for physical training, but I have not found much evidence of official army interest in this approach to premilitary preparation. Schleicher stated in 1932 that, in principle, labor service activity had to be kept separate from Wehrsport: RK, Cabinet Protocol (i.e., record of cabinet discussion), 18 Jun 32, 3598/D790242.

training program under the foreign and domestic conditions of 1929, although later this kind of program was to develop strongly.[52]

A third mode of training was that of the so-called "militia," which was the name for a plan developed by Erich von Manstein and others for giving Wehrsport graduates a short period of actual peacetime military training, making them into trained reservists. There were precedents of sorts for this in the Krümper system and, more recently, in the temporary volunteers" (*Zeitfreiwillige*) trained during the 1923 Ruhr crisis, but no training of this kind was attempted in the later 1920s. Groener spoke to Brüning in early March 1930 about youth training and "a people's army on the Swiss model." In May 1930, a Foreign Ministry official attended an army war exercise and during his many convivial talks with high-ranking officers he learned of their hope that the army of the future would include a standing force of 200,000 to 300,000 men, along with "a militia of several 100,000 citizens, who would be called up for exercises at certain intervals." The Reichswehr officers did recognize clearly that this program could not be considered under existing political conditions at home and abroad.[53]

The militia scheme was in fact a complicated deception, one that has deceived most commentators right up to the present.[54] A program

[52] Groener's 16 Apr 30 directive on the tasks of the Reichswehr (to be discussed below) stated that the conditions of personnel mobilization depended on "the date of beginning, the nature, and the extent of the training outside the Reich army," implying that none of these things had yet been settled. In this statement, Groener may have also been referring to plans for a militia (also discussed below); although the militia plans came to provide for training *within* the army, this may not yet have been decided in early 1930. See ML, PG 34072, Reichswehrminister 14/30, att., 16 Apr 30, reel 30/917. On nature of external training and relation to Wehrsport: OKW, Wi/IF 5.535, Chef HL TA 153/33, atts. from Finance Ministry, 28 Feb 33, 73/744786, 744792; Wi/IF 5.498, TA 251/31, 20 Mar 31, 109/837591; TA 784/31, 24 Jul 31, 109/837600; T2 III A 1207/31, 24 Nov 31, 110/837779-80; Hans-Jürgen Rautenberg, *Deutsche Rüstungspolitik vom Beginn der Genfer Abrüstungskonferenz bis zur Wiedereinführung der allgemeinen Wehrpflicht 1932-1935* (Doctoral diss., University of Bonn, 1973), pp. 219-220, 233. On Rautenberg's interpretation, cf. note 61 below.

[53] ML, Shelf 5939, A II a 12 and att., 15 Jan 32, reel 17/762-763; Brüning, *Memoiren*, pp. 159-160; Schäffer Diary, 17 Jun 32; AA, Memo by Schlimpert, 3 Jun 30, K6/K000480. A Major Mummentey also contributed to the militia plan; Mummentey proposed four-month service, Manstein three-month. Henceforth, where the word "militia" is used in this study in reference to official German military planning, it is to be understood as referring, not to a separate "'citizen army' as distinguished from a body of mercenaries or professional soldiers" (as in the pertinent OED definition), but to the training of men for relatively short periods (three months to a year) within regular units.

[54] Historians often write of various plans for a militia, stemming from Groener, Schleicher, and Reichenau, supposedly differing in method of recruitment, but all involving an entity separate from the professional army. See, for example, the relatively recent, in many ways brilliant, essay of Andreas Hillgruber, "Militarismus am Ende der Weimarer Republik," in his *Grossmachtpolitik und Militarismus im 20. Jahrhundert* (Düsseldorf: Droste, 1974), pp. 37-51, esp. p. 44. So far as I can deter-

to train reserves in peacetime could not be concealed entirely, and it would obviously run counter to existing treaty provisions; the promoters hoped, however, to make it appear innocuous, and even to gain acceptance for it at the coming World Disarmament Conference. In 1932, foreign governments were told that Germany wanted to form a militia on the Swiss model; service in it would be for three months, its projected strength was usually given out as 40,000, and its purpose would be "the maintenance of internal order as well as frontier and coastal protection." Sometimes the militia would be described as a volunteer force, sometimes as a force raised by conscription; in Germany it was presented as a less-expensive substitute for the professional army forced on the country in 1919—an appealing point when the depression had led to painful budget cuts in other areas.[55] With such small numbers and service so short, the force would surely not appear threatening to anyone.

But in actuality, this "militia on the Swiss model" amounted to a Krümper-Roon system. Unlike the Swiss army, the men in the projected German militia were to be trained in, and in case of war integrated with, an enlarged regular army. As with the legendary Krümper, the short service would make it possible to train a large number of men in a nominally small force. Indeed, since military experts calculated strength in terms of "average daily strength," or the total number of man-days in the year divided by 365, then 160,000 men (for example) could have been given three-month training in a year, and have been reported as 40,000. Each German three-month trainee would count as only one fourth of a man in average daily strength, while French conscripts, who served a year, would each count as one; thus Germany could have accepted a lower strength figure than France, and still have trained more reserves.[56] In prospective practice, the number

---

mine from contemporary documents, the militia was always seen in the army as expansion personnel for (and at latest after mid-1932 was planned to be trained in) regular army units. See also Appendix.

[55] A British diplomat was informed in 1931 of a desire for a militia force training about two weeks each year: *DBFP*, 2/III, No. 218; ML, Shelf 5939, Memo by Frohwein, 24 Sep 31, reel 17/438-444. A small-size militia was mentioned to an American assistant military attaché early in March 1932 (Georges Castellan, "Le Réarmement clandestin du Reich" [doctoral thesis, the Sorbonne, 1952], p. 28, note VI), through diplomatic channels to the British and Americans (according to the German record) in Bessinge in April 1932 (see Chapter Three), and to the French in August 1932 (see Chapter Four). Schleicher used the economy argument to Bernhard von Bülow before mid-April 1931 (Schäffer Diary, 17 Apr 31), in addressing officers in June 1932 (Vogelsang, *Reichswehr*, Annex 28), and in his 26 Jul 32 broadcast (see Chapter Four).

[56] See also Dülffer, p. 154, and MI, Shelf 5939, A II a 12 and att., 15 Jan 32, reel 17/762-763; the latter document makes the militia a separate organization with its own cadre, but I believe this to be a misunderstanding, probably due to naval

of German trainees would be somewhat restricted by the number of Wehrsport graduates available. Moreover, actual Reichswehr plans were not simply to put a large number of men through a three-month basic-training course, but rather to give the recruits three-month training in the ranks of regular units and then retain some of them to augment those units. Thus, aside from those who had three-month training only, the number of cadre and specialists would be enlarged. Plans were developed by November 1932 to induct 85,000 men each year from 1934 on, of which about 50,000 would serve only three months, while the rest would go on to serve three years or more. Although regulars and militia would total over 200,000, the average daily strength would have been just over 150,000.[57] Apparently the hope was that outside observers would construe this as the old Reichswehr, slightly augmented by 10,000 men, plus a 40,000-man militia force. But by 1938, it would have been possible to mobilize about 450,000 trained men.

The militia scheme meshed with a new mobilization plan that Manstein had proposed in late 1929. This plan would make the listing of potential reserves, the acceleration of rearmament, and the training of young men still more urgent than before. On April 28, 1930, after considerable deliberation within the Reichswehr Ministry, General Heye adopted this plan, which directed that the army prepare to mobilize twenty-one divisions on a territorial basis, with the regular personnel uniformly diluted by volunteer reservists, roughly in the proportion of two reservists for each regular. In contrast to previous arrangements, all divisions of the mobilized army, known as the "field army," were thus to have the same proportion of regulars and reservists, and so far as possible, the same armament. Until the rearmament program was farther advanced, there would necessarily remain some differences in armament between the field-army divisions in the East, which would have preference, and those in the West; for a time, four

---

unfamiliarity with army mobilization plans. Cf. ML, Shelf 5939, Reichswehrminister 134/31, 10 Apr 31, reel 17/535-536, in which Groener said he aimed for a defense organization providing "a short service militia" alongside "a sufficient number of officers and long service men for cadre and training"; also AA, "Militärische Grundgedanken über die deutschen Forderungen," [July 1931], 9095/H221692, given to Frohwein by Col. Kurt Schönheinz. On the possibility of actually training more men than France with a lower nominal strength: ML, Shelf 5993, Chef HL TA 16/31, 8 Aug 31, att., reel 5/1066.

[57] See esp. BA-MA, RH 15/49, TA 737/32, 7 Nov 32, and Annex 6; II H 228, T2 549/32, 15 Jul 32. Castellan, *Réarmement clandestin*, pp. 83-85, 334, 392-393, 407, quotes extensively from TA 737/32 and its Annex 6. A primary reason for training three-month men in regular units rather than in separate training battalions was the greater ease of mobilizing slightly trained men within the organization in which they had been trained.

of the field-army divisions to be formed in the West could only be armed in part. No doubt because army leaders wanted to provide for all the modern auxiliary services, the total number of men mobilized was to be not 300,000 but 450,000, without counting Grenzschutz, air forces, construction troops, and a "reinforcement" army (*Ergänzungsheer*), presumably made up of older or less thoroughly trained elements. Thus the requirements for reservists were expanding just when the potential reservists of the past, the veterans of the 1914-1918 war, were becoming overage. This scheme was to come into full effect on April 1, 1933, but it seems to have become the basis for preparations after its official adoption in April 1930. The existing four regular divisions in the West were given an option as to the date (up until April 1, 1933) when they would adopt the new mobilization procedure; apparently the three existing divisions in the East were to go over to it as soon as possible.[58] Probably the First Division, in East Prussia, had already virtually reached this stage of preparation, since it had long since listed potential reservists.

Manstein's plans for a militia and for mobilization carried quite a freight of hopes and implications. In the late 1920s, sober staff officers had maintained that, within the near future, Germany would only be able to mobilize fifteen or sixteen divisions, at least at the start of a conflict. Although a number of officers, and particularly Werner von Blomberg, then Chief of the Truppenamt, believed that Germany should prepare to fight a war on two fronts, it would be almost impossible to fight even Poland alone with sixteen divisions. But Manstein seemed to demonstrate that twenty-one divisions or more were attainable after all.[59] No doubt most staff officers wanted to believe this. Manstein's planning was a tour de force, however, and it strained the limits of what was practical, at least under the conditions then existing.

Militarily, Manstein's plans confirmed the army's commitment to a full national mobilization of materiel and manpower. And by adopt-

---

[58] Erich von Manstein, *Aus einem Soldatenleben 1887-1939* (Bonn: Athenäum-Verlag, 1958), pp. 110-116; Burkhart Mueller-Hillebrand, *Das Heer 1933-1945*, 3 vols. (Darmstadt: Mittler, 1954-1969), I, 18; BA-MA, II H 288, TA 453/30, 28 Apr 30, and TA 454/30, 28 Apr 30. In 1924, Joachim von Stülpnagel had already favored the expansion of the regular divisions over the creation of a separate, second-string force with its own cadre: Stülpnagel Papers, N5/10, Lecture to Reichswehr officers, February 1924.

[59] Carsten, *Reichswehr and Politics*, pp. 272-273; Post, pp. 152-157, 195, 199-201; Geyer, "Zweite Rüstungsprogramm," p. 128 (incl. notes 26, 28); Michael Geyer, "Militär, Rüstung und Aussenpolitik: Aspekte militärischer Revisionspolitik in der Zwischenkriegszeit," in Manfred Funke, ed., *Hitler, Deutschland und die Mächte* (Düsseldorf: Droste, 1976), p. 241. Post and Geyer hold that Blomberg was removed from the Truppenamt because his two-front thesis, involving defense against France in the West, was unacceptable to Schleicher.

ing these plans, the army rejected General von Seeckt's conception, that of leaving heavy fighting to a small, élite, professional force, supported by a separate, second-line, genuinely militia style organization. Presumably even the third-line reinforcement army would be used for replacements. Theoretically, the government retained the option of using the "march-ready" regular army alone, either for curbing domestic disorder, or for meeting incursions by partisans or by fleeing foreign troops. But the regulars were too valuable as cadre to be thrown away as ordinary private soldiers in a fight against a foreign enemy with a vast numerical superiority, and the reduction of this margin of superiority was the main object of the plans. Even a domestic use of the regular army alone might alienate some potential trainees, disrupt training activities, and risk the confusion of shifting from an initial domestic deployment of the regular army to a mobilization of the field army to meet a foreign enemy.[60]

A striking—and politically fateful—feature of the Manstein plans was their dependence on Wehrsport. Wehrsport preparation was essential in the first place for the success of the three-month militia-training program; without Wehrsport preparation, three months would be quite insufficient. But also, and more immediately, as long as reserves had not been accumulated through the militia—and that would take years, even after international approval was obtained— the mobilization of Manstein's twenty-one-division field army would require the induction of large numbers of volunteers, who would have to be given rudimentary training before they went into action. Beginning in 1931, the army conducted trials in ultra-short-term training in its own ranks. As an example, a young officer received the assignment in April 1932 of taking a company two-thirds composed of new Reichswehr recruits and preparing it for defense duties in two weeks; the task was of course exactly the one he would have in the event of mobilization. For this kind of training, prior Wehrsport preparation was particularly desirable.[61]

[60] OKH, H1/326, TA 275/30, 28 Apr 30, 395/6398217-21. On "reinforcement" army: BA-MA, RH 15/49, Att. 6 (pp. 9-10) to TA 737/32, 7 Nov 32. On use of "march-ready" regular army: ML, PG 34072, Reichswehrminister 14/30, 16 Apr 30, reel 30/913-914. The eventuality of fleeing troops had occurred in 1920, during the Polish-Soviet war. On the disruption that might follow on internal use of regulars, see Hammerstein's comment of early 1932 (quoted later in this chapter, in Section 9): Vogelsang, "Neue Dokumente," p. 422.

[61] On emergency training: OKW, Wi/IF 5.498, TA 628/31 and att., 17 Sep 31, 110/837637-52; BA-MA, II H 226, Chef TA 140/30, 15 Mar 30; II H 285, TA 50/31, 31 Jan 31; Hans Meier-Welcker, "Aus dem Briefwechsel zweier junger offiziere des Reichsheeres 1930-1938," *MGM*, 14 (1973), 74 and note. Contrary to Rautenberg, *Rüstungspolitik*, p. 220, TA 628/31 and att. refer to the emergency or ultra-short-term training of new recruits within the army.

Political Implications of Reichswehr Plans—45

The dependence on Wehrsport was politically fateful because a large-scale Wehrsport program would in practice mean dependence on the Wehrverbände. It is true that Reichswehr leaders weighed the possibility of official, perhaps compulsory, Wehrsport programs, and the Reichswehr would later establish an organization for at least officially guiding and supporting the Wehrsport work of the Wehrverbände.[62] But under existing foreign and domestic conditions, only the Wehrverbände could quickly provide significant numbers of willing trainees and Wehrsport instructors, at relatively low cost and with some security against undesirable disclosures. Landesschutz officers had always looked largely to the Wehrverbände for veterans or willing youths for the Grenzschutz, or for emergency reinforcements (*Marschfreiwillige*) to make the regular army over-strength in a domestic crisis, and programs for training outside the army in nonborder areas inevitably drew upon the military enthusiasts who were also trusty Wehrverbände members. When it was a question of finding young men for mobilization in the field army, the army was bound to rely on those defense-minded youths who had undergone Wehrsport or other training in the Wehrverbände. And indeed, while waiting for the militia system to be authorized, the army began to expand its covert external military training program, and this again led to ties with the paramilitary groups. The Chief of the Truppenamt would note in March 1933 that, under existing conditions, the bulk of the "war volunteers" for filling out the twenty-one divisions were to come from the Wehrverbände.[63] Most officers favored the Stahlhelm and its affiliates over

[62] See Stülpnagel Papers, N5/20, ltr. to Hasse, 26 Jun 25; OKW, Wi/IF 5.498, ltr., Groener to Wirth, 1 Aug 30, 109/837584-88; Brüning, *Memoiren*, pp. 554-555. On the official "roof" organization for guiding and supplementing the Verbände work: Vogelsang, *Reichswehr*, pp. 137, 160, 231-232, 285-287; Thilo Vogelsang, "Der Chef des Ausbildungswesens (Chef AW)," in *Gutachten des Instituts für Zeitgeschichte*, II (1966), 146-147.

[63] The chief of the Truppenamt, General Wilhelm Adam, made his report to General von Blomberg in mid-March 1933: National Archives, Nuremberg Trial 10 (Krupp Trial), Defense Exhibit 62, Affidavit quoting document, 5 Mar 48; see also Chapter Seven, Section 3. Probably many recruits for the regular Reichswehr itself had come from the Wehrverbände; suspicion of this was part of the SPD complaint at the end of 1926. On the pursuit of Wehrsport by private organizations and its role in the preparation of volunteers for mobilization, see also OKW, Wi/IF 5.498, TA 784/31, 24 Jul 31, 109/837595-600; TA 570/32, 25 Jul 32, 109/837571-72. BA-MA, RH 15/49, TA 737/32, 7 Nov 32, Annex 6, says that militia volunteers were to be drawn from the "segment ready and willing to support defense [*wehrwilligen Teil*]" of the population, a euphemistic way of referring to the right-wing Wehrverbände. The same document stated that paramilitary training was to be an obligatory precondition for three-month militia training, and pointed to the "financial and material" obstacles to training all fit men of military age. This document added that internal changes "(for example, elimination of the Wehrverbände)" might make the (official) training of all fit youth extremely desirable, and if external

other organizations, but after 1930, the Nazi SA would mushroom, attracting large numbers of young eighteen- to twenty-year-old men, the very people the army wanted as Wehrsport trainees. To renounce such a repository of defense-mindedness was more than the army could have borne. As a result, preparations for mobilizing the field army would require the cooperation of some of the same paramilitary groups that the regular army might otherwise be asked to suppress. Aside from military sympathies with the right, military technical considerations would later lead the army to seek an integration of the Nazis into the official system.

In early 1930, staff officers may not have realized that an expanded Wehrsport program would lead to dependence on the Wehrverbände, and they could hardly have foretold the growth of the Nazi SA. They saw that the emergency mobilization plans might be risky militarily, but General Heye wrote: "It is necessary—on psychological and pedagogical grounds, as well as with a view to the later future—to give the long-service peacetime army the high goal of a stronger Wehrmacht."[64] Despite risks and problems, the idea of national mobilization followed naturally from a refusal to be content with a Versailles army and a state of defenselessness.

Indeed, Heye's reference to "the later future" suggests more than defensive goals, and we have seen how, in 1927, he urged that Germany should prepare to throw her weight into the scales in conflicts between other governments. This was in fact one objective of the 1930 plans. On April 16, 1930, a directive from Groener to the Chiefs of the Army and Navy Commands reviewed the possible tasks of the Reichswehr, and envisioned the possibility, not only of using the twenty-one-division army to conduct a successful defensive war, but also of employing it—under favorable diplomatic conditions—to intervene at the behest of other powers whose support would benefit Germany, or even to wage war on Germany's "own free initiative." The document did rec-

---

conditions permitted that, "a new situation would be created." In other words, Wehrverbände preparation was a key element under existing conditions, although official preparation, if permitted by other powers, could replace that and permit the training of the whole population. See also *DBFP*, 2/III, No. 113 (p. 141). On use of *Marschfreiwillige*: ML, PG 34097, TA 25/33, 18 Jan 33, reel 31/923; WKK, WK VII/2670, Marschbereitschaft of I/I.R. 19, May 1931, 55/257. OKW, Wi/IF 5.335, Att. to Finance Ministry order of 20 Feb 33, 73/794792, shows that external training had been authorized by the guidelines of 26 Apr 29; WKK, WK VII/2670, WKK VII Ib 711/32, 9 Jun 32, 55/477, refers to a steady increase in the importance of external training.

[64] OKH, H1/326, TA 275/30, 28 Apr 30, 395/6398217-21. In November 1935, a lecturer at the War Academy observed that it was fortunate this system had not been put to the test (OKH, H1/324, Lecture by Capt. Scherff, 12 Nov 35, 395/6397884-85).

ognize that even a successful defense of Germany presupposed an enemy heavily engaged elsewhere, or timely foreign intervention on Germany's behalf, and that such a defense would also require carrying out personnel and supply preparations that had not so far been completed. But despite the warnings of German diplomats, Reichswehr leaders still hoped, in the face of the realities of the late twenties and early thirties, to secure a favorable diplomatic constellation, and to use armed force to support an "active" policy.[65] The idea of joining others to gain their support could lead readily to an old German idea, which Hitler would revive, of developing armaments to become more *bündnisfähig*, more attractive as an ally.

5. BRÜNING'S ECONOMIC AND POLITICAL IDEAS

Brüning was an austere and courageous bachelor in his mid-forties. His memoirs suggest that, as a scholarly, middle-class Catholic, he felt himself condemned to remain forever an outsider to the Prussian Protestant aristocracy, and that his response was to make himself, in some ways, more Prussian than the Junkers. Despite physical disabilties, he had gained a commission and commanded a machine-gun section during the war. During the French occupation of the Ruhr in 1923, Brüning, on behalf of the Ministry of Labor, collaborated with the Reichswehr in guiding passive resistance. He was elected to the Reichstag in 1924, where in the years that followed he immersed himself in financial problems, including those of reparations.[66]

In 1929 and 1930, reparations became the center of national debate. In return for the evacuation of the Rhineland, Germany accepted the Young Plan; this was designed by its sponsors to be a final settlement

---

[65] ML, PG 34072, Reichswehrminister 14/30, att., 16 Apr 30, reel 30/917. Post, who first called attention to this document, discusses it in his pp. 197-198, 231-234, 237, 328, 333, 341. On the more sober appraisal of the diplomatic situation by the diplomats, see Post, esp. Chapter Six. A military analysis of early 1929 had concluded that in general, either an independent war with a great power or, given existing alliances, a war with a small state was unthinkable, and that Germany could only have a chance of success in a war in a highly favorable political constellation and in alliance with other powers, conditions not foreseeable at that time; Germany might nevertheless wish to, and have to, fight Poland (presumably without hope of success): Geyer, "Zweite Rüstungsprogramm," p. 128 and note 26; Geyer, "Militär, Rüstung und Aussenpolitik," p. 241. The basic conclusions were not too different from those of the Groener directive, but the outlook in early 1929—before the Manstein militia and mobilization plans—was much less optimistic. In view of the plans and decisions of 1930, I do not think that this 1929 analysis can have been (as Geyer suggests) the military-political study referred to in December 1931 in Doc. No. 2 of his "Zweite Rüstungsprogramm" (p. 146 and note 169).

[66] Gottfried Treviranus, *Das Ende von Weimar* (Düsseldorf: Econ, 1968), pp. 28-33, 57-58; Brüning, *Memoiren*, pp. 90-95, 108, 116, 129.

of the reparations levied against Germany by the Treaty of Versailles, transforming them in large part into a regular debt to private creditors to be paid on a fixed long-term schedule. When the new settlement was proposed, it evoked bitter denunciations from the German right, which charged that the German nation was being placed in serfdom for generations. Alfred Hugenberg, the leader of the DNVP, and a notoriously obstreperous and narrow-minded individual, waged a propaganda campaign against the reparations "tribute," and his Nazi rivals on the far right reaped much of the political benefit. Reparations were widely thought to explain Germany's economic problems, as a monstrous inflation was now succeeded by a ferocious depression. The impression began to spread that the nation was undergoing a crisis which could only be resolved by the introduction of a more authoritarian system; among other things, such a system might eliminate the duality of Reich and Länder, or, in other words, curb the power of Prussian Social Democracy.[67]

As a trained economist and a conservative politician, Brüning had definite views about the financial policy to be followed. According to his memoirs, he believed that a heavy influx of foreign capital had produced a false prosperity and a false impression of Germany's ability to pay reparations. Past governments had depended on foreign loans rather than really balancing the budget, and this had made Germany vulnerable to financial pressure from abroad. Brüning's own economic orthodoxy and love of Old Prussian austerity were such that he probably would have wanted a balanced budget and deflation in any case, but he also thought that such policies would convince foreign lenders that Germany was making an honest effort to pay reparations from her own resources, and demonstrate the impracticality of large-scale intergovernmental payments, not only for Germany but also for the world economy. German, and then world, prices would fall, and a severe financial crisis would follow, but even this would serve German ends, leading British and American private creditors to use their influence to end reparations and debts. Brüning's calculation worked, in that Anglo-American anxiety over private debts and the security of

[67] The most comprehensive survey of the sociopolitical crisis is in Bracher, *Auflösung*, first section ("Probleme der Machtstruktur"); see now also his *Die deutsche Diktatur* (Cologne: Kiepenheuer & Witsch, 1969). Eyck, II, Chapters 6-8, and Vogelsang, *Reichswehr*, pp. 65-71, are other notable analyses. Maier, *Recasting*, pp. 45, 250-252, 358-364, 445, 482-483, 513, 515, points out that different social groups had actually fared differently in the great inflation, with many middle class elements suffering less than they believed, and also that reparations were not strictly speaking the cause of inflation. But (as Maier says on his pp. 250, 358n) most German experts blamed reparations, and surely this was a widespread belief.

banks, enhanced by a deliberately alarmist declaration by Brüning's government, encouraged President Herbert Hoover to propose, in June 1931, a one-year moratorium on intergovernmental war debts and reparations. Brüning still had the task of making the moratorium permanent, but he thought it might be possible to link up an ending of reparations with German arms demands. His idea was that the way to enable the administration in Washington to annul the American war debts, and thus release the European Allies from their principal need for reparations, was for the Europeans to reach agreement on a disarmament convention (meaning the relative disarmament of France vis-à-vis Germany)—thus giving Hoover something positive to show "the emotional American people."[68]

British and American political leaders thought that Heinrich Brüning was an apostle of financial responsibility and political moderation. His memoirs, along with recent research, now cast him in a somewhat different light. For example, he writes that he separately told Hitler, Hugenberg, and the Reichswehr chieftains that he wanted sharp opposition from the right at this stage so that he could exploit this in diplomacy, evidently by making himself appear as the moderate German to whom concessions had best be made; he told Hitler that he hoped to collaborate later with the right, including the Nazis, on a change in the constitution, and that personally he hoped to bring about a monarchist restoration. In the pursuit of reparations revision, Brüning and his officials deliberately sought to encourage apprehension among foreign creditors, and to make the most of Germany's economic difficulties. Brüning's deflationary policies and his calculated incitement of a reparations crisis in June 1931 doubtless contributed to the severity of the depression, not only in Germany, but throughout the world—and although this was probably no part of his calculation, the depression was to affect seriously the ability of other countries to unite and to resist German political and military initiatives. Brüning was playing for political stakes which he regarded as more important than economic well-being. In his memoirs, he speaks proudly of the second standstill agreement, which barred the withdrawal of foreign credits from Germany, as putting his country in such a position that

[68] Brüning, *Memoiren*, pp. 138-139, 192-194, 222-223, 228-232, 274, 377, 401, 412-413, 431, 433-435, 437, 491-492, 556-557. See further Wolfgang J. Helbich, "Between Stresemann and Hitler: The Foreign Policy of the Brüning Government," *World Politics*, 12 (1959), 22-44, and the same author's *Die Reparationen in der Ära Brüning* (Berlin: Colloquium, 1962), also Edward W. Bennett, *Germany and the Diplomacy of the Financial Crisis, 1931* (Cambridge, Mass.: Harvard University Press, 1962), and Werner Link, *Die amerikanische Stabilisierungspolitik in Deutschland 1921-1932* (Düsseldorf: Droste, 1970).

if the Treaty of Versailles were not revised as it wished, it could "push over, like Samson, the pillars on which prosperity and order for a generation depended."[69]

In domestic affairs, Brüning was a thoroughgoing conservative, and an advocate of limiting the powers of parliament and of the Länder. He also relied on the presidential decree power under Article 48. Nevertheless, he avoided an open confrontation with the left, and tried to achieve his aims by budgetary, bureaucratic, and procedural tactics. There was a contrast between this approach, which avoided a constitutional struggle, and that involved in conducting a nonparty government ruling purely by presidential decree, the other procedure Schleicher had considered. The election of September 1930 was a blow to Brüning's and Schleicher's hopes of wooing major portions of the right away from Hugenberg. Instead of a strong and responsible right developing under the leadership of Treviranus and similar figures, the radical-rightist Nazi party suddenly increased its number of seats from 12 to 107. At this stage, the Bendlerstrasse, disturbed by Nazi influence within the army's ranks, and not yet expecting to pay a political price for Wehrsport members, had no interest in sharing power with the Nazis. The thoughts of Schleicher and Groener turned instead with no great hesitation to the idea of governing without the Reichstag. If parliament became unworkable, this could only increase the Reichswehr's influence, and Groener (in a speech at autumn maneuvers) said that not one move could be taken henceforth in German politics without the Reichswehr playing a role.[70]

It turned out, however, that the parliamentary system was still not completely unworkable after all, as the Social Democrats decided that their wisest course was to "tolerate" the Brüning cabinet, lest it be replaced by something worse. Brüning made little concession to the SPD leaders in return, beyond keeping in touch with them and keeping them generally informed, but he found them more reasonable and reliable than the gentry on the right, and the SPD was very valuable to him personally. Its support, and Brüning's consequent ability to govern to a significant extent through the Reichstag, helped him in his efforts to convince foreign governments, particularly the British and American, that he represented the liberal and constitutional ele-

---

[69] Brüning, *Memoiren*, pp. 192-194, 373, 379, 485, 491-492; see also Bennett, *Diplomacy*, pp. 7-8, 114-116, 118-122. Brüning was particularly successful in winning over the American Secretary of State, Henry L. Stimson; see Link, *Stabilisierungspolitik*, pp. 518-522.

[70] Hermann Pünder, *Politik in der Reichskanzlei: Aufzeichnungen aus den Jahren 1929-1932*, Thilo Vogelsang, ed. (No. 3 of the *Schriftenreihe der Vierteljahrshefte für Zeitgeschichte*) (Stuttgart: Deutsche Verlags-Anstalt, 1961), p. 79; Vogelsang, *Reichswehr*, p. 95, Annex 6.

ments in Germany, as opposed to the extreme nationalists. Perhaps even more important, although Brüning makes no direct acknowledgment of this in his memoirs, Socialist support saved him from complete dependence on President von Hindenburg and the Reichswehr.

Neither Hindenburg nor the Reichswehr seems to have objected at first to Brüning's being tolerated by the Social Democrats. But the presidential household and the army leadership gradually became convinced that the stream of history was moving to the right, and they became impatient with Brüning's apparent unwillingness to come to terms with rightist elements. Beyond this, Schleicher at least seems to have suspected, by August and September 1931, that Brüning was considering an open alliance with the Social Democrats, including the detested Prussian party leaders. Later, in the last months of the Brüning government, the general decided that Brüning, and now Groener too, would never weed out the antidefense elements in government administration and overcome the power of the Land governments, with their indifference to defense interests.[71] In the light of the long conflict between the army and the Socialists, it is clear that Schleicher wanted a revamping of government personnel and attitudes from top to bottom. But there were also inherent conflicts between the army's developing plans and the natural policy of any moderately conservative government, such as Brüning was trying to conduct. These differences involved strategy at the Disarmament Conference, preparations for the second armament program, and conditions for premilitary training. After Brüning's fall, Schleicher would seek to initiate changes in all these areas.

6. THE QUESTION OF DISARMAMENT CONFERENCE TACTICS

Once the Disarmament Conference finally met in Geneva, it would evidently have an effect on Germany's power status. But although Brüning had a "grand design" dealing with disarmament as well as reparations, he usually devoted little direct attention to disarmament and rearmament, being mainly preoccupied with the more immediate (and to him, more familiar) matters of reparations policy, domestic economic and financial problems, and political infighting. Brüning argued that Germany could only rise to a level of equality if other nations disarmed to some degree, since economically and financially she would never be able to catch up with their existing levels of armament. Although he expressed the hope that the West European powers might be obliged to disarm in order to inspire the American public

[71] See Vogelsang, *Reichswehr*, pp. 124-131, 160-180, 185, 190-191, also the same author's *Schleicher*, pp. 67-68.

and induce the United States to forgo war debts, Schleicher was skeptical of the feasibility of linking disarmament with reparations, and Brüning may not always have believed in the possibility himself; on at least one occasion, in January 1932, he accepted the idea that there was no serious chance for an American remission of war debts.[72]

The election results of September 1930 naturally incited the German government to adopt more forceful language on the disarmament question, but the government also had to keep reparations policy in mind. General Kurt von Hammerstein-Equord, the new Chief of the Army Command, complained at a meeting at the end of October that Germany lacked the most elementary security, and that the illegality of German armament had bad effects on both foreign and domestic politics; domestically, because (owing to the secrecy) the efforts of the armed services to improve defense capabilities were not sufficiently recognized by the German people. He indicated that, if grand policy required, the situation could be endured for one more year, but he considered it extremely unsatisfactory. Nevertheless, he agreed with civilian leaders that Germany should avoid calling for rearmament. Brüning pointed out that world opinion would now accept stronger words than it would have six months before, but also that it might be necessary in the coming year to seek an international investigation of Germany's capacity to pay reparations, and that the investigators would probably object to German arms expenditures. Julius Curtius, the Foreign Minister, affirmed his support of illegal rearmament, but he observed that 1931 would probably be dedicated to the reparations question, while the armament question could perhaps be placed in the forefront in 1932. It already appeared that the Disarmament Conference would not meet before late 1931 at best, and in fact the League Council (in January 1931) set the date for February 2, 1932.[73]

This interval gave the German government plenty of time to consider the strategy to be followed. The basis for any German position was that the Allies had committed themselves in 1919 to disarm as well: the preface to Part V, the section of the Versailles treaty limiting armament, stated that German rearmament was being ordained in or-

[72] Brüning, *Memoiren*, pp. 194, 228-229, 431, 433; also Wilhelm Vernekohl and Rudolf Morsey, eds., *Heinrich Brüning: Reden und Aufsätze eines deutschen Staatsmanns* (Münster: Verlag Regensburg, 1968), pp. 171-176; RK, Note on discussion at Reichskanzlei, 5 Jan 32, 9242/E650037.
[73] RK, Record of chiefs' meeting, 30 Oct 30, 3642/D810816-20; *DBFP*, 2/I, No. 331; United States, Department of State, *Foreign Relations of the United States* (Washington, D.C.: G.P.O., 1862-), 1931, I, 476 (series referred to hereafter as *FRUS*); Arnold J. Toynbee et al., *Survey of International Affairs* (Oxford: Oxford University Press, 1925-1971), 1931, p. 279 (series referred to hereafter as *SIA*). The October 30 discussion took place under the immediate impact of a critical resolution from the Reichstag Foreign Affairs Committee (*DBFP*, 2/I, No. 335n).

der "to render possible the initiation of a general limitation of the armaments of all nations"; Article 8 of the League Covenant called for "the reduction of national armaments to the lowest point consistent with national safety"; and in replying on June 16, 1919, to German comments on the Versailles treaty text, the Allied powers had declared that the general disarmament made possible by German disarmament "will be one of the first duties of the League of Nations to promote."[74] The Germans concerned with the question did not believe that the other powers would in fact disarm to the level prescribed for Germany, and they inferred that such a failure would entitle Germany to rearm. German policy-makers seem to have recognized, almost instinctively, the diplomatic importance of how proposals are framed—the desirability, that is, of actually begging the question by so stating it that it can only be answered in one way. All German officials agreed on the way the disarmament—or rearmament—question should be posed. Rather than ask for rearmament, they would ask for "equality of rights," which could be interpreted in a number of ways: e.g., that Germany must be treated generally on terms of equality with other powers; that German arms and the arms of other countries must all be limited in the same convention for the same length of time; that Germany should be permitted whatever weapons were permitted others, that is, qualitative equality; or that, quantitatively, German armament and effectives should be on a par with those of, say, France. At a minimum, one thing was always meant by equality of rights: Germany could not remain the only major nation subject to limitation. German soldiers and diplomats agreed on demanding equality of rights in this sense. They concurred in the demand for equality of status, meaning that there should be an end to Part V, so that Germany would have the same right to determine the shape of her forces as other governments, and possess the same weapons under the same convention. There was agreement that German strategy should begin by demanding the disarmament of others. But the diplomats did not agree to demands, such as for quantitative equality, which would reveal an intent to rearm. On such demands, differences would arise repeatedly from 1931 until October 1933.

On March 16, 1931, under the guise of propaganda proposals, Groener sent to Curtius a statement of what Germany's goal and strategy should be at the Disarmament Conference. This statement maintained that disarmament restrictions applying to Germany alone (i.e., Part V of the Treaty of Versailles) must disappear, that the disarmament convention must permit Germany the defense organization she

[74] The pertinent part of the note of 16 Jun 19 is contained in *FRUS, Paris Peace Conference 1919*, VI, 954.

desired, that the one-sided demilitarization of the Rhineland must be eliminated, and in particular that Germany must have parity of security, meaning at a minimum parity in forces with France, at whatever level might be established for the French army. No halfway solution was acceptable, as this would mean a voluntary reacceptance by Germany of one-sided restrictions, under which the country would then be bound for the future.[75] Curiously, Groener and Schleicher appear to have said nothing about these demands when they met with Brüning and Curtius two days later; they did not then question an assertion by the chief permanent official of the Foreign Ministry, Bernhard von Bülow, that legal inferiority (*Deklassierung*) was more onerous than the actual limitation of effectives. On concrete objectives, they spoke only of shortening the term of service, reducing (!) the number of professional soldiers, and training a militia alongside them, while their suggestion that an annual budget of 800 million would serve better than 700 million gave Brüning a deceptively modest picture of their probable future demands. They seem, indeed, to have tried to shelter him from full knowledge of their plans. Brüning himself, at this meeting, supported the view that Germany should "uphold as long as possible the thesis of disarmament and only let the German wishes [for rearmament] be perceived at the last possible moment."[76]

Groener's statement on Germany's goal and strategy was not left unchallenged, however. Curtius replied to it in a letter of March 23, expressing "essential" agreement but urging that quantitative demands be deferred until later, that the question of the demilitarized zone be likewise postponed, and that there should be no early disclosure of the policy Germany would follow in the event of a failure of the Disarmament Conference.[77] Foreign Ministry officials, such as

[75] The statement of goals and strategy (*Ziel und Wege Deutschlands auf der Abrüstungskonferenz*) is in Vogelsang, *Reichswehr*, Annex 15; as Dülffer points out (p. 154n), Vogelsang's dating of the document is incorrect. Cf. ML, Shelf 5939, Reichswehrminister 111/31, 16 Mar 31, and att., reel 17/523-526.

[76] RK, Memo on discussion of disarmament questions, 19 Mar 31, 3642/D811005-09; cf. Dülffer, pp. 153-154. While Groener and Schleicher evidently hesitated to describe the full scope and cost of Reichswehr plans, the Foreign Ministry representatives were probably preoccupied, on their side, with immediate diplomatic questions. Curtius needed this meeting to authorize a statement on German disarmament policy that he was to give to the British ambassador the same day, along with an acceptance of a British invitation to him and Brüning to pay a visit to England. See *DBFP*, 2/I, Nos. 355, 356; AA, Memo by Curtius, 19 Mar 31, 3154/D666573-76; Bennett, *Diplomacy*, pp. 87-89. Important cabinet deliberations on proceeding with the customs union with Austria would also take place later on the eighteenth (Bennett, *Diplomacy*, pp. 54-56); Brüning was lukewarm about this plan, and Curtius and Bülow may have thought that this was not the time to embark on a full-scale debate over Groener's statement of position.

[77] ML, Shelf 5939, ltr., Curtius to Groener, 23 Mar 31, reel 17/531-533. On Groener's rejoinder, see note 81 below.

Albert Frohwein, Ernst von Weizsäcker, Karl Schwendemann, and State Secretary Bernhard von Bülow would continue to uphold these positions.[78] Bülow, a nephew of the imperial chancellor of the same name, son of the commander of the Second Army in the August 1914 advance, and a Brüning appointee, was in practice the chief civilian policy-maker on disarmament questions. He was a strong nationalist, like Brüning, but he also resembled Brüning in his belief in diplomatic finesse and maneuver. He was opposed to surprise moves in diplomacy, and to flat statements of nonnegotiable demands. Holding that the main objective was to get rid of the discrimination of Part V, he thought that it was unnecessary to press for an early recognition of parity, a course that would make Germany appear to be seeking rearmament; the initial convention would only last about five years, and anyway, Germany would have little money to spare for rearmament during that period.[79] Fundamentally, Bülow and his officials believed that Germany could not afford to flout foreign opinion.

On the other hand, the army officers involved with the disarmament question, principally General Werner von Blomberg and Colonel (soon General) Kurt Schönheinz, were haunted by the fear that a new convention might give permanent sanction to an inferior military status for Germany, particularly as to strength in effectives. As matters stood, German spokesmen could claim that the Treaty of Versailles was a "Diktat" and that the ex-Allied governments were failing to honor their promise to disarm. But if Germany freely signed a new general agreement, these arguments would collapse; in order to get a German signature on a convention perpetuating a fundamental inferiority for Germany, other powers might offer tempting concessions on particular points. Blomberg and especially Schönheinz insisted that Germany must demand parity from an early stage and refuse any partial solutions; German negotiators should not directly ask for increases, but simply assert a right to whatever level of national security France insisted on for herself. More generally, they believed that a recognition of Germany's full equality of rights should be a precondition for negotiation; thus they wanted to predetermine all the later practi-

---

[78] Other expositions of Foreign Ministry views (aside from Bülow's) in 1931 are in ML, Shelf 5939, ltr., Curtius to Groener, 6 May 31, reel 17/540-541; Memo of discussion, 2 May 31, reel 17/566-573; Draft delegation guidelines, [18 Dec 31], reel 17/733-737; AA, Memo by Weizsäcker, 2 Mar 31, 4604/E191096-104. For later statements of views, see esp. Chapter Ten, Section 4.

[79] For early statements of Bülow's views, aside from their expression at the 18 Mar 31 meeting, see AA, Record of meeting, 29 Aug 31, 4604/E191203-06; Memo by Bülow, 7 Jan 32, 4604/E191251-52. He frankly described many of his ideas to French ambassadors Pierre de Margerie and André François-Poncet: AA, Memo by Bülow, 30 Jan 31, 3154/D666439-41; RK, Memo by Bülow, 25 Feb 32, 3642/D811614-20.

cal decisions. This policy might lead to severe strains at the start of the conference, but they thought it preferable to face this, rather than to settle for less. Military thinking was probably influenced by regrets that Germany, by signing the Treaty of Locarno, had freely reconfirmed the demilitarization of the Rhineland; there was probably also an awareness of the bitter struggles over parity that had characterized naval treaty negotiations between other states. At the same time, unless Germany soon obtained recognition of her right to larger forces, the army's plans for an unconcealable amount of rearmament and expansion might lead to severe problems, perhaps to foreign intervention, while Germany was still relatively disarmed. Even Blomberg did not advocate an open demand for German rearmament, lest Germany be blamed for a collapse of the conference.[80]

The conflict of military and diplomatic views was partly kept within bounds by agreement on the fundamental aim of ending the Versailles system, as well as on some lesser matters: at the March 18 meeting, for example, Bülow and Schleicher both favored cautious soundings with the French, rather than leaving the disclosure of German desires to the latest possible moment. The conflict was also confined because the army was pretending not to plan any serious rearmament, and would not reveal to the diplomats its more far-reaching plans, as for example to prepare to mobilize a field army of 450,000 men, plus Grenzschutz. And Groener did not at this stage rigidly uphold the military view on negotiating tactics.[81] Despite all, the two ministries did jointly launch a formidable propaganda campaign. This had the support of almost all respectable German opinion: a list of professors and others attend-

[80] The statement of goals and strategy forwarded by Groener on 16 Mar 31 (see note 75) was prepared in Schönheinz's office, the League of Nations Section (Group Army) (Völkerbundsabteilung Gruppe Heer); most of the Reichswehr position statements on arms questions were drafted in this office. Blomberg was to be the chief military delegate at the conference, as well as divisional commander in East Prussia, until he became Hitler's Minister of Defense. The basic Reichswehr position dated back at least to 1926: *ADAP*, B/I/1, No. 144 (esp. p. 347). Other pre-conference statements of Blomberg's and Schönheinz's views: ML, Shelf 5939, Memo of discussion, 2 May 31, reel 17/566-567; Note for the files by Schönheinz, n.d. [probably June 1931], reel 17/542-543; Memo of discussion, 27 Oct 31, reel 17/693-695; Draft guidelines, 23 Dec 31, reel 17/740-742; Shelf 5993. Chef HL, TA 16/31, 8 Aug 31, and att., reel 5/1063-1067; OKH, H27/10, Memo of discussion, 27 Oct 31, 324/6315982-84; AA, Memo by Frohwein and atts. (probably by Schönheinz), 25 Jul 31, 9095/H221687-94; Record of meeting, 29 Aug 31, 4604/E191204-05; Memo by Frohwein, 17 Nov 31, 9095/H221926-31. From their July 1931 talk, Albert Frohwein concluded that Schönheinz wanted to block any German acceptance of a new voluntary arms convention. But Schönheinz was more intransigent than most Reichswehr leaders. On Blomberg, see the two 27 Oct 31 memoranda.

[81] Curtius, in his letter of 23 Mar 31, asked for information on what Groener had described as "the defense organization striven for by Germany." Groener replied that he wanted "a defense organization which, along with a sufficient number of officers and long-service men for cadre and training, allows for a short-service militia, and is equipped with the materiel which is allowed to other states"; Groener

ing a Reichswehr-Foreign Ministry briefing, in preparation for lecturing abroad on Germany's views on disarmament, including such varied academic personalities as Arnold Bergsträsser (University of Heidelberg), Fritz Berber (Hochschule für Politik, Berlin), Moritz Bonn (Handelshochschule, Berlin), Willy Hellpach (University of Heidelberg), Otto Hoetzsch (University of Berlin), Hajo Holborn (Hochschule für Politik, Berlin), Ernst Jäckh (Hochschule für Politik, Berlin), Herbert Kraus (University of Göttingen), and Arnold Wolfers (Hochschule für Politik, Berlin).[82] A few years later, Bonn, Holborn, Jäckh, and Wolfers would be emigrants, and Kraus would be dismissed for political reasons.

In December 1931, Brüning chose Rudolf Nadolny, then ambassador to Turkey, to head the German delegation to the conference. In his memoirs, Brüning says that he made this choice, in the face of Bülow's misgivings, because he thought that Nadolny's "Eastern" personality, i.e., his Ostelbian outspokenness and aggressiveness, would reassure the Reichswehr that he was a "strong man." The instructions finally worked out for the delegation in January 1932 no doubt increased the need for reassurance, as they followed Bülow's wishes as opposed to the army's. In regard to parity, the instructions specified only that "the necessary steps through which the final balancing of armaments [*Rüstungsausgleich*] will be reached must be set forth either in the convention or simultaneously with it."[83]

If the issue of strategy at the Disarmament Conference had not been settled before, it was in effect foreclosed when Brüning decided, early in January 1932, to seek a half-year postponement of the Lausanne Conference, the conference of Young Plan signatories, which was supposed to take up the discussion of reparations in the middle of that month. Brüning's desire for postponement resulted in part from his deep involvement at the time with prolonging Hindenburg's tenure in the presidency, and with other domestic political problems.[84] Also,

---

conceded that the quantitative issue should remain in the background, but stressed the idea of parity of security: ML, Shelf 5939, Reichswehrminister 134/31, 10 Apr 31, reel 17/534-536. Schönheinz informed Frohwein in July 1931 that the army wanted a three- to six-month militia and an average (peacetime) strength of 240,000 (AA, 9095/H221688, H221692-93), but the Reichswehr did not provide details until October 1932, and then for the Foreign Minister's eyes only.

[82] RK, Memo by Weizsäcker, 11 Feb 31, 3642/D810988-93; Memo by Mackensen, 28 Mar 31, 3642/D811022-31; Reichskanzlei memo, 9 Apr 31, 3642/D810998; AA, Memo by Bülow, 30 Jan 31, 3154/D666439-41; Memo by Weizsäcker, 2 Mar 31, 4604/E191101; ML, Shelf 5939, Memo of discussion, 30 Mar 31, reel 17/555-557; Memo of discussion, 30 Apr 31, reel 17/566-573; ltrs., Frohwein to Reichswehr Ministry, 10 and 27 Jul 31, reel 17/635-639.

[83] Brüning, *Memoiren*, p. 492; RK, Cabinet Protocol, 15 Jan 32, and annex, 3598/D789052-56, D789066-69. For background, see AA, Record of meeting, 6 Jan 32, 9095/H122121-28.

[84] Brüning, *Memoiren*, pp. 497-498.

he believed that a delay until summer would enable Germany to get a permanent settlement of reparations, whereas an immediate meeting could only produce agreement on a two- to five-year suspension. Such a suspension was particularly undesirable since, after this interval, Germany would have emerged at least partly from the depression, and so would have lost her "trump card"—the argument that she was simply unable to pay. Brüning obtained the delay by publicly affirming that Germany's financial and economic position made the resumption of political reparation payments impossible; the resulting uproar, especially in France, led to postponement.[85] But if the Reparations Conference was to be postponed, it would be rash indeed to make sharp demands in the interim at the Disarmament Conference, destroying the belief abroad in Brüning's moderation. In fact, Brüning remarked to his cabinet in May that it had been tactically right not to precipitate a premature crisis on disarmament questions, and that the necessity to prepare for reparations negotiations must always be kept in mind.[86]

Schleicher seems to have accepted the delay at first. When addressing divisional commanders in January 1932, he stated that there was com-

[85] The prospect of French and German elections in the spring created a poor atmosphere for negotiation, and the Germans expected that, by summer, the growing depression in France (where it had only just begun) would make the French more receptive to economic arguments against reparations. Brüning was aware that a delay would mean leaving the United States and war debts out of the picture, as by summer Congress would have adjourned and the presidential election campaign would have begun, but the American Dr. O.M.W. Sprague, an adviser to the Bank of England, counseled him (and also the British Foreign Office) that Europeans should give up hope of an American remission of war debts. Brüning suspected (rightly) that the British and French governments were about to propose a two- to five-year moratorium, so he warned the British ambassador, Sir Horace Rumbold, that Germany would not accept an interim solution, and that he would declare Germany unable to pay reparations now or in the foreseeable future. Probably with the connivance of interested parties in either London or Berlin, Reuters published the substance of his statement to Rumbold in a flat-footed form that provoked a sharp reaction in Paris. Brüning then issued an already-drafted public statement; this contained the slightly more diplomatic language described in my text. The opening session of the conference was put off at first for a week, and ultimately it proved impossible to do anything but postpone the conference until June, as Brüning wished. See RK, Notes on meetings, 5, 6 Jan 32; Memo by Pünder, 8 Jan 32; ltrs., Pünder to Bülow, 6, 8 Jan 32, 9242/E650032-71; AA, tel., London to Foreign Ministry, 4 Jan 32, 3243/D730182-83; Memo by Bülow, 8 Jan 32, 4618/E196617; Schäffer Diary, 4, 7, 9 Jan 32; Pünder, *Politik in der Reichskanzlei*, pp. 110-112; DBFP, 2/III, Chapter 1, passim, and Appendix I; Public Record Office, London, Foreign Office Papers (FO 371), Record of interview with Sprague, 5 Jan 32, C125/29/62; Minutes by Nichols and Sargent, 11 Jan 32, C248/29/62; tel., Rumbold to F.O., 11 Jan 32, and Minutes by Nichols, Sargent, and Vansittart, 11-12 Jan 32, C276/29/62; Sir Horace Rumbold Papers (private possession), ltr., Rumbold to Vansittart, 20 Jan 32; ltr., Rumbold to Anthony Rumbold, 24 Jan 32; *FRUS, 1932*, I, 638-640; Brüning, *Memoiren*, pp. 497-500.

[86] RK, Cabinet Protocol, 2 May 32, 3598/D789852-53.

plete agreement between the Foreign Ministry and the Reichswehr Ministry, and endorsed the strategy of demanding the disarmament of others rather than stating that Germany wanted to rearm. The military, however, were to become restive. There was a quarrel between civilian and army delegates in Geneva in February, and in March, Blomberg, the chief military member of the delegation, complained to Schleicher that the Foreign Ministry (as opposed to Nadolny) did not want to negotiate equality of rights until the end of the conference. By June, Schleicher himself would become highly impatient to establish the principle of full equality, and would refer in this connection to the particularly urgent question of the militia.[87]

In April 1932, Brüning met with the American Secretary of State and the British Prime Minister at Bessinge, near Geneva, and obtained, as he thought, their agreement to German equality of rights (this meeting is discussed more fully in Chapter Three). But even after the Bessinge meeting, Brüning's ideas involved a long delay, and heavy reliance on others. As he indicated to the Reichstag Foreign Affairs Committee on May 24, there could probably be no settlement linking war debts and disarmament until at least December 1932, after the American elections, and a settlement then depended on President Hoover's being reelected and becoming able to act. Whoever was elected in the United States, Brüning hoped for a positive result by March 1933, probably because April 1, 1933, was the date set for the initiation of new mobilization, training, and rearmament plans. At best, however, his hopes depended on the cooperation of all the former Entente powers, and in particular on an evocation of American generosity on debts by French military disarmament. That could not be expected unless German rearmament plans were kept carefully concealed.[88]

7. THE QUESTION OF THE COST OF ARMS

To the frictions connected with the Disarmament Conference were added frustrations arising from rearmament planning. There had been

---

[87] Vogelsang, "Neue Dokumente," pp. 416-417; ML, Shelf 5940, Report [by Freyberg], 9 Feb 32, reel 18/326-329; Schleicher Papers, N42/91, tel., Blomberg to Schleicher, 2 Mar 32; AA, ltr., Schleicher to Bülow, 9 Jun 32, 7360/E535897. The diplomats were later said to have claimed, in February, that they had won the military over to their viewpoint: France, Ministère des affaires étrangères, Commission de publication des documents relatifs aux origines de la guerre 1939-1945, *Documents diplomatiques français (1932-1939)* (Paris: Imprimerie nationale, 1963–), First Series, Vol. I, No. 205 (this publication hereafter *DDF*).

[88] Vernekohl and Morsey, pp. 172-174. President Hoover also hoped to establish a link between debts and disarmament, in the sense that there would be no sympathy for debt reduction without disarmament, but he refused to make any commitment to reduce debts. See *FRUS, 1932*, I, 190, 198.

hopes that rearmament would soon be well launched. In September 1930, the Ordnance Office referred to Truppenamt plans "after the now-completed evacuation of the Rhineland to take measures outside the limits of the [Treaty of Versailles] to improve our defense capabilities," and suggested the elimination of various restraints on arms production. A Truppenamt memorandum of June 1931, on the concealment of aircraft testing, expressed the belief that "on the basis of the 1932 Disarmament Conference, a positive decision on the question of a military air force for Germany is to be expected." As far as legal and diplomatic obstacles were concerned, the Truppenamt seems to have thought that there would soon be few or no restrictions on German rearmament. The 1931 financial crisis compelled the army to practice some economies, but in view of the political climate and Hindenburg's backing, it could escape serious cuts, and it maintained its secret arms expenditures. The plans for rearmament preparation in the 1933-1938 second armament program were drawn up in preliminary form in January 1931, and estimates were revised and reworked over succeeding months, being ready for final consideration by the chiefs of the offices of the Army Command in December of that year.[89]

The office chiefs found, however, that they faced painful decisions. The most modest program considered was the so-called "Six-Week Program," designed to ensure that the army could be supplied beforehand for six weeks of combat. Over the five years from 1933 to 1938, the program would cost RM 484 million, or just under 100 million marks each year. The funds were to be raised (as they had been for the first program) from over-appropriations for overt items in the Reichswehr budget, and to a small extent from the budgets of other ministries. This sum was considered a very scant minimum, but it already severely strained the German budget. This situation led the arms experts to comment that armament was primarily a financial problem.[90]

A second alternative was the "Billion Program [*Milliarde-Pro-*

[89] OKW, Wi/IF 5.3049, H Wa A 611/30, 5 Sep 30, 399/1651123-25; Wi/IF 5.509, TA 562/31, 12 Jun 31, 113/841862; Wi/IF 5.499, TA 957/30,* 30 Sep 30, 110/837868-73; TA 240/31, 28 Feb and 26 Jun 31, 110/838023-24; TA 43/32,* 15 Jan 32, 110/837914-15; Schäffer Diary, 29 Sep 31. The documents with asterisks (*) in this note (and the three following) are largely reproduced in Geyer, "Zweite Rüstungsprogramm," in which see also p. 130 and esp. Doc. 2. On economies, also Schleicher Papers, N42/1, Memorandum of Ministeramt for briefing of commanders, [August 1931]; part of this document is in Vogelsang, *Reichswehr*, Annex 11.

[90] Mueller-Hillebrand, I, 19; OKW, Wi/IF 5.499, TA 43/32,* 15 Jan 32, 110/837917-20; Geyer, "Zweite Rüstungsprogramm," pp. 128, 132, Doc. 2. A later memorandum stated that six to twelve months would have been needed before further supplies could have been provided for more than sixteen divisions: Wi/IF 5.406, Wa Wi 1770/34 and att., 15 Sep 34, 92/817730-34.

gramm]," involving the expenditure during the second armament period of roughly one billion marks over and above regular expenses. By this means, the "first armament goal" could be attained in 1938. This program, although twice as costly as the Six-Week Program, was still regarded as inadequate by military leaders. A January 1932 memorandum recorded the views of the Army Command office chiefs, led by the current Chief of the Truppenamt, General Wilhelm Adam: "Neither the Billion Program nor still less the Six-Week Program represent a *militarily* acceptable goal, not even a minimal goal. Such can only be: a combat-effective [*kampfkräftiges*], operationally viable army (including air forces) with modern equipment and adequate replacement of materiel. Regard for personnel and materiel compel for the time being an adherence to 21 divisions + corresponding air force + Grenzschutz." The memorandum stated that the operational details of an adequate minimum of equipment and supply for such an army —i.e., for a third alternative program—would be worked out by the Truppenamt. It recognized that, "from the current state of affairs," such an alternative did not come into consideration for the second, 1933-1938, armament period, and that the immediate aim must be to launch a modified version of the Six-Week Program. The office chiefs urged, however, that the Billion Program had to be demanded "by the soldier" as the next intermediate objective, leaving the responsibility for rejection to the cabinet. On January 23, General Kurt von Hammerstein authorized a modified form of the Six-Week Program; even with that, the uncertain financial situation made it necessary to establish priorities, in order to close the most serious gaps first.[91]

Cabinet records—which usually did not register discussions on sensitive military questions, at least in the versions we have—do not show whether that body took up the question of the Billion Program. The cabinet did, however, consider the possibility of launching work-creation programs to deal with the unprecedented number of unemployed, and in April a letter to Brüning, prepared for Groener's signature in a section of Schleicher's office, argued that the work-creation program should include a provision for spending one billion marks for arms over a five-year period. Whether propter or merely post hoc, ordnance

[91] OKW, Wi/IF 5.499, TA 43/32,* 15 Jan 32, 110/837915-21; TA 78/32, 25 Jan 32, 110/837924-25; T2 III notes on presentation before Chef HL,* 23 Jan 32, 110/838122-26. Cf. the German edition of F. L. Carsten's study: *Reichswehr und Politik 1918-1933* (Cologne: Kiepenheuer & Witsch, 1964), p. 401. OKW, Wi/IF 5.3682, Rüstungsausschuss 5547/32, 23 Apr 32, 436/1300325-29, elucidates the various alternatives. Completion dates for the "first armament goal" under both programs were indicated in a draft and a final letter from Groener to Brüning appealing for the Billion Program (BA-MA, RH 15/40, Undated and Reichswehrminister W. 111/32,* 13 Apr 32).

officers were informed a few days later that Chancellor Brüning might make special funds available in July after the Reparations Conference. The extra expenditures, which were supposed to reduce unemployment and provide orders for idle factories, were to be concealed as much as possible "out of regard for Geneva." Brüning may have been trying at this juncture to appease Schleicher, although he claims in his memoirs that the use of work-creation funds for armament was Schleicher's idea and that (excepting for small sums for the militia) he opposed it while negotiations were pending in Geneva. At all events, when the ordnance experts heard that there was a possibility of receiving more funds, they reacted like hungry prisoners set loose at a banquet table. They reported that there would be no problem in working out how to spend the money, and they proposed a speedup, so that the Six-Week Program would be completed in 1932 and 1933 and the Billion Program in 1934 and 1935. All that was needed for the time being was an authorization from Hammerstein. Then a further program would follow; its details, which would offer "the possibility of freer decision" as to requirements, had yet to be worked out, but the objective seems to have been that of the office chiefs' third alternative. The Ordnance Office protested, however, when it learned that increased funds could only become available after the Reparations Conference. That office had been developing the argument that a bankruptcy of the principal arms manufacturers, which was threatened by the depression, would endanger German defenses, and in May its chief, General Alfred von Vollard-Bockelberg, sent a memorandum to Schleicher's Ministerial Office, maintaining that financial aid had to be given in some cases "even if considerations of internal, foreign, and economic policy must be subordinated." The view was gaining ground that the Reichswehr should set its own financial requirements. But in late May, the Ministerial Office took the position that there was no prospect for large new sums for the time being, "pending clarification of the situation." One thing requiring clarification was presumably the political fate of the Brüning government.[92]

8. The Question of Wehrsport and the Wehrverbände

A greater problem than either tactics at the Disarmament Conference or the cost of actual armament production was that involving

[92] BA-MA, RH 15/40, Reichswehrminister W. 111/32,* 13 Apr 32; Brüning, *Memoiren*, pp. 572-573; OKW, Wi/IF 5.3682, Wa Wi 322/32, 20 Apr 32, 436/1300340-41; Rüstungsausschuss 5547/32, 23 Apr 32, 436/1300324-28; H Wa A 366/32, 26 Apr 32, 436/1300336-37; Wehramt 5569/32, 24 May 32, 436/1300307; OKH, H15/103, Chef HWA 90/32,* 23 May 32, 146/6118619-22; W.L 130/32, 17 Feb 32, 146/6118702-04; Phelps, "Groener Dokumenten," p. 1021; Geyer, "Zweite Rüstungsprogramm," pp. 131-133.

plans for premilitary training or Wehrsport. Groener had supported such programs for some time, and, as we have seen, he had mentioned them to Brüning in March 1930. On August 1 of that year, Groener wrote to the Minister of Interior of the Reich, Joseph Wirth, explaining that he opposed any "play-soldiering" that was useless and dangerous from the foreign-policy standpoint, but favored the "furthering of efforts whereby youth should be made fit to bear arms." The solution, as he saw it, was for the state to provide help and guidance. Indeed, the Reichswehr already considered it the task of the Ministry of Interior to support this kind of activity, as that ministry had supposedly assumed the immediate responsibility. In addressing the divisional commanders in October 1930, Schleicher complained that after the sport associations had been turned over to the Ministry of Interior, that ministry had failed to guide the associations in making people fit to bear arms. Some of the funds for this had even gone to support antidefense propaganda, while "poisonous antidefense elements" in that ministry were circumventing the intentions of even the most prodefense ministers. Groener (Schleicher said) had asked that the Interior Ministry get Defense Ministry approval before giving money to Verbände.[93]

Any plans for developing Wehrsport would raise the question of the army's relations, not only with nonmilitary sport associations, but also with the Wehrverbände. Neither Groener nor Schleicher showed enthusiasm for the Wehrverbände at this stage. Schleicher told the division commanders that experience showed that these Verbände were not usable, that they wanted to stand in the limelight without assuming responsibility. He also said, however, that every individual "national man" was welcome for Landesschutz work. This implied tolerance for individual Nazis in Landesschutz, a tolerance which might seem somewhat surprising, considering that enlistment in the Reichswehr was supposedly closed to anyone who could not swear to uphold the constitution, and that members of the Reichswehr were expected to refrain from all political activity. In July 1929, the Reichswehr leadership had ordered the discharge of Nazi-organized civilian employees of the army and navy. But the Grenzschutz (and the rest of Landesschutz preparations) had always depended on the Wehrverbände for recruits, and while reliance had earlier been mainly on the Stahlhelm, the Nazi SA had now become an important potential source of men. A certain problem arose, in that Hitler had forbidden

[93] OKW, Wi/IF 5.498, Memo, Groener to Wirth, 1 Aug 30, 109/837584-88; Vogelsang, "Neue Dokumente," pp. 405, 423. Before World War I, Groener had written an article on the army as a school for innoculating young Germans against the pacifistic and anarchic lures of socialism: Helmut Haeussler, *General Groener and the Imperial German Army* (Madison: State Hist. Soc. of Wisconsin, 1962), p. 44.

his followers to serve in the forces of the republic, but his edict seems not to have been strictly followed. The Nazi electoral success in September 1930 deeply impressed the government with the strength of this ultranationalist crusade, and German officers saw it as the manifestation of an unstoppable youth movement. They themselves obviously found the SA part of the supposed "youth movement" irresistible: it seemed to mean that now, suddenly, after years of fear that the military spirit would die out, German youth was returning en masse to military ideals. Now there might be alternatives to working through recalcitrant officials. It was symptomatic that in December, Schleicher told the cabinet that Nazis had to be used in the Grenzschutz, although not in closed (purely Nazi) formations. Brüning and the cabinet agreed, although the Chancellor thought that Nazis should still be barred from enlistment in the Reichswehr proper.[94]

The Reichswehr soon took steps to bypass the Ministry of Interior and to promote Wehrsport activity through the Wehrverbände. Groener's August 1930 letter to Wirth had resulted from complaints, which Wirth had forwarded, against paramilitary activity on the part of radical rightist students. If the Reichswehr had not already begun secretly to back such activity—and it may have been doing so—it was backing it by March 1931. As an intermediary cover device, the Reichswehr used "working groups," particularly two named the Geländesport-Verbände-Arbeitsgemeinschaft (GVA) and the Akademische Wissenschaftliche Arbeitsamt (AWA). Subsequently, the AWA took over all the work with the student groups, including the National-Sozialistische Deutsche Studentenbund, and the GVA sponsored training by other, more general organizations. Providing channels for material support, program guidance, and instructor training for the Verbände, these working groups served as a pilot model for the official institute for youth training, or *Jugendertüchtigung*, which the Reichswehr leaders hoped to establish. The training the working groups sponsored had a more definite military cast than that formerly given by the sport associations, and could hardly have escaped exposure. The German public was in fact to learn in early 1932 that a so-called "Deutsche Volkssportverein," actually a part of the Berlin SA, had

[94] Carsten, *Reichswehr and Politics*, pp. 59, 94-95, 224-232, 264-272, 310, 317, 350-355; Vogelsang, *Reichswehr*, pp. 58, 62, 99n, 116-120; RK, Cabinet Protocol, 19 Dec 30, 3598/D785532. On reaction to September 1930 elections: Bennett, *Diplomacy*, pp. 11-13; *DBFP*, 2/I, No. 324n. Although Vogelsang (in *Reichswehr*, pp.. 58-59, 118) says that Hitler's ban was apparently widely adhered to, Heinrich Bennecke, in *Hitler und die SA* (Munich: Günter Olzog, 1962), pp. 157-158, and *Die Reichswehr und der "Röhm-Putsch"* (Munich: Günter Olzog, 1964), p. 13, states that the ban was only for cover, and not serious; Bennecke himself was an SA leader, specializing in student affairs. See also Otto Braun, pp. 268-269.

received material and financial support from the GVA, and had been training for some time at the Reichswehr drill grounds at Döberitz.[95]

The effort to promote an enhanced Wehrsport program must have gained urgency from the experiments the Reichswehr was conducting in ultra-short-term training after induction. As Hammerstein had affirmed in March 1930, "As long as the creation of reserves in peacetime is not possible, the expansion of the army in an emergency can in part only be accomplished through the induction of short-term trainees." Experiments in ultra-short-term training in the spring of 1931 had shown that regulars from line units could train men to do the simplest defense tasks in seven days, and to participate in short attacks after two weeks. But the experiments also brought a practical realization that "our mobilized units under existing conditions will *at best* look roughly like the rapidly trained experimental units," evidently no sight to gladden a military heart. It was noted that prior training in small arms would be a great advantage, as would refresher courses, and that there were reasons for concern over the "moral firmness" or morale of new recruits.[96] Premilitary training and indoctrination thus appeared more than ever desirable. Schleicher presumably had this need in mind when, in March 1931, he told the Stabschef (commander) of the SA, Ernst Röhm, that every individual should have the right to serve if he rejected the violent overthrow of the constitutional state and institutions. As Hitler himself now said that he would come to power legally, a loyal Nazi could presumably meet Schleicher's criterion. Schleicher also mended his relations with the Stahlhelm, which in fact received a subsidy for Wehrsport training in 1931.[97]

In October 1931, an opportunity occurred to advance Wehrsport and other training more officially. The failure of the Austro-German customs union forced Brüning to drop his Foreign Minister, Curtius, from the cabinet, and Brüning also decided to dispense with Wirth,

---

[95] OKW, Wi/IF 5.498, TA 251/31, 20 Mar 31, 109/837589-91; TA 784/31, 24 Jul 31, 109/837595-600; Vogelsang, *Reichswehr*, pp. 161-162, 231, 286; Otto Braun, p. 269. The GVA and AWA were headed respectively by a retired general named Vogt and a certified engineer named Schwab; the GVA was also called the "General Vogt Arbeitsgemeinschaft." The AWA also encouraged students to study subjects related to national defense. TA 784/31 (and also TA 570/32, 25 Jul 32, 109/837570-71) indicate that the new Wehrsport training was to build on the old, and was to prepare men to expand the army in case of mobilization.

[96] BA-MA, II H 226, Chef TA, 140/30, 15 Mar 30; II H 285, TA 50/31, 31 Jan 31; OKW, Wi/IF 5.498, TA 628/31, 17 Sep 31, 110/837637-52.

[97] Vogelsang, *Reichswehr*, pp. 118-119, also 130; Volker R. Berghahn, *Der Stahlhelm: Bund der Frontsoldaten 1918-1935* (Düsseldorf: Droste, 1966), pp. 193-195. The Stahlhelm's subsidy was reported as only RM 40,000; one must suspect that it received more. There are a number of references to payments to Wehrverbände (see e.g. Vogelsang, "Neue Dokumente," p. 416; Otto Braun, pp. 268-269), but few hard figures have come to light.

who was unpopular on the right. Schleicher managed to use this occasion to obtain for Groener the Ministry of Interior, in addition to his post as Minister of Defense. Groener and Schleicher saw in this combination of offices a chance to overcome official foot-dragging in the Ministry of Interior and to organize under that ministry an institute to direct the Wehrverbände toward Wehrsport activity, and away from politics. In presenting the new cabinet's program to the Reichstag, Brüning stated that the most important task of the government would be to awaken, especially among youth, the will to undergo self-discipline and to make sacrifices for the Fatherland. Soon after this, and also after Hitler had charged that the new cabinet was turning the Reichswehr into an internal police force, Groener issued a press statement on his plans, making a transparent bid for cooperation from the Wehrverbände. He promised to support efforts "to get the youth off the streets, train it in discipline and order, prepare it physically, and make it psychologically ready to bear arms." The military and police under his control were strong enough so that wide scope could be allowed the activity of the "constructive forces" within the people. "Youth needs ideals," he concluded, and "should have the freedom to live for them, if they are not directed against the state, but toward Germany's future."[98]

Such public utterances were of course kept vague and general, so as not to increase domestic resistance or arouse foreign suspicions. The actual scope of the army's plans emerges more clearly from the incomplete record of a series of briefings given on November 24 and 25, 1931, to staff officers from the seven divisions. One briefing surveyed the general objectives of defense policy, as follows:

> Point on program of present Reich Chancellor: Strengthening *of Germany's defense power as precondition for stronger foreign policy.*
>
> To this end, serious, planned efforts:
>
> 1) for appropriate *psychological influencing* of the German people and shaping of its will.
>
> Propaganda, permeation of schools (statistics on educational level of recruits).
>
> 2) for physical training of the people for defense [*zur körperlichen Wehrertüchtigung des Volkes*] (see declarations of Reich Chancellor and Reich Defense Minister). New initiative by Reich intended.

[98] Vogelsang, *Reichswehr*, pp. 131-137; Vogelsang, "Neue Dokumente," pp. 416, 422, 423; Castellan, *Réarmement clandestin*, p. 41; Brüning, *Memoiren*, pp. 427, 554-555; *Frankfurter Zeitung*, 14 Oct 31 (2. Morgenblatt), 20 Oct 31 (2. Morgenblatt); *Völkischer Beobachter*, 16 Oct 31 (Bayernausgabe).

3) for drawing in all elements of the state in preparations for national defense.

Most important step: *formation of the Reich Defense Council.*
. . .

The Reich Defense Council, an old objective in the Reichswehr Ministry, was to comprise all the appropriate national agencies, including the German central bank, the Reichsbank, as well as the ministries of Finance, Post, Economics, Food, Transport, and Interior. Under the last ministry, the briefing stated, there was to be a "Working Reich Office for Youth Training [*Bearbeitende Reichsstelle für Jugendertüchtigung*]"; presumably the "new initiative" in the field of physical training consisted in the formation of this office.[99] These proposals reveal an intent to indoctrinate and train the German people, and also to compel the other German government departments to do their bit, under Reichswehr guidance, for preparedness.

Groener's press statement had hinted at a new official acceptance of the Wehrverbände, and the divisional staff officers learned at the November briefings that an order had been prepared for permitting Nazi enlistments in the regular army. There was no dearth of non-Nazi recruits for the few regular vacancies, and probably there was no strong desire to admit Nazis as regulars. Apparently the order was meant to serve as an official indication, to civil authorities and Nazis alike, that the Nazis were now accepted for any defense work, and in particular, it was meant to overcome resistance to Nazi participation in training outside the army. The briefer who revealed the new order suggested that civil officials would only cooperate smoothly when they became sincerely convinced (*innerlich durchdrungen*) of the value of Landesschutz work. And he added: "*Relation to the NSDAP* of decisive significance. Relation of the party to the armed forces and attitude of the Länder to the party vital question for training outside the army." The briefer stated further: "*NSDAP* not yet declared legal. Merely admission of party members in Reich army will be authorized. Before officially clear that NSDAP no revolutionary party, cessation of resistance of civil authorities on questions of participants, unpaid volunteer instructors, etc., not to be expected."[100]

These statements implied that there should be a sort of governmental *Gleichschaltung*, a forced acceptance of the Nazis, in the interests of defense training. It was not only, of course, that officials should ac-

[99] OKW, Wi/IF 5.498, T2 III A 1207/31, att., 24 Nov 31, 110/837781-83. On Jugendertüchtigung project: Groener Papers, Caro-Oehme questionnaire, 12 Jan 33, reel 25/Stück 224; Vogelsang, "Neue Dokumente," p. 416; "6," "Jugendertüchtigung," *Militärwochenblatt*, 117 (1932-1933), no. 29 (4 Feb 33), 959-961.

[100] OKW, Wi/IF 5.498, T2 III A 1207/31, att., 25 Nov 31, 110/837784-86.

cept Nazi involvement in training, but also that they should assist in preventing disclosure of that training. The statements also indicate that, so far, Prussian (and other) civil officials were continuing to resist Reichswehr plans. Reichswehr leaders presumably realized that an authorization of Nazi participation in Wehrsport, as well as in external training, would arouse further opposition. But Schleicher had been negotiating with Hitler, and Schleicher and Groener seem to have thought that the Nazi leader could be induced to cooperate. Cooperation from the SA, and a broader acceptance of the SA by others, would certainly be necessary if either external training or Wehrsport were to develop within the near future. The Reichswehr made its contribution on January 29, 1932, by issuing the order permitting Nazi enlistments.[101]

Historians have usually attributed such actions as the admission of Nazis to the army to Schleicher's political ambitions, his tactics, and his basic political shortsightedness. There was indeed something in Schleicher's nature that made him like to play the "gray eminence," and he manipulated his fellow men in much the way he habitually played with a glass menagerie on his otherwise barren desk. Counterespionage was one of his official responsibilities; wiretapping, cryptanalysis, and espionage were used by him as weapons of domestic politics.[102] Certainly he was insensitive to the danger National Socialism posed to civilized society, an insensitivity shared by most of the officer

---

[101] The same briefer (Capt. Buschenhagen) who discussed relations with civil officials in connection with external training was also scheduled to give a later briefing on the further development of Jugendertüchtigung (OKW, Wi/IF, 5.498, TA 1151/31, 10 Nov 31, 110/837769-71); there seems to be no surviving legible record of this latter talk on the T-77 films. On Schleicher-Hitler negotiations and resulting beliefs: Carsten, Reichswehr and Politics, pp. 333-335; Vogelsang, Reichswehr, pp. 136-142. Text of 29 January order: ML, Shelf 5916, Reichswehrminister 600/32, reel 11/1145-1146. The order was revealed in the press a few days later (Frankfurter Zeitung, 6 Feb 32, [1. Morgenblatt]); a Communist resolution against it was passed in the Reichstag on 26 Feb 32, the Nazis being absent (Vogelsang, "Neue Dokumente," p. 422n). But the Reichswehr continued to regard it as valid (see also Vogelsang, Reichswehr, Annex 32).

[102] On Schleicher and his glass menagerie: Treviranus, pp. 114, 282-286. On wiretapping and espionage: Wheeler-Bennett, Nemesis of Power, pp. 184, 283; Brüning, Memoiren, pp. 395-398; Magnus Freiherr von Braun, Weg durch vier Zeitepochen (Limburg a. d. Lahn: C. A. Starke, 1965), pp. 257-258; Günther Gereke, Ich war königlich-preussischer Landrat ([E.] Berlin: Union Verlag, 1970), p. 210. Schleicher Papers, N42/33, Note on Curtius and relations with France (January 1930), appears to be a record of conversations between Curtius and the French ambassador, obtained either from a telephone tap, a deciphered report, or an agent. The Schleicher Papers contain several items indicating the use of intercepted diplomatic correspondence, including that of France and Italy, as a source of information on German political figures: N42/23, 32—"VN" material. See also Institut für Zeitgeschichte, ED-86, Ferdinand von Bredow Papers, Vols. 1-5, Orientation notes [Kurzorientierungen], passim, and Vol. 6, VN 17, 11.30/D, 3 Nov 30, and att., 17 Mar 31.

corps. But the first concern of Schleicher and other officers was national defense, and it must be borne in mind that they looked at the Nazis first of all not as a political movement, but as a potential source of reserves and a badly needed adjunct to a system of military training. The acceptance of Nazis for training did not mean an uncritical admiration for the top men in the party, or for their policies, and in fact Schleicher thought that, aside from directing the training work of the paramilitary groups, the Reichswehr Ministry could use financial contributions to make them desist from political activity. The briefing on relations with civil authorities had specified that there should continue to be a ban on the recruiting of open enemies of the state, and that participants must be approved by the local authorities.[103]

Whatever we may think of the political or military judgment of Reichswehr leaders, we must recognize that, from their perspective, there were real dangers from abroad, and therefore real reasons for pressing their programs. In the past, the Soviet Union had been a deterrent to any Polish attack on Germany. With the Manchurian crisis, however, the USSR had to reckon with a possible conflict with Japan, and even before that, Moscow had begun discussions with Poland and France aimed at the conclusion of nonaggression pacts.[104] These circumstances and the internal strains of the first Five-Year Plan made it appear that Germany could not count on Soviet aid. In February, German military intelligence, the Abwehr, conveyed a warning from no less a source than the League High Commissioner in Danzig, the Italian Count Manfredi Gravina, that this diplomatic situation, together with internal dissension in Germany, was encouraging thoughts of an invasion among Polish nationalists. France was incensed over Brüning's January statement that reparations could not be resumed, and from Paris came indications of the calling up of reservists and the transport of supplies to the frontier. At the end of February, Hammerstein gave credence to a report that France would reply to a German refusal to pay reparations by taking the question to the Court of International Justice at the Hague; then, if that body ruled against Germany (as it had done in the case of the Austro-German customs-union proposal), France would deputize Poland and Czechoslovakia to take armed sanctions against the Reich. This danger seems farfetched now (and seemed so to Bülow at the time), but Hammerstein believed in it. He emotionally told his generals that

---

[103] Vogelsang, "Neue Dokumente," p. 416; OKW, Wi/IF 5.498, T2 III A 1207/31, 24 Nov 31, 110/837785.
[104] The Polish treaty was initialed on 26 Jan 32; the French treaty was substantially completed on 10 Aug 31; both were held up by difficulties in Rumanian-Soviet negotiations: see DDF, 1/II, No. 29.

while the army could in a pinch give way to a French invasion, on grounds of overwhelming French superiority, a failure to resist a Polish and Czech invasion would mean that "we have lost our right to exist."[105] Precisely because German military leaders believed in seizing favorable opportunities for taking the initiative, it was easy for them to believe that other nations might act against Germany while she was weak.

9. BRÜNING'S OVERTHROW

Such policies as the admission of Nazis to the Reichswehr increased the sense of insecurity among moderate and prorepublican elements. The Social Democrats could see that the Reich government had begun to tolerate and work with the Nazis while showing distrust of the Reichsbanner. They were disturbed in October 1931 when the Reich Interior Ministry took no action after an SA mass parade in Braunschweig resulted in bloody disorders. When the Ministry of Interior dismissed an expert who had specialized in watching right radical formations, this was interpreted by some observers as a concession to the Nazis. Republican critics of the government became bitter in December when that ministry saw no cause for alarm in the discovery of documents—at Boxheim, in Hessen—which seemed to the critics to prove that the Nazis planned to carry out a coup. The continuation of SA involvement in street fighting after the new year began, particularly in connection with the several election campaigns, and the uncovering of new evidence of plans for putsches, caused concern among

[105] On the Soviets, General Adam wrote in March 1933: "Even if we pursue a resolute Eastern policy, Russia would not be able to help us at present. Aside from [her Far] Eastern involvement, in my opinion she is not at present capable of fighting a war [*kriegsfähig*]" (Krupp Trial, Defense Exhibit 62). Polish threat: ML, PG 49031, Abw 447, 18 Feb 32, and note of 14 Nov 33, reel 35/469-473; Vogelsang, "Neue Dokumente," pp. 420, 425; Institut für Zeitgeschichte, ED-1, Curt Liebmann MS Notes, pp. 176-177 (typed series); OKH, H24/6, ltr., Bülow to Hammerstein, 14 Mar 32, 248/6227978-82. Hammerstein described his source as "certain" or "good," but it is possible that the French or Poles played false information into German hands. All Germans concerned seem to have recognized that France was unlikely unilaterally to take military action, which would have directly violated the Locarno agreements, but French preparations could have fitted in with a concerted Polish-Czech attack, diverting German forces from the East. On French preparations: AA, Dispatches, Hoesch to Ministry, 22, 29 Jan 32, K936/K241027-31, K241033-36; *DBFP*, 2/III, No. 78; AA, tel., Bülow to Nadolny, 10 Feb 32, K936/K241059; Captain B. H. Liddell Hart, *The Memoirs of Captain Liddell Hart*, 2 vols. (London: Cassell, 1965), I, 190-191; National Archives, State Department Papers (SDP), Warrington Dawson reports, 29 Feb, 29 Mar 32, 751.62/180, 183. See also Geoffrey Warner, *Pierre Laval and the Eclipse of France* (London: Eyre & Spottiswoode, 1968), pp. 51-53, and Neville Waites, "The Depression Years," in N. Waites, ed., *Troubled Neighbours* (London: Weidenfeld & Nicolson, 1971), pp. 136-137. A study based on French archives would be very helpful here.

the Land governments, not only in Prussia, but also in many other Länder. In March the Länder took restrictive steps on their own initiative, and at the end of the month the Bavarian government suggested that the SA be banned, informing the Reich and five interested Land governments that it would outlaw the organization on its own if the Reich did not act.[106]

The mounting dissatisfaction of the Socialist and moderate civilians was duplicated, but in the reverse sense, among the military. Despite a few changes made at the Reich Ministry of Interior, other old officials remained, and the Wehrsport project still did not progress satisfactorily. Hermann Dietrich, the Minister of Finance, was holding up funds for Wehrsport until the presidential election on April 10, apparently because he suspected that the Nazis would use the money to support Hitler's campaign against Hindenburg. Schleicher and Groener themselves had a large stake in Hindenburg's reelection, and until this was accomplished, they had to restrain their impatience with the republican politicians who were backing the old man's candidacy. But Schleicher intended to mince no words after April 10.[107]

Reichswehr leaders believed themselves to be nonpartisan. Hammerstein rejected the idea of arming the SA or any other Wehrverband, as that would mean taking political sides, alienating part of the population, diverting stocks of weapons, and delaying military training: "Every rifle will be used against [the] foreign enemy. Disrupting of mob[ilization (preparations)] for domestic pol[itical] purposes makes mob[ilization] 'employability' [*Verwendbarkeit*] impossible for 2 years." A reduction of the power of the Wehrverbände was obviously needed. But for the future militia as well as for Wehrsport generally, it was important to capture and use the personnel of the Verbände through the projected youth training institute, rather than simply disbanding the Verbände and alienating their members. Least of all did the army want to lose the young men of the SA, who belonged to a movement that might win an absolute majority in the next Reichstag election.[108] To Groener and Schleicher, the Prussian government seemed overeager to use any material it could find against the Nazis, and the demands of the Land governments also appeared as attempts to place

---

[106] Vogelsang, *Reichswehr*, pp. 136-141, 161-166; Bracher, *Auflösung*, pp. 482-486.

[107] Vogelsang, "Neue Dokumente," pp. 422-424; Groener Papers, Caro-Oehme questionnaire, 12 Jan 33, reel 25/Stück 224; Gordon A. Craig, "Quellen zur neuesten Geschichte: Briefe Schleichers an Groener," *Welt als Geschichte*, 11 (1951), 130.

[108] Vogelsang, "Neue Dokumente," pp. 420, 422, 423. The reference to using every rifle against a foreign enemy indicates that "mob[ilization] 'employability' " referred to the availability and employability of volunteers from outside the Reichswehr for mobilization into an expanded field army. If the army got entangled domestically, the members of both the groups it opposed and those it supported would be for a time unavailable for training or eventual external use, and the vital work of making further cadre out of the first class of trainees would be postponed: see note 50.

the Reich government under pressure. It could even be suspected that those who sought to place restrictions on the Nazis were trying to cripple Nazi preparations for the coming Land elections, or were looking for a pretext to cancel those elections. Later, when he had fallen out with Schleicher, Groener claimed that he had initiated the ban of the SA, implicitly for political reasons. At the time, however, the Minister of Defense and Interior tried to evade a ban, and explicitly for military reasons, for the sake of the militia scheme. He only agreed to a ban after all the major Land governments had made it clear that they would otherwise act without him. Then he made a pretense of independent initiative to hide the fact that he was yielding to dictation.[109]

This action was to place Groener at odds with the Reichswehr, as represented by Schleicher, and thereby was to lead to his and also Brüning's downfall. The main elements of the story are well established.[110] Schleicher and Hammerstein at first agreed with the ban, perhaps because of reports that the Nazis planned to seize Grenzschutz arms and would even refuse to join in the defense of the eastern provinces in the event of a Polish invasion. But even when agreeing, Schleicher still wanted to win the "good parts" of the SA over, and as he considered the matter further, thinking of the Wehrsport program and not of the danger on the domestic political scene, it seemed to him that the proper action was to send Hitler an ultimatum demanding a reconstruction of the SA, say within a week. This reconstruction could indeed help smooth the path for the new Wehrsport program, while the contrary course would alienate all of the SA, encourage the foot-dragging of civil officials, and cripple Wehrsport and militia plans. Groener, however, could not readily change the stance he had adopted, and he remained committed to declaring a complete ban at an early date. Brüning had some doubts as to the ban, but he was loyal to Groener, and after the Minister of Justice raised legal objections to Schleicher's proposal, the Chancellor was won over to the action.

Schleicher resented the overruling of vital defense considerations, and in his eyes the issue became one of whether or not the Land governments, or unsympathetic elements in the Reich government, would

[109] Craig, "Briefe Schleichers an Groener," p. 130; Vogelsang, "Neue Dokumente," pp. 416, 424; Vogelsang, *Reichswehr*, pp. 146-147, 162-168, Annex 21; also Waldemar Besson, *Württemberg und die deutsche Staatskrise 1928-1933* (Stuttgart: Deutsche Verlags-Anstalt, 1959), Annex 10; O. Jacobsen, *Erich Marcks*, p. 51.

[110] On the last two months of Brüning's cabinet, surveyed below, see Vogelsang, *Reichswehr*, pp. 167-202; Sauer, "Mobilmachung," pp. 705-706; Bracher, *Auflösung*, pp. 486-526; Brüning, *Memoiren*, pp. 538-603; Conze, "Politischen Entscheidungen," p. 236; Schäffer Diary, 2, 7, 17 Jun 32. Groener told Schäffer that Schleicher at first intended to retain him, and to take Brüning's place himself.

be permitted to obstruct national defense. He also became personally hostile to Brüning after a private meeting at which Brüning (according to his memoirs) challenged the General to assume the chancellorship and lectured him on the lack of resolution of Prussian generals, including Schleicher himself.[111] Although Brüning and Groener, by using every pressure tactic at their command, obtained Hindenburg's signature to the decree of a ban on April 13, the President's dislike of the measure, which seemed to him a partisan antiright action, was encouraged by Schleicher, who did what he could to inspire objections to this "one-sided" measure from within the armed forces. Ex-officer advocates of preparedness also protested loudly.[112] Hindenburg was led to send an inquiry to Groener, asking whether the Reichsbanner should not also be banned. Groener and Brüning did not consider a ban of the Reichsbanner justified, and they also considered it impolitic, for reasons of both foreign and domestic policy. Their reaction tended to reinforce Hindenburg's longstanding suspicion that Brüning had a weakness for the left, and was refusing to recognize the overwhelming trend of opinion toward the right. The President resented rather than appreciated Brüning's success in winning his own reelection with votes from the center and left, and on the morrow of that event it was difficult to dissuade him from telling the press that he would only retain the Brüning cabinet provisionally. The position of Brüning's cabinet was further weakened when the Land elections of April 24, involving more than three fourths of the German people, produced sweeping gains for the Nazis, and in particular ended that Socialist dominance of the Prussian parliament that had been a mainstay of the German republic.

Brüning succeeded in winning Reichstag votes on budgetary questions in early May, but his minority cabinet served at the pleasure of the President and the Reichswehr, and did not base itself upon the Reichstag. Although he had maintained good relations with the Social Democrats, he probably could not have formed an anti-Hindenburg majority cabinet, and he lacked the will to try. In this situation, Schleicher, by threatening resignation by himself and other generals,

[111] On Schleicher's probable views, see Wohlfeil and Matuschka, pp. 295, 297; see also O. Jacobsen, *Erich Marcks*, p. 51, for the "Schleicher line." Brüning, who describes his talk with Schleicher at length and in almost purple prose, says it took place on the night of 2-3 May (*Memoiren*, pp. 575-580); Treviranus (p. 292) dates the quarrel at the beginning of January.

[112] On Schleicher's activity, see Dülffer, pp. 127-128; Carsten, *Reichswehr and Politics*, pp. 341-348. Examples of external criticism: ML, Levetzow Papers, Manifesto of the Vereinigte Vaterländischen Verbänden Deutschlands, 21 Apr 32, reel 26/800-803; [George] Soldan, "Zur Auflösung der SA und SS," *Deutsche Wehr*, 5 (1932), 293-295. The latter argued that the Nazi organizations were for use against a foreign enemy, not for revolution at home.

was able to force Groener's resignation as Defense Minister. The presidential household wanted Groener to resign as Minister of Interior as well, a demand that Brüning loyally rejected, making his own threat to resign. Hindenburg did not for the moment force the issue, but he still wanted Groener's complete removal from office, and his State Secretary, Otto Meissner, together with Schleicher, began to plan a complete reconstruction of the cabinet, drawing on the old plan for a presidial cabinet without party ties. The opposition of Ostelbian landowners to the cabinet's land-settlement plans served to strengthen Hindenburg's resolve, but his impatience with Brüning and the military reaction against the SA ban would probably have caused him to act anyway.[113] On May 29, he refused to sign any more of Brüning's emergency decrees, and Brüning resigned the next day.

For months before his downfall, Brüning had told the President that he had no objection to a realignment of the cabinet toward the right, or even to being replaced, but that such steps should not be taken while diplomatic negotiations were still in progress on reparations and disarmament. He asked in effect to be allowed to maintain a republican, moderate façade until these two foreign-policy questions were settled, after which an old-fashioned authoritarian monarchy could be introduced. Hindenburg, and also Schleicher, on the other hand, did not question Brüning's diplomatic abilities, and proposals were made until the last moment that he should stay on as foreign minister. But they had come to see Brüning's tenure as chancellor as an obstacle to authoritarian rule, the elimination of Socialist influence, and an accommodation with the Nazis.

There is some record of Schleicher's political and military reasons for desiring Brüning's removal. Just after that event, in a speech before officers and in a contribution to Meissner's official account of the cabinet change, Schleicher expressed the ideas that the Nazis now represented a majority of the population, that they were furthering the development of military spirit (*Wehrwille*), that their objectives were therefore similar to the Reichswehr's, that it was a contradiction for the Minister of Defense to ban such a "defense organization" as the SA, and that "the Grenzschutz would be greatly weakened by the banishment of the young National Socialists who constituted the main contingent." Schleicher also indicated his intention to reorganize the army soon without too much regard for the Treaty of Versailles, and he suggested that it had been necessary to remove Brüning's cabinet

[113] Cf. Vogelsang, *Reichswehr*, pp. 193-196; Bracher, *Auflösung*, pp. 511-517. The contemporary testimony of Brüning and Groener strongly emphasized Schleicher's role, with scarcely a mention of agrarian intrigues: Schäffer Diary, 7, 17 Jun 32; *FRUS, 1932*, II, 293-294.

in order to avoid a situation in which the army would have had to fire on the Nazis. Some of these statements may have been bids for the support of the Nazis or of young officers, but Schleicher's allusion to the importance of young Nazis for the Grenzschutz probably expressed a serious concern, not only for the Grenzschutz, but also for both Wehrsport and training outside the army. As noted earlier, the Reichswehr used the Grenzschutz as cover for Landesschutz activities, the Grenzschutz being widely known and accepted in Germany. Training outside the army was increasing in the spring of 1932, doubtless with Nazi participation. The young men of the SA were probably not yet in fact the main contingent in the Grenzschutz, but they were the army's main hope under existing conditions for voluntary Landesschutz preparations and for filling out the twenty-one divisions in the event of war.[114] To the extent, however, that national defense depended on the SA, it was vulnerable. The members of the SA were more loyal to their movement than to the existing state, which they identified with "the system"; they were also more closely controlled by their private paramilitary organization. In reaching out for reserve forces in the nation at large, the Reichswehr was making itself dependent on outside elements.

THE RESOLVE to revise the peace treaties affected every aspect of Weimar foreign policy, and much of domestic politics too. Such issues as the Corridor, Anschluss with Austria, and reparations kept the flame of resentment alive, and we have seen how far Brüning was willing to go in the interests of reparations revision. The revision or overthrow of the restrictions on armament was crucial, in German eyes, to the revision of other treaty provisions. And this in itself tells us something about the extent of German revisionism.

German officers were prepared to use armed force in international politics—not only for the defense of Germany, but also for "national liberation," and for the offensive. They hoped to find a situation in which Germany could intervene and gain useful allies in conflicts between other countries, or advantageously wage war on her "own free initiative." Presumably, continued growth in armed strength would eventually make it less difficult to find situations in which Germany could take the initiative. It was not that German officers were thirsting for blood; but they were completely imbued with the doctrine that one cannot control a situation and succeed without taking the offensive. And they believed that this principle applied to politics as well as to

---

[114] Schleicher Papers, N42/91, ltr., Schleicher to Meissner, 9 Jun 32; Vogelsang, *Reichswehr*, Annex 25, also pp. 157-160 (on SA in Grenzschutz); WKK, WK VII/2670, WKK VII Ib 711/32, 9 Jun 32, 55/477.

warfare.[115] From their point of view, he who does not take the offensive will probably find himself the victim of someone else's attack, and in fact they lived in fear that Germany's neighbors, especially Poland, would attack while Germany was still weak.

German civilian policy-makers were also strong revisionists. German diplomats refused to rule out war with Poland, cabinet members listened without demur to Heye's talk of deliberate intervention, and Brüning (according to an army briefing) believed in the strengthening of German armed force to make possible a stronger foreign policy. Most of the civilians, including Brüning, had had military indoctrination, and they, too, believed that by seizing the initiative, they could impose on an opponent the "rules of action" (*das Gesetz des Handelns*) that he would have to follow.[116] Despite the experience of 1914-1918, they still believed war could be an instrument of policy. Brüning and the diplomats had, however, more awareness of foreign opposition than the soldiers had, and they were more anxious to avoid diplomatic isolation. They had more hope of securing revision with the consent of other powers, rather than in defiance of them.

The honored status of the military, their access to the President, and the urgency of their demand for rearmament threatened to outweigh diplomatic caution, and the military did overwhelm the domestic opposition. The power of German revisionism ensured priority for external policy. But in Germany's situation, this tended to translate into a priority for military policy, and that had domestic impact. Army leaders believed that Germany's future, not to speak of their own careers, depended on solving soon the problem of training reserves; even delay might mean a permanent loss of power. And the solution of the reserve problem seemed to require an elimination of Socialist opposi-

[115] The Reichswehr's Field Service Regulations—I. D.V.Pl. 487-Führung und Gefecht der verbundenen Waffen (F.u.G.), 1921—stated: "The attack alone imposes the rules [of action] on the enemy [*Der Angriff allein schreibt dem Feinde das Gesetz vor*].... The defense is only justified against an overwhelmingly superior enemy, and to make possible an attack at another place or at a later time": Jehuda L. Wallach, *Das Dogma der Vernichtungsschlacht: Die Lehren von Clausewitz und Schlieffen und ihre Wirkungen in zwei Weltkriegen* (Frankfurt: Bernard & Graefe, 1967), p. 332. A political expression of this general outlook was the last sentence of Schleicher's radio broadcast of 26 Jun 32, in which he advocated, for Germany's government, the "greatest soldierly virtues: courage, the ability to make decisions, and readiness to accept responsibility": Vogelsang, *Schleicher*, p. 81.

[116] This would be strikingly exemplified when, in a cabinet discussion of domestic politics, Franz von Papen exclaimed, "The motto of the government must be: act, act, act": RK, Cabinet Protocol, 15 Aug 32, 3598/D790546. Brüning indicates in his *Memoiren* his resentment of military criticism of his "hesitancy," and disparages the exaggerated stress of German officers on "action," while trying to show his own readiness to take the offensive: pp. 200-201, 211, 400-401, 489, 579-580. See also Albrecht Mendelssohn-Bartholdy, *The War and German Society* (New Haven: Yale University Press, 1937), p. 41, which points out Brüning's own "decisionism."

tion, and an alliance with the "defense-minded" Nazis, although no senior officer probably thought yet of subordinating the army to Hitler. More generally, the Reichswehr needed a compliant government to serve its human, material, and financial needs in preparing for mobilization.

The significance of mobilization planning has not usually been recognized. Most accounts of this period speak of the small size of the existing Reichswehr and its limited stock of armaments. It seems true that the number of men actually serving in the army proper did not, in 1932, exceed the treaty limits. But the 100,000-man army constituted only the core of the force that German planners intended to put into the field in case of war. Already there were secret stocks of arms, a secret Landesschutz or mobilization organization, covert ties with the Wehrverbände, a framework for a Grenzschutz organization. Beyond these things, the planners hoped to provide soon for increased arms production and the peacetime training of reserves within the army. Existing conditions counted for little, as compared with plans for the future. Furthermore, mobilization planning could logically lead beyond such matters as arms and training programs and to a totalitarian control of the whole society. The army was preparing again for war in a world that had moved far beyond the comparative simplicity of Scharnhorst's or Roon's times. The task was not impossible, but it would lead to a kind of regime that neither Scharnhorst nor Roon had ever dreamed of.[117]

[117] On the significance of mobilization planning, see also Rautenberg, pp. 238-239; Geyer, "Zweite Rüstungsprogramm," pp. 125-126, 136. Aside from documents cited earlier (note 37), an interesting internal Ordnance Office memo of 11 Apr 32 called for a break with the "1914 idea" that war was simply the affair of soldiers, for preparations by all ministries for the eventuality of mobilization, and for a Reich Defense Council; it also hinted at the need for measures to combat pacifism and defeatism: OKH, H15/103, Leiter des Beschaffungswesens 18/32, 11 Apr 32, 146/6118665-67. Evidently officers might be susceptible to Adolf Hitler's ideas on a total mobilization of opinion as well as of resources.

CHAPTER TWO

Perceptions and Preoccupations of France and Britain

1. WESTERN DETECTION OF GERMAN REARMAMENT

If historians find that German military preparations began early, then they will naturally expect an early Western reaction. Or if they find that the preparations began later, then they will expect the reaction to come later. In the 1940s, when it was taken practically for granted that Adolf Hitler had long planned and prepared for world conquest, Western appeasement seemed obviously culpable. On the other hand, when A.J.P. Taylor published in 1961 his revisionist study, *The Origins of the Second World War*, his partial exculpation of Western leaders for their concessions to Hitler went hand in hand with a description of Hitler as a feckless dreamer who did not prepare for a general war. But the implicit interaction between German and Western policy commands less attention when historians concern themselves mainly with one country. Thus, some other British writers have subsequently taken up the case for the "appeasers" at the same time that German historians have begun to stress the early development of Hitler's plans and armament preparations.[1] In the present study, however, we are again concerned with both German and Western policy. We have seen that German military planning, even well before Hitler, had the objective of rebuilding German military strength in order to further an actively revisionist policy—including by war, if conditions were favorable. The emphasis was on the training of young men, and this had political consequences which might have signaled to foreign observers that something was afoot. How well, then, was the German objective perceived by Western governments? If they did not take cognizance of German plans, was there any excuse? What *were* their guiding concerns?

[1] References here will be to the American edition: A.J.P. Taylor, *The Origins of the Second World War* (New York: Atheneum, 1961). The second British edition (London: Hamish Hamilton, 1963, and Harmondsworth, Middlesex: Penguin, 1964), not generally available in the United States, includes an important new essay, "Second Thoughts." Aside from his view of Hitler, much of Taylor's thesis has won acceptance, and I strongly agree with his view that Nazi foreign policy restated the older problem of Germany's place in Europe. Other studies of British policy are discussed in this chapter, Section 5, below. On the literature on Hitler's plans: Klaus Hildebrand, "Hitlers Ort in der Geschichte des preussisch-deutschen Nationalstaates," *HZ*, 217 (1973), 584-632.

It was no secret outside Germany that there were Germans who wanted their country to regain its military power. President von Hindenburg, the former field marshal, now usually wore civilian clothes, but he still personified Germany's military past. The newspapers of the 1920s and early 1930s frequently reported large-scale rallies of the Stahlhelm, or plans for new pocket battleships. From September 1930 on, every German election seemed to bring new strength to the militant National Socialist party, whose uniformed storm troopers appeared to supplement the regular army and the more conventional paramilitary groups. Less frequently, there were press reports of illicit German armament, most notably in the *Manchester Guardian* of December 3, 1926 (republished in the German *Vorwärts*, the organ of the Social Democrats, on December 5), which Scheidemann used in his speech of December 16. Carl von Ossietzky's *Weltbühne* made its exposés, as did the French press.[2]

The British and French governments naturally collected much larger quantities of information. The Allies had foreseen that Germany might not willingly disarm as prescribed by the Treaty of Versailles, and in the early 1920s the Inter-Allied Military Control Commission had carried out sweeping inspections to ascertain the degree of German compliance. This supervision had evoked bitter resentment among German officers. One comprehensive Control Commission report in 1925 had listed numerous cases in which the treaty was being violated. The French press published stories, based on this report, which described German attempts to obstruct the commission's inspections, the capacity of Germany to produce war material, and the ability of the Reichswehr "army of cadres" to draw reserves from short-term volunteers, auxiliary police, and paramilitary organizations. Another, final report was being drawn up as the commission prepared to disband, and this survey still listed many violations. But the Germans issued decrees that appeared to meet some of the worst complaints, and in the political interests of closing off a source of friction, the Allies proceeded to dissolve the commission on January 31, 1927. In its place, a few experts were appointed to verify that the remaining discrepancies were removed. These experts lacked clear authority to inspect, and although they jointly reported three years later their doubts as to German compliance, mere doubts could not carry much weight. German officials had some apprehensions after the experts'

[2] In the Weimar period, the *Guardian*'s correspondent in Berlin was Frederick Voigt, whose reports showed unusual penetration into German military thinking. On Voigt, see David G. O. Ayerst, *Guardian: Biography of a Newspaper* (London: Collins, 1971), pp. 501-513. On Ossietzky and the *Weltbühne*, Istvan Deak, *Weimar Germany's Left-Wing Intellectuals* (Berkeley: University of California Press, 1968), esp. Chapters 8 and 13.

comments were forwarded to the League in March 1931, but although the joint statement was too restrained to suit some of the French, the British government was inclined to ignore it.[3]

In the early 1930s, the British and French governments each had, of course, its own current intelligence on German violations. For the professional intelligence analyst, there are many other sources of information besides the reports of secret agents. The most obvious and incontrovertible indicator of illegal German armament was the German budget. To judge by the budget, the cost of military equipment was strangely higher in Germany than elsewhere. Indeed, German military expenses obviously exceeded the amounts required to maintain the strengths authorized by the treaty; presumably the excess amounts represented excess stocks of authorized weapons, or more likely, expenditures for other—forbidden—items. Field Marshal Sir George Milne, Chief of the Imperial General Staff, signed a report in February 1930 noting that if only the authorized number of replacements was made, the sum allotted for German trench mortars would make them cost over RM 100,000 apiece, twice the cost of a field gun, or 120 times the cost of a British trench mortar. The British military attaché in Berlin, Colonel J. H. Marshall-Cornwall, pieced together much information, and he and the financial attaché reported in July 1931 that despite the depression-induced fall in prices, for example of fodder, budget allotments for German army purchases of these items did not decline correspondingly. In February 1932, the French Deuxième Bureau, which evaluated intelligence for the French General Staff, informed the Premier, André Tardieu, that for a national army one-fifth the size of the Prussian army of 1913, Germany was spending two-fifths as much money, even allowing for changes in the cost of living. The French also discovered without great difficulty that some military expenses were being met from the budgets of other ministries:

---

[3] Gatzke, *Stresemann*, pp. 15-16, 19-23, 26-28, 31-36, 44-45, 64-66, 70-71, 92-93; Castellan, *Réarmement clandestin*, pp. 537-540; Jon Jacobsen, *Locarno Diplomacy* (Princeton: Princeton University Press, 1972), pp. 91-97; John P. Fox, "Britain and the Inter-Allied Military Commission of Control, 1925-1926," *Journal of Contemporary History* (*JCH*), 4 (1969), no. 2, 143-164. The Control Commission's work is studied at length in Salewski, *Entwaffnung*, which concludes (pp. 376-378) that it had not been possible to deceive the Allied officers, but that Allied politicians had different priorities. He suggests, however, that Germany could not have been kept disarmed against her will. For 1925 report: BA-MA, II H 272, Heeresfriedenskommission 379/[25], att. 5 Jun 25; file II H 272 contains abundant material showing the German view of the commission and efforts to deceive it. Summary of 1927 report: ML, Shelf 5993, Sachverständiger der Reichsregierung 96/27, att., 26 Aug 27, reel 5/729-828. Ambassadors' 1930 report: *DBFP*, 2/I, No. 357. Concern of the German government: ML, Shelf 5993, Circular tel. from Bülow and att., 11 Apr 31, reel 5/1045-1053; Briefing note, 22 Apr 31, reel 5/1033-1034.

for example, the Ministry of Interior was observed to support historical and geographical services on behalf of the Reichswehr.⁴

Putting together scraps of information from newspapers, Reichstag debates, on-the-spot observations, photographs, directories, purloined documents, and agent reporting, the French and British gleaned other information. For example, both observed that German generals were retiring at an early age, and the French surmised that a reserve of such officers was being built up. Both British and French analysts were well aware that arms were being manufactured in countries such as Sweden and the Netherlands by firms linked financially and technically with Germany, thus maintaining German technical expertise. The British and French both knew that German aviators had been trained in Russia, although the Deuxième Bureau had been at first skeptical of Polish intelligence reports on the subject, while Marshall-Cornwall reported prematurely that the practice had apparently ceased. Such matters as the construction of new barracks and the failure to dispose of of old ones inevitably attracted attention, as did mechanization of transport and the acquisition of modern technical equipment. In all, many violations of the Treaty of Versailles were noted; the French in particular compiled an enormous "dossier" of them, and frequently hinted that they were about to divulge its contents. Sometimes certain violations were candidly admitted; in December 1931, General von Hammerstein told Colonel Marshall-Cornwall:

> It is true that we are hard at work mechanizing our army, and improving technical equipment in certain ways. If this is contrary to the treaty, it merely shows on what petty and ungenerous lines the military clauses were drafted. We consider that we have not only the right to motorize our army, but also to equip it with an aviation unit and other defensive arms which are denied to us—though not with any offensive intentions. Of course, we are accused of training officers as pilots in Russia, but, if that is so, it is again the fault of the Versailles Treaty.⁵

Hammerstein's remarks, for all their apparent candor, also pointed to a weakness in Western intelligence: despite the various treaty vio-

⁴ *DBFP*, 2/I, Appendix II; 2/II, No. 186; Castellan, *Réarmement clandestin*, pp. 57, 60; Martin Gilbert, *Sir Horace Rumbold* (London: Heinemann, 1973), pp. 321-322.

⁵ Castellan, *Réarmement clandestin*, pp. 131, 189, 192-195, 267, 274-284, 405; *DBFP*, Ia/V, No. 268, att.; 2/I, Appendix II; 2/II, Appendix IV; 2/III, Appendix IV; 2/IV, No. 57. A summary of some of the material in the French dossier is given in *DDF*, 1/IV, No. 65, Annex. Marshall-Cornwall's estimate that aviation training in Russia had ceased was actually dated the same day as his (presumably later) talk with Hammerstein.

lations, the nature of German intentions remained unclear to Western analysts. Hammerstein could attribute the violations simply to the petty and inflexible nature of the restrictions, and it was difficult to contest his assertions. The British and French general staffs had very little authentic information at this time on the long-term plans of the Reichswehr. The British obtained a German document which forecast that a regrouping of the powers would permit a new German mobilization, but this did not lead them anywhere. The first French awareness (in March 1932) of definite German plans for a militia seems to have come from an American assistant military attaché in Berlin, who was informed by the Germans themselves. The Deuxième Bureau apparently never gained knowledge of the first armament plan, nor any detailed information on the second. The greatest independent success of the French seems to have been their acquisition of an early document on the preparations of the Landesschutz organization for mobilization. The Poles, who had 200 officers working on intelligence collection, as opposed to 60 for France, were somewhat more successful. Many important German documents dating from 1932, including plans for a new peacetime army, with three-year enlistments and a militia for training reserves, were obtained by Polish intelligence. The principal agent in obtaining this material was probably Captain Jerzy Sosnowski, an ex-cavalryman with a flair for seducing the wives and daughters of German officers, particularly those who were also employed as secretaries in the Bendlerstrasse. The fact that their Reichswehr pay was low was also a help in recruitment, although the suspicious rise in their standard of living eventually gave them away. In late 1933 and early 1934, the Poles suddenly became more forthcoming and gave the French some of their documents. By that time, however, the French also had other, more up-to-date information.[6]

In the absence of conclusive evidence of German aims, some intelligence officers made shrewd observations. In March 1930, General Tournès, then French military attaché in Berlin, estimated that Germany would soon seek to change the twelve-year voluntary-enlistment

---

[6] *DBFP*, Ia/V, No. 268, att.; Castellan, *Réarmement clandestin*, pp. 186n, 390-391; also Castellan's 1952 doctoral thesis, "Le Réarmement clandestin du Reich," pp. 26-29, Note VI, and pp. 110-112, Note XIV. On the Sosnowski case: Michael Alexander Graf Soltikow, *Rittmeister Sosnowski* (Hamburg: Verlag der Sternbücher, 1954), and Bernard Newman, *The Sosnowski Affair* (London: Werner Laurie, 1954); there are contemporary references in the Bundesarchiv, Koblenz, Z Sg 101/28, Brammer Collection, Information report, 18 Feb 35 and Z Sg 110/1, Traub Collection, ltr. from Kurt Metger, 18 Feb 35. (The Z Sg files contain private reports and letters by journalists on current events and Propaganda Ministry instructions.) Two of Sosnowski's German accomplices, Benita von Berg (or von Falkenhayn) and Renate von Natzmer, were beheaded in 1935; he was exchanged. French intelligence on Germany in 1934 as reflected in *DDF*, 1/V, No. 227; 1/VI, No. 40.

system. French intelligence noted in 1930 and 1931 the problems the Germans faced with the lack of young trained reserves. Colonel Marshall-Cornwall, referring to equipment and technique as well as organization, wrote in December 1931 of evident signs that "a change of system is not only meditated but is being actively prepared." A British War Office study of March 1, 1932, commented that "the German army is departing to an increasing degree from the type envisaged by the treaty."[7]

The dearth of authentic information on intentions led nevertheless to weak or distorted conclusions, either on the part of the report writers or on that of their readers. There was the laissez-aller view, which Hammerstein had tried to foster in Marshall-Cornwall's mind, and which was most common among British and American observers; this reading of the situation was that the violations were unimportant, simply a byproduct of the pettiness of Versailles, and of the even more detailed restrictions laid down by the Inter-Allied Control Commission in the early 1920s. The American military attaché commented in 1931, "In any case I feel that the sooner these galling and childish restrictions are replaced by some larger-minded and mutually acceptable convention, the better it will be for the peace of Europe." Those who felt this way could argue that the Reichswehr was still far from able to cope with a major attack. A War Office (MI-3) memorandum noted in January 1932 that violations were increasing in armament, equipment, training, and organization, but it added, "On the other hand, no attempts have been made at important violations, such as conscription, or an increase in peace effectives, which might foreshadow aggressive tendencies." This conclusion was accurate as to fact, but of little value as a prediction.[8] And there were deeper reasons for avoiding predictions.

In contrast, French reports occasionally overinterpreted the evidence to show an aggressive intent. An October 1931 report on industrial mobilization was filled with conclusions that stressed the obvious, or else presented speculations as certainties:

> If the present activity of German industry in producing war material is relatively reduced, its possibilities of extension are immense. . . . There is no doubt that at the present time, German industry is

[7] Castellan, *Réarmement clandestin*, pp. 403-404, 533n-534n; *DBFP*, 2/II, Appendix IV; 2/III, Appendix IV.

[8] National Archives, U.S. Army Military Attaché Reports, Berlin Report No. 11,784, 17 Nov 31; FO 371, War Office ltr. of 6 Jan 32 and att., C211/211/18. The U.S. military attaché reporting from Berlin was of very little value, failing to report on such developments as Umbau or the Jugendertüchtigung program, even though the American officers were aware of German plans (cf. Chapter One, note 55, above; *DBFP*, 2/IV, No. 165).

in a state to respond (on condition of being provided with certain raw materials) to all the requirements of the armed nation in time of war . . . several essential industries are suited for immediate or nearly immediate transformation into war industries: the chemical, automotive, aeronautical, and electrical industries. . . . Moreover, it is certain that Germany is actively preparing for industrial mobilization. . . . The Reichsverband der Deutschen Industrie, which has its headquarters in Berlin in a building next to the Reichswehr Ministry, brings together, under the authority of the most important chiefs of industry, the principal branches of the German economic system, directing and coordinating their activity. This powerful association is admirably placed for studying and preparing industrial mobilization and it seems indeed, in effect, that the RVDI plays an important role in this matter. There is no doubt that it maintains close relations with the government and notably with the Reichswehr Ministry.

Despite its suspicious proximity to the Reichswehr Ministry, there seems to be no evidence that the RVDI was actually engaged in the planning of industrial mobilization. Probably most of the French staff officers who read this report accepted its conclusions without question. Ultimately, however, it would be desirable to provide information, if only indirectly and in a bowdlerized version, which could carry conviction in more skeptical quarters, both in other countries and in French political circles.[9]

It was common practice to disparage the quality of French intelligence, which on the whole was really more sound and professional than most people supposed. An incident in 1931, when French officers were arrested for openly photographing antiaircraft equipment at the artillery barracks in Königsberg, was regarded by Marshall-Cornwall as an example of tactlessness, not to say petty snooping. General von Blomberg later commented, however, that the French officer responsible had not been stupid; he had been doing his job. The Deuxième Bureau tried to reach the public through articles by officers, sometimes under pseudonyms, charging the Germans with rearmament, but these articles, which could not reveal secret sources, and which tended to seem alarmist in tone, made little impression outside of France; both the British and American military attachés in Berlin belittled their

[9] Castellan, *Réarmement clandestin*, pp. 262-263; see also *DBFP*, Ia/V, No. 268, att. Carroll, *Design*, pp. 68-69, indicates that it was the "Statistical Society" (Stega) that sponsored industrial mobilization preparations, and that, according to Gen. Thomas, some industrialists did not cooperate, including I. G. Farben chairman, Carl Duisberg, who headed the RVDI during much of this period.

value.[10] The reliability of French intelligence was sometimes questioned in France itself. Once, when the German ambassador suggested to Premier Edouard Herriot that he was being misinformed on the Wehrsport program by "interested parties," Herriot replied, not that the information he received officially (for the most part, presumably, from the Deuxième Bureau) was correct, but that he had other, personal sources of information. In April 1933, another French premier, Edouard Daladier, told American diplomats that he had "most reliable" information that Krupp's was building heavy artillery; as recorded by Allen W. Dulles—the future chief of the Central Intelligence Agency—Daladier indicated that his information came from German sources opposed to the Hitler regime, "not the type of military espionage sources that are so unreliable." Such comments took no account of the fact that the reports of the Deuxième Bureau were based on a sifting of information from many sources; false information is usually weeded out by this means, and important insights may emerge. In 1932, General Maurice Gamelin knew that the Germans intended to treble the Reichswehr on mobilization. Contrary to what British and American observers usually thought, the (admittedly incomplete) record does not indicate that the Deuxième Bureau seriously exaggerated the scale of German preparations, as now known from German sources.[11] But, as noted, the French army seems to have lacked information on the armament plans and on the Manstein plans for mobilization and a militia. And the British and French had almost no hard intelligence on the ultimate problem of German intentions.

2. THE SPECTER OF GENERAL VON SEECKT

In the absence of secret evidence, it was natural to try to unravel the mystery by looking for overt statements on German plans and strategy. As it happened, an abundance of material was available in the published writings of General Hans von Seeckt, who had resigned

[10] *DBFP*, 2/II, Appendix IV; National Archives, Microfilm T-84, Blomberg Manuscript Memoirs, IV (EAP 21-a-14/30c—no frame numbers), serial 202; U. S. Military Attaché Reports, Berlin Nos. D-12, 602, 16 Jan 33; D-13264, 24 Jan 34; also, in the same collection, for indications of official inspiration of articles published in France, Paris Reports 18,073-W, 16 Jan 32; 18,149-W, 15 Feb 32; 18,210-W, 5 Mar 32; 19,951-W, 8 Dec 33.

[11] RK, disp., Köster to Ministry, 18 Nov 32, 3642/D812989-90; SDP, Memo on conversation with Daladier, 21 Apr 33, 500.A15A4 General Committee/321; *DDF*, 1/I, No. 286; note also 1/IV, No. 65. On Herriot, see also Bredow Papers, Orientation note, 6 Oct 32. Dulles wrote the report on Daladier's views, and it may have been Dulles himself who considered military espionage sources "so unreliable." But skepticism about the French military was very much in character for Daladier. On German refusal of promise not to fight, see Chapter One, note 16.

as Chief of the Army Command in 1926. Monocled, stiff-backed, and laconic, Seeckt impressed foreign observers as being the very model of a Prussian military thinker. As leader of the army, he had accepted the idea that the land forces should be expanded three- or ninefold in the event of war; indeed, it is possible that he may have devised the plan. Even in 1919, however, he distinguished sharply between professional and conscript elements, and discounted the latter.[12] As time passed, the aristocratic Seeckt came to regard the 100,000-man professional army as the model for a self-contained élite force, capable of delivering an instant attack. As he envisioned future developments, there would also be militialike elements comprising large numbers of men who had undergone short-term training, although these would only perform defensive tasks. The age of the mass army, Seeckt thought, was over.[13] Seeckt has sometimes been regarded as the father of the Blitzkrieg, and it may be that the German employment of armor in World War II owed something to his ideas.[14]

But Seeckt's ideas were not dominant within the Reichswehr in the later 1920s and the early 1930s. While he was still in office, there was dissatisfaction: General Otto Hasse commented in 1924 that Seeckt wanted "a model army of 100,000 without regard for mobilization," and Joachim von Stülpnagel thought that under Seeckt's guidance, the army was being prepared for war with Poland while shutting its eyes to France, competing with the navy, and overlooking air development. We have seen that in April 1930, the German army committed itself to a mobilization policy completely opposed to the idea of an élite force, choosing instead to dilute every regular division by large numbers of reserves. Later that year, after Seeckt had entered politics, he astonished the Reichswehr Ministry by stating in an article that the limitation of weapons was not a very important matter, as the rapid progress of technique made large accumulations of arms uneconomic and of doubtful military value; this argument was promptly used by a French diplomat to support financial limitation instead of the limitation of arms stores. Even Seeckt's faithful fol-

[12] Rabenau, pp. 474-475; National Archives, Microfilm T-132, Seeckt Papers, Memo, Seeckt to Groener, 17 Feb 19, reel 21/Stück 126.

[13] Hans von Seeckt, *Gedanken eines Soldaten* (Berlin: Verlag für Kulturpolitik, 1929), pp. 81-100; cf. Rabenau, pp. 613-619. One might suspect that Seeckt's writings were deliberate deception, but in his *Gedanken* (p. 81), Seeckt warned explicitly that his observations were only personal opinion and lacked any official stature, including such as might be inferred from his past career.

[14] B. H. Liddell Hart, *The German Generals Talk* (New York: Morrow, 1948), pp. 10-11, 14; Irving M. Gibson, "Maginot and Liddell Hart: The Doctrine of Defense," in Edward Meade Earle, ed., *Makers of Modern Strategy: Military Thought from Machiavelli to Hitler* (Princeton: Princeton University Press, 1943), p. 369.

lower, Friedrich von Rabenau, reluctantly conceded in his 1940 biography that the general had been ready to settle for less than maximum rearmament.¹⁵ Seeckt's immediate successors hardly foresaw the full development that took place under Hitler, but they did turn directly to preparations for the mobilization of all of Germany's resources in materiel and manpower. Instead of consoling themselves with the idea that the day of the mass army was over, they tried to see what could be done to regain a mass army for Germany.

In Britain and especially in France, however, it was assumed that Seeckt's ideas still inspired the German army. For one thing, Seeckt was virtually the only available authority. But beyond this, the military beauty of Seeckt's ideas lay to a large extent in the eyes of their British and French beholders. The idea of a long-service professional force of modest dimensions fitted in with British military tradition and preference. It had been Prime Minister David Lloyd George, in fact, who had insisted on having Part V prescribe a long-service volunteer system for the Reichswehr, his expressed aim being to avoid any necessity for a large-scale army in Britain itself, such as would require peacetime conscription. Marshal Foch had argued that a long-service army would provide cadres for mobilization, but he had been overruled, and had then insisted on a limit of 100,000. Although British politicians in 1919 had seen the voluntary system as an antimilitarist safeguard, Seeckt's view of the possible employment of a professional army coincided later with the interest of British military thinkers in highly mobile armored forces.¹⁶ It also suited the predilections of those French professional officers who argued that long-service elements were the backbone of any army, and in the case of Charles de Gaulle, that a professional army was superior to an army of millions of mobilized reserves. Admiration for Seeckt's analysis thus had self-congratulatory overtones in certain military quarters in both Britain and France. At the same time, civilian critics of the military on the

¹⁵ Meier-Welcker, *Seeckt*, pp. 537-538, 604-605; Waldemar Erfurth, *Die Geschichte des deutschen Generalstabes von 1918 bis 1945* (Göttingen: Musterschmidt, 1957), pp. 79-80; Vogelsang, *Reichswehr*, pp. 96-97; Rabenau, pp. 612-613.

¹⁶ On Lloyd George's policy and Foch's views: *FRUS, Paris Peace Conference, 1919*, IV, 217-219, 263-264. On British army theorists: Robin Higham, *The Military Intellectuals in Britain: 1918-1939* (New Brunswick, N.J.: Rutgers University Press, 1966), pp. 67-116. Although the British had a German draft manual stating that the Reichswehr was an army of cadres, Sir George Milne believed in December 1928 that the Reichswehr was following Seeckt's doctrine (*DBFP*, Ia/V, No. 268), and the following month Milne urged all British army officers to study an article by Seeckt (Liddell Hart, *Memoirs*, I, 159); in 1932, Milne and others stressed that the Germans were applying Seeckt's ideas; see Chapter Five, Section 5. Belief in Seeckt's influence has persisted: see, e.g., Michael Salewski, "Zur deutschen Sicherheitspolitik in der Spätzeit der Weimarer Republik," *VfZ*, 22 (1974), 134-136.

French left traditionally regarded a professional army as a danger to democracy at home; opinion was therefore divided as to whether and to what degree a Seeckt-style army was a model for France. French governmental spokesmen argued that Seeckt's Reichswehr might attack France before she could mobilize, and they used this as an argument against the disarmament of France. The official leaders of the French army—Marshal Henri-Philippe Pétain, Generals Maxime Weygand and Maurice Gamelin—still followed Foch, regarding the Reichswehr, correctly, as a source of cadres for mobilizing a larger army. All French observers tended to agree that, in German hands, the professional Reichswehr was a formidable threat, but most of them overlooked the greater danger of the mobilized nation. When the German proposals for a militia for Germany came to light, this disclosure was widely taken as a confirmation that Seeckt's ideas were being followed. In view of Seeckt's theory, however, the militia was not considered dangerous in itself, in the way that the German professional soldiers were, and indeed the French government was to advocate a militia army, with a minimum of professionals, for Germany. This would have given Germany the same military system that the French left considered desirable for France.[17]

[17] On French military concerns and preoccupation with Seeckt, see Judith M. Hughes, *To the Maginot Line* (Cambridge, Mass.: Harvard University Press, 1971), pp. 82, 193-194, 209-211; Jacobson, *Locarno Diplomacy*, pp. 111-112; Gibson, "Maginot," p. 370n; Philip C. F. Bankwitz, *Maxime Weygand and Civil-Military Relations in Modern France* (Cambridge, Mass.: Harvard University Press, 1967), pp. 42-43; Jeffrey Albert Gunsberg, "'Vaincre ou Mourir': The French High Command and the Defeat of France, 1919–May 1940" (Ph.D. diss., Duke University, 1974), pp. 26, 37. In Castellan, *Réarmement clandestin*, references to Seeckt or his ideas appear on pp. 28, 72, 73, 108, 256, 426, 447-449, 459-460, 461-463, 465, 473-474, 488-489, 491-492, 494, 511, 513, 527, reflecting the interest of French military circles, shown in Castellan's documentation. The interest of politicians and diplomats is shown in *DDF*, 1/I, Nos. 120, 126, 166, 172, 197, 244, 250, 272, 286, 315; in the last document, the ambassador in Berlin, André François-Poncet, expressed some reservations as to Seeckt's influence. The army leaders took the Foch view in a top-level meeting (*DDF*, 1/I, No. 286), but they did not argue with civilian statements of the Seeckt interpretation, and seem not to have been clear on the form the expansion would take. An analysis by General Baratier (in *DDF*, 1/I, No. 166), stressing the supposed Seeckt influence on current German military policy, was given to Premier Herriot, who drew on it for his 25 Sep 32 speech: FO 371, ltrs., Tyrrell to Simon and Heywood to Tyrrell, 29 Sep 32, C8261/211/18. (At this time Baratier was attached to the French Foreign Ministry.) The opposing French doctrines of professional and national armies are studied in Richard D. Challener, *The French Theory of the Nation in Arms, 1866-1939* (New York: Russell & Russell, 1965) and Volker Wieland, *Zur Problematik der französischen Militärpolitik und Militärdoktrin in der Zeit zwischen den Weltkriegen* (Boppard: Harald Boldt, 1973). On de Gaulle and his forerunners, see esp. Challener, pp. 245-251. De Gaulle (like Seeckt) was not so much concerned with using armor as with developing an élite army of professionals: Volker Wieland, "Pigeaud versus Velpry: Zur Diskussion über Motorisierung und Mechanisierung nach dem ersten Weltkrieg," *MGM*, 17 (1975), 49-51.

3. Distortion through British Hopes and Fears

Attitudes on professional versus militia armies probably affected French judgments more than British, but another consideration distorted the British perception of the German army. This distorting influence was the overwhelming British desire to find harmonious solutions to European problems. We will see presently why this interest was so strong; what is pertinent here is that this interest seems to have affected the use of intelligence. Everyone knows that it is tempting to report what people will want to hear, and everyone agrees that this temptation, if it means distortion, should be resisted. But there are many reasons, aside from fear of displeasing superiors, why officials do not resist the temptation. One is the reporter's own preference. Another is that an unwelcome report requires more evidence, a tighter argument, a preparation to fend off criticism. A third reason, most potent in the British case, is that writing or relaying a report is not simply an academic exercise: to report is to act, and possibly to affect later developments. A report of foreign military preparations might lead to preparations on one's own side, and to resignation to the eventual likelihood of war. British officials in 1932 seem to have hesitated before making such reports, with a potential effect so contrary to their own hopes.

In most governments, a few offices or individuals do have an interest, or perhaps a fixation, that encourages them to circulate opposing views and to raise awkward questions. In the British government, however, the reporting of German violations was discouraged in the mid-twenties, in the interests of reconciliation,[18] and in the early thirties virtually no one wished to sound the alarm on Germany, even if he had disturbing information. In February 1931, the highly perceptive ambassador, Sir Horace Rumbold, reported that "most Germans have no affection for disarmament in the abstract," that a war with Poland would be highly popular in Germany, and that senior German officers were very concerned over the loss of conscription, more so than over

[18] Brig. Gen. J. H. Morgan, a British member of the Inter-Allied Control Commission until 1923, did not believe that the Germans had disarmed. His introduction to the volume he published later, *Assize of Arms* (New York: Oxford University Press, 1946), shows the pressure within the British government of the mid-1920s against revelations of German rearmament, and suggests, despite disclaimers, that he was then officially discouraged from publishing his book. A second volume, reportedly more pointed than the first, has not been published yet; see Fox, "Britain and the Inter-Allied Military Commission," p. 144 and note. In 1924, the army general staff argued that French fears were justified, and that the German danger would become acute in ten years, but the Foreign Office dismissed these views as unconstructive: Schuker, pp. 254-255.

their lack of materiel. But along with these accurate insights, Rumbold also assumed that the Reichswehr was controllable by the Reich and Prussian civil governments; perhaps he was influenced by his conviction that the British government should trust and support Brüning. British military observers did note at the end of 1931 the increasing political power of the German army, along with that army's closer relations with the Nazis, and the military potential of the Wehrverbände. Colonel Marshall-Cornwall pointed out that it would be difficult to arrest the Reichswehr's development once it was launched, and he suggested that German infractions of Part V might justify a protest. But he also doubted that such a protest could be enforced, and both the War Office and Foreign Office opposed any such diplomatic step, especially on the eve of the Disarmament Conference. With this decision, made within the two departments, the whole matter was put aside, without any apparent effort to consider how Britain might meet the possible danger. In late May 1932, the Berlin embassy received an account of how military considerations had led to Groener's downfall, but the embassy failed to link this to Brüning's overthrow, and embassy analysis attributed that overthrow almost entirely to the intrigues of the Ostelbian landowners.[19]

Although Rumbold's and Marshall-Cornwall's reports, included in the published British documents, suggest that the British government was aware of German violations, it was the ministers who really made decisions in that government, and it is not clear that in late 1931 or early 1932 any of them, aside from the Foreign Secretary and War Secretary in their separate departmental roles, were seeing and considering such reports.[20] In March 1932, the ministers did consider an overall review of the situation by the chiefs of staff of the three services, but this review centered almost entirely on the Japanese danger in the Far East.[21] It appears that the cabinet only began to have inklings of

[19] *DBFP*, 2/I, No. 352; 2/II, Appendix IV; 2/III, Nos. 113-116, 118-122, 127; FO 371, War Office ltr. of 6 Jan 32 and att., C211/211/18; War Office ltr. of 13 Jan 32, C451/211/18; Minute by P. Nichols, 25 Jan 32; rpt. by Col. Waterhouse (Mil. Att., Paris), 13 Jan 32; disp., F.O. to Rumbold, 3 Feb 32, C735/211/18. On Rumbold's desire to support Brüning: Gilbert, *Rumbold*, p. 344. Rumbold was absent from Berlin when Brüning fell. The British feared that Brüning's fall might cause a failure of the Lausanne Conference: *DBFP*, 2/III, Nos. 123, 124, 127.

[20] The War Secretary (Lord Hailsham) reportedly reviewed one of the War Office memos commenting on a Marshall-Cornwall report: FO 371, Minute by Nichols, 13 Jan 32, C211/211/18. Another W.O. memo on German infractions (in *DBFP*, 2/III, Appendix IV) was said to have been shown to the Foreign Secretary (Sir John Simon): FO 371, ltr., Nichols to Carr, 16 Mar 32, C1696/211/18. Circulated in the F.O. as a confidential print, this pointed to many problems, including the Verbände, and then concluded that German preparations were mainly defensive.

[21] Public Record Office, London, Cabinet Records, Cabinet Papers (Cab 24), Vol.

a German danger in the fall of 1932, and at that time Field Marshal Milne assured them that the Germans, in the near future and also in 1938, would only be able to mobilize fourteen infantry divisions—not the twenty-one divisions that General Gamelin correctly estimated. Milne expected that, in 1938, Germany would still be at the mercy of France or Poland, and would still be far from having the army of Seeckt, the goal toward which Milne believed the Reichswehr was striving.[22]

What British officials and ministers wanted to do was not to discover and discuss a German rearmament program, but to find ways of winning German acceptance for a disarmament convention. A Foreign Office memorandum assumed in April 1931 that "a good deal will have to be done for [the Germans] in the direction of equality before we can bring them into line," and it added, in justification, "The failure of the Disarmament Conference would have incalculable consequences for Europe and the League." A three-party committee on disarmament concluded, among other points, that the conference should consider the circumstances of each country in arriving at its armament level, that offensive power should be limited, that the methods of Part V had been generally effective (although budgetary limitation was needed as well), and that there should be no difference in the principles applied to different nations. Aside from budgetary limitation (which Britain did not in fact pursue), these points tallied with the overt German position. The main implication of this committee report was that a reduction in the arms of France would actually promote security.[23] A whole system of British hopes was based on the assumption that Germany was not rearming, and some influential citizens even thought it dangerous to suggest otherwise. In April 1934, Lord Allen of Hurtwood was to write, "I incline to think it would be a mistake to seem to be on the side of France about the secret rearming of Germany under the *Versailles Treaty*. To do that means, however carefully we put it, that we appear to re-endorse that wicked treaty and to justify the evil policies of France toward world reconciliation during the past ten years."[24]

229, CP 104 (32) (1082-B, 23 Feb 32); Committee of Imperial Defence Minutes (Cab 2), Vol. 5, 255th meeting, 22 Mar 32; Conclusions of Cabinet Meetings (Cab 23), Vol. 70, meeting of 23 Mar 32, 19(32)2. The "Conclusions" are minutes.

[22] Cab 23/72, meetings of 30 Sep 32, 49(32)—; 11 Oct 32, 50(32)5; 31 Oct 32, 56(32)3; Cab 24/233, CP 326 (32), 29 Sep 32; Cab 24/234, CP 362 (32), 28 Oct 32.

[23] *DBFP*, 2/III, No. 208; Cab 24/227, CP 19 (32), 27 Nov 31.

[24] Martin Gilbert, ed., *Britain and Germany between the Wars* (London: Longmans, Green, 1964), pp. 59-60.

4. France: The Quest for a British Alliance

The French and British views of German rearmament were only parts of their outlooks on European and domestic problems. Anyone's perception of information naturally tends to be influenced by his preoccupations, and French and British preoccupations were very different. Each nation had its own view of what had happened in the past, and what lessons should be drawn from it. The French, for their part, often said that Germany had invaded France three times within a century. The real point of this statement was not that Germany had been three times the aggressor, which appeared doubtful to British and other non-French students of history, but that the Germans had three times penetrated deeply into French territory, and France alone had not been able to expel them. Non-French observers often believed after 1919 that France had regained the power position she had held in the days of Louis XIV and Napoleon. There was talk of French hegemony, and of a need to balance the overwhelming power of France. But the French did not boast of their power. They never forgot that Germany was still a nation of nearly 70 million, as against 40 million Frenchmen. They realized that German industry had a productive capacity several times that of their own. Deep down, they were afraid, not triumphant. They were ready to believe, as the British were not, that Germany was rearming.

These concerns and beliefs might conceivably have led to punitive or preventive action; a certain justification for such action could have been found in German derelictions in the matters of arms and reparations. But although there were frequent rumors that the French were considering punitive steps, such as a reoccupation of the Rhineland, and although plans did exist for limited incursions, the French government had more or less unwittingly reduced its own capability to take even the latter kind of action. Changes during the 1920s in the organization of the French army, especially successive shortenings in the term of conscript service, reduced the number of trained men on active service, and rendered that army unable to move rapidly into Germany; extensive measures of partial mobilization (the establishment of the *couverture*) would first be required. Despite the growing threat from Germany, the left-of-center governments of 1932-1934, facing a severe budgetary crisis, would reduce defense expenditures further and resist any proposals to restore longer service; financial weakness in itself also tended to rule out punitive action. French military strength would foreseeably decline sharply in the later 1930s, the "lean years," as the low birth rate of the World War I years would result in a halving of the number of conscripts in each new annual class. Anyway, the Maginot line of fortifications in the East, begun in 1930, appeared to make a French offensive action a dangerous and unnecessary adventure, impossible to justify to the war-weary French public.[25] Such action was also likely to be a diplomatic disaster. In theory, and assuming that Germany had not committed aggression elsewhere, the Treaty of Locarno of 1925 bound Britain and Italy to aid Germany against a French attack as well as to aid France against a German one; in practice these countries might not give Germany any effective military support, but they would otherwise use their influence against France, should she advance into German territory. Even before Locarno, the entry into the Ruhr in 1923 had brought France into acute financial difficulties and a painful diplomatic isolation.

Isolation, if it lasted, would make a defense of France impossible, while firm alliances should discourage any German attack. Although the French government showed some indifference to foreign opinion in the early 1920s and again in 1931, the first concern of French diplomats between the wars, and especially after the Ruhr experience, was usually the avoidance of isolation and the winning of reliable guarantees from strong allies. The victory of 1918, they believed, had been largely due to French success in gaining the support of other powers. This belief was well founded. France had been bled white by October 1916, by which point, midway through the time-span of the war, she had suffered 66 percent of her total wartime casualties. The world heard little about it, but mutinies had crippled the French army in May and June of 1917.[26] Victory could not have been achieved without the other allies, in particular without the heavy sacrifices of the Russians in the first half and of the British in the second half of the war, along with the smaller but psychologically very important contribution of the United States in 1918.

In the 1920s, France was unable to secure comparable alliances. She had ties, it was true, with Poland and the Little Entente (Czechoslovakia, Rumania, and Yugoslavia), but instead of increasing French

---

[25] France, Assemblée nationale, Commission chargée d'enquêter sur les événements survenus en France de 1933 à 1944, *Rapport fait au nom de la commission chargée d'enquêter sur les événements survenus en France de 1933 à 1944*, 2 vols. (Paris: Presses universitaires de France, [1952]), I, 39-42, 47-50, 65; Paul-Emile Tournoux, *Haut Commandement: Gouvernement et défense des frontières du Nord et de l'Est, 1919-1939* (Paris: Nouvelles Editions latines, 1960), Annex I; Bankwitz, *Weygand*, pp. 84-85; Gunsberg, pp. 24-29. Mrs. Hughes's book is a study of this whole development.

[26] Winston S. Churchill, *The World Crisis*, 4 vols., 2nd ed. (New York: Charles Scribner's Sons, 1955), III, 300 (in Appendix I); C.R.M.F. Cruttwell, *A History of the Great War, 1914-1918*, 2nd ed. (Oxford: Oxford University Press, 1936), pp. 413-417.

strength, these increased French responsibilities. René Massigli, the leading French career diplomat concerned with disarmament questions, once (in February 1930) summarized French preoccupations to a British official:

> Since 1922, France had been constantly straining for some international agreement which would give her the only guarantee worth having, namely some form of international protection in the event of her being again subjected to a flagrant attack. While agreeing that the various peace pacts all had a certain value in calming world opinion, he pointed out that there was nothing either in the Covenant of the League or in the Briand-Kellogg Pact which really afforded such a guarantee even to a Power which was the innocent victim of the most flagrant violation of the Peace Pact. . . . Locarno was of the utmost value as far as Germany's western frontier was concerned: however, it definitely left open the question of Germany's eastern frontier so that if trouble arose between Poland and Germany at any moment, France might be forced in, but the guarantors of Locarno [Britain and Italy] would retain their liberty of action.[27]

Although Massigli spoke initially of a guarantee against a flagrant attack on France, he evidently wanted a guarantee for the small allies of France as well. He apparently believed—quite rightly—that Germany was more likely to launch a war over an Eastern issue than any other, and while France was committed to intervene in such a case, Britain and Italy might well stand aloof in this all-too-conceivable eventuality. Actually, France herself might also feel unable to act without British support, leaving Poland and the Little Entente a prey to Germany. As A.J.P. Taylor has put it in a characteristic paradox, "France could act offensively, to aid Poland and Czechoslovakia, only with British support; but this support would be given only if she acted defensively, to protect herself, not distant countries in Eastern Europe."[28]

It was not even entirely clear that Britain would defend France. In 1919, President Woodrow Wilson and Prime Minister Lloyd George —in order to induce France to forgo a permanent occupation of the Rhineland—had offered treaties guaranteeing France against German aggression. But the United States Senate failed to ratify, or even consider, the Franco-American treaty, and following this failure, the British declared the corresponding Franco-British treaty to be null and void. Because the British did not want to involve themselves in a con-

[27] *DBFP*, 2/I, No. 150.
[28] Taylor, *Origins*, p. 38.

flict in which the United States was neutral, and indeed did not want to enter another conflict in any case, they remained reluctant to ally themselves directly and unequivocally with France. In 1922, the British offered a promise to help defend France itself against attack, but they rejected French counterproposals for a guarantee of the whole peace settlement, backed by a convention between the general staffs. The French Foreign Ministry still seems to have thought that the British might agree to a wider commitment under the League. Pacts like the Geneva Protocol of 1924 would have involved Britain in such a commitment, but even though some British enthusiasts for the League supported these proposals, they never had a chance of acceptance by London, particularly in view of the opposition of the British dominions and the United States.[29] Although the British had promised at Locarno to assist France if she were invaded, the sense of the Locarno Treaty was to make British action depend upon the British view of the situation when conflict arose. Aside from the lack of an Eastern commitment, Britain's responsibility to both France and Germany in the West prevented any preliminary military planning with France. As the French well realized, joint general staff planning provides a much stronger guarantee of action than a promise of support in case of attack. When war breaks out, the identity of the attacker is often a highly debatable question; joint planning not only means timely preparation, it also creates an identity of interest and leads to a common view of the rights and wrongs.[30]

Evidence of German rearmament strengthened the already existing wish of French diplomats for firm commitments of support from other nations, especially Britain. But while Germany was secretly violating the restrictions on her armament, she was publicly demanding that France disarm in the way that she herself had, invoking the inspired slogan of "equality of rights." As the 1932 Disarmament Conference approached, this German demand was gaining support in many other countries, including the United States and Great Britain, where many people assumed, naturally enough, that the main function of the conference was to disarm "the heavily armed states," meaning particularly France. The Quai d'Orsay, the French Foreign Ministry, privately urged the British government, in July 1931, not to encourage the Germans in their demands for equality; the Quai argued that the

[29] F. S. Northedge, *The Troubled Giant: Britain and the Great Powers, 1916-1939* (London: London School of Economics and G. Bell, 1966), Chapter 9; Arnold Wolfers, *Britain and France between Two Wars* (New York: W. W. Norton, 1966), pp. 78-80 (incl. n. 7), 88, 344-347 (incl. notes 18, 19, 21, and 25); David D. Burks, "The United States and the Geneva Protocol of 1924: 'A New Holy Alliance?'" *AHR*, 64 (1958-1959), 891-905.

[30] See also Taylor, *Origins*, pp. 54-55; Wolfers, pp. 171, 259.

Germans had no legal right to equal treatment in armaments, and that, due to German violations of Part V, the preconditions for general disarmament specified in the preface to Part V had not been met. France would therefore not concede equality. If, however, the Germans felt confident of British support, they would leave the conference and rearm when their equality was not admitted. It would be well, the French suggested, for the British to warn the Germans if they did not support such actions, so as to lead the German government to moderate its demands and exert a restraining influence on the German press.[31]

Fundamentally, however, the French could only compete with a slogan like equality of rights by promoting a different principle, a different definition of the question under negotiation: the principle they had long supported and would rely on now was security. Their public memorandum of July 15, 1931, argued that the reduction of national armaments foreshadowed in the League Covenant presupposed the development of a system of common action and the maintenance of disarmament in the states disarmed. It pointed out that, although there had been little movement toward a system of common action—i.e., a system that would guarantee French security and the Versailles settlement—France had already reduced her armed strength. In particular, the gradual reduction of the military service obligation from three years in 1921 to one year in 1928 had meant a loss in the military effectiveness of the French soldier, as well as a reduction in the number of trained men available for home defense by 42 percent, or by 60 percent if the "expeditionary force" was sent to meet emergencies in the colonies. Without promising any further disarmament, the memorandum stated that reduction in armament required confidence, and that this could only be provided if a system of mutual security were organized for all countries.[32]

This particular memorandum was less a bid for support than a rationale for refusing to consider any reduction in French forces; Foreign Minister Aristide Briand told Curtius that it represented a Ministry of War position, rather than a position of the Foreign Ministry. Thanks to the devaluation of the franc in 1928, the French were in an unusually strong financial position in 1931, and the right-of-center governments of Pierre Laval and André Tardieu could be relatively indifferent to foreign opinion—including British opinion, as the British were financially weak. By imposing conditions on their financial assistance, the French forced the Austrian government to abandon the project for an Austro-German customs union in the sum-

[31] *DBFP*, 2/III, No. 214.
[32] *DBFP*, 2/III, No. 213.

mer of 1931, in spite of a gesture by the Bank of England in support of the Austrians; the French and British were also very much at odds over reparations and war debts.[33] Behind this, many on the French right nursed a resentment of seeming British blindness and faithlessness, and there was a current in French opinion which believed that disarmament was aimed directly at France. On November 26, 1931, a Parisian audience jeered at speakers before an international disarmament rally at the Trocadéro, finally bringing the session to an inglorious and disorderly end; the police took only a tardy interest in checking the violence of the far-rightist Leagues, which were mainly responsible, and there was some suspicion that the government had wittingly tolerated the disturbances. The demonstrators showed special hostility to Alanson B. Houghton, a former American ambassador to Germany and Great Britain, who had recently suggested that if France would not reduce her armaments, other governments should proceed to reach an agreement without her—thus seeming to dictate disarmament under threat of isolation. As a British historian has written, a "disarmament" that meant an increase in German relative strength was, to the French, "a mad enterprise" at a time when extreme nationalism was advancing in an unstable Germany.[34]

On the eve of the Disarmament Conference, French experts drew up two alternative plans for their government, both in the tradition of advocating a stronger League. One, the maximum plan, would have completely replaced national standing armies by international standing forces controlled by the League. The other, the minimum plan, proposed that each nation designate certain units to form an international police force under League command, which could be supplemented in case of aggression by further contingents from existing national standing armies; the latter plan also suggested the transfer of military aircraft and heavy land and sea armaments to the League Council. In the event that the maximum plan was adopted, the individual nations were to retain only militia forces (land, air, and naval contingents to provide border defense pending the arrival of international forces), specialized colonial forces for nations with overseas territories, and police. These militias would include only a few professionals, mainly as instructors. With both maximum and minimum plans, the French experts also called for the internationalization of civil aviation and the transfer to the League of all heavy, long-range

[33] AA, Memo by Curtius, 22 Jul 31, 3154/D666724; Bennett, *Diplomacy*, pp. 7, 82-85, 171, 204-218, 226-228, 252-258, Appendix I.
[34] *SIA, 1931*, pp. 285-286; Michel Soulié, *La Vie politique d'Edouard Herriot* (Paris: Colin, 1962), pp. 341-342; R.A.C. Parker, *Europe 1919-45* (New York: Delacorte, 1970), p. 248.

bombers; they further suggested the conclusion of regional pacts of mutual assistance, and of provisions for inspection and sanctions. Either plan would have involved Great Britain in firmer commitments, while, regarding the United States, both plans raised the question "of the guarantees to be obtained from states not belonging to the League, so that the system envisioned might be able to function." The proposal of militia armies, contained in the maximum plan, would have generalized the kind of army that, under pressure from the left, had already been largely introduced into France.[35] Germany could then have shortened her term of service and begun the accumulation of reserves, although the plan was largely intended to strip her of her highly trained professional forces.

The maximum and minimum plans were both approved on January 8, 1932 by French ministers, acting as the Supreme Council of National Defense (Conseil supérieur de la Défense nationale). Nevertheless, General Gamelin, the Chief of the General Staff, stated in the name of the Minister of War that although the French delegation at the Disarmament Conference was entitled to present the maximum plan, that plan appeared to be utopian under existing world conditions, and it therefore implied a profound (and prior) change in those conditions. The views of the Minister of War, André Maginot, were not by themselves necessarily binding, even though they may have acquired a certain testamental force from the fact that he had died the previous day, having suffered food poisoning from eating bad oysters. Nevertheless, his successor, André Tardieu (who would become Premier in February), offered to the Disarmament Conference only an elaboration of the minimum plan, and the maximum plan was shelved and kept secret for the time being. It was probably General Weygand, Inspector General of the army and, as such, designated to command it in the event of war, who actually inspired Maginot's comment.[36] In later deliberations, Weygand would emerge as the principal opponent of the maximum plan, regarding it as fatal for the military effectiveness of the French army. As the maximum proposal would have virtually ended the professional element that Weygand regarded as both heart and spine of the army, his opposition is comprehensible. Expert military opinion may have carried extra weight in January 1932, when relations with Germany were particularly

[35] *DDF*, 1/I, No. 244, incl. Annex I; Challener, pp. 181-183.

[36] *DDF*, 1/I, Nos. 244 (incl. Annex I), 250; Bankwitz, *Weygand*, pp. 49-57, 66-67; *SIA, 1932*, pp. 196-199; Georges and Edouard Bonnefous, *Histoire politique de la troisième république*, 7 vols. (Paris: Presses universitaires de France, 1956-1967), V (1962), 105, 109-110.

strained over reparations. Tardieu did not always support Weygand, but he respected the general's military judgment, and probably shared his skepticism about militia armies, a proposal with evident leftist antecedents. Tardieu's proposals to the conference in fact involved no clear-cut commitment to any disarmament at all. A French historian, Jacques Chastenet, has commented that it is not certain Tardieu believed his plan could be adopted.[37] Such doubts would indeed have been in order, in view of the differing British and American perspective.

Chastenet may also be suggesting, however, that Tardieu's plan was largely tactical, and not a serious proposal. And indeed, when French leaders offered proposals like the Tardieu plan, involving a strong League and no disarmament, they seem to have sought in part to forestall domestic charges of inactivity, and in part to protect the French diplomatic and power position. Germany was rearming anyway, and an Anglo-American-style agreement would not affect that, except to legalize it, but such an agreement *would* disarm France and further weaken her vis-à-vis Germany.[38] How could she protect herself? A plan like Tardieu's would gain support from French allies in Eastern Europe and from some other medium-sized or small nations. And if the plan were rejected, at least France could always claim that she had made her constructive internationalist offer, failing which she would stand on the letter of the Treaty of Versailles.

Small-power support was not an optimum goal for France. The ideal, even for Laval and Tardieu, would have been to stand before the world as a full-fledged partner with Britain, laying down a common policy that would defend French interests.[39] But far worse than a position supported by small powers would be involvement in a conclave of the major powers, in which France would rank along with Germany and Italy, while the British assumed the role of mediator or initiator. This mise-en-scène would deprive France of her status as a leader and victor, and alienate her East European allies. If such a meeting of four or five powers took place, France would in fact probably stand alone, with Germany and Italy hostile to her, and Britain (and the United States, if participating) professedly impartial and actually following the line of least commitment. As the definition of questions does much to decide their answers, so does the

---

[37] Jacques Chastenet, *Histoire de la troisième république*, 7 vols. (Paris: Hachette, 1952-1963), VI (1962), 31-32.

[38] See also Parker, *Europe 1919-45*, p. 264.

[39] *DBFP*, 2/III, Nos. 88, 91, 92, and 236 show the interest of Laval and Tardieu in the appearance and reality of an Anglo-French entente. Tardieu seems to have thought that he could best gain an entente by hard bargaining.

choice of the forum of discussion. Unfortunately for France, bilateral association with her held the least attraction for the British, and a four- or five-power conclave the most. Professions of international principles provided France with a means to thwart such meetings.

To appeal to international sentiment was more than a tactic for evading disarmament or five-power meetings, however. At the beginning of 1932, the League of Nations enjoyed high prestige, its principles were widely professed, and it could provide valuable support to a power interested in maintaining the status quo, like France. The French certainly did not want to fight another war, but international institutions, combined with other methods, might serve to curb Germany without war. This would not have seemed at the time to be mere wishful thinking. At the start of 1932, the political and economic weakness of France, which would soon emerge and last through the 1930s, still lay in the future. The recent past had witnessed displays of French power. In opposing the 1931 Austro-German proposal for a customs union, France had used not only financial pressure, but also appeals to the League Council and the International Court of Justice. A development of international institutions might make it possible to employ inspection and economic sanctions, and arbitration might be made compulsory in all cases. Ultimately, an international order could replace a system of deterrent alliances, in the name of higher principles. Probably some French officials believed in such proposals, even if the hard-edged Tardieu did not. Although such ideas may look like a false front to critical outsiders, they are apt to be taken seriously by their proponents, who cannot understand the outsiders' resistance. It would seem that the French must have had some belief in such proposals to have offered them so many times, in spite of British and American opposition. And if the proposals could have appealed to world opinion, they might have had some success.

It has sometimes been assumed, in the 1930s and since, that there was no way to stop Germany without armed intervention, which neither France nor Britain was ready to undertake. It may also appear that "world opinion" was of no significance except in the electoral calculations of Anglo-American politicians. But actually, world opinion was very important, as Hitler recognized in 1933 and after. A convincing exposure of German rearmament plans would have had a devastating effect, reaching even into Germany itself. When these plans went unexposed, and German arguments for equality of rights were accepted, Western policy-makers could find no support for pressures short of war, and indeed their own resolution and will to form a united front were sapped.[40]

5. BRITAIN: THE REPUDIATION OF THE BALANCE OF POWER

While the French tended to regard enmity as the inevitable characteristic of their relations with Germany, and wanted to organize a solid alliance in opposition to that country, it was the British who chiefly determined the response to German actions, and they had different views. They thought that the 1914-1918 conflict with Germany might have been avoided, had it not been for prewar alliances and alignments. They now hoped to eliminate enmity and conflict by a process of mutual accommodation, thus obviating any pretext for new alliances. The reconciliation of former enemies was the original purpose of appeasement, a word that was an active part of the British political vocabulary long before Hitler became chancellor. Leaders in London thought that conferences of the powers, under their own leadership, provided an excellent means of promoting appeasement. Later, after the Germans had begun to press their claims more vigorously, Prime Minister James Ramsay MacDonald pointed out that Great Britain had put forth great efforts to assist Germany in international affairs.[41] This was certainly true, even if Britain had not acted entirely unselfishly.

As already mentioned, the cabinet, functioning as a committee, customarily made the final decisions on British foreign and defense policy. Most ministers had no experience in diplomacy and only slight understanding of foreign or military problems, and though Foreign Office officials tended to sympathize with France, the ministers usually saw things from the perspective of domestic or British Commonwealth politics. Of course, the cabinet system had its merits: decisions once taken did have full authority, including that of a disciplined parliamentary majority. Indeed, decisions would usually be assured of public support—as they were usually in line with public opinion.

Given the dominance of the cabinet in British policy making, the

---

[40] The point was made by a German writer, who argued on the eve of war that the British lion was still very strong, but caged by its inability to use power except for "ethical" (i.e., publicly acceptable) ends, particularly those acceptable to the dominions. Thus, although the Munich settlement had been contrary to the traditional British interest in the balance of power, the German case could not be resisted because it was "morally unassailable." See C[arl] E[rdmann] Graf Pückler, *Wie stark ist England?* (Leipzig: W. Goldmann, 1939), pp. 202-207.

[41] Cab 23/81, meeting of 20 Mar 35, 16(35)—.

sensitivity of cabinet members to public opinion, and their personal accessibility to leaders of opinion like Lord Cecil of Chelwood, President of the League of Nations Union, or Geoffrey Dawson, Editor of the *Times*, there could hardly be a serious divergence between public and actual policy. Parliament usually had little tolerance for ministers who tried to deceive it. Ministers, and officials too, shared the opinions common within the educated public. The diplomatic officials of the Foreign Office joined in the public enthusiasm for disarmament, and officials in the service departments were at one with the public in their dislike of military alliances or peacekeeping forces. Further, the cabinet system brought the full weight of Treasury views to bear on foreign and defense policy, and these views usually coincided with the outlook of the financial community of the City of London.

This constituted a very different background for policy formation from that in Berlin. There was no room here for elaborate military or diplomatic deceptions, no place for a doctrine of seizing the initiative. There was little room even for departmental professionalism. The British resisted specialization, and refused to place their trust in specialists—except in finance. The government could not consider any policy that it could not explain to the public with a reasonable hope of acceptance. Of course, British ministers did not in fact always tell Parliament and the public the truth. But their most common deception was to try to make their foreign policy appear as a "bold" and farsighted application of high principles when actually it was a series of hasty makeshifts. Whenever diplomatic dangers appeared, the British government would produce new proposals, largely designed to shore up public hopes of peace, and to fend off accusations of inactivity. In all, the British method of making policy did not encourage careful attention to the awkward and ugly problems of power.

Power has little moral inspiration to it, and a government's search for it can cost a citizen money, and even his life. Quests for national power often appear to stem from quests for personal power on the part of individuals, or for a preferred status on the part of special interests. Many consider a hankering for power to be an unattractive and unhealthy trait. But in any case, like it or not, a demand for power affects everyone. The presence on the scene of an unsatisfied, power-conscious nation means that, by a sort of political Gresham's law, other nations will also, sooner or later, be impelled to turn away from more constructive pursuits and concern themselves with power. In the interests of self-preservation, they will probably try to combine their forces to offset and balance the dissatisfied, expansive state.

A.J.P. Taylor has suggested that the tendency to form a balance of power can indeed function as the political equivalent of the self-operating laws of economics, and he has noted that prior to 1914, the weaker powers had tended to join together, "almost unconsciously," against the aggressor. In this sense, the tendency to form a balance of power can be stated as a *descriptive social law*; the tendency results from natural impulses toward self-defense. Such a social law is not inescapable, however, as men can make conscious decisions for or against obeying it. Thus they may also either consider the balance of power to be a *norm* to be sought, a *prescriptive rule* or *principle*— or they may reject such a normative principle. Taylor says that "the Allies" made such a rejection after World War I; W. M. Jordan states, rather more specifically, that Britain rejected the principle.[42]

In the first sentence of a study of eve-of-war diplomacy, Keith Middlemas has elegantly stated the case for the balance of power as a prescriptive rule: "So long as a balance of power exists in the world, British foreign policy in 1937-9 will be a matter of controversy." In other words, freedom (and not only, of course, to debate the events of 1937-1939) depends on there being no single dominating nation. But historians who believe in the balance of power as a norm seem to be on shakier ground when they argue that the principle still influenced British policy-makers between the wars. Middlemas and W. N. Medlicott have pointed to a secret Foreign Office memorandum from 1926 (now officially published) as evidence of, among other things, a belief in the prescriptive rule of balance. The memorandum stated: "The maintenance of the balance of power and the preservation of the *status quo* have been our guiding lights for many decades and will so continue."[43] But another (not yet published) document shows that different attitudes had penetrated even the Foreign Office by the early 1930s. Sir Robert Vansittart, the Permanent Under-Secretary of State, referred in a cabinet paper at the end of 1931 to the

---

[42] A.J.P. Taylor, *The Struggle for Mastery in Europe, 1848-1918* (Oxford: Oxford University Press, 1954), pp. xx, xxi; Taylor, *Origins*, pp. 30-31, 135-136; W. M. Jordan, *Great Britain, France and the German Problem, 1918-1939* (London: Oxford University Press, 1943), pp. 2, 4-5. Martin Wight, in "The Balance of Power," in Herbert Butterfield and M. Wight, eds., *Diplomatic Investigations* (Cambridge, Mass.: Harvard University Press, 1968), pp. 149-175, distinguishes nine different usages of "the balance of power," including that it is a norm, that it involves a special role to be played by one power, and that it is an inherent tendency (Taylor's sense), a sort of sociological law.

[43] Keith Middlemas, *Diplomacy of Illusion* (London: Weidenfeld & Nicolson, 1972), pp. 1, 17-18; W. N. Medlicott, *British Foreign Policy Since Versailles, 1919-1963* (London: Methuen, 1968), pp. xvi-xviii, 78-80. Medlicott also says, however, that the government seemed to believe in the 1930s "in the possibility of 'appeasement'" (p. xix), and Middlemas says policy makers hovered uneasily between policies of a free hand and a power balance (p. 17).

danger of "the old—and vicious—'balance of power,'" adding, "We have learned by experience what that is apt to mean," evidently a reference to the 1914-1918 war. A Foreign Office memorandum of August 1932 stated that the aim of British policy since the war had been to work, both for humanitarian and material motives, "for the maintenance and consolidation of peace, for appeasement in Europe and the bringing together of the two great protagonists, France and Germany, and finally for the gradual reinstatement of Germany on a footing of equality in the community of Nations."[44] This statement suggests a concern for atmospherics rather than for power, and a susceptibility to German appeals. Certainly the cabinet itself, so far as its own records show, paid little conscious heed in the early 1930s to the balance-of-power principle.

Of course, a lack of conscious belief in the principle of balance would not necessarily prevent governments from tending in practice to balance each other. Arnold Wolfers, in his still indispensable *Britain and France between Two Wars* (1940), argues cautiously that the British government actually continued, instinctively if not avowedly and intentionally, to follow a policy of balance. A supporter of Weimar Germany's policies, he finds it natural that British policy in the 1920s should have sought a balance against the French, and he suggests it was only at a relatively late date that German might reached such a dangerous level as to oblige the British to throw all of their power into the scales against that country instead.[45] Wolfers's argument implies more the working of social law than the conscious pursuit of a principle, and what happened in 1939 does seem to show that the law still—in the final analysis—worked. British opinion itself then accepted a popularized version of the balance of power: that Hitler was too powerful and had to "be stopped." It has recently been suggested that this final change resulted merely from an orchestrated propaganda campaign, but for people at the time it was a recognition of a law of survival, that which Taylor's weaker powers had recognized before 1914.[46]

[44] Cab 24/227, CP 4 (32), 1 Jan 32; FO 371, Memo, 26 Aug 32, C7359/211/18.
[45] Wolfers, pp. 211, 220, 233, 242-253, 307-308, 380, 389-390.
[46] The idea of the propaganda campaign is stressed in Dietrich Aigner, *Das Ringen um England* (Munich: Bechtle Verlag, 1969). It also appears in the study of Maurice Cowling, *The Impact of Hitler* (Cambridge: Cambridge University Press, 1975); see, e.g., p. 398. Cowling is correct in stating that it is far from obvious now that Hitler wanted to destroy the British Empire (p. 8), but even if this had been (as he says and I doubt) "not obvious then," Hitler's personal assurances or even wishes would have provided scant security to a Britain next to a Europe governed by a militarized, expansionist Germany. Would Hitler have tolerated a Labour government, Jewish ministers, or an uncensored BBC in London? Also, his aim was not to spare Britain from fighting a war, but to bring her into war on Germany's side.

Still, the prolonged refusal to avow the principle made a difference. The organizers of a power balance hope to deter aggression, but they accept the possibility that war may result instead; they still prefer fighting with allied support to fighting alone or surrendering. The chief reason why the British refused to think in balance-of-power terms was (it seems) that they refused to accept this risk of war. They evaded considering the alternatives of fighting alone or surrendering by hoping to evade involvement in war on the continent. In spite of League Covenant provisions for armed sanctions, in spite of the Locarno obligation to defend the existing Franco-Belgian-German border, and in spite of regular involvement in continental questions, Sir John Simon, the Secretary of State for Foreign Affairs from 1931 to 1935, recorded in cabinet papers his hope that England would be able to stay out of a war in continental Europe. He evidently believed that his fellow ministers shared this hope, which tended to become a premise. A natural result of this attitude was that, as Field Marshal Milne, Chief of the Imperial General Staff, reported in October 1932, the British army was in no condition to give any early help to France if an obligation arose under the Treaty of Locarno.[47] We have already noted the British reluctance to look at disagreeable intelligence. Toward war, British policy-makers seemed to adopt the ironic German motto: "What ought not to be, cannot be."[48]

There is no modern parallel elsewhere to such an attitude within a government. The positions of British leaders stand in need of explanation, and recognition of this need has resulted in a considerable literature; mention has already been made of some writers who have discussed British policy in terms of the balance of power. Some other historians have concentrated on the personalities of the leaders themselves, and on the immoral or moral ideas that inspired them. A. L. Rowse, in *All Souls and Appeasement*, and Martin Gilbert and Richard Gott, in *The Appeasers*, have sharply denounced the British leaders, largely shapers of opinion, who made concessions to an aggressive, racist, unscrupulous Hitler. In *The Roots of Appeasement*, on the

---

[47] Cab 24/233, CP 347 (32), 17 Oct 32; Cab 24/239, CP 52 (33), 28 Feb 33; Cab 24/234, CP 362 (32), 28 Oct 32. Some Foreign Office officials argued that Britain was already committed to Europe by her geographical position and the extent of her economic relations with the continent (Cab 24/225, CP 301 [31], 26 Nov 31), but their suggestion of further commitments was rejected; see below.

[48] Günter Wollstein specifically applies this motto (*was nicht sein darf, auch nicht sein kann*) to a comment by Sir Eric Phipps, who said that Britain could not judge Hitler solely by *Mein Kampf*, as that would logically necessitate a preventive war: *Vom Weimarer Revisionismus zu Hitler* (Bonn: Verlag Wissenschaftliches Archiv, 1973), p. 235. Neville Chamberlain reacted like Phipps when he read an anti-Hitler book in 1937: Keith Feiling, *The Life of Neville Chamberlain*, 2nd ed. (London: Macmillan, 1970), p. 328.

other hand, Gilbert praises those pre-Hitler appeasers who called for a reconciliation with the one-time German enemy. These writers express a demand for fundamental decency and honesty, in the best traditions of English life, and they are well qualified to understand the debates of the 1930s, which took place largely in the same terms as those they use. They do not, however, examine the social, economic, and institutional factors in British policy, or judge that policy in the context of the policies of other governments.[49] Still other writers, dealing with the later 1930s, have sought to explain British policy in terms of a capitalist plot to turn an anti-Bolshevik Germany eastward against the Soviet Union, or as the form that British party conflict took, the expression or concealment of class conflict in Britain.[50] Whatever merit these latter theses may have, they can scarcely shed much light on 1932 and 1933. Soviet-German official relations remained friendly until the end of 1933, and foreign policy only became an active issue in British electoral politics three months before that. Since British leaders were already trying to propitiate the Germans well before this, there must be other explanations, which probably also have some bearing on the later period. To do justice to British leaders, in the senses of both understanding and criticism, a review of the foundations of their foreign and military policies seems needed.

6. OTHER NATIONS IN BRITISH EYES

British policy toward other governments was influenced by attitudes, sentimental and practical, toward other nations. Not long after the war was over, British sympathies began to turn in the Germans' favor, a trend that was reinforced by what seemed to Britons to be the vengefulness and rapacity of the French, especially in the episode of the French occupation of the Ruhr in 1923. The British were not entirely motivated by disinterested magnanimity, of course. The financiers of the City were anxious to see Germany return to financial normalcy, and the Bank of England, directed by its governor, Montagu Norman, gave crucial aid in the stabilization of the mark. British industry also wanted to resume normal business relationships with German counterparts.

---

[49] For adverse criticism, see Bernd Jürgen Wendt, *Economic Appeasement* (Düsseldorf: Bertelsmann Universitätsverlag, 1971), p. 18, and Robert P. Shay, Jr., *British Rearmament in the Thirties* (Princeton: Princeton University Press, 1977), p. 8. Northedge, *Troubled Giant*, and Keith Robbins, *Munich 1938* (London: Cassell, 1968) take a less personalized approach.

[50] Andrew Rothstein, *The Munich Conspiracy* (London: Lawrence & Wishart, 1958) is representative of the former view, which has been widely circulated in Soviet and East European writing. The latter thesis is that of Maurice Cowling's *Impact of Hitler.*

In one case, when the British arms and machine manufacturers, Vickers, sought to revive prewar ties with the firm of Friedrich Krupp of Essen, Krupp's replied that, aside from business relations, it was desirable to develop better (i.e., more friendly) reporting on Germany in the British press. Vickers cooperated by sending a free-lance reporter over, and Margarete Gärtner, an accomplished propagandist, conducted him around. The experience and the favorable articles that resulted gave her the idea of establishing a special bureau under the auspices of German industry to encourage visits from England and America, and assist in the production of books and articles by the visitors. By 1932, Fräulein Gärtner was in a position to place articles in the London press, or to have questions asked in Parliament. She was no Nazi herself, but the Nazis later imitated her methods and exploited some of her old acquaintances.[51]

But it would be a mistake to think that, either in the 1920s or the 1930s, British opinion was simply manipulated by German hands. At the start of the thirties, a spontaneous wave of Germanophilia seemed to wash over England and the United States. Perhaps it was partly the piquant appeal of the lost and condemned cause, partly a reaction against the orthodox hatred preached during the war. This was the period when Christopher Isherwood, Stephen Spender, W. H. Auden, and Wyndham Lewis, as well as Sinclair Lewis, Dorothy Thompson, Thomas Wolfe, and Katherine Anne Porter went to Berlin to live. Berlin was felt to be alive, contemporary, free, a place where things were happening in politics, literature, and the arts; Germany was ahead of the rest of the world in music and architecture. People who stayed home could read in translation the novels of Thomas Mann, or as they usually preferred, of Vicki Baum and Erich Kästner. Those who respected technical achievement could admire the Graf Zeppelin, the DO-X ten-motored airplane, the speed of the *Bremen* and the *Europa* and of Dr. Kruckenberg's propeller-driven railcar. Others could watch Marlene Dietrich on the screen or foxtrot to the strains of "Auf Wiedersehen." Germany became fashionable, a matter that can have political significance.

Sympathy for Germany often meant antipathy for France. French policy in the early 1920s had often appeared from London to be clearly expansionist, as in the French encouragement of Rhenish separatism. In 1923, when relations were very strained over the Ruhr occupation, Warren Fisher, the chief permanent official of the Treas-

---

[51] Fräulein Gärtner's experience is described at length in her memoirs, *Botschafterin des guten Willens* (Bonn: Athenäum, 1955). Her work, and much else on unofficial Anglo-German relations, is covered in D. C. Watt, *Personalities and Policies* (London: Longmans, 1965), Essay 6.

ury, was arguing that the French were acting "illegally and madly" and that France considered herself in a position to dictate to Britain.[52] Before the crisis was over, the British government had decided on an ambitious program for a fifty-two-squadron air force, this size being required to achieve parity in the air with France. Execution of the scheme was later slowed down as relations improved, as the Foreign Office insisted that the French would not fight a war with Britain, and as a return to prewar monetary parity imposed economies, but a 1931 general-staff assessment still characterized France as the dominant power whose arms kept Europe in tension. Many British officers would probably have agreed with Colonel Marshall-Cornwall, who told a group of German officers in May 1930 that he hoped British cannon would not in future be fired against German troops, but side by side with them.[53]

On the antimartial side, British internationalists regarded Part V of the Treaty of Versailles, which disarmed Germany, as only the first step toward general disarmament, and they felt that it was up to the French to make the largest contribution. In the view of many Englishmen, general disarmament would remove one of the foremost causes of international friction, and render the coercion the French were seeking unnecessary. It had been the British who, at the peace conference, had originated the idea of an implicit Allied pledge to disarm.[54] It appeared to many British observers that the French were trying to evade this promise, and that in so doing, they were blocking the restoration of harmony in Europe, so essential from the English standpoint.

Experiences with the French in 1931, including the French attitudes toward a naval agreement and reparations as well as the French memorandum on disarmament of July 15, created an unusual amount of ill-feeling. In London, French policies were thought to have aggra-

[52] Keith Middlemas and John Barnes, *Baldwin* (London: Weidenfeld & Nicolson, 1969), pp. 183-184. On British attitudes generally toward France, see John C. Cairns, "A Nation of Shopkeepers in Search of a Suitable France," *AHR*, 79 (1974), 710-743.

[53] Basil Collier, *The Defence of the United Kingdom* (London: H.M.S.O., 1957), pp. 14-19, 25; Sir Charles Webster and Noble Frankland, *The Strategic Air Offensive against Germany, 1939-1945*, 4 vols. (London: H.M.S.O., 1961), I, 54, 57; BA-MA, II H 226, Der Kommandeur der Artillerieschule Ia 107/30, 9 May 30. Foreign Office view: *DBFP*, Ia/I, Appendix (p. 850). In 1932, Sir Maurice Hankey used RAF estimates of the number of bombs the French could drop on London in the first days of war—estimates intended to show the need for a bigger air force—to argue to King George V the need to abolish bombing: Stephen Roskill, *Hankey: Man of Secrets*, 3 vols. (New York: Walker, 1970-1974), III, 42.

[54] Paul Birdsall, *Versailles Twenty Years After* (New York: Raynal & Hitchcock, 1941), p. 169; *FRUS, Paris Peace Conference, 1919*, V, 299; VI, 354, 365-367 (cf. 363, 954-955).

vated the problems of the shaky British financial structure, leading to the fall of the Labour government in August and the British departure from the gold exchange standard and prewar parity in September 1931. The momentary financial power of France was regarded as an almost unbearable domination. MacDonald agreed in August with Henry L. Stimson, the American Secretary of State, that British opinion was swinging away from France in her efforts "to secure her hegemony as against Germany." Vansittart, the chief permanent official in the Foreign Office, usually sympathized with France, but 1931 experiences overcame this tendency, and influenced his views in particular with respect to the coming Disarmament Conference. Although in May 1931 he had written that rearmament, or at least "undisarmament," was the first aim of German policy, he stated in his end-of-the-year cabinet paper that, as opposed to the French case on disarmament,

> the German case is on every ground of morality and equity exceedingly strong, . . . indeed, almost unanswerable—and it happens largely to coincide with our own interests. We desire disarmament —or the end of over-armament—and we do not desire a perpetual [French] hegemony. This is recognized not only by the Cabinet, but by an overwhelming majority in the public opinion of this country.[55]

Two subordinate Foreign Office officials had suggested that one solution to the vicious circle of economic and political insecurity would be for Britain to undertake guarantees similar to those in the Geneva Protocol. But the cabinet on December 15 had recorded its aversion to any guarantee, over and above Locarno, whereby British forces might conceivably become engaged in a continental war even over Germany's eastern border. Vansittart supported this refusal in his memorandum. The cabinet also took the view that Germany had "strong moral backing for her claim to the principle of equality." They thought that the solution of the Franco-German controversy must be some sort of compromise, while they refused to contribute any commitment to facilitate such an agreement. They would certainly resist any proposal which, like the Tardieu plan, would put military forces and heavy armaments, including parts of the Royal Navy, at the disposal of an international command under the League

---

[55] *FRUS, 1931*, I, 515; Cab 24/225, CP 317 (31), 10 Dec 31 (encl. 1 of May 1931); Cab 24/227, CP 4 (32), 1 Jan 32. Ramsay MacDonald's own view of the French was characterized by resentment and suspicion: David Marquand, *Ramsay MacDonald* (London: Jonathan Cape, 1977), pp. 716-717; Public Record Office, Ramsay MacDonald Papers (PRO 30/69; hereafter, MDP), 8/1, Diary, 18 Jan, 7 Apr, 1 May, 21 Jun, 3 Dec, 5 Dec [32], 9 Mar, 3 Jun [33].

Council; the sanctions and inspection foreseen in that plan also went against the British grain.[56]

Along with attitudes of sympathy toward Germany and suspicion toward France, the British showed a concern to maintain—or perhaps one should say establish—good relations with the United States. As Milne noted in 1932, with a slight hint of disapproval, "successive British governments have declared that a good understanding with [the United States] is an axiom of our foreign policy." This certainly did not mean blindness to American shortcomings, and sometimes American actions aroused considerable frustration and resentment. There had always been sharp dissatisfaction in some quarters, such as the Royal Navy, and there were serious suggestions that Britain should cease tailoring its policies to suit American ideas on war debts, disarmament, and Japan. The almost complete lack of mutual comprehension on questions of debts and of international trade and monetary policy induced despair over the possibility of good relations, particularly after the summer of 1933. The idea spread that Americans were very long on talk and very short on action.[57] Yet in the last analysis, Britain always came down, so far as possible, on the side of trying to please the United States.

Various explanations, not at all mutually exclusive, can be advanced for this. First, America could be either very helpful or a serious problem for Britain herself. World War I had shown that it would be difficult for Britain to enforce a strict blockade against an unwilling United States. Presumably, British leaders recognized that American industry would be a highly desirable resource for supporting any major British war effort, not to speak of eventual American financial aid or direct military support. More immediately, a naval construction race with the United States would have been an immense burden on the British Treasury, while the war debt problem for a time made it seem worthwhile to propitiate American opinion in the hope that the debts could be scaled down or eliminated by agreement.[58] Second, there was the influence of the British Commonwealth. The senior

[56] Question of commitment: Cab 23/69, meeting of 15 Dec 31 (PM), 91(32)2; Cab 24/225, CP 301 (31), 26 Nov 31; Cab 24/227, CP 4 (32), 1 Jan 32. On resistance to Tardieu plan: *SIA, 1932*, p. 200; FO 371, Minute by Howard Smith, 9 Feb 32; ltr., Hodsoll (CID) to Cadogan, 9 Feb 32, W1564/1466/98; Minute by MacDonald, 22 Apr 32, W4550/10/98.

[57] Milne statement: Cab 24/234, CP 362 (32), 28 Oct 32; he was interested in improving relations with Japan. On frictions, see Watt, *Personalities and Policies*, esp. pp. 39-44; Roskill, *Hankey*, III, 29-30, 34-35, 37; *DBFP*, 1a/I, Appendix (p. 878); some expressions of annoyance are reported in Chapters Three, Eight, and Nine below.

[58] On difficulty of blockade: Cab 24/225, CP 301 (31), 26 Nov 31; Baldwin recognized this in a speech in November 1934: Watt, *Personalities and Policies*, p. 43. Some British officials believed that the American refusal to agree to British cruiser

dominion, Canada, occupied a strategic and economic position that rendered her incapable of seriously opposing the United States, and indeed she was quasi-American in her view of international affairs. The Canadians, especially the French Canadians, did not want to be bound by any automatic commitments in Europe, and they were apprehensive of Japanese expansion in the Pacific. South Africa usually supported Canada in her isolationism, and for the sake of commonwealth unity, the United Kingdom had to heed Canadian wishes.[59] Indeed, there was strong isolationist sentiment in Great Britain, and many Englishmen regarded the English Channel as a cultural barrier oceanic in width. Dominion and American resistance to European entanglements provided England with a welcome excuse for her own refusal of commitments to France. Third, if Britain could enlist American backing for British proposals, and could speak for the whole English-speaking world, this would strengthen her own hand. Sir John Simon hinted at the point in the fall of 1932, when urging the cabinet to take a position on German claims for equality of rights: "The influence of our National Government, backed by American sympathy and approval, is vital to the maintenance of peace in Europe." In October 1933, it would be the United States that would influence Simon. But in any case, despite American isolationism, the United States was seen as a decisive force in world affairs, and never more so than after President Hoover's 1931 moratorium proposal. Foreign Office experts stated in November 1931: "No final settlement of any major international issue can be reached without the approval of the United States of America," and an economic committee observed in a May 1933 report: "It is of the utmost importance, with regard to every phase of world affairs, to develop the cordial co-operation of Great Britain and the United States."[60]

---

demands stemmed from a wish to be able to overcome British interference with American commerce in wartime: W. N. Medlicott, *The Economic Blockade*, 2 vols. (London: H.M.S.O., 1952-1959), I, 11. U.S. industrial support had been crucially important in World War I (see Correlli Barnett, *The Collapse of British Power* [New York: Morrow, 1972], pp. 12, 14, 82, 85, 308), although it is not clear that the need for such support in future was recognized in ministerial circles, which, as we have seen, were inclined not to think too much about war dangers. On this and on the financial relief provided by the naval agreements, see Northedge, *Troubled Giant*, pp. 280, 289-290, 345; Roskill, *Hankey*, III, 31, 74-75. On British concern for U.S. opinion in connection with war debts: Cab 24/225, CP 301 (31), 26 Nov 31; FO 371, tel., Simon to MacDonald, 17 Jul 32; ltr., Simon to MacDonald, 18 Jul 32, W8178/10/90 (sic:—/98).

[59] Canadian and other dominion influence: Watt, *Personalities and Policies*, pp. 146-158. Canadians have sometimes complained that the United Kingdom has sacrificed their interests to American interests.

[60] Simon: Cab 24/233, CP 247 (32), 17 Oct 32. Stamp Committee: Cab 24/241, CP 131 (33), 16 May 33. Hoover's action: Bennett, *Diplomacy*, Chapter 5.

Although there were some anti-American sentiments and utterances, it seems obvious that the prevailing trend of policy and opinion was pro-American. After all, when the British government made an early and unfavorable settlement of war debts, when it conceded parity to the American navy, when it gave up the number of cruisers that professional naval opinion considered necessary, these actions were taken in expectation of public acceptance. Conservative, "realist" considerations were outweighed by the popular desire and financial need for harmony with the United States. The British public emerging from the Great War (as it was called) supported the ideals President Woodrow Wilson had publicized, and this public seems to have wanted the sense of security that American friendship could afford. It could be argued that Britain was unconsciously seeking a balance of power through American support. But the United States was noncommital, and could not be counted on explicitly. Pro-American attitudes seem to have stemmed more from sentiment than calculation. There was the often criticized but certainly influential belief in a "transatlantic Anglo-Saxon family," an idea more often avowed in England than in America. The heyday of racial Anglo-Saxonism had passed, but ties nevertheless existed in a number of common traditions and attitudes, and (the familiar epigram notwithstanding) in a common language with a continuing common literature. And despite all taboos, frictions, and Washingtonian traditions, the actual conduct of American foreign policy in the 1930s suggests that many Americans, especially in élite circles, also felt themselves to be kindred with the English.[61]

[61] Watt, in *Personalities and Policies*, pp. 19-99, devotes three essays to Anglo-American relations, mainly with respect to the British élite. Medlicott, *British Foreign Policy*, pp. 29-31, deplores British policy at the Washington Conference of 1921-1922, but indicates the trend of British opinion by speaking of "the immediate emotional satisfaction of a dramatic act of friendship and temporary pacification" afforded by that policy. Stephen Roskill (*Naval Policy between the Wars*, Vol. I, *The Period of Anglo-American Antagonism, 1919-1929* [New York: Walker, 1968], p. 25) says the Admiralty as well as the British Navy League accepted that an Anglo-American war was unthinkable, even if American navalists did not. The arguments of Corelli Barnett against the "myth" of Anglo-American friendship (esp. pp. 258-263) concede at least the strength of belief in the myth. As to American policy, Hoover's two bombshell messages, on intergovernmental payments in June 1931, and on disarmament in June 1932, were both launched after prior consultation with MacDonald, but not (effectively) with French leaders. Henry L. Stimson once wrote MacDonald that Hoover had said, "Tell MacDonald that I believe that . . . civilization . . . can only be saved by the cooperation of Anglo-Saxons; we cannot count on the other races": MDP, 2/12, ltr., Stimson to MacDonald, 27 Jan 32. Norman Davis and Stimson in 1932 spoke without hesitation of "Anglo-Saxon" instinctive aversion to advance commitments: Nancy Harvison Hooker, ed., *The Moffat Papers* (Cambridge, Mass.: Harvard University Press, 1956), pp. 50-51; *FRUS, 1932*, I, 139; SDP, Memo of conversation with French chargé, 16 Sep 32, 500.A15A4 Steering Committee/56. Roosevelt disappointed MacDonald in June and July 1933, but sought to consult Chamberlain on a major initiative in January 1938. Both the French and the Germans habitually bracketed the British and Americans together.

7. ENGLAND'S MILITARY DECLINE

Along with sometimes superficial attitudes toward other nations, British foreign and defense policy had roots in English tradition, experience, and conditions. Just as Germany had certain military traditions stemming from Scharnhorst and Roon, Britain too had its military traditions, and these were not forgotten within the services. Great Britain is an island, and, paradoxically, modern military organization and technique worked in some ways to increase British insularity and insulation from the continent. From 1727 to 1868, the annual Mutiny Acts, required for the maintenance of a standing army, usually described the army's function as "the preservation of the balance of power in Europe."[62] Before the mid-nineteenth century, Britain had frequently participated in warfare on the continent, both by subsidizing or hiring the service of continental troops, and by sending its own contingents of professional soldiers. The perfecting of Prussian universal military service by 1870, the copying of this system by other continental countries, and the development of mass armies eliminated mercenaries and helped to make subsidies impractical; further, universal service on the continent meant that the British professional forces were completely outclassed in scale. In the Waterloo campaign of 1815, British regulars and a directly hired Anglo-German Legion had numbered 42,000, and had been a major force in themselves. They had been supported, moreover, by a British-subsidized Prussian army of 116,000. But at Mons in August 1914, a British Expeditionary Force of 70,000 was a minor contingent. The British army could only have remained a major military factor, by continental standards, if it had adopted universal training. Some British officers, notably Lord Roberts, advocated this, but since the channel and the navy made isolation seem possible, this idea failed to win wide support.[63]

Indeed, a powerful school of thought, to which Winston Churchill early adhered, argued well before 1914 that a "servile imitation of continental operations" was not the British way in warfare.[64] In the First World War, over two and a half million volunteers enlisted—even though many enrolled under considerable social pressure—be-

---

[62] Wight, "Balance of Power," p. 161.

[63] Factual data: Archibald Becke in *Encyclopedia Britannica*, 14th ed., 24 vols. (Chicago: William Benton, 1958), under heading "Waterloo Campaign"; Sir Charles [K.] Webster, *The Foreign Policy of Castlereagh*, 2 vols. (London: G. Bell & Sons, 1950), I, 133; Barbara Tuchman, *The Guns of August* (New York: Macmillan, 1962), p. 252. On the conscription movement: Michael Howard, *The Continental Commitment* (London: Temple Smith, 1972), pp. 37-39.

[64] See Trumbull Higgins, *Winston Churchill and the Second Front* (New York: Oxford University Press, 1957), p. 49; Michael Howard, *The British Way in Warfare: A Reappraisal* (London: Jonathan Cape, 1975), pp. 9-10; Howard, *Continental Commitment*, p. 41.

fore the government turned to conscription in 1916.[65] Even before the first full-scale British engagement in the Battle of the Somme, some cabinet ministers were arguing that England should have her army stand pat and rely on economic and financial pressures to defeat Germany—a strategic doctrine that would become familiar in the 1930s. Churchill and Lloyd George advocated attempts to strike at the flank of the Central Powers, i.e., at some place other than on the Western front. A month after the Somme attack commenced, Churchill had launched a polemic against it within cabinet circles.[66] Later, as prime minister, Lloyd George maneuvered to prevent further major attacks on the main front without actually venturing to forbid them.

British strategy in the First World War became a highly controversial subject, and the debate itself has affected history. There was much in the planning and execution of the Somme and Passchendaele attacks of 1916 and 1917 that is open to criticism: the lack of surprise, in particular. British casualties on the opening day of the Somme battle were more than five times the British and American losses on D-Day in 1944. The technique of prolonged preliminary bombardment failed to silence the German machine guns on the Somme, and in Flanders it made the battlefield impassable. Tanks were introduced in small numbers on the Somme when they could only produce a decisive effect in quantity. Sir Douglas Haig, the British commander in France, was clearly overoptimistic in his estimates of what could be achieved with the forces at his disposal; sometimes he made poor choices of subordinates, and he was also stubborn. But for all his faults, Haig was probably better qualified than any other British general of the time, and his determination to fight on, a characteristic linked to his optimism and stubbornness, kept the Allies from being defeated and eventually enabled them to succeed.[67]

[65] Lord Hankey (Sir Maurice Hankey), *The Supreme Command, 1914-1918*, 2 vols. (London: Allen & Unwin, 1961), II, 477n.

[66] Maurice V. Brett and Oliver, Viscount Esher, eds., *Journal and Letters of Reginald, Viscount Esher*, 4 vols. (London: Ivor Nicholson & Watson, 1934-1938), IV, 1-3, 14-15, 52; Churchill, *World Crisis*, III, 187-194. Although out of office, Churchill had been given access to top secret records; see also Hankey, II, 505. After the war, however, Churchill advocated a Swiss militia scheme to retain conscription: Robin Higham, *Armed Forces in Peacetime* (London: Foulis, 1962), p. 92.

[67] Somme first-day casualties: James E. Edmonds et al., *History of the Great War: Military Operations, France and Belgium, 1916*, 2 vols. (each in two parts) (London: Macmillan, 1932-1938), I, 483. D-Day casualties: Chester Wilmot, *The Struggle for Europe* (New York: Harpers, 1952), p. 293n. There is a vast literature on Haig and the controversial battles, notably including Robert Blake, ed., *The Private Papers of Douglas Haig* (London: Eyre & Spottiswood, 1952); John Terraine, *Douglas Haig, The Educated Soldier* (London: Hutchinson, 1963); Sir John Marshall-Cornwall, *Haig as Military Commander* (New York: Crane, Russak, 1973); Basil H. Liddell Hart, *History of the First World War* (London: Cassell, 1970); Leon Wolff, *In

Since the point is often forgotten in discussions of British strategy, it is important to recall what was at stake. There were clear indications from 1915 to 1918 that German war aims included control of Belgium and the annexation of further lands from France, and as long as the Germans still occupied Belgium and northeastern France, a negotiated peace could only have led to the success of such plans. Such a peace would have amounted to a German victory, and would have meant German domination in Europe. The Allies could only defeat Germany by evicting the Germans and in particular by convincing German military leaders that their forces could not win. Inevitably, if very tragically, it took repeated offensives, involving a great deal of resolution and bloodshed, to implant this conviction in the minds of men like Erich Ludendorff, leading a professionally schooled officer corps and millions of well-trained men. The task was all the more costly where it fell to an army that had been largely improvised during the course of the war. But at the end it was mainly the tanks and infantry of Britain, Australia, and Canada (though preceded by a successful Franco-American counterattack) that broke Ludendorff's will; the lesson was only underlined when the American army, long and ominously expanding, finally began major operations.[68] The cost of the war for England was awful, but this was the cost of defeating Germany, and if Britain lost fewer lives in the Second World War, this was mainly because others, primarily the Russians, sacrificed many more.

Flanders Fields (New York: Viking, 1958); Churchill's *World Crisis*; and Hankey's *Supreme Command*. On the broad issues, see Howard, *Continental Commitment*, pp. 57-58, and *British Way*, pp. 18-20.

[68] German expansionism was expressed most explicitly by the middle-class and conservative parties and pressure groups, but the government never clearly disowned claims in Belgium, as the Allies always asked, until the autumn of 1918. Relevant studies include Hans W. Gatzke, *Germany's Drive to the West (Drang nach Westen): A Study of Germany's Western War Aims during the First World War* (Baltimore: Johns Hopkins University Press, 1950); Frank P. Chambers, *The War Behind the War, 1914-1918: A History of the Political and Civilian Fronts* (New York: Harcourt, Brace, 1939); Gerhard Ritter, *Staatskunst und Kriegshandwerk: Das Problem des "Militarismus" in Deutschland*, 4 vols. (Munich: R. Oldenbourg, 1954-1968), Vols. III and IV; and the well-known volume of Fritz Fischer, *Griff nach der Weltmacht: Der Kriegszielpolitik des kaiserlichen Deutschland 1914-18*, 3rd ed. (Düsseldorf: Droste, 1964). The British perception of the issues at the time of the war is shown, for example, by a statement Prime Minister Asquith made in Parliament in February 1916: "We shall never sheath the sword, which we have not lightly drawn, until Belgium recovers in full measure all, and more than all, which she has sacrificed; until France is adequately secured against the menace of agression; until the rights of the smaller nationalities of Europe are placed upon an unassailable foundation; and until the military dominance of Prussia is wholly and finally destroyed" (cited from Chambers, p. 257). For all the wartime rhetoric, the problem was genuine. For the German recognition of defeat, see Ritter, IV, 388-414; Fischer, pp. 846-853.

If Germany had continued to possess a major short-service army after 1919, the British might perhaps have maintained a major army as well, giving up the tradition of the small, insular, long-service force. But as we have seen, Lloyd George successfully insisted on a long-service army for Germany, and thereby enabled his countrymen to escape peacetime conscription. We begin to see how strong some of the reasons were for British observers to minimize German infractions of Part V: all British military policy after 1919, and much of Britain's financial and political policy as well, was predicated on the assumption that the German army had ceased to be a significant force. British generals now thought less about European warfare than they had in Edwardian days, and the army was soon reduced below its prewar size. The public did not perceive it as a bastion of national pride, like the navy, or as technically and socially progressive, like the air force. The cultivation of professional staff officers had never been a strong point of the British army, and such military intellectuals as appeared turned to developing doctrines on armor and to rationalizing minimum forces as "the British traditional method in warfare."[69] Implicitly, the bloodshed of modern full-scale warfare would be left for others. For the chiefs of the services, thought focused on the all-red, worldwide Empire; the navy was largely concerned to maintain imperial communications, the army and the air force to intervene against poorly armed tribesmen, although the airmen also developed theories of strategic bombing. One deterrent to continental thinking among the generals may have been the probability that, in another war alongside France, a British force on the continent would fall under the control of a French supreme commander. Another deterrent was certainly the extreme unpopularity of continental involvement, and a third, the related lack of financial support.

8. ENGLAND'S SPIRITUAL DISARMAMENT

While England disarmed materially, it also disarmed spiritually. The contrast with Germany is striking. There were German critics of war and wartime policies, at least until the last phase of the Weimar Republic, but they were outsiders, strongly opposed by respectable society and the political élite, and for the most part, their views had only narrow circulation. In England, dislike of military commitments and distrust of military leadership were authoritatively, if unintentionally, fostered by men who had themselves held leading wartime positions. The war had given rise to a number of hard-fought policy

[69] See Howard, *British Way*, pp. 6-8, 19-20.

controversies, and after the armistice, these quarrels came out into the open as the contestants and their supporters took up their pens and, still more, as the British mass-circulation press seized on their revelations. With some of these disclosures, the intention may have been to show that the protagonists had taken prudent and far-sighted measures to make sure the country was prepared. But much of the public, having believed during the war that they were fighting mainly to honor a commitment to Belgium and to defeat German aggression, were later shocked to learn that secret military arrangements, not known at the time to the cabinet at large, had in effect made Britain an ally of France from 1905 on.[70] Many Englishmen made up their minds that no such links should be forged again. Indeed, this resolve was probably the strongest political idea affecting British foreign policy between the wars.

Aside from the issue of British involvement in the war, the nation chewed over and over the experience of the Great War itself, and this discussion colored everyone's thinking. In this regard, one of the first issues refought before the public was whether British forces should have been concentrated on the Western front, or used in flanking moves in other theaters. Churchill's flanking strategy had been rather discredited by the fiasco of the Dardanelles, but his capacity to write vividly and eloquently now enabled him to weaken the position of his opponents. His use of casualty statistics, comparing British losses with German losses, and arguing that most Allied offensives had been costly failures, was particularly effective, and he made the public more aware of the total human cost of the war. Churchill did not attack Haig, but others did. In 1930, Captain B. H. Liddell Hart, Britain's best-known military writer, published a revisionist history of the war, and Lloyd George (with Liddell Hart's assistance) added a biting indictment of military stupidity in a series of volumes that began to appear in 1933.[71]

[70] The published biography and diaries of Sir Henry Wilson (Sir Charles Caldwell, *Field Marshal Sir Henry Wilson: His Life and Diaries*, 2 vols. [New York: Scribner's, 1927]) provided new detail on the origins of British involvement in the war, as did also books by Richard Burdon Haldane, Lord Haldane (*Before the War* [London: Cassell, 1920]), Field Marshal Sir William Robertson (*Soldiers and Statesmen, 1914-1918*, 2 vols. [London: Cassell, 1926]), and Churchill's *The World Crisis*. Sir Edward Grey had revealed the commitments to the French in Parliament when war broke out, but they were not widely discussed until the 1920s.

[71] Churchill, *World Crisis*, esp. III, Chapter 2, pp. 188-194, and Appendix I; B. H. Liddell Hart, *The Real War, 1914-1918* (London: Faber, 1930); Liddell Hart (*Memoirs*, I, 357-375; David Lloyd George, *War Memoirs*, 6 vols. (Boston: Little, Brown, 1933-1937). Sir Charles Oman, in "The German Losses on the Somme, July-December 1916," in Lord Sydenham of Combe et al., *"The World Crisis" by Winston Churchill: A Criticism* (London: Hutchinson, [1928]), pp. 40-65, challenged Churchill's discussion of the German losses, and the data given later in the German official

Thus the military direction of the war came under persuasive criticism, which made it appear that lives had been wantonly thrown away. The antiwar writings of disillusioned front-line veterans strongly reinforced this last impression.[72] Perhaps an American, whose country has had no comparable losses since the Civil War, is unqualified to comment on this literature. A British historian, Correlli Barnett, has recently argued that the antiwar writers, such as Siegfried Sassoon and Robert Graves, were unrepresentative of the British army as a whole, that the mud and blood of trench life was less of a shock to lower-class Englishmen who had lived at home in squalor and even danger, that the impression of a "lost generation" stemmed from the exceptionally heavy losses of upper-class, public-school-educated subalterns, and that the proportion of United Kingdom war dead to total population was little more than half that of France; he could have added that it was also little more than half that of Germany.[73] But however this may be, the English élite and English leaders themselves felt keenly the loss of relatives and friends, and unlike the situation in

---

history (Kriegsgeschichtliche Forschungsanstalt des Heeres, *Der Weltkrieg, 1914 bis 1918*, 14 vols. [Berlin: Mittler, 1925-1944], Vols. X [1936] and XI [1938]) were much closer to Oman's estimates than to Churchill's. Exact, comparable casualty losses will probably never be established. It is somewhat ironic to see Churchill, in his Second World War memoirs, castigate the members of the Oxford Union and their mentors for passing the famous resolution in 1933 against fighting for King and Country, when he himself contributed to bringing about this state of mind: Winston S. Churchill, *The Second World War*, 6 vols. (Boston: Houghton Mifflin, 1948-1953), I, 85.

[72] These include Richard Aldington, *Death of a Hero* (London: Chatto & Windus, 1929); Edmund Blunden, *Undertones of War* (Garden City, N.Y.: Doubleday Doran, 1929); Robert Graves, *Goodbye to All That*, rev. ed. (London: Cassell, 1957); Charles Edward Montague, *Disenchantment* (London: Chatto & Windus, 1922); R[alph] H[all] Mottram, *The Spanish Farm Trilogy, 1914-1918* (London: Chatto & Windus, 1927); Siegfried Sassoon, *The Memoirs of George Sherston* (New York: Literary Guild, 1937); and Robert Cedric Sherriff, *Journey's End: A Play in Three Acts* (New York: Brentano's, 1929), as well as the American Ernest Hemingway's *A Farewell to Arms* (New York: Scribner's, 1953) and the German Erich Maria Remarque's *All Quiet on the Western Front*, A. W. Wheen, trans. (Boston: Little, Brown, 1929).

[73] Correlli Barnett, pp. 425-435. I am indebted to Barnett's fascinating work, but doubt his suggestion (p. 428) that the officers who were lost were no loss to the nation's capacity to revive industry, win markets, solve great social problems, or "dominate the course of world affairs." I calculate a proportion of war dead to population of about 1:30 for France, 1:33 for Germany, 1:61 for Great Britain and Northern Ireland, and 1:65 for the 1914 United Kingdom including all of Ireland. The ratio for the American Civil War (both sides combined) was 1:60, or possibly higher. Brüning's friend Treviranus, a former naval officer, translated *Goodbye to All That*: Robert Graves, *Strich Drunter!*, trans. G. R. Treviranus (Berlin: Transmare, [1930]). However, for Treviranus, German publication of the book did not represent a protest against the war, but an affirmation (as he wrote in his translator's preface, p. 9) of "the comradeship of front-line soldiers," healing "the blood sacrifice which was not given in vain."

Germany and France, this feeling was rendered more acute by a belief that the suffering and loss had been unnecessary.

The British recoil against the war was extended to much of the peace settlement by John Maynard Keynes's *Economic Consequences of the Peace* (1919), which has been described as "a brilliantly destructive tract exemplifying at once both the worst and the best qualities of the British liberal intellectual tradition."[74] Thanks largely to Keynes, British opinion soon concluded that the reparations laid down in the peace settlement should be abolished. The general impression Keynes gave of deviousness and blind selfishness at Versailles was to be reinforced by the later speeches and writings of Lloyd George, the British member of the 1919 Big Four. It seemed that the various peace treaties had only set the stage for more Balkan-style quarrels—over the minorities of Czechoslovakia, Yugoslavia, and Rumania, over the boundaries of Greece, over Upper Silesia, over the Tyrol, over the separation of Austria from Germany, and in particular over the Polish Corridor. In no way did the British want to become involved in another war over such questions. In the initial postwar period, Lloyd George had nearly involved Britain in a war with the Turkish Nationalists, who had refused to accept the Treaty of Sèvres. The three largest dominions, Canada, South Africa, and Australia, had proven reluctant to promise help, and the British Conservatives had rebelled against Lloyd George's leadership, causing his downfall. This event made a lasting impression on political leaders, and encouraged thoughts of imperial isolation.[75] With regard to Germany, the idea took root that the territorial boundaries of the Versailles Treaty should be peacefully revised. In August 1931, Prime Minister Ramsay MacDonald gave Secretary Stimson his frank opinion that "the treaty must be amended eventually"; he indicated that this should not be done by force, and that it was necessary first to free France from her fear of isolation, but after that the next step was "to get the Versailles Treaty up for discussion."[76]

One part of the peace settlement stood apart, however, in the eyes of most British observers: this was the League Covenant. Lord Cecil's

---

[74] By Watt in *Personalities and Policies*, p. 120. See also Franklin Reid Gannon, *The British Press and Germany, 1936-1939* (Oxford: Oxford University Press, 1971), p. 7.

[75] See Martin Gilbert, *The Roots of Appeasement* (New York: New American Library, 1966), Chapter 9; Gilbert, *Rumbold*, Chapter 13; Middlemas and Barnes, pp. 110-124. Andrew Bonar Law, Lloyd George's Canadian-born Conservative successor, declared during the crisis that unless France supported Britain in Turkey, Britain would have to follow the American example and withdraw into imperial isolation (p. 114).

[76] *FRUS, 1931*, I, 516. In April 1933, MacDonald wrote in his diary, "A Peace Treaty of Conquest has always a secret & undisclosed article, a brief one: the date of the consequential war": MDP, 8/1, Diary, 5 Apr [33].

League of Nations Union lobbied powerfully in support of the League, and the Labour party always emphasized its desire for a stronger League. Cecil himself was a Conservative, and aside from the most rigid of the old-school imperialists, most Conservatives avowed that they too supported the "right kind" of League, one that would settle disputes by conciliation, not coercion. Indeed, most of Labour held to roughly the same conception. But despite the fact that most of the nation was opposed to the idea of a League "with teeth," the British supporters of the League were extremely serious about the League, in the way that evangelists are serious. What most British internationalists thought necessary was not coercive institutions, but a sort of conversion, the adoption on everyone's part of a whole new outlook. Somewhat as other prophets or reformers had appeared at dark moments, President Wilson had appeared at the end of the Great War with his Fourteen Points and, especially, his proposal for a League of Nations. His gospel struck particularly fertile soil in England, for the good reason that most of the ideas had already been circulating there. The desire to apply universal moral principles to politics had deep roots in British liberal tradition, as manifested for example in the pro-Boer protest, in William Gladstone's revulsion against the Turks, in William Wilberforce's movement to abolish slavery, and indeed in the Great Rebellion of the seventeenth century.[77] Despite Wilson, the United States had refused to join the League, but Englishmen could hope that the United States might cooperate with the League, or perhaps even join it at a later date, provided it was kept noncoercive. For the British, the League served the psychologically important function of justifying the sacrifice of the war: if the war could bring about the establishment of a society of all nations, the substitution of cooperation and understanding for selfish opportunism and hatred, then it would not have been fought in vain. This idea was probably what made support for the League a majority position in British political life.

9. ENGLAND'S ECONOMIC ORDEAL

In British traditions, interests, and ideals, the leaders of the country had many reasons to shun the balance-of-power doctrine, and to

[77] On the early League of Nations movement and Wilson: Lawrence W. Martin, *Peace Without Victory: Woodrow Wilson and the British Liberals* (New Haven: Yale University Press, 1958), pp. 58, 62, 95, 103-104, 108-111, 114-115, 150-151; George W. Egerton, "The Lloyd George Government and the Creation of the League of Nations," *AHR*, 79 (1974), 419-444; Arthur Marwick, *The Deluge: British Society and the First World War* (Boston: Little, Brown, 1965), pp. 214-217, 309-310; Arno J. Mayer, *Political Origins of the New Diplomacy, 1917-1918* (New Haven: Yale University Press, 1959), pp. 45, 53-58.

shudder at the prospect of another war. Explanations of British attitudes have tended, increasingly, to stress the trauma of the Great War, and that experience unquestionably had a tremendous effect. But underlying British foreign policy and every other aspect of British life was the problem of how Britain, with its material and psychological investment in the nineteenth-century world, could adjust to a new world of advancing technology, trade barriers, and relatively self-sufficient economies. The difficulties, sharply aggravated by the 1914-1918 war, were staggering, and it was hard for British leaders to conceive of how Britain could afford to prepare for another war, let alone sustain one. Insular traditions and revulsion from war, powerful as they were, probably did less to make Britain militarily weak than financial difficulties did.

Financial counsel tended to assume the same authority in England as military counsel in Germany, and British Treasury officials, like German officers, had certain fixed principles, which were shared by the political chiefs. In London, financial policy-makers believed in the Gladstonian doctrines of peace, governmental retrenchment, and the reduction of public debt—as had been done in the nineteenth century with the debt from the Napoleonic Wars. To the credit of the British Gladstonians, they did not tolerate the covert swindling of domestic war loan creditors through large-scale inflation, as happened in other major European countries. The British government also made a point of reaching an early settlement of its external war debt to the United States—while the French and Italian governments achieved much more lenient settlements by delay. This British policy of attempting honestly to pay the costs of war meant that, in 1929, about one quarter of the total of public expenditure (including local as well as national) went to meet interest payments on the national debt; in 1913, the figure had been six percent.[78] Considering other inescapable expenses, this naturally placed narrow limits on defense spending.

Up until September 1931, debt and budgetary problems were aggravated by the deflationary return to the prewar parity of the pound (at $4.86), decided on in principle in 1918 and finally carried out in 1925. Keynes and other economists have debated the importance of this policy in hampering British exports and maintaining high rates of unemployment.[79] But whatever the economic effect of the return to

[78] B.W.E. Alford, *Depression and Recovery? British Economic Growth, 1918-1939* (London: Macmillan, 1970), p. 65.
[79] R. F. Harrod, *The Life of John Maynard Keynes* (London: Macmillan, 1951), pp. 338-362; Alford, Chapters 3 and 4. Keynes's criticism was expressed esp. in a pamphlet entitled *The Economic Consequences of Mr. Churchill* (London: V. & L. Woolf, 1925), a title playing on his own earlier critique of the peace settlement and on Churchill's policy as Chancellor of the Exchequer.

prewar parity, it clearly led to cuts in spending on the armed services. On the recommendation of its Finance Committee, and without serious consideration of the diplomatic or strategic problems involved, the War Cabinet decided in August 1919 that the services need expect no war for ten years, and that the defense budget should be reduced as far as possible to prewar levels. By 1922, the reduction was nearly achieved in the case of the army, and in that year the Geddes Report pressed for further reductions (the "Geddes Axe"). The return to prewar parity with the dollar appears to have caused the application of a brake to naval and air force planning in 1925; it was Churchill, as Chancellor of the Exchequer, who applied this brake, and in 1928, at his urging, the assumed period of ten years of peace was made to advance automatically each day, until the cabinet should decide otherwise.[80]

Certainly economic conditions made it no easier to sustain the currency and to carry the load of debt and taxation, and some of these conditions pointed up the penalties of war. The basic industries—textiles, coal mining, iron and steel, shipbuilding—had become victims of obsolescence, the postwar slump, and a wartime loss of foreign markets to competitors such as Japan and the United States.[81] Between the wars, Britain's imports of food and raw material were increasingly purchased by income from overseas services and investments, rather than exports. Although the automobile and light industries thrived, little seemed to get done to revive the stagnant older industries that still dominated the economy, or to find adequate substitutes to gain foreign exchange and reduce the unemployment rate, which from 1923 to

[80] N. H. Gibbs, *Rearmament Policy* (in Sir James Butler, ed., *History of the Second World War: Grand Strategy*, vol. I) (London: H.M.S.O., 1976), pp. 3-6, 44-64; Peter Dennis, *Decision by Default: Peacetime Conscription and British Defence, 1919-1939* (London: Routledge & Kegan Paul, 1972), pp. 10-13, 18-19; Higham, *Armed Forces*, pp. 68-69, 76-77, 85-86, 95, 123-127, 130, 163-165, 248, 263, Appendix II; Middlemas and Barnes, Chapter 13 (esp. p. 327) and pp. 759-760; Correlli Barnett, pp. 277-278; Shay, p. 19.

[81] For a survey of these industries, see Derek H. Aldcroft, *The Inter-War Economy: Britain, 1919-1939* (New York: Columbia University Press, 1970), Chapter 5. British exports to South America fell over one third between 1913 and 1923, largely due to American competition, and the export of cotton piece goods to India declined 53 percent in the same period; this last was particularly serious because the prewar trade surplus with India had served to settle about half of Britain's trade debts. See Alan S. Milward, *The Economic Effects of the Two World Wars on Britain* (London: Macmillan, 1970), pp. 50-51; Alford, p. 60. Figures in Sidney Pollard, *The Development of the British Economy, 1914-1950* (London: Edward Arnold, 1962), p. 121, show that cotton piece-good exports to India had fallen 90 percent (from the 1909-1913 average) by 1939. David S. Landes, *The Unbound Prometheus* (Cambridge: Cambridge University Press, 1959), pp. 368-369, 451-453, 467-477, describes the problems in the cotton and especially the steel industries.

1939 never fell below 10 percent, and which reached 22.5 percent in 1932. The average annual rate of growth in domestic output was little more than 2 percent between 1924 and 1937. Many explanations have been given for this state of affairs, including social pressures at all levels against enterprise and effort, educational institutions that failed to train engineers and managers, and the continued investment of capital overseas that might have been invested at home. But in the old industries, innovation and new investment could also aggravate unemployment by reducing the number of hands needed to supply an already shrinking export market.[82]

One might have expected these economic problems to have serious political and social consequences. The 1925 return to prewar parity, along with obsolescence in the coal industry and foreign competition, forced wage cuts in the mining industry, and this led to the General Strike and Coal Strike of 1926. Unemployment was a continuing worry, and the 1931 financial crisis brought a reduction in the dole and the application of the means test to dole recipients, a source of class contention. A flagrant concentration of wealth remained, as one percent of those over 25 in England and Wales still owned 55 percent of the total capital in 1936-1938. On the other hand, the wealthy could sense they were losing ground: their wealth had less value than before the war, and income tax and supertax, which reduced the highest incomes by one twelfth in 1914, cut them in half in 1925. Yet the prevailing attitude, on the part of rich and poor alike, was that things might be much worse, and that any change would endanger existing advantages. Few wanted a circulation of élites or a rationalization of working practices. The guiding rule, at home as abroad, was not to raise awkward, indiscreet questions, and not to rock the boat. Britons did not want to risk a Bolshevik revolution or a German-style inflation, and in fact, partly because they had enjoyed no real boom, they did avoid the acutely traumatic depression experienced in Germany

---

[82] Aldcroft, and also Harry W. Richardson, *Economic Recovery in Britain, 1932-9* (London: Weidenfeld & Nicolson, 1967), take a more positive view of the British economy, stressing the new industries and comparing the growth rate with past British records, but cf. the criticism in Alford, pp. 15-29. On the causes of difficulty, see Correlli Barnett, pp. 485-495; Alford, Chapter 4; Landes's pp. 316-358, although on the pre-1914 period, give an excellent summary of the persisting British problems. One stagnant industry was armament, and the difficulty of maintaining an arms industry without significant arms orders was under cabinet consideration in early 1933, in connection with the question of an arms embargo against Japan: Roskill, *Hankey*, III, 72-75; Cab 24/238, CP 42 (33), 18 Feb 33; CP 48 (33), 24 Feb 33; Cab 24/239, CP 64 (33), 13 Mar 33; CP 94 (33), 7 Apr 33. On the contribution of increased productivity to unemployment: Aldcroft, p. 135.

and the United States after 1929. Social stability was secured, but Britain did not keep pace with other countries.[83]

The political leaders of the period, Stanley Baldwin and Ramsay MacDonald, symbolized rather than solved Britain's problems. Baldwin, the paternal, tradition-loving Conservative leader, had an inherited stake in an old-established iron and steel concern, a business which slumped badly after the war. When Baldwin predicted that another war would end "the civilisation of the ages," he was probably thinking not only of aerial bombardment, a special bugbear of his, but also in part of the blow such a conflict would give to the ailing economy and to social and political institutions. MacDonald, the Labour (later National Labour) leader of humble origin, had bravely upheld pacifist principles through the war, but his second (1929-1931) Labour government could not cope with unemployment, and he was becoming notorious for the woolly imprecision of his public utterances. He specialized in the cultivation of good relations with the United States—a policy favored by most of the nation, but necessitated in particular by Britain's financial weakness and her desire for relief from war debts.[84] Both Baldwin and MacDonald were centrists, concerned primarily to smooth over domestic differences. After the financial and political crisis of 1931, another, tougher political leader would begin to emerge, the dour, hawk-faced Neville Chamberlain. He would move Britain toward recovery through economic nationalism, and thus all the more toward disengagement from the continent. But all three men shared the prevalent feeling that the country was skating on thin ice, and that no risks should be taken.

[83] Kathleen M. Langley, "The Distribution of Capital in Private Hands in 1936-38 and 1946-47 (Part II)," *Bulletin of the Oxford University Institute of Statistics*, 13 (1951), 46-47. Where the number of persons with a gross income over £3000 had been 32,500 in 1913-1914, those with an income equivalent in purchasing power had fallen in 1924-1925 to 24,000: A. L. Bowley, *Some Economic Consequences of the Great War* (London: Thornton Butterworth, [1930]), p. 138. On Britain's relatively mild depression, Aldcroft, pp. 40-43; Richardson, pp. 15-20. On the prevailing attitude, see esp. Parker, *Europe 1919-45*, Chapter 7.

[84] On Baldwin: Cab 24/239, CP 52 (33), 28 Feb 33; Middlemas and Barnes, pp. 731-733, 735-736, 937. Baldwin coined the phrase, made famous by Keynes's *Economic Consequences of the Peace*, about "hard-faced men who looked as if they had done well out of the war," and in 1919 he freely and anonymously *gave* his country £120,000, in part, apparently, so as not to feel himself to be one of these men: see Middlemas and Barnes, pp. 72-73, 260, 529. On MacDonald: Marquand, esp. pp. 693-696, 714-715, 733-737; [Robert], Lord Vansittart, *The Mist Procession* (London: Hutchinson, 1958), esp. pp. 318, 385-391, 403. MacDonald's diaries show that he was painfully aware of his declining powers, and also that he was still more lucid than his public speeches suggested. He felt able to work with Americans (see e.g. MDP, 8/1, Diary, 1 May, 16 Oct [32]; 13 Jun [33]), although often disappointed or irritated with them (MDP, 8/1, Diary, 21 Jun, 23 Nov [32]; 4, 7 Apr, 16 May, 13 Jun, 10 Jul [33]).

10. CHAMBERLAIN AND THE TREASURY AS MILITARY PLANNERS

For the British, the financial and political crisis of 1931, Arnold Toynbee's *Annus Terribilis*, was the climactic event of the years between Versailles and Munich, an event that made an avoidance of commitments not only desirable, but essential. The immediate cause was a panic among investors, especially foreign short-term lenders. They were alarmed by a preceding financial crisis in Austria and Germany, and by a British budget unbalanced by unemployment benefits. If German governments had to enjoy the support of the army, a British government needed the support of the financial community. MacDonald and his Chancellor of the Exchequer parted from most of their Labour colleagues and joined with opposition leaders in forming a coalition government, so as to muster a national consensus for economies, particularly in the dole, and show foreign lenders that Britain could live within her means. When pay cuts led to protests in the Royal Navy, the symbolic foundation of British power, even the new government could not prevent further withdrawals of gold and a departure from the gold standard. Henceforth, the government felt obliged to display the purest orthodoxy in finance.[85]

Within a year, an economic recovery had begun, but financial confidence was still considered crucial. For the sake of confidence and principle, British leaders worked to make permanent the Hoover moratorium on war debts and reparations, and at the Lausanne Conference they were to achieve a practical end of reparations, while ignoring indications that the United States would not agree to end the debts. At the time Brüning fell, the British concern was that this event might interfere with success at Lausanne; this helps to explain why they overlooked the role of the German military in Brüning's removal. Neville Chamberlain, as the new Chancellor of the Exchequer, pursued another policy goal by introducing a permanent and general tariff, with preferences for the dominions. This may not in practice have had all the empire-buttressing results that Chamberlain hoped for, but it gave some apparent substance to the idea of a commonwealth independent economically, and also politically, of the European continent.

[85] The 1931 politico-financial crisis is covered in detail in R. Bassett, *Nineteen Thirty-One: Political Crisis* (New York: St. Martin's, 1958) and in Robert Skidelsky, *Politicians and the Slump: The Labour Government of 1929-1931* (London: Macmillan, 1967). Toynbee's contemporary commentary, in *SIA, 1931*, Part I, Section 1, shows the impact on the British upper and middle classes. See also Landes, pp. 380-382, 397-398; H. V. Hodson, "Nemesis: The Financial Outcome of the Post-War Years," *SIA, 1931*, pp. 161-242; Charles P. Kindleberger, *The World in Depression, 1929-1939* (Berkeley: University of California Press, 1973), pp. 157-163; A. J. Youngson, *The British Economy, 1920-1957* (Cambridge, Mass.: Harvard University Press, 1960), pp. 80-86.

Otherwise, Britain's situation and Treasury policy continued to dictate budgetary conservatism. Until early 1933, the meager supply of gold and gold equivalents held by Britain's Exchange Equalization Account hardly sufficed to maintain the level of the pound at moments of weakness. And only strict policies of retrenchment made it possible to convert the domestic War Loan from 5 percent to 3½ percent without evoking fears of inflation.[86] Free trade, limping since the war, was now gone for good, but at home Gladstonian policies were more fervently pursued than ever.

Neville Chamberlain's nature and background equipped him well to reorganize an overextended, nearly bankrupt enterprise. He had early experienced failure and lived it down, and, scrupulously honest, he had learned to despise big spenders and big talkers. If, later on, he failed to be shocked by Hitler, this was perhaps because he had a low opinion of most political leaders, from those in the Kremlin to many of his associates in the cabinet. If he had an ideology, it was a belief in the sustenance and revival of business and trade within the British Empire. He favored social expenditures, but had no sympathy for the idea of reviving the economy through rearmament, and he thought that England could not afford a full-scale rearmament.[87]

Chamberlain had not been Chancellor for many months before he and the Treasury were called on to take a stand on defense. The chiefs of staff annual review of March 1932 pointed out that serious deficiencies existed in all three services, and urged that the ten-year rule should be revoked. Regarding the army, the review stated:

> 25. For major military liabilities, such as might arise under the Covenant of the League of Nations or the Treaty of Locarno, we are but ill-prepared. In 1914 we intervened on the Continent with six well-equipped and well-trained Divisions within the first month of the War. If today we committed our Expeditionary force to a Continental campaign in response to our liabilities under the Pact of Locarno, its contribution during the first month would be limited to one Division, and during the first four months to three divisions,

---

[86] On the British reaction to Brüning's fall, see Section 3 above and note 19. On post-crisis economic policies, Youngson, pp. 86-95; Richardson, pp. 182-202, 207-231; Kindleberger, pp. 179-181; H. W. Arndt, *The Economic Lessons of the Nineteen-Thirties* (London: Oxford University Press, 1944), pp. 121-134.

[87] On Chamberlain generally, Feiling, *Life of Chamberlain*, and Iain Macleod, *Neville Chamberlain* (New York: Atheneum, 1962). For another view, see also A.J.P. Taylor's comments in *English History, 1914-1945* (Oxford: Oxford University Press, 1965), esp. pp. 205-206, 255-258, 404-406, 414-416. Shay, and also F[rancis] A. Coghlan, "Armaments, Economic Policy, and Appeasement: Background to British Foreign Policy, 1931-1937," *History*, 57 (1972), 205-216, bring out Chamberlain's key role in defense decisions.

arriving piecemeal; and except for the moral effect of its presence on the Continent it would have little effect on the fortunes of the campaign.

The chiefs of staff placed their main stress not on European dangers, however, but on the sudden emergence of the Japanese threat in the Far East, where Britain was also completely unready; "the position [the annual review stated] is about as bad as it could be."[88] The navy was the service most involved, and the admirals may have hoped that the coincidence of a new National Government with the Manchurian crisis would give them a chance at last to obtain a long-sought improvement of the base at Singapore, and to press for a restoration of cruiser strength from the reduction dictated by the London Naval Treaty of 1930.

But Chamberlain presented a characteristic protest from the Treasury against any attempt to increase expenditures: this said that Japan was probably as much influenced by British financial and economic difficulties as by British military weakness. The memorandum stated:

> For some years past, and at the present time more than ever, the position and future of this country depend on the recovery and maintenance of sound finances and a healthy trading position. Without these we cannot provide resources for Imperial or national defence. What we need above all is a period of recuperation, diminishing taxes, increased trade and employment. . . .

The Treasury argued that Britain's public debt was disproportionate to that of any other country: "It is not therefore surprising if our people are anxious to avoid heavy expenditure on armaments; and that such is the attitude of the nation is undoubted." The assumption of the priority of finance is startling to a present-day reader. But even more striking was the Treasury's assumption that it was competent to judge foreign attitudes and defense needs. The paper argued that the British inability to fulfill military commitments, although involving some diplomatic disadvantages, "would for the most part only involve danger in the event of war with France. France is the only Power that constitutes a threat at one and the same time to our home defence (especially in the air), to our sea-borne communications in the narrow seas and to some extent in the oceans, and to certain of our coast defences at home and abroad." France, however, would be deterred (in the Treasury memorandum's words) by:

[88] Cab 24/229, CP 104 (32) (1082-B, 23 Feb 32); Gibbs, *Rearmament Policy*, pp. 78-80; Shay, pp. 22-23. The chiefs of staff had stated in 1925 that they could "only take note" of continental commitments, no forces being available for honoring them: Howard, *Continental Commitment*, p. 94.

(a) The age-long experience of the French nation in war and their known horror of a war on two flanks.
(b) The permanent Teutonic threat to their eastern frontier.
(c) The minor threat on their Italian frontier.
(d) The complete control of the Mediterranean and the eastern Atlantic by the naval forces of Great Britain.
There are also important economic deterrents.

Although Vansittart had lately complained that France had achieved virtual hegemony, and although the army general staff had spoken of French arms keeping Europe in tension, the Treasury's concern over a French danger was hardly shared by the Foreign Office or the Chiefs of Staff and it probably reflected only the intense Francophobia in financial circles at this time. Perhaps there was an element of reductio ad absurdum here, an attempt to ridicule military fears. But the suggestions illustrate the Treasury's tendency to claim that financial experts were the only experts who counted, even in other fields—a claim then often accepted. The Treasury paper concluded with a submission "that today financial and economic risks are by far the most serious and urgent that the country has to face and that other risks must be run until the country has had time and opportunity to recuperate and our financial condition to improve."[89]

Subsequent events would show a continued domination of British policy by financial considerations. The cabinet, greatly daring, did cancel the ten-year rule in March 1932, but in view of Chamberlain's position, it did so with the proviso that this not be taken to justify an increase in expenditure without regard for "the serious financial and economic situation that still obtains." There was no immediate move to increase expenditures, the government being in any case bound by an arms truce at least until November 1932, and foreseeably until the

---

[89] Cab 24/229, CP 105 (32) (1087-B, 11 Mar 32); Gibbs, *Rearmament Policy*, pp. 79-82; Shay, pp. 23-24. There have been charges that Sir Warren Fisher, for many years Permanent Under-Secretary at the Treasury, was an appeaser, opposed to rearmament. For the mid-thirties, these charges are refuted by Watt in *Personalities and Policies*, Essay 5, and by DRC records. But the Treasury, doubtless with Fisher's participation, did impose drastic disarmament earlier, and although Fisher backed rearmament in the 1930s, he had no regard for expert service opinion. See also Coghlan, 212n; Shay, pp. 186, 188, 220. Shay shows (pp. 74 and note, and 284) that Fisher began to lose influence in 1935, when he supported the idea of a defense loan, opposed by Neville Chamberlain and by other Treasury officials. Fisher's hostility toward France and the U.S. verged on the hysteric, and this March 1932 memo was probably his work. Chamberlain himself was not anti-French, and in the fall of 1933 he even favored closer discussions with France—in the interests of avoiding a duplication of military preparations: Cab 2/6 (1), 261st meeting, 9 Nov 33; Public Record Office, Cabinet Records, Cabinet Committee Records (Cab 27), Disarmament Conference 1932—Ministerial Committee (DC [M] 32, in vols. 504-511), Vol. 505, meeting of 30 Nov 33.

Disarmament Conference disbanded. Indeed, except for a decision in October 1932 to proceed with work on the base at Singapore, long-term planning was postponed until the fall of 1933. The chiefs of staff still gave priority to the Far East in their 1933 review, dated October 12 of that year; they stated, "We believe Germany already to have started that rearmament which, in the nature of things was sooner or later to be expected," but they did not anticipate an early and deliberate disturbance of the peace by her.[90]

If the British discounted German military preparations, one major reason was an assumption that Germany had the same financial limitations as were binding Britain. In September 1932, Simon maintained that if Germany were given equality of status, it was improbable "that Germany would seek to increase her armaments materially—for one thing she could not afford it." Field Marshal Milne advised that German budget restrictions would not permit the Reichswehr to accumulate expensive reserves of materiel. Even when Vansittart began stressing the German danger in 1933, he sought to make ministerial flesh creep by arguing that financial stringency would force Germany to concentrate on those relatively inexpensive weapons, also the weapons most fearsome to England, the airplane and the submarine.[91] As the precepts of German generals made them suspect the Poles of a desire to seize the military initiative, so the financial inhibitions in the minds of British policy-makers led them to suppose that the same restraints operated equally in Germany.

MANY considerations restrained the British government from following the principle of the balance of power. Nevertheless, it would not seem fair to say of 1932 and 1933, as Paul W. Schroeder has said of the 1938 Munich settlement, that a British policy of appeasement was massively overdetermined.[92] Indeed, as already mentioned, the British did eventually go over to a balance-of-power policy, in 1939. They did this, not because the problems of British weakness, Dominion and American isolationism, and distrust of France had been solved, but because the menace of German domination had become too clear to be ignored; resistance would certainly prove after 1945 to have a high cost, but not as high as submission would have had. A nation did not have

[90] Cab 23/70, meeting of 23 Mar 32, 19(32)2; Cab 23/72, meeting of 11 Oct 32, 50(32)9; Cab 24/244, CP 264 (33) (1113-B, 12 Oct 33).
[91] Cab 24/233, CP 305 (32), 15 Sep 32; Cab 24/234, CP 302 (32), 28 Oct 32; Cab 24/239, CP 52 (32), 28 Feb 33.
[92] Paul W. Schroeder, "Munich and the British Tradition," *Historical Journal*, 19 (1976), 242. Schroeder has also stated that World War I was massively overdetermined: "World War I as Galloping Gertie: A Reply to Joachim Remak," *JMH*, 44 (1972), 320.

to be strong to adopt a balance-of-power policy; it did have to look steadily and clearly at the dangers.

The possibility of a policy of balance always existed, and indeed some forces militated in favor of such a policy. English idealism could well be repelled by German militarism and, especially, by Nazi brutality. The Foreign Office understood and largely sympathized with French concerns. Air power meant an end of British isolation from the continent, and the potential threat of a German air force was early recognized. At the same time, British air strategy and air technology did not fall under the heavy hand of tradition, which elsewhere had such influence on British life. Although economic weakness was felt to dictate a policy of caution, the British economy did begin to regain strength after mid-1932, and in practice, the later rearmament programs (contrary to Gladstonian doctrine) would actually further recovery. And in fact, there were times in 1932, and even more in 1933, when it seemed that Britain might join in opposing Germany.

But in Britain's situation, unlike that of France, it was always easy to find reasons, even good reasons, for standing aside. A basic question was whether, in the face of economic preoccupations, nostalgic hopes for isolation, plausible German demands for "equality" and self-determination, and annoyance with French rigidity, British policy-makers could recognize in time the danger that lay in Germany's potential power and militant expansionism. Also, if they recognized it, could they convey an understanding of it to the British public? Thus the problem, in a broad sense, was one of finding, interpreting, and disseminating intelligence. Even more broadly, the problem was one of willingness to face a hard, unpleasant reality.

CHAPTER THREE

The United States and the Legends of Lost Opportunity

1. THE NEW PEACE CONFERENCE

Ironically, the problem of German rearmament came to a head just as the nations finally gathered to consider the limitation and reduction of armament. The World Disarmament Conference had its origin in Point 4 of Wilson's Fourteen Points and in the implied Allied promise to disarm, contained in the preamble to Part V of the Treaty of Versailles, in Article 8 of the Covenant, and in the Allied letter of June 16, 1919. Preparations had begun in 1920 for a general agreement to reduce the armed strength of those nations not disarmed by the postwar treaties, and these preparations had taken a somewhat more active turn in 1926, as one result of the Locarno agreements. Progress, however, was still far from rapid, a fact that evoked increasing resentment and suspicion in Germany. A Preparatory Disarmament Commission convened irregularly from May 1926 until the end of 1930, and it finally produced a draft disarmament convention for consideration at a general conference. Early in 1931, the League Council scheduled the sessions to begin on February 2, 1932, and a little later it appointed Arthur Henderson, at that time the British Foreign Secretary, to be chairman.

Although the conference met in Geneva, in association with the League of Nations, it was not simply a League affair; the United States and the Soviet Union took part, although neither nation was a League member, and in all, fifty-nine states sent delegates. Arnold Toynbee wrote in 1933 that the Conference "was recognized to be the most important international gathering since the Peace Conference of Paris." And Secretary of State Stimson stated in 1931, 1932, and again in 1933 that he considered the Disarmament Conference to be essentially a European peace conference. Although by saying this Stimson was denying American responsibility for the fate of the gathering, he was also bearing witness to its expected importance.[1]

---

[1] *SIA, 1932*, p. 194; Robert H. Ferrell, *American Diplomacy in the Great Depression* (New Haven: Yale University Press, 1957), pp. 200-201; Yale University Library, New Haven, Conn., Henry L. Stimson MS Diary, entry for 15 Apr 32; Edgar B. Nixon, ed., *Franklin D. Roosevelt and Foreign Affairs*, 3 vols. (Cambridge, Mass.: Harvard University Press, Belknap Press, 1969), I, 56. Nadolny, in a speech on 18 Feb 32, stated rhetorically that he regarded the conference as a peace conference:

Stimson's was an apt description. For Germany and for the Western governments too, the sessions in Geneva seemed to offer a chance to make some major corrections that each thought were needed in the peace treaties. The Disarmament Conference was also like a peace conference in that the various parties had different and conflicting objectives. Indeed, the conference was not so much a meeting on the subject of disarmament as an arena for conflicting subjects of debate.

One way of describing the situation is to say that, in effect, at least three conferences took place in Geneva. First, the German government maintained that the gathering was a Conference on Equality of Rights, a meeting to give Germany the same right to national defense as the other powers continued to demand for themselves. The German Foreign Ministry and the Reichswehr agreed, as we have seen, in demanding equality of status, meaning that German armament should be limited only by the same convention as governed the armament of other powers, and not by Part V of the Treaty of Versailles. Thus Germany sharply rejected Article 53 of the Preparatory Commission's draft convention, which would have retained existing arms limitations in force. The equality principle could pave the way for German rearmament, since once the conference agreed to equality of status and the elimination of Part V, there would be no logical basis for giving Germany's material claims any different treatment from those of France, and the practical, detailed questions would have to be resolved on a basis of equality. The attitude of Englishmen and Americans also helped; the idea of equality appealed to their belief in fair play, while in their hope for a harmonious conclusion, they were inclined to overlook any indications of German rearmament. But if it once became completely obvious that Germany was trying to rearm, then from her point of view the conference would no longer serve any useful purpose; the concealment afforded by the equality doctrine would be lost, and the French might then use the conference to expose Germany's rearmament, even organizing opposition against her.

Second, the French government wanted to make the sessions a Conference on Security. Hard experience may have made the French less hopeful than the Germans that they could impose their definition of the agenda. But French policy continuously sought to "organize the peace" with new international guarantees of security, and aside from its intrinsic merits, this policy gave France an important bargaining counter. If other countries were thinking of pressing the principle of equality, or of calling for drastic cuts in (French) arms, France could

---

League of Nations, Conference for the Reduction and Limitation of Armaments, *Records of the Conference*, Series A, *Verbatim Records of Plenary Meetings*, Vol. I, 146.

threaten to ask for a decision on her proposals. If French plans were rejected, that would be unfortunate, but France could then at least cling to her existing arms status and blame the "Anglo-Saxons" for refusing guarantees, or the Germans for seeking rearmament. The French government always had an interest in exposing German rearmament, if that could be done in a way which would carry conviction in Britain, the United States, and elsewhere. But in early 1932, Tardieu may have had another strategy in mind, one of offering the Germans modifications of Part V restrictions, or even a form of equality of rights, to entice them into entering a French-style security system.[2] Even a mere show of German interest would strengthen the French position vis-à-vis the British.

Third, the British and American governments believed that this was in truth a Conference on Disarmament—meaning specifically the disarmament of France, and generally the elimination of European resentments and tensions. With a few exceptions, such as Sir Maurice Hankey in London and, in some moods, Secretary Stimson in Washington, British and American leaders and officials shared the public belief in both countries that disarmament agreements would prevent war. And indeed the case for disarmament agreements was overwhelming—as long as no party assumed that the agreement had ended the use of arms altogether, and provided that all parties could actually be held to observance. Britain and the United States had themselves, with Japan, reached two successive agreements limiting their navies, and this experience seemed to show that arms limitation was feasible; it also, along with economies in land armament, gave the British and especially the Americans a somewhat smug belief that they had already made their disarmament contribution. (Actually, the naval agreements had served largely to legitimize American naval equality with Britain and superiority over Japan; in Japan, the agreements were much resented, while if Britons outside the Royal Navy accepted them, it was for the sake of economy and smooth relations with the United States, as well as out of a belief in disarmament.) The British and Americans seldom reflected that the construction of unauthorized warships is easier to detect than that of gun barrels, and the British, in particular, did not like the idea of inspection, which among other things would reveal the near-absence of arms in England.[3] Inspection and sanctions were both French ideas, distasteful to Englishmen who preferred to rely on trust.

[2] See AA, tel., Nadolny to Brüning and Bülow, 26 Feb 32, 7360/E534972-73; tel., Hoesch to Ministry, 27 Feb 32, 7360/E534984; tel., Nadolny to Ministry, 29 Feb 32, 7360/E535013; tel., Nadolny to Ministry, 17 Mar 32, 7360/E535055.
[3] See, e.g., Cab 24/227, CP 4 (32), 1 Jan 32; CP 19 (32), 27 Nov 31.

Although Germany and the Western powers (not to mention the Soviet Union, Italy, and other nations) had sharply divergent conceptions of the aim of the conference, one major consideration worked to keep the conference from flying apart: A principal British objective was, precisely, to avert a breakdown and promote confidence, and this meant that London had more interest in keeping negotiations going than even in disarmament itself. Sometimes British officials tried to allay French fears, but more of the time, and more significantly, they looked actively for ways to satisfy the German demand for equality of rights. 1932 and 1933, roughly the period of the Disarmament Conference, can be seen as two cycles, in each of which the British began by trying to meet German demands, later were driven into reassuring France, and then, finally, to preserve an apparent hope of agreement, again called for concessions to Germany. Being the squeakiest wheel, Germany got the most grease.

As the conference began, the British had two general ideas for meeting German demands and promoting agreement. One was to hold out hope of full equality later by putting the still-needed sections of Part V in a new general arms convention of limited duration, dispensing with Part V.[4] The other general idea was that of qualitative disarmament. Qualitative disarmament meant the outlawing of certain categories of arms, or the limiting of weapons to a maximum tonnage or caliber. Such action, it was thought, would hamper aggression. Part V, of course, already applied qualitative restrictions on Germany: absolute prohibitions in the case of aircraft, tanks, and submarines, size restrictions in the case of artillery and battleships. If the new convention applied similar restrictions to all countries, this would greatly reduce the apparent discrimination against Germany, and meet, at least to some extent, the German demand for equality in the types of arms permitted, although the British did not as yet admit that, where the other powers did not accept the full qualitative limits of Part V, Germany would also be entitled to disregard those limits. In one more specific area, too, British officials, probably unwittingly, recommended acceptance of a German idea: Vansittart and Allen Leeper, a Foreign Office expert on disarmament, argued that Germany might well be permitted to have a militia on the Swiss model, though her standing army should be reduced in that case.[5] They of course supposed that a German militia and a standing army would be separate entities.

[4] This idea was already contained in the April 1931 memorandum referred to in Chapter Two, Section 3: *DBFP*, 2/III, No. 208 (8 Apr 31).
[5] Simon's ideas are discussed more fully below. On Swiss militia: Cab 24/227, CP 4 (32), 1 Jan 32; CP 19 (32), 27 Nov 31; FO 371, Memo by A. Leeper, 11 Jan 32, W11790/1466/98.

In February and March, the Disarmament Conference accomplished almost nothing beyond inaugural speeches and committee meetings. This had been expected by Vansittart in London and by Bülow and Schleicher in Berlin; they had all thought that little could be done until May, when parliamentary elections in France were expected to produce a more accommodating government in Paris.[6] But it was difficult for the British government simply to stand pat. For one thing, Tardieu actively pushed his plan. In order to be first in the field, he used an irregular means to get his proposal publicized before Simon made the opening speech: Tardieu presented his plan at a preliminary meeting for the election of conference officers. Thereafter he sought in various ways to bring about a debate on it. Since the British government did not like that plan, a debate would force them to define their own position, and this prospect became imminent when the conference decided that, upon reconvening on April 11 after a recess, it would sit continuously to reach decisions on basic principles.[7] For another thing, the British and American advocates of disarmament naturally thought that something should be happening. In Britain, ministers had to deal with a public demand for destroying the instruments of war, and since the British government believed that a disarmament agreement would make the American public more willing to cancel war debts, the British cabinet was anxious to satisfy both British and American opinion.[8]

Probably no one in London was more alive to these considerations than Sir John Simon, who had become Foreign Secretary in November 1931. A noted lawyer and Liberal leader, tall, and benevolent in expression, Simon had believed in August 1914 that Britain should remain neutral, and he had submitted his resignation as minister, although he had soon withdrawn it. At the end of 1915, he had resigned again, more lastingly, in protest against conscription.[9] He seems to have resolved, by 1932, not to make another such sacrifice for principle. Rather than taking stands, he now searched for compromises and hankered for personal approval. In trying not to antagonize anyone, he inspired distrust. He smarted under discourtesies, and apparently resented Tardieu's maneuver to present the French plan first. Later on, Simon showed anxiety lest Tardieu either secure approval for his plan, or withdraw from the conference. On the other hand, when Brüning addressed the conference and received demonstrative

---

[6] Cab 24/277, CP 4 (32), 1 Jan 32; Schäffer Diary, 17 Apr 31; Vogelsang, "Neue Dokumente," p. 416.

[7] *SIA, 1932*, pp. 196-200, 211-216.

[8] On British concern with U.S. opinion: Cab 24/229, CP 119 (32), 11 Apr 32, encl.; Cab 24/230, CP 164 (32), 26 May 32.

[9] [Max Aitken], Lord Beaverbrook, *Politicians and the War, 1914-1916* (London: Oldbourne, 1960), pp. 20-22, 247; Roskill, *Hankey*, I, 240.

applause, lasting noticeably longer than that accorded to Tardieu or to Simon himself, Simon was one of those who applauded to the end.[10]

Simon's own speech stressed the idea of qualitative disarmament and, in diplomatic conversation, he frequently referred to this and to the idea of placing the restrictions on Germany in the general convention to be concluded by the conference. He also stated repeatedly that the British government would not undertake new commitments to France. On the other hand, whether out of a wish to throw some sop to the French, or to maintain some check on the Germans, or simply as a matter of strict legal logic, he held that the continued validity of Part V did not depend on an execution of disarmament by the former Allies; for one thing, he argued, who could define the degree of disarmament that would be required to fulfill the alleged condition? In early April, the ministerial committee on disarmament (and then the cabinet) approved Simon's views on qualitative disarmament, on transferring restrictions on Germany to a general convention, and on rejecting any new engagements in the name of security. When it was pointed out in the committee that the destruction of the French plan would make the French more utterly negative, this led one member to comment that there were really only two courses open: to submit to France or to isolate France. The committee agreed that in any case Britain could not contemplate submission to France. Simon and his colleagues were very alert against further continental commitments. They hoped, however, that the United States would support their overall position.[11]

[10] Simon's autobiography, *Retrospect* (London: Hutchinson, 1952), is notorious for its lack of information. Edward Barton Segel, "Sir John Simon and British Foreign Policy: The Diplomacy of Disarmament in the Early 1930's" (Ph.D. diss., University of California, 1969) is an excellent study, which uses Simon's (now unavailable) personal papers, held by the present Lord Simon. On Simon's personality, see also Anthony Eden, Lord Avon, *The Eden Memoirs: Facing the Dictators* (London: Cassell, 1962), pp. 26, 28-30, 46, 175, 219-220; A. L. Rowse, *All Souls and Appeasement* (London: Macmillan, 1961), pp. 15-17; Liddell Hart, *Memoirs*, I, 205; Paul Schmidt, *Statist auf diplomatischer Bühne 1923-45* (Frankfurt: Athenäum, 1964), pp. 232, 295-296; Cowling, *Impact of Hitler*, pp. 74-76. Attitude toward Tardieu and his plan: *Records of Conference*, Series A, I, 58; Cab 24/228, CP 78 (32), 20 Feb 32; CP 89 (32), 1 Mar 32; Public Record Office, Foreign Office Private Collections (FO 800), Simon Papers (in Vols. 285-291), Vol. 286, ltrs. to MacDonald, 16, 18 Apr 32. Applause: *Frankfurter Zeitung*, 10 Feb 32 (2. Morgenblatt).

[11] *Records of Conference*, Series A, I, 55-59; DBFP, 2/III, Nos. 235, 236, 239; AA, tel., Nadolny to Ministry, 29 Feb 32, 7360/E535011-12; Cab 24/228, CP 78 (32), 20 Feb 32; Cab 24/229, CP 110 (32), 21 Mar 32; CP 119 (32) (DC [M] 32, 2nd cons., 5 Apr 32, and Annexes A and B); CP 125 (32), 31 Mar 32; Cab 23/71, meeting of 13 Apr 32, 21(32)2; Isaac Alteras, "The Geneva Disarmament Conference: The German Case" (Ph.D. diss., City University of New York, 1971), p. 35. The British chiefs of staff contested the theory of hampering aggression by limiting "offensive" weapons, holding that it was impossible to distinguish offensive from defensive arms (see Cab 24/227, CP 19 (32) [CID Sub-Committee report, 27 Nov 31, citing

2. THE UNITED STATES: IDEALS AND INTERESTS

While the French were preoccupied with trying to build alliances against Germany, and the British were preoccupied with evading any such ties, for the United States, the problems associated with a possible German revival hardly seemed to exist. The country was practically invulnerable as long as no single aggressive power controlled Europe's industrial and human resources, and in 1932 no one thought of that as an imminent possibility. The first interest of the American public was in domestic affairs, and at this juncture that public's attention turned increasingly to the depression, which promised to be the central issue in the 1932 presidential campaign. Nevertheless, the United States did concern itself actively with the disarmament question. Aside from its leading role in the Washington and London naval treaties of 1922 and 1930, the American government had participated in an unsuccessful three-power disarmament conference in Geneva in 1927, and in the prolonged labors of the Preparatory Commission. Outside the area of naval limitation, it was not so much national interest as the domestic politics of idealism that shaped American policy on disarmament.

The idea of an American mission to reform world affairs, championed by Woodrow Wilson, had not disappeared with the end of World War I. Although the American Senate had refused to assume the commitments involved in the League of Nations and the Treaty of Versailles, and had ignored the Franco-American treaty of guarantee, it is often forgotten that a majority of the Senate membership (if not the necessary two-thirds majority) approved of the Treaty of Versailles either as it stood or with minor reservations; with more flexibility, Wilson might have secured a League membership as meaningful as that practiced by the British. Interest in international peace efforts seemed to decline during the 1920s, but an active minority still supported such efforts, and party leaders continued to call for disarmament, arbitration agreements, and conciliation, presumably in the belief that such measures for peace enjoyed broad public support. Members of the peace movement showed their political weight in 1927 and 1928 by pushing an initially indifferent administration into concluding the Kellogg-Briand pact, outlawing war. Various political leaders, including both President Hoover and the future Democratic

---

CID paper 1064-B]), but their views seem to have been ignored. On Germany's legal demand for Allied disarmament, a 1926 Foreign Office paper had stated that she had "not only a good case in equity, but the letter of the Treaty of Versailles" on her side: *DBFP*, Ia/I, Appendix (p. 857). For another legal objection to the claim that Part V could lapse through Allied non-disarmament, see Chapter Five, Section 3.

candidate and President, Franklin D. Roosevelt, had themselves once supported Wilson, and despite their own nationalist lapses, they knew the political potency of appeals for peace. In 1932, American internationalists believed that disarmament was the next step to be taken, and they were still a political force to reckon with, even if a diminishing one.[12]

Along with Wilsonian idealism and its political exploitation, other considerations also led to American involvement abroad. It has been plausibly argued that the United States was nationalist rather than isolationist in the 1920s.[13] American commercial interests influenced policy in Latin America, the Far East, and Africa. Japanese expansionism and American racialism fostered fears of a war with Japan. As Tardieu pointed out, the United States had increased its arms expenditures (for the most part, naval expenditures) more than any other power,[14] and Secretary Stimson vigorously opposed any reduction in American naval strength. Stimson also hoped to give more substance to the Kellogg pact by developing, unilaterally, an unwritten custom of consulting with other nations when problems arose. In early 1932, he was trying, in consultation with the British, to discourage Japanese aggression in China, even though he had little means of pressure at his disposal beyond bluff.

Both Britain and France wanted to encourage American consultation and cooperation in dealing with world problems. But the war debts of the ex-Allies to the United States caused friction, while private loans to Germany gave Americans a material interest in helping that country. Republican tariffs made the repayment of both war debts and private debts more difficult. Hoover attempted in 1931 to save German private credit with his moratorium on intergovernmental war debts and reparations, and even so the payment of German private short-term debts also stopped in 1931, only resuming under restraints thereafter, while German long-term debts were later repaid only in part. The French greatly resented Hoover's proposal at the time he made it, as it meant for them a loss in reparations

---

[12] Jean-Baptiste Duroselle, *From Wilson to Roosevelt*, Nancy Lyman Roelker, trans. (Cambridge, Mass.: Harvard University Press, 1963), pp. 110-115, 121-125; Selig Adler, *The Uncertain Giant: 1921-1941: American Foreign Policy between the Wars* (New York: Macmillan, 1965), pp. 25-42, 87-92, 125-126, 148-149; Charles A. Beard, *American Foreign Policy in the Making, 1932-1940: A Study in Responsibilities* (New Haven: Yale University Press, 1946), Chapter 4; Robert H. Ferrell, *Diplomacy in the Great Depression*, Chapter 2. On political influence in 1932, see Hooker, *Moffat Papers*, pp. 56-57, 60.

[13] Duroselle, *Wilson to Roosevelt*, pp. 134-135; William Appleman Williams, in *The Tragedy of American Diplomacy* (New York: Dell, 1962), Chapter 4, states the thesis in terms, rather, of economic imperialism.

[14] *FRUS, 1932*, I, 37.

greater than their savings from war debt remission, whereas they had nothing significant to lose on German private debt. The United States also disappointed France in the matter of consultation itself. When Premier Laval visited Washington in October 1931, he proposed an actual consultation pact; Hoover replied that this was a political impossibility, and indeed the Americans told Laval that security could best be obtained by disarmament and a revision of the Polish Corridor.¹⁵ When, from December 1932 on, the French and later the British ceased payment on their war debts, this in turn evoked American resentment. Along with congressional exposés, the depression, and the growing danger of war in Europe, the default on war debts would make American isolationism more rampant in the middle and late thirties.

At no time between the wars did the doctrine of the balance of power win significant acceptance in the United States. Looking back in 1953, Robert E. Osgood argued that American diplomacy until World War II had been marred by an overemphasis on ideals and a failure to recognize and act on Americans' real interests. This failure caused an oscillation between periods of crusading fervor and bitter disillusion. Osgood traces the absence of realism to American traditions and new aspirations, but also, no doubt rightly, to the American feeling of security, to the general confidence that America was safe from foreign attack. Thus the spur of fear, which might have encouraged a balance-of-power doctrine, was usually lacking. Before and during World War I, some American leaders, including some of Wilson's closest advisers, feared the growing power of Germany, and the threat of a German victory to American security. Yet, when war came they found it more expedient to argue for American preparedness and intervention on grounds of morality, honor, and neutral rights. The American public, and President Wilson himself, were never convinced that there was a German threat to America's own security. Osgood points out that, to judge by the Treaty of Brest-Litovsk, a German victory would have been highly inimical to American ideals and interests. But that treaty came after the United States had entered the war, and it was, as John Wheeler-Bennett called it in 1938, the "forgotten peace."¹⁶ Few people reflected that America's past century of security had resulted from a European balance and a strong British

---

15 *DBFP*, 2/II, Nos. 280, 288; Ferrell, *Diplomacy in Great Depression*, pp. 199-203.
16 Robert E. Osgood, *Ideals and Self-Interest in America's Foreign Relations: The Great Transformation of the Twentieth Century* (Chicago: University of Chicago Press, 1953), esp. pp. 17-20, 78-84, 114-176, 180, 201-202, 205-208, 262-263; John W. Wheeler-Bennett, *Brest-Litovsk: The Forgotten Peace* (London: Macmillan, 1938). Cf. Ernest R. May, *The World War and American Isolation, 1914-1917* (Cambridge, Mass.: Harvard University Press, 1959), pp. 436-437.

fleet, or that Britain might now become unable and unwilling to uphold such a balance in a more unstable Europe. No evidence appeared after the war to show that there had been a direct German threat to the United States, and the idea that there had been a German menace now seemed a figment of war propaganda.

To a greater degree than in England, involvement in the war came in retrospect to appear as unnecessary, and the motives of those responsible as suspect. The exposure of munitions profiteers and Allied propagandists reached its climax later, but from early on, populist politicians and newspaper owners could raise the bogey of foreign entanglements any time the United States government showed symptoms of a new interest in European affairs. These men could argue, without being contradicted, that the only war danger was that which might arise from a foolish American involvement in foreign quarrels. The government fell prey to contradictory impulses. On the one hand, internationalist pressure groups and financial and business interests encouraged the government to take an active part in world affairs. Leaders like Hoover and, later, Roosevelt sometimes also thought that they could enhance their own political following and ensure their place in history by taking internationalist initiatives in such areas as disarmament. But on the other hand, the supporters of internationalism appeared to be far less imposing and far less potentially numerous than the forces opposing entanglement.[17] Hoover and Roosevelt tended to retreat quickly if there was any complaint in Congress or in the press of partiality for Britain and France, or of willingness to embroil the United States in foreign problems. In this way, Roosevelt would manage to be the first Western leader to accept a fait accompli by Hitler.

In fairness to Hoover, he did have a genuine interest in disarmament. A Quaker and a lifelong admirer of Woodrow Wilson, he sincerely believed in working for peace. He did not look to "idea men" for proposals, but worked out his own ideas on disarmament. At a congress of the International Chamber of Commerce in May 1931, Hoover had expounded the political danger and also the economic wastefulness of armaments. At one point he said, "Of all the proposals for the economic rehabilitation of the world, I know of none which

---

[17] On isolationism and internationalism: Adler, *Uncertain Giant*, Chapters 1-6. Ernest R. May points out that no more than 15 percent of the general public is particularly interested in foreign affairs, and that the composition may vary; moreover, in 1933, this foreign affairs public was fragmented into numerous small groups: *"Lessons" of the Past: The Use and Misuse of History in American Foreign Policy* (New York: Oxford University Press, 1973), pp. 146-147. This would indicate that the political support for the League, disarmament, or a consultative pledge was strictly limited.

compares in necessity or importance with the successful result of that [the Disarmament] Conference." His lack of acquaintance with modern methods of preparing for war was betrayed by the remark in his speech that the vast world forces of five and a half million men under arms and twenty million more in reserves "still are to be demobilized." Someone more familiar with military organization would have known that reserves are by definition demobilized until recalled to the colors in an emergency; at the time Hoover was speaking, there was not a nation in the world whose reserves were mobilized. But Hoover had made his reputation during World War I by organizing overseas relief and administering food supplies, not as a soldier. He had a sharp antipathy for the French, but ironically, he used the same argument as they and the British did, asserting in his speech that American forces had already been reduced, and not mentioning the rise of American naval power.[18]

During the summer of 1931, Hoover worked out the idea that the minimum land forces required by any country for domestic security might be computed, using 100,000 men for the German population as a standard; these forces might be considered irreducible, but all further contingents needed for defense against foreign attack were a relative matter, i.e., the strength required for them was a function of the strength of similar contingents in neighboring countries. Hoover believed that all the defense contingents could then be reduced simultaneously.[19] His approach fitted in with German demands, in that his plan would have ended the special regime of Part V. It would also have reduced French forces relative to German, and it might even have lent some support to a German demand for an increase; if Germany's quota for domestic security was 100,000, presumably she needed a supplement for defense from foreign attack.[20] But although Hoover tended to sympathize with Germany, his plan was not designed, as Simon's proposals were, to win German agreement by accommodating the principle of equality of rights. Hoover did nothing so mundane as to try to accommodate the demands of other nations. He tried, rather, to construct a plan that would stand by itself as an ultimate

---

[18] *FRUS, 1931*, I, 493-495.
[19] *FRUS, 1932*, I, 2-3, 29, 59n; Stimson MS Diary, 5 Jan 32.
[20] General von Blomberg thought that the Hoover ideas offered a solution, assuming that Germany was given equal treatment, including a defense contingent adequate in the light of Germany's situation: AA, Memo by Blomberg, 17 Jun 32, 7360/E535457. Schönheinz complained, however, that the police figure for Germany was too low, that reserves were not counted, and that the limiting character of the plan would tend to prevent increases in the German defense contingent: AA, Memo by Schönheinz, 4 Jun 32, 3154/D667348-49. In January 1933, Nadolny urged in Berlin that the Hoover plan offered a basis for a claim for a national security contingent: AA, Memo by [Nadolny], [Jan 33], 3154/D668415-19.

solution to the disarmament problem. He was the only Western leader in office who proposed disarming the rest of the world to something like the German level. At first, he was willing merely to have his ideas advanced anonymously as a possible contribution. The initial American statement at the conference by the acting chairman of the delegation, Hugh Gibson, outlined the Hoover scheme in moderate terms and without reference to Hoover's name, also making other suggestions. As the election drew nearer, however, Hoover's desire to publicize a "Hoover Plan" would grow stronger; he would also become more interested in developing a counter to European demands for war debt cancellation.[21]

Although Secretary Stimson was the nominal head of the American delegation to the conference, the effective leaders in Geneva were Gibson, who had taken part in the Preparatory Commission, and Norman Davis, a leading Wilsonian Democrat and Wall Street lawyer. Gibson and Davis pressed hard to make a success of the conference. Disarmament, for them, was a noble ideal, and they could further their own careers by helping bring it about. For practical reasons, and because Congress was reluctant to appropriate money for the delegation, Gibson and Davis, in contrast to the officials of other countries, would take a particular interest in getting prompt, highly visible results. Sometimes they would bend the truth to further negotiations. Their approach to diplomacy was eclectic. As Stimson did, they believed in confidential, high-level conversations, but like Hoover, they also believed in issuing high-minded pronunciamentos.

3. CONFLICT BETWEEN THE "ANGLO-SAXONS" AND THE FRENCH

The United States ambassador to France, Walter Edge, thought the delegates whom he saw pass through Paris were destined for disillusion: "Our delegation seemed to be wholly innocent of any real appreciation of the pent-up passions and practical antagonisms which were latent in Europe. . . . I sensed that trouble lay ahead and was fearful that it would be the kind of trouble which would further strain Franco-American relations."[22] Senator Claude Swanson of Virginia, a leading delegate, termed the French proposals not only "an unreality," but also unrighteous. Favorably impressed by Brüning's

[21] *FRUS, 1932*, I, 28. On Hoover's political motivation: Harvard University Library, Cambridge, Mass., Jay Pierrepont Moffat Manuscript Diary, entry for 22 Apr 32; Stimson MS Diary, 18, 19, 20 Jun 32. Stimson opposed publicity on the Hoover formula in early April, and in June he tried to discourage Hoover from announcing his plan: *FRUS, 1932*, I, 71, 182-185; Stimson MS Diary, 18 Jun 32.

[22] Walter E. Edge, *A Jerseyman's Journal: Fifty Years of American Business and Politics* (Princeton: Princeton University Press, 1948), p. 212.

"frankness" in stating that Germany could not resume reparations, the delegates proposed to Washington on January 31 that Gibson's opening speech begin with remarks encouraging "the questioning even of such fundamental law as the peace treaties which it has previously been considered heretical even to question." Stimson himself had earlier helped to inspire the delegation to think along such lines, but events in recent weeks—principally, no doubt, his efforts to condemn Japanese treaty violations—now led him to veto any statement countenancing "the repudiation of agreements as a principle"; he was also wary of appearing to take a position with regard to European quarrels, "at this stage at least."[23]

After the conference recessed for three weeks on March 19, the American delegation concluded that the United States would have to assume an attitude of leadership in order to lift the proceedings out of the doldrums. Norman Davis made a hasty trip back to Washington to convey this feeling, and to offer two suggestions for action. One was to make a dramatic proposal in a plenary session; the other was for Secretary Stimson, the titular head of the delegation, to travel to Europe and negotiate directly and privately with European leaders.[24] Both approaches were adopted.

For the dramatic proposal, the delegation had originally planned a speech pointing up Hoover's scheme for reducing effectives. After Davis had left for Washington, however, Gibson sent a message proposing that the speech also discuss qualitative reduction, emphasizing the idea that the abolition of "aggressive" weapons would eliminate the possibility of invasion, making other countries feel as secure as the United States. Simon, the Italians, and others had already advocated qualitative reduction in the opening sessions of the conference; in fact, Gibson himself had mentioned the idea in his own initial speech. But Gibson and the delegation now apparently became intrigued by this corollary thought that an abolition of offensive weapons would provide security in itself—more security than such "paper engagements" as the French plan, and without any American commitment. As Gibson took the view that "the primary problem of the reduction of arms rests with France to retrench," he was in effect saying that France, the nation most concerned for security, should find it by destroying her own tanks and heavy guns. After some cabling back and forth, Washington, largely on Stimson's recommendation, decided that the speech should not mention Hoover's proposal, but

---

[23] Hugh R. Wilson, *Diplomat Between the Wars* (New York: Longmans, Green, 1941), pp. 268-269; *FRUS, 1932,* I, 16-17, 19; Stimson MS Diary, 5 Jan 32.

[24] Hooker, *Moffat Papers,* pp. 58-60; Stimson MS Diary, 29 Mar 32; *FRUS, 1932,* I, 59.

concentrate on this "security through disarmament" argument; it was also decided that the speech should make no reference to capital ships, largely on the grounds that it was the turn of the land powers to disarm. Although the French delegation had earlier claimed the privilege of speaking first when the conference resumed its meetings on April 11, Gibson arranged to be the first speaker.[25]

Gibson's remarks in themselves amounted to a flat contradiction of French ideas. In omitting any reference to naval armaments, Gibson seemed now, from a French viewpoint, to be saying that disarmament was for others, not the United States. Moreover, the timing of the speech made it look like an attempt to forestall discussion of the French plan—which indeed it probably was. When shown the speech shortly before its delivery, Tardieu—a hypertensive individual already strained by his election campaign and the failure of proposals he had recently made for a Danubian Confederation—completely lost his temper. At first he said that he would accuse the United States of trying to leave France defenseless, and demand that the Americans sink their battleships; the American proposal, he ranted, was a very bad joke. Gibson remarked afterwards that Tardieu had almost bitten him. Tardieu's public speech of reply was somewhat more restrained, but after he returned to Paris, he asked Edge (probably with the 1931 Hoover moratorium also in mind) when the United States government would learn that it could not force France to make concessions she was unwilling to make: "He was through. He would not go back to Geneva. The Americans and British could hold their Disarmament Conference by themselves if they wished."[26]

Tardieu suspected that Gibson's speech had been prepared in coordination with the British, and he obviously thought that the two countries were trying to prevent discussion of the French plan. Gibson

[25] SDP, tels., Carr to Delegation, 18 Mar 32, and Gibson to Dept., 19 Mar 32, 500.A15A4/927, 928; *FRUS, 1932*, I, 29, 59-62, 65, 69-76; Hooker, *Moffat Papers*, pp. 59, 61, 64-65. General Douglas MacArthur, the American chief of staff, supported the abolition of military aircraft, tanks, and heavy artillery; aircraft in particular absorbed too much of the meager army budget: Hooker, *Moffat Papers*, pp. 63-64; *FRUS, 1932*, I, 65. But he had forebodings about German rearmament, and privately advised French army leaders not to disarm: SDP, Extracts from ltr., MacArthur to [?McBride?], 15 Oct 32, 862.00/2855½; *DDF*, 1/I, No. 250; [Maurice] Gamelin, *Servir*, 3 vols. (Paris: Plon, 1946-1947), II, 78-79.

[26] *FRUS, 1932*, I, 72, 74-85; Stimson MS Diary, 9 Apr 32, SDP, Transatlantic phone call, 11 Apr 32, 500.A15A4/993½, League of Nations, Conference for the Reduction and Limitation of Armaments, *Records of the Conference*, Series B, *Minutes of the General Commission*, Vol. I, 50-55; Edge, pp. 213-214. Tardieu was paralyzed by a stroke in 1939 and may have had one as early as 1915; see Rudolph Binion, *Defeated Leaders: The Political Fate of Caillaux, Jouvenel, and Tardieu* (New York: Columbia University Press, 1960), pp. 251, 336. On Tardieu, also Joseph Paul-Boncour, *Entre Deux Guerres*, 3 vols. (Paris: Plon, 1945-1946), II, 213-216.

denied that there had been any collaboration, and Stimson rebuffed Simon's efforts to enlist him against the French scheme, but of course there was a general affinity in the British and American opposition to fixed commitments and a super-League, as well as in the idea that France should disarm. Simon, at any rate, definitely desired to stave off a debate on the French plan. He did not, however, seek qualitative disarmament merely as a means of reducing tension and evading French demands for security. He also saw it as a bridge to meeting German demands for equality, and on April 18, in Geneva, he explained to Brüning, as he had earlier to Nadolny, his ideas on this and on including the restrictions on Germany in a general convention. He told Brüning he feared that if the French plan came to a vote before qualitative disarmament, a negative vote would give France an excuse to refuse to disarm, while a positive vote would not be accepted by Italy, Britain, or the United States, and would rule out qualitative reductions. To forestall a decision on the French plan, Simon proposed on April 20 a simplified resolution providing for the prohibition of offensive weapons. Tardieu rushed back to Geneva, and between visits with American and German leaders, he presumably helped to guide the French and associated delegations in their successful effort to press the British to "clarify" their resolution. The resolution was changed to provide that offensive weapons might be either prohibited or internationalized, thereby making it reconcilable with the French plan. But the British were annoyed to have to make this change, which kept the French plan in the running and blunted their demand for French disarmament; some British officials thought that a showdown with the French had only been postponed.[27]

The friction between the United States and Britain on the one hand, and France on the other, did not provide a promising background for the other American move, the visit of Secretary Stimson to Geneva for high-level private negotiations. Although Tardieu had the reputation of being a rigid supporter of the treaties—he had in fact been one of the principal French authors of them—Davis had reported that the French leader had had a change of heart, and now wished to reach a more satisfactory relationship with Germany; such unfounded hopefulness was characteristic of Davis, whom a high State

[27] *FRUS, 1932*, I, 97-103; *SIA, 1932*, pp. 218n, 219-220; AA, tels., Bülow to Ministry, 18, 21 Apr 32, 7360/E535147-48, E535168-69; tel., Nadolny to Ministry, 22 Apr 32, 7360/E535175-76; FO 800/286, Simon Papers, ltrs. to MacDonald, 16, 18 Apr 32; FO 371, tels., Patteson (Geneva) to F.O., and minutes by C. Howard Smith and G. M[ounsey], 21-24 Apr 32, W4549/— and W4606/1466/98. If Gibson did not coordinate his speech with the British, it is still possible that British—and Italian—diplomats had exerted earlier and more subtle influence; see also Edge, pp. 213-214; *FRUS, 1932*, I, 67-68.

Department official was later to call "the king of the optimists." Stimson's main interest in going to Europe may have been to discuss the Far East with the British, but Davis's advice had doubtless influenced his decision. As Stimson told Tardieu on his way through Paris, "Now that I arrive I find you and Hugh Gibson pulling at each other's hair and I feel as if I had made a mistake."[28]

German diplomats usually welcomed friction among the Western powers, and they must have been pleased to learn from Simon that he and, as he told them, the American delegation too, were opposed to treating Germany as subject to the Treaty of Versailles instead of to the projected disarmament convention.[29] But Germany also had in any case to reckon with France. If Tardieu stormed out of the conference, his next move would hardly be to agree to German rearmament. More likely, it might be some kind of preventive action. As long as Germany had not regained her own strength, she had an interest in reaching some understanding with France, or at least in encouraging the French to hope for some understanding.[30]

Brüning and his Foreign Ministry did seek to negotiate privately with the French, over reparations as well as disarmament, and Brüning was probably aware that Schleicher, too, was interested in private discussions with the French on disarmament; the general actually extended his own feelers in May, if not before. Yet diplomatic discussions in Geneva and Paris with Tardieu, as well as in Berlin with Poncet, did not produce much result in February and March. The French asked the Germans to "say where the shoe pinched" with regard to Part V, and the Germans refused to do this as it amounted to adopting the French perspective, recognizing the overall validity of Part V, and putting themselves in the invidious position of asking to rearm; the Germans preferred to say that it was up to the French

---

[28] *FRUS, 1932*, I, 62-63; Hooker, *Moffat Papers*, p. 59; AA, tel., Prittwitz to Ministry, 7 Apr 32, 7360/E535105; Stimson MS Diary, 29, 31 Mar, 15 Apr 32; Arnold A. Offner, *American Appeasement: United States Foreign Policy and Germany, 1933-1938* (Cambridge, Mass.: Harvard University Press, 1969), p. 21.

[29] AA, tel., Nadolny to Ministry, 29 Feb 32, 7360/E535011-12; *DBFP*, 2/III, No. 239. Simon's letter to MacDonald of 18 Apr 32 (FO 800/286, Simon Papers) shows Simon describing this position to Brüning, and the latter as much interested and disclaiming any wish to rearm.

[30] Brüning advised Simon to consult the French before offering his resolution, lest this give Tardieu an occasion for breaking off negotiations before his election: AA, tel., Bülow to Ministry, 18 Apr 32, 7360/E535147-48; FO 800/286, Simon Papers, ltr. to MacDonald, 18 Apr 32. Simon apparently did not consult the French, but later, to explain his modification of his resolution, he told members of the British delegation that Brüning, although favoring the resolution, had been most anxious to avoid a French withdrawal, which in turn would compel a German denunciation of Part V (*DBFP*, 2/III, No. 240n). Simon's and Bülow's records of the April 18 Simon-Brüning discussion do not mention any warning of an eventual German denunciation in case of a breakdown; German military and conservative opinion would indeed have expected one, however.

to declare how much they would disarm. The French also hinted that the Germans could obtain some hitherto forbidden arms if they accepted the Tardieu plan. The Germans agreed to review the French plan and see what their position would be with respect to it, on the assumption that equal rights were conceded, but when the French ambassador in Berlin, André François-Poncet, inquired two weeks later, he was told that the analysis would not be completed for another three weeks. One reason for keeping the French waiting for this review may have been to deter Tardieu from delivering the speech he frequently threatened to make on German illegal rearmament. Tardieu at any rate did not deliver such a speech. He may also have been restrained from disclosures by having learned at the end of February that Brüning hoped to meet him in Geneva after the German presidential elections.[31] Brüning, Tardieu, MacDonald, and Stimson were all in fact to converge on Geneva in mid-April.

4. THE EPISODE OF THE BESSINGE MEETING

The high-level meetings held at Geneva (or Bessinge) on April 26 and planned for April 29 have been set down in histories and memoirs as the most hopeful moment in interwar disarmament negotiation, the time when a settlement of European quarrels was within reach; the failure of Tardieu to take part and accept Brüning's modest demands has been described as the "April Tragedy."[32] In the light of the evidence now available, this picture can no longer be accepted. Those who did take part seem to have reached tentative agreement on the British idea of using the convention to supplant Part V, but even

[31] Schleicher Papers, N42/34, ltr., Regendanz to Schleicher, 12 May 32; AA, tel., Nadolny to Ministry, 5 Feb 32, 7360/E534936-37; tel., Nadolny to Ministry, 26 Feb 32, 7360/E534972-73; ltr., Bülow to Nadolny, 27 Feb 32, K957/K251164-65; ltr., Bülow to Hoesch, 27 Feb 32, K957/K251166-69; tel., Hoesch to Ministry, 27 Feb 32, 7360/E534982-86; tel., Nadolny to Ministry, 29 Feb 32, 7360/E535013-14; Memo by Bülow, 29 Feb 32, 7360/E534995; tels., Hoesch to Ministry, 1 Mar 32, 7360/E535025-26 and E535028-31; Memo by Bülow, 1 Mar 32, 7360/E535033; tel., Nadolny to Ministry, 17 Mar 32, 7360/E535055-57; RK, Memo by Bülow, 25 Feb 32, 3642/D811614-20; Extract from ltr., Nadolny to Bülow, 12 Mar 32, 3642/D811929-30; Memo by Bülow, 15 Mar 32, 3642/D811934-35. On the tactical refusal to say "where the shoe pinched," also AA, ltr., Neurath to Nadolny, 20 Jun 32, 3154/D667391-97.

[32] See John W. Wheeler-Bennett, *The Pipe Dream of Peace: The Story of the Collapse of Disarmament* (New York: W. Morrow, 1935), pp. 26-34; Carl Loosli-Usteri, *Geschichte der Konferenz für die Herabsetzung und die Begrenzung der Rüstungen, 1932-1934: Ein politischer Weltspiegel* (Zurich: Polygraphischer Verlag, 1940), p. 101; Wilson, *Diplomat*, p. 271; Liddell Hart, *Memoirs*, I, 193-195. Cf. the observations of Christoph M. Kimmich, *Germany and the League of Nations* (Chicago: University of Chicago Press, 1976), p. 163. Internal evidence suggests that Wheeler-Bennett's account of the meeting stemmed from the German record, and in fact he let his government know that he had learned about the Bessinge talks from Brüning (FO 371, Memo by J. W. W[heeler]-B[ennett], 22 Nov 32, W13253/1466/98). See also Alteras, p. 53n.

without Tardieu, those involved were too divergent in aim to reach an honest understanding on German armament. Brüning's objectives and Tardieu's situation and concerns have both been misjudged. And even at the time, the German and Anglo-American participants drew different conclusions from the meetings; this difference would help lead in a few months to a diplomatic crisis.

The leaders who met in Geneva in April did not convene in a single-minded resolve to hold a secret conclave on disarmament, nor were they brought together simply by Stimson's presence. As just noted, Brüning had planned at the end of February to meet with Tardieu, and it seems that when the Chancellor went to Geneva, his main goal was to reach understandings with France on reparations and armaments. In the absence of Tardieu, Brüning spent his first few days largely in discussing reparations with the British. Tardieu, on his part, evidently wanted to stave off isolation. He could not spare much time from his election campaign, and when he did come briefly on the twenty-first and twenty-second, separate meetings with Stimson and Brüning did not produce much result.[33]

Stimson, as noted, had come to Europe largely to talk with the British on Far Eastern problems, and with some further hope that Tardieu was ready for sweeping concessions on disarmament. When he met with Tardieu on his way through Paris on April 15, Tardieu was cordial, but clearly as rigid in his outlook as ever. As it turned out, Stimson's greatest preoccupation in Geneva was to counter suggestions that American unwillingness to join in a consultative pact was responsible for the British refusal to give hard-and-fast security commitments to France. In private talks with Simon and MacDonald (who had advanced his arrival from the twenty-fifth to the twenty-first to meet Stimson's schedule), Stimson made an issue of these suggestions; MacDonald, becoming "very intense," denied "with the utmost seriousness" that his government had instigated them. MacDonald and Simon conceded to Stimson that, irrespective of American policy, British opinion would not consent to any new continental commitments, and that Stimson had in practice shown a readiness to consult over the Far Eastern crisis.[34]

[33] AA, Memo by Bülow, 29 Feb 32, 7360/E535995; Memo by Bülow, 1 Mar 32, 7360/E535033; tels., Bülow to Ministry, 21, 22 Apr 32, 7360/E535168-69 and E535173; Brüning, *Memoiren*, pp. 498, 545-546; RK, "Material for Reich Chancellor's trip to Geneva in April 1932," undated, 3642/D811917-19.

[34] Stimson MS Diary, 15, 16, 19, 20-23, 27 Apr 32; SDP, tels., Dept. to Atherton, 1 Apr 32, and Atherton to Dept., 2 Apr 32, 500.A15A4/957, 958; *DBFP*, 2/III, No. 240. Simon told Stimson that British opinion was even more opposed to a British commitment to France than it had been two years earlier (Stimson MS Diary, 20 Apr 32).

Along with their other aims, each of the major participants undoubtedly hoped to advance his own goals in the disarmament question. The American delegation had arranged for Stimson to occupy a beautifully furnished and situated villa outside Geneva named Bessinge; this provided a center at which the prime ministers and foreign ministers could gather, away from the crowd of minor diplomats and enthusiasts regularly attending the conference. Before leaving Geneva on April 22, Tardieu separately told both Stimson and Brüning that he would return on April 26. On April 23, MacDonald suggested to Stimson that they should try to get Brüning and Tardieu to negotiate together at Bessinge in their presence, preferably on the twenty-sixth; Stimson agreed, provided that MacDonald would make the arrangements.[35]

Tardieu failed to appear again in Geneva on April 26, or indeed before his fall from office. There have been suggestions that he was unwilling to discuss Brüning's proposals on the eve of an election, or that his ambassador in Berlin, François-Poncet, warned him on the basis of information from Schleicher that Brüning would soon fall and that therefore there was no point in negotiating with him.[36] The detailed circumstances indicate no such deliberate obstruction, but rather a series of misunderstandings and mishaps. When Tardieu promised to return to Geneva on April 26, he seems to have meant that he would leave Paris that night, arriving on the morning of the twenty-seventh and staying until the twenty-ninth. He planned to spend the twenty-sixth in Paris, after devoting the previous three days to campaigning in his rural constituency (Belfort campagne). Then, by the morning of the twenty-fifth, he decided not to try to return to Geneva, unless that should be urgently necessary. One reason for this change was that the conference steering committee, the Bureau, had recommended a postponement of plenary sessions; another reason was probably a decision to deliver a major speech in Belfort on April 28. The British did not attempt until the twenty-fifth to inform

---

[35] Moffat MS Diary, 7 Apr 32; Stimson MS Diary, 16, 22, 23 Apr 32; Brüning, *Memoiren*, p. 556; AA, tel., Bülow to Ministry, 22 Apr 32, 7360/E535173; *DBFP*, 2/III, Nos. 103, 240; FO 371, Minute by Cadogan, 26 Apr 32, W4949/10/98.

[36] See Wheeler-Bennett, *Pipe Dream*, p. 33. François-Poncet, in *Souvenirs d'une ambassade à Berlin: septembre 1931-octobre 1938* (Paris: Flammarion, 1946), pp. 34, 36-38, 41, maintains that Brüning's position was materially weakened by the Nazi success in the Prussian election of April 24, but denies that he urged Tardieu not to negotiate with the Chancellor. Brüning's *Memoiren*, pp. 562-564, imply that Tardieu deliberately evaded a meeting. Other books suggesting or implying that Tardieu was blameworthy include A. C. Temperley, *The Whispering Gallery of Europe* (London: Collins, 1938), pp. 202, 205; Werner, Freiherr von Rheinbaben, *Viermal Deutschland* (Berlin: Argon Verlag, 1954), pp. 267-268; Northedge, *Troubled Giant*, p. 371; Liddell Hart, *Memoirs*, I, 195; Wolfers, p. 85.

Tardieu of the four-power meeting proposed for the morning of the twenty-sixth, and Tardieu, who was busy visiting remote villages in his district, reportedly did not receive the message until late in the evening of the twenty-fifth, when about to leave for Paris. He was now completely exhausted, and it seems understandable that he did not change all his arrangements and try to reach Geneva by the next morning. He arrived in Paris on the morning of the twenty-sixth and agreed that day to return to Geneva on the twenty-ninth, which would have involved leaving Belfort at 1 a.m. that day. On the twenty-seventh, however, he came down with the grippe and acute laryngitis; Ambassador Edge went to Tardieu's sickbed that day to get his signature on a double taxation treaty, and afterwards informed Stimson that the French Premier was indeed "quite seriously ill with a temperature." His April 28 speech was read for him in Belfort, and as late as May 4, Leopold von Hoesch, the German ambassador in Paris, reported that he had not been able to see the Premier, who was now said to be suffering not merely from grippe but from general overfatigue. In view of Tardieu's activities in the preceding weeks—negotiation in London on Danubian Confederation on April 4, major speeches on April 6 and 16, two hectic visits to Geneva (April 11-12 and 21-22), and a fierce electoral campaign, in which, within a period of sixty hours, he reportedly addressed meetings in his constituency for over forty-five hours—this is quite credible.[37]

Nevertheless, Tardieu may well have found his campaign involvement and illness to be timely. It was not that he did not want to meet Brüning; he had already done so, and had told the German Chancellor that he would return and confer with him again, although his main reason for returning was probably to speak for France in plenary sessions. If Tardieu received reports that Brüning might fall from office, this probably did not make much difference to him; for him, unlike the British and Americans, there was probably little to choose between one German leader and another. A more likely difficulty for Tardieu lay in MacDonald's plan for four-power talks with British and American participation.[38] Tardieu's experience over the last two

[37] *Le Temps* (Paris), 23-28 Apr 32; FO 371, Minute by Cadogan, 26 Apr 32, W4949/10/98; *FRUS, 1932*, I, 103, 112; Stimson MS Diary, 27 Apr 32; Edge, p. 216; François-Poncet, pp. 34, 41; Bonnefous, V, 116, 118; AA, tel., Hoesch to Ministry, 4 May 32, 2406/D510423; *Manchester Guardian*, 28 Apr 32.

[38] Tardieu and Brüning agreed to postpone public consideration of the French plan and equality of rights until after the French elections, but they expected to discuss disarmament and reparations privately with each other on 26 April: AA, tel., Bülow to Ministry, 22 Apr 32, 7360/E535173. One of Brüning's staff later told an American diplomat that Brüning had established good working relations with Tardieu, so that Tardieu's electoral defeat was unfortunate: SDP, disp., Sackett to Dept., 11 May 32, encl., 751.62/185. Werner von Rheinbaben, while blaming Tar-

weeks with the "Anglo-Saxons" must have convinced him that they would support the Germans against him and completely reject his plan. On April 22, MacDonald had frankly told one of Tardieu's associates that Britain considered the French plan unworkable. Indeed, MacDonald may have proposed four-power meetings partly in order to guard against a Franco-German understanding in support of the French plan. With an apprehension that now seems grossly exaggerated, the British were worried lest the French get their plan before the conference, succeed in avoiding all disarmament, and perhaps even change the whole character of the League.[39] But the composition MacDonald planned for the private meetings would safeguard British interests. Up until April 1933, at least, he consistently favored great-power meetings of the chiefs of four or five governments. This approach placed him in a position of leadership, enabled him to monitor what the French and the others were doing, and indeed offered openings for modifying the Treaty of Versailles. The French, on the other hand, consistently disliked this kind of meeting. If Tardieu had not really been sick, he might well have developed a diplomatic illness.

On learning that Tardieu would not appear on the twenty-sixth, MacDonald proposed that he and Stimson go ahead and meet with Brüning anyway. As Brüning describes it in retrospect, this meeting at Bessinge was the moment he had been waiting for. He had just returned after leaving Geneva briefly to vote in the Prussian election of April 24, and he had arranged to travel back as far as Basel with Groener, who now explained to Brüning the full scope of the Reichswehr's plans for mobilizing an integrated army. Groener showed Brüning that, thanks to the decline in the French birthrate during World War I, it was possible for Germany, "with seemingly modest demands and by clever use of the advantages conferred on us by 12-year service, to achieve an actual superiority over France in the course of the next five years." After this interval, Germany would be able,

---

dieu, has put his finger on the critical point: "Tardieu did not want to put himself in the situation of being forced into a compromise by the British and American Foreign Ministers": Comment on Deist, "Internationale und nationale Aspekte," pp. 94-95. See also *Le Temps* (Paris), 26, 28 Apr 32. For the 29 April meeting, MacDonald planned that he and Stimson should first have separate meetings with Brüning and Tardieu (Stimson MS Diary, 27 Apr 32; AA, Record of interview with MacDonald, 1 May 32, 7360/E535229); this would have given him and Stimson even more the status of arbiters.

[39] FO 371, Minute by MacDonald, 22 Apr 32, W4550/10/98; Minutes by C. H. Smith and G. M[ounsey], 24 Apr 32, W4549/1466/98; Record of delegation meeting, 23 Apr 32, and minute by Leeper, 29 Apr. 32, W4761/1466/98; Cab 27/505, DC (M) 32, meeting of 21 Mar 32; AA, Record of interview with MacDonald, 1 May 32, 7360/E535228-31.

by using militia and former Reichswehr personnel, to mobilize "700,000 hard-hitting infantrymen and 300,000 specialists, superior to those of all other armies, for the technical arms." Brüning had already warmly supported Groener's plans for compulsory Wehrsport, of which he writes: "They were intended to make it possible in the course of five years to muster a people's army with a military striking power such as we had never possessed before the war. It was the consummation of Scharnhorst's ideas, taking account of the experience acquired since his time." Brüning also discussed negotiating tactics with State Secretary von Bülow, and on these tactics he comments:

> It was a matter of achieving something which was psychologically extremely difficult, that is, to win the other powers simultaneously for three demands which seemed to contradict each other: elimination of reparations, disarmament of the others, and rearmament for us. I was now reaching the point where with one false move, or an incomplete understanding of the psyche of the negotiating partner, all the French propaganda could successfully revive again, particularly in the United States—that propaganda which tried to prove that we only sought the elimination of reparations in order to make money available for our own rearmament. If the French succeeded in gaining acceptance for this conception in Britain and America, then the great two-year-long approach march, accomplished with endless sacrifices, would have been in vain.[40]

In view of the esteem in which Brüning has been held by people devoted to disarmament, these remarks make startling reading. It should be added, however, that Brüning's memoirs appear to exaggerate his zeal and skill in pursuing nationalist goals, in an attempt to show that he could have restored Germany to the position of 1914 if only he had been left in office a little longer. His account of the actual discussion at Bessinge strains our credulity to show that he was a master of Anglo-American psychology, and that he gained complete acceptance for German goals. The contemporary German and American records of the meeting must be regarded as far more reliable sources.[41]

---

[40] Brüning, *Memoiren*, pp. 546, 552-557. That MacDonald proposed to go ahead with the meeting: FO 371, Minute by Cadogan, 26 Apr 32, W4949/10/98. Thanks to the existence of a Prussian exclave near Lake Constance, Brüning did not have to return to North Germany to vote.

[41] Brüning's account of the meeting is in *Memoiren*, pp. 557-563. The fullest contemporary German record is in AA, Memo by Bülow, 26 Apr 32, 3154/D667202-04; see also AA, tel., Bülow to Ministry, 26 Apr 32, 7360/E535192-93. The American record is in *FRUS, 1932*, I, 108-112. There apparently is no British official record, and there was none, even in August-September 1932: MDP, 2/12, ltr., Sir M. Hankey to MacDonald, [22 Aug 32]; FO 371, Memo by R.M.A. Hankey, 5 Sep 32,

At the April 26 meeting, Brüning and Bülow were the German participants, MacDonald and Lord Londonderry the British, and Stimson, Gibson, and Davis the Americans. The contemporary official records agree that MacDonald began by stating his conception of the problem: Germany wanted to be freed from the military clauses of Versailles; Britain and America would, however, be opposed to Germany's starting a new race in armament. Brüning maintained in reply that Germany really wanted disarmament and did not wish to start an arms race. He and Bülow stated that Germany was prepared to have her arms remain at their existing level—or in the German record, "essentially" (*im wesentlichen*) at their existing level—provided only that she should be freed from certain unreasonable hardships arising from Part V and the regulations of the Inter-Allied Control Commission; for example, the rules governing munitions manufacture, which caused unnecessary expense, should be eliminated,[42] and twelve-year enlistments should be shortened to six or eight years, while retaining the voluntary enlistment system. According to the American record, Brüning said he would not suggest two- or three-year service, as the French would regard this as designed to amass reserves. The German report, by Bülow, indicates that "a militia on the Swiss model" was proposed, and that when Davis asked about the size, Brüning said it would not be larger than the Reichswehr; if they did mention this, the Germans presumably did not mention that the militia they had in mind, combined with premilitary training, would provide trained reserves for mobilization.[43] The participants discussed the possibility of having Part V superseded by the convention, and the British and Americans seemed to agree, as Simon had indicated to Nadolny in February. But MacDonald pointed out that France would be concerned over Germany becoming free at the end of the convention (or in the American version: over whether or not Germany would be free). France might therefore refuse to limit her own arms. Brüning argued that it was necessary to leave open the possibility of later German rearmament in order to bring pressure on the French to disarm,

---

C7946/211/18. MacDonald did make some general comments on the meeting to dominion representatives on 29 April: MDP, 1/501, Note of meeting with Dominion Delegates, 29 Apr 32. See also note 47.

[42] The German army's Ordnance Office was very interested in a relaxation of these restrictions (for example, on the export of arms) which kept unit production costs high, and thereby obstructed stockpiling and industrial mobilization preparations. See Geyer, "Zweite Rüstungsprogramm," pp. 132-133.

[43] The American record does not mention the Swiss-style militia proposal, and Stimson denied five months later that a militia was mentioned (SDP, Memo of conversation between Stimson and Henry, 8 Sep 32, 763.72119 Military Clauses/27), although the detail about Davis's question makes the German account seem plausible. See also Link, *Stabilisierungspolitik*, p. 519.

but no definite conclusion seems to have been reached on this question of the possibility of increases after the first convention.[44]

Despite the degree of mutual confirmation, certain differences between the German and American records reveal a wide divergence in outlook and aim. Bülow's record tries to minimize Brüning's readiness to accept restrictions in order to show the Chancellor's resolute support of the national cause, while the American report tends to maximize that readiness, so as to suggest that the Germans were willing to remain disarmed, making an agreement possible. For example, according to Bülow, Brüning referred to German willingness to remain "essentially" at the existing level of forces, while in the American report, the Germans were ready to limit German forces "to what they are" and to accept a virtual transfer of Part V into the convention. If the Germans did in fact raise the subject of the "militia on the Swiss model," the Americans apparently did not realize that this had any significance; they apparently tended to overlook anything that might suggest a German desire to rearm—they did not want to believe that. On the other hand, Bülow made the most of the readiness of the British and Americans to see Part V replaced, and to accept minor actual adjustments; this conceded equality of rights in a theoretical sense, and left an opening for a broader equality. There is no American counterpart to Bülow's statement, "The Americans and British declared themselves in agreement with our demand for equality of rights in the sense of a supersession of Part V [of the] T. of V. [Treaty of Versailles] and [also declared] our standpoint to be reasonable and justified." Bülow reports that MacDonald rejected any extension of new British guarantees to France, and that both the British and Americans were very cool toward France; the American record betrays this attitude only by two references to the French "so-called plan."

At the time of the April 26 meeting, the participants thought that this was merely a preliminary to more important sessions with Tardieu, and they did not suppose that they were reaching decisions. On May 2, Brüning told his cabinet that he could not report on the negotiations "at this stage." Privately he said that the disarmament ques-

[44] The French and Italians apparently received private information on the talks. The German delegation reported stories in the Geneva press that Brüning had demanded an increase in the Reichswehr in the discussions, and it commented that this "obviously goes back to a leak from the French delegation, which like Grandi received a record of the discussion. The reports, which are tendentiously distorted, have been denied [i.e., by the German delegation]" (AA, Memo by Völckers, 28 Apr 32, 7360/E535196). A note in the French diplomatic documents (*DDF*, 1/I, No. 142n) states that American delegates provided a French delegate with information on the meeting; the record of this delegate (Louis Aubert) contained a reference to the question of German rearmament, "to which MM. Stimson and MacDonald had declared themselves very definitely opposed."

tion could turn out well, perhaps with a recognition of Germany's formal equality, but he immediately added that financial reasons by themselves would compel her to make very limited use of it.[45] This did not suggest a decisive and unqualified success in the talks. Years later, however, Brüning would write in his memoirs: "In two and a half hours, at the first attempt, there had been success in putting across all the demands of the Reichswehr one hundred percent with all the great powers excepting France." Even in June, Schleicher wrote, "The overwhelming majority of Conference participants have unconditionally recognized Germany's right to equal rights in national security." And Bülow was to tell François-Poncet in August that "the other great powers" had accepted "the main lines of the German thesis."[46] Implicitly, Bülow was referring to Bessinge, and by August "the German thesis" clearly meant not only equality of rights in principle, but also a certain measure of German rearmament.

That German rearmament had been agreed to at Bessinge was not the recollection of Stimson and MacDonald. Stimson strenuously denied that he had condoned rearmament at all, and there is no reason to suppose that he intended to condone it. MacDonald's case is more complicated. In September, the Prime Minister denied that he had encouraged the Germans to press their demand, and in October he claimed to have been wary about rearmament. He also noted on a telegram from Berlin that "the German claim" (in this case, probably, to replace Part V) had only been discussed "informally and without decision." Informality had certainly been the order of the day as far as MacDonald himself was concerned; he had not even bothered to record at the time what had happened. But in various statements after the Bessinge meeting, MacDonald also indicated a belief that the conferees had accepted "equality" or the replacement of Part V, and the idea that Germany would remain for a time, voluntarily, at her existing strength, becoming as free as any other power at the next conference.[47] Evidently, he considered that the Bessinge participants had

[45] RK, Cabinet Protocol, 2 May 32, 3598/D789852; Schäffer Diary, 2 May 32. The German naval delegation understood on May 7 that there would be further high-level talks after mid-May: ML, Shelf 5939, ltr., [Freyberg] to Raeder, 7 May 32, reel 17/1009-1010.

[46] Brüning, *Memoiren*, p. 563; AA, ltr., Schleicher to Bülow, 23 Jun 32, 7360/E535911; *DDF*, 1/I, No. 115. See also AA, Memo by Bülow, 2 Jun 32, 7360/E535360.

[47] On August-September 1932 repercussions: *DBFP*, 2/III, Nos. 173, 174, 241; 2/IV, Nos. 51, 56, 104n, 110, 136n, 217; *FRUS, 1932*, I, 421; FO 371, Memo by R.M.A. Hankey, 5 Sep 32, C7946/211/18. On 20 July, Stimson told the German ambassador in Washington that Nadolny's current actions were "a reversal of the attitude which Brüning had taken with me in Geneva last April in which he said he would be perfectly satisfied to leave the present figures of armament inserted in the new treaty, only making a statement of reservation to the effect that this was done voluntarily by Germany" (SDP, Memo of conversation between Stimson and Pritt-

accepted the policy which Simon had been advocating, and which the ministerial committee on disarmament had approved in early April: this policy was that the first convention should contain Germany's old limits, and that she should be free, after the convention expired in seven to ten years, to seek higher figures in a second convention. Simon had already described this position to Brüning in their conversation on April 18. Neither Simon nor MacDonald wanted to agree to German rearmament, but they did want to meet a demand for equality of rights in principle, and they could hold out a prospect of arms freedom later. This, and Stimson's acquiescence, was enough to encourage the Germans to think that they had Anglo-American agreement, and that only French consent was lacking.

5. AMERICAN EFFORTS FOR FURTHER TALKS

After the April 26 meeting, Stimson cabled to Washington that Brüning had been "more conciliatory toward making a reasonable compromise with the French on their fundamental issues than we had anticipated."[48] Since the Americans believed that Brüning had renounced any rearmament and had shown himself ready for compromise, the next step, in their eyes, was to arrange further informal meetings, this time bringing in the French. Stimson himself was not prepared to delay his departure beyond May to attend such meetings, even though urged by Hoover to do so, and an electoral defeat for Tardieu and the necessity of a glaucoma operation for MacDonald soon created further obstacles. A new French government could not be formed before the new parliament assembled on June 1, although

---

witz, 21 Jul 32, 500.A15A4/1357). On lack of British record, see note 41. MacDonald's failure to make a record may be partly explained by the fact that, on doctor's orders, he was taking two or three hours off from work each afternoon: MDP, 8/1, Diary, 1 May [32]. He told the dominion delegates he thought Germany would be prepared to agree to "a measure of disarmament . . . which would not in itself repeat specifically the German subordination as set out in Part V." Germany might insert "in the present convention figures which France could accept and then when the Treaty came to be revised in due course, say in 10 years, Germany would be free to go into the negotiations on an equal basis with everyone else": MDP, 1/501, Note of meeting, 29 Apr 32. In September, MacDonald wrote letters to Lord Cecil (13 Sep 32) and, in particular, to Simon (22 Sep 32), recalling what had happened: British Museum, London, Cecil of Chelwood Papers, 817A/51081; Segel, "Sir John Simon and British Foreign Policy," pp. 296-298 (text of ltr. from the Simon personal papers). MacDonald thought that Tardieu too had been ready to accept the "line" taken at Bessinge, and had really been prevented from coming by illness: Segel, "Sir John Simon and British Foreign Policy," p. 269; *DBFP*, 2/III, No. 175; AA, Record of interview with MacDonald, 1 May 32, 7360/E535229.

[48] *FRUS, 1932*, I, 112. Stimson still believed in 1947 that Brüning had made an offer "which, if accepted, might well have avoided World War II": Link, *Stabilisierungspolitik*, pp. 519-520.

it was probable that Edouard Herriot, the leader of the Radical Socialists, would head it, while MacDonald's convalescence was likely to last into June. But Stimson directed Gibson and Davis to travel to London and Paris in an effort to promote private discussions, and they took up this task with characteristic energy.[49]

In London, they found Simon anxious to expose the impracticality of the French plan, and MacDonald (in a nursing home) dubious about the capacity of the next French government to negotiate seriously, at least until after the Lausanne Conference. What most impressed the visitors was an exposition by Stanley Baldwin of some radical ideas he had in mind for disarmament, involving the total abolition of military aircraft, capital ships, heavy guns and tanks, and subsidies for civil aircraft, as well as a drastic reduction of land effectives. Baldwin told Davis and Gibson that, although a proposal to abolish battleships would shock both American and British opinion, such a step would be the price of real relief. The prohibitive cost of these vessels doomed them to early disappearance anyway, and it would be wise to use their abolition to bring about a general arms reduction. Baldwin appears merely to have been throwing out some ideas; of these ideas, only the abolition of military aircraft, a matter close to Baldwin's heart, seems to have been seriously discussed by the British cabinet. But reports from Gibson and Davis of Baldwin's supposed program, particularly the proposal to abolish battleships, produced a deep—and highly unfavorable—impression on Secretary Stimson, though Gibson and Davis themselves were greatly encouraged.[50]

In Paris, Gibson and Davis talked with Tardieu, whose cabinet was now acting in a caretaker capacity, and who still nursed a hope that Herriot would need his collaboration in two or three months. He asserted that Herriot's views were more like his own than their political

[49] *FRUS, 1932*, I, 112-115, 117; AA, tel., Nadoluy to Ministry, 29 Apr 32, 7360/E535204-05; tel., Neurath to Brüning, 14 May 32, 7360/E535327; Marquand, pp. 693-694. MacDonald's other eye had been operated on in January, and he had not been able to return to work until March.

[50] *FRUS, 1932*, I, 117-119, 121-126; Stimson MS Diary, 18, 22 May 32; Moffat MS Diary, 31 May 32. On Baldwin's ideas, see also Cab 23/71, meeting of 4 May 32, 26(32)2; meeting of 7 Jun 32, 31(32)1; Middlemas and Barnes, pp. 732-733. This seems to be one of a number of cases in which the enthusiasm of Davis and Gibson for disarmament led them to suggest a greater readiness for disarmament (in this case, on the part of the British government) than actually existed; on this case, see esp. *FRUS, 1932*, I, 142-144, 145-150, 157. But Gibson did make a slight effort to qualify his first report: *FRUS, 1932*, I, 131. Finally, hearing from Gibson that the British cabinet was about to consider Baldwin's plan, Stimson directed Ambassador Andrew Mellon to tell Simon and Baldwin that it was impossible for the U.S. to consider the abolition of capital ships; Baldwin and Simon, "rather astonished" at the "formality and gravity" of Mellon's démarche, assured him that only personal ideas had been expressed, and no view of the British government: *FRUS, 1932*, I, 157-168; Stimson MS Diary, 7, 8 Jun 32; Moffat MS Diary, 7, 9 Jun 32; FO 800/287, Simon Papers, ltr. to Lindsay, 8 Jun 32.

opposition in the preceding year would suggest; when Davis received an invitation to lunch with Herriot, he consulted Tardieu, who said that he had no objections whatever, and that such a meeting might be very useful.[51]

On May 22, Davis and Hugh Wilson, the United States minister in Berne and an alternate delegate to the Disarmament Conference, were received by Herriot at the Restaurant Carillon in Herriot's home city of Lyons, where he was mayor. Herriot's first term as premier had come in 1924 and 1925, when he had liquidated the occupation of the Ruhr, recognized the Soviet Union, and tried without success to obtain a military guarantee from MacDonald and (later) Sir Austen Chamberlain. A *normalien* and *docteur-ès-lettres*, as well as at one time a lycée professor, Herriot was a member of the French intellectual élite; he had studied French, German, and Greek philosophy, and had written and lectured on early-nineteenth-century literature. The tradition he honored, however, was not really so much that of Descartes as that of Sarah Bernhardt. He was extremely sensitive to any slight. He could produce tears on demand, and like Bernhardt, he frequently threatened to leave the stage, confident that there would be pleas for his return. But (again like Bernhardt) there were reasons for such pleas; he was an eloquent supporter of Western liberal ideas, and a man of integrity. He was very human, and susceptible to appeals to his humanity, but he also had a considerable tactical skill in politics and diplomacy.[52]

Herriot was, furthermore, a gourmand of massive dimensions, and we may imagine that his party had a large and excellent lunch at the Carillon. The conversation ranged widely over Franco-American relations, debts and reparations, the Gustav Stresemann letters, and disarmament. Some of the late German foreign minister's papers, revealing that he had always aimed at revision, had lately been published in a French magazine, and Herriot commented sharply on this; he apparently did not mention that he himself was especially piqued because Stresemann had described him as a "jellyfish." Herriot remarked that the German always regarded the man he was dealing with either as master or servant; it was extraordinarily difficult, he said, to deal with people in whom you could not have confidence. Somewhat pointedly, he emphasized several times that the United States, England, and France should work together. On disarmament, the Americans urged

[51] *FRUS, 1932,* I, 127-128, 132.
[52] On Herriot, see Soulié, passim, also Peter J. Larmour, *The French Radical Party in the 1930's* (Stanford: Stanford University Press, 1964), pp. 50-51, 122-125; Schuker, pp. 126-127, 232-236. For 1932, at least, I do not share Schuker's view (e.g., pp. 235-236, 238-245, 299-300) that Herriot was an incompetent diplomat.

that a limited interim agreement would reduce expenditures. This was probably a deliberate attempt to play on one of Herriot's major concerns: the French budget was falling seriously out of balance. Herriot said he would have to reserve his position until he had studied the problem, but he added that "the idea was sympathetic to him."[53]

Davis then described the talks at Bessinge, and his discussions in Paris and London. At Bessinge, Davis said, MacDonald had been "emphatic on the question that no new German competition in armaments could be envisaged and that equality of armament could not now be considered," while Brüning had been open to the idea of restating the Versailles disarmament clauses in the new convention, "with a footnote to the effect that Germany voluntarily accepted this obligation." Davis (according to the American memorandum for the record) urged that Herriot and MacDonald, "knowing and appreciating each other," should both come to Geneva a few days before the Lausanne Conference to work on disarmament. Herriot was agreeable, provided that calendar problems could be overcome, and he suggested that, if necessary, the Lausanne Conference might be postponed a few days for this purpose.

Later, a legend would arise that Brüning had been on the brink of reaching a disarmament agreement when he was dismissed by Hindenburg at the end of May. This legend seems to have stemmed partly from an American distortion of Herriot's views, which first appeared in the cable report from Geneva to Washington on the Carillon lunch. The cable reported that Herriot had described Brüning as a good man, although Herriot wondered whether Brüning could bind Germany; the cable also stated that Davis and Wilson had suggested Brüning, as well as MacDonald, as a participant in pre-Lausanne talks, and that Herriot "thought well of the idea." Now, these additional references to Brüning were not contained in the very full memorandum for the record on the Carillon lunch, which the cable otherwise summarized. Nor does Herriot's own brief note on his talk with Davis reflect such ideas.[54] Actually, the French leader was very interested in having early and direct Franco-British discussions, not only on disarmament but even more on reparations. But as the memorandum for the record shows, and as he also indicated to the British ambassador to Paris, Herriot's aim was to establish solidarity with the British (and also the

[53] Record of the meeting: *FRUS, 1932*, I, 132-139; Edouard Herriot, *Jadis*, 2 vols. (Paris: Flammarion, 1948-1952), II, 293. On Herriot's abiding pique over Stresemann's comment: *FRUS, 1932*, I, 477, also SDP, Excerpts from confidential statement to certain British and American correspondents, 28 Sep 32, 500.A15A4/1483½. Stresemann's diaries also embarrassed Herriot by reporting his indiscretions to the German leader on internal French policy differences: Schuker, p. 374 and note.
[54] SDP, tel., Gibson to Dept., 23 May 32, 500.A15A4/1066; Herriot, *Jadis*, II, 293.

Americans), as opposed to the Germans. Herriot habitually tried to believe that he shared common interests and feelings with the "Anglo-Saxons" generally, and with MacDonald in particular. In mid-June, Herriot expressed very clearly to MacDonald and Simon his opposition to including Germany in disarmament discussions by the principal powers; "her inclusion would in his view amount to some sort of recognition of her claim to equality of rights." If Davis and Wilson did suggest, on May 22, the inclusion of Brüning, it seems probable that Herriot did not hear, misunderstood, or did not reflect on it. And the British do not seem to have been advised of the idea of a Mac-Donald-Herriot-Brüning meeting.[55]

It looks as though Davis and Gibson talked things over after Davis had returned to Geneva from Lyons, and as though they decided that if Herriot was ready for any high-level meetings, it should be possible to bring about the four-power encounter which failed to take place at Bessinge in April. The American delegates were anxious to present some minimum agreement to the public as soon as possible, and Gibson in particular may have glimpsed an opportunity to settle everything at what would now be called a summit meeting. At all events, Davis and Gibson developed the idea of sending an unofficial emissary to Berlin with a message to induce Brüning to "complete the group" (as Gibson put it) at Geneva at the appropriate time, and also to warn him of the unfortunate impact of the Stresemann letters. Officials in the State Department were lukewarm about this procedure, wondering, among other things, what was so new and startling in the disarmament situation, and there was some delay in carrying out the plan. Frederic Sackett, the American ambassador in Berlin, had received a letter from Gibson by the thirtieth, but (or so Bülow informed Nadolny) the ambassador did not communicate it; Bülow surmised that this was because Brüning resigned as chancellor that morning, and because the message was too much aimed at Brüning personally to be given to anyone else.[56]

Brüning claimed later that Sackett showed him the letter, just before his last meeting with Hindenburg on May 30, and that it indicated that Herriot, as Premier, had accepted the German disarmament formula already (according to Brüning) accepted by Britain, the United States, and Italy in April; the President's entourage did not, however, allow Brüning time to explain the situation to Hindenburg. Appar-

[55] FO 371, Memo by Tyrrell, 25 May 32, W6020/221/98; *DBFP*, 2/III, No. 241; *FRUS, 1932*, I, 142, 145.

[56] *FRUS, 1932*, I, 139-140, 142; Moffat MS Diary, 24 May 32; AA, tel., Bülow to Nadolny, 31 May 32, 7360/E535349. Bülow's statement to Nadolny does not entirely rule out the possibility that the message was shown to Brüning, but not to him, Bülow; see next note.

ently, Sackett at some point did give Brüning, or Brüning's friend Treviranus, at least an oral account of the message. There seems to be no trace of the letter in either German or American official records. Gibson probably wrote in it something along the lines of what he told Nadolny on May 30: that at the proposed meeting, which would follow preliminary discussions of the basic questions between delegations, the heads of government could "in case of necessity make the authoritative decision. Herriot will probably also be ready for this."[57] Perhaps mainly owing to what he had heard of Baldwin's ideas, Gibson had at this time a wildly optimistic impression of the readiness of governments to disarm themselves.[58] But whatever Gibson told Nadolny, and whatever his letter to Sackett said, it was not the case that Herriot was ready to concede equality of rights, let alone an increase in German arms; this is patent from the later course of negotiations. Also, Herriot did not become premier until four days after Brüning's fall. May 30, 1932, was not in fact an occasion when a great opportunity was lost. But through Gibson's zeal and overoptimism, Brüning had been given a misleading impression, one that would heighten Brüning's conviction of the tragic character of his own downfall.

[57] Brüning gave one version (A) of the story in "Ein Brief," *DR*, 8 (1947), 10-11; another (B) in a "Zeugenschrifttum" (witness's report) for the Institut für Zeitgeschichte, Munich, of 3 Dec 51 (ZS 20); and a third (C) in his *Memoiren*, pp. 601-602. The version given here is that of (A). Wilhelm Deist, in "Brüning, Herriot und die Abrüstungsgespräche von Bessinge, 1932," *VfZ*, 5 (1957), 265-272, criticized the (A) version and pointed out that Herriot was not even premier in May. Brüning's post-Deist version (C) does not claim that Herriot accepted his Bessinge position, but rather that Gibson and Herriot discussed German disarmament and reparations demands, and that Herriot recognized the need to keep Brüning in office. Version (B), which was not intended for publication, makes the astounding claim that the United States had compelled Herriot to accept the German disarmament proposals. Nadolny's report of his talk with Gibson is in AA, tel., Nadolny to Bülow, 30 May 32, 7360/E535346-48; Nadolny reported Gibson as saying that Herriot had expressed readiness for energetic disarmament action for budgetary reasons, and that he (Gibson) had sent a letter to Sackett for use in informing Brüning. Treviranus's memoirs (pp. 319-320) say Sackett gave Brüning an oral message from Gibson that Herriot was ready to give up French opposition to German equality of rights in armaments, and "only had to win over two more cabinet members"; Gibson urged Brüning to come to Geneva, where he could expect a quick settlement. Treviranus says that Brüning told Sackett that he had to resign that morning, and suggested that Bülow be informed, but Sackett refused because the message was aimed at Brüning personally. Treviranus notes (p. 413, n. 119) that he confirmed the content of the message with Sackett the next day, and with Gibson in 1943. But at the time, on 30 May 32, Treviranus reportedly told Kuno, Graf Westarp, that the message was from Hoover, on reparations: Westarp Memo, 1 Jun 32, in W[erner] Co[nze], "Zum Sturz Brünings," *VfZ*, 1 (1953), 285. This 30 May 32 version presumably represents an account from Brüning, although at third hand, and it suggests either that Brüning may have misunderstood the message under the strain of the moment, or that Sackett did not pass the message to Brüning at that time, and that Brüning guessed at its content.

[58] On Gibson's overoptimistic view of the prospects for the conference: *FRUS, 1932*, I, 145-150. See note 50 above.

6. NEGOTIATIONS WITHOUT GERMANY

That downfall did end for a time the American efforts to bring the Germans into a high-level meeting. Brüning, then and later, was entirely right on one point: he was the one German leader who had the full confidence of the British and Americans. His successor and Schleicher's new nominee, Franz von Papen, was a peculiarly inept choice from this standpoint. A wealthy Catholic aristocrat and former cavalry officer, Papen had been military attaché in Washington in 1914 and 1915, and after coming under public suspicion of fomenting sabotage and other conspiratorial acts, he had been declared persona non grata. On his way home, the British seized his private records and used them as a basis for a "white book," purporting to expose Papen's sabotage and espionage activity. Papen's reputation was such that the first British report from Berlin on his candidacy described him as "notorious," while Secretary Stimson told the British ambassador in Washington that Papen was a man who "if proposed as ambassador here would unhesitatingly be refused." The new Chancellor, who spoke flawless French and whose wife came from a partly French family in the Saar, took a particular interest in the improvement of Franco-German relations; nevertheless, his political and religious orientation did not commend him to Herriot, who was a thorough republican and an old opponent of the French church. As to Schleicher's reputation with the Americans, Brüning, in his first bitterness over his removal, told Ambassador Sackett that he (Brüning) had been removed through military intrigue alone, and that reports of disagreement between him and Hindenburg over "the new economic proposals" (i.e., land settlement in the East) were subterfuge; Sackett concluded that "the personnel of the new cabinet is strongly indicative of a military dictatorship in close cooperation with nationalist groups having monarchial sympathies. . . ." It was widely noted that almost all the members of the new cabinet were nobles, and the new leadership was generally regarded as a "Junker" government.[59] On June 5, Gibson reported a feeling in Geneva that opinion, particularly in France, Britain, and the United States, had become reunited "as to the necessity of main-

[59] Franz von Papen, *Der Wahrheit eine Gasse* (Munich: Paul List, 1952), pp. 53-78; May, *World War*, pp. 164-165; *FRUS, 1932*, II, 293-296; *DBFP*, 2/III, Nos. 116, 117, 127, 129; Herriot, *Jadis*, II, 315. Brüning had earlier spoken of military intrigues to a British friend, Major Church: MDP, 1/301, Note by Sir F. Leith-Ross, [ca. 24 May 32]. On Brüning's bitterness, also Schäffer Diary, 7 Jun 32; Co[nze], "Zum Sturz Brünings," 288. For other negative reactions: Stimson MS Diary, 1 Jun 32; University of Birmingham Library, Birmingham, England, Austen Chamberlain Papers, ltr. to Hilda Chamberlain, 5 Jun 32, AC 5/11/566; Harry Graf Kessler, *Tagebücher 1918-1937* (Frankfurt: Insel Verlag, 1961), p. 670. See also Link, *Stabilisierungspolitik*, pp. 523-524.

taining a rigid regime and attitude until it is seen what line Germany is going to take." And he added, "If progress can be achieved among the other powers they can speak with one voice in telling Germany she must take what she can get as a first step toward equality." Gibson, recently the chief proponent of Brüning's inclusion, now expected and hoped for a series of conversations without German participation.[60]

Aside from the change of government in Berlin, other, possibly more important, reasons soon appeared for proceeding with private talks without the Germans. One reason was the thought that some nominal disarmament might be agreed on without Germany, and another was the need for both Britain and France to avoid a sharp conflict over disarmament while the Lausanne Conference was meeting on reparations. MacDonald and Simon do not seem to have shared in the new rigidity described by Gibson, and they still hoped eventually to bring Germany into discussions. But public opinion in Britain and the United States would be highly incensed if the conference did not produce soon some actual disarmament. In order to get things moving, and in pursuit of his idea of qualitative disarmament, Simon suggested to the French that the major powers, excepting Germany, meet to draw up a list of weapons which should be abolished or limited in size; such a discussion would be easier without Germany, and her exclusion could be explained as owing to the fact that she was already disarmed. If, however, the French wanted to take up the broader issues of internationalization, then in that case the Germans would have to take part. The new Herriot government, on its side, refused at this stage to take part in private discussions with the Germans on equality of rights; to do so might imply or lead to acceptance of that doctrine. Herriot recognized, if the British did not, that to accept equality at this point would not only impair Part V, but also prejudge the substantive arrangements in an eventual disarmament convention. The other powers should first decide, free of German pressure, what disarmament they could undertake, and only discuss equality after that. Joseph Paul-Boncour, Herriot's new Defense Minister and an old Geneva hand, told the British and Americans that France was ready to negotiate without Germany on specific reduction. But he intimated that if other powers wanted larger reductions, then French ideas of internationalization would also have to be discussed.[61]

[60] *FRUS, 1932*, I, 152-153.
[61] *DBFP*, 2/III, Nos. 241, 245; Herriot, *Jadis*, II, 316; *FRUS, 1932*, I, 169-170, 174-177. According to British records, the American delegates agreed with them on seeking a transfer of Part V provisions to the convention: *DBFP*, 2/III, No. 244; FO 371, tel., Simon to F.O., 19 Jan 32, W7010/1466/98.

Thus the French and British each held the other in check, and while neither the British nor the French ministers admitted to reducing their aims,[62] perhaps they were not entirely unhappy with this situation. The British government was committed by Treasury policy to seek a complete cancellation of reparations at the Lausanne Conference, and British leaders presumably recognized that the French would not consent to that if they were also pressed to the limit on disarmament and German equality.[63] For them, reparations cancellation had priority. The new French government probably did not want a showdown on Tardieu's plan, an event that would have led to a sharp public split with Britain and the United States and, no doubt, to a rejection, all on the eve of Lausanne. Despite all Franco-British differences, the leaders and diplomats of the two countries were more familiar with each other, and (after the departure of Tardieu and Brüning) got on better with each other than with their German counterparts. For a part of June and July, the British largely dismissed Germany's equality of rights from their minds, something they would do again in the summer of 1933.

Yet another reason for ignoring German views was absorption with a new American proposal. On June 19, representatives of Britain, France, and the United States began the informal private talks suggested by Simon, and at French suggestion, these talks concentrated on specific reductions. Only three days later, the American government jarred the delicate balance between the French, British, and American positions by announcing, with all publicity, a variation of Hoover's plan. Previously, the plan had received little public attention, and the quantity of the reductions had been left open for diplomatic negotiation. Now Hoover proposed to cut the number of battleships by one third, and of cruisers and aircraft carriers by one fourth, and he also, in application of his ideas for dividing land forces into domestic security and defense elements, called for a one-third reduction in all defense contingents. Moreover, he proposed to ban chemical warfare and to abolish tanks, large mobile guns, and military aircraft except naval observation planes.[64]

[62] *DBFP*, 2/III, Nos. 242, 244; *FRUS, 1932*, I, 169-170, 174-175; FO 371, tel., Simon to F.O., 19 Jun 32, W7048/1466/98.

[63] Simon had thought to compensate the French for reparations cancellation by arranging a political "truce" for a set period of years on the raising of new issues, but neither the French nor the Germans thought much of the idea: *DBFP*, 2/III, Nos. 124, 128 (and encl.), 132, 133, 134, 136 (Annex III). MacDonald told Gibson and Davis that Chamberlain had induced the cabinet (over MacDonald's opposition) to form a common front with other debtors on the war-debt question, and that this policy was now required in order to compensate France for the loss of reparations: SDP, tel., Gibson to Dept., 20 Jun 32, 462.00R296A1/179.

[64] *DBFP*, 2/III, No. 245; *FRUS, 1932*, I, 178-182.

This whole proposal seems to have been partly set in motion by Stanley Baldwin's suggestion in May (reported by Gibson and Davis) that battleships might be abolished entirely. Where Stimson was simply opposed to the elimination of American battleships, Hoover saw Baldwin's ideas as a bid for disarmament laurels that he might have to compete with, and he wanted to get a proposal of his own out before the British announced one. Hoover also wanted to press for European disarmament to forestall and discourage the claim by Europeans (after ending reparations at Lausanne) that they could not afford to pay war debts. He had no desire for a discussion of the war debt issue during the presidential election campaign, since although he opposed cancellation, as did most of the American public, his 1931 moratorium and his discussions with Premier Laval had paved the way for a cancellation demand, based on the termination of reparations. He was anxious, moreover, as the election campaign moved into full swing, to do something spectacular to convince the American public that he was capable of taking a far-sighted, liberal initiative. And there was certainly an element of Quaker idealism in his proposal. Hoover gave little or no consideration to the three-power discussions when he decided to announce his disarmament plan. He and Stimson hoped that the conference would simply adopt a resolution approving the plan in principle, and then adjourn for six months, putting disarmament and reparations on ice until after the election.[65]

An early acceptance was hardly likely, however, considering the radical nature of the proposals, and the scant amount of prior consultation with other powers. There were hostile reactions within both the British and French governments. Simon fumed at the (as he put it) "peremptory summons of [the] General Commission to receive orders from Washington," just when private conversations were getting launched. As to substance, Hoover's proposal for naval reduction, especially the reduction in cruisers, was the biggest surprise in his announcement, and it was completely unacceptable to the Royal Navy.[66] On July 7,

[65] *FRUS, 1932*, I, 185-186, 189-191, 212-214; Stimson MS Diary, 22, 24, 25 May, 7, 18, 19, 20, 21 Jun 32; Hooker, *Moffat Papers*, pp. 72-75. On Hoover's lack of thought of the private discussions, see esp. *FRUS, 1932*, I, 205-206. Stimson told Davis that Hoover's real objective was to use the proposal to give the conference a good reason for adjournment: Library of Congress, Norman Davis Papers, Memos of phone conversations, 21 Jun 32 (1:20 a.m.), 22 Jun 32 (1:00 a.m.). This may have been more Stimson's idea than Hoover's, however.

[66] AA, tel., Nadolny to Ministry, 23 Jun 32, 7360/E535475-76; *DBFP*, 2/III, Nos. 250, 263; FO 371, Minutes by Howard Smith, Craigie, and Vansittart, 22 Jun 32, W7107/1466/98. The British cabinet decided on 24 June to proceed with their announcement, and Canadian and Australian concurrence was obtained by July 1: Cab 23/71, meeting of 24 Jun 32, 38(32)2; meeting of 27 Jun 32, 39(32)1; FO 371, Minutes by A.E.H. Wiggin and Howard Smith, 30 Jun and 1 Jul 32, W7419/10/98.

the British issued a statement which, while largely agreeing with Hoover on land armament, conveyed their navy's opposition to the American approach to naval limitation. While the American navy was less interested in numbers than in retaining large ships with a long cruising range, British naval experts believed that the London Naval Treaty had already reduced their forces below the number needed to ensure communications with their widely scattered possessions and with the dominions. So, instead of Hoover's numerical reduction, the British proposed a reduction in the size of ships.[67]

The French found an unpleasant resemblance between the new plan and Hoover's unilateral bombshell of a year earlier, proposing a moratorium on war debts and reparations, and Herriot issued a sarcastic press statement on June 25. Their ambassador in Washington, the poet Paul Claudel, was angry over Stimson's failure to inform him, and he advised Paris that Hoover's proposals would not be ratified by the Senate. On substance, however, the French were less critical than the British. The Herriot government planned anyhow to make some reductions in effectives for budgetary reasons, although nothing like the proportion Hoover called for; given time, it might do something in return for counter-concessions, and in this connection, the French suggested that the United States might undertake in some way to consult with other governments in case of a conflict. But in the tripartite talks, Boncour, using the same lever as before, emphasized that if the Hoover plan was to be debated, the French plan would have to be debated also.[68]

In their messages home, Gibson and Davis implied that they would press hard for comprehensive disarmament as proposed by Hoover, in contrast to a limited agreement on a few specifics. Nevertheless, a close examination of these messages and of British records shows that Gibson and Davis realized by June 23 that they could not obtain an early conference vote adopting Hoover's plan. Gibson's response was to propose a resolution that would allude generally to Hoover's prin-

[67] *DBFP*, 2/III, Appendix VI; Medlicott, *British Foreign Policy*, pp. 76, 91-92. On the Royal Navy's problems with disarmament and the London Treaty, see Gibbs, *Rearmament Policy*, pp. 6-31.

[68] *FRUS, 1932*, I, 223, 225-228, 234-236, 243-246, 249-251, 253-255; AA, tel., Prittwitz to Ministry, 23 Jun 32, 3154/D667435; SDP, tel., Gibson to Dept., 30 Jun 32, 500.A15A4/1212; Stimson MS Diary, 29 Jun 32; *DBFP*, 2/III, Nos. 252, 257. Boncour told Weygand on 8 June that the total budget would have to be cut by 500 million francs, and Weygand agreed to a reduction of the army budget by 300 million, on condition that this was not taken from the allotment for materiel: [Maxime] Weygand, *Mémoires*, 3 vols. (Paris: Flammarion, 1950-1957), II, 386. An October 1932 document states that a reduction of approximately 40,000 men was planned for budgetary reasons: *DDF*, 1/I, No. 244. Weygand stated in October that the Hoover plan, as it stood, would reduce French effectives by at least 120,000 men: *DDF*, 1/I, No. 286 (p. 624).

ciples, announce the limited agreed results of the continuing tripartite talks, and ordain the study, during a conference adjournment, of Hoover's practical proposals, including those for naval reduction. With the references to Hoover's plan further watered down, this procedure was acceptable to Simon and Boncour, particularly as it presented some semblance of progress without involving a recognition of equality of rights. A reservation of "political questions" for later decision protected the French position, and it was hoped that the Germans would also regard this as protecting their position on equality of rights. Although Hoover had doubtless hoped for more, he was too wrapped up in other concerns to raise serious objections, and Stimson was more interested in protecting American naval strength and in obtaining an adjournment than in Hoover's scheme—although he could not refrain from telling Gibson that the final resolution, as it emerged after lengthy negotiation on July 20, was "not much more than eyewash."[69]

Possibly the German Foreign Ministry might, by itself, have accepted such a postponement of the equality-of-rights issue, but the Papen government, under the influence of General von Schleicher and the Reichswehr, would not. Germany had had no voice in framing the resolution, and Herriot, when approached by the Germans with suggested amendments, was very unresponsive. His position was that equality of rights could only be considered after a convention had been worked out, as only then could it be seen if the convention would be a suitable substitute for the clauses of Part V.[70] As the French premier had already indicated to the British and Americans, he wanted to keep the doctrine of equality from influencing the convention; no doubt he also feared the political reaction in France to a recognition of equality.

IN JUNE AND JULY, the preoccupation of the British and American governments with their own proposals and their own direct concerns worked against German interests. The "Anglo-Saxons" had only accepted equality of rights at Bessinge as a Platonic matter of status; they thought that such equality might be largely achieved, for exam-

[69] *FRUS, 1932*, I, 225-228, 233-234, 253-258, 259, 263-267, 284-289, 291-305; *DBFP*, 2/III, Nos. 252, 257; Stimson MS Diary, 5, 25 Jul 32; Davis Papers, Memos of transatlantic phone conversations, 1, 20 Jul 32; Moffat MS Diary, 24, 27, [29] Jun, 2, 8, 9, 18, 20 Jul 32. According to Moffat, Stimson told *Davis* that the resolution was "bellywash"; according to *FRUS, 1932*, I, 302, Stimson told *Gibson* it was "not much." The British record seems to show that Gibson and Davis were discussing alternatives for the conference treatment of the Hoover plan in the private conversations with the British and French, something they concealed from Washington.

[70] AA, tel., Hoesch to Ministry, 15 Jul 32, 9284/H254360-61; tel., Nadolny to Ministry, 20 Jul 32, 3154/D667477.

ple, by putting the 1919 restrictions into a new convention. Their fear of a discussion of the French plan now enabled Herriot and Boncour to block further consideration of equality of rights. Yet overall, the policy of the British and Americans had provided support for Germany, not only in their somewhat misunderstood endorsement of equality of rights at Bessinge, but in their advocacy of qualitative (and with Hoover, also quantitative) disarmament. The demand for qualitative disarmament pressured France to dispose of her heavy weapons, and if she refused to do this, as was probable, the principle of qualitative equality would tend to justify the acquisition by Germany of the arms France retained. As a matter of fact, although the Germans did not know it, a British Foreign Office memorandum had suggested just before Brüning's downfall that there was no reason in equity for not allowing Germany the weapons retained by the French, even though she should not "of course" be permitted a quantitative increase in guns or effectives or tonnage.[71] In saying this, the staff of the Foreign Office was already proposing to concede what was to appear in the autumn, in British eyes, as the major new demand of the Papen-Schleicher government.

[71] Cab 24/230, CP 164 (32) (Encl. II, 26 May 32).

CHAPTER FOUR

General von Schleicher's New Approach

1. MILITARY DOMINANCE IN THE NEW CABINET

Although Franz von Papen was a poor choice as chancellor from the standpoint of retaining the confidence of foreign governments, this had not been a major consideration when General von Schleicher recommended him to Hindenburg. Nor had it much mattered that Papen had little standing in his own party, the Catholic Center. Indeed, the very fact that Papen was an outsider in his own party as well as independent of Alfred Hugenberg, the obstinate and uncontrollable leader of the DNVP, may have been part of his attraction; for these reasons, and because of his wealth, party considerations would presumably not influence him. Schleicher was aiming now for a cabinet of wellborn experts, standing above parties, a reversion to prewar practice that he and others had long had in mind. Papen was no doubt also attractive as an old opponent of the Prussian Social Democrats, and, even more, as a declared advocate of bringing the Nazi movement closer to the state. But probably Papen's greatest merits, in Schleicher's eyes, were that the ex-cavalry officer and gentleman jockey, whom Schleicher had known before the war at the general staff school, was sympathetic to Reichswehr aims and was also (or so Schleicher believed) someone whom he could guide and control. As Schleicher himself put it, Papen was not to be the head of the new government, but rather a "hat" for it. Schleicher, who now took the position of Reichswehr Minister, obviously considered that he himself was to be the leader of the government, and he meant to see that there would be no more obstacles to strengthening national defense. Though observers noted the high proportion of titled aristocrats in the new government, and called it a "cabinet of barons," it was also a cabinet of ex-officers and ex-reserve officers.[1] The onset of the Papen-Schleicher government marked, indeed, a new stage in the primacy of military policy, and in devotion to the doctrine of seizing the initiative.

[1] On Schleicher's motives for choosing Papen, see Bracher, *Auflösung*, pp. 518-519, including 519n; Carsten, *Reichswehr and Politics*, pp. 364-365. Thomas Trumpp, *Franz von Papen, der preussisch-deutsche Dualismus und die NSDAP in Preussen* (Doctoral diss., University of Tübingen, [1963?]) shows Papen's desire to bring the Nazis into government. Schleicher's belief that this was his cabinet was especially evident at its first meeting: RK, Cabinet Protocol, 2 Jun 32, 3598/D790074-78. Ex-officer makeup of cabinet: Magnus von Braun, p. 226.

As was to become clearer during the summer, dominance over official policy—even if one had it—did not guarantee political domination within the country. Schleicher himself did not expect to succeed over the long term without popular support. While freeing himself from Brüning and Groener, from the moderates and also the Socialists, he courted the right, and especially the National Socialists. He had a double interest in the Nazi party: he hoped to harness that movement's national fervor and popularity for the government by inducing the Nazi leaders to become junior partners in that government, and he hoped to bring the SA into the developing scheme for promoting Wehrsport through a youth-training organization. In order to propitiate the Nazis and achieve these two ends, the Papen government scheduled new Reichstag elections for July 31, in line with Nazi desires, and it replaced the ban on the SA with a decree putting that and the other Wehrverbände under the supervision of the Ministry of Interior.[2] These concessions have been hard to understand in retrospect, particularly since Hitler refused to give any more in return than an oral promise to "tolerate" the government until the election. In view of the growing Nazi following, the election would obviously increase Nazi strength in the Reichstag; this would hardly make Hitler easier to deal with afterwards. The impenitent attitude of the Nazis made it equally obvious that a removal of the ban on the SA would result in a resumption of their violence—as immediately happened. Apparently, however, Papen and Schleicher assumed that a strongly nationalist movement would naturally support an authoritarian government led by aristocrats and high officials, and that precise parliamentary majorities would no longer count for very much. The Nazi leaders would be tamed when given a modest share in the responsibilities of power, while the SA would likewise settle down when given some regular premilitary training to occupy its time. And premilitary training and the silencing of leftist opposition were of course militarily desirable in themselves. Taken all together, the wave of popular nationalism was too good a thing to be left unexploited.

Even though the new cabinet was nominally nonpartisan, it had its own constituency to serve. It would naturally protect the landowners and industrial leaders. Beyond this, Schleicher believed that the new government would have to adopt a strongly nationalist foreign policy to win popular support on the right.[3] Above all, it had to champion Germany's claim to equality of rights. Schleicher needed to satisfy not

---

[2] Vogelsang, *Reichswehr*, pp. 200, 205, 209, 216-217; Bracher, *Auflösung*, pp. 545-552.

[3] See AA, Record of chiefs' meeting, 4 Jun 32, 3642/D811887; RK, Cabinet Protocols, 7 Jul, 15 Aug 32, 3598/D790351, D790547-48.

only the conservative leanings, but also the professional concerns and interests of Reichswehr officers, upon whom his own position ultimately depended. Their immediate concern, and his also, was to strengthen Germany's armed forces. The Truppenamt had presented Hammerstein on March 2 with initial proposals for an enlarged peacetime army, and staff officers had since been perfecting these plans. It was vital to speed the development of Wehrsport training, and the militia was, as Schleicher put it, a question "particularly urgent for us."[4] The officers responsible for material rearmament and industrial mobilization preparations were also, as we have seen, clamoring for more funds. Some of these demands would be hard to meet.

Even though he faced problems, Schleicher, from his earliest days as a cadet, had distinguished himself by his self-assurance and his ability to influence others. As with many politicians, promoters, and salesmen, self-assurance and influence were not for him the consequences of a real situation, but rather the creative sources of a fictitious reality, one that could only be upheld with further displays of confidence. Schleicher's health was unstable, and Brüning suggests that, because the general believed his days to be numbered, he was therefore anxious to act quickly. In mid-June, Schleicher appeared very worn out (though shrewd) to one observer, while a week later he gave another an impression of vigor, physical strength, and unusual penetration; fluctuations in his health probably had much to do with occasional fluctuations in his resolve. He was often arrogant, and prone to indiscreet boasting at social gatherings.[5] He had little in common with the Helmuth von Moltke of Bismarck's day, who had believed in the virtues of silence, but Schleicher adhered to the general staff tradition of seizing the initiative, and he looked forward to substituting action for what he regarded as Brüning's hesitation.

Schleicher had little direct contact with foreign diplomats, such as might have led him to be cautious about exposing German aims.

---

[4] I have not found the March proposals (TA 72/32, 2 Mar 32), but they are referred to in BA-MA, RH 15/19, TA 737/32, 7 Nov 32; see also Castellan, *Réarmement clandestin*, p. 83. On the militia: AA, ltr., Schleicher to Bülow, 9 Jun 32, 7360/E535897; see further below.

[5] Vogelsang, *Schleicher*, esp. pp. 10, 17 (citing an acute portrait by Col. Albrecht von Thaer from 1918), 101; Brüning, *Memoiren*, pp. 452-453, 580, 581; Wheeler-Bennett, *Nemesis of Power*, pp. 182-183, 237-238n. Hans Rudolf Berndorff, in *General zwischen Ost und West* (Hamburg: Hoffman & Campe, 1951), pp. 12-13, relates that Schleicher, as a new boy at the cadet school at Plön, impressed his fellow students with references to a fictitious ancestor, "Count Schlei." Vogelsang (*Schleicher*, p. 101) says Schleicher suffered from anemia; Brüning (*Memoiren*, p. 580) says he had liver trouble. General MacArthur, who toured Europe in the summer of 1932, heard that Schleicher had both (SDP, Extracts from ltr., MacArthur to [?McBride?], 15 Oct 32, 862.00/2855½). Observers: Vogelsang, *Reichswehr*, Annex 28; U.S. Military Attaché Report, Berlin No. 12,255, 28 Jun 32.

Perhaps partly on the basis of exaggerated reports of what had been achieved at Bessinge, and partly due to a habit of counting unhatched chickens, he overestimated the willingness of foreign governments to accept German rearmament. Foreign Ministry officials were aware that the British and Americans, at Bessinge, had at most only accepted theoretical equality, not rearmament. But they apparently did not make this clear to the new Reichswehr Minister.[6]

Since Britain, the United States, and also Italy had supposedly recognized Germany's equality of rights, Germany's next step was evidently to negotiate with France on the theoretical and practical issues of armament. Papen's interest in improving Franco-German relations may have been an additional reason, in Schleicher's eyes, for choosing him as chancellor. In May, a private intermediary had approached the French ambassador, André François-Poncet, on Schleicher's behalf, attempting to find out what special French wishes would have to be met in order to secure a French recognition of Germany's theoretical equality of rights. An idea had long been circulating that the best way to settle the Franco-German problem would be to have German military leaders talk directly with their French counterparts. The German advocates of such talks expected that the French military would prefer a certain amount of German rearmament to a restriction of French armament along the lines proposed by such interlopers as Sir John Simon or Herbert Hoover.[7]

On the Foreign Ministry side, Constantin von Neurath, the new

---

[6] Treviranus, p. 241; AA, ltr., Schleicher to Bülow, 23 Jun 32, 7360/E535911; Vogelsang, *Reichswehr*, Annex 28. The Foreign Ministry recognized that the British and Americans would not (knowingly) accept German rearmament (AA, Draft guidelines, [prior to 2 Jun 32], 7360/E535374), but Bülow encouraged Schleicher's optimism on 4 June by saying that if French agreement could be gained, "one could unquestionably expect that no difficulties would be raised in any other quarter against a satisfactory continuation of the Conference" (RK, Record of chiefs' meeting, 4 Jun 32, 3642/D811886-87).

[7] AA, Memo, Bülow to Neurath, 12 Jul 32, 4619/E197476; Reichswehrminister 231/32, 26 Aug 32, 7474/H184054; Schleicher Papers, N42/34, ltr., Regendanz to Schleicher, 12 May 32; Schäffer Diary, 19 Jan 31; Brüning, *Memoiren*, p. 555. The July proposal of [Major Erich] Marcks for talks on equality and security (Bredow Papers, Orientation note, 25 Jul 32) may have been a follow-up to the approach to Poncet. Geyer (in "Militär, Rüstung und Aussenpolitik," p. 243) attributes to Major Marcks an elaborate plan for unifying the German and French economies and especially rail systems, thus barring rapid, secret mobilization and reassuring the French. But the "Marks" who advocated this plan (to Hans Schäffer and his Ullstein associates: Schäffer Diary, 23-24 Jun 32) was apparently a Consul S[alomon] Marx, an old collaborator of Stresemann's; see Fernand L'Huillier, *Dialogues franco-allemandes, 1925-1933* (Strasbourg: Diffusions Ophrys, 1971), pp. 113-114; there is a whole file on S. Marx in AA, Serial 5720: Abteilung Pol II: Po 2C Frankreich, Die Bemühungen des S. Marx zur deutsch-französischen Verständigung, Aug 1931-Jun 1932. For the rather different views of Major Marcks, see his article, "Frankreichs Sicherheit," *Zeitschrift für Politik*, 21 (1931-1932), 830-839.

Foreign Minister and most recently ambassador in London, had not been much involved so far in disarmament discussions. But State Secretary Bernhard von Bülow suggested that, since little progress had been made at Geneva, Germany should use the occasion of the Lausanne Conference on reparations to try to secure French acceptance of equality of rights. The Foreign Ministry drew up guidelines arguing that a special agreement with France was desirable, particularly since without such an agreement, the British and Americans, who did not want German rearmament, would also become suspicious, and all three powers would oppose Germany's desired adjustments. In Bülow's eyes, there was a good basis for getting French concurrence, since Germany would in any case be financially unable to undertake extensive rearmament during the next few years, and she could afford to offer France special agreements limiting her numerical strengths in men and weapons for, say, a five-year period. He assumed that France would recognize Germany's equality of rights in order to secure these agreements.[8]

At a ministerial discussion on June 4, Schleicher agreed with Bülow's basic ideas, provided the German public was not given the impression that Germany was forgoing rearmament.[9] Difficulties soon developed, however. Schleicher stressed to Bülow on June 9 that his position was to demand "*complete* equality of rights," limited only de facto by a temporary agreement with the French on quantities; he also said at this time that the Foreign Ministry's guidelines were specifically inadequate "with regard to the question, particularly urgent for us, of the militia." Bülow then added a preface to the guidelines, entitled "The *Hidden* German Goal [*Das* interne *deutsche Ziel*]," which included a summary reference to the German desire for a certain increase in personnel strength, a shortening of the term of service, freedom in organization, and a militia. But Bülow also wrote Schleicher privately, saying that with regard to the special urgency of the militia question, "I wish to point out that we cannot count on winning approval of a German militia in a short period by way of the Disarmament Conference, since even if the disarmament convention is concluded this year, it will only go into effect after ratification, which experience shows takes many months."[10]

Such arguments as this could only make Schleicher more impatient

---

[8] AA, ltr., Bülow to Neurath, 2 Jun 32, 7360/E535360-61; Draft guidelines [prior to 2 Jun 32], 7360/E535368-74.

[9] RK, Record of chiefs' meeting, 4 Jun 32, 3642/D811886-89; AA, Memo by Bülow, 4 Jun 32, 7360/E535363; ltr., Bülow to Nadolny, 6 Jun 32, 7360/E535364.

[10] AA, ltr., Schleicher to Bülow, 9 Jun 32, 7360/E535897; Guidelines for disarmament discussion at Lausanne, [10 Jun 32], 7360/E535883-84; ltr., Bülow to Schleicher, 10 Jun 32, 7360/E535895-96.

with the Disarmament Conference and with the conduct of German policy by the Foreign Ministry. He had not formed the new cabinet with the idea of accepting long delays in rearmament. He told a group of young officers in this period that diplomatic restraints on armament "no longer count for so much, and under present conditions it is no longer necessary to peer anxiously at the provisions of the Treaty of Versailles." The Bendlerstrasse was beginning to claim the sole right to control disarmament policy. When a report appeared on June 10 that MacDonald was proposing a ten- to fifteen-year freeze on armaments in case the conference failed, the Reichswehr Ministry, without consulting the Foreign Ministry, advised the press that "an arms holiday of long duration was absolutely unacceptable, since it signified a further anchoring of the status quo and a prolongation of the special treatment of Germany, and thus stood in complete opposition to our demand for equality of rights." And General von Blomberg, the chief military delegate, returned to Geneva dissatisfied with the state of cooperation between the delegation and civilian officials in the capital. The military delegation asked the Reichswehr Ministry whether Schleicher himself should not come to the Lausanne Conference to negotiate equality of rights there.[11]

Before this query was received, Schleicher dealt with the problem in another way, by laying down the law on disarmament policy generally, both for Lausanne and Geneva. The Reichswehr Ministry on June 14 put the finishing touches on a new paper, now entitled (with evident borrowing from Bülow's latest draft) "The Hidden German Goal at the Disarmament Conference." This document listed some nine minimum adjustments that should be immediately conceded to Germany for the period of the first convention, including the reduction of the period of regular service to something varying between three and twelve years; a "modest provision" of technical services, heavy artillery, aircraft, and an experimental battalion for tanks; the establishment of a militia with three-month service "for domestic security and border guard purposes, as well as for coastal defense"; and a "rounding out" of the army to about 160,000 men and of the navy to 27,000 men. Obviously much of the problem between diplomats

[11] Vogelsang, *Reichswehr*, Annex 28; AA, Memo by Völckers (with marginal note by Koepke), 10 Jun 32, 7360/E535377-78; National Archives, Microfilm T-120, Völkerbundsabteilung Gruppe Marine documents (hereafter NA-VGM), F VII a 11, ltr., Freyberg to Carls, 14 Jun 32, 7792/E565361-63; tel., Schönheinz to Reichswehr Ministry-VGH, 14 Jun 32, 7792/E565364-65. Freyberg suggested in his letter that Admiral Raeder, Chief of the Naval Command, might also come. Raeder replied on 15 June that Schleicher thought the new, clear guidelines removed any danger that the delegates to Lausanne would be too yielding, and that Schleicher and Raeder himself did not consider their own presence necessary (ML, Shelf 5939, reel 17/1071).

and soldiers had arisen from the former's ignorance of the latter's intentions, and this part of the paper gave the Foreign Ministry a better idea of Reichswehr desires than it had had before. This was by no means a complete idea, however; in particular, the real function of the so-called militia—to provide reserves—was not discussed, and it was not pointed out that the mobilized army would be much larger than 160,000 men.[12]

The Reichswehr paper not only stated the goals of the army and navy, but also tried to indicate the form the disarmament treaty should take. It prescribed that specific armament figures should be inserted in the convention showing Germany's full "equality of rights to security," which was defined for the army as "parity with France or at least parity with Poland + Czechoslovakia," and for the navy as parity with France or Italy. Alongside these figures, other significantly lower figures would be entered, representing what Germany would actually confine herself to for the course of a first convention, lasting at most no more than five years. The conclusion of the paper stated: "If the foregoing minimum demand of Germany is not met, then the interests of the armed forces require a breaking off of negotiations, which must then necessarily lead to a development of the armed forces in the sense of the foregoing demands." In other words, the Reichswehr planned to rearm whether or not the Disarmament Conference agreed.

This was not the first time that Reichswehr officers had proposed to claim a numerical strength equal to that of France, but in contrast with the situation under Brüning, the army as personified by Schleicher now believed that its wishes in arms questions were decisive, and outweighed diplomatic considerations. According to naval records, Papen and Neurath saw the Reichswehr guidelines on June 13 and approved them.[13] Bülow, however, wrote to Schleicher from Lausanne

[12] AA, [Reichswehr Ministry], "The *Hidden* German Goal . . . ," [14 Jun 32], 7360/E535898-901. After the Reichswehr's program (which may have been drawn up well before) had reportedly been approved by Papen and Neurath on 13 June, Schleicher made minor revisions: ML, Shelf 5939, Notes, 13, 14 Jun 32, reel 17/1094, 1098. Freyberg had suggested on 9 June the desirability of giving civilian delegates more information: ML, Shelf 5939, ltr., Freyberg to Raeder, 9 Jun 32, reel 17/1058. One might ask how far the Foreign Ministry officials guessed the army's real intentions; not very far, to judge by Bülow's attitude in October (see Chapter Five, Section 5).

[13] ML, Shelf 5939, Note, 14 Jun 32, reel 17/1098. In a letter from Neurath to Nadolny of 20 Jun 32, the Foreign Minister stated that the question of demanding specific figures was reserved for later decision (AA, 3154/D667392). This would seem to indicate that, although he had been informed, Neurath did not yet regard the Reichswehr paper as official policy. This letter appears, however, to have been drafted by Bülow: see handwriting and changes on copy on AA, 7360/E535459-70.

on June 16, objecting to the proposal to demand explicit numerical parity with France, and commenting that this idea "has always been turned down," implying the right of the Foreign Ministry to have the last word in such matters. It would be better, Bülow said, to leave the question of the proper relative strengths of the powers to a later disarmament conference, when the German position would be much stronger. Schleicher replied that once lower figures were accepted in the first convention, it would be difficult to get them changed, and that he believed the German position would never be stronger than at the present conference, since "the overwhelming majority of the conference participants have unconditionally recognized Germany's right to equality of rights in national security." Therefore, said Schleicher, himself claiming the status of final arbiter, he was unable to take account of Bülow's hesitations.[14]

2. THE DISAPPOINTMENTS OF LAUSANNE

Although Papen seems to have regarded a reparations settlement as secondary to armament and other questions, the Germans did not after all have much success in discussing nonfinancial problems with the French at Lausanne. The new Chancellor met Herriot, the new Premier, for the first time informally on June 16, and advanced some very far-reaching ideas, including economic reconstruction, a Franco-German alliance against Communism, an Eastern Locarno, and also discussions between the two general staffs. Similarly, at another meeting with Herriot on June 24, Papen again proposed a discussion between the general staffs, limiting the practical application of the principle of equality of rights. So far, Papen tried to present these ideas as though they were German concessions, offered to compensate France for the loss of reparations. Herriot, who was Foreign Minister as well as Premier, at first showed interest in the German Chancellor's proposals, but he became convinced by the end of June that Papen was merely maneuvering, and was not sincere, while MacDonald, when he heard of these talks, strongly disapproved, fearing a Franco-German deal behind Britain's back.[15] Aside from Papen's activities, Bülow

[14] AA, ltr., Bülow to Schleicher, 16 Jun 32, 7360/E535904-06; ltr., Schleicher to Bülow, 23 Jun 32, 7360/E535909-13.

[15] Papen, *Wahrheit*, pp. 198-206; Herriot, *Jadis*, II, 322, 338; *DBFP*, 2/III, Nos. 148, 149, 150; Waites, "Depression Years," pp. 138-139; Thilo Vogelsang, "Papen und das aussenpolitische Erbe Brünings: Die Lausanner Konferenz 1932," in Carsten Peter Claussen, ed., *Neue Perspektiven aus Wirtschaft und Recht* (Berlin: Duncker & Humblot, 1966), pp. 495-500. On Papen's presentation of his ideas as concessions, see Herriot's statement in *DBFP*, 2/III, No. 175 (p. 385). Weygand (in August) was opposed to general-staff talks with the Germans (Castellan, *Réarmement clan-*

spoke with a French diplomat, suggesting direct agreements between France and Germany for establishing the proportionate military strengths of the two countries; according to the French record, Bülow stated that Germany in no way sought actual equality with France, either in men or in arms. No progress was made toward equality of rights, however, and another discussion at Geneva between Nadolny and René Massigli of the French delegation led back to the old impasse over which nation should state its requirements first.[16]

As far as reparations were concerned, the key issue at the Lausanne Conference was whether Germany would make a final payment in recognition of her legal obligation, or simply stand on her claim that she was unable to pay anything. Actually, Brüning had expected to make a final payment, and had discussed this with the British. The German delegation now intended to claim at first an inability to pay, and then fall back on a final payment—hoping to get something in return. Neurath privately indicated to MacDonald on June 22 that a small "recognition payment" might be possible on certain conditions, including a satisfaction of Germany's "modest wishes" in the disarmament question. Papen gave similar indications to MacDonald on June 27. On the next day, at an Anglo-French-German meeting, Neurath clearly defined German conditions for a final payment: a complete cancellation of reparations (i.e., of the reparations system), an international monetary settlement, and "a settlement of disarmament on the basis of equality of rights." An arms agreement was now clearly a German desideratum, rather than a German concession. But the ardent German quest for equality of rights only undermined the German reparations position. Herriot triumphantly seized on the new German proposal as an admission that Germany could pay after all, and he maintained that this destroyed the whole German argument on reparations: "The Germans were in effect saying that they could not pay, but that if they could have satisfaction on disarmament they would pay." For Herriot, there was now no need to offer concessions to Germany, certainly not on disarmament, and the only things remaining to be negotiated were the size of the final payment and some provision for the possibility that the United States might refuse to reduce war debts. On June 29, he told Papen that military questions would have to be discussed at Geneva, and although on Papen's plead-

---

destin, p. 544). MacDonald's 5 July offer of a Franco-British exchange of information on any proposals for revision (see below) evidently stemmed in part from suspicions of Papen's "piecemeal approaches" to Herriot.

[16] *DDF*, 1/I, Nos. 46 (Annexes II, III), 127; AA, Memo by Nadolny, 16 Jun 32, 7360/E535451-55; Memo by Neurath, 20 Jun 32, 3154/D667391-98. Bülow told the French diplomat (André Lefebvre de Laboulaye) that it was not Schleicher but Hammerstein and Blomberg who had advocated military talks to him.

ing Herriot finally said that he would consult his cabinet, he did refuse thereafter to consider disarmament matters in Lausanne.[17]

Toward the end of the Lausanne Conference, on July 5, MacDonald as conference chairman tried to accommodate German and French political desires and to bridge the gap between German financial offers and French financial demands by advancing a new draft agreement. As bait for the Germans, this included a six-power consultative pact proposal (based on an idea of Papen's), an effective striking of the war guilt article (Article 231, the legal rationale for reparations), and a promise to seek a disarmament settlement "beneficial and equitable for all." Privately, he told the German delegates, rather unclearly, that this latter language would mean "release from conditions of force majeure" (meaning presumably from Part V), along the lines he had discussed with Brüning at Bessinge. The Germans were encouraged by this, but they wanted to see their projected equality spelled out more explicitly, and indeed they apparently still hoped for direct negotiations with France. Adopting an idea worked out in their delegation in Geneva, they suggested a draft formula by which the powers would affirm that the principle of equality of rights should find its application in the future disarmament convention, and that they would enter into negotiations without delay in order to work out the practical problems. Along with his private proposals to the Germans, MacDonald also privately offered enticement to the French, in this case reading a declaration that proposed a mutual undertaking by Britain and France, linked to the six-power consultative pact, to consult with each other over any German proposals for revising the peace settlement; aside from promoting a financial compromise, MacDonald evidently wanted to discourage further Papen-Herriot discussions. The idea of an Anglo-French undertaking impressed Herriot favorably, but on the same evening he rejected all other political (i.e., nonreparation) proposals, and he adamantly maintained that disarmament would have to be discussed at Geneva. On July 6, Herriot told the British and Germans that "he could not incorporate in any text at Lausanne a phrase which would suggest that he was recognizing

[17] Brüning, *Memoiren*, pp. 496-497, 530-531; Schäffer Diary, 4 Jan 32; Lutz Graf Schwerin von Krosigk, *Staatsbankrott* (Göttingen: Musterschmidt, 1974), pp. 119, 125, 126; RK, Memo by Pünder, 16 Jan 32, 9242/E650128-30; ltr., Pünder to Carl Bergmann, 20 Jan 32, 9242/E650133-34; Memo by Neurath, 22 Jun 32, 3650/D812558-59; Memo of discussion between Papen and Herriot, 29 Jun 32, 3650/D812659-60; AA, ltr., Bülow to Paul Scheffer, 13 May 32, 4617/E192916-17; *DBFP*, 2/III, Nos. 149, 150, 151; Helbich, *Reparationen*, p. 90; Vogelsang, "Papen," pp. 488-489, 501-502, 504-505. Krosigk's account (also in his *Memoiren* [Stuttgart: Seewald Verlag, 1977]) badly confuses the sequence of events at Lausanne.

the principle of equality of status. . . . He considered that even a reference to an equitable settlement was equivocal."[18]

Papen and Bülow made one further attempt at Lausanne to win an acceptance of equality of rights, at a meeting on the morning of July 7 with Herriot and Joseph Paul-Boncour, the head of the French disarmament delegation and minister of war. Although the Germans had modified their draft formula, removing an explicit statement that equality of rights would be applied in the convention, the new formula reintroduced the same idea indirectly by saying that the powers would enter into contact in order to work out without delay a solution to the practical problems arising from the principle of equality of rights. Herriot and Boncour stated immediately that this new formula was unacceptable, as it attempted to settle in an incidental way a major question which had been raised at Geneva; indeed, Herriot became more and more outspoken in refusing to settle any political questions at Lausanne, and the French record of the meeting suggests that he overawed and virtually routed the German participants. Aside from the termination of reparations, the Germans obtained in the final act of the Lausanne Conference only a general declaration expressing the hope that this agreement would be followed by others in both the economic and political spheres.[19]

The British did offer one slight consolation a few days after Lausanne. At the end of the Conference, Herriot wanted to pursue MacDonald's July 5 suggestion of a Franco-British promise to report any German proposals to each other, and British leaders did not feel able to refuse him an oral assurance; indeed, they wanted to be kept informed themselves. But immediately after returning to London, MacDonald and Sir John Simon moved to attenuate this secret oral commitment to France by proposing a larger, overt declaration, open also to Italy, Belgium, and Germany, under which the adherents would seek, with respect to disarmament, a "beneficial and equitable" solu-

---

[18] *DBFP*, 2/III, Nos. 144 (Annex), 172 (including Annex III), 173, 174, 175 (including Note), 177, 179; Papen, *Wahrheit*, pp. 202-203; ML, Shelf 5940, Diary, 6 Ju[l] 32, reel 18/555; Soulié, p. 366; Waites, "Depression Years," pp. 137-139. Where the German consultative pact proposal had called for the discussion of matters concerning all signatories, the British proposal—probably to discourage bilateral Franco-German agreements—called for consultation on any matter concerning two or more signatories. Herriot rejected the British consultative pact with the comment that it was contrary to League practice and principle, and would exclude interested smaller powers (*DBFP*, 2/III, No. 175 [Note]). Krosigk claims that Herriot was willing to declare Art. 231 void, but not on German demand: *Staatsbankrott*, p. 126; *Memoiren*, p. 145.

[19] *DDF*, 1/I, No. 68, Annex II; AA, Circular dispatch, 15 Jul 32, 7360/E535552-53; University of Birmingham Library, Neville Chamberlain Papers, Lausanne Diary, 7 Jul 32, NC 2/16.

tion. Aside from diluting their association with France, the British apparently meant to give Germany some token consolation on equality and on Papen's consultative pact idea, and to create an instrument for convening great-power meetings. Although Herriot acceded to the British proposal on July 13, and the British-drafted declaration was published that day, the French Premier turned the British maneuver against its authors by inviting France's smaller allies to join in as well; this meant that the pact need not function as a revisionary great-power forum in which France would be isolated. Herriot also insisted on obtaining a British assurance that the term "equitable" was "not intended expressly to recognize any particular political claim," thus avoiding any implicit recognition of the German claim for equality of rights. Despite British intentions to give Germany some satisfaction, the German press regarded this initially Anglo-French accord with suspicion, and officials in Berlin adopted at first an attitude of reserve. We shall examine the significance of this agreement in Anglo-French relations in the next chapter.[20]

The German failure to gain French agreement on equality of rights at Lausanne was embarrassing to Papen, the more so as the attempt to win concessions by agreeing to a final payment had not been kept secret from the German public. Although the rest of the world believed that Germany had gained tremendously by the termination of reparations for a highly contingent promise to pay a capital sum only 20 percent more than the reparations installments paid in the single year of 1929—a capital sum that was never in fact to be paid—the German cabinet was greatly disappointed that Germany had had to agree to a final payment without gaining any political compensation.[21]

3. SCHLEICHER INTERVENES PERSONALLY

Schleicher had a particularly strong desire to settle the equality-of-rights issue and to carry out army reorganization and expansion, and other circumstances combined with the Lausanne results to bring these

[20] *DBFP*, 2/III, Nos. 184, 189, 190; *DDF*, 1/I, Nos. 16 (including Annexes), 30. Documentation on the Anglo-French negotiations is cited in Chapter Five, notes 3-10 and 13-17. The explanations given to the Germans by the French and British were widely divergent: AA, tels., Hoesch to Ministry, 15, 20 Jul 32, 7360/E535739-44 and E.535616-19; disp., Bernstorff to Ministry, 14 Jul 32, 7360/E535737-38; tel., Bernstorff to Ministry, 18 Jul 32, 7360/E535745-46. Of these documents, the 20 July telegram from Hoesch (E535616-19) reported particularly significant statements by Ralph Wigram; see Chapter Five, Section 1, below. On the German reaction to the proposal: AA, tel., Neurath to Bernstorff, 16 Jul 32, 7360/E535739-44; *DDF*, 1/I, Nos. 24, 28.

[21] RK, Cabinet Protocols, 7 and 11 (11:35 a.m.) Jul 32, 3598/D790348-56 and D790360-75.

matters to a climax for him. He had heard from Blomberg in Geneva that Papen was unlikely to get French recognition for equality of rights in Lausanne, and that the adjournment resolution of the Disarmament Conference would only show meager results. The conference would probably reject any proposal made to it for a full recognition of equality of rights, and Blomberg assumed (or professed to assume) that in that case Germany would have to withdraw permanently from the conference. Therefore, Germany should only make such a proposal if she were prepared to withdraw. He suggested, however, that alternatively, if the German government was not ready to leave the conference, it might point out the failure of the proceedings so far, and try to secure a resolution instructing governments to reach agreement on the question of equality of rights before the conference resumed. In any case, the military delegates urgently desired a clarification of Germany's equality of rights before any further negotiation in Geneva. Blomberg argued that as long as equality of rights was not recognized, German military proposals could only consist of unbelievable demands that other nations disarm to German levels; this meant that the military delegates could not reach any practical agreements.[22]

At about this time, probably on July 7 or 8, Schleicher also learned of some Foreign Ministry aspersions on the Reichswehr's "Hidden German Goal" guidelines. Bülow had written to Nadolny, describing these unwelcome guidelines as a sort of "Christmas list" (*Wunschzettel*) of the Reichswehr Ministry's wishes, and had asked Nadolny to take a position on the matter. Nadolny then replied that until his instructions had been changed by a cabinet decision, with his own participation, he could only regard the Reichswehr's paper as representing merely the wishes of one department. But Nadolny included in his letter a reference to Bülow's use of the term "Christmas list," and he gave a copy of the letter to Blomberg. The phrase then made its way to the Bendlerstrasse, presumably from Blomberg.[23]

[22] NA-VGM, F VII a 11, ltr., Blomberg to Schleicher, 6 Jul 32, 7792/E565383-84; ltr., Freyberg to Raeder, 7 Jul 32, 7792/E565378-82; ML, Shelf 5939, tel., Blomberg to Schleicher, [12 Jul 32], reel 17/1202-1203; *DDF*, 1/I, No. 46, Annex II. For Blomberg's recollections of this controversy, see Blomberg MS Memoirs, IV. Dülffer (p. 176) writes that Blomberg telegraphed Schleicher on 6 July that "Papen had to put through 'equality of rights' in Lausanne under all conditions; later, i.e., probably, after acceptance of a German final payment, such a demand would no longer be possible." Untypically for him, Dülffer does not give a source reference for this, and I have not been able to find such a telegram in the NA-VGM or U of M-ML material, where his related sources can be found. Generally, he seems to me to exaggerate Blomberg's intransigence.

[23] AA, ltr., Nadolny to Bülow, 2 Jul 32, 7360/E535921-24; ltr., Bülow to Nadolny, 28 Jun 32, 7360/E535914-15; ltr., Bülow to Schleicher, 6 Jul 32, 7360/E535929-32; Memo by Völckers, 9 Jul 32, 9284/H254268. Nadolny was notoriously jealous of his

The Reichswehr Minister was already troubled, as he showed at a cabinet meeting on July 7, by Herriot's refusal to discuss equality at Lausanne, and by the unwillingness of the French at Geneva to mention equality of rights in the adjournment resolution. Apparently, Schleicher exploded when he heard that the Reichswehr guidelines had been called a mere departmental "Christmas list." His answer was a teletype message to Blomberg, sent on July 8:

> For present cabinet, only the guidelines "The German Goal at the Disarmament Conference" of 14 June, drafted by the Reichswehr Ministry and recognized by the Reich cabinet, are binding. Calling the ambassador [Nadolny] in for the fundamental discussion on these guidelines was expressly proposed by me, but was turned down by Foreign Ministry.
>
> Please inform Ambassador Nadolny of foregoing. Foreign Ministry has been informed. Orientation of Reich Chancellor has begun.

It was true that Bülow had declined a suggestion by Schleicher to bring Nadolny to Berlin for consultation, although that had been before Bülow had known of the scope of Schleicher's demands; incidentally, informing Nadolny of this was well calculated to encourage existing friction between the Foreign Ministry and the delegation leader. But although (according to naval records) Papen and Neurath had approved the guidelines, there had certainly not been any consultation of the full cabinet: Foreign Ministry officials checked and made sure of that fact.[24] Schleicher's message amounted to an assertion that he alone would determine German policy on armament questions, and then as necessary inform the cabinet.

On the same day, in reply to Blomberg's report that an inadequate adjournment resolution was being drafted, Schleicher fired off another message to Geneva:

> A resolution is only acceptable to Germany if it corresponds in every respect to our demands on the basis of the T. of V. [Treaty of Versailles] or in another way brings *full* equality of rights. Every other resolution is to be sharply rejected. Considering importance, final German position requires agreement of cabinet.

---

prerogatives; see AA, ltr., Koepke to Bülow, 9 Apr 32, 4617/E195622. It is hard to tell whether, in passing the "Christmas list" reference on to Blomberg, Nadolny was acting to embarrass Bülow—Nadolny heartily distrusted his diplomatic colleagues—or was merely being clumsy, something he was quite capable of.

[24] RK, Cabinet Protocol, 7 Jul 32, 3598/D790350-51; AA, Copy of teletype, Schleicher to Blomberg, VGH 105, 8 Jul 32, 9284/H254266; ltr., Schleicher to Bülow, 9 Jun 32, 7360/E535897; Memo by Völckers (with marginal note by Bülow), 9 Jul 32,

This amounted to a warning to the Foreign Ministry, the delegation, and also Franz von Papen not to accept a resolution unsatisfactory to the Reichswehr. The Reichswehr Ministry sent copies of both messages, after transmission, to the Foreign Ministry and the Reich Chancellery, in the latter case with a pointed request that Papen's attention be drawn to the record of the cabinet meeting of July 7, held while Papen was in Lausanne, "at which a particularly firm position on the disarmament and equality of rights question at Geneva had been declared to be absolutely necessary, after people had had to find themselves ready for concessions at Lausanne." It was Schleicher who had voiced that sentiment, and at another cabinet meeting, on July 11, he said he would not remain one day longer in office if the cabinet decided to depart in the slightest respect from the demand for equality of rights.[25] Schleicher's second teletype message did not so much reply to Blomberg as express a determination not to be betrayed by Papen and the diplomats.

Bülow was afraid Schleicher would propose to the cabinet that Germany force an immediate decision on equality of rights and leave the conference if her demands were not met.[26] But this does not seem to have been Reichswehr policy, at least at this stage. The military in Berlin and Geneva now thought that an attempt to secure a resolution on equality of rights would only court a public defeat, and that the best policy would be simply to announce that German patience was at an end and that Germany would only cooperate in future on the basis of equal rights. The German government could declare itself ready for negotiation on equality during the conference recess, while indicating that "we must reserve all further steps" if equality was not conceded in these negotiations.[27] Thus Reichswehr officers did not favor an immediate showdown, but did favor a sort of ultimatum, making an implied threat to rearm if equality was not conceded during the recess.

Meanwhile, the German civilian negotiators along Lake Leman

9284/H254268; Note for file, 12 Jul 32, 9284/H254271; Schleicher Papers, N42/91, ltr., Bülow to Schleicher, 10 Jun 32. On Papen's and Neurath's approval of the guidelines, see notes 12 and 13 above.

[25] AA, Copy of teletype, Schleicher to Blomberg, VGH 106, 8 Jul 32, 9284/H254267; Memo by Völckers, 9 Jul 32, 9284/H254269; RK, Cover ltr. from Obstfelder, 8 Jul 32, 3642/D811841; Cabinet Protocols, 7, 11 (11:35 a.m.) Jul 32, 3598/D790350-51 and D790368-71.

[26] AA, Memo, Bülow to Neurath, 12 Jul 32, 4619/E197476-77; tel., Bülow to Nadolny, 12 Jul 32, 9284/H254292.

[27] AA, Memo by Völckers (on call from Obstfelder), 9 Jul 32, 9284/H254269-70; tel., Blomberg to Schleicher, 13 Jul 32, 9284/H254328; ML, Shelf 5939, tel., Blomberg to Schleicher, [12 Jul 32], reel 17/1202.

still doggedly hoped to reach an accommodation with the French. Following the disastrous July 7 meeting with Herriot and Boncour, Neurath, Nadolny, and probably Papen agreed in Lausanne that Germany would have to issue a declaration on her position at adjournment time, but they also decided that, before the adjournment, further efforts should be made in Geneva to get the French to accept some statement on equality. In contrast with Reichswehr views, Nadolny believed the resolution might be amended to embody both German and French objectives in the respective areas of equality and security.[28]

After Papen, Neurath, and Bülow had returned to Berlin, the cabinet convened at Schleicher's request on July 12, mainly to clarify policy on equality of rights—and probably also to make an honest statement out of Schleicher's claim to have cabinet approval for the "Hidden German Goal."[29] The official record of the meeting says only that Neurath announced he had reached agreement with Schleicher, but apparently the record was abbreviated in the interests of secrecy. Other documents indicate or state that: (1) the cabinet probably endorsed the Reichswehr Ministry's "Hidden German Goal," or at least generally accepted the rearmament proposed in it; (2) the cabinet agreed that Germany should vote against any resolution not conceding equality of rights; (3) Neurath proposed to say that a "completely changed situation" would arise if the principle of equality was not settled before the conference reconvened; and (4) the cabinet approved a proposal by Neurath for an immediate further effort to obtain French agreement to discuss equality of rights during the recess. Schleicher seems now to have accepted further efforts at amending the conference resolution as preparing a clear basis for further negotiations between governments during the recess.[30]

One aftermath of this cabinet meeting was a conference on July 14 between Schleicher and Generals von Hammerstein and Adam, at

[28] AA, tel., Neurath to Nadolny, 13 Jul 32, 9284/H254304-05; tel., Nadolny to Ministry, 14 Jul 32, 9284/H254308; tel., Nadolny to Hoesch, 10 Jul 32, 7360/E535513-14; tel., Nadolny to Hoesch, 10 Jul 32, 7360/E535515.

[29] AA, tel., Bülow to Nadolny, 12 Jul 32, 9284/H254292. Just before the meeting, Bülow warned Neurath that other countries would not understand a German withdrawal from the conference over equality of rights, and would certainly not accept such equality before adjournment; neither would they be ready to establish minimum strengths for each army: AA, Memos (2), Bülow to Neurath, 12 Jul 32, 4619/E197476-79.

[30] Official record: RK, Cabinet Protocol, 12 Jul 32, 3598/D790385. On the points apparently discussed: (1) BA-MA, II H 228, TA 562/32, 14 Jul 32; (2) ML, Shelf 5939, Marginal note, 12 Jul [32], on copy of VGH 106, 8 Jul 32, reel 17/1174; AA, tel., Nadolny to Neurath, 13 Jul 32, 9284/H254342; (3) AA, ltr., Schleicher to Neurath, 15 Jul 32, 9284/H254355; (4) AA, tel., Neurath to Nadolny, 13 Jul 32, 9284/H254304; Cover note, Bülow to Neurath, 12 Jul 32, 4619/E197475; Memo by Völckers, 13 Jul 32, 9284/H254302. The 13 July memo by Völckers also indicates Schleicher's attitude.

which the army command's proposals for a new peacetime army were presented to and approved by Schleicher. The recorded result was: "The secret armament goal in the course of the second armament period up until 1938—independently of the results of Geneva—is the 'Minimal Demands for the Disarmament Convention (1st Stage).'" These "Minimal Demands," which were said to have been "demanded by the Reichswehr Minister and resolved on by the cabinet," were presumably those of the "Hidden Goal" guidelines. Schleicher accordingly authorized the first steps in reorganization to be taken on October 1, 1932, with the introduction of 200 additional officer candidates and the shortening of new enlistments, where possible, to three years. On April 1, 1933, army strength was to be increased by 15 percent, bolstering established units and providing for modern specialists, and on that date staff officers hoped also to start training the militia at 50 percent of eventual strength. Militia training was to take place in the ordinary operational units, and not in the special training battalions used hitherto by the army. The regular army and army air force were to number 142,000, and additionally, 50,000 militiamen and 105,500 Grenzschutz men were to be trained each year. Longer-serving personnel (three fourths of whom would now serve three years) would in future be selected from the militia. The three-year men were indeed pivotal to the plan, as they increased the proportion of cadre to militiamen both during the training period and in the event of mobilization, supplied needed specialists, and provided a sort of glue to hold together what would otherwise be a rather disjointed conglomerate.[31]

In material armament, Schleicher had already, as new minister, indicated his sympathy and support for the Billion Program. It was not yet clear, however, where the Reichswehr could get more money.

---

[31] BA-MA, II H 228, TA 562/32, 14 Jul 32; T2 549/32, 15 Jul 32; II H 296, Leiter der Heeresausbildungsabt. 509/32, 28 Jul 32. According to a later naval report, the army in 1932 in fact enlisted more than double their usual number (160) of officer candidates, namely 340: ML, PG 34176, Memo by Saalwächter, 13 Dec [32], reel 27/349. Bredow Papers, Orientation note, 15 Sep 32, indicates that the prospective shortening of service caused problems, as existing Reichswehr personnel feared early discharge. The difference between the figure of 160,000 in the "Hidden Goal" and the 142,000 proposed on July 15 might be partially explained by the inclusion of the Grenzschutz (i.e., its average daily strength) in the former. But the fact is that figures in official strength proposals were frequently changed, and sometimes there even appear to be internal inconsistencies in the same document. Obviously, the planned increase in the number of regulars and regular reserves, through an augmented number of three-year men, would greatly assist in the training and eventual mobilization of the militia. Another later document (ML, Shelf 5939, Memo by Schönheinz, 18 Mar 33, reel 18/83) indicated that if Germany could not have three-year men, as well as short-service militia and career soldiers, she would not be able to have a cadre superior to those of other continental states.

The end of the Lausanne Conference meant that Germany no longer had to make a show of financial conservatism to impress her reparations creditors, but a balanced budget was still a totem at home, and inflation a taboo. A significant work-creation program, such as might include sizable arms expenditure, could only be financed by some form of inflation; one proposal for financing work creation was to issue some sort of parallel currency or scrip. But Papen and his financial advisers preferred what were, or seemed to be, more conservative ways of stimulating the economy, especially that of tax relief for private industry. Some time remained, in any case, until the next budget year began on April 1, 1933.[32]

Other old obstacles to Reichswehr plans were now in the process of being cleared away. After the removal of the ban on the SA, the storm troopers had become more violent than ever, but this did not seem to disturb the Reichswehr leadership, and in fact the Reichswehr was considering the establishment of liaison officers to maintain touch with the Nazi formations. This interest in the SA gained impetus from the impending establishment of an official body for promoting Wehrsport, the Reich Institute for Youth Training (Reichskuratorium für Jugendertüchtigung), under the retired general Edwin von Stülpnagel; a trial training course was launched on July 15. The cabinet was also resolved, with Schleicher one of the moving spirits, to take a firm hand with press criticism, and the ministers decided at the end of June to ask the Prussian government to ban *Vorwärts*, the Social Democratic newspaper, for five days; the paper had published a cartoon suggesting that funds cut from payments to handicapped veterans were being used to provide uniforms for National Socialists.[33]

The Prussian government, once the great impediment for Groener and Schleicher, could now no longer oppose Reich toleration of the Nazi SA, question the preparations for premilitary training and a militia, or afford protection to the liberal and socialist press. The April 24 election had left the Otto Braun/Carl Severing government with only minority support, and it remained in office only because the new

[32] OKW, Wi/IF, 5.3682, Minister Amt 271/32, 18 Jun 32, 436/1300275-76; Wehramt 5783/32, 28 Jun 32, 436/1300247-49; Dülffer, pp. 220-227; Geyer, "Zweite Rüstungsprogramm," 133, 162; Gerhard Schulz, "Die Anfänge des totalitären Massnahmenstaates," in Bracher, Sauer, Schulz, *Die nationalsozialistische Machtergreifung*, pp. 658-660; Dieter Petzina, "Hauptprobleme der deutschen Wirtschaftspolitik 1932/33," *VfZ*, 15 (1967), 19-26. The Papen policy was to issue tax-credit certificates that corporate taxpayers could use in future for tax payments, and meanwhile could borrow on: Petzina, "Hauptprobleme," 22-23. At this stage, army leaders still gave some recognition to the necessity of government economy: see BA-MA, II H 228, TA 562/32, 14 Jul 32; TA 633/32, 30 Aug 32.

[33] Vogelsang, *Reichswehr*, pp. 229-232, 285-287; RK, Cabinet Protocols, 14, 25 Jun 32, 3598/D790228-29 and D790280-81.

Prussian parliament could not produce a majority for any other cabinet. Other Länder, particularly Bavaria, tried to oppose the lifting of the SA ban, but the Prussian cabinet did not feel able to resist the wishes of the new Reich government, and indeed the new Reich Minister of the Interior, Freiherr Wilhelm von Gayl, found the Prussian leaders surprisingly cooperative. They recognized, as did the press, that the Papen government was likely soon to place a Reich-appointed commissar in control of the Prussian state, and in fact Papen and his colleagues had planned from the outset to eliminate the dualism of Prussia and the Reich. The Reich cabinet debated this action on July 11 and 12, and decided that the step should be taken on July 20. Throughout these discussions, Schleicher strongly supported the action, maintaining that military force could be used to overcome any resistance, and opposing a suggestion that the Prussian government be given prior warning in the form of an ultimatum. On the appointed day, with Otto Braun absent on extended sick leave, Severing and two other Prussian ministers were summoned to the Reich Chancellery, ostensibly on other business, and were then informed that they were removed from office. Schleicher apparently thought (correctly) that Albert Grzesinski, at this time chief of police in the Berlin area, might try to organize resistance; General Gerd von Rundstedt, the regional military commander, abruptly informed Grzesinski of his ouster. The Prussian government's resistance was broken without a struggle.[34]

Thus a few days in mid-July witnessed major decisions. The cabinet approved the reconstruction of the German armed forces, and Schleicher and the army command moved to carry out plans to this end regardless of developments in Geneva. The cabinet also agreed to the installation of a Reich commissar in charge of the Prussian state administration, a decision that among other things eliminated for good an old (if now much weakened) obstruction to a German military revival. Further, the ministers acquiesced in the idea of declaring that a "completely changed situation" would arise if the equality-of-rights question was not settled before the Disarmament Conference reconvened. This meant, in effect, that if other govern-

[34] Vogelsang, *Reichswehr*, pp. 211-221, 238-248; Bracher, *Auflösung*, pp. 582-591; Trumpp, pp. 128-143; Ehni, pp. 256-264; *DBFP*, 2/III, No. 120; 2/IV, Nos. 5, 6; RK, Cabinet Protocols, 21, 25 Jun, 11 (4:30 p.m.), 12, 13, 16, 20 Jul 32, 3598/D790251, D790277-78, D790377-82, D790385-88, D790396-97, D790410, D790416-18. A newsman reported to his editors on 15 June that, for the Reichswehr Ministry, the political situation was dominated by the Prussian problem, which would be solved "without taking too much account of the constitution"; the Reichswehr Ministry also complained of continuing resistance within the Reich Ministry of Interior: Brammer Collection, Z Sg 101/25. On Grzesinski: Richard Breitman, "On German Social Democracy and General Schleicher," *CEH*, 9 (1976), 356.

188—General von Schleicher's New Approach

ments did not by that time concede to Germany a degree of rearmament, Germany would probably withdraw from the conference.

4. SHOULD GERMANY OPENLY THREATEN REARMAMENT?

One matter had not been clearly decided, however: whether Germany should publicly state that she would proceed to rearm if equality of rights was not recognized by the time the conference resumed. The original instructions for the delegation, drafted by the Foreign Ministry and approved in January, had foreseen a need to intimate to other countries that the nonfulfillment of German demands and a consequent collapse of the conference would "present us with grave decisions," but these instructions had immediately added: "It must however always remain unclear what conclusions we will draw in case of such a collapse." Germany, the instructions explained, should avoid isolation, and instead isolate France, although to France, too, the readiness for agreement should remain discernible.[35] Schleicher now proposed a different policy.

Following the July 12 cabinet meeting, the Foreign Ministry instructed Ambassador von Hoesch to continue to urge the French to discuss the practical application of equality of rights, leaving to his discretion whether he would mention the threat of a German declaration.[36] The flexible wording of this instruction made Blomberg and Nadolny suspect that the Foreign Ministry was backing away from making a declaration and from insisting on a recognition of equality by the end of the recess.[37] After receiving Blomberg's complaint, Schleicher advised Neurath that, in his view, if the conference resolution did not grant full equality, Germany should declare "very precisely" that she would not participate further in the conference unless diplomatic negotiations could bring about a recognition, at the first session in the next phase of meetings in October, of Germany's full equality of rights. But beyond this, under all circumstances, Nadolny's declaration should leave the foreign delegates in no doubt that, without such an acceptance, "Germany will regard the precondition for

[35] RK, Cabinet Protocol, 15 Jan 32, attachment, 3598/D789068-69. Nadolny expounded the same ideas before the ministers at this meeting (D789053).

[36] AA, tel., Neurath to Hoesch and Nadolny, 12 Jul 32, 9284/H254294-99. Bülow probably drafted this message; see AA, Cover note by Bülow, 12 Jul 32, 4619/E197475.

[37] AA, tel., Nadolny to Neurath, 13 Jul 32, 9284/H254342; Copy of teletype, Blomberg to Schleicher, 13 Jul [32], 9284/H254328-29; tel., Nadolny to Ministry, 14 Jul 32, 9284/H254308; Memo by Völckers, 13 Jul 32, 9284/H254302; ML, Shelf 5940, Diary, 13, 14, 16 Jul 32, reel 18/570-573, 581.

the maintenance of the military provisions of Versailles as unfulfilled, and herself as released from them, that she will leave the Conference and take the measures which in her judgment are necessary for the national security of the country." In other words, Schleicher proposed an open threat to rearm. Indeed, he stated to Neurath: "Our point of view and the conclusions that result from it for us must be presented thus sharply and clearly before the whole world in order that our manifestation in Geneva produce a success in domestic politics along with the diplomatic effect."[38]

Neurath was a wealthy nobleman from Württemberg, who on all occasions exuded an air of calm and self-confidence. That he was really so self-assured is doubtful, in view of his pronounced tendency to adjust his own remarks to please whomever he was addressing. But even Neurath could not let Schleicher's rigid and yet reckless policy prevail unchallenged; to do so would have destroyed Neurath's own position as head of the Foreign Ministry, and might have made Germany a pariah. He sent a pacifying message to Geneva, justifying the language Nadolny and Blomberg had complained of, and he submitted this to Schleicher, along with a cover letter that claimed there was no real difference of view between soldiers and diplomats. But the letter also went on to state:

> I do not consider it possible, if negotiations on the equality of rights question do not in the next weeks lead to a result satisfactory to us, to give immediately just on that basis alone a declaration stating that we withdraw from the Conference, that we consider ourselves no longer bound by Part V of the Treaty of Versailles, and have full freedom in this respect; that is politically not acceptable. For the time being we must at least await the next developments before making final decisions in this direction. Besides, the second phase of the Conference will presumably not begin in October, but only later.[39]

In spite of Schleicher's power within the Papen government, Neurath had the authority of diplomatic experience behind him, he apparently was supported by Papen, and he had the option of taking up the question with President von Hindenburg. Neurath had accepted his post only after an urgent personal appeal from the President, who had already asked him to take the office on earlier occasions, and only

[38] AA, ltr., Schleicher to Neurath, 14 Jul 32, 9284/H254325-27.
[39] RK, tel., Neurath to Nadolny, 15 Jul 32, 3642/D811816-17; AA, Cover ltr., Neurath to Schleicher, 14 Jul 32, 9284/H254330-31. On Neurath generally: John L. Heineman, "Constantin von Neurath as Foreign Minister, 1932-1935" (Ph.D. diss., Cornell University, 1965).

on Hindenburg's assurance of full support.[40] Schleicher now, on July 15, backed down a step—for the moment—replying to Neurath that the minimum to be done was to say, as Neurath had proposed to the cabinet, that a completely changed situation would arise if equality of rights was not settled when the conference reopened, and that in that case Germany would not hesitate to draw the necessary conclusions.[41] But Schleicher would soon reestablish his position in another way.

Notwithstanding Hoesch's efforts in Paris, and also Nadolny's in Geneva, Herriot (as noted in the last chapter) was unresponsive, and he still refused to discuss German equality of rights during the recess. The French Premier maintained that it was illogical to try to say whether the disarmament convention would apply to Germany when it was not yet known what the convention would contain, and whether it could appropriately be substituted for Part V. He told Nadolny that if he granted equality of rights, "this would amount in practice to a rearmament of Germany, and he saw no possibility of expecting such a concession from France." Meanwhile the French, British, and Americans reached agreement on the final text of the adjournment resolution.[42]

From the viewpoints of both the German Foreign Ministry and the Reichswehr, there was now no alternative to making a strong declaration and voting against the resolution. Neurath and Schleicher agreed on a declaration demanding an early recognition of the equality of all states in the matter of national security and with respect to the application of all provisions of the convention. The declaration stated that Germany was ready for immediate negotiations on the application of the principle of equality, but concluded: "The German government must however point out even now that it cannot hold out the prospect of its further cooperation if a satisfactory clarification of this point, which is decisive for Germany, is not achieved by the time the work of the Conference recommences." This definite threat not

---

[40] Papen's interest in negotiating with the French tended to align him with the Foreign Ministry, and indeed, in conversation with Poncet on 13 July, he reportedly omitted (as Neurath and Bülow did not) any threat of diplomatic consequences if Herriot did not agree to discuss equality during the recess (*DDF*, 1/I, No. 14). On Neurath's acceptance of his post: *DBFP*, 2/III, No. 120; 2/IV, No. 113; Heineman, "Neurath," pp. 85-86, 89-91; Bracher, *Auflösung*, p. 533.

[41] AA, ltr., Schleicher to Neurath, 15 Jul 32, 9284/H254355-56. Schleicher's retreat was still very assertive in tone; he conceded that it was not yet possible to judge "in what form we will express the point of view and demand laid down by me."

[42] AA, tel., Hoesch to Ministry, 15 Jul 32, 9284/H254360-63; tel., Nadolny to Ministry, 19 Jul 32, 3154/D667477-78. According to the diary of the naval delegation, Herriot was completely intransigent with Nadolny, said he had lost the confidence he had formerly had in the Germans, and referred to Stresemann's memoirs: ML, Shelf 5940, Diary, 20 Jul 32, reel 18/587.

to return was inserted by Neurath personally, presumably to please Schleicher.[43] Nadolny delivered the declaration at the conference on July 22, and together with the Soviet Union, Germany voted the next day against the adjournment resolution. By means of the declaration, Germany had issued an implied ultimatum.

5. SCHLEICHER OUTFLANKS THE FOREIGN MINISTRY

Bülow had pointed out to Neurath that most other governments were much more interested in disarming France than in the question of equality of rights, and that if Germany demonstratively withdrew from the conference over the adjournment resolution, other countries would not understand this step.[44] Even Germany's threat not to return surprised and rather baffled the other powers, whose attention had been mainly directed elsewhere. The French and British had been largely engrossed with the Lausanne Conference and with subsequent and conflicting efforts to promulgate a consultative declaration, while the American delegates had concentrated on efforts to make the adjournment resolution resemble the Hoover proposal. If we can judge by Massigli's reports, the French delegation in Geneva took a condescending rather than a serious view of Nadolny's activities, and German records suggest that Herriot and especially Paul-Boncour may have seen in the German bid for equality a chance either to bargain vis-à-vis both the Germans and the British for recognition of security, or to make Berlin appear responsible for a failure of the conference.[45] Herriot's adjournment speech conceded that all countries had a right to security, and he and Boncour took the German declaration calmly, privately expressing the hope that there would be negotiations on both security and equality before the conference reconvened. The French

---

[43] AA, Text of declaration, with amendments in Neurath's hand, and cover ltr. to Schleicher, 20 Jul 32, 3154/D667493-95. Blomberg's MS Memoirs suggest that "the diplomats" wanted to vote for the resolution, believing that it took care of equality of rights. This may have been true of some lesser delegation members, but it was untrue of Neurath, Nadolny (see AA, tel., Nadolny to Bülow, 15 Jul 32, 3154/D667173), and even Bülow (see AA, Memo by Bülow, 12 Jul 32, 4619/E197476-77), and probably only reflects Blomberg's warped recollection. Neurath, in a letter to Schleicher of 20 July (AA, 9284/H254393), described the delegation's drafts for the final declaration as "not clear and powerful enough," but the delegation had actually offered both a strong and a moderate draft (AA, tel., Nadolny to Ministry, 20 Jul 32, 3154/D667487-90); Neurath may have been trying to ingratiate himself with Schleicher at Nadolny's expense.

[44] AA, Memo, Bülow to Neurath, 12 Jul 32, 4619/E197477.

[45] *DDF*, 1/I, Nos. 5, 15, 55; AA, tel., Nadolny to Ministry, 10 Jul 32, 7360/E535513-14; tel., Hoesch to Ministry, 12 Jul 32, 9284/H254286-89; tel., Hoesch to Ministry, 15 Jul 32, 9284/H254360-63 (in which Herriot said it might be simplest if Germany did make a declaration against the resolution); tel., Nadolny to Ministry, 20 Jul 32, 3154/D667477-78.

do not seem to have been very conscious, at this point, of the implied German threat to rearm.[46]

The British government, by comparison, was more disturbed when it learned, in the last days before adjournment, that Germany might vote against the resolution; the concern, however, was mainly with appearances. Sir John Simon had temporarily lost interest in equality of rights, and now he wanted to win approval for a resolution which —without further weakening British defenses or involving Britain in any new commitments—would give the British public the impression of some progress toward disarmament. British ministers and officials had tried to spread the idea, and had themselves tried to believe, that the Lausanne Conference had ended the miseries of the postwar era, and that a new age of peace and prosperity was about to commence. At the same time, Simon feared that any amendment in favor of equality would lead the French to raise their demand for security. He made last-minute efforts to dissuade the Germans from their negative vote, giving assurances that Germany would get equality if she did not rush things, and he finally urged the French to consider a supplementary statement to appease the Germans. After the voting, Simon tried to preserve a semblance of harmony by proposing to Nadolny an immediate German adherence to the new Anglo-French consultative declaration of July 13.[47] Actually, the German government had by this time decided on adherence, and it did adhere to the consultative declaration on July 25. It could hardly refuse to associate itself with a general promise to negotiate when it was pressing for negotiation on equality of rights, and in fact, in informing the British government of Germany's accession, Count Albrecht von Bernstorff said that Germany intended to start negotiations on disarmament in the near future.[48]

[46] AA, tels., Nadolny to Ministry, 22, 23 Jul 32, 3154/D667508-12; see also *Le Temps* (Paris), 24 Jul 32.

[47] *DBFP*, 2/III, Nos. 188, 189, 265-269; *DDF*, 1/I, No. 55; AA, tel., Nadolny to Ministry, 20 Jul 32, 9284/H254395-96; tel., Nadolny to Ministry, 21 Jul 32, 3154/D667500; Memo by Neurath, 22 Jul 32, 3154/D667505-06; Memo by Nadolny, 23 Jul 32, 3154/D667517-18; tels., Nadolny to Ministry, 22, 23 Jul 32, 3154/D667508-12. On hopes for a new era: Roskill, *Hankey*, III, 50; AA, tel., Hoesch to Ministry, 20 Jul 32, 7360/E535616-19; FO 371, Memo by Wigram, 7 Jul 32, C6330/5810/18; *Times* (London), 9 Jul (Lausanne story), 13 Jul (ministers' statements) 32.

[48] AA, tel., Neurath to Paris embassy, 20 Jul 32, 7360/E535751-53; tel., Neurath to London embassy, 16 Jul 32, 7360/E535741-44; FO 371, Memo by O'Malley, 25 Jul 32, C6458/5920/62. Neurath smilingly told Ambassador Sackett that there had been no problem about the German accession, as "the disparity between the French and British conceptions of the scope and meaning of the pact, and the unwieldly (sic) development which it had assumed as a result of the adherence of so many small Powers, had, as a practical matter, in his opinion, rendered the pact meaningless" (*FRUS, 1932*, I, 700).

The only government to take immediate and serious notice of the German threat not to return was the American, and that was because it misunderstood the threat: Gibson reported on July 20 that Germany would leave the conference if she did not get recognition of equality of rights in the adjournment resolution, implying that this step would be taken permanently forthwith. Secretary Stimson had a deeply ingrained dislike for German militarism, stemming originally from an incident in his childhood: during a celebration in Berlin of the return of victorious troops from the Franco-Prussian War, he and his pregnant mother had been pushed off the sidewalk on Unter den Linden by a drunken Prussian officer, and his father had consequently abandoned his medical studies in that city, moving the family to Paris. Stimson thought Brüning represented a different Germany, but all his old antipathy was aroused when he heard that the German army was assuming police powers in Prussia, and (from Gibson) that Germany was threatening to leave the conference over the resolution. Misinformed on the terms of Nadolny's equality demand, Stimson concluded (in a larger sense correctly) that the issue was German rearmament. As he recorded, on July 21 he told the German ambassador in Washington, Friedrich Wilhelm von Prittwitz und Gaffron, that this was "very bad work on the part of Nadolny" and "was a reversal of the attitude which Brüning had taken with me in Geneva last April in which he said he would be perfectly satisfied to have the present figures of armament inserted in the new treaty, only making a statement of reservation to the effect that this was done voluntarily by Germany."[49]

General von Schleicher was soon to provide a better foundation for Stimson's criticism, and to make the British and, particularly, the French more conscious of the possibility of German rearmament. He was dissatisfied with the declaration Nadolny had made, and he decided to indicate Reichswehr intentions more clearly. On July 26, he broadcast a radio address arguing that, although some Germans had not recognized the fact, Germany was sadly lacking in that security of which France, the "strongest military power in the world," was

[49] SDP, tel., Gibson to Dept., 20 Jul 32, 500.A15A4/1351; Memo of conversation (on Disarmament Conference), 21 Jul 32, 500.A15A4/1357; Memo of conversation (on action in Prussia), 21 Jul 32, 862.00/2805; DBFP, 2/III, No. 269; Stimson MS Diary, 22, 25 Jul 32; Moffat MS Diary, 22 Jul 32; FRUS, *1932*, I, 309, 315-316; AA, tel., Prittwitz to Ministry, 21 Jul 32, 3154/D667503-04. On Stimson's childhood experience and suspicion of German militarism: Brüning, *Memoiren*, pp. 344-345; ML, Shelf 5939, Memo on Stimson-Groener-Curtius discussion, 10 Aug 31, reel 17/645-649. Cf. Treviranus, p. 241. It was customary in Prussia for civilians to yield the sidewalk to officers: Hans-Ulrich Wehler, *Deutsche Kaiserreich 1871-1918* (Göttingen: Vandenhoeck & Ruprecht, 1973), p. 159.

hypocritically demanding more for herself. "How can Germany obtain security?" Schleicher asked, and then gave his own answer:

> Theoretically, there are two ways:
> 1. The other powers could disarm to Germany's level, as they are legally and morally bound to do. After the course taken by the Disarmament Conference to date, few people in the world believe in such a miracle. . . .
> 2. We can attain this security if we so organize our armed forces—by reorganization, not enlargement [*umbauen, nicht ausbauen*]—that they would give at least a certain degree of security, and I wish, in connection with the German declaration at Geneva, to leave no doubt that we shall take this course if full security and equality of rights are further withheld from us.

Although Schleicher stressed the unwillingness of France to disarm, he also told a story about an expert who, when asked when a battleship was a defensive weapon, replied: "When it flies the British or American flag." Although the broadcast was no doubt aimed in part at the domestic audience, the general had definitely intended foreign governments to take his speech seriously. Still believing Germany had broad support abroad, he suggested to Papen that the German government make clear to France and the other powers its ability to force a resolution on equal rights through the League Assembly in September; he also complained that the Foreign Ministry seemed to have made the French believe that his remarks were merely intended for internal German consumption.[50] Schleicher was circumventing the Foreign Ministry's opposition to open threats of rearmament, and acting to bring the issue to a head.

Schleicher need not have feared that the French were taking his broadcast lightly. Although Ambassador François-Poncet had been ready to believe that the Papen-Schleicher government was well disposed toward France, Poncet now raised the question, to the Quai d'Orsay, of whether Germany wanted to reach a disarmament agreement or instead to make an understanding impossible, so as to have a pretext to rearm freely. The French would have to decide whether they would discuss equality of rights, whether they would agree to conversations during the vacation period, whether they would permit modifications in the treaty, and what they would do if Germany proceeded to flagrantly violate the treaty on a large scale. Poncet himself

[50] John W. Wheeler-Bennett et al., eds., *Documents on International Affairs* (London: Oxford University Press, 1928-1973), 1932, pp. 184-185; RK, ltr., Schleicher to Papen, 2 Aug 32, 3617/D800776-78. Schleicher was also unhappy because "the cabinet" had not been consulted about Neurath's adherence to the consultative pact.

believed it would be better to try to reach a new agreement, rather than to adopt a rigid position.[51]

Herriot, Poncet's chief, was less pragmatic. The Premier and Foreign Minister was angered by Schleicher's allegation that France was hypocritical in demanding more security, and dismayed that the general had overlooked his concession in his last speech at Geneva that all countries had a right to security. When he saw Ambassador von Hoesch on July 28, Herriot argued that if equality of rights was conceded, there must be some compensation for France in the field of security. He complained that Germany had, as usual, immediately followed up one success in revising the Versailles treaty, the elimination of reparations, by demanding another revision, concerning disarmament. Regarding Franco-German relations, Herriot observed darkly that he saw many things coming, and could only conclude that the situation threatened to become more and more difficult. Hoesch gave the reply, not very reassuring, that once reparations and disarmament were settled, the remaining problems, aside from that of the Saar, did not involve a direct opposition between German and French interests. Further grounds for French apprehension were provided by the July 31 German election, in which the Nazis obtained 230 seats in the Reichstag, more than double the number of seats they had won in 1930. On August 1, the French cabinet resolved to maintain French opposition to any German rearmament.[52]

Schleicher's radio speech proved to be no isolated incident. He made another statement, on August 8, to the *New York Times*, saying, "the German people have waited thirteen years for their due and they can wait no longer." In what was probably a calculated indiscretion, Major Erich Marcks, who was in charge of Reichswehr press relations, told a French visitor that the *Umbau* or reorganization of the Reichswehr would certainly be carried out, and that France would undoubtedly "be confronted with a fait accompli"; the change, Marcks said, would involve primarily a reduction in the duration of military service, and he added that many soldiers had indeed already been discharged before the expiration of their term of service. Marcks also said that if an emergency should make it necessary to send troops into the de-

[51] *DDF*, 1/I, Nos. 65, 76, 77. On Poncet's belief in reaching an understanding, see esp. *DDF*, 1/I, Nos. 77 and 125 (Annex), also SDP, disp., Sackett to Dept., 23 Aug 32, encl. (by Kliefoth), 19 Aug 32, 763.72119 Military Clauses/9; tel., Sackett to Dept., 30 Aug 32, 763.72119 Military Clauses/4.

[52] *DDF*, 1/I, No. 72; AA, tel., Hoesch to Ministry, 28 Jul 32, 7360/E535766-69. Possibly referring to this conversation, the French later complained to the British that Hoesch had openly told them, in studiedly offensive terms, that Germany would pursue a series of demands: *DBFP*, 2/IV, No. 90. On general reaction, also *Le Temps* (Paris), 28 Jul 32.

militarized zone, the French "would be informed beforehand"; he did not say French permission would be sought. The French deputy military attaché in Berlin (Commandant de Laforest de Divonne) relayed Marcks's statements to Paris; he also reported hearing from other military attachés in Berlin that the term of service would be shortened, and some of these observers also expected hitherto forbidden arms to be acquired. Herriot expressed to an American diplomat his anxiety over such reports of German rearmament. The efforts of Papen and Schleicher to bring Hitler into the German cabinet (which failed on August 13) and a wave of Nazi murders in East Prussia and Silesia also made alarming impressions. On August 14, a rally took place at Pirmasens, in the Palatinate, at which one of the principal speakers (according to French press reports, denied by Bülow) spoke of lost territories, including Alsace-Lorraine. In a speech in Metz on the same day, Herriot spoke with concern of "the appeals to violence, the panegyrics of war" coming from Germany and Italy.[53]

6. Berlin Plans a Meeting

Along with the increased—in part, deliberately increased—tension, rumors began to spring up in Berlin of talks either pending or already taking place between the French and the Germans. According to one story Colonel Chapouilly, the principal French military attaché, had made a hurried trip to Paris with what were thought to be written German proposals. Another story, circulated by the United Press, stated that the two governments were on the verge of starting negotiations, and that these would begin when Poncet returned to Berlin from France; this last report seems actually to have stemmed from statements by Papen and Neurath to H. V. Kaltenborn of the Columbia Broadcasting Company. It was, of course, the German government that was interested in having the talks, and probably this leak was also meant to ensure that the talks would take place in Berlin. The Reichswehr apparently wanted to have its own representative at the parleys, a participation that would be easier to bring about in Berlin, and General von Blomberg feared a Paris discussion might lead to a return to the conference without a determination by the cabinet (i.e., by Schleicher) that actual equality was assured. The Foreign Ministry itself also wanted to negotiate in Berlin, largely because it believed—

[53] *New York Times*, 8 Aug 32; *DDF*, 1/I, Nos. 100 (Annexes), 125 (Annex); Castellan, *Réarmement clandestin*, pp. 77-79; SDP, tel., Armour to Dept., 13 Aug 32, 751.62/197; Moffat MS Diary, 16 Aug 32; Soulié, pp. 380, 383-384. Marcks, who had worked closely with Schleicher for years, was son and namesake of the historian Erich Marcks; shortly after this he took charge of the Press Section of the Reich government: Vogelsang, *Reichswehr*, pp. 268n, 283.

correctly—that Poncet was particularly interested in reaching an understanding.⁵⁴

To forestall any possible misunderstanding by the French government—and probably also to forestall objections or new antics from the Bendlerstrasse—the Foreign Ministry drafted by mid-August a résumé of the German position to be handed to Poncet and later communicated to other governments. Neurath gave this draft to Schleicher for his review, and after some difficult interdepartmental negotiation, an agreed version was approved by the two ministers on August 27. The résumé began by restating the German demand for equal status. Since Germany had the same right to security as any other state, the initial disarmament convention had to prepare the way for "the necessary adjustment of armament [*Rüstungsausgleich*]." The paper distinguished three aspects of such a convention: (1) its legal form; (2) its duration; and (3) its material content. Germany insisted on being treated in the same way as other states with respect to the first two aspects, which meant that Part V would have to be superseded by the convention. Regarding the third aspect, however, there was room for negotiation. For the period of the first convention, Germany (the paper said) would accept less than the armament status her national security really required, in the expectation that other states would go much further in disarmament in the second convention. Meanwhile, she had to claim the right to possess all categories of weapons not prohibited to all, and had to have the same right as others to make organizational changes, including a shortening of the regular-enlistment term and the training of a special militia "for the purpose of maintaining internal order as well as frontier and coastal protection." Germany would be ready to discuss any concrete French plans that would establish security in a uniform manner for all states.⁵⁵

The Reichswehr showed its determination to rearm in the changes

---

⁵⁴ *Times* (London), 19 Aug 32; SDP, disp., Sackett to Dept., 23 Aug 32, enclosing memo by Wuest, 19 Aug 32, 763.72119 Military Clauses/9; disp., Edge to Dept., 25 Aug 32, 763.72119 Military Clauses/11; *FRUS, 1932*, I, 416-417; Moffat MS Diary, 16, 17 Aug 32; *DDF*, 1/I, No. 115n; *DBFP*, 2/IV, No. 45; OKH, H27/10, ltr., Blomberg to Hammerstein, 19 Aug 32, 324/6315884-86; AA, Cover ltr., Neurath to Schleicher, 15 Aug 32, 7474/H184014-17. On Poncet, see note 51.

⁵⁵ AA, Cover ltr., Neurath to Schleicher, 15 Aug 32, 7474/H184014-16; First Foreign Ministry draft for résumé, 7474/H184018-25; Final text, as given to Poncet and sent to embassies, 29-30 Aug 32, 7474/H184112-18. The intervening frames contain other drafts and memos exchanged between the two ministries on what should be said. Date of final approval: AA, ltr., Frohwein to Schönheinz, 30 Aug 32, 7474/H184148. The British Foreign Office translated *Rüstungsausgleich* as "equalisation of armament" (*DBFP*, 2/IV, No. 62), while the French embassy in Berlin—perhaps out of regard for Herriot's sensitivities—chose the word *ajustement* (*DDF*, 1/I, No. 128). That the French translation was made in the embassy: AA, tel., Neurath to Paris embassy, 29 Aug 32, 3154/D667561.

it dictated in the Foreign Ministry's original draft. A provision was added that the militia would be based on compulsory service, while a promise to keep arms expenditures at the level of recent years was reduced to an observation that account would naturally be taken of the German financial situation. Out of regard for British and American sensibilities, the Foreign Ministry tried unsuccessfully to resist these changes. Also, Schleicher ruled out a proposal for a special agreement with France, which would have dealt with certain points previously covered in Part V but not included in the general convention. The general now maintained that this kind of agreement, which had formed the center of German proposals in June, could no longer be considered, as there was no longer any question of a Franco-German military agreement (such as Papen had proposed); further, such a separate agreement would arouse the suspicion of other governments, and place Germany under French tutelage and inspection. It was also made clear to the Foreign Ministry that Schleicher would not agree to German participation in minor interim conference activities—particularly the meeting of the steering committee of the conference, the Bureau, scheduled for the week beginning September 19—until the negotiations with the French had been satisfactorily concluded.[56]

Schleicher's view that a special military arrangement with France was now out of the question seems to have reflected not only French refusals but also the idea that threats would be more effective than pleas with France. Perhaps he did not want the résumé to be attractive to the French; such episodes as his July 26 speech or the Marcks indiscretions on Umbau plans and the demilitarized zone suggest he may have wanted to provoke a refusal to negotiate. The general apparently believed (as in his suggestion of seeking a League Assembly decision on equal rights) that France was diplomatically isolated, and that Germany's claims were widely supported by other governments; either (he seems to have thought) the French would yield to pressure, or they would refuse discussion and put themselves in the wrong. He had informed Admiral Erich Raeder by August 15 that the question of equal rights had to be settled within the next three months.[57] He was not averse to gambling, and he calculated that if no solution was

---

[56] AA, Memo by Frohwein, 25 Aug 32, 7474/H184029-34; Reichswehr Ministry's "General Observations on the Discussion of the Equality of Rights Question," [received by Foreign Ministry, 23 Aug 32], 7474/H184035-36; RWM 231/32, 26 Aug 32, 7474/H184054. Geyer ("Militär, Rüstung und Aussenpolitik," pp. 241-244) contrasts a readiness of Schleicher to negotiate an arms agreement, esp. with France, with a refusal by "professional military men" like Blomberg, but these positions were not consistently held by either Schleicher or Blomberg.

[57] NA-VGM, F VII a 11, B Nr M 2069/32, 15 Aug 32, 7792/E565410; M 2302/32, 30 Aug 32, 7792/E565415.

reached permitting a German return to the conference, Germany could simply begin rearming without a convention. At this point, Schleicher probably would have preferred to see the conference break down, in the expectation that this would obviate further dickering and delay.

As might be guessed, the general's incautious and aggressive attitude meshed badly with Herriot's psychology, and with the actual position of the French government. It was not that Herriot could not be frightened—he seemed unabashedly fearful—but the effects of his fear were not in the German interest. If he were only moderately pressed, and not made unduly apprehensive, Herriot might be ready for a compromise. In early August he began to develop the idea, which was to become his policy, of accepting equality within a convention that also guaranteed security. He later told an English visitor that he had obtained cabinet approval for an acceptance, in principle, of equality of status, although his ministers remained opposed to German rearmament and hoped Germany would undertake not to rearm.[58] Herriot admitted that the former allies were theoretically obligated to disarm, but only in the context of an organized security system.[59] In terms of practical domestic politics, he would need a tranquil atmosphere for any concession, and he would also need to achieve some apparent counterbalancing success, presumably in the security field, just as he had achieved a nominal success at Lausanne in getting German agreement to a final payment.[60] If, however, the Germans appeared menacing, Herriot would personally take alarm and draw back. He was quite aware that a mere discussion of equality would prejudice later arrangements, and a discussion under threat would be intolerable. At the same time, however, German threats would improve Herriot's diplomatic position by encouraging anti-German and pro-French sentiment in Britain and the United States. Here another matter came into play: neither Schleicher nor the German Foreign Ministry were aware that the consultative declaration of July 13 stemmed from a secret British promise to inform France of any proposals for revision—really, of any danger—emanating from Germany, and that Herriot had made the same promise in return to the British. MacDonald had told Her-

[58] Soulié, p. 383; House of Lords Record Office, London, Viscount Samuel Papers, ltr., Samuel to [Simon], 31 Aug 32, A/88/1; Cab 27/505, DC (M) 32, meeting of 15 Sep 32.
[59] Soulié, p. 383; DDF, 1/I, Nos. 250 (pp. 478, 480), 272 (p. 566).
[60] Herriot had clearly indicated to Hoesch a wish for compensation, including the establishment of an international security system: AA, tel., Hoesch to Ministry, 29 Jul 32, 7360/E535769. Nadolny and Hoesch had also earlier discussed compensation in the area of security with the French: AA, tel., Nadolny to Paris and Ministry, 10 Jul 32, 7360/E535513-14; tel., Hoesch to Ministry, 12 Jul 32, 9284/H254288; tel., Nadolny to Ministry, 23 Jul 32, 3154/D667511-12; Rudolf Nadolny, *Mein Beitrag* (Wiesbaden: Limes Verlag, 1955), pp. 124-125.

riot that the Foreign Office would later inform and warn the Germans of the British promise,[61] but this warning was apparently never delivered. MacDonald's private and oral promise of consultation gave Herriot an excellent reason both to keep the British informed and to avoid private bilateral discussions with the Germans. Precisely because a satisfactory written confirmation of the British promise was lacking, Herriot probably thought it all the more important to confirm the promise by practical usage at the earliest opportunity.

By the end of the third week in August, Herriot, disturbed over Schleicher's statements and over the attaché reports from Berlin, was ready to invoke this Anglo-French agreement. He directed his ambassador to London, Aimé Joseph de Fleuriau, to disclose some of the attaché information to the British (without revealing the source), to refer to the Anglo-French agreement to consult, and to ask for the views of Sir John Simon. On the day that Fleuriau carried out this directive, August 22, Alexis Léger, Director of Political and Commercial Affairs and soon to be the chief permanent official at the Quai d'Orsay, also spoke with the British chargé d'affaires in Paris, Ronald Campbell—in Campbell's words, "in a great state of nervousness and alarm." Léger deplored the Schleicher pronouncements and the unending series of German demands, and said that the only way to meet this situation was for the British and French governments to unite in saying to the Germans, "Thus far and no farther." He later separately told Campbell and Norman Armour of the American embassy that the situation was more disquieting than at any time since the war. Léger probably helped to turn Herriot, if any help was needed, away from the idea of bilateral Franco-German negotiations.[62]

French apprehension and the French tendency to seek British support could only intensify after Bülow informally told Poncet on August 23 that he would soon be called in to discuss equality of rights, involving the replacement of Part V and (according to Poncet's record) the concession to Germany of samples of hitherto forbidden weapons. Time pressed, as the Bureau of the Conference would convene on September 20, and Germany would not participate in the meeting unless she had had a discussion with France (or according to Bülow's record, a decision in principle) on equality of rights. Bülow also said the Reichswehr was planning to establish a militia of 40,000 men, serving for three months. Poncet observed that this was

---

[61] *DBFP*, 2/III, No. 184.

[62] *DBFP*, 2/IV, Nos. 46-51, 54; SDP, disp., Edge to Dept., 30 Aug 32, encl. by Armour, 27 Aug 32, 763.72119 Military Clauses/12. On Léger's views, also AA, tel., Hoesch to Ministry, 27 Aug 32, 3154/D667547-49; *FRUS, 1932*, I, 417-419. Campbell privately described Léger's statements as an "almost hysterical outburst" (FO 371, ltr., Campbell to Sargent, 23 Aug 32, C7197/211/18).

the first he had heard of the project, and that the historical analogies it evoked (i.e., of the Krümper system during the Napoleonic Wars) would arouse active suspicion in France.[63] This was indeed the case. In Paris, General Weygand protested the prospective negotiations. The Quai d'Orsay jumped to the conclusion that four militia contingents would be trained each year, making an annual total of 160,000, or a total in five years of 800,000; the French spread this word to the British and Czech, and later the American, governments. On August 25, Herriot sent word to Poncet, and also to the London and Prague governments, that although France recognized her obligation to disarm, Germany should find her equality of rights in the convention finally concluded in Geneva, and in return for a recognition of the "equality of duties." He would be ready for a conversation after September 1, but France could not consent to increases in German arms in the form of samples of forbidden weapons, nor to increases in German effectives, and if the claim to train 160,000 militiamen a year was maintained, France would have to take that problem "to Geneva"—i.e., to the League of Nations.[64]

Actually, the French misconstrued the danger; they should have remembered Roon as well as the legend of Scharnhorst's Krümper. The Germans would indeed soon be planning to induct almost 100,000 a year instead of 40,000, but more than a third of the inductees were to go on to serve three years or longer. The goal was not simply to form an additional second-string conscript army, but to be able to mobilize a single integrated force.[65] The Germans were not, however, inclined to reveal their actual plans, or to consider modifying their predetermined proposals, while in any case, the French were now committing themselves to a refusal to discuss German material rearmament. The projected Franco-German discussions could therefore hardly come to anything. The German government put its plan into opera-

---

[63] *DDF*, 1/I, Nos. 115, 125 (Annex); AA, Memo by Bülow, 23 Aug 32, 4602/E190810-15. It was important, in diplomatic etiquette, that the implied ultimatum date of 20 September be only implied, not explicit. Fleuriau told Simon, respecting this same deadline, that France would not tolerate an ultimatum: *DBFP*, 2/IV, No. 79. On reference to samples, see Chapter Five, note 47.

[64] Castellan, *Réarmement clandestin*, pp. 543-544; *DDF*, 1/I, Nos. 120 (incl. note), 166; SDP, Memo of conversation between Castle and Henry, 1 Sep 32, 763.72119 Military Clauses/41. The French military attaché in Berlin later reported, as did Poncet, Schleicher's explanation at the August 29 meeting that there would be only one militia contingent each year: Castellan, *Réarmement clandestin*, pp. 80-81; *DDF*, 1/I, No. 154.

[65] Projections in October were for a total annual contingent of 90,000 to 110,000; for the most part there was to be one contingent per year, but in some branches of service two (AA, ltr., Schleicher to Neurath, with atts., 15 Oct 32, 3154/D671232-42). The eventual total was set in November at 85,000 (BA-MA, RH 15/49, TA 737/32, 7 Nov 32, and Annex 6).

tion on August 29, calling Poncet in for a meeting with Neurath and also Schleicher, handing him the résumé, and discussing various aspects of the question with him. Their tactics ran aground, however, on the French position, which was approved by the French cabinet council on September 1, and which received its formal statement in a note of reply of September 11. This position was to refuse bilateral negotiations on what would constitute rearmament, using the arguments that this had nothing to do with the conference, and that it would affect the interests of many nations. Herriot had never much wanted to discuss equality of rights, and now he seemed to have a chance, thanks to the German demands, to turn the issue against Germany.[66]

7. GERMANY APPARENTLY ALONE

There had evidently been little understanding in Berlin of how prejudicial and alarming the proposed negotiation would appear to Herriot, and also of how vulnerable Germany was making herself to accusations of rearmament.[67] The German leadership, so bent on returning at home to the authoritarian Reich of pre-1914 days, might have profited from studying Bismarck's experience in the "war-in-sight" crisis of 1875, when the French leaked out German talk of preventive war and evoked an unlikely collaboration of Russia and Britain in their own support.[68] As it was, German policy reflected, not any careful review of the diplomatic situation, but Schleicher's highhanded methods, his desire to win over the German right, and his interest in early rearmament.

The German government also had not foreseen that its diplomatic initiative would be publicized. Poncet had agreed that the August 29 discussion should be kept secret for as long as possible. But Havas, the French news agency, reported on August 31 the arrival of "a note

[66] *DDF*, 1/I, Nos. 125 (Annex, note), 137 (Annex I), 167, 169; AA, tel., Neurath to Paris embassy, 29 Aug 32, 3154/D667643-50; Soulié, p. 386. The council of ministers (the cabinet plus the President of the Republic) approved the position on 3 Sep 32. Eduard Beneš and François-Poncet later noted that France had astutely turned the tables on Germany (*DDF*, 1/I, Nos. 150, 154), but Beneš warned Herriot not to antagonize Anglo-American opinion by appearing too rigid.

[67] A few officials may have foreseen problems. A Foreign Ministry official noted that the proposal for a special agreement, which the Reichswehr now vetoed, was "the most important point for France": AA, Reichswehr draft for résumé, [ca. 20 Aug 32], 7474/H184072.

[68] William L. Langer, *European Alliances and Alignments, 1871-1890*, 2nd ed. (New York: Knopf, 1950), pp. 44-55; Andreas Hillgruber, "Die Krieg-in-Sicht Krise 1875—Wegscheide der Politik der europäischen Grossmächte in der späten Bismarckzeit," in Ernst Schuler, ed., *Gedenkschrift Martin Göhring: Studien zur europäischen Geschichte* (Wiesbaden: Franz Steiner, 1968), pp. 239-253.

from the German government concerning claims connected with the reorganization of the Reichswehr." Herriot had been visiting the English Channel Islands on the French warship *Minotaure*; a Havas representative in Guernsey learned that an important message was coming, and Herriot's chef de cabinet brought it out by boat to the *Minotaure*, waving it in the air for all to see. According to Herriot's later explanation, he could not deny, surrounded as he was by newsmen, that he had received important news from Germany. Herriot had often made indiscreet statements to the press, the French had a reputation of using press leaks, and German officials now found it hard to believe that this leakage and the subsequent excitement of the French press were really uninspired. Even the description of the résumé as a note annoyed them, because it made the paper appear more like a formal demand. Schleicher, however, also contributed to the uproar with further provocative public statements, reported in the Italian *Il Resto del Carlino* of August 31 and in the *Heimatdienst* of September 1. On September 6, while attending maneuvers in East Prussia, he made his strongest statements yet, reportedly telling a newspaperman, "I can only assure you that Germany will in any case—yes, in any case—carry out such measures as are necessary for her national defense. We cannot submit any longer to being treated as a second-class power." On the same day, Neurath felt compelled to publish the résumé to dispose of what he considered to be French distortions of Germany's intentions, but the résumé itself, when read against the background of Schleicher's statements, exposed the Achilles' heel of the German case, the intent to increase German armed strength.[69]

The initial reaction in London and Washington to the German démarche also came as a surprise to Berlin. As described earlier, the Germans had understood from the April 26 meeting at Bessinge that

[69] AA, Neurath to Paris embassy, 29 Aug 32, 3154/D667561; tels. and memos on press indiscretion, 20 Aug-1 Sep 32, 7474/H184160, H184163-64, H184179, H184197-98, H184202-03, H184205; François-Poncet, p. 52; *DDF*, 1/I, Nos. 137 (incl. Annexes), 149; *Times* (London), 1 Sep 32; *DBFP*, 2/IV, Nos. 71, 75. Eric Phipps wrote of Herriot in 1924: "Not only . . . has he nothing up his sleeve, but he has no sleeve. His whole attitude conveys the impression of . . . a man laying all his cards on the table. He has, of course, the defects of his qualities, for he fails to pick up the cards when the press come into the room" (Schuker, p. 233). In this case, the press chief at the Quai told an American diplomat that France had deliberately leaked the German "aide mémoire" to avoid falling into the trap of Franco-German military talks, and that Poncet's support of Franco-German talks was suspect to French civilian leaders because of his ties with the Comité des Forges (SDP, Ferdinand Mayer Geneva diary, 15 Oct 32, 500.A15A4/1469½). Schleicher later told Bülow that his 6 Sep 32 statement only referred to East Prussia, and that he had not in fact said that Germany would assume her freedom to rearm (AA, ltr., Bülow to Neurath, 24 Sep 32, 3154/D671863-65); Bülow afterward argued accordingly (Schäffer Diary, 18 Oct 32). Thilo Vogelsang has used the term "Achilles heel" to describe the place of rearmament in German diplomacy in this period (*Reichswehr*, p. 297).

the British and Americans agreed to equality of rights in the sense that Part V should be replaced, and that they regarded the German position as "reasonable and justified." MacDonald and Stimson had indeed been generally sympathetic to a transfer of restrictions from Part V to the disarmament convention, but no specific formula on equality of rights had been worked out, and although adjustments in the restrictions on German arms had been discussed, the British and American participants had not understood that the Germans were proposing anything in the nature of rearmament. Yet Germany was now presenting demands to France on the assumption that only the French stood opposed to her Umbau program. Indeed, the German tactics only made sense on the assumption that France was isolated.

When Bülow talked with Poncet on August 23, he explained (as Poncet reported to Paris) that Germany was now addressing herself to France because she had met with no objections from the other great powers when she had described the main elements of her thesis to them. (According to Bülow's own record, he told Poncet that "the other powers concerned have already given their agreement to our demands.") The French then naturally made inquiries in London as to whether Bülow's assertion was correct, all the more since the assertion seemed to refer to some quite recent German soundings with other governments, such as the British should have reported to the French government under the consultative declaration of July 13, and particularly under the secret oral assurances from MacDonald which had preceded that declaration. Simon told Fleuriau on August 26 that the British government had received no specific information such as Bülow had given Poncet; in the spring there had been general and informal talks at Geneva between Brüning, Stimson, MacDonald, and also Tardieu, but "this matter had never taken definite shape."[70]

The British government was anxious to allay French suspicions, and British leaders were disturbed at the German renewal of controversy, threatening economic recovery and the settlement of war debts. Preparations had begun for an international Economic Conference, probably to be held in London, and at this juncture, Simon thought it might convene within a few months.[71] Also, the British distrusted any negotiations between Germany and France from which they were excluded. Hoping in vain to discourage the Germans from pressing their claim at this time, Simon spoke sharply to Count von Bernstorff on

[70] *DDF*, 1/I, Nos. 115, 125 (Annex), 129n; *DBFP*, 2/IV, No. 51; AA, Memo by Bülow, 23 Aug 32, 4602/E190814-15.

[71] See *FRUS, 1932*, I, 808-823; also note the reference to the Economic Conference in the 19 September British statement (*DBFP*, 2/IV, No. 92).

August 29, stating that the exchange of views at Bessinge did not constitute British assent to equality of rights, and warning that negotiations between France and Germany "by themselves" were likely to lead to a clash. Simon's denial, indeed, was stronger than was strictly justified, considering the extent to which the British government, with himself in the lead, had been ready in the spring to recognize German legal equality.[72]

Berlin reacted with pain and genuine surprise. The German Foreign Ministry sent documentary material to London to support its claim that the British and Americans had given their agreement. Bülow insisted to the British ambassador, Sir Horace Rumbold, that MacDonald and Stimson had conceded (*zugestanden*) equality of rights at Bessinge, and Neurath suggested that Simon did not know the facts because he personally had been absent from both Bessinge and Lausanne. The German Foreign Minister did not see why Simon feared a Franco-German crisis: "Considering the moderate character and sound foundations of our demands, that would only be possible if the French take an intransigent position." But the German rebuttal had little effect: Simon and MacDonald had both departed for holidays in Scotland, and the Foreign Office, which lacked any documentation of its own on the Bessinge discussion, preferred to pass on to the problem of what to do next.[73]

Other governments, too, rejected the German claim to have the agreement of all but France. The Italian government apparently denied on September 1 that it had consented to German claims, and although Mussolini, who had recently decided to follow a more militant and revisionist line, did give his backing to the principle of equality of rights the next day, he advised the German government to be moderate in asking for arms, and not to withdraw from the conference. The Belgian Foreign Minister assured the French that his government

---

[72] AA, tel., Bernstorff to Bülow, 29 Aug 32, 3154/D667566-69; Memo by Dieckhoff, 31 Aug 32, 7474/H184158-59; *DBFP*, 2/IV, No. 52; *DDF*, 1/I, No. 129. According to Dieckhoff's 31 August memo (see also Roskill, *Hankey*, III, 58-59), Simon and MacDonald interrupted their vacations to deal with these matters, and also consulted Lord Tyrrell, the British ambassador to France, before talking with Bernstorff. Sir Maurice Hankey recorded in October that he had urged Simon and MacDonald in late July and August to take the initiative and try to dissuade the Germans from raising equality of rights, but Simon and the Foreign Office had failed to act before the Germans did (Roskill, *Hankey*, III, 53-54, 56-59). See next chapter.

[73] AA, tel., Frohwein to London embassy, 31 Aug 32, 7474/H184156-57; Memo by Bülow, 30 Aug 32, 7474/H184145; tel., Neurath to London, 31 Aug 32, 7474/H184127-32; tel., Bernstorff to Ministry, 1 Sep 32, 3154/D667579; *DDF*, 1/I, No. 142. The content of the Bülow record of the April meeting, which was sent to Bernstorff "for your personal information," seems not to have been communicated to the British government; cf. FO 371, Memo by R.M.A. Hankey, 5 Sep 32, C7946/211/18.

had not been consulted.[74] As for the United States, its diplomatic spokesmen, including Secretary Stimson himself, repeated Stimson's affirmation in July that there had been no acceptance of German rearmament at Bessinge. Stimson was concerned for the sanctity of treaties. Furthermore, he had stated on August 8, in a major speech before the Council of Foreign Relations, that a threat of war would inevitably bring about consultation between the United States and other powers. There was a strong belief in the State Department that Germany was putting herself in the wrong by threatening to withdraw from the Disarmament Conference, and Stimson in particular felt a surge of moral indignation. He noted in his diary: "The Germans are getting heady. The old Prussian spirit is coming up, and now we have a new very dangerous sore spot in the world."[75]

IF THE British and American governments denied that they had approved German policy, this was mainly because now, in contrast to the situation in April, the Germans were perceived as seeking a measure of rearmament. This change was much more due to Schleicher's speeches and his influence on policy statements than it was to French appeals for support. As the German Foreign Ministry had recognized at an earlier stage, revelations of the German intent to rearm tended to reduce the isolation of France, and to isolate Germany instead. The plan of getting the French alone in a corner, and then either frightening them into accepting German rearmament or else placing the blame for a collapse of the conference on their shoulders, did not seem to be working out. Foreign condemnation of German policy appeared to reach a climax when the British government published a statement of its views on September 19; this scolded Germany for launching a new controversy so soon after Lausanne, and contested her fundamental claim that a failure of other nations to disarm entitled her to abrogate the disarmament clauses of the Treaty of Versailles. Although the statement also recommended, in rather involved language, that Part V should be replaced at Geneva by a single convention "binding upon all," this implied recognition of German legal equality passed largely

[74] SDP, Memo of conversation between Castle and De Martino, 1 Sep 32, 763.72119 Military Clauses/18; AA, tel., Schubert to Ministry, 2 Sep 32, 3154/D667582-85; *DDF*, 1/I, Nos. 133, 141.

[75] *FRUS, 1932*, I, 417-426, 575-583; SDP, Memo of conversation between Castle and Henry, 1 Sep 32, 763.72119 Military Clauses/41; Memo of conversation between Stimson and Henry, 8 Sep 32, 763.72119 Military Clauses/27; AA, tel., Leitner to Ministry, 31 Aug 32, 3154/D667576-77; tel., Leitner to Ministry, 2 Sep 32, 3154/D667587; *DDF*, 1/I, Nos. 132, 136, 152; *DBFP*, 2/IV, Nos. 60, 66, 74, 88; Moffat MS Diary, 30 Aug, 1, 2 Sep 32; Stimson MS Diary, 7 Sep 32.

unnoticed.⁷⁶ Schleicher had undertaken to direct Germany's disarmament diplomacy, and his direction had been too brusque, exposing too much of his intention to rearm. He had thought that all the major powers except France supported Germany's "right to equality of rights in national security," but they would not, it now appeared, support an open German rearmament.

Schleicher had also hoped his policy in the armament question would make the cabinet popular at home. This plan, too, now seemed to fail. In October, Hitler would denounce the government's "senseless policy of manifestations," which (he charged) had clumsily led to a united front against Germany, and which in particular, through (supposed) naval demands, had antagonized the British, with whom good relations were essential for Germany's future. Many on the left naturally criticized Schleicher's policy. Nevertheless, foreign opposition did spur on the many Germans who felt that somehow, if not by Schleicher's methods, Germany had to regain her former power. Some thought that German policy should be more drastic yet. François-Poncet, in a perceptive dispatch, noted that German press organs close to the Reichswehr were now advocating withdrawal from the League, and he observed that, although this idea had not been pressed hard and had not won general acceptance, it would acquire great force if (as some French sources had suggested) the League, at French instigation, should order an investigation of German armament.⁷⁷

⁷⁶ *DBFP*, 2/IV, No. 92. The statement and its background are discussed more fully in the following chapter.

⁷⁷ See *Völkischer Beobachter*, 21 Oct 32 (Bayernausgabe); *DBFP*, 2/IV, Nos. 97, 101 (encl.), 102, 104, 105, 106, 113; *DDF*, 1/I, No. 205. Hitler asserted in an "open letter," apparently wrongly, that the Papen government had told other powers it wanted a 300,000-man army and large battleships. Poncet also commented, as did Rumbold (*DBFP*, 2/IV, Nos. 96, 113), on the divergence of views between the Wilhelmstrasse and Bendlerstrasse. The *Manchester Guardian*'s correspondent (Frederick Voigt), who had more working-class and leftist contacts, reported in the 5 Sep 32 issue that ordinary Germans had no enthusiasm for militarism, and that some of them were denouncing official German policy.

CHAPTER FIVE

Britain Intervenes to Prevent a Break

1. THE ACTUAL SIGNIFICANCE OF THE CONSULTATIVE DECLARATION

Despite their disapproval of German rearmament, neither the British nor the United States government would in fact side with France in the coming months. The position of the American government can be quickly explained. Americans did not see the European balance of power as their concern. And of all times, the period of an election campaign was the poorest for a policy of European involvement—all the more if that policy seemed to mean supporting France against Germany. There was no "French vote" in the United States, and there was a German vote. Stimson himself never ventured to give the French public support. His backing through diplomatic channels was sharply undercut on September 20, when Hoover, who distrusted the French anyway, and who feared the loss of German-American support, insisted on declaring that the United States took no position respecting Germany's equality of rights, and that that was "solely a European question." The most Stimson could do was to turn the end of Hoover's announcement into an appeal to Germany to return to the conference. By October 20, Stimson had decided that it was best, while preserving close relations with France, to "string along with the German demands," so as to avoid driving Germany over to the side of Japan in the coming League Council debate on the Manchurian problem.[1]

In the case of the British government, the basic policy has been less transparent. Although British leaders were wary after the experience of 1914 of assuming any commitment toward France, and were interested in smoothing over every kind of political or economic disturbance, Simon had ceased during the summer of 1932 to press the case

---

[1] See also below, in Section 3. Hoover statement: *FRUS, 1932*, I, 439-440, 442; Stimson MS Diary, 16, 20 Sep 32; Moffat MS Diary, 20 Sep 32. Stimson indicated to the French chargé on 20 Sep 32 that he agreed with the strictures in the British 19 Sep 32 statement, but could not say so publicly: *DDF*, 1/I, No. 198; Stimson MS Diary, 20 Sep 32. October position: *FRUS, 1932*, I, 467-468; Link, *Stabilisierungspolitik*, pp. 532-533. The Hoover statement caused embarrassment for the State Department, and Bülow took the occasion (according to his own record) to tell the American ambassador in Berlin that "no pretty words or vague promises" from America would bring Germany back to Geneva without a recognition of equality of rights: Moffat MS Diary, 20, 22 Sep 32; *FRUS, 1932*, I, 439-443; RK, Memos by Bülow, 21, 24 Sep 32, 3642/D811718-21, D811715-17.

for German legal equality. Britain and France together had sponsored a consultative declaration appearing on July 13, and the British public statement of September 19 suggested that Britain had now accepted the French case. To the extent that historians and others have dealt at all with the July 13 declaration or the September 19 statement, they have usually seen these pronouncements as evidence of an Anglo-French entente.[2] But far from providing evidence of a pro-French policy, the detailed stories of the July 13 declaration and of the September 19 statement show the full extent and nature of the actual divergence between Britain and France. As we shall see, the British hoped that the July 13 Anglo-French consultative declaration would actually forestall alliances and alignments and promote mediation and timely revision, while the September 19 statement originated in a plan to entice the Germans to the Disarmament Conference with French concessions. Neither the declaration nor the statement reflected a concern for the power balance. Both were encouraged by financial anxieties and, still more, by apprehensions over public opinion. In Britain, internal problems and politics had primacy.

MacDonald had undoubtedly intended, by his various political proposals at Lausanne on July 5, to broaden and further the reparations negotiations. Although from his viewpoint the six-power pact and other elements might serve the long-term goal of reconciliation and revision, his immediate aim was privately to satisfy the Germans with half-promises of support on equality of rights, and privately to entice the French with suggestions that Britain and France make a special agreement to inform each other of any new German proposals for treaty revision. MacDonald read a declaration to Herriot which stated that, as a part of its duties under the six-power agreement, Britain would refuse to reach any conclusions regarding German proposals for treaty modification before discussing the matter with France. The French government was to give a similar undertaking. In his declaration, MacDonald stated: "The object of this understanding would be to protect both governments against the dangers of piecemeal approaches by the German government, the development of which no government can at present foresee." The declaration also suggested close contact at the Disarmament Conference and in preparation for the coming World Economic Conference, and the avoidance of tariff discrimination between the two countries. There was no suggestion to the French that

[2] Views on consultative declaration and 19 Sep 32 statement: *SIA, 1932*, pp. 115, 263-265; Bracher, *Auflösung*, p. 557; Dülffer, p. 179 (citing Wilhelm Deist, *Die Haltung der Westmächte gegenüber Deutschland während der Abrüstungskonferenz 1932/33* [Doctoral diss., University of Freiburg, 1956], pp. 140-141); Papen, *Wahrheit*, pp. 213-214, 231; Northedge, *Troubled Giant*, p. 373; Waites, "Depression Years," pp. 137-139, 141-142.

Britain would make any corresponding declaration to Germany, or to any other power. Behind this overall statement to the French, there was clearly a desire to give the impression of a special Anglo-French understanding, and also a serious wish to forestall any further German-French bilateral negotiations of the kind Papen had been attempting. But MacDonald thought of this as an understanding within a six-power agreement, i.e., within a forum of major powers.[3]

Later that day, July 5, relations between MacDonald and Herriot suddenly fell under severe strain. MacDonald described to the French the German proposals on equality of rights and an end to the war-guilt clause. He recorded in his diary:

> As I proceeded with the report I felt a strange change in the atmosphere. Herriot shrugged shoulders; Germain-Martin [the French Finance Minister] did the same and raised eyebrows & expanded palms as well. When I finished the storm of emotion burst. Honour, buying German agreement to pay 'compensation' by selling French women & children (!) & so on flashed like fireworks & rumbled like heavy artillery. Herriot would never trust Germans & so on, & there was more than a suspicion that they believed that we had negotiated with the Germans behind French backs.

The Belgians told the British afterward that this was indeed the French suspicion, and one Belgian delegate believed that the French would settle for a lower payment if the political-conditions difficulty were eliminated.[4]

Herriot's indignation may have been partly feigned, but to the British Prime Minister, for a moment, the conference seemed hopeless. In the middle of that night, MacDonald sent Herriot an emotionally worded note, asking for a frank personal talk. On the morning of the sixth, MacDonald reportedly told Herriot he was mistaken in thinking that the British delegation agreed with the German conditions. Good relations were restored, apparently on terms favorable to Herriot.[5] At the start of this private meeting, Herriot handed over a writ-

---

[3] *DBFP*, 2/III, No. 172, incl. Annex II. On MacDonald's ideas, see Cab 23/72, meeting of 12 Jul 32, 44(32)6; FO 371, Minute by MacDonald, 19 Jul 32, C6165/5920/62; AA, tel., Hoesch to Ministry, 20 Jul 32, 7360/E535616-18; these latter documents are discussed in the text below.

[4] Marquand, pp. 722-723; Neville Chamberlain Papers, Lausanne diary, 5 Jul 32, NC 2/16; Soulié, p. 366; *DBFP*, 2/III, Nos. 175, 177.

[5] Marquand, p. 723; MDP, 8/1, Pocket diary, 6 Jul [32]; Soulié, p. 366; *DBFP*, 2/III, No. 177 (and note). Soulié says that Herriot and MacDonald agreed, with respect to the Gentlemen's Agreement (see below), that the satisfaction of *all* European creditors was required for ratification; this was a French desideratum. MacDonald's diary says he "straightened things out" with Herriot. After their talk, Mac-

ten refusal to agree to a six-power pact, along with a French translation of the statement on Franco-British consultation, which the British Prime Minister had read to him the morning before; this translation omitted previous references to the six-power pact, and it bore the heading (in French), "Declaration to be made by M. MacDonald."[6] In effect, Herriot now asked MacDonald to repeat the British offer to consult with France regarding German approaches for revision, even though France would not agree to the six-power pact. We have no record of what was actually said between the two men on this. Apparently MacDonald did not immediately commit himself, but considering that he was anxious to repair his relations with Herriot, and in view of his position two days later, it is likely he at least indicated some personal willingness, and promised to discuss the matter with Simon.[7] Simon visited MacDonald that day (July 6), and on the morning of the seventh, Simon told Ralph Wigram (the first secretary of the Paris embassy, temporarily detailed to Geneva) that despite the lack of a six-power consultative pact, he was thinking of reaffirming MacDonald's declaration to Herriot about German proposals for revision.[8] Simon and Herriot also conferred on the subject during the evening of the seventh.[9]

The outcome was that, at the final Anglo-French discussion at Lausanne, on July 8, MacDonald and Simon solemnly repeated the promise to consult with France regarding any German proposals. MacDonald said that, even without the "truce" (the six-power pact) he had wanted, his government would work for "the most amiable understanding between France and Great Britain," and he specifically re-endorsed the assurance he had given on July 5. Simon affirmed that this was "a promise and would be kept," adding that a confirmatory

---

Donald said that Herriot had gone away "quite mollified" and that the Lausanne prospects were much improved (Neville Chamberlain Papers, Lausanne diary, 6 Jun 32, NC 2/16; *FRUS, 1932*, I, 685).

[6] *DBFP*, 2/III, No. 175, Note; FO 800/287, Simon Papers, French text with note by Wigram, [6 Jul 32].

[7] That MacDonald did not immediately commit himself seems to be shown by Herriot's reported query to Simon on the evening of the 7th (*DBFP*, 2/III, No. 184; cf. Soulié, p. 368).

[8] FO 371, Minute by Wigram, 7 Jul 32, C6330/5810/18. Wigram's minute also suggested that a declaration should be made to the Italians too, similar to that proposed toward France, so as to "make it more difficult for them to intrigue with the Germans" over treaty revision. That Simon visited MacDonald: *FRUS, 1932*, I, 271-272.

[9] *DBFP*, 2/III, No. 184. According to Soulié's biography, based on Herriot's papers, Simon that evening accepted Herriot's reservations in connection with the Anglo-French *accord de confiance* (p. 368), meaning presumably the elimination of references to the six-power-pact idea and to an "equitable" disarmament settlement.

dispatch would be sent to Lord Tyrrell (the British ambassador in Paris) for communication to Herriot. The only reservations MacDonald made were that nothing should be said of this immediately to the Germans, and that the parties should not give the impression "that the French and British governments had made in Lausanne a new alliance against Europe."[10]

There is a familiar saying among lawyers, "Hard cases make bad law." Perhaps there is a principle in international politics that commitments given in haste and under pressure make bad diplomacy. In this case, at any rate, MacDonald and Simon rapidly began to qualify their promise to establish a special consultative relation with France.[11] There are, of course, some justifications for their shift of ground. These British leaders had given their promise in a state of exhaustion; MacDonald later wrote in his diary that he had had to close the conference "under powders and dopes." They had been alarmed over Papen's maneuvers, and Herriot seems to have played on their apprehensions. They had wanted another kind of agreement, a consultative pact among several powers, enabling them to mediate between France and Germany. And they evidently feared the reaction in the cabinet and in political circles in Britain to a bilateral Anglo-French understanding. Public coolness in Britain toward France was evident in Parliament on July 11 and 12 in attacks on the "Gentlemen's Agreement" reached at Lausanne. By this undertaking, France, Britain, Italy, and Belgium had made their ratification of the Lausanne agreement conditional upon the settlement of their own war debts to the United States. Parliamentary critics held that this common-front agreement was likely to hurt Britain's own chances of getting a favorable debt settlement from the Americans. On the afternoon of July 12, Lloyd George, referring to the Gentlemen's Agreement, assailed the government with charges of secret arrangements with France, an ominous accusation for any British cabinet of the interwar period to face.[12]

---

[10] *DBFP*, 2/III, No. 184. MacDonald spoke imprecisely of proceeding without "the political truce," but it had been the six-power pact that had been referred to in his original declaration, and which had constituted its presupposition.

[11] They did for a time uphold the promise in the sense of passing on information. As Simon showed in discussions with Fleuriau and Bernstorff at the end of August, he and MacDonald regarded themselves at that time as bound to inform the French of any German proposals.

[12] "Powders and dopes": MDP, 8/1, Diary, 13 Jul [32]; on his exhaustion, also MDP, 8/1, Pocket diary, 6, 9 Jul [32]; Neville Chamberlain Papers, Lausanne diary, 7 Jul 32, NC 2/16. The MacDonald-Simon apprehension as to the cabinet reaction seems evident in their explanations to that body on 12 July; MacDonald stressed that his 5 July statement on informing France had been an "illustration" of the way the six-power pact should work, and Simon stated that an arrangement was now being offered to Herriot which "would not appear as an Anglo-French agreement"; if Herriot wanted to amend this, further consultation would be needed

Even before most of this criticism was made, however, Simon acted to nullify any exclusive bilateral link with France and to turn the arrangement into a revival of the multimember consultative pact. Herriot had been anxious to have something he could allude to in the French parliament—perhaps he also wanted to get the British committed before their ardor cooled. At his urging, Simon had the promised dispatch ready to send to Tyrrell by the early evening of the eleventh, only a few hours after his own return to London from Geneva. This dispatch itself contained a statement of intent to consult, but only in the generalized form of denouncing any attempt by any European power to bring about changes affecting the rights of another power without consulting that power. Simon personally took pains to eliminate from the dispatch all reference to MacDonald's special assurance to France. Moreover, the dispatch was communicated only orally to Herriot. For eventual public consumption, the British also forwarded a draft Anglo-French declaration, which repeated nonspecific pledges to exchange views, keep in close contact on disarmament and economic problems, and avoid tariff discrimination, and added a promise to cooperate to find a "beneficial and equitable" solution to the disarmament question. As far as these points went, they were satisfactory enough to Herriot, except for the reference to an "equitable" solution, which he had consistently opposed. But the British also watered down the Anglo-French character of this declaration by proposing to communicate it also to the Italian, Belgian, and German governments before publication, inviting them to adhere to it. Herriot, who badly needed a public statement to ease the reception of the Lausanne accords in his parliament, was obliged to concur, subject to a clarification that the term "equitable" did not refer to equality of rights, and on the express understanding on his part that the declaration referred to "the relevant proceedings at Lausanne," that is, to the British statements of July 3 and 8. The new declaration was published on July 13, but there was hardly a trace of any special bilateral Anglo-French understanding, and indeed it was somewhat difficult for outsiders to understand why the declaration was being made.[13]

(Cab 23/72, meeting of 12 Jul 32, 44[32]6). On parliamentary and public attitudes, *Times* (London), 12, 13 Jul 32; *FRUS, 1932*, I, 687-691; AA, Memo by Bülow (on phone call from Bernstorff), 12 Jul 32, 9242/E650350.

[13] FO 371, tel., Simon to Tyrrell (no. 123), 11 Jul 32 (8:25 p.m.), C5891/5810/18; *DBFP*, 2/III, No. 189. As noted in Chapter Four, Herriot now obtained an explicit assurance that "equitable" did not constitute a recognition of the German claim; see *DBFP*, 2/III, No. 190. Simon had left Geneva the evening of the tenth (*Times* [London], 11 Jul 32; *DDF*, 1/I, No. 5) and was sending Tyrrell a telegram from London by 2:00 p.m. on the eleventh (FO 371, C5810/5810/18). He probably went by rail overnight to Paris, and then flew to London. On Simon's editing of his dispatch to Tyrrell to eliminate reference to MacDonald's proposal: FO 371, Minute

The need to explain and justify the agreement led MacDonald to produce a minute for the Foreign Office, in which he presented the agreement as a revival of his original proposal for a six-power pact. He had wanted such a pact, he said, to prevent the formation of different camps among the powers, and he went on to say that there could be "no real progress towards the establishment of peace and the creation of the peace mind . . . until France abandons its search for security by means of military alliances or guarantees." France would not cooperate with England, however, unless French security requirements were somehow met. MacDonald thought the need could be satisfied through an agreement to consult on problems at an early stage:

> If, instead of saying to France "We will give you a free hand in working out alone your diplomatic policy and then support you with arms if you get yourselves into a mess," we were to guarantee that, so soon as questions of international import begin to occupy the minds of Foreign Offices, they will be discussed in a friendly and helpful spirit, we thereby give France its required protection and, at the same time, secure ourselves against being made to pay bills which we have never incurred.

In contrast to some tendencies, he said, the pact would mean that "the more responsible Powers will have the initiative"; small powers would be allowed to join, as otherwise they would be hostile, but they should be confined to a minor role. He made only a slight passing reference to the Anglo-French bilateral agreement, describing it as a French counterproposal to the six-power pact, and he suggested that the French had not noticed that the pact proposed on July 13 represented a reversion to the original six-power pact proposal. But the minute also indicated MacDonald's real motives by stating that the French at Lausanne had seemed to want some special recognition to enable them "to swallow the Reparations proposals we were putting up."[14] Thus, the July 5 and July 8 assurances had been given, at least in large part, to further the Lausanne negotiations.

---

by Simon, 11 Aug 32, C6666/5920/62. In reporting Simon's announcement of the agreement, the *Manchester Guardian* (14 Jul 32) said: "Indeed, the House [of Commons] was at a loss to know what it all meant." *Le Figaro* (Paris), hostile to Herriot, commented (14 Jul 32): ". . . quel brouillard, quelle purée de pois! De quoi s'agit-il?"

[14] FO 371, Minute by MacDonald, 19 Jul 32, C6165/5920/62. The division of Europe into two camps was a consistent worry of MacDonald's: MDP, 8/1, Diary, 18 Jan [32], 13 Apr, 17 Nov [33]. Another minute (of 22 July) under this file number, by Orme Sargent, attempted to summarize MacDonald's thoughts. Letting the small powers join had hardly been MacDonald's own original intention; Herriot had forced it on him.

Britain Intervenes to Prevent a Break—215

A multilateral agreement to consult could at least be defended publicly in England, and British officials seem to have genuinely hoped it might prove effective. But a separate bilateral agreement with France was evidently not considered defensible; for MacDonald and Simon, it was something to be swept under the rug. There is definite evidence of such concealment, going beyond the rationalizations in MacDonald's minute. As the consultative declaration of July 13 had led to many questions, a circular dispatch was drafted in the Foreign Office to explain its origins and purpose to British diplomatic missions, the dominions, and perhaps the cabinet. This draft dispatch quoted MacDonald's July 5 declaration to the French, and noted that Herriot had accepted the public declaration of July 13 on the understanding that it referred to MacDonald's secret declaration of July 5 and the assurances of July 8. Simon, however, expressed serious misgivings about this dispatch in a minute. He did not like the circulation which would be given to the July 5 declaration. After first referring to this as a "declaration" and then crossing this word out and substituting "conversation," he argued that (as MacDonald had stated in a cabinet meeting) this had been only an illustration of the method proposed for the six-power pact: ". . . it would be a serious error to instil in the minds of the Dominions, our missions abroad, & members of our Cabinet that it is really after all at bottom ["at bottom" inserted by caret] a Franco-British assurance. In the Cabinet, the P. M. has already explained that it is not so. [The following sentence was squeezed in, evidently as an afterthought:] Italy's resentment would be redoubled." Obviously, Simon was trying to push the Lausanne promises down an Orwellian memory hole. His wishes prevailed, and a different dispatch was sent, an anodyne version of MacDonald's rationale for a six-power pact.[15]

Herriot had no doubt been immensely encouraged by the promises given to him on July 8, and he must have been deeply disappointed by the actual dispatch and draft announcement communicated to him by Tyrrell. We have seen that he insisted on a clarification of the term "equitable," and that he circumvented the British attempt to erect a great-power directorate by encouraging France's small allies

---

[15] FO 371, Draft dispatch, minutes by O'Malley, 4 Aug 32, and by Simon, 11 Aug 32, C6666/5920/62. Herriot's understanding regarding "the relevant proceedings at Lausanne" had been recognized earlier in the Foreign Office as referring to the 5 and 8 July statements: see FO 371, Minutes by Nichols, 28 Jul 32, and by Simon, 4 Aug 32 (which did not question Nichols' interpretation), C6327/5810/18. In his 11 August minute, Simon professed to believe that Herriot was contented with his 13 July dispatch. The circular dispatch finally sent on 23 Aug 32 (C7119/5920/62) was apparently based on Sargent's summary of MacDonald's minute (C6165/5920/62; see note 14 above).

to join as well. But nevertheless, he chose to regard Simon's July 11 dispatch as, fundamentally, a confirmation of the British oral promises, so drafted (as MacDonald had urged on July 8) as not to give outsiders the impression of an "alliance against Europe." With Herriot, the will to believe was apparently at least as strong as Simon's will to forget. The French Premier wanted to believe that the old entente had been renewed, and he wanted to be able to make statements in his cabinet, his parliament, and elsewhere, conveying this idea. Tyrrell reported to London that he had communicated the July 11 dispatch orally, while Herriot "took careful note, point by point, of my oral observations, and read over to me the note which he made, which conformed exactly to your dispatch." Indeed, Herriot seems to have taken down the content of the dispatch word for word, and his notes now became, in his hands, an aide-mémoire, ostensibly from the British embassy, giving him at least something in the form of a written British statement for France alone, something he could show his cabinet. He made as much of the agreement as he could in public statements. And he continued to cling to the earlier oral assurances of special Anglo-French consultation, appealing to them in the months that followed.[16]

Herriot always inclined toward a pro-British policy, and would not willingly give up his belief in the promises which had been made; he did not, of course, know that Simon was pretending these promises had not been given. MacDonald sent Herriot a letter on July 13 with a warning hint that the understanding they had reached was not after all "some alliance or other." Herriot replied with a warm and eloquent message, in which he said, "More than ever, it is necessary to accustom Great Britain and France to work together, not only in their common interest and for their mutual support in the difficulties of the present, but for all peoples." Herriot was reassured when a mutual friend informed him that MacDonald had written her, "I am so anxious to make great friendship with the French." This, Herriot thought, was an answer to those in France who doubted MacDonald's sincerity.[17] MacDonald may have written that he wanted friendship with France, but he did not want another entente, and he wanted to entrust the

[16] *DBFP*, 2/III, No. 190; *DDF*, 1/I, No. 16; Soulié, p. 370. Herriot seems to have taken verbatim notes, as he had also done at Lausanne (Schmidt, *Statist*, p. 243). In statements Herriot later made to the press (*Times* [London], 14 Jul 32), he spoke of the desire of Britain and France to pool their efforts, work, and experiments, and to promote understanding among the peoples of Europe, and "the organization of peace." He told his Chamber of Deputies that the Entente Cordiale had been resurrected (*New York Times*, 14 Jul 32).

[17] *DDF*, 1/I, Nos. 17, 20 (and note); Soulié, pp. 370-371. The statement by MacDonald, sent to Herriot by Princess Marthe Bibesco, and quoted by the *DDF* (from Herriot's papers) in English, seems to have been mangled somewhere en route.

direction of European affairs to a small group of powers, a procedure to which France was opposed. He would revert to this idea in the succeeding months.

Despite assurances of an intent to consult interested parties, a principal aim of a directorate of powers was always to circumvent the objections of lesser states to changes affecting their interests. While Herriot was writing to MacDonald about Anglo-French cooperation "for all peoples," Ralph Wigram, who had drafted the July 13 consultative declaration, was outlining very different ideas to a German diplomat in Paris.[18] One objective of the declaration, Wigram said, had been to show the United States that European governments were trying to settle political problems as well as reparations; although he did not elaborate, the British leaders were indeed afraid that a failure of Europe to disarm might make the United States unwilling to cancel war debts.[19] Another objective (Wigram continued) had been the "pacification" of Europe. MacDonald was far from wanting a special Anglo-French agreement, but he had decided that success could only be achieved if France were "harnessed to the cart" first:

> MacDonald and Simon had come to the conclusion that events would indeed lead readily [*geläufig*] to the revision of untenable provisions of the peace treaty. At one moment there was thought of bringing an explicit reference to Article 19 [the "revision article" of the Covenant] into the text, but this was abandoned because the French would have seen it as an incitement to Germany to seek the revision of certain treaty provisions. But the thought of revision is anything but excluded by the agreement. Wigram repeatedly stressed that MacDonald's basic view was that, just as now with the liquidation of the reparations question at Lausanne, the force of events would in itself gradually lead to revision in other areas. But for that, a calming of the European situation and, above all, the establishment of relations of confidence between France and Germany was necessary. This was particularly true in connection with the Polish question.

Wigram apparently thought Poland was next on the German list, and he commented that the Herriot government might "be inclined to exercise pressure on Poland in favor of a compliant attitude toward Germany." Wigram's observations foreshadowed some of the directions

[18] Wigram's account, which follows here, is from AA, tel., Hoesch to Ministry, 20 Jul 32, 7360/E535616-18. Wigram's minute of 7 July (C6330/5810/18) indicates his involvement in the drafting of the various declarations. See also note 23 below.
[19] See FO 371, tel., Simon to MacDonald, 17 Jul 32; ltr., Simon to MacDonald, 18 Jul 32, W8178/10/90.

218—Britain Intervenes to Prevent a Break

British policy would later take, including the acceptance of Mussolini's four-power pact proposal in 1933 and the exercise of pressure on the Czechs to accept the Munich agreement in 1938.

By means of these remarks, made on July 20, Wigram was evidently doing his best to make the declaration attractive to the Germans, in order to encourage them to adhere to it. Still, Wigram was no doubt accurate as to MacDonald's views; as noted earlier, MacDonald had told Stimson in August 1931 that, after overcoming French fears, the next step was "to get the Versailles Treaty up for discussion." The British believed, of course, that such revision could be part and parcel of a process of reconciliation and elimination of differences. Orme Sargent, the key official in London concerned with the consultative pact, was anxious for Germany to adhere, as he feared the pact would otherwise lead to exactly the division of Europe into two camps which Britain wanted to avoid. Also, Sargent was annoyed when the French—without first consulting Great Britain—extended specific invitations to the smaller European states (including the Irish Free State); he thought the French were trying to use the pact against Germany. He might have been a little crestfallen had he known of the comment to Berlin of Count Bernstorff, who concluded after several discussions with Sargent that the consultative declaration was obviously "a typical product of the unclear thought patterns of MacDonald, which aim merely at the creation of atmosphere."[20]

2. THE ACTUAL BACKGROUND OF THE SEPTEMBER 19 DECLARATION

Nadolny's July 22 declaration and, more clearly, Schleicher's July 26 broadcast soon showed that equality of rights, not the Polish Corridor, was the immediate problem. British leaders would probably have preferred to deal with the Corridor. In the long term, a revision of the Corridor might remove a source of resentment—at least on the part of the German beneficiaries, the more powerful party—while a contention over arms would tend to feed on itself. More immediately, the equality-of-rights issue would presumably set off a sharp and direct Franco-German controversy, reducing the chances that the European

[20] FO 371, tel., Wellesley to Simon, 21 Jul 32, and minute by Sargent, 22 Jul 32, C6212/5920/62; Minute by Sargent, 22 Jul 32, C6165/5920/62; Memo by Sargent, 22 Jul 32, C6254/5920/62; Draft dispatch by O'Malley, Aug 32, C6666/5920/62; DDF, 1/I, No. 75; AA, tel., Bernstorff to Ministry, 18 Jul 32, 7360/E535746. In December 1932, MacDonald wrote: "Disarmament as tried now at Geneva is either futile or a danger unless the Treaty of Versailles is changed" (underlined in pencil, probably by MacDonald): MDP, 8/1, Pocket diary, 27 Dec [32]. In July 1933, MacDonald regarded the four-power pact, as finally concluded, as like his own proposals at Lausanne for a consultative pact: MDP, 8/1, Diary, 19 Jul [33]; 1/6, ltr. to Grandi, 20 Jul 33.

great powers might act in concert, and preventing a post-Lausanne revival of economic confidence. Furthermore, an equality-of-rights controversy would raise problems at home for a British government already exposed to domestic attack over its disarmament policy. Simon himself admitted to MacDonald, "No one pretends that the resolution [the conference recess resolution he had supported] was a great achievement...." On July 23, a leading article in the *Times* faulted the resolution for its failure "to mitigate [the] anomalous inequality of status, which is one of the sources of political unrest in Europe"; on July 28, another *Times* leader praised Schleicher's forthrightness and said that governments had to make good the conference's failure to deal with equality of rights. Liberal opinion denounced the resolution, with Lloyd George saying that the Allies could not protest against German rearmament, since they had failed to disarm. The *Manchester Guardian* concluded that Britain had to choose between German rearmament ("the long road to ruin"), a refusal of German demands (which would intensify grievances), and a beginning of disarmament by powers other than Germany (obviously the right choice). Aside from his concern over existing criticism, MacDonald worried over the intentions of Arthur Henderson, the leader of the Labour party, who was also in a good position, as the President of the Disarmament Conference, to denounce the government's failure to push disarmament.[21]

Officials in Whitehall did believe, however, that the German demand could be dealt with in the spirit (as they interpreted it) of the public consultative declaration, i.e., by direct discussion between a few great powers, with Britain acting as mediator. When the Germans (on July 25) notified the British of their adherence to the consultative declaration, adding that they intended soon to start negotiations with the principal powers concerned with disarmament, the Foreign Office informed the French embassy, just as Herriot would have wished.[22] Sir Maurice Hankey had discussed with Wigram the possible value of the consultative declaration for handling both economic and disarmament questions, and Hankey urged MacDonald and Simon to use this declaration to deal with the equality-of-rights demand; if the Germans could not be dissuaded from pushing this demand at this

[21] FO 800/287, Simon Papers, ltr., Simon to MacDonald, 28 Jul 32; *Times* (London), 23, 28 Jul 32; *News Chronicle* (London), 3, 6 Aug, 5 Sep 32; *Manchester Guardian*, 2 Sep 32; Segel, "Sir John Simon and British Foreign Policy," pp. 175, 265. The *Guardian*'s Berlin correspondent warned that the Reichswehr was militarizing the Verbände for a reserve, aiming at a restoration of the "nation in arms" (issues of 29 Aug, 2, 5, 6 Sep 32). Such material was used editorially (esp. 9 Sep) to urge the need for reduction by the "great armed powers."

[22] FO 371, Minute by O'Malley, 1 Aug 32, C6458/5920/62.

time, Britain should try to show them how to raise the matter without antagonizing France. When the French, in late August, began to express alarm over German rearmament plans, London assumed the French were appealing, or were about to appeal, for British participation in discussions with Germany under the consultative understanding. Foreign Office officials expected that Britain would shortly have to take a position, and they offered various suggestions.[23]

The Foreign Office of this period encouraged staff members to contribute to a collective judgment through minutes and memoranda.[24] Most of the professional staff was sympathetic, by British lights, with France, and the younger members also supported disarmament. Certain officials, particularly Allen Leeper, proposed (on August 26) that the German demands be met so far as possible by an offer of qualitative disarmament toward the German level. (Quantitative problems, including the number of effectives, would be reserved for discussion at the conference.) Qualitatively, Germany would then be permitted to have any weapons allowed to other countries, as long as this did not produce a total increase in effectives, guns, or tonnage. Of Leeper's specific suggestions—banning submarines, limiting the number of aircraft, and reducing the size of artillery, warships, and tanks—most would have been much harder for France to accept than for Britain. But from London, the French government appeared ready to concede some equality of rights, and Leeper thought that France might agree to his program if Britain would stand firmly beside her on it, refusing any further concession and declaring, "Thus far and no farther"— an allusion to Alexis Léger's outburst to Ronald Campbell of August 22.[25]

This program would have furthered disarmament, but to say the least, it would have been hard to negotiate. The French government

[23] Roskill, *Hankey*, III, 51, 53-54, 56-58; FO 371, Minute by Sargent, 24 Aug 32, C7178/211/18; Foreign Office memo, 26 Aug 32, C7359/211/18. There seem to be chronological errors in Hankey's diary account of 3 Oct 32, reproduced by Roskill. Wigram himself wrote Simon on 26 July on the possible usefulness of the consultative agreement, esp. since it had reassured the French (FO 800/287, Simon Papers).

[24] See Robert Peyton Gilpin, in "Christian Names vs. Führerpolitik" (Ph.D. diss., Duke University, 1972), p. 11.

[25] FO 371, Foreign Office memo, 26 Aug 32, C7359/211/18. Leeper and Orme Sargent wrote this memo, and R. C. Craigie and Alexander Cadogan made minor changes: FO 371, Minute and note by A. Leeper, 25, 26 Aug 32, W9391/10/98; also see Minute by Leeper, 24 Aug 32, C7178/211/18. On supposed French readiness in August-September for concessions, see *DBFP*, 2/IV, Nos. 46, 51, 54, 59, 64, 69, also Sir Herbert Samuel's reports on his talks with Herriot in the Channel Islands at the end of August: Samuel Papers, ltr., Samuel to [Simon], 31 Aug 32, A/88/1; FO 371, Note by Simon, 4 Sep 32, C7436/211/18; Cab 27/505, DC (M) 32, meeting of 15 Sep 32. The French record (see *DDF*, 1/I, Nos. 120, 126, 127, 132; Soulié, pp. 383, 386) would indicate that the French never decided to concede equality except at the conclusion of a convention also guaranteeing security.

and the British service departments would have resisted the proposed limitations in arms. And the British cabinet would surely have wanted to know in just what sense Britain was supposed to stand beside France on this program. Moreover, it was not even possible at the moment to convene the cabinet, since most of the important members were scattered on vacation, trying to recover the energies they had squandered so prodigally in Lausanne and Geneva, or in Ottawa, at the conference on Commonwealth trade.[26]

Meanwhile, the days were passing. Simon's interview with Bernstorff on August 29 came much too late to discourage a German broaching of the equality issue. The Germans made their démarche with Poncet on the same day, and the French would no doubt soon produce some reply. Neurath and Bülow told Sir Horace Rumbold they hoped they might gain an acceptance in principle of equality of rights before the Bureau meeting on September 20; in the light of other statements, this implied that Germany would not participate in the Bureau's work without an acceptance. If the Germans did not attend the Bureau, this might, it seemed, foreshadow the failure of the conference. When Simon's officials argued that a policy decision by the cabinet was urgently needed, he noted down his agreement, but he added that "an urgent decision is at this moment easier to demand than to get. Would it not be well to begin by deciding if we can agree to the German points (1) (2) & (3)?"—a reference to the German demand, in their August 29 résumé, that the convention express their equality in its legal form, its duration, and its material content. Thus, in order to forestall a collapse of the conference and the domestic maelstrom that would follow such an event, Simon already inclined toward accepting the German definition of the problem, and indeed toward meeting their demands.[27]

Another Foreign Office official, Owen O'Malley, soon advanced a proposal more in line with Simon's ideas. O'Malley believed that if Germany did not gain sufficient satisfaction to attend the September 20 Bureau meeting, she would soon denounce Part V unilaterally. He gathered from Rumbold's reports, however, that a French concession of equality in principle might suffice for the moment to secure German attendance. To bring the French to make this concession, he thought Britain should threaten to express public sympathy with the German point of view. If this threat did not move the French, then

[26] Leeper himself referred to the problem of the service departments in a minute of 24 Aug 32 (C7178/211/18). On absence of ministers: *DDF*, 1/I, No. 155. Simon and MacDonald did return briefly to London at the end of August: AA, Memo by Dieckhoff, 31 Aug 32, 7474/H184158-59.
[27] *DBFP*, 2/IV, Nos. 52, 53, 55, 56; FO 371, Minutes by Sargent, A. Leeper, Wellesley, Eden, Simon, 1 Sep 32, C7393/211/18.

Britain might state publicly her own independent position. This public statement would say that Britain "viewed with horror" any unilateral breach in Part V, and favored general disarmament rather than German rearmament, but it would also accept the replacement of Part V by a general convention which, while still weighted against Germany, "would give her more elasticity within existing maxima than she now enjoys."[28] Where the Leeper program sought to carry out real disarmament downward, inducing France to join in by offering her assurances, O'Malley concentrated on the immediate aim of bringing Germany back to the conference, using the stick of threatened nonsupport on France, and offering the carrot of "more elasticity" to Germany. In O'Malley's view, getting Germany back was desperately urgent; in a note for MacDonald, he wrote: "It is generally agreed that, if nothing is done and if the Bureau meets on September 20 with the Germans sulking in their tents, the Disarmament Conference is doomed and a dangerous situation will arise deeply prejudicial to the economic restoration of Europe, on which we would wish to see all thoughts concentrated."[29]

Events were still moving too rapidly for London. On the same day that O'Malley made his proposal, September 6, the French—in keeping with Herriot's desire for the fullest kind of consultation with, and involvement of, MacDonald—asked for British comments on a statement of their own position. The French position was that while they were ready to discuss disarmament, the German demands, amounting to rearmament, should be treated separately from the Disarmament Conference; also, the French held that there was no legal foundation for a claim to equality of rights in armament, including a replacement of Part V. By this time, crisis pressures were affecting judgment in the Foreign Office, and officials there jumped to the conclusion that the French were proposing action through the League Council and the International Court at the Hague, and threatening not to ratify the Lausanne agreements. O'Malley described the legalistic French statement of views as "a hopeless document." He, Leeper, and E. H.

---

[28] FO 371, Minute by O'Malley, 6 Sep 32, C7465/211/18. O'Malley's idea of the German Foreign Ministry position was based on *DBFP*, 2/IV, Nos. 55, 56, 67. O'Malley admitted it was doubtful that his proposed public statement would, if made, bring the Germans back, but he evidently thought that anything which might be worth trying. O'Malley was considered something of a maverick in the Foreign Office, largely because he was less pro-French than most officials; see Thomas H. Keene, "The Foreign Office and the Making of British Foreign Policy, 1929-1935" (Ph.D. diss., Emory University, 1974), p. 398n.

[29] Note for MacDonald by O'Malley: FO 371, 7 Sep 32, C7529/211/18. Departmental critics pointed out that Germany might interpret "elasticity" in a generous way; FO 371, Minutes by E. H. Carr and Leeper, 6 Sep 32, C7465/211/18; also memo by Cadogan, 12 Sep 32, C7728/211/18.

Carr agreed that the Bureau meeting should be postponed for a month, and that, in the interim, four-cornered discussions should be held with the French, Germans, and Italians.[30]

Meanwhile, the French, not receiving any replies from London, seemed to assume that they now had full British support, or perhaps that they could maneuver the British into giving it. On September 8, Herriot sent to London a draft of his note of reply to Germany, having earlier explained that he would send the note to Berlin as soon as he knew it had British approval. The note itself stated, very diplomatically, the same general position communicated to the British on September 6. Soon thereafter, word leaked to the press in Paris that the note had been submitted to London. This compelled Simon to issue a denial of any British responsibility for the French note.[31]

While Herriot was attempting to practice his kind of consultation, MacDonald and Simon tried to invoke their kind of consultation. In MacDonald's view, a four-power meeting (as suggested by the Foreign Office) would fulfill the purpose of the July consultative pact. At such a meeting, the Germans could be pressed to act more carefully and the French could be urged to face the changed conditions since Versailles; the "position at Bessing[e]" could provide a point of departure. He and Simon now discreetly and tentatively urged Herriot not to send his note, but instead to propose a postponement of the Bureau meeting and a four-power discussion in the interim. But although the idea of four- (or five-) power discussion was later to determine the outcome of the crisis, on this offering it did not succeed. The French Premier declined to delay transmitting his note to the German government, supposedly on the ground that the preparation of the note had already become public knowledge, and the note was in fact delivered in Berlin on September 11. Herriot also sent word to London that he opposed any postponement of the Bureau meeting, saying that the whole conference apparatus should continue its scheduled work without the Germans. Privately, he told the Czech minister in

[30] *DDF*, 1/I, No. 146 (Annexes); *DBFP*, 2/IV, No. 76 (encl.); FO 371, Memo by Carr, 5 Sep 32, C7576/211/18; Minute by Leeper, 6 Sep 32, C7465/211/18; Joint minute by O'Malley and Leeper, [6 Sep 32], and minutes by O'Malley and Wellesley, 6 Sep 32, C7508/211/18; Note for MacDonald by O'Malley, 7 Sep 32, C7529/211/18. The officials were almost as frustrated with the impossibility of getting decisions from their own government as with the French; both Simon and MacDonald were apparently away for the first week in September, MacDonald returning late on the fifth and Simon on the seventh (*News Chronicle* [London], 6 Sep 32; *Daily Telegraph* [London], 9 Sep 32). Originally, Simon had intended to be away for the first half of September (*DDF*, 1/I, No. 135; AA, tel., Bernstorff to Ministry, 29 Aug 32, 3154/D667569).

[31] *DBFP*, 2/IV, Nos. 69, 79, 80; *DDF*, 1/I, Nos. 160, 169. Simon and also MacDonald (MDP, 8/1, Diary, 10 Sep [32]) had understood that Herriot was not asking for approval or comment.

Paris that he was firmly against the British idea of a four-power conference, in which Germany and Italy would oppose France, while Britain would play the referee; he was determined to follow a Geneva policy in Geneva. But Herriot apparently felt unable to express explicit opposition to the British themselves, and Ambassador de Fleuriau in London left Simon with the impression that such a conference might be possible at a later date.[32]

By sending their reply on to Germany, the French confirmed they would not make any immediate concession, such as might have induced the Germans to attend the Bureau meeting. But Simon had become (as he had reportedly put it) "most anxious there should be no impression of an Anglo-French front to Germany," and he still hoped for negotiations on equality of rights.[33] Apparently to establish Britain's independent position and her readiness to mediate, he directed O'Malley to draft a public statement for the British government. O'Malley used the ideas he had advanced on September 6, and completed a first version by the twelfth; this was eventually to become the September 19 statement. But the statement could now hardly induce German attendance at the Bureau, and on the thirteenth Simon learned that the Germans would in fact not appear there.[34] Somehow, the world continued to revolve on its axis.

Although the proposed British statement might still eventually lead the French, the Germans, and also the Italians and Americans to negotiate on equality of rights, the immediate aim now became even more to show the world and the British public that the British government had a definite policy and was not responsible for the diffi-

[32] MDP, 1/504, Note by MacDonald, [ca. 9 Apr 32]; 8/1, Diary, 9 Sep [32]; *DBFP*, 2/IV, Nos. 80, 82, 83, 84n, 86, 91; *DDF*, 1/I, Nos. 158, 160, 161, 163, 167, 170, 174; National Archives, Microfilm T-120, German-captured Czech documents (hereafter referred to as Czech docs.), tel., Osusky (Paris) to Beneš, 9 Sep 32, 2376/D497256. The German résumé of 29 Aug 32 had appeared in the press of 7 September; the French reply appeared on 14 Sep 32. Herriot hinted at his opposition to a four-power meeting with a remark about "collective pressure on France" (*DDF*, 1/I, No. 163), but Fleuriau weakened this hint in the version of Herriot's position he gave to the British (*DBFP*, 2/IV, No. 86). Probably Herriot thought that open opposition might antagonize the British, and that it was unnecessary, as he could evade their proposal without it. He followed similar tactics in early October; see below.

[33] SDP, tel., Mellon to Dept., 10 Sep 32, 763.72119 Military Clauses/30.

[34] *DBFP*, 2/IV, Nos. 83, 87; SDP, tel., Mellon to Dept., 10 Sep 32, 763.72119 Military Clauses/30; FO 371, Minutes by Cadogan, 12 Sep 32, and by O'Malley, 14 Sep 32, and undated draft memo by O'Malley, C7728/211/18; AA, tel., Bernstorff to Ministry, 13 Sep 32, 3154/D667663. Since the draft memo was completed by Monday, 12 Sep 32, Simon had quite likely ordered its preparation on Saturday. If this is correct, Simon gave the instructions to O'Malley under the immediate impact of the French decision to send their reply to Berlin, and well before the German refusal to attend the Bureau.

culties in Geneva. In explaining the statement to the ministerial committee on disarmament, Simon said "it seemed to him extremely possible that the Disarmament Conference might fail." He argued that if the British government "made no expression of opinion at all, we might easily find ourselves in a very equivocal position," whereas "if, after . . . this statement, Germany still refused to return to Geneva, then the blame would inevitably rest with her."[35] Simon's greatest concern was now evidently the growing criticism at home of inaction, and of failure to meet Germany's just claim to equal status by carrying out Allied (especially French) disarmament. The tone of public discussion at this time is illustrated by the *News Chronicle*'s argument (on September 9) that, despite Schleicher's tactlessness, "the [German] demand is logically irresistible: what is asked, as has been pointed out again and again, is not equality of armaments but equality of status. Our government cannot possibly deny the justice of this claim. What is it doing—what is it even proposing to do—to further it?"[36]

Aside from meeting domestic criticism, Simon intended that the government's statement offer Germany some satisfaction. He told the disarmament committee that the Germans might be gratified by it, though he did not expect "that France would very much like the substance of [the] draft." But modifications obscured that substance and made the statement more abrupt; they eliminated, among other things, an explicit bid to Germany to return to the conference.[37] Even O'Malley himself apparently had little real appreciation of German sensitivities, while he believed that a "bedside manner" was desirable in dealing with the French.[38] And British leaders were very unhappy over the new German clamor for equality of rights, which endangered the prospects for economic stability and carried the threat of an arms

---

[35] Cab 24/233, CP 305 (32), 15 Sep 32; Cab 27/505, DC (M) 32, meeting of 15 Sep 32.

[36] See *Manchester Guardian*, 2, 9 Sep 32; *News Chronicle* (London), 5, 9, 13, 15 Sep 32.

[37] Originally, the statement proposed a "restricted and partial reorganization by way of transition," such as "would meet the immediate requirements of Germany on the one hand, and on the other fall short of threatening in any degree whatever the security of other Powers." This was dropped, probably because Britain was not ready to state in specific terms what this happy solution might be. Sir Herbert Samuel and J. H. Thomas suggested the deletion of the explicit invitation to return, on the grounds that the invitation would almost certainly be refused, and thus invite a snub, which Thomas thought the French would then seek to exploit. Simon had the impression—apparently as a legal deduction—that the Germans did not disagree with his legal argument that Part V was unaffected by an Allied failure to disarm. See FO 371, Draft text, and minute by Cadogan, 12 Sep 32, C7728/211/18; Cab 27/505, DC (M) 32, meeting of 15 Sep 32.

[38] See Cab 27/505, DC (M) 32, meeting of 15 Sep 32; FO 371, Memo by O'Malley, 24 Sep 32, C8251/211/18.

race, and they were inclined to read the Germans a lesson. MacDonald wrote Lord Cecil on September 13 that while France had passed up opportunities to reach an adjustment, Germany had handled her widely accepted case in such a way as to arouse the maximum of fear in Europe, appearing to return to pre-1914 militarism; he thought the outlook deplorable. In a letter to Lord Tyrrell, Simon took the view that "in the last resort the only way of restraining her [Germany] from excesses is to show that the solid and informed opinion of the world is against her *methods*." Nevertheless, Simon also maintained to Tyrrell—and in a paper for the cabinet—that France and "all of us" had to choose "between an agreement which Germany (grudgingly) accepts for the next five or ten years and a disregard by Germany of agreements altogether." He ruled out as impractical any third alternative, "such as that Germany should be held to the Treaty and 'not allowed any variation of it.' "³⁹

3. THE SEPTEMBER 19 STATEMENT MISFIRES

The final memorandum or statement of views, published on September 19, did not read like an invitation to the Germans to negotiate. It did propose that the arms clauses of the treaty, unless modified by agreement, should reappear in the convention, which would then become "the effective obligation binding upon all." This meant that Germany would be included within the convention, and it implied that she would be free of any restriction when the convention expired, excepting such limits as she might then choose to accept; Britain was actually conceding equality in two of the three senses demanded in the German résumé of August 29, those of legal form and duration. Indeed, the British memorandum also indicated in an incidental way the possibility of an "agreed adjustment" in the convention itself, a step toward meeting the third German demand. But the tone of the memorandum made the British government appear to side strongly with France against Germany. One characteristic statement ran: "In view of Germany's economic difficulties, the initiation of acute controversy in the political field at this moment must be accounted unwise. And, in view of the concessions so recently granted to Germany by her creditors, it must be accounted particularly untimely." Although Germany had been disarmed with the declared object of making general disarmament possible, this did not entitle her to abro-

---

³⁹ Cecil Papers, 817A/51081, ltr., MacDonald to Cecil, 13 Sep 32; FO 371, ltr., Simon to Tyrrell, 15 Sep 32, C7796/211/18; Cab 24/233, CP 305 (32), 15 Sep 32. See also MDP, 8/1, Diary, 9 Sep, 11, 16 Oct [32]; 1/504, ltr., MacDonald to Cecil, 15 Sep 32.

gate Part V: "To state what the object or aim of a stipulation is is a very different thing from making the successful fulfilment of that object the condition of the stipulation. . . . The correct position under the Treaty of Versailles is that Part V is still binding and can only cease to be binding by agreement." An agreed settlement could not "be attained by peremptory challenge or by withdrawal from deliberations which are about to be resumed," but only "by patient discussion through the medium of conference between the States concerned."[40]

The German Foreign Ministry did not entirely overlook the concessions made on equality of rights, but officials in the Wilhelmstrasse felt that the memorandum represented a regrettable retreat from what MacDonald's earlier statements and British press comment had led them to expect, and they were especially annoyed at the lecturing tone of the document. From the perspective of the Berlin government, there was no evident necessity for Britain to take any position, let alone to scold like a governess. Both Bülow and Schleicher commented that the tenor of the memorandum illustrated in itself the absence of a German equality of rights, and Bülow told Rumbold that Germany had not received this sort of communication since the time of Poincaré and the invasion of the Ruhr. It seemed to the German government very unjust to be blamed for suddenly raising a new issue and then for not negotiating when they had been trying unsuccessfully for some time to launch negotiations on equality of rights. Official anger was doubtless increased by the fact that the British memorandum appeared immediately in the press and came as a total surprise to the German public, which had heard little of Simon's August 29 warning, and which had been led to believe that the British stood on the German side. Rumbold wrote to his son: "There was an explosion of resentment—especially in official circles—and I am told that the great Schleicher foamed at the mouth."[41]

The British memorandum probably helped bring home to the German government the failure of its initial tactics; it seems to have discouraged Schleicher's tendency to defy, even provoke, French opposition, and the experience may have made Germany more ready in December to accept an agreement without explicit provision for practical adjustments. But contrary to O'Malley's original hope, the paper

---

[40] *DBFP*, 2/IV, No. 92. Like the 29 August German résumé, this paper was not strictly speaking a "note"; it was not formally communicated to Germany. It has been variously dated; I use here the date of publication.

[41] RK, Circular tel., 19 Sep 32, 3642/D811650-52; Memo by Bülow, 21 Sep 32, 3642/D811730-32; Memo by Bülow, 22 Sep 32, 3642/D811733-34; Cabinet Protocol, 19 Sep 32, 3598/D790778-79; AA, Extract from memo by Bülow, 20 Sep 32, 3154/D667712-13; *DBFP*, 2/IV, Nos. 97, 104, 105, 106, 109, 110, 113; *DDF*, 1/I, No. 205; Rumbold Papers, ltr., Rumbold to Anthony Rumbold, 25 Sep 32.

did not of course produce any immediate change in the German position with respect to the Bureau and the conference. Governments never openly bow their heads to foreign criticism and say they made a mistake, and utterances of this kind scarcely provide an overture to friendly discussion. Before the memorandum appeared, however, Bülow and Neurath had already begun very discreetly to indicate that they hoped for negotiations at the coming meetings of the League Council and Assembly, beginning on September 23.[42] They believed in quiet, unpublicized conversations, even if Schleicher, and now the British and French too, seemed to prefer public speeches, press leaks, and declarations. And the very sharpness of the British memorandum was to turn to the German advantage, because the reaction made Simon feel he had gone too far.

The memorandum was well received in some quarters, it is true. As Lord Tyrrell brought the paper into Herriot's office, the ambassador exclaimed: "Good news!" Herriot accepted the memorandum in this spirit, and he did not perceive that the British now proposed Germany should be as free as any other state at the end of the first convention. French officials, who had been very unhappy with the British, now generally took the statement as a gain for France, and the permanent chief at the Quai d'Orsay, Philippe Berthelot, ascribed it in large part to American influence. In Washington, Stimson commented to the British chargé d'affaires that the paper was obviously "the work of an able lawyer," meaning Simon. He only wondered if it was not a little too diplomatic "to make an impression on German psychology"; he had found "that stern, blunt methods were effective in bringing the Germans to terms and in extinguishing excessive pretensions when more conciliatory methods only prompted further demands. It was a case of the yellow streak in the composition of the bully." But Stimson's bold talk rang hollow when Hoover issued his statement that Germany's equality of rights was "solely a European question."[43]

To the British cabinet, the most important reaction to Simon's statement of views was that in England itself, and this was tepid at

[42] AA, tel., Bernstorff to Neurath, 13 Sep 32, 3154/D667663; tel., Hoesch to Ministry, 17 Sep 32, 3154/D667693-94; *DDF*, 1/I, No. 189. Neurath's own report of his meeting with Poncet omits the mention Poncet gives (in last reference) of discussion of a possible meeting in Geneva: AA, Memo by Neurath, 16 Sep 32, 3154/D667687-88.

[43] Soulié, pp. 388-389; *DDF*, 1/I, Nos. 193, 198, 203, 205, 248; 1/II, No. 4; *DBFP*, 2/IV, Nos. 98, 100, 156; Czech docs., tels., Osusky (Paris) to Foreign Ministry, 14, 19 Sep 32, 2376/D497261, D497263; Stimson MS Diary, 18, 20 Sep 32; Moffat MS Diary, 20, 22 Sep 32; *FRUS, 1932*, I, 433, 439-440, 442. On Herriot's misunderstanding, see note 72 below. *Le Temps* (Paris), 20 Sep 32, likewise inferred that, according to the British statement, Part V remained in full force, with or without a convention, except as modified by agreement, and it concluded that a convention would mean a free German confirmation of the Versailles clauses.

best. The American embassy noted that the statement received "a curiously apathetic reception in the London press." Some conservative and moderate organs supported the statement, but the Liberal press —the *Manchester Guardian, News Chronicle,* and *New Statesman*— and also the *Spectator* thought that it did not do justice to the German claim to equality of rights. The *Times* not only regarded that claim with sympathy, but unlike the liberal papers, it had also implied that Germany might well be allowed some increase in arms. On September 19 the *Times* commented (correctly) that the statement had apparently been written before the German refusal to attend the Bureau meeting, and it observed that the chances for agreement were now reduced. On September 22 another leading article was somewhat kinder as to the statement, but regarded it as only the first step, and called on the former Allies to give favorable consideration to German demands. Five days later the *Times* was regretting that Herriot had made no advance toward the German viewpoint, and it argued that the German determination to have a militia—"for it appears to be a determination, and who could prevent its execution?"—was probably part of the quest for equality of status. Contrary to an impression Bernstorff had, the professionals of the Foreign Office did not surreptitiously inspire the *Times* to take a more sympathetic view of German claims than Simon had taken; on the contrary, they were trying without success to instill a little more skepticism on Germany into that paper.[44]

Aside from newspaper criticism, MacDonald also had to deal with letters and prepare for deputations from the League of Nations Union and from a Church of England group led by the Dean of Chichester; both groups wanted the government to commit itself to a policy of drastic disarmament. Privately, Thomas Jones, a close unofficial adviser to Stanley Baldwin, suggested to Baldwin that the Foreign Office had been outwitted by the French, and he asked why Britain was afraid of France. J.C.C. Davidson, a former chairman of the Conservative party and a good friend of Baldwin's, proposed to him that Simon should turn the Foreign Office over to MacDonald and go to the Home Office. MacDonald himself began to complain that Simon's manner

[44] SDP, disp., Mellon to Dept., 22 Sep 32, 763.72119 Military Clauses/139; *Manchester Guardian,* 19, 26 Sep 32; *News Chronicle* (London), 19, 20, 24 Sep 32; James H. Thompson, "Great Britain and the World Disarmament Conference, 1932-1934" (Ph.D. diss., University of North Carolina, Chapel Hill, 1961), pp. 129-133; *Times* (London), 23, 28 Jul, 5, 19, 22, 27, 30 Sep 32 (also ltrs. to editor, 12, 13 Oct 32); AA, tels., Bernstorff to Ministry, 20, 22 Sep 32, 3154/D667714, D667723; FO 371, Minutes by Vansittart and R[eginald] Leeper, 30 Sep, 3, 4 Oct 32, C8206/211/18; Minutes by Sargent and R. M. [?], 3 Oct 32, C8263/211/18; Minutes by Vansittart and R. Leeper, 4 Oct 32, C8438/211/18.

230—Britain Intervenes to Prevent a Break

did not promote confidence. The positions of both MacDonald and Simon were shaky at this juncture, since much of the Liberal contingent in the National Government was just deciding to withdraw in protest over the Ottawa agreements. With the government majority even more overwhelmingly Conservative than before, MacDonald feared that he would become a mere figurehead, or as he put it, "a limpet in office." Baldwin assured him of his continuing support, but aside from Baldwin's sense of loyalty and lack of ambition, the chief reason for retaining MacDonald and Simon was their presumed appeal to the liberal, reformist public at large. Much of this appeal depended in turn on their ability to maintain hopes for disarmament and peace.[45]

4. A NEW BRITISH CONFERENCE PROPOSAL

Simon—surprised, baffled, and perturbed at the feeling in Germany and at home—left Geneva to attend the League Assembly meetings there and to try to engage in some quiet diplomacy. As the British memorandum had retained from O'Malley's original concept a cryptic reference to the possibility of an "agreed adjustment," provided there was no increase in the "sum total" of German strength, Simon now decided to put more stress on this idea of adjustment. On the morning of September 23, he suggested to Joseph Paul-Boncour, the French Defense Minister and League delegate, the possibility of conceding to the Germans samples of the weapons forbidden to them under Part V. Boncour expressed grave reservations and pointed out that French policy opposed any such concessions. Nevertheless, that afternoon, Simon told an American diplomat, Hugh Wilson, that he did not rule out letting the Germans have samples of the weapons possessed by others. In the evening, Simon sought out Neurath at his hotel, and after the German Foreign Minister voiced his dissatisfaction over the British memorandum—in terms that appear much stronger in Neurath's own record than in Simon's—Simon stressed the indication in that memorandum that Part V should be superseded. Then the British minister said that he wondered, and had meant to express this thought in the memorandum, whether Germany might be given a full choice

[45] FO 371, Minutes, ltrs., and memos respecting League of Nations Union and Church deputations, 28 Sep-10 Oct 32, W10844/130/98, W10953/130/98, W11047/130/98; Roskill, *Hankey*, III, 60; Cecil Papers, 817A/51081, ltrs., Cecil to MacDonald, 11, 14 Sep 32; Thomas Jones, *A Diary with Letters, 1931-1950* (London: Oxford University Press, 1954), pp. 55-57, 63-64; Robert Rhodes James, ed., *Memoirs of a Conservative: J.C.C. Davidson's Memoirs and Papers, 1910-37* (London: Weidenfeld & Nicolson, 1969), pp. 379-380; MDP, 8/1, Diary, 22 Sep, 23 Oct [32]; Herbert Louis, Viscount Samuel, *Memoirs* (London: Cresset, 1945), pp. 228-229; Middlemas and Barnes, pp. 685-687, 734; Marquand, pp. 724-737.

of arms, with the concession of new arms compensated by reductions in other sorts of arms Germany already possessed. Simon later told Wilson that he had concluded Germany would not reenter discussions or sign an eventual treaty unless she were permitted samples of the arms permitted to others.[46]

The proposal of samples amounted to a discreet formulation of the German demand for qualitative equality, well suited to make that demand more acceptable to advocates of disarmament. Although Simon was to recall later (in October 1933) that it was Neurath who suggested the possibility of samples, it was probably Simon who raised the matter with Neurath, as Simon himself had just discussed the concept with others. The term "samples" may have been a euphemism devised by François-Poncet to forestall resistance in Paris to qualitative equality; Simon at any rate had fastened on it as a possible solution.[47] Actually, the idea had been implicit in the Foreign Office memorandum of May 26, 1932, which suggested allowing Germany weapons permitted to France, without an overall increase. The question of samples was to become a source of animosity in October 1933, when Simon, having thought Germany would be content with modest numbers, and now himself opposing even this, discovered that she actually wanted significant quantitities.[48]

In 1932, in any case, the idea of samples constituted the thin end of the wedge for an admission of the third German demand, that Germany obtain all weapons not generally prohibited—and also have the same right as other states to shape her defense system. Upon his return to London, Simon circulated a paper to the cabinet stating that,

[46] FO 800/287, Simon Papers, ltrs. to J. L. Garvin and J. A. Spender, 20 Sep 32; DDF, 1/I, No. 208; FRUS, 1932, I, 437, 444-445, 446-447; DBFP, 2/IV, No. 111; RK, Cabinet Protocol, 29 Sep 32, and tel., Neurath to Foreign Ministry, 23 Sep 32, 3598/D790844-46, D790859-62. Simon inquired through Wilson as to Stimson's views on the three German demands, plus quantitative equality (which Simon himself rejected). Stimson, who distrusted Simon's firmness vis-à-vis Germany, considered sending a trenchant legal critique of the German demands, but finally decided it would be wisest, especially in view of the 20 Sep 32 Hoover statement, to send only a short reply opposing qualitative equality in battleships and submarines, but otherwise largely agreeing: Moffat MS Diary, 26, 27, 30 Sep 32; FRUS, 1932, I, 444-445, 447, 449-450.

[47] 1933 recollections: DBFP, 2/V, No. 443. Simon's first accounts of the 23 Sep 32 discussion (DBFP, 2/IV, No. 111; Cab 24/233, CP 323 [32], 27 Sep 32) rather suggest that Neurath mentioned samples first, but do not say so specifically. Neurath's own report on the meeting, which was shown to the German cabinet (see note 46) did not mention samples. Poncet did report Bülow as mentioning samples in their conversation of 23 August, and this version was relayed to the British: see DDF, 1/I, Nos. 115, 125 (Annex); DBFP, 2/IV, No. 49. Bülow's record of the conversation says that Poncet had already used the word *Ansätzen* (in this context, "beginnings") to refer to the arms desired: AA, Memo by Bülow, 23 Aug 32, 4602/E190813.

[48] Cab 24/230, CP 164 (32) (encl. II); DBFP, 2/V, Nos. 419, 421, 422, 423, 431, 434, 443.

according to Neurath, Germany would be content with samples. Simon asked his fellow ministers to decide whether, if a choice became necessary, they preferred "Germany's continued abstention from the Conference and all that involves to the future of disarmament and the peace of the world," or "agreeing that Germany may have a range of choice between weapons permitted to others, provided that this does not involve, during the period of the Convention, a substantial increase in her equipment as a whole."[49]

Meanwhile, O'Malley had written an intradepartmental memorandum, developing the ideas he had advanced earlier. He proposed an agreement between the powers on replacing Part V and allowing Germany certain changes (a few tanks, a general staff [!], a reduced period of enlistment, a militia), while still leaving her in a position of hopeless military inferiority to France. If Britain could determine that a Franco-German compromise was possible, the next step would be to hold direct, highly private, round-table discussions between the British, French, Germans, and Italians. The French would doubtless make difficulties, and they would have to be asked in the clearest way to choose between German breaches in the Treaty of Versailles, against which there could be no strong British reaction in view of British opinion, and an agreement with Germany that would leave her almost as powerless as she was, and the violation of which would turn British opinion against her.[50]

Sir Robert Vansittart forwarded this memorandum approvingly to Simon, minuting:

> As I have said before, and I rather think this is your own view, a mere theoretical recognition of Germany's equal status can solve nothing. It does not go far enough to meet Germany who wants a practical application of such treatment in the reconstruction of her fighting forces, and will not return to the Conference unless she can be assured of getting this in the near future.
>
> It would seem that only complete frankness with France, Germany, and Italy can solve the present crisis, which, if unsolved, must have disastrous consequences. The first necessary step appears to be an urgent and vigorous discussion with M. Herriot facing him with the necessity as we see it of offering Germany definite assurances of practical satisfaction on the concrete points of her demand. So far as possible this should be by other countries discarding weapons forbidden to Germany: in some cases, the alternative course of

[49] Cab 24/233, CP 323 (32), 27 Sep 32.
[50] FO 371, Memo by O'Malley, 24 Sep 32, C8251/211/18.

allowing Germany weapons hitherto forbidden will be necessary. But this should only be the last alternative under proper safeguards. . . . Should M. Herriot be unwilling to agree to a concrete offer of this kind to Germany he should be faced with the alternative of Germany tearing up the Treaty and the British government washing their hands of the matter.

Vansittart suggested an approach to Herriot, the Italian government, and Papen in turn. The German Chancellor, for his part, could be confronted with the alternatives of a generous offer "or of being left alone, with the opinion of his own country much divided, to face the uncompromising opposition of all other European States to German schemes of aggressive rearmament." If the cabinet approved Simon's suggestion of allowing Germany "a range of choice between weapons permitted to others," then (Vansittart noted) O'Malley's and his suggestions could be put into effect.[51]

Both France and Germany now appeared from London to be aggravating the difficulties. Simon, while in Geneva, had apparently hoped for a time to hold private talks together with Neurath and Herriot.[52] Herriot, however, went to deliver a sharp speech at Gramat (a small town in southwest France) on the twenty-fifth, which was answered both by cutting remarks in the German press—Theodor Wolff of the *Berliner Tageblatt* hinted that he had been drunk—and by a public statement from Papen. If the thin-skinned Herriot had ever been ready for private talks, this response must have eliminated such readiness. When he came to Geneva thereafter, Neurath reportedly indicated willingness to meet him, but Herriot delayed his reply, and Neurath then left Geneva on the twenty-eighth.[53] A hastily prepared review of the situation, drawn up by Simon on the basis of contributions from Sir Maurice Hankey and, to some extent, from Vansittart, showed some recognition that Germany was moving to rearm and that French fears had to be faced. But it also seemed necessary to have a policy which could be defended vis-à-vis the advocates of disarmament, and as Sargent minuted, "the direct intervention of H.M.G. with an invi-

---

[51] FO 371, Minute by Vansittart, 27 Sep 32, C8251/211/18; Minute by Vansittart, 28 Sep 32, W10844/130/98.

[52] See AA, Guideline for briefing by Neurath, 15 Nov 32, 3154/D672151-52; also *DBFP*, 2/IV, No. 112 (antepenultimate paragraph); RK, tel., Neurath to Foreign Ministry, 23 Sep 32, 3598/D790861-62.

[53] Soulié, pp. 389-390; Loosli-Usteri, p. 190; *DDF*, 1/I, Nos. 215, 224; *DBFP*, 2/IV, No. 114. On Herriot's reaction to German criticism, also RK, Memo by Bülow, 28 Sep 32, 3642/D811657; SDP, Notes [by Sackett] on declaration by Herriot, 28 Sep 32, 550.A15A4/1483½. Neurath's departure: RK, Cabinet Protocol, 29 Sep 32, 3598/D790845; Memo by Bülow, 28 Sep 32, 3642/D811657.

234—Britain Intervenes to Prevent a Break

tation to consider concrete proposals would seem in present circs. the only effective way of putting an end to these provocative and sterile disputations."[54]

When the cabinet met on September 30, the Foreign Secretary stressed the seriousness of the situation. The ministers agreed that British policy should aim at the return of Germany to the conference, and they also accepted Simon's suggestion that, because continued German absence would be so serious, they should not rule out the possibility of conceding hitherto forbidden weapons to Germany. To get things moving, the cabinet took up the O'Malley idea of a great-power meeting, and approved a proposal to invite the French, German, and Italian governments, on the basis of the consultative declaration of July 13, to send representatives to a meeting in London on October 11. The invitations were officially issued on October 3 and 4, along with a suggestion to the French that the intervening period might be used for private Franco-British discussions. This last had the appearance of a gesture toward France, but there was no consultation with France before the announcement, and the background indicates that the purpose of the private discussions, in British eyes, was to put pressure on the French to make concessions. Simon told American diplomats that he had invoked the (multilateral) consultative understanding for a London meeting to prevent France and Germany from refusing to attend. Sir Maurice Hankey was dubious about the French attending, or their contributing anything if they did, but he noted in his diary that "our government have 'got the wind up' badly owing to the threats of the pacifists, Bishops and Free Churches etc. over the coming failure of the Disarmament Conference, which has been certain from the first."[55]

After Bülow learned on October 4 of this proposed meeting in London, he was jubilant, and with good grounds.[56] Although the original German plan to deal with France in isolation had failed, a critical gain had now been won in another way. Now there would be a chance for German diplomats to gain a recognition of equality of rights through negotiation, so as to forestall the risky course, favored by

[54] FO 371, Minutes by A. Leeper, C. Howard Smith, Vansittart, 28 Sep 32, W10844/130/98; Minutes by R.M.A. Hankey, Sargent, Vansittart, 29-30 Sep 32, C8206/211/18; Cab 24/233, CP 326 (32), 29 Sep 32. Although Vansittart participated in preparing the review showing the German intent to rearm, Hankey claimed to have taken the initiative on this (Roskill, *Hankey*, III, 59-60), and in view of Vansittart's minute of 27 Sep 32, quoted above, it does indeed seem likely that the Under Secretary merely chimed in.

[55] Cab 23/72, meeting of 30 Sep 32, 49(32)—; *DBFP*, 2/IV, Nos. 115, 116, 124; *FRUS, 1932*, I, 450; Roskill, *Hankey*, III, 60.

[56] See *DDF*, 1/I, No. 224; also *DBFP*, 2/IV, No. 131; RK, Memo by Bülow, 4 Oct 32, 3642/D812804-06.

some German officers,[57] of simply going ahead with rearmament and negotiating later. The format of a four- or five-power meeting, especially with Italian participation, would be favorable to German interests and unfavorable to those of France; for this reason, and others, Herriot was to resist the proposal. Bülow might have been even happier had he known that the British cabinet had opened its collective mind to the idea of conceding samples of all weapons; this enhanced Germany's prospects for achieving practical changes, not only in weapons but also in organization. If (as in O'Malley's latest paper) the British did not stick at the Germans' having a few tanks, they would also hardly refuse proposals for a militia. Of course, they still maintained that adjustments should not increase the sum of German armed strength. Definite opposition to German rearmament did exist in the Foreign Office, and for that matter, in the British cabinet. But from the British point of view, an overriding objective was to try to get the Germans back to the conference to accept an agreed limitation. After that, perhaps some disarmament might still be accomplished, i.e., on the part of the French. Since they gave priority to securing a German return, the British were not likely to stand firm against "minor" increases in German strength.

5. Current Reichswehr Plans and British Assessments

The German proposals for qualitative equality, in arms and in the right to reorganize, were no minor affair, however. We have seen that the German general staff was anxious to obtain reserves of materiel and men for mobilization. German generals realized that if something were not done soon to provide trained manpower, Germany would long remain unable to use threats of war, or war itself, as an instrument of policy. On the other hand, if the German army could train young men, who had had Wehrsport preparation, for three months, and retain a third of them for three years of regular service, it would be able in a few years to mobilize a major force. A concession of qualitative equality in arms would also permit a significant increase in German strength, as it would facilitate the development of prototypes, of preparations for mass production, and even of large stocks; any arms detected could then be explained away as one of the authorized number.

We have direct testimony from just this part of the fall of 1932, not only on actual Reichswehr plans, but also on the significance of these plans, as seen from a critical perspective. Spurred by word of the pro-

---

[57] *DBFP*, 2/IV, No. 165 (incl. encl.); NA-VGM, F VII a 11, Memo by Schönheinz, 1 Dec 32, 7792/E565449-52.

posed four-power meeting, Bülow wrote to Schleicher on October 4, telling him of the British invitation and asking for more information about the Reichswehr's program, so as to be better prepared for the prospective discussions in London. Bülow's letter stated, almost naively: ". . . it will be essential that we give the opponents a rather clear picture of our intentions, so that they cannot maintain that we want to rearm, and so that they actually know what they have to deal with." Schleicher did not agree that details could be discussed in London, and apparently he did not reply to Bülow, but on October 15 he did provide Neurath with information on Reichswehr plans, strictly for the Foreign Minister's own background. The material forwarded by Schleicher disclosed that, at a minimum and irrespective of any reductions by other powers, the army wanted nine heavy artillery units, twenty-two air squadrons, and an experimental tank battalion. Regular strength was to increase to 145,000 men, and for half the regulars the basic term of service would be reduced over a three-year period to either—this had not yet been decided—two or three years. If regular service were two years, there would be 300,000 reserves from this source in five years, and if service were three years, there would be 200,000 reserves; significantly, the army refused to provide figures beyond a five-year period. The militia would not be a special unit outside the army, but was to be trained within army ranks, and regular recruits would be drawn from it; with one militia contingent per year for most arms, 110,000 men would enter the militia each year in the event of two-year regular service, or 90,000 in the event of three-year service. The militia would in itself train 160,000 men in five years (aside from those who went on to the regular army) in the event of two-year training, or 250,000 in the three-year eventuality. The Reichswehr did not want to describe those who merely went through the militia as trained reservists, as the French supposedly did not regard men with less than six months' training as reservists. But this was merely a pettifogging technicality: the actual reserve status of these men was apparent from the intention to give them a further two weeks of training every three years "in the course of the 15-year period of general service obligation." The navy, for its part, wanted to build five battleships of 25,000 tons each, one aircraft carrier, 35,000 tons of submarines, and twenty-one naval air squadrons, to name only the additions in the then-forbidden categories.[58]

[58] AA, ltr., Bülow to Schleicher, 4 Oct 32, 7360/E535936-41; List of questions for Schleicher, [12 Oct 32], 3154/D672029-30; ltr., Schleicher to Neurath, with atts., 15 Dec 32, 3154/D671232-42; RK, Cabinet Protocol, 7 Oct 32, 3598/D790958-60. The information provided by Schleicher seems to have been hastily collected from more than one office, and to be not always consistent, with itself or with other sources. Both the preliminary program approved in July (BA-MA, II H 228, T2 549/32, 15

Bülow had supposed that the Reichswehr's "Umbau" plans would not involve a perceptible amount of rearmament, and when Neurath let him study the information from Schleicher, the State Secretary was taken aback. In a note to Neurath in his own hand, he pointed out that the increase in regulars and reduction in regular service would completely change the character of the army, making it resemble the prewar national service army, only with a larger proportion of non-commissioned officers. Foreign propagandists would add up the regulars and militia, together with a projected annual total of Grenzschutz trainees of 143,000, and would arrive at a total of around 385,000, nearly four times the number of troops permitted under the Treaty of Versailles. At least 500,000 rifles and carbines would be needed, as opposed to the 102,000 authorized in the Treaty. Bülow calculated that in fifteen years the regulars and the regular and militia reserves would produce an army of over 1,000,000. The naval demands for battleships also disturbed him greatly; he thought that Germany could not compete with British naval power again, and that demands for such ships would arouse the opposition of Britain and the United States. Finally, Bülow believed that the program was beyond Germany's financial means. Neurath wrote a marginal comment at the end of Bülow's note: "Completely correct (*Durchaus richtig*)." There seems to be no evidence, however, that Bülow's objections had any serious result.[59]

If Bülow's reaction confirms that the German proposals involved major rearmament, his previous attitude shows that even a high German official could have illusions on this subject. And British Foreign Office officials were more likely to be misled than Bülow had been. Their thoughts about the German military tended to concentrate on the less important question of forbidden weapons.[60] In this area, mi-

---

Jul 32) and the later program issued in November (BA-MA, RH 15/49, TA 737/32, 7 Nov 32) simply provided three-year service for three-fourths of the regular force. The furnishing of the information was pursuant to an agreement between Schleicher and Neurath of 12 Oct 32 regarding the conduct of any four-power talks; see Dülffer, pp. 254-255. Dülffer suggests that the naval proposals, which went beyond actual naval plans, were *Spielmaterial*, i.e., exaggerated for negotiating purposes; the navy also had alternate, minimal demands. The army proposals, however, were only slightly larger than the plans approved in November.

[59] AA, Note by Bülow, [16 Oct 32], 3154/D671228-31. Neurath did discuss the future military budget with the Finance Minister (Lutz, Graf Schwerin von Krosigk), who said that the budget could not be increased in the next years, and who agreed that there was no use in seeking international acceptance for sums larger than Germany could actually expend (AA, Note by Neurath, 15 Nov 32, 7474/H185474-76). A discussion with both Krosigk and Schleicher had been planned (AA, Unsigned memo, 9 Nov 32, 3177/D683451-55), but I have found no record that it took place.

[60] Vansittart suggested on 29 Sep 32 that Germany might drop her demand for a shorter service period if she could get samples of all weapons not generally prohibited: FO 371, Minute by Vansittart, C8363/211/18.

nor violations had been reported for years, and British diplomats had become skeptical as to the importance of any reported German military preparations, believing that the French always exaggerated German transgressions. In a briefing note for MacDonald, O'Malley wrote that "Germany has not, at least on any appreciable scale, violated any international obligations." When, in August, Ambassador de Fleuriau passed on information that the implementation of the Umbau program had begun, Sargent, the principal Foreign Office expert on Germany, checked with the British mission in Berlin, expecting a contrary report. He got it: Rumbold seized an opportunity to ask Neurath himself, and Neurath—naturally—gave an emphatic denial. Simon and his officials had little grasp of such matters as reserves; he asked Ronald Campbell (while passing through Paris on September 27) if the French would consider a reduction in the Reichswehr term of service as rearmament. Campbell replied that they would, but did not recall until later that the chief French objection was that this would increase the number of Germans who would receive military training. As far as civilian officials were concerned, there was some truth in a comment Tardieu had made to Brüning in Geneva in April, that "the Americans and British didn't understand anything about land armaments."[61]

For that matter, advice from the War Office did little to clarify the danger. The Army Council commented in early September that in spite of "numerous and systematic minor breaches of the military clauses," Germany was not capable of waging even a successful defensive war, and that "even if the German proposals were accepted in full, and Germany obtained a long-service army 100,000 strong, provided with all modern armament, backed by a short-service conscript militia trained for three months, the German Army would still be in a position of absolute military inferiority in comparison with the French army." The council did not believe that a conscript militia trained for three months could be an effective instrument in an aggressive war, or that it would enable Germany to build up a trained reserve comparable to that of the French. The council did predict, however, that even if a compromise could be reached, it would not satisfy Germany

---

[61] FO 371, Minute by Sargent, 24 Aug 32, C7178/211/18; Note by O'Malley, 7 Sep 32, C7529/211/18; ltr., Campbell to Vansittart, 28 Sep 32, C8363/211/18; *DBFP*, 2/IV, No. 55; AA, tel., Bülow to Ministry, 21 Apr 32, 7360/E535169. Although Neurath denied that steps toward rearmament had begun, one of the leaders of a propaganda organization for national security had told a Foreign Ministry official, apparently on the basis of inside information, that the Umbau program had been launched; Neurath had raised this matter with Schleicher a few days before he spoke with Rumbold (AA, ltr., Neurath to Schleicher, 27 Aug 32, and encl., 3154/D667550-52).

for long, and they concluded resignedly: "Germany cannot be tied down in perpetuity and yet the facts of population and industrial efficiency must inevitably weigh down the scales against France in the future."[62]

After a German decree appeared on September 14, officially establishing the Reich Institute for Youth Training, MacDonald showed some apprehension that the Germans were engaged in something serious. Probably Sir Maurice Hankey influenced the Prime Minister in his concern. As noted, Sir Maurice had contributed to a report that called attention to German armament plans. But Hankey's views also had a certain distortion, for the report referred to the view of "some professional quarters" that the militia was intended to maintain internal order and to replace casualties in a very highly trained striking force. The Berlin embassy and Hankey's son Robert (a Foreign Office official) had just drawn attention to an article by General von Seeckt envisioning such a role for the militia.[63]

One confirmed believer in Seeckt's influence was the Chief of the Imperial General Staff, Field Marshal Sir George Milne, who himself looked like a somewhat more amiable Seeckt. In the course of a lengthy memorandum circulated to the cabinet at the end of October 1932 (referred to in Chapter Two), Milne reviewed Seeckt's ideas, saying that they "undoubtedly" represented current German aims. Milne wrote that it was Seeckt's concept of a professional force able to strike on the first day of war which "excites the fears of the French and Polish General Staffs." He argued that every German government for years past had worked on a definite program for the destruction of the Treaty of Versailles, and that it was evident to both Germany and Poland that the questions of the Corridor and of Upper Silesia could only be settled by force. Milne pointed to the establishment of a thinly disguised general staff and staff college, to evidence of increased armament and mechanization, to the likelihood that the German general staff had discussed military cooperation with their Soviet counterparts

---

[62] FO 371, ltr., A. T. Widders (War Office) to Vansittart, 9 Sep 32, C7649/211/18. Some of this spirit of resignation was already evident in 1928: *DBFP*, Ia/V, No. 268.

[63] MDP, 1/504, ltr., MacDonald to Cecil, 15 Sep 32; FO 371, Minute by R.M.A. Hankey, 15 Sep 32, C7800/211/18; tel., Rumbold to F.O. (no. 198), 16 Sep 32, C7826/211/18; tel., Rumbold to F.O. (no. 199), 16 Sep 32, and minute by R.M.A. Hankey, 19 Sep 32, C7829/211/18; Minute by R.M.A. Hankey, 23 Sep 32, C7988/211/18; disp., Rumbold to F.O., 19 Sep 32 (enclosing memo by Thorne), C8043/211/18; Cab 24/233, CP 326 (32), 29 Sep 32. Sir Maurice Hankey thought the Germans were so attached to their élite army that a leveling down by the Allies might be more convenient to them than a leveling up on their own part; he did however notice that they were showing an intent to reduce the term of service so as to gain trained reserves usable as NCOs, develop token specimens as models for production, and launch national registration: see Roskill, *Hankey*, III, 59-60.

(the Soviets were a bête noire of Milne's), and to excessive German defense expenditures, which were probably being spent primarily on the accumulation of materiel reserves for eventual mobilization.[64]

Milne himself showed less concern, however, over German intentions than over Soviet designs on India and in the Far East, where he suggested that better relations with Japan would help to contain the Soviet threat. He evidently regarded Germany as much more a French than a British problem, and in any case, he believed (as noted earlier) that even with an acceptance of German demands and the establishment of a militia, Germany would still only be able to mobilize fourteen infantry and three cavalry divisions in 1938, and would still at that date be at the mercy of France or Poland. Seeckt's ideal army would be finally established only at the end of a further stage, when with favorable military alliances Germany might have a good chance of defeating France and her existing allies. These allies did not include Britain, which might, in Milne's view, choose to intervene later on either side.

Three of Milne's conclusions must have suited British political leaders very well. One was that the critical period of German rearmament would only come after 1938. Another conclusion stated (as Simon and the *Times* also held) that there was "no practical means of holding Germany bound indefinitely to the Treaty or of preventing her advance towards her goal." The third was that Britain "ought to keep clear of entanglements and make no promise of military support to any group of Powers." The British services were not in any case in a position to act on such promises, and Milne wanted to concentrate on the consolidation of the Empire and its communications, which entailed an acceptance of and an adjustment to the dislike of the dominions for European entanglements. From either a popular disarmament perspective or an imperial military point of view, it was most convenient to minimize the importance of German demands, accept the impossibility of resisting revision, and avoid any firm commitments. If British leaders occasionally saw that the Germans intended to rearm, they were also impressed by the sympathy of the British public for the German case, they were influenced by the demand in élite circles and in the Foreign Office itself for a British disarmament plan,

[64] *DBFP*, 2/I, Appendix II; Cab 24/234, CP 362 (32), 28 Oct 32. On Milne's appearance, Liddell Hart, *Memoirs*, I, 70 and illus. facing p. 99. The Milne memorandum has already been referred to in Chapter Two, Section 3. Roskill states (*Hankey*, III, 54) that Sir M. Hankey coached Milne in October on the preparation of a memo for the cabinet, presumably this one. Milne was stressing in 1927 the Soviet threat to India: Howard, *Continental Commitment*, pp. 90-91.

and they preferred on the whole to believe that the German proposals were relatively innocuous.

State Secretary von Bülow, in spite of his disapproval of the Reichswehr's program, patriotically helped to deceive the British government and public in an episode that illustrates how the German government was able to exploit the desire of high-minded Englishmen to promote disarmament. John W. Wheeler-Bennett, then editor of the *Bulletin of International News* of the Royal Institute of International Affairs, had received information from Brüning on the April 26 conversation at Bessinge, and when the British, French, and Germans were at loggerheads at the end of September, it occurred to him that he might clarify the situation by drafting a statement of what, as he understood it, the Germans wanted and were ready to give, then clearing it with top German officials and circulating it. He talked on two separate visits with Papen, Erwin Planck (state secretary in the Reich Chancellery), and also Bülow; Bülow actually guided the others in their response, regarding this as a chance to propagate a more favorable impression of German aims. The Wheeler-Bennett formula, as finally amended, approved in Berlin, and published in a letter from Wheeler-Bennett to the *Times* on November 15, covered eight points; among these were a "partial reduction" in the period of service in the Reichswehr, a militia of not more than half the size of the Reichswehr (i.e., 50,000 men), an acceptance of existing strength levels for the period between the first and second Disarmament Conferences, a claim to "token" equality in arms not limited by the convention, and an agreement not to increase the "normal average" German military budget. Bülow's correspondence with Planck reveals that, on each of these five points specifically described, the Germans either deliberately misled Wheeler-Bennett, or kept him incompletely informed, or knowingly left him with a misunderstanding. Bülow's cynicism became more clear-cut after he had learned of Schleicher's intentions. Rather than setting Wheeler-Bennett straight on some matters, Bülow thought it better to be content with "a few slight retouches in order to strengthen our propaganda." In the case of the militia, Bülow advised Planck that although Reichswehr wishes went beyond 50,000, "such complicated things" as average daily strengths could not now be given to the public; it was tactically best to leave the formula as it was "and declare later that he [Wheeler-Bennett] misunderstood us." Wheeler-Bennett's formulae had a generally favorable reception in the Foreign Office, where Vansittart noted that one draft was "not too unpromising" if (as Wheeler-Bennett stated) Papen had approved it. Wheeler-Bennett also left a copy of his *Times* letter, with an explanatory memorandum,

242—Britain Intervenes to Prevent a Break

with MacDonald's private secretary, Nevile Butler, who passed it on to the Prime Minister. We do not know what impression this information made on MacDonald, but he was definitely impressed on his first reading of statements from Neurath, Schleicher, and Papen that appeared in the British press; he understood these statements to represent a recession from the German position of August 29, and he said that the Foreign Office should be on to them "like a knife." The publication of these statements appears to have been arranged by Margarete Gärtner, whose propaganda activities we have noted (Chapter Two).[65]

6. FRANCE BIDS FOR INTERNATIONAL SUPPORT

If Bülow exulted over the British proposal for a four-power meeting in London, French leaders did not. When Bülow cheerfully informed Poncet of the proposed meeting, the French ambassador exclaimed: "What will Edouard say?"[66] Edouard Herriot had in fact already reacted, having learned of the proposal on October 2, through press leaks and a loose tongue in Geneva. He had immediately concluded that the meeting should be held, if anywhere, in Geneva, with Czech, Polish, and Belgian participation. But he was generally upset, and also hurt. On October 3, Tyrrell officially informed Herriot of the proposal for a meeting, adding (according to the French record) that the British mediation would be based on a recognition of equality of rights with-

[65] FO 371, Draft formula by Wheeler-Bennett, 1 Oct 32, ltr., Cadogan to C. Howard Smith, 9 Oct 32, minutes by Leeper, Smith, G. M[ounsey], Vansittart, 19-20 Oct 32, initialed by Simon, 23 Oct 32, C8686/211/18; Memo by Wheeler-Bennett, 22 Nov 32, minute by Leeper, 3 Dec 32, W13253/1466/98; RK, ltr., Papen to Bülow, 6 Oct 32, 3642/D812875-76; ltrs., Bülow to Planck, 5, 27 Oct 32, 3642/D812873-74, D812889-91; ltrs., Planck to Bülow, 26 Oct 32, 3642/D812866; ltr., Planck to Wheeler-Bennett, 31 Oct 32, 3642/D812892-94; ltr., Wheeler-Bennett to Planck, 19 Oct 32, 3642/D812895; MDP, 1/504, Note by Neville Butler, 3 Dec 32. Wheeler-Bennett initially hoped, as did Papen, to interest the French in the formula. Perhaps with this in mind, Wheeler-Bennett gave the formula to the British delegation in Geneva; Cadogan, however, sent it to the Foreign Office. For a relatively recent account by Wheeler-Bennett himself, see his *Knaves, Fools, and Heroes* (London: Macmillan, 1974), pp. 63-65; when writing this, he seems to have been unaware of the British and German official files on his proposal. On MacDonald's reaction to German statements: FO 371, Minutes by Sargent, Howard Smith, Vansittart, 8 Nov 32, newspaper clippings, 7 Nov 32, W12532/130/98; *DBFP*, 2/IV, No. 206; *DDF*, 1/II, No. 60. Vansittart's minute records a belief that MacDonald later came to think "that the Germans were only trying to give themselves a better appearance and were not honest." But the *DBFP* and *DDF* documents show that he considered the articles possibly significant. For a probable reference to the placement of the articles: Gärtner, *Botschafterin*, p. 249.

[66] RK, Memo by Bülow, 4 Oct 32, 3642/D812809-10; *DDF*, 1/I, No. 224. In his conversation with Bülow, Poncet remarked that "people in Paris have had about enough of these improvised invitations of MacDonald's." MacDonald had made such proposals in July 1931, January 1932, and April 1932.

out an increase in armament, on a reduction by stages on the part of the heavily armed powers, and on Germany's right to have arms of all categories, but in smaller quantities than those of the former Allies. Herriot told Tyrrell that the British proposal was "a great satisfaction for Germany and a humiliation for France." He also emphasized that the absence of prior consultation with France was not in accordance with the understanding he had arrived at with MacDonald in Lausanne. To Simon, passing through Paris on October 4, Herriot described the British proposal as "most imprudent."[67]

We can well imagine that the British proposal came as a shock to Herriot. Since the consultative agreement in July, he had tried to believe that he had established a special relationship with England. There had indeed been warning signals, particularly MacDonald's and Simon's suggestion, in early September, that Herriot propose a four-power meeting. But the British statement of September 19 appears to have swept away any doubts on Herriot's part about British sympathies. Thanks to the German demands and their exploitation by France, it had appeared from Paris that Germany was isolated.[68] Herriot's recent strategy had been to urge the continuation (through the Bureau) of the work of the conference without Germany; eventually, the conference could proceed without German participation to write a convention, which Germany could then either accept or reject. If she accepted it, she would agree to terms satisfactory to France. If she rejected it, the blame for blocking disarmament would fall on her, and she would still be legally bound by Part V. And the convention, if rejected by Germany and accepted by the other powers, would then tend to unite the latter in opposition to Germany. Of course, to perform its intended function, the convention would have to appear to the non-German world as reasonable and equitable, but Herriot presumably expected that the experts of France and her allies could produce a draft which would have this appearance and also protect French interests. Herriot had in fact established a committee on September 10 to draft new instructions for the French disarmament delegation.[69]

While speeding the development of a French plan, Herriot had to

[67] *DBFP*, 2/IV, Nos. 123 (and note), 134, 135, 141; *FRUS, 1932*, I, 454-456; Soulié, p. 391; *DDF*, 1/I, Nos. 226, 228 (and note), 229; FO 371, ltr., Tyrrell to Simon, 6 Oct 32, minute by Sargent, 7 Oct 32, C8517/211/18.

[68] On previous French expectations and shock at the British move, from a German perspective, see AA, tel., Forster to Ministry, 7 Oct 32, 3154/D667869-71. The observations in *DDF*, 1/I, Nos. 244 (pp. 439n, 441-443) and 250 (pp. 477-480) are revealing, and yet, due to the context, probably an understatement.

[69] See *DDF*, 1/I, Nos. 163, 175, 180, 244 (p. 439n, pp. 442-443), 250 (p. 477), 273 (p. 577); *DBFP*, 2/IV, No. 134. Herriot did not, however, want to assume responsibility for blocking an adjournment of the Bureau: *DDF*, 1/I, Nos. 180n, 186.

deal immediately with the British proposal for a meeting of the powers. As he did not want the meeting, but was reluctant to offend MacDonald with a flat refusal, the German dislike for the Geneva milieu offered the French Premier an escape from a dilemma. On October 6 he formally agreed to attend a meeting of the kind proposed by Britain, but asked that, "since these conversations can only have an unofficial and preliminary character," they be held in Geneva, so as to be "within the framework of the League, which alone is qualified to take definitive decisions with the participation of all interested parties."[70] Although Herriot no longer asked that the lesser powers be included in the talks proposed by the British, a meeting in Geneva would be very awkward for the Berlin government, as it would look as though Germany were returning to the conference without achieving her announced goal. To gain British agreement to this shift from London to Geneva, and also to convince the British that France intended to disarm, Herriot took up the idea of direct bilateral conversations with MacDonald. The French Premier came to London on October 12, bringing a preliminary draft of the French plan with him. After looking the draft over, however, MacDonald said that he was not ready to discuss it, and he indicated that he preferred to concentrate on plans for a four-power meeting. Herriot did not press for a discussion of the French plan. It was more important in the immediate situation to get the British to agree to the proposed change of venue to Geneva. Franco-British agreement on this point was announced on October 14, and although Bülow had said that the place of meeting was not a fundamental issue, Neurath now declared that Geneva was unsatisfactory to his government as a meeting place.[71]

The London conversations must have strengthened Herriot's desire to get the French plan out as soon as possible. While he argued there that the powers should demand that the Germans make a formal statement of their specific demands, MacDonald doubted the wisdom of further formal exchanges, and wanted to explore the German aims in

[70] *DBFP*, 2/IV, Nos. 140 (encl.), 141; *DDF*, 1/I, Nos. 228, 229.
[71] *DBFP*, 2/IV, Nos. 141 (incl. notes), 152-157; Soulié, pp. 392-393; RK, Memo by Bülow, 7 Oct 32, 3642/D812825. No official French record of the London talks is available. The information on MacDonald's reception of the French plan is indirect, obtained from Bernstorff from an unnamed source: *FRUS, 1932*, I, 463; AA, tel., Bernstorff to Ministry, 14 Oct 32, 3154/D667967-68. The British had already received a secret report on the French plan, although this was not circulated on the grounds that it might be misleading (it was in fact fairly accurate): FO 371, F.O. minute, 11 Oct 32, W11402/1466/98. The British had no serious objection to a change of venue, and the cabinet had agreed to this before Herriot came: Cab 23/72, meeting of 11 Oct 32, 50(32)5. The German Foreign Ministry claimed later that the British had been warned just before Herriot's visit that Geneva would be unacceptable (AA, Circular tel., 15 Oct 32, 3154/D672059).

personal conversations at a four- or five-power gathering. At the final session, Simon (as Vansittart had recommended) cross-examined Herriot in an effort to lead him to accept British views on equality of rights. Although the visitor welcomed British statements that they opposed German rearmament, and was happy to assure them that he too wanted to disarm, he was clearly less comfortable with their idea that Part V should be replaced by a new arrangement with the same duration for all. If steps of this kind were contemplated, it was all the more essential from the French point of view to come forth with a plan to restrain the Germans.[72]

The French solution was their *plan constructif* (constructive plan) for both disarmament and security. The moving spirit behind this plan was the Minister of Defense, Joseph Paul-Boncour. Boncour was a highly mannered orator, with chiseled features, thin lips, and a striking snow-white mane, a former Socialist who cultivated a resemblance to Robespierre. He had figured in the French contingent in Geneva for years, and had become the chief French delegate to the League after the death, in March 1932, of Aristide Briand. A rigid supporter of organized security, he was steeped in the tradition of French internationalism.[73] Boncour's memoirs suggest that the constructive plan was a new departure, and a carefully conceived French answer to the Hoover proposals. Actually, the basis of the new plan was that maximum plan which Tardieu had passed over in January, well before the Hoover proposals were publicized. Boncour does appear to have been an initiator of the original maximum plan, and one may suspect that he was seeking a chance to revive it.[74] But the new French scheme might have taken a long time to emerge, had it not been for supplementary proposals from Eduard Beneš, and above all,

[72] *DBFP*, 2/IV, Nos. 152, 153, 156; *DDF*, 1/I, No. 243 (and note). The British, particularly Allen Leeper, thought that they had gotten Herriot's agreement to a replacement of Part V and a common duration (*DBFP*, 2/IV, No. 156; FO 371, Memo and minute by Leeper, 14 Oct 32, minutes by Howard Smith, Vansittart, Wigram, 14-15 Oct 32, W11778/1466/98). But it seems to me that Herriot's remarks at the final session in London (see *DBFP*, 2/IV, No. 156), though not including an explicit refusal, really showed he had *not* accepted the supersession of Part V. Herriot had earlier failed to recognize that under the 19 September memorandum, Part V would be superseded; see *DDF*, 1/I, Nos. 240, 248.

[73] Chastenet, VI, 51-52; Bonnefous, V, 133-135; Joseph Paul-Boncour, *Entre Deux Guerres*, 3 vols. (Paris: Plon, 1945-1946), II, 271.

[74] For Boncour's account: Paul-Boncour, II, 220-231. The inspiration from the maximum plan is shown in *DDF*, 1/I, Nos. 244 (incl. Annex 1) and 250. The final plan is published in *DDF*, 1/I, No. 331. Boncour had long presided over the Study Committee of the Supreme Council of National Defense, and in this capacity he had presumably helped develop the maximum plan of January 1932. In 1924, he had prepared a plan for Herriot for vesting control of German arms in the League: Schuker, pp. 341-343.

for the shock of the British suggestion of a four-power meeting. One gets the impression that the news of the British proposal led the French government to pull together hastily a plausible alternative to a British adaptation of the German demands.[75]

It will be recalled that the maximum plan involved replacing national standing armies by international standing forces controlled by the League. The only forces individual nations would still retain would be militias, colonial troops, and police. The League forces would consist of professional élite troops, but within the national militias there would be virtually no professional elements except instructors. The militia rank and file would be recruited by compulsory national service, which was now expected to run for nine months.[76] French planners also specified that the changeover to the proposed system should be made carefully by stages, so that at no time would any state be threatened or defenseless.[77] In the context of the fall of 1932, the maximum plan appeared to have the important added virtues that it would meet the German demands for equality, and specifically for a conscript militia, while eliminating the élite professional Reichswehr, and tying Germany down in a system of mutual security, including provisions for annual inspection. Deep down, French leaders seem to have doubted that the French army, already approaching the character of a militia, could cope with the Reichswehr. Bringing the German army under the same system would, however, make the continent safe for military democracy.[78]

There was one major modification to the original maximum plan, however. Because the British refused to accept any new responsibilities, Boncour, by April 1932, had decided that France could content herself with a security system embracing only continental European nations, at least for guarantees of direct military assistance.[79] Eduard Beneš

[75] Gamelin writes (in his memoirs, II, 77) that, with Weygand's agreement, he brought Boncour a first set of proposals; "then the matter dragged on" until October 14, when Boncour's military cabinet completed its "constructive plan"—meaning, presumably, the basic report (no finished plan) in *DDF*, 1/I, No. 244. On the sudden scramble of the French government, see *DDF*, 1/I, Nos. 228 (and note), 240 (and note), 244 (notes, p. 439); *DBFP*, 2/IV, No. 134. Annexes 2-4 of *DDF*, 1/I, No. 244, are pertinent, in that they do *not* give any indication that the experts were working on a new overall plan.

[76] Duration of training: *DDF*, 1/I, Nos. 260 (p. 520), 272 (pp. 569-570), 286 (pp. 625-626).

[77] *DDF*, 1/I, Nos. 244 (p. 445), 260 (p. 522), 331 (p. 716).

[78] *DDF*, 1/I, No. 260 (p. 519); Challener, pp. 181, 183.

[79] Leeper and Vansittart welcomed this conclusion, which they had already reached themselves: FO 371, Memo by MacDonald, 22 Apr 32, minutes by A. Leeper, Howard Smith, Vansittart, 5-9 May 32, W4550/10/98; disp. Tyrrell to Simon, 18 May 32, minute by Leeper, 20 May 32, W5723/22/98; Memo by Leeper, 11 Jan 32, W11790/1466/98; Cab 24/227, CP 4 (32), 1 Jan 32.

elaborated the same idea in a plan he devised in September 1932, and this seems to have done much to inspire the new French plan.[80] Beneš, and now the French plan, proposed three concentric circles of responsibility. The strictest obligation for immediate armed assistance would, if necessary, be limited to an innermost circle of the continental members of the League; a simple majority of the League Council could decide that these powers should act against an attack or invasion. Other members of the League, including Great Britain, would form the next circle; they would continue to be bound by their League obligations, particularly their obligation under Article 16 to take sanctions against aggression, and by any other existing commitments, such as Locarno in the British case. Beyond these first and second circles of states, a third circle of non-League members—principally the United States—would, in the event of a violation of the Kellogg-Briand pact, consult about measures to be taken, isolate the aggressor economically and financially, and refuse to recognize any de facto results of aggression.[81] Secretary Stimson's doctrine of nonrecognition of treaty violations (in connection with the Manchurian question) and his public acceptance of the idea of consultation (in his August 8, 1932, speech) encouraged the French to hope that the United States would agree to this program. This was very important, since American acceptance would presumably forestall or weaken the objections that Britain might otherwise raise to taking effective action under Article 16. Actually, the French hoped that in spite of everything, the British might be inveigled into the inner circle.[82]

Although neither Herriot nor Boncour had an adequate idea of the American and British repugnance for even the slightest additional commitment, Herriot did have some doubts about the practicality of international contingents, and he and others wondered if it might be wiser to propose a modification of the Hoover plan. That plan could bar German rearmament, and its acceptance would presumably bring American support; Herriot now had more confidence in American policy than in British. But Boncour tenaciously held to his conception, and his adroit arguments and his confidence in the established doctrine swept such doubts aside.[83] In any case, Herriot and Boncour agreed on the need to avoid isolation by presenting positive proposals, and on the need for governmental economy, including in defense spending, where, politically speaking, cuts were reputedly easy to

[80] *DDF*, 1/I, Nos. 240, 244n, 248, 250, 286 (p. 618); FO 371, tel., Cadogan to F.O., 12 Oct 32, W11237/1466/98.
[81] *DDF*, 1/I, No. 331.
[82] *DDF*, 1/I, Nos. 255, 286 (pp. 618-619).
[83] See *DDF*, 1/I, Nos. 250 (pp. 477, 479, 480-481, 485, 487-488), 266 (p. 535), 268 (p. 549), 273 (pp. 577-584), 286 (pp. 624, 626-627).

make.[84] They also agreed that General von Seeckt's ideas were guiding the Reichswehr's current plans. Their obsession with Seeckt's influence —it was nothing less than an obsession—was supported by some expert opinion, but it also coincided with their left-republican suspicions of the military professionals of their own country.[85] They did not see that they were conceding what Seeckt's successors most desired: the opportunity to accumulate trained reserves again. As Simon's idea of samples was an opening wedge for German material rearmament, the French proposal for standardized short-service forces was an opening wedge for increased German effectives in peacetime and for a mass mobilization in time of war.

Boncour's ideas and personality clashed sharply with those of General Maxime Weygand, who also participated in the advisory committee and in the Supreme Council of National Defense session that formally reviewed the plan. Weygand outspokenly opposed Boncour's proposals for international contingents and national militias. He doubted that the plan would weaken the Germans. Although the Reichswehr would supposedly be dissolved, he pointed out that each nation contributing a contingent to the League—and Germany would be one of them—would have a professional force, "in a word a Reichswehr, perhaps reduced in effectives and certainly strongly reinforced in materiel." If long-service men were discharged from the Reichswehr, they would not for that reason forget their training. Weygand also pointed out that the Germans wanted a militia to provide themselves with reserves; he and General Maurice Gamelin noted that German military tastes and paramilitary organizations would make possible a fruitful pre- and postmilitary training, which France would not have. Weygand's main concern, however, was with the effect of the proposed plan on the French army. He believed that the army of France had

[84] On fear of isolation, see further below, including note 87. Weygand reports that the national-defense sector ultimately absorbed 51 percent of the reductions in the 1933 budget: Weygand, II, 386-387, 397; Bankwitz, *Weygand*, pp. 87-89. It has been shown, however, that reductions in economic and social expenditure were actually far greater than in defense: Robert Frankenstein, "A propos des aspects financiers du réarmement français (1935-1939)," *Revue d'histoire de la deuxième guerre mondiale*, [26] (1976), no. 102, 8-9.

[85] On concern over Seecktian plans, see *DBFP*, 2/IV, No. 152; *DDF*, 1/I, Nos. 120, 126, 197, 244 (p. 442), 250 (pp. 478, 479), 272 (p. 567), 286 (pp. 616, 627); Castellan, *Réarmement clandestin*, pp. 426, 447; Paul-Boncour, II, 228; SDP, disp., Edge to Dept., 20 Sep 32, 763.72119 Military Clauses/138; AA, tel., Hoesch to Ministry, 20 Oct 32, 3154/D667999. Herriot told Hoesch that "his conviction that the German fight for equality of rights was a matter of carrying out Seeckt's army reform plans was so deeply rooted and so fixed in him through a continuous study of the material that he had to describe it as unalterable." Schönheinz drew up an outline for the Wilhelmstrasse of the differences between Seeckt's proposals and current German demands: AA, Reichswehrminister 322/32, 8 Nov 32, 7474/H185601-03. On left republican suspicions, Bankwitz, *Weygand*, pp. 31-33, 60, 65, 68-69, 90-91.

already been reduced in numbers and length of training to the minimum or below, and he and the general staff argued that France would be unable to come to the aid of Belgium or the allies of France in Eastern Europe. It would be necessary (the general staff stated) to extend the Maginot line along the Belgian frontier. Weygand predicted that those parts of the plan providing for security and commitments would be eliminated, and that only those parts would be retained which weakened the French forces. He doubted that the other powers would undertake the proposed guarantees, or honor them if they undertook them. He maintained that the military aspects of the plan should be weighed by the nation's highest military body, the Supreme War Council (Conseil supérieur de la Guerre).[86]

Diplomatic considerations, however, outweighed the objections raised by Weygand. To Herriot, the conclusive argument for advancing the new plan was not that it would satisfy the Germans or guarantee security; it was that, unless France showed she was not obstructive, none of the other powers would side with her, either immediately, when the Germans proceeded to rearm without permission, or later, when a Franco-German war broke out. As Herriot said, "The great apprehension of the government is that, one day, France might find herself alone, confronted by Germany." Boncour himself admitted that the Germans were likely to reject his plan, and in fact he had aimed to make a proposal which would put those refusing it in a bad light, although he also thought it should be—and was—a plan that France could live with if it did win acceptance. If the Germans did reject the plan, the French leaders hoped this would make other nations recognize Germany's intention to rearm and lead them to sympathize with France. So Weygand was overruled, and the general scheme was officially adopted on October 28.[87] Herriot spoke to the Chamber of Deputies on that same day, describing the plan in very general terms, and he won a strong vote of confidence. Boncour gave a more complete account of the plan to the Bureau of the Disarmament Conference on November 4, and the final text was published in full on November 14.

[86] Weygand was Inspector-General and Vice-President of the Supreme War Council (Conseil supérieur de la Guerre), and although he would only assume command in the event of war, he had more moral authority than the Chief of Staff, General Gamelin. On his views and those of the General Staff on the Boncour plan, see Weygand, II, 390-391; *DDF*, 1/I, Nos. 250, 260, 268, 272, 273, 286 (esp. pp. 620-622); Jacques Minart, *Le Drame du désarmement français (ses aspects politiques et techniques)* (Paris: Le Nef de Paris, 1959), pp. 30-32; Gamelin, II, 77-78.

[87] *DDF*, 1/I, Nos. 244 (esp. p. 439, n. 3), 250 (esp. pp. 477-478), 260, 272, 273 (esp. p. 577), 286 (esp. p. 616); Czech docs., Memo of conversation between Beneš and Herriot, 11 Nov 32, 2376/D497273-74. See also Parker, *Europe 1919-1945*, pp. 248-249. On the seriousness of the plan, cf. Hughes, pp. 233-238.

250—Britain Intervenes to Prevent a Break

7. "Anglo-Saxon" Alternatives

As an effort to win the support of the Americans and British, the French plan was a gross miscalculation. It attempted to provide either a general security system, with varying responsibilities for all nations, or failing that, a means of aligning governments alongside France, in common opposition to Germany. But the British and American governments did not want to assume responsibilities or form alignments; they wanted to bring about some kind of disarmament agreement, including Germany, that would, without cost to them, satisfy public opinion in their own countries.

The French took a great interest in the American reaction, in itself and because it was likely to influence the British position.[88] Although Norman Davis had said to MacDonald that plans such as the French were "useless," Davis at first responded cordially to an explanation by Herriot of the French plan. But even at the start, Davis had to point out to Herriot that Stimson's August 8 statements on consultation could not be incorporated in a disarmament treaty, and Stimson himself soon warned the French ambassador in Washington that this would be quite impossible. As Jay Pierrepont Moffat (Chief of the Division of Western European Affairs in the State Department) studied the full plan, he became convinced it was "a reincarnation of the Protocol of 1924 and the Tardieu Plan." Noting the complication and difficulty of negotiating the French plan, and the reluctance of Congress to vote more funds to support the American disarmament delegation, Davis began to elaborate ideas of his own for quickly concluding an interim convention, supposedly to give the powers more time to study such far-reaching plans as the French were making. Though Davis sometimes described his plan as a mere registration of the results already obtained—or as the Geneva correspondent of the *Manchester Guardian* put it, "what are humorously called the results already obtained"—his proposal involved such measures as an acceptance of a standard pattern for continental armies, scrapping large tanks and possibly heavy mobile guns, the abolition of aerial bombing and chemical warfare, and the completion of the London Naval Treaty by French and Italian accessions. Davis thought such an agreement would make it possible to say that the conference had accomplished something, and he seems to have had a particular personal interest in being

[88] The French interest in the American response is esp. clear in the Czech docs., Memos of conversation between Beneš and Herriot, Berthelot, and Massigli, 17, 23 Nov 32, 2376/D497281, D497283, D497285-86. MacDonald told Beneš on 12 Nov 32, not very frankly, that there would be no difficulty about British acceptance of the consultative provisions of the French plan if the United States accepted them: 2376/D497303; also *DDF*, 1/I, No. 330.

able to make such a claim. After the Democratic electoral victory of November 8, Davis hoped to become the next Secretary of State, but because he had no personal claim on Franklin D. Roosevelt (he had favored the nomination of Newton D. Baker), Davis's main hope for preferment lay in achieving, quickly, a major diplomatic coup.[89]

As for the British, they had been continuing, during the preparation of the French plan, with their own consideration of how to bring Germany back into talks on disarmament by meeting her demand for equality. As before, two approaches seemed possible. One was for Britain and the other former Allies to disarm to the German level; the other was to allow Germany a modification of her arms status. Neither approach made any allowance for security commitments, such as those in the French plan, and both were intended to avert the division of Europe into two camps. But disarmament downwards, the Leeper approach, cut against the conservative grain; it might weaken imperial defenses. When the cabinet discussed, on October 11, the idea of forgoing the arms denied to Germany, the ministers decided there should be no such proposal until they had deliberated further. Such a step would be a very serious one, although it was pointed out that such a proposal had almost no chance of universal acceptance.[90]

The official supporters of disarmament downwards now made their most serious effort. Allen Leeper believed that Herriot, during his London visit, had agreed to the replacement of Part V and the control of the armaments of all nations by a single convention with a common duration, as well as to disarmament by stages toward the German level in preference to letting the Germans increase their arms. Leeper now suggested that Britain, France, and Italy could propose to Germany a single convention with a uniform duration, disarmament on their own part, and no German rearmament; all four powers could then confer in some Swiss locale other than Geneva. Ralph Wigram and Vansittart urged the necessity of taking prompt action, and Anthony Eden (then Parliamentary Under-Secretary at the Foreign Office) joined with Van-

---

[89] MDP, 8/1, Diary, 11 Oct [32]; *FRUS, 1932*, I, 348-350, 398-404; David Papers, Draft cable to Walter Lippmann, [ca. 26 Jun 32]; ltr., Davis to Sumner Welles, 8 Dec 32; ltr., Davis to Castle, 20 Dec 32; ltr., Davis to Gibson, 31 Dec 32; Moffat MS Diary, entries of 15 Nov 32, 1 Mar 33; *DDF*, 1/I, No. 258; I/II, Nos. 5, 13; *Manchester Guardian*, 5 Dec 32. Massigli indicated at the final French discussion of the draft *plan constructif* that the part of the plan dealing with the obligations of the outermost ring of powers had been drafted by Davis (*DDF*, 1/I, No. 286 [p. 618]; see also FO 371, Memo by Wigram, 30 Oct 32, W11936/10/98). As regards the version attached to this discussion (*DDF*, No. 286, Annex, *Avis* No. 2, II, A, 1), this authorship seems credible, despite Davis's denial that he had given advice (*FRUS, 1932*, I, 481-482). But the wording was changed in the version finally published (*DDF*, 1/I, No. 331), and made more compulsory and automatic.

[90] Cab 23/72, meeting of 11 Oct 32, 50(32)5.

sittart and Simon in concocting a memorandum over Simon's name to secure cabinet agreement.[91]

This paper pulled out all stops, saying that although Germany was probably rearming, only a "bold course" would divert public sympathy from Germany and bring opinion to bear against German rearmament. Also, only a "bold declaration" would avert a collapse of the conference. If Germany did not return to the conference and the conference collapsed, Germany's neighbors would not disarm.

> A Conference which began by aiming at general disarmament will end in increased armaments and the probability of a renewal of war. Our own country may keep out of it, but even so, we cannot expect our own burdens in the matter of armaments to be reduced, and they will probably have to be increased. And a new war in Europe will shatter economic recovery.

Criticism in Parliament and in the press would become more intense, the goodwill of the United States would be prejudiced, and in the end Germany's claim to equality would prevail "in a really effective form." The paper advised that the government should accept a common convention with a common duration, and also concede to Germany all arms permitted to other states. To keep this from leading to German rearmament, the memorandum urged a declaration that *"we are now prepared to accept the principle, if other states will do the like, that we content ourselves with those kinds of arms to which Germany is now limited"* (emphasis in original). This was the full Leeper position, although Simon believed (as Leeper himself also did) that the specific restrictions on German arms might be altered without permitting an overall increase in strength. On October 19, Simon presented this position to the cabinet, pointing to the strong public demand for a British lead.[92]

While Foreign Office officials argued that it was essential to forestall and outbid the French plan, the army and air force counterattacked against the idea of a qualitative disarmament to the German level. Briefed by Sir Maurice Hankey, Lord Hailsham, the War Secretary, spoke strongly against the proposal in a cabinet committee. Field Marshal Milne contended that disarmament to the German level would be disastrous, as Germany would continue her covert preparations and would have a flying start in five years. The War Office also raised specific arguments against the abolition of tanks and heavy

[91] FO 371, Memo and minute by A. Leeper, 14 Oct 32, minutes by Howard Smith, Vansittart, Wigram, 14-15 Oct 32, W11778/1466/98; Minute by Leeper, [ca. 18 Oct 32], W11474/1466/98; Eden, p. 26.

[92] Cab 24/233, CP 347 (32), 17 Oct 32; Cab 23/72, meeting of 19 Oct 32, 53(32)1.

guns, and against an attempt to apply Part V-type restrictions, designed to provide only for the maintenance of German internal security, to the army of a colonial power. The Air Ministry naturally objected to any proposal for eliminating military aircraft or internationalizing civil aviation. MacDonald appears not to have liked the idea of further British disarmament, especially in the air.[93] In any case, the emergence of the French plan made it apparent by the end of October that the French would not join in any London-drafted three-power declaration, so what now came under consideration was a unilateral statement, for which Britain would bear full responsibility. Simon had retreated from the Leeper position by October 28, when he rephrased his formula to say that all states should be content with the same kinds of arms—no longer necessarily the kinds prescribed for Germany in the Versailles Treaty—and this was the kind of qualitative equality the British eventually proposed. Thus the British government declined to accept for itself the kind of arms scrapping urged by disarmament enthusiasts—a refusal that was almost certainly fortunate, since our present knowledge of German resentments, plans, and objectives shows that a Western disarmament would hardly have affected German policy. But if the Germans were not granted qualitative equality, and if the other powers did not disarm, no defensible basis remained on which to deny weapons to Germany.[94]

As in September, the main objectives of the ministers were to get Germany back to the conference and to make a favorable impression on the British public. Simon stressed to the cabinet, on October 31, the need for decisions in order to anticipate the French plan and also to be ready to negotiate with Neurath, who would soon be coming to Geneva for a League Council discussion of the Manchurian problem. In the Foreign Office, information about the French plan was still incomplete, but the plan would apparently call for further British commitments, while not satisfying Germany. Simon indicated his reaction and perspective in a brief minute: "I cannot understand *why* the Fr[ench] plan should bring the Germans back." When Beneš visited MacDonald, the latter also stressed the primary importance of securing Germany's return, as opposed to the elaboration of plans. On the other hand, British officials believed that Germany should

[93] Foreign Office views: FO 371, Memo by Cadogan, 26 Oct 32, W12080/1466/98; Memo by Wigram, 30 Oct 32, minutes by Leeper and Vansittart, 31 Oct 32, W11936/10/98. Military views: Roskill, *Hankey*, III, 61; Cab 24/234, CP 362 (32), 28 Oct 32; CP 369 (32), 29 Oct 32; CP 373 (32), 1 Nov 32; CP 384 (32), 7 Nov 32. MacDonald's views: *DBFP*, 2/IV, Nos. 152, 153, 156; Roskill, *Hankey*, III 63. Aside from bringing an early draft of the French plan to London, Herriot permitted Wigram on 29 Oct 32 to see a version of the plan corresponding to *DDF*, 1/I, Annex, *Avis* No. 2.

[94] Cab 24/234, CP 360 (32), 28 Oct 32; see also *DBFP*, 2/IV, Appendix I (I).

also make some positive contribution, they wished to reassure France without actually accepting French security proposals, and apparently they felt there should be some restraint to balance the concession of qualitative equality. Therefore, the Foreign Office proposed and the cabinet agreed that the grant of equality of rights should be contingent on the securing of "satisfactory guarantees that Germany will not upset the tranquillity of Europe." This was a reference to the idea of a "political truce," a proposal that Vansittart had advocated in 1931, and that Simon had briefly advanced just before the Lausanne Conference. A cabinet committee worked out a policy statement, the cabinet approved it on November 9, and Simon unveiled it in the House of Commons the next day. In short, the British ministers took their position before the full French plan was known, but in some measure to undercut it and outbid it—especially in the eyes of the British public. This time the British statement was not passed to Herriot beforehand, as had been done with the September 19 memorandum.[95]

The British policy, stated by Simon in the Commons, was that Part V should be superseded by a single convention binding on all, that the duration should be the same for Germany as for other states, and that Germany should have qualitative equality; the stages and means of the application of this last should be discussed by the conference, with Germany again taking part. Simon opposed any increase in strength in the name of equality, and his acceptance of the three points on equality of rights was subject to the condition that "the European states should join in a solemn affirmation that they will not in any circumstances resolve any present or future differences between themselves by resorting to force." Without saying so explicitly, the British were apparently thinking here primarily of the Polish Corridor; they had never succeeded in getting a satisfactory German renunciation of a warlike solution to that question. Simon paid his respects to what

[95] Cab 23/72, meeting of 31 Oct 32 (a.m.), 56(32)3; meeting of 31 Oct 32 (p.m.), 57(32)—; meeting of 9 Nov 32, 60(32)1; Cab 24/234, CP 360 (32), 28 Oct 32; FO 371, Memo by Wigram, 30 Oct 32, minutes by Cadogan and Leeper, 31 Oct 32, W11936/10/98; Minute by Simon, 6 Nov 32, W12125/1466/98; Czech docs., Memo of conversation between Beneš and MacDonald, 12 Nov 32, 2376/D497302-06; MDP, 8/1, Diary, 13 Nov [32]; *DBFP*, 2/IV, Appendix I. On the immediate motives behind the "political truce," see also FO 371, Minutes by R.M.A. Hankey and Vansittart, 22, 26 Sep 32, C7908/211/18; Memo by Sargent, 4 Oct 32, C8438/211/18; Memo by Cadogan, 26 Oct 32, W12080/1466/98; ltr., Simon to Vansittart, 12 Nov 32, W12571/1466/98; AA, tel., Hoesch to Ministry, 14 Nov 32, 3154/D668189. On the earlier background of the truce idea: Cab 24/227, CP 4 (32), 1 Jan 32; *DBFP*, 2/III, Nos. 128, 134. On lack of consultation with France: AA, tel., Hoesch to Neurath and Bülow, 11 Nov 32, 3154/D668161; tel., Hoesch to Ministry, 12 Nov 32, 3154/D672114. The French, Italian, and Belgian governments were briefly informed on 9 Nov 32 that Britain would concede equality of rights in principle, but would leave the discussion of details for Geneva: *DBFP*, 2/IV, No. 169.

was known of the French proposal, and claimed he was not trying to set up a rival to "anybody else's plan," but it was obvious that the British approach to equality of rights would likely conflict with French ideas. In a further speech at Geneva on November 17, Simon associated his concession of equality of rights with specific suggestions, somewhat along the lines of Leeper's August proposals, for reducing the size or numbers of battleships, artillery, tanks, and bombers. All these proposals would probably be unwelcome to France, although Simon also advanced the theory that German overall strength should be kept at the same level, as by offsetting the introduction of a militia and of shorter regular service by reductions in regular strength.[96] Simon and MacDonald hoped, however, that a German agreement not to use force would weaken the French case for security guarantees, that preliminary talks could be held during the League Council meeting on Manchuria, and that these talks might blossom into a four- or five-power conference to recognize German equality.[97]

8. Conference Plans and Apprehensions

The September 19 statement and the October 4 proposal for a four-power meeting had failed to induce the Germans to negotiate. London was now anxious to see whether Germany would accept the latest British proposal. Leopold von Hoesch, freshly transferred as ambassador from Paris to London, attended the whole debate in the House of Commons on Simon's proposal, and reacted most favorably. In his first excitement, he told Bülow by telephone that Simon had offered full recognition of the German demands. But Bülow, who was acting in Neurath's absence, seems to have been cool, indeed suspicious, and he stated that Germany had to ask not merely for token specimens of the arms retained by others, but for the same proportion of them in relation to effectives. Germany could join others in a "no force" declaration, but could not confirm the status quo, or make a unilateral commitment ahead of other nations. Simon wanted very much to get immediate German agreement to the "no force" declaration, without the Germans awaiting the agreement of other governments, but Bülow did not yield to Rumbold's pleas on this, which Simon and Vansittart duplicated through Hoesch. (Bülow explained to Hoesch that he was

[96] *Documents on International Affairs, 1932*, pp. 209-217; *DBFP*, 2/IV, Nos. 170, 183. *DBFP*, 2/IV, Appendix I seems to show the association of the "no force" declaration with the Corridor question. See also Chapter One, note 16.

[97] Czech docs., Memo of conversation between Beneš and MacDonald, 12 Nov 32, 2376/D497304; *DDF*, 1/I, No. 330. See also AA, tels. (2), Hoesch to Ministry, 4, 13 Nov. 32, 7360/E536070, 3154/D672110-15; *DBFP*, 2/IV, No. 170 (p. 265); FO 371, Minute by Sargent, 8 Nov 32, W12532/130/98.

concerned lest the German government appear to accept, even if only indirectly, the Eastern borders.) When Neurath returned on the fifteenth, he backed Bülow up, if in a more friendly way. Yet, along with their hesitations, Neurath and also Bülow expressed the hope that problems could be worked out in the coming talks in Geneva on the fringes of the Manchurian debate.[98]

Bülow and Neurath had in fact a difficult task, one of making German policy appear acceptable abroad, and at the same time of satisfying Reichswehr and nationalist sentiment at home. The official German position was that Germany would only return to the Disarmament Conference in Geneva if the other powers conceded her equality of rights; Schleicher, as we have seen, had stressed that this should be *full* equality. Germany had also declined to enter into talks in Geneva, in the shadow of the conference. But by Simon's latest declarations, Britain had now virtually agreed to equality of rights in the terms of the German August 29 résumé, even though the British opposed rearmament and thought that material details should be worked out at the conference. It was scarcely possible to ask publicly for more than London had conceded. The French plan did not attempt so clearly to accommodate German demands, and it contained many provisions objectionable to Germany, but it represented a first French retreat from Part V, and it actually advocated short-service conscript forces for all European states, in this regard conceding more than the British. German officials still wanted an explicit French recognition of equality of rights, but although they thought that a stiff position had been fruitful for Germany,[99] they could hardly refuse now to enter into private talks in Geneva under cover of the League Council's Manchurian debate. Indeed, the accommodating attitude of the British provided an apparent chance to bring pressure on France in such talks. And the diplomats of Germany themselves wanted to return to the conference.[100]

The Reichswehr, however, had already decided in July to proceed with its own plans, "independently of the results of Geneva." In fact, General von Hammerstein had just issued final orders on November 7

---

[98] AA, tel., Hoesch to Ministry, 11 Nov 32, 3154/D668156-60; Memos (2) by Bülow, 11 Nov 32, 3154/D672093-97; Unsigned memo for Marcks, 11 Nov 32, 3154/D672381-83; tel., Koepke to Paris, 12 Nov 32, 3154/D672106-09; tel., Hoesch to Ministry, 12 Nov 32, 3154/D672114-15; tel., Bülow to London embassy, 14 Nov 32, 3154/D668169-71; Memo by Bülow, 15 Nov 32, 3154/D668194; RK, Memo by Bülow, 13 Nov 32, 3642/D812941-44; *DBFP*, 2/IV, Nos. 171, 173-175, 178, 179; FO 371, ltr., Simon to Vansittart, 12 Nov 32, W12571/1466/98.

[99] RK, ltr., Schwarz to Planck, 15 Nov 32, 3642/D812971; *DDF*, 1/I, No. 287.

[100] For Foreign Ministry outlook, see esp. AA, Unsigned memo, 11 Nov 32, 3154/D672381-83; Briefing notes (by Frohwein), 15 Nov 32, 3154/D672144-52; Circular dispatch, 22 Nov 32, 3154/D672171-83.

for reorganizing the army, and these were explained to officers from the divisions on November 11 and 12, a procedure which tended to make them more irrevocable. These plans were similar to one of the alternative schemes passed to Neurath in October: three fourths of the members of the regular army were to serve for three years, the rest of the regulars for twelve years, and 85,000 men (instead of 90,000) were eventually to be inducted for training each year, including those who would only serve for three months and those who would go on to enter the regular army. Although all inductees were initially to be volunteers, the planners clearly intended to introduce conscription later. Diplomatic problems might well arise from the plan to begin increasing inductions on April 1, 1933, something difficult to conceal. Although the militia program would not operate at full scale until 1934, these increased inductions represented a pilot program that would permit army leaders to test the new procedure of training men for three months within line organizations rather than in training battalions, and enable them to prepare cadre for the later influx. Hammerstein's covering directive stated: "In connection with the still-pending disarmament negotiations, the *preparation* for the first reorganization period must be kept secret." The directive stressed the need for precautions "through which the initiation of the measures will be *guaranteed*, but which will also achieve the result that foreign powers will not be able to *demonstrate* that we are rearming."[101] Schleicher had told Admiral Raeder in mid-August that equality of rights would have to be settled in three months, and now, after three months, the Reichswehr was proceeding as if the question had been settled. The Reichswehr seemed to assume that disarmament negotiations would fail before April 1, or that, if somehow an agreement were reached, Germany could conceal its activities that did not coincide with the agreement. Evidently matters would be much simpler for the army if the Disarmament Conference did not resume.

Probably Neurath and Bülow did not know how far the Reichswehr had already committed itself. Schleicher had spoken in the cabinet in early October as though German defense decisions were awaiting the result of the Disarmament Conference, and as though Germany had nothing to fear from French disclosures.[102] The plans Schleicher revealed to Neurath in mid-October had been still unsettled. But Schleicher's prestige was associated with the demand for equality of

---

[101] Emphasis in original. BA-MA, RH 15/49, Chef HL 6296/32, 7 Nov 32; TA 737/32, 7 Nov 32 (incl. Annexes 5 and 6); Wehramt 6003/32, 2 Sep 32; OKW, Wi/IF 5.498, TA 381/32, 29 Oct 32 (incl. Annex), 110/837756-58. On July decisions, see Chapter Four, Section 3.

[102] RK, Cabinet Protocol, 7 Oct 32, 3598/D790959-60. Neurath had heard, however, that Umbau had begun; see note 61.

rights, and he would obviously scrutinize closely any agreement reached on this. Neurath now explained to the cabinet that he had to go to Geneva to attend not only the League Council debate on Manchuria, but also discussions on certain Polish questions affecting Germans in Poland, a matter of greater German interest. It would be impossible for him to avoid discussing disarmament while he was there, but he would uphold the German refusal to return to the conference without a recognition of equality. His briefing notes played down the likelihood of agreement, and emphasized that he would not commit himself to anything without the agreement of his colleagues, and "in particular, Herr von Schleicher."[103]

The German Foreign Ministry had hoped that Herriot would come to Geneva for the League Council meeting, as Simon and Neurath were doing. The Germans apparently thought that, with the help of Britain, Italy, and the United States, they might then quickly wring an explicit recognition of equality of rights out of the French Premier and Foreign Minister. They doubtless believed that Herriot would be easier to deal with than the inflexible Paul-Boncour, who would otherwise represent France. Herriot, however, showed no inclination to rush to Geneva. Instead, picking up a passing suggestion of bilateral talks from Roland Köster, the new German ambassador in Paris, Herriot said he would be ready, following on a settlement of equality of rights (and doubtless of security), to consider a personal discussion between two high military officers, French and German, on German military wishes. In reviving this idea, which Papen had broached the previous summer, Herriot may have been encouraged by other, unofficial feelers from emissaries claiming to speak for Papen and Schleicher. And in the wake of Simon's November 17 speech at Geneva, in which the British Government seemed (as Massigli put it) "to persist more than ever . . . in assuming the position of arbiter" between Germany and France, Herriot may perhaps have thought seriously for a moment of seeking a German entente to reinsure his country against British perfidy—even though this would have been a new departure for the French leader most devoted to good relations with the "Anglo-Saxons." Whatever Herriot had in mind, the German Foreign Ministry was very wary, although it did not rule out the possibility of military talks later, if the French, like the British, would abandon their resistance to equality of rights.[104]

---

[103] AA, Briefing notes (by Frohwein), 15 Nov 32, 3154/D672144-52; RK, Cabinet Protocol, 17 Nov 32, 3598/D791182-83.

[104] AA, tel., Koepke to Paris embassy, 12 Nov 32, 3154/D672106; tel., Bülow to Neurath (at Geneva), 21 Nov 32, 3154/D672169-70; RK, disp., Koester to Foreign Ministry, 18 Nov 32, 3642/D812991; *DBFP*, 2/IV, No. 185. Bülow was also wary in

In the private diplomatic conversations that began on November 21 on the fringe of the council meeting, Simon made evident his readiness to do whatever was needed to bring about a five-power meeting and a German return to the conference. This attitude naturally encouraged Neurath and alarmed the French. According to Neurath's report to Berlin, Simon promised him that if Herriot did not agree to Simon's own formula conceding equality of rights, as the Italians and Americans were doing, Simon would conclude that the responsibility for the failure of the conference lay with France. To Neurath, this opened up glorious prospects, and he commented to Bülow: "I believe that, to begin with at least, we have achieved that which we intended, namely the formation of a front of the three principal powers against France." It soon became apparent that, contrary to Neurath's first impression, Simon would not insist that the French agree even before the five powers met. Nevertheless, Neurath still seems to have reasoned that the outlook for forming a common front against France was favorable. Disregarding the standard German position that recognition of equality had to precede negotiation, Neurath told a group of German correspondents he would be ready to take part in unofficial talks without seeking preliminary assurances.[105]

From the French, Simon sought acceptance of his formula and of a five-power meeting in Geneva; this last, he argued, would be the meeting that Herriot had "pressed for" in October. Herriot had, of course, only pressed for a Geneva meeting then because he had expected the Germans to refuse it, but much as he might have liked to, the French Premier could hardly decline to attend such a meeting now, especially as he had recently sent word to MacDonald that he would come to Geneva when it was necessary. Still, Herriot and Boncour could and did refuse to agree to equality of rights in abstraction from considerations of security and disarmament. When Herriot agreed to come to a five-power conference, he stressed (to Tyrrell) that his cabinet had formally decided to stand firm on linking equality of rights with disarmament and security.[106]

French suspicions were aroused by Neurath's declared readiness to

---

deciding not to inform the Reichswehr Ministry at this stage of Herriot's new proposal. On unofficial feelers, see Czech docs., Memo of conversation between Beneš and Herriot, 11 Nov 32, 2376/D497275-76; tel., Ibl (Paris) to Foreign Ministry, 12 Nov 32, 2376/D497277. Massigli view: *DDF*, 1/II, No. 6.

[105] AA, tels., Neurath to Bülow, 22, 24 Nov 32, 3154/D668265-67, D668278; *DBFP*, 2/IV, Nos. 188, 196, 198, 199; *DDF*, 1/II, No. 33.

[106] *DBFP*, 2/IV, Nos. 187, 189-194, 197; Czech docs., Memos by Beneš on conversations with Herriot and Massigli, 17, 23, Nov 32, 2376/D497281-82, D497285; *DDF*, 1/II, Nos. 23, 25, 33, 42, 51; Pompeo Aloisi, *Journal (23 juillet 1932-14 juin 1936)*, Maurice Vaussard, trans., Mario Toscano, ed. (Paris: Plon, 1957), pp. 22-23.

enter talks without preliminary assurances, and these suspicions deepened when Neurath told two journalists that Simon had promised to support the recognition of equality of rights, and had indicated that the Reich had everything to gain and nothing to lose from five-power talks. The British tried to reassure the French, but with very indifferent success. In confirming to the French government that Neurath had agreed to come "without seeking to exact previous assurances" (or in another version, "specific assurances")—language the French transcribed and translated as "without seeking exact precise assurances"—Simon unwittingly encouraged suspicions that *general* assurances had been given. Herriot pointedly observed to Tyrrell that Simon had "certainly not failed" to inform Neurath of the French insistence on discussing equality only within the framework of the French plan—a case in which "certainly" indicated considerable uncertainty. When Davis traveled to Paris to try to interest Herriot in his scheme for an interim convention, Herriot gushed out his anxiety over the coming five-power meetings, and at the end of one interview in his office he pushed the astonished Davis into his own chair, saying that "if only Mr. Davis would sit in that chair for a while, with the responsibility for millions of French people on his shoulders, he would then understand his [Herriot's] preoccupation in the face of Germany." Herriot also expressed his fears, especially of isolation, to Tyrrell, saying that Germany intended to use the British to squeeze the French, and if this failed, to separate the British from the French; Germany stood to gain either way. MacDonald responded by telling the ambassador to inform Herriot that "the danger pointed out by the Premier is very much on his [MacDonald's] mind, that he will go to Geneva preoccupied by this, and that he will do everything in his power to avoid this danger. He does not intend to fall into the trap which the Germans may set."[107]

There was now concern in certain German quarters also, arising from the possible conflict between disarmament talks or agreements and the Reichswehr's plans. Shortly after Simon's statement in Parliament on November 10, an officer in the Reichswehr Ministry told the British military attaché in Berlin that the British acceptance of the

[107] *DDF*, 1/II, Nos. 33 (and note), 34, 39, 42, 51, 55; *DBFP*, 2/IV, Nos. 198, 199, 201, 202; *FRUS, 1932*, I, 476-481, 484. MacDonald was indeed concerned to avoid a Franco-British rift. When he heard that Hungarian leaders had compared the French plan unfavorably with the British proposals, and had stated that Reichswehr authorities took the same view, MacDonald commented that there appeared to be "a move to concentrate on us in order to divide us from the French. Do not encourage this kind of language or expectation. Something might usefully be said to show that we are aware of possibility of this": FO 371, Draft ltr., Howard Smith to Chilston, 1 Dec 32, W13031/1466/98.

German right to equality seemed to be balanced by an unmistakable refusal to permit that right to be put into execution. General (as he now was) Kurt Schönheinz, head of the League of Nations Section (Group Army) and the moving spirit in the Bendlerstrasse in advancing army aims in the disarmament question, gave a briefing to Schleicher and General Adam, Chief of the Truppenamt, on December 3, and expressed his anxiety that Neurath, by agreeing to Simon's proposals, might secure equality of rights in principle, but foreswear rearmament in practice. It says much for the atmosphere within the German government that Schönheinz drew on an Abwehr agent report from Geneva to show that Neurath had accepted British proposals too readily. Schönheinz also gave Schleicher a long memorandum elaborating on his view of the dangers. Germany (the memorandum said) would be asked to state expressly that she would not achieve an equalization of armaments by rearmament, and indeed some German officials had already made too many statements along this line. (Orally, Adam named Bülow in this connection.) There had been a truce on new arms construction during the conference, and its extension would now be sought, which would bring Germany, with her April 1, 1933, measures, into a difficult situation, i.e., one in which she might appear as a violator of the truce. (Schönheinz did not suggest that a postponement of the measures could be considered.) A negotiation on the basis of Simon's practical proposals would militate particularly against German claims in the areas of personnel and air forces.[108]

Schönheinz inferred that recognition of German equality would now scarcely lead to a four- or five-power substantive conference in which practical decisions could be reached. Instead, Germany would be forced to come back to the Disarmament Conference. And if she returned to the conference for a mere moral equality, she would have thrown away her best card (the equality issue) and would become the scapegoat if she later refused to accept an unfavorable treaty. "Militarily," Schönheinz wrote, "there is no interest in reviving the bogged-down conference by our return and providing the whipping-boy our enemies are waiting for in order to evade disarmament themselves." Schönheinz could only envision one set of conditions under which Germany would enjoy a favorable situation at a resumed conference: this would be if she secured beforehand "legal and practical" as well as moral equality. In that case the level of German armament in the

---

[108] *DBFP*, 2/IV, No. 175; NA-VGM, FVII a II. Note on briefing of Reichswehr Minister, 3 Dec 32, Abwehr report, 28 Nov 32, memo by Schönheinz, 1 Dec 32, 7792/E565444-52. Adam cited statements Bülow had made to the Italian ambassador; these had probably become known to the army through the decryption of Italian diplomatic messages. See Chapter One, note 102.

first period would result from "voluntary German concessions and not from German demands."

Schleicher assured Schönheinz that there could be no question of backing down from the German (i.e., the Reichswehr) position. He also agreed that it was advantageous not to pull other people's chestnuts out of the fire by returning to the conference. He indicated, however, that he had confidence in Neurath, and although he said that he would clarify the situation at the cabinet meeting that day, there is no record he did so. As would appear on other occasions in the following weeks, he now seemed to be seeking publicity rather than secrecy on German military plans, saying that propaganda in the press on Umbau should stress those measures, such as the shortening of the term of service, which the French had declared to be disarmament. Generally, Schleicher professed to support Schönheinz's objectives, while at the same time displaying more optimism about the situation.[109] Actually, he had just become chancellor, and would now be compelled to put the best face on the situation. Certainly he had no interest at this moment in provoking international tension, and the circumstances of his accession weakened his position vis-à-vis Neurath.

9. THE FIVE-POWER CONFERENCE, DECEMBER 6-11, 1932

MacDonald and Simon arrived in Geneva for five-power meetings on December 2, and Herriot came on December 3. Neurath only appeared late on December 5, having been held up by the negotiations through which Schleicher replaced Papen; Neurath himself retained his post, as did most of the other members of the former Papen cabinet. During the interval before Neurath's arrival, MacDonald and Simon talked with Boncour and Herriot, and Davis attempted to interest the others in his plan for an interim convention. Herriot continued to maintain that the other powers should begin drawing up a convention without Germany, which she could then accept or reject. MacDonald spoke to the French of his concern over German divisive maneuvers, but in his diary he vented his annoyance with French policy:

> ... [Saw] Herriot, Boncour & Massigli & it was evident they did not want any conversations with Germany, but just to get us to stand

[109] NA-VGM, F VII a 11, Note on briefing of Reichswehr Minister, 3 Dec 32, 7792/E565444-45; RK, Cabinet Protocol, 3 Dec 32 (12:30 and 12:45 p.m. meetings), 3598/D791242-47. It is possible that, as with army matters, a cabinet discussion on disarmament questions might have gone unrecorded. But the protocols of the meetings of 7 and 14 December give extensive coverage to disarmament discussion (3598/D791250-56, D791265-68), and the content of the former suggests that there had not been any recent consideration of the subject.

firm with them in telling Germany she was bound by treaty & that nothing could be done. They desired to put pistols to Germany's head & ask them to give us in writing straight away what they meant by "equal rights" & so on.—in fact to do what they wd. have done the day after a great victory.—Most hopeless mentality & shows how France can drive Germany into war again.

For MacDonald, the first essential was to get Germany back to the conference.[110]

At first Davis seemed surprisingly successful in his high-pressure salesmanship on behalf of an interim convention. He won the support of the Italians for his plan, which tended to uphold their claim to naval parity without increasing their actual arms expenditures. The British and French had little enthusiasm for the Davis proposal, but apparently they thought it wisest—particularly at a time when their war debts were under discussion—initially to give him his head, and let him discover for himself the impossibility of reaching agreement on his plan within a few days. There was also the consideration that he might be the next Secretary of State. But Boncour and Herriot insisted that Davis not disclose to the Germans the specific disarmament measures he had in mind, as the Germans would then regard these measures as something already conceded, while nothing would have been obtained on the score of security.[111]

The critical discussions began when Neurath appeared for the first time at the meeting of the powers on December 6. After Davis had expounded a version of his plan, leaving out the detailed disarmament proposals as the French desired, Neurath observed in his most diplomatic way that he seemed to be confronted with a new plan. He maintained that in November, all except the French had agreed both that Simon's proposals could form the basis of discussion, and that the French should be asked to agree to proceed on that same basis. Herriot then plunged in with a statement of the basic French position, as approved by his cabinet:

> France concedes that the aim of the Conference is to grant equality of rights to Germany and other powers disarmed by treaty within

---

[110] *DBFP*, 2/IV, Nos. 204, 206, 208; *DDF*, 1/II, Nos. 60, 67; *FRUS, 1932*, I, 489-490; Cab 23/73, meeting of 13 Dec 32, 66(32)1; MDP, 8/1, Pocket diary, 3, 5 Dec [32]. The 3 Dec [32] entry also shows, however, an interest in blaming Germany for any breakdown of the conference.

[111] Aloisi, pp. 24, 26-27; *DBFP*, 2/IV, Nos. 202, 205, 207-209; *DDF*, 1/II, Nos. 67, 68; *FRUS, 1932*, I, 489-490. A memo by Cadogan of 7 Dec 32 (FO 371, W13537/1466/98) indicates a British expectation that Davis would fail. The British delegation hoped that it would be objections from others, and not themselves, that would provide a stumbling block for the Davis plan (FO 371, Minute by Leeper, 7 Dec 32, W13709/1466/98).

a regime which will provide security for all nations, as for France herself.[112]

This brought the conferees squarely up to the terms on which equality of rights and security should be recognized and a German return to the conference arranged. It appeared to MacDonald that the French and Germans were on the verge of agreement when Davis practically advised Neurath not to agree. Davis thought that the main topic was disarmament, and he tried to tempt Neurath into demanding immediate disarmament measures, which he seems to have naively supposed Neurath wanted. But Neurath, who had received a number of reports on the disarmament provisions in the Davis scheme, was even less attracted by that scheme than MacDonald and Herriot; for him, it would have constituted a cheap and very inferior substitute for equality of rights. He wanted a recognition of equality without delay, and he did not want that equality to depend on the achievement of security by means of the French plan—especially since MacDonald and Davis made it clear that their governments would not extend their security commitments.[113]

In any case, Neurath did not have full powers to conclude an agreement. It was agreed that Neurath would submit a version of the French formula to his colleagues in Berlin, as a statement from all the other powers. Neurath informed Bülow the same night that he was quite impressed by Herriot's newly stated position, with its explicit recognition of equality of rights. He commented that such a declaration, if it had been made in July, would probably have obviated the German departure from the conference; it was important not to let this

[112] DBFP, 2/IV, No. 210; FRUS, 1932, I, 492-494; DDF, 1/II, No. 71; RK, Cabinet Protocol, 7 Dec 32, 3598/D791250-53; tel., Neurath to Foreign Ministry (no. 48), 6 Dec 32, 3642/D813015-17. The text given here is my translation of the French version of Herriot's statement, as given in French and German sources. In British and American sources, words corresponding to "le but de la conférence" were omitted.

[113] MDP, 8/1, Pocket diary, 6 Dec [32]; 1/513, ltr. to Baldwin, 9 Dec 32; FO 371, Memo by Cadogan, 7 Dec 32, W13537/1466/98; ltr., Michael Wright to A. Leeper, 7 Dec 32, W13597/1466/98; DBFP, 2/IV, Nos. 210, 211; DDF, 1/II, Nos. 71, 72; FRUS, 1932, I, 492-493, 497; SDP, tel., Davis to Dept., 6 Dec 32, 500.A15A4 Steering Committee/221; AA, tel., Prittwitz to Neurath, 1 Dec 32, 7360/E536110-12; tel., Weizsäcker to Ministry, 3 Dec 32, 3154/D668316-20; tel., Weizsäcker to Bülow, 4 Dec 32, 7360/E536114-15. Davis believed afterwards that an opportunity was lost. Years later, in a letter to Raymond Leslie Buell, 18 Jun 40 (Davis Papers), Davis wrote: "I do feel somewhat resentful myself when I think of how hard I worked to try to get a disarmament agreement—how nearly we came to an agreement on two different occasions before Hitler came to power—and which in my judgment would have prevented him from coming to power, but it failed because the French refused to make some slight concessions which were only sensible and fair." Presumably one of the two occasions was when Tardieu failed to come to the Bessinge meetings, and the other was in December 1932.

concession slip away by making a purely negative answer. At the same time, Herriot's proposal could not be simply accepted as it was. Neurath suggested that the cabinet consider accepting the French formula and adding a German interpretation saying that equality of rights should not only be accepted in principle, but should be applied in practice in all respects.[114]

In reporting to the German cabinet the next day (December 7), Bülow went even farther, and tried at first to induce the ministers simply to accept the French formula without an interpretation. He argued that it was scarcely possible to obtain an improvement, that a German refusal would provoke a storm in the foreign press, and that the French formula went beyond those formerly proposed by Germany at Lausanne and Geneva. The problem, he said, was largely whether German public opinion could be asked to accept the French formula, even though Germany's "clear wishes"—presumably the specific wishes for a militia and arms—were not fulfilled.[115]

But Schleicher, now Chancellor, did not accept Bülow's recommendation. He remarked that the British, the Americans, and probably the Italians only wanted to grant equality on the condition that no use be made of it for five years; thus it was only a matter of a moral concession. Even worse, the French still wanted to leave it up to the conference to decide how much of a claim Germany had to equality. In his view, the Herriot formula offered no more than had been offered before the conference adjourned. After discussion, however, Schleicher became somewhat more amenable, and he summed up his final views by saying Germany would welcome a statement that general security was to be the goal of the conference. But

> on the other hand, the thesis could not be accepted that our claim to equality of rights should likewise be only a goal of the Conference. If equality of rights were given us under conditions, this too would be no significant advance; instead, under all circumstances it is to be arranged that the definitive treaty must contain the recognition of our claim to equality of rights.

In other words, Schleicher would return to the conference only if he obtained now an unconditional and irrevocable concession of equality. He did not, however, press for "practical" equality, or for specific concessions in men or arms. Bülow now fell back on the idea of a

---

[114] *DBFP*, 2/IV, Nos. 210, 211; SDP, tel., Davis to Dept., 6 Dec 32, 500.A15A4 Steering Committee/221; RK, tel., Neurath to Bülow, 6 Dec 32, 3642/D813014; tel., Neurath to Ministry, 6 Dec 32, 3642/D813015-17.

[115] RK, Cabinet Protocol, 7 Dec 32, 3598/D791248-53. On 10 December, Bülow sent materials to the Reich Chancellery to support the contention that the Herriot formula represented a significant advance (RK, 3642/D813018-19).

German "interpretation" of the French formula, as suggested by Neurath in his message from Geneva. But Schleicher thought it better to explicitly ask the other powers to confirm "that the recognition of our claim to equality of rights will form a significant part of the treaty." Accordingly, Bülow told Neurath confirmation must be obtained that equality of rights was not merely a goal of the conference, but the basis of its negotiations.[116]

At a brief meeting of the representatives of the powers on December 8, Neurath asked for a clarification that equality of status was "to receive practical effect in the future convention in every respect," and that it would be the starting point for future conference discussions respecting "the disarmed states." Boncour, in turn, wanted to know exactly what was meant by "equality of status in every respect," and Neurath provided on December 9 a detailed list of points, much of it taken from the August 29 résumé; he implied that the list might later be extended.[117] To the others, Neurath's list seemed to anticipate the work of the conference and to foreshadow still further demands. When Neurath tried to prepare a statement combining the Herriot statement with a German interpretation, Herriot commented that a document containing two opposing theses, one stating that equality of rights was a "point to be reached," and the other, that it was a "point of departure," would mean nothing.[118]

MacDonald and Herriot had both left Geneva after the five-power meetings on December 6 to confer in Paris on war debts with their finance ministers. Herriot then became entangled in the final struggles of his collapsing cabinet, and he never returned to Geneva as Premier. MacDonald did return on December 9, and he guided the five powers back to the question of Germany's return to the conference. Davis suggested that a joint declaration be drafted, and by the afternoon of December 9, MacDonald (and probably Simon) had worked out privately what was to be the major modification in Herriot's formula, one which conceded the demand Schleicher had made

---

[116] RK, Cabinet Protocol, 7 Dec 32, 3598/D791254-56; tel., Bülow to Neurath, 7 Dec 32, 3642/D813003-04. Bülow's telegram painted a deceptively rosy picture of the cabinet's reaction to the Herriot formula and the proposal of a unilateral German statement, and Schleicher, who read the message, had Bülow called to account (RK, Note by Planck, 8 Dec 32, 3642/D813002). Bülow explained that his message had been framed as positively as possible "in view of the coming struggle over the responsibility for a breakdown," i.e., presumably, with an eye to eventual "White Book" publication, and that Neurath had been told the true situation by telephone.

[117] DBFP, 2/IV, No. 215; DDF, 1/II, No. 82; FRUS, 1932, I, 498-499; RK, tels., Neurath to Ministry, 8 Dec 32, 3642/D813005-06.

[118] DBFP, 2/IV, Nos. 215-217; DDF, 1/II, Nos. 82-84, 88, 89; RK, Memo by Bülow, 10 Dec 32, 3642/D813024-25; tel., Neurath to Foreign Ministry, 10 Dec 32, 3642/D813028-30.

in Berlin. Instead of saying that "one of the *aims* of the Conference" was to accord Germany equality in a system of security for all, the MacDonald formulation had the British, French, and Italian governments declare that "one of the *principles* that should *guide* the Conference" was to accord equality in a system of security, and that "this principle should find itself *embodied*" in the disarmament scheme agreed on by the conference. (Italics supplied.) Aside from the British, apparently only Davis and his aide, Allen Dulles, knew on the ninth of this formulation, which MacDonald held in reserve.[119] There seems to have been no thought of consultation with the French; the Lausanne promises of July 8 were now a dead letter. Neurath, still ignorant of this formula, predicted to Bülow that the negotiations would break down, and that Germany would not return to the conference, but also that Germany would be in a position to put the blame on France, and that an anti-French front would probably be formed.[120]

MacDonald met on the morning of the tenth with the second-ranking members of the other delegations, supposedly to draft an agreed statement. MacDonald, the first among unequals, presented the others with a ready-made declaration, combining the language he had worked out the day before with some of Simon's November formula. The key first paragraph (actually written in the form of two one-sentence subparagraphs) ran:

> 1. The Governments of Great Britain, France, and Italy declare that one of the principles that should guide the Conference on Disarmament should be the grant to Germany, and to the other disarmed Powers, of equality of rights in a system which would provide security for all nations, and that this principle should find itself embodied in the convention containing the conclusions of the Disarmament Conference.
>
> This principle implies a single convention for all States, lasting for the same period, and permitting the same qualities of arms, but leaves open for discussion all questions of stages and quantities, it being clearly understood that the object of the Disarmament Conference must be to bring about the maximum of positive disarmament that can be generally agreed—not to authorize in the name of equality any increase in armed strength.

[119] MDP, 8/1, Pocket diary, 9 Dec [32]; *FRUS, 1932*, I, 499-500, 523-524.

[120] RK, Memo by Bülow, 10 Dec 32, 3642/D813024-25; see also tel., Neurath to Ministry, 10 Dec 32, 3642/D813028-30. Neurath's poor-mouthing as to the prospects of the meetings was not new, and may have been a stance adopted to show Berlin that he was hard-headed, and immune to the internationalist enthusiasms of Geneva. See also AA, Notes for report by Neurath, 15 Nov 32, 3154/D672152; RK, Memo by Bülow, 8 Dec 32, 3642/D813000-01.

The second numbered paragraph stated that on this basis, Germany would return to the conference and join in a no-force declaration, and the third paragraph, that all the powers were prepared to work without delay for a convention involving substantial disarmament. Blank space was left under a Roman numeral II for practical recommendations to the conference, if there was time left over to work them out, but everyone except perhaps Davis and his advisers knew that this was most unlikely.[121]

Massigli (according to the German record of this morning meeting) took some exception to the first sentence of paragraph 1, which made equality of rights a guiding principle rather than a goal, but the main debate, pursued in the afternoon by the principal delegates, centered on the second sentence. This incorporated, Simonized so to speak, the three points first stated in the German résumé of August 29. In their preoccupation with this section, the French tended to overlook the variance between the first sentence and the original Herriot formula; Boncour said at one point that he accepted the whole draft declaration, except for the second sentence. Finally, after tea, Davis produced an abbreviated draft of the second sentence, omitting reference to stages, and also to duration and qualitative equality. Implicitly as compensation for the last two omissions, Neurath asked that there also be nothing said about not authorizing an increase in arms. Although they had originated the stipulation on nonrearmament, which had constituted an important part of Simon's public statements in November, MacDonald and Simon now supported Neurath's request, presumably because they considered the reaching of an agreement, and the return of the Germans to the conference, to be more urgent. After more debate, the second sentence was further reduced to state only that the arms limitations of all states would be contained in the convention, while details were up to the conference to decide. Boncour raised some last-minute objections to the change from "aim" to "guiding principle" in the first sentence, but he and Neurath agreed to refer the revised draft to their governments.[122]

[121] *FRUS, 1932*, I, 501, 525-526. A Davis telegram to the State Department on the night of 10 December admitted that discussion of the equality-of-rights formula had taken so long that there would be little or no chance to discuss substantive disarmament, although it still noted with a touch of hope that talks would resume the next day (SDP, 500.A15A4 Steering Committee/243). A final short meeting was held on the morning of the eleventh, largely "to give satisfaction to M. Davis" (*DDF*, 1/II, No. 95; *FRUS, 1932*, I, 505-508; *DBFP*, 2/IV, No. 219). This resulted only in general oral assurance of mutual cooperation.

[122] RK, tel., Neurath to Foreign Ministry, 11 Dec 32, 3642/D813033-36; *FRUS, 1932*, I, 501-505; *DDF*, 1/II, Nos. 92-93; *DBFP*, 2/IV, No. 218; MDP, 8/1, Pocket diary, 10 Dec [32]. French officials may have tried to conceal their defeat. Boncour's final objections to the substitution of "principle" for "aim" were not reported in the (generally very detailed) French record.

Although Paul-Boncour expressed his doubts to Herriot, and the latter had misgivings of his own, the French were impelled to approve the draft. It would have been hard for any French premier to take the responsibility for a rejection, a step that probably would have placed France in isolation for some time. The five-power conclave had a built-in bias against France, but it was too late to remedy that now. Moreover, at this juncture Herriot's other plans virtually forced him to accept. The Herriot cabinet had been threatened for some time by the inability of its supporters to agree on a way to cope with the financial deficit, and it was not expected to survive much longer in any case. At Herriot's meeting with MacDonald and Neville Chamberlain on war debts on December 8, the British had indicated that they intended to pay the sum due to the United States on December 15. Herriot then decided to urge the same course to the French parliament. He knew—indeed he shared—the resentment of most French deputies at the idea that France should make further payments when, as a delayed aftermath of Hoover's moratorium initiative of June 1931, the Germans had ceased to pay reparations. He probably expected to be overthrown on this issue. But by calling on the French parliament to approve the payment, he could express his personal respect for the sanctity of contracts ("C'est cela la France," he was to say), he could testify to the value he placed on solidarity with the British and, even more, on avoiding a break with the United States, and he could in this way give his tenure in office a worthy conclusion. In Herriot's decision, sincere conviction was mingled with the spirit of Bernhardt. The drama he had now decided to perform reached its theatrical climax with his final speech in the Chamber, in the small hours of December 14, in which he called for a unity of the free against the dictators, and likened himself to the imprisoned Socrates as an advocate of obedience even to unjust laws. Three nights earlier, when presented with the draft five-power agreement from Geneva, the only course consistent with his chosen role—and with his genuine belief in American friendship and the Entente Cordiale—was to accept the draft five-power statement hammered out by MacDonald.[123]

Neurath telephoned Bülow that the German goal in the negotiations had been reached: it had been recognized, he said with some exaggeration, that the principle of equality of rights should find immediate practical application.[124] Everything had been achieved that

[123] Soulié, pp. 407-417; *DDF*, 1/II, No. 79; Chastenet, VI, 43-45. In his diary, Herriot also compared himself with Petrouchka (Soulié, pp. 416-417).

[124] According to a German report, MacDonald explained on the morning of the tenth that the use of "embodied" meant the practical realization of the equality of rights in the convention, and that this language could not be interpreted as a purely theoretical recognition of equality: RK, tel., Neurath to Ministry, 11 Dec 32, 3642/D813033. This report is not confirmed elsewhere, however.

could be, and there was no longer any purpose in remaining outside the conference. He said that reaching an agreement had been very difficult, and that until that noon (December 10, when he had first seen the MacDonald variation of the Herriot formula), he had not really believed agreement would be reached. When Bülow got in touch with Schleicher by phone in Berlin, the Chancellor was favorably impressed by the draft, although he decided to withhold his decision until the next morning; he did not think it necessary to convoke his cabinet, in view of Neurath's position and the cabinet discussion of the seventh. Then at 9:30 a.m. on Sunday, December 11, Schleicher told Bülow of his acceptance: "The expression 'embodied,' in the first paragraph, seemed to him especially useful, since this was probably the strongest expression known to the English language for anchoring the principle of equality of rights."[125]

THE "equality in a system of security" agreement has usually been regarded as a major success for Germany, and this is on the whole correct. By the change from "aim" to "guiding principle," equality had been made a basis of negotiation, and by the provision that the principle should be "embodied" in the convention, a pledge had been given that equality of rights would be expressed in the terms of the first disarmament convention, the one now to be negotiated. Germany could claim that this five-power agreement already granted equality of rights irrevocably. The overthrow of Part V seemed assured by the fact that, even in the final draft of the second sentence, it was specified that the "limitations of the armaments of all states" should be included in the disarmament convention. Moreover, Neurath had avoided making any promise not to rearm, such as Simon's proposals had originally provided for. And since the British were inclined to accept qualitative equality, and the French were willing to allow Germany a short-service conscript force, even if with other restrictions, it seemed that Germany was now in a position to secure significant practical alleviations in the first convention. If other powers refused to concede specific steps toward equality, the Germans could presumably justify unilateral action by charging a failure to honor the agreement. In a radio broadcast on December 15, Schleicher warned that the outcome of the disarmament question would have "decisive significance" for future German participation in the League.[126]

[125] AA, Memo by Bülow, 12 Dec 32, 7465/H178529-30. MacDonald evidently assumed that the agreement was acceptable to the Germans, as he noted after the French acceptance that the meetings had succeeded: MDP, 8/1, Pocket diary, 10 Dec [32].

[126] *Vossische Zeitung* (Berlin), 16 Dec 32 (a.m.).

The German success was not complete, however. The Germans had not in fact obtained explicit parity with France, an endorsement of "equality in every respect," or specific agreement to shorter regular service, freedom in arms manufacture, or the organization of a militia. Although Neurath had talked much of "practical" equality, and had even claimed at the end to have already secured it, it had not been won as yet. Foreign acceptance of practical rearmament would depend on negotiations when the conference resumed, and these promised to be difficult. The Berlin *Börsenzeitung*, which usually expressed the military view on political questions, asked if the Reich had not accepted the shadow of theoretical equality in place of the substance of practical military equality. In briefing army generals in mid-December, Schleicher defended the new agreement against its critics, saying, "We have a completely free hand for that which we want to demand as 'equality of rights.' " He admitted, however, that there was a danger that matters might be postponed and delayed; if so, "this would force us to assume 'freedom to rearm' ['*Rüst. Freiheit*']."[127] Such action might indeed be "forced," considering that Hammerstein had already issued orders for proceeding with rearmament, regardless of the results of negotiation. From a military viewpoint, negotiation was likely to lead to embarrassment, and it would have been better to avoid reentering the Disarmament Conference. We will consider in the next chapter why Schleicher agreed to reentry.

The officials of the French government may have felt they had been worsted; certainly they showed even more aversion after this to the idea of a great-power conference.[128] But they did not openly admit defeat. On the morrow of the agreement, *Le Temps*, which often spoke for the government, claimed that the agreed text safeguarded, in spirit and letter, the main point of the French position, and that Germany was now returning to the conference without gaining the firm prior commitment she had sought. A circular dispatch from the Quai d'Orsay stated that the first sentence of the agreement was "almost identical" with Herriot's original formula, linking the grant of equality of rights with a regime providing security for all nations. As a matter of fact, German diplomats discovered to their irritation that, compared with the English text, the language of the French text of the agreement was indeed less forceful in changing Herriot's formula and making equality of rights a guiding principle to be embodied in the convention. The Quai was also pleased to observe that Simon's

---

[127] *Börsenzeitung*: see *DDF*, 1/I, No. 109. Joachim von Stülpnagel had taken a leading position on this paper after leaving the army. Schleicher: Vogelsang, "Neue Dokumente," pp. 428-429.

[128] See *DDF*, 1/II, Nos. 356, 358, and esp. 364.

no-force declaration had been accepted, and might apply to such questions as that of the Corridor, and Massigli promptly arranged with Arthur Henderson that the French plan should be the first topic on the agenda when the conference resumed.[129] The five-power agreement did indeed give a general recognition to the French doctrine of security, even though, as with equality of rights, it remained to be seen what the cash value of this recognition would be.

One specific result of the December 1932 agreement was, however, beyond debate: Germany would now return to the conference. This had been a concrete British aim since September, an aim in which neither Paris nor Berlin had been nearly as interested. By staving off an overt breakdown, British policy also succeeded in fending off public blame for an inadequate disarmament policy. British policy did not succeed in obtaining a renunciation of rearmament. In any case, the Reichswehr had resolved to proceed with its rearmament program without regard for disarmament agreements. Also, British policy did not reduce German dissatisfaction, the aim of appeasement, or enable peaceful German forces to exert more influence.[130] Those forces were by now routed and dispersed. On the whole, the German return probably served to perpetuate illusions in England about the German government's willingness to forgo rearmament. It was, to be sure, desirable to continue negotiating, to maintain channels of communication. Continued negotiation would in fact hamper German rearmament, although it would also provide cover for it. But British leaders appeared unable to view German political forces and intentions realistically. And a bad precedent was set when Britain, after at first apparently resisting concessions, then proceeded to make them under pressure.

[129] *Le Temps* (Paris), 12 Dec 32; *DDF*, 1/II, Nos. 99, 148. The French text translated "should guide the conference" as "devrait servir de guide à la conférence," and "should find itself embodied" as "devrait trouver son expression." German diplomats in Geneva induced Henderson to declare the English text (which alone had been signed) to be governing, but the French delegation maintained that it was only on the basis of their translation that the French government had been able to agree. Bülow then advised the German delegation to repeat to the French that the English text was definitive, but he hinted that it was desirable not to pursue the controversy, "which can produce no useful result." See AA, tel., Weizsäcker to Ministry, 13 Dec 32, 3154/D668367; Note by Frohwein, 14 Dec 32, 7465/H178581; tel., Weizsäcker to Ministry, 15 Dec 32, 7465/H178619-20; tel., Bülow to Weizsäcker, [17 Dec 32?], 7465/H178638. See also note 122 above.

[130] As suggested in FO 371, Memo by Sargent, 3 Oct 32, C8603/211/18; Memo by Cadogan, 26 Oct 32, W12080/1466/98; Cab 24/233, CP 347 (32), 17 Oct 32.

CHAPTER SIX

Schleicher Reaches a Dead End

1. DOMESTIC POLITICS AND THE OTT WAR GAME

Diplomatically, Schleicher had had a real success. Of course, problems remained. Approval had not been won for the army's more specific aims. Even when claiming a triumph at the cabinet meeting of December 14, Neurath had to point out that there was unanimous opposition abroad to letting Germany rearm.[1] The discussion of specifics at Geneva might lead to further disclosures and alarms, such as had followed the attempt to negotiate with the French in August. Moreover, the Reichswehr was already beginning to carry out its plans, and in 1933 this implementation would foreseeably become harder to conceal.[2] Yet Germany had achieved her symbolic goal of having her equality recognized and her legal goal of weakening the Treaty of Versailles, and this would presumably have practical consequences.

In domestic politics, Schleicher had been far less successful.[3] He had incurred the lasting distrust of the Socialists, particularly by the removal of the Prussian Social Democratic ministers, but he had still not been able to "lead the Nazis over to the state." This was not for lack of effort on Schleicher's part, nor even for lack of willingness on Hitler's, provided that the latter's very high terms were met. During the July election campaign, Nazi propaganda, which sharply attacked the rest of the Papen cabinet, did not attack Schleicher. On the election day, July 31, the Nazis won 230 seats (out of 608), a plurality and the largest number they were to win in a free election, although not a majority. This encouraged Hitler to claim for himself the chancellorship in a presidial cabinet, as well as six Reich and Prussian ministries for his followers, including the Prussian Ministry of Interior with its control of the Prussian police. If such demands seem dangerous in retrospect, critical observers could see that they were dangerous then, too: the Nazis were showing their propensity for

[1] RK, Cabinet Protocol, 14 Dec 32, 3598/D791265-67.
[2] On 14 Jul 32, Schleicher approved plans for increasing the number of officer candidates and limiting enlistments so far as possible to three years, beginning both steps on 1 Oct 32; see Chapter Four, note 31. On 1933 problems, see Chapter Five, Section 8.
[3] On domestic politics during the Papen-Schleicher period, see generally Bracher, *Auflösung*, Part 2, Chapters 6-9; Vogelsang, *Reichswehr*, Chapters 3-5; Alan Bullock, *Hitler: A Study in Tyranny*, rev. ed. (New York: Harper & Row, 1962), Chapter 4; Eyck, II, Chapters 12-14.

violence in such events as the "Bloody Sunday" at Altona on July 17, when fifteen people were killed, the shooting and bombing of political opponents in Königsberg on July 31, and the Potempa murder of August 9, when five Nazis beat and kicked a Communist to death in the presence of his mother, an action that Hitler later condoned. Nevertheless, after meeting with Hitler, Schleicher concluded that it would be appropriate for the Nazi chieftain to become chancellor.[4]

Hindenburg, however, thwarted the plans of Schleicher and Hitler. He had no wish to appoint Hitler, the "Austrian corporal," and he intended to retain Papen, for whom he had taken a great liking. At a meeting with the President on August 13, Hitler asked for leadership with full powers for himself and his party, but the old man refused, saying that he could not "assume responsibility before God, his conscience, and his country for turning the whole of governmental power over to one party, a party, moreover, which was one-sidedly biased against those who differed with it." Hindenburg went on to lecture Hitler on political proprieties, and warned him expressly against any illegal behavior. After this, Schleicher seems to have changed his mind, at least for the moment, deciding that after all Hindenburg's refusal was wise.[5]

The Papen government took pains to publicize Hitler's humiliation at Hindenburg's hands, and the hostility between the Nazis and the cabinet became much sharper. The Nazis enjoyed a measure of revenge a month later when a Nazi leader, Hermann Göring, presiding over the new Reichstag, pretended to overlook an order from Hindenburg for dissolution, and pushed through a vote of no confidence against Papen. The vote had no official standing, but the devastating tally of 512 to 42 underlined the fact that Papen was opposed by nine tenths of the nation. The dissolution was followed by further elections on November 6. This vote registered a Nazi recession to 196 seats (out of 584), but the Communist strength simultaneously rose to 100 seats, so that, as after the July 31 vote, more than 50 percent of the seats belonged to these two radical parties. Some signs suggested that the difference between them might be diminishing. Just before the November election, the Berlin Nazis had backed, as had the Communists, a violent strike of Berlin transport workers against a wage reduction. The catastrophic depression, with its wage cuts and its millions of unemployed, could presumably help to fuel more widespread disorders.

The Papen government had originally been formed, not to combat

---

[4] See esp. Vogelsang, *Reichswehr*, pp. 256-258; Bracher, *Auflösung*, pp. 612-614. On violence, cf. Bennecke, *Hitler und die SA*, pp. 190, 194, 197-198.

[5] See also Thilo Vogelsang, "Zur Politik Schleichers gegenüber der NSDAP 1932," *VfZ*, 6 (1958), 86-101, 108-110.

the Nazis, but to work with them, guide them, and tame them by giving them secondary responsibility or, in the case of the SA rank and file, military training. Before coming to power, Papen himself had advocated bringing the Nazis into governmental responsibility. But the plans for integrating the Nazis had suffered a setback when Hindenburg refused in August to make Hitler chancellor with full powers and Hitler refused to accept a lesser appointment. After the November election, the inclination of Papen and Gayl, the Minister of Interior, was to try to govern without party backing aside from the DNVP, relying on Hindenburg's support and the strength of the Reichswehr. They had developed ideas of resorting eventually to a constitutional "reform," involving the establishment of an upper house appointed by the President, a lower house based on a suffrage weighted in favor of heads of families and veterans, and a permanent elimination of the dualism of Prussia and the Reich. Papen had in mind a defiance of the Reichstag, following the precedent set by Bismarck in his conflict with the Prussian parliament in the 1860s.[6] But most of the ministers continued to hope that somehow the Nazis could be enticed into governmental responsibility. While Papen and Gayl argued in cabinet meetings that Nazi claims were dangerous, other ministers expressed hopes for a revival of the Harzburg front (a fleeting 1931 union of the Nazis, DNVP, and other rightists) or for a government based on conservative and nationalist forces. It was pointed out that the Nazis were not yet cooperating extensively with the Communists, but some ministers feared that the Nazis, or part of them, might eventually go over to the Communists if the Nazi party was not brought into the government. Although correspondence between Hindenburg and Hitler had proven fruitless, Schleicher now took up negotiations with the Nazi leader—and also with other elements.[7]

The army had its own reasons to wish for a reconciliation of the Nazis with the government. Grenzschutz and external training were developing, and for these, as noted earlier, Nazi cooperation was considered "vital." In mid-October, the Reichswehr promulgated new

[6] Papen asked Joachim von Stülpnagel if he would serve as Defense Minister in a "conflict cabinet," warning Stülpnagel that this might involve "the use of live ammunition": Vogelsang, *Reichswehr*, p. 325 and note. Stülpnagel agreed in principle.

[7] RK, Cabinet Protocols, 3 Nov, 9 Nov (6:00 p.m.), 25 Nov 32, 3598/D791121, D791124, D791164-66, D791168-71, D791210-14; National Archives, Microfilm T-84, Miscellaneous German Records Collection, Rathmannsdorfer Hauschronik (includes Lutz Graf Schwerin von Krosigk's diary entries for November 1932–February 1933), MS Diary, 27 Nov 32, 4 Dec 32, reel 426 (no frame numbers); Vogelsang, *Reichswehr*, pp. 312-330. After Hitler came to power, Schleicher maintained that he had sought all along to bring the Nazis into the government: Vogelsang, "Zur Politik Schleichers," pp. 88-90.

guidelines for Landesschutz, replacing those which the Braun-Severing government had so long contested. These guidelines expanded the Grenzschutz area and relaxed the restrictions on Grenzschutz training and on external training for leaders and specialists (and young replacements for them) in other areas. In the border lands, the population which was "capable and ready to serve"—characteristics of Wehrverbände members—could now be trained with arms in the open country. As Ferdinand von Bredow, Schleicher's top assistant, put it, the aim of the guidelines was to provide "expanded Grenzschutz" (i.e., expanded Grenzschutz training) in the border areas and "existing Grenzschutz" in the rest of the country.[8] Beyond the guidelines, we have seen that mobilization plans depended on the induction, in time of war, of Wehrverbände members into the field army. In a 1932 transport exercise, for example, the army had explored the possibility of concentrating the whole field army, about 400,000 men, in Eastern Pomerania for an attack on "the enemy's" north wing.[9] Such a concentration would only be possible with the participation and support of a large number of Wehrverbände men, including from the SA. An important instrument for preparing would-be soldiers for ultra-short-term training during mobilization, or for the projected peacetime militia training, was the new Reich Institute for Youth Training (Reichskuratorium für Jugendertüchtigung), which made its public debut on September 14. Schleicher's circle in the Reichswehr Ministry wanted to secure trainees for the Institute's program from all the (non-Communist) Wehrverbände, bringing in the largely Socialist Reichsbanner; probably they wanted a political balance to the SA and a split in Socialist ranks, as well as the maximum number of recruits. Karl Höltermann, the Reichsbanner leader, was willing, hoping for food and boots for his men, but old suspicions within the army, and now even more within the SPD, obstructed collaboration with this Verband, and made the success of the program depend all the more on the SA. In November, 714 trainees in the Institute's Wehrsport courses, the largest contingent from any organization, came from the SA and Hitler Youth, while 549 came from the Stahlhelm and 25 (in Bavaria only) from the Reichsbanner. Some army officers expressed a willingness to fire on the Nazis if ordered to do so, but there is no sign that Schleicher thought seriously of ceasing to work with the SA, and in-

[8] OKW, Wi/IF, 5.335, Reichsminister der Finanz, G II RV—, 20 Feb 33, and att., 73/794782-92; Bredow Papers, Orientation notes, 30 Jul, 12 Aug, 3 Oct 32; Lecture notes, 2 Dec 32. On "vital" need for Nazi cooperation, see Chapter One, Section 8.

[9] ML, PG 34095, ltr., Henning to B. H., 2 Jul 32, reel 31/650-652; A II a 2669/32, 10 Oct 32, reel 31/653; TA 739/32, 17 Oct 32, reel 31/649.

deed, developing military preparations presupposed friendly relations with that organization.[10]

The Reichswehr's friendliness toward the Nazis encouraged the latter to hope for army tolerance for an NSDAP attack against the left and republican institutions, and one must suspect that the army's interest in the SA was regarded in Nazi party circles as a lever for forcing acceptance of Nazi political demands.[11] Such hopes were premature, but the military outlook did have a tilt against the left and toward the Nazis. This is evident in the actual scenario, hitherto unpublished, of a famous "war game" exercise in dealing with domestic disorder.

In mid-November, Schleicher agreed to the proposal of one of his staff officers, Major Eugen Ott, to conduct a war game to test the Reich government's ability to cope with a general civil emergency. As Ott recalled it after World War II, the game, conducted on November 25 and 26, assumed "the worst possible case," one in which the army would have to deal with "the terrorism of right and left," and also with radical Polish elements, who would take advantage of a momentary German weakness to attack East Prussia. The Nazis (the SA) were supposed to provide most of the reinforcements for mobilizing a border protection force (Grenzschutz) in East Prussia, but if the Nazis were in revolt against the government, the regulars might simultaneously have to deal with civil unrest and defend the border, "a very dangerous test of their discipline." It would not be possible to spare any reinforcements for East Prussia from the rest of the Reich, and indeed conditions would be critical there too, especially with striking dockworkers in Hamburg and "Communist terror" in the Ruhr. The Technische Nothilfe, an organization designed to use civilian volun-

---

[10] The strong interest of the Reichswehr leadership in Wehrsport emerges frequently in the Bredow Papers; on the Reich Institute, see Orientation notes of 6, 15 Jul, 4, 25 Aug, 17 Sep, 12 Oct, 14, 19 Nov 32, also ltr. to Schleicher, 28 Oct, and Remarks at chiefs' meeting, 11 Nov 32. On the relations of Reichsbanner and Reichswehr: Karl Rohe, *Das Reichsbanner Schwarz Rot Gold* (Düsseldorf: Droste, 1966), pp. 372, 444-446, 448-452; Breitman, "Social Democracy and Schleicher," pp. 360-362; Vogelsang, "Neue Dokumente," 430; Bredow Papers, Orientation notes, 3 Aug, 6 Sep, 25 Nov, 6, 19 Dec 32; Briefing note, 6 Dec 32. November training figures: Vogelsang, *Reichswehr*, p. 286n. The courses, still in development, enrolled a total of 2150 men. Willingness to fire: Carsten, *Reichswehr and Politics*, pp. 374-375. Bredow's notes for Schleicher show friction primarily with the Stahlhelm, and at the end, increasingly cordial relations with the SA: Bredow Papers, Orientation notes, 2 Jul, 12 Aug, 14 Sep, 3 Oct, 6, 19, 21 Dec 32, 14, 24 Jan 33; ltr. to Schleicher, 25 Oct 32; Remarks at chiefs' meeting, 14 Nov 32.

[11] See Vogelsang, *Reichswehr*, pp. 251-252, 330 and n. 1569, Annexes 31 and 32; also Hitler's ltr. to Reichenau, described in Section 4 of this chapter. The SPD secured minutes of an alleged Reichswehr-Nazi conference on 24 July on possible joint action against "leftist" moves: Breitman, "Social Democracy and Schleicher," p. 355.

teers to maintain essential services in an emergency, lacked equipment and many of its members were Nazis, who would presumably refuse to serve under the conditions postulated. Participants in the Ott exercise included civil servants (Prussian as well as Reich) and Technische Nothilfe functionaries, and each of the seven divisions sent representatives, in five cases their chiefs of staff.[12] (Although Ott does not say so, Col. Walther von Reichenau, who was chief of staff of the First Division in Königsberg, may have influenced the conclusions as to the situation in East Prussia; he had reportedly for some time been in touch with the Nazis.)[13] According to Ott's later testimony, the exercise led to the conclusion that, in a state of emergency, "the forces of order of the Reich and the Länder in no way sufficed to maintain order against the National Socialists and the Communists and to protect the borders. It was therefore the duty of the Reichswehr Minister to prevent the Reich government from having recourse to a state of military emergency." The representatives of the divisions asked Ott to make it clear to Schleicher that the political crisis should be overcome without resorting to the Reichswehr.[14]

Ott's postwar recollections are not entirely accurate, however. They have given the impression that the war game—and thus official thinking—was aimed primarily at a possible Communist-Nazi combination, such as had formed during the Berlin transport strike. Actually, the starting hypothesis for the war game was rather of a general strike of all the left, similar to that which had defeated the Kapp putsch in 1920. A document shows that the postulated problem was as follows:

[12] On Ott testimony, see note 14 below. War game participation: Conze, "Politischen Entscheidungen," p. 243. Sohn-Rethel says (*Ökonomie und Klassenstruktur*, p. 194) that the Technische Nothilfe had been notorious strikebreakers in the 1920s; "one can describe them, like the *Baltikumer*, as one of the seedbeds of Nazism."

[13] Reichenau's contacts with the Nazis: Westarp memo in Co[nze], "Zum Sturz Brünings," p. 284; Hermann Foertsch, *Schuld und Verhängnis: Die Fritsch-Krise im Frühjahr 1938 als Wendepunkt in der Geschichte der nationalsozialistischen Zeit* (Stuttgart: Deutsche Verlags-Anstalt, 1951), pp. 24, 31-32; Thilo Vogelsang, ed., "Hitlers Brief an Reichenau vom 4. Dezember 1932," *VfZ*, 7 (1959), 430. I have no specific evidence that Reichenau personally took part in the Ott exercise.

[14] On the war game: Bracher, *Auflösung*, pp. 674-675; Vogelsang, *Reichswehr*, pp. 316, 326-327, Annex 38; Vogelsang, "Neue Dokumente," p. 427 and note; Edgar Röhricht, *Pflicht und Gewissen: Erinnerungen eines deutschen Generals* (Stuttgart: Kohlhammer, 1965), p. 14. Ott testimony: G[eorges] Castellan, "Von Schleicher, Von Papen et l'avènement d'Hitler," *Cahiers d'histoire de la guerre*, no. 1 (Jan 1949), 23-25; the version in Papen, *Wahrheit*, pp. 247-249, and Schüddekopf, pp. 374-375, is less complete. Rainer Wohlfeil and Hans Dollinger, *Die Deutsche Reichswehr—Bilder-Dokumente-Texte zur Geschichte des Hunderttausend-Mann-Heeres 1919-1933* (Frankfurt: Bernard & Graefe, 1972), p. 244, has a photocopy of Reichswehrminister 549/32, 18 Nov 32, ordering the war game. According to Ott, the proposal for the game stemmed from a group headed by Carl Schmitt, defending the Reich government in the case of Prussia vs. Reich before the Supreme Court arising from the 20 July action (Conze, "Politischen Entscheidungen," p. 243n).

The Reich cabinet has issued strict new provisions for maintaining law and order this winter. In particular, strikes in vitally important enterprises are forbidden under threat of penal sanctions. A strong protest has arisen against this, inspired by the SPD and KPD [Communist Party of Germany]. Political work stoppages have flared up, supported locally by National Socialists. In various places, especially in the Ruhr and Central Germany, there have been bloody clashes with the police.

In the night of 23-24 November, the SPD and KPD have issued a call for a general strike. The strike call has at first, on 24 November, only been partly heeded. There is clearly a lively internal controversy going on within the Reich leadership of the NSDAP over whether to join in the strike. The press of the Center and the Bavarian People's Party oppose the strike, but demand the immediate resignation of the Reich cabinet.

Thus the imagined situation presupposed a malign unity between the two "Marxist" parties, and a division of the Nazis into "good elements" and troublemakers. This casts a revealing light on the deep distrust that had developed between officers and officials on the one hand and the SPD on the other, a distrust outweighing sporadic tactical efforts to cultivate better relations. The final results of this actual war game are not shown in the available documents, although a fragmentary record of the notional "developments" in one area suggests that the Nazis remained simply an uncertain quantity. The documents do show that the planners did not throw up their hands, but proceeded to refine their preparations, including those for the employment of the Technische Nothilfe, in spite of its supposed infiltration by the Nazis. They also testify to concern that Polish border incursions and a closing of the Corridor might lead to "more serious developments," and show that care was taken to keep domestic security measures from interfering with an eventual mobilization of the field army.[15]

[15] ML, PG 34097, W 308/32, 22 Nov 32, reel 31/902-903; see also Fritz Arndt, "Vorbereitungen der Reichswehr für den militärischen Ausnahmezustand," *Zeitschrift für Militärgeschichte*, 4 (1965), 195-203; ML, PG 34097, Chef ML B AIIa 20385/32, 3 Dec 32, reel 31/890-892; TA 25/33, 18 Jan 33, reel 31/922-924; PG 34095, TA 2569/32, 12 Dec 32, reel 31/655-664. Several of Arndt's documents are also in ML, PG 34097. Other references to a general strike are contained in Vogelsang, *Reichswehr*, Annex 38; Vogelsang, "Neue Dokumente," p. 427. A potential conflict between emergency plans for internal unrest and eventual mobilization against a foreign enemy arose in the case of Kreis Arnswalde, in civil affairs a part of the province of Brandenburg, and militarily a part of Wehrkreis II (Hq. Stettin, in the province of Pomerania). It was asked whether difficulties would arise for Landesschutz preparations if Kreis Arnswalde (in line with its civil affiliation) was tempo-

The Reichswehr leadership, then, had not been thinking of the Nazis as the major problem. Indeed, when Schleicher, at a November 26 cabinet meeting, commented on the idea of using formations like the SA as auxiliaries to help restore domestic order, he assumed they would stand on the Reichswehr's side; his objection was that they would "in no case strengthen the apparatus."[16] But Schleicher's last efforts to negotiate with Hitler broke down on December 1, and now he had to face the prospect of another Papen government, which would mean a conflict with the Nazis as well as with the left. When Hindenburg decided to reappoint Papen, and when Papen outlined a program of "struggle," of dismissing the Reichstag for an unconstitutional six months, promoting "reforms," and ruling by force, Schleicher thought this impractical and dangerous. He wanted to make further efforts to bring in the Nazis, or a segment of them, and also to reduce tensions with the center and the left. No doubt he shared the aversion of most German officers to seeing the army employed as a domestic police force; that was the role which the Treaty of Versailles had tried to assign to the army and from which the army had been working to escape.[17] As in the spring, he did not want a conflict with the Nazis, the largest source of manpower for national defense.

When the cabinet ministers convened on the morning of December 2, most of them agreed with Schleicher in opposing another Papen government. Neurath spoke against such a course, and Lutz, Graf Schwerin von Krosigk, the Minister of Finance, particularly stressed that a Papen cabinet would destroy any hope of economic recovery. Papen did not give up, however, and Otto Meissner, the President's State Secretary, supported Papen with doubts that Hindenburg would change his mind. Schleicher then seized an opportunity to bring Ott in to report to the cabinet on plans for dealing with resistance. Presumably to reinforce Schleicher's argument against a policy of struggle, Ott expanded on some of the war-game results to paint as black a picture as possible of the dangers. In contrast to the actual war game, and more in line with his postwar recollections, he now projected a

---

rarily subordinated, in a domestic emergency, to Wehrkreis III (Hq. Berlin, including Provinz Brandenburg). The Reichswehr decided to transfer the Kreis permanently to Wehrkreis III. See Arndt, Docs. 10 and 11.

[16] RK, Cabinet Protocol, 25 Nov 32, 3598/D791214. Cf. Andreas Dorpalen, *Hindenburg and the Weimar Republic* (Princeton: Princeton University Press, 1964), p. 388n. In December 1932, the use of closed Verbände as auxiliary police—a Nazi aim attained in February 1933—was explicitly barred: Arndt, Docs. 10 and 11.

[17] Aside from the general question of the army's mission, note Hammerstein's objections to an upsetting of mobilization plans for internal purposes (Chapter One, Section 9 and note 108), also Bracher, *Auflösung*, pp. 661-662, incl. n. 39.

general strike by virtually all elements, and he definitely included the Nazis among the strikers. Thus the Reichswehr would have to face nine tenths of the population as the defenders of a thin upper crust. Among the Wehrverbände and other organized groups, only the Stahlhelm, at best, would support the government, and the Reichswehr would have particular difficulty dealing with sabotage and passive resistance. The troops had to obey orders, and would do so, but they hoped they might be spared this trial. Even Papen was taken aback, and he wiped his eyes frequently with a handkerchief. Schleicher seems to have feared that Ott was overdoing it, and he said that such a "worst case" need not actually arise, but the presentation had the desired effect, and Papen's subsequent report to Hindenburg led the old man to change his mind. The reported judgment of the Reichswehr was decisive for the President. He sadly agreed that Schleicher should "try his luck" as Chancellor. Schleicher had professed reluctance to take the post, but now he was willing, and there was no other alternative to Papen in sight—except Hitler himself.[18]

The story of the Ott war game has led to retrospective comments that the army was politically too sympathetic with the Nazis to fight them, or that Schleicher and his cohorts had neglected their duty to prepare to put down domestic unrest, or that Schleicher wanted to take Papen's place. Each of these comments contains at least some truth. But another lesson can also be drawn: that the army had already opted to work with the Nazis—in the Grenzschutz, in mobilization preparations generally, and now also in the Technische Nothilfe —and that any elements standing in the way of this collaboration, whether they were Prussian Socialists or romantic reactionaries like Papen and Gayl, had to be thrust aside. The war gamesters had confined their deliberations to the employment of the "ready to march" Reichswehr, along with the police and other civil organizations. Army officers would never have discussed national mobilization problems with the civil officials who participated in the war game. But it had proven impossible to exclude Grenzschutz problems from consideration, and behind these problems lay the potential need for a twenty-one-division mobilization, which could not be carried out in the face of public hostility, particularly in the face of Nazi hostility. The prospect of fighting the Nazis fell as a pailful of cold water upon the army's aspirations.

[18] See Papen, *Wahrheit*, pp. 246-250; Krosigk MS Diary, 27 Nov, 4 Dec 32; RK, Cabinet Protocols 2, 3 (12:45 p.m.) Dec 32, 3598/D791241, D791245; Vogelsang, "Zur Politik Schleichers," pp. 105-107, 110-113; Vogelsang, "Neue Dokumente," pp. 426-427; Magnus von Braun, p. 231; Krosigk, *Staatsbankrott*, pp. 150-153; Vogelsang, *Reichswehr*, pp. 332-334, Annex 38.

2. SCHLEICHER'S POLITICAL AND DIPLOMATIC PROBLEMS

Despite Hindenburg's words to Papen, Schleicher did not have much luck. Since Hitler had just refused, as before, to accept any cabinet post less than that of chancellor with full powers, Schleicher now tried to entice other Nazi leaders into collaboration, in effect trying to steal Hitler's followers away from him. Schleicher hoped in particular that Gregor Strasser, a Nazi leader with whom he was in close contact, would help in this maneuver. But perhaps because he saw the Nazi movement mainly as a collection of potential soldiers, Schleicher did not understand the strength of Hitler's messianic leadership. Other party leaders remained faithful to Hitler, and Strasser resigned his party position and took his family on a holiday to Italy. This meant that Schleicher's goal of taming the Nazis was farther away than ever. He had already tried to muster other support through the Socialist-linked trade unions and Reichsbanner, and he extended feelers to the SPD itself, but the latter distrusted him and barred collaboration by its affiliates. Otto Braun found Schleicher unwilling to write off his hopes of Nazi backing.[19] At the same time, Schleicher's soundings on the left, and his "socialist" proposals, such as for subsidized land settlement in the East, alarmed conservative industrialists and landowners. Without political support, Schleicher, like Papen before him, could not carry on for long unless Hindenburg would give him full backing to govern without parliament and against the constitution.

Even before the Strasser fiasco, Schleicher's actual power was reduced by the manner of his accession. One of the wellsprings of his influence had been his access to Hindenburg. This was particularly important for his standing within the army itself. Now, however, Hindenburg only accepted grudgingly the replacement of Papen, who had become his favorite, and Papen, who began to resent bitterly Schleicher's maneuvers, remained in close contact with the President. The President's son, who lived with his father, had also become an enemy of Schleicher's. This was not an auspicious situation for the new Chancellor. Furthermore, the other ministers had developed positions of power, each in his own area, and also collectively. Hindenburg respected their opinions. As with Hindenburg, their acceptance of Schleicher rested mainly on their belief that he was the man best qual-

---

[19] On efforts to restore contact between the SPD and Schleicher: Breitman, "Social Democracy and Schleicher," 352-378; Gerard Braunthal, "Trade Unions During the Rise of Nazism," *Journal of Central European Affairs*, 15 (1955-1956), 343-346; Rohe, pp. 444-446, 448-452; Otto Braun, pp. 436-439; Arnold Brecht, "Die Auflösung der Weimarer Republik und die politische Wissenschaft," *Zeitschrift für Politik*, 2 (1955), 303, 306 and the same author's *Federalism and Regionalism in Germany* (New York: Russell & Russell, 1971), p. 119.

ified to win wider support, and they hoped in particular that he would win backing from the Nazis. If he could not bring this about, his standing in their eyes would decline. And Schleicher did not have any party to back him up and share his responsibility, not even the DNVP, which had generally supported Papen.

It is difficult, however, to feel very sorry for Schleicher. Whatever Hindenburg's shortcomings were, the old man apparently had some inkling of the threat a Hitler government might pose to civilized society. But Schleicher and the other military leaders worried little about such matters, and in fact they themselves had been moving in the direction of totalitarianism. In mid-October, as part of the Landesschutz guidelines, Schleicher secured a cabinet affirmation that all national and Länder officials had a duty to support his national defense measures; he apparently regarded this as an "enabling act" for any defense policy decision he might make. The new guidelines directed civil officials to assist Reichswehr efforts to register "the whole population capable of military service according to age, occupation, and training," mainly by providing census and tax data. They also strengthened legal sanctions against the betrayal of military secrets and against treason.[20] Beyond defense policy properly speaking, the Reichswehr had for some time been combatting pacifist films and publications, and trying to punish those who disclosed German treaty violations. In October 1932, Schleicher sought unsuccessfully to have two hostile periodicals charged with defamation of the state: one had carried articles saying that reparations were justified and that Germany had sold arms abroad in violation of the Treaty of Versailles, while the other had favorably reviewed a book (by a French colonel) charging Germany with treaty violations, and had published a letter arguing that Germany had started the 1914-1918 war and that certain German circles were working for another war.[21] While trying to curb these publications, the Reichswehr subsidized one Berlin newspaper, the *Tägliche Rundschau*, and financial backing was considered for the *Deutsche Allgemeine Zeitung* and the *Frankfurter Zeitung*. At a December 1932 meeting with divisional commanders, Ott explained on Schleicher's behalf various plans for "leading youth to the state"; teachers were to be in-

[20] RK, Cabinet Protocol, 14 Oct 32, 3598/D790987; OKW, Wi/IF 5.335, Reichsminister der Finanz, G II RV—, 20 Feb 33, and att., 73/794782-83, 794789-90, 794792. Schleicher invoked the cabinet's resolution in January to excuse himself from submitting Reichswehr Umbau plans to the cabinet; see below.

[21] *FRUS, 1931*, II, 310; Deak, *Left-Wing Intellectuals*, pp. 189-192; Schleicher Papers, N42/22, Reichswehrminister 423/32, 21 Oct 32; RK, Cabinet Protocol, 28 Oct 32, 3598/D791035-36; Bredow Papers, ltrs. to Schleicher, 21, 25, 28 Oct 32. The cabinet rejected the Reichswehr's proposal, presented by Bredow, for a decree to deal with these publications, Neurath saying that this would not be appropriate in the current diplomatic situation.

doctrinated and new teaching materials introduced, replacing those current in Prussia, which were lacking in "heroism." In a radio broadcast on December 15, describing his governmental program, Schleicher urged that education should arouse love of homeland and a sense of union with the German *Volk* and state; he also warned against an overemphasis on intellectualism and said that the renewal of the Volk would come from below, not above.[22] As Umbau measures began to get under way, the Reichswehr became more concerned to avoid disclosures and to encourage defense-mindedness and patriotism.

Schleicher continued to present himself as an advocate of German rearmament, apparently trying to satisfy military and rightist opinion. We have seen him trying to dispel skepticism among the generals on the December 11 agreement, claiming to have gained full equality of rights, while indicating that he would assume a freedom to rearm in the event of any delay in Geneva. Indeed, at a dinner given by Major Marcks on January 13, Schleicher assured a conservative German journalist that he favored authoritarian government, and then went on to say that he intended to introduce conscription on April 1 in the form of a militia with six-month service; further, he would then begin the provision of heavy weapons. If, as he expected, the Disarmament Conference decided by the end of March to take a long recess, Germany would be free to act. According to the journalist, this last statement made Bernhard von Bülow of the Foreign Ministry, who was present, turn "as pale as a tablecloth."[23] At about this time, the general also told a Reuters correspondent that he planned to train 300,000 militia men each year. The British military attaché heard from German officers that an annual figure of 300,000 was a minimum for the militia, which should preferably be trained for six months in two successive contingents of 150,000. In a series of public statements, Schleicher indicated that he intended to introduce universal military service. By comparison, the Hammerstein directives of November had been much more limited, providing that a three-month militia program, initially on a voluntary basis, would only come into full operation on April 1,

[22] Walter Struve, *Elites against Democracy: Leadership Ideals in Bourgeois Political Thought in Germany, 1890-1933* (Princeton: Princeton University Press, 1973), pp. 372-373; Bredow Papers, Orientation notes, 25 Jun, 11, 17 Nov 32; Vogelsang, "Neue Dokumente," p. 429. Schleicher speech: *Frankfurter Zeitung*, 16 Dec 32 (2. Morgenblatt); *Vossische Zeitung* (Berlin), 16 Dec 32 (a.m.).

[23] Vogelsang, "Neue Dokumente," p. 428; Brammer Collection, Z Sg 101/26, Information report, 14 Jan 33, pp. 41-47. Someone has written "1934" in pencil on the Brammer Collection typewritten record of Schleicher's statement about 1 April, but everything about the context makes it appear that he was referring at the time to 1 Apr 33. Dülffer notes (p. 237n) that the "1934" was probably a later addition, but also that a naval document, A I 517/33 of 2 Feb 33 (in ML, PG 34176, reel 27/439-441), indicates that both army and navy were supposed to begin militia training with the 1934 budget. On actual Reichswehr plans, see further below, including note 37.

1934. Conceivably, Schleicher may have wanted longer service with conscription in order to make himself less dependent on Nazi cooperation in defense activity. He may have thought that such a program would now be diplomatically feasible, since the French plan conceded nine- to twelve-month conscript service and, presumably, an increase in German effectives. But the French, for their part, now inferred with annoyance that the Germans planned to build up a large-scale militia while still retaining their élite professional Reichswehr.[24]

Within the German army, the pressure continued for rearmament, without regard for French or other foreign views. Although the French plan offered Germany a chance to have a large conscript militia, it sought a drastic reduction in the number of German professional soldiers, and while German military leaders did not in fact want to retain a professional élite force, as Seeckt's writings suggested, they did want a large cadre of long-service men and specialists to shepherd and support the militiamen in training and in battle; indeed, this was a prerequisite for the militia program. The army's staff officers had planned for months to increase the number of regulars from April 1933 on, a step which would be obstructed by the French plan. To a German military observer like General Schönheinz, the French plan was from the start an object of suspicion.[25] In early January, the Reichswehr Ministry submitted a draft—quite possibly by Schönheinz himself—for new instructions for the disarmament delegation, and this draft gave short shrift to the French proposals for military organization; it

[24] *DBFP*, 2/IV, No. 276 (incl. encl.); BA-MA, RH 15/49, TA 737/32, 7 Nov 32, Annex 6. Schleicher had told Schönheinz on 3 Dec 32 that propaganda should make the most of those measures which the French had now declared to be disarmament or defensive: NA-VGM, F VII a 11, Note by Schönheinz, 3 Dec 32, 7792/E565445. The attaché's report concluded that the militia would come into being in 1933 on the basis of universal service. Schleicher's public statements were in his radio broadcast of 15 December, in a statement at the New Year's reception given by President von Hindenburg, and in a short speech to the Kyffhäuser Bund on 15 Jan 33: *Vossische Zeitung* (Berlin), 16 Dec 32 (p.m.), 16 Jan 33 (p.m.). The German press devoted much attention to the militia idea, especially in January 1933; this press discussion was reviewed in Günter Nickolaus, *Die Milizfrage in Deutschland von 1848 bis 1933* (Berlin: Junker & Dünnhaupt, 1933), pp. 159-169. After Schleicher's fall, an anonymous book, *Die deutsche Miliz der Zukunft: Eine Frage von entscheidender Bedeutung für das deutsche Volk* (Berlin: E. S. Mittler & Sohn, 1933), appeared from a military press; it advocated six-month service, and seems to have been intended to support Schleicher's ideas. (The Library of Congress identifies the author as Friedrich Immanuel, who was a retired officer and veteran writer on military subjects.) Kurt Hesse, the author of *Feldherr Psychologos*, also wrote on the militia in 1933 (according to Sauer, "Mobilmachung," p. 725n; see also Nickolaus, pp. 147-148n). French conclusions: *DDF*, 1/II, Nos. 173 (and note), 184, 212; AA, Memos by Nadolny and Bülow, 18 Jan 33, 7360/E536142-43, E536146. Boncour seems to have been particularly concerned.

[25] OKW, OKW 845, Notes by Schönheinz, 5 Nov 32, 793/5522490-91. Schönheinz described the French plan as designed to "destroy" the Reichswehr. Bredow and Schleicher were wary of proposals for fewer professionals and a voluntary militia: Bredow Papers, Orientation notes (with marginal note by Schlcicher), 21 Jan 33.

286—Schleicher Reaches a Dead End

stated: "The military system of the Herriot plan does not correspond to German interests." The Reichswehr draft also displayed great hostility to any proposals for inspection such as the French had advanced, and it argued that inspection would be more effective in Germany, and more dangerous for her, than in the case of other countries.[26]

Those who prepared this draft clearly had little patience with the discussions in Geneva. In their view, any further questioning of the principle of equality of rights should be met by a renunciation of further participation in the conference. In explaining their opposition to a prolongation of the arms truce, they indicated the major reason for their impatience:

> In view of preparatory steps toward the reorganization of the Reichswehr, to be introduced on April 1, 1933, Germany cannot accept a new compulsory restriction. With respect to April 1, Germany must however press with all means for an acceleration of the work of the Conference. A prolongation by other states poses for us the question of another withdrawal from the Conference. A conclusion by the end of March must be striven for, otherwise foreign policy difficulties can arise in connection with April 1, 1933.

The draft guidelines concluded with an ominous statement: "In the negotiations, account must be taken of the possibility that according to experience to date, the execution of the emergency program for increasing German security can only be realized outside the limits of a convention."

This constituted a forthright expression of Reichswehr aims, in the tradition of the March 16, 1931, statement of German goals and strategy at the Disarmament Conference, and the "Hidden German Goal" guidelines of June 14, 1932. The impatience was justified, if one accepted the premise that new steps toward rearmament on April 1 were both imperative and impossible to conceal. Schleicher's statements to the generals and at Marcks's dinner party seemed to show that he shared the belief that new steps on April 1 were imperative. And concealment would be particularly difficult if Schleicher was going to introduce large-scale militia training on that date; such a program could hardly be covert, as recruitment for it would involve the compromising complications of physical examinations, government pay (including for dependents), and leave from work.[27]

[26] AA, Reichswehr draft for guidelines [Jan 33], 7616/H188327-28/10. The neighboring states' "far-reaching intelligence net left over from the period of inter-allied control" was one worry.

[27] On the problems of concealing a militia program: Wohlfeil and Matuschka, p. 208; OKW, Wi/IF 5.701, Reichswehrminister, TA 421/33, 22 Mar 33, 177/912300-01; ML, F VIII a 11, Minister-Amt 464/33, 6 Sep 33, reel 54/229-232.

Schleicher, however, now failed to take up the cudgels for an early decision at Geneva. When the official instructions for the delegation were finally worked out in negotiations between Nadolny and the Reichswehr, they incorporated certain points from the Reichswehr draft, but they omitted all threats to leave the conference, and all allusion to any deadline existing on April 1. According to these official guidelines, the basic aims were to eliminate completely the one-sided restrictions of the Treaty of Versailles, to disarm the highly armed states as far as possible, and to establish a military power balance that would ensure the defense of the Reich. These guidelines concluded by repeating the instructions of the previous year to leave uncertain what Germany would do if the conference collapsed, and to bear in mind the need to avoid diplomatic isolation. A covering letter from Neurath, dated January 19, enjoined the delegation to work for a positive outcome to the conference, in the form of a convention legalizing "the armament reorganization measures [*Umrüstungsmassnahmen*] necessary for us"; this endorsed the measures, to be sure, but it did not suggest much urgency. Neurath stated that the guidelines had the approval of the Reich government—and hence, presumably, of Schleicher.[28] Schönheinz had claimed that Schleicher supported a refusal to extend the arms truce beyond February 28, but on January 16 Neurath recorded an oral assurance from Schleicher that there must be some mistake about this, and that he (Schleicher) completely agreed with Neurath that such a refusal would be a tactical error; Schleicher added that, in general, the tactical treatment of all these questions was entirely up to the Foreign Ministry. Schleicher also backed Neurath in another case. Nadolny had suggested Germany might claim a right to rearmament on the grounds that her forces, according to President Hoover's formulation, only sufficed for domestic security, and so did not provide the national security to which she was entitled under Article 8 of the Covenant; Neurath objected that to advance this claim would reveal Germany's intention to rearm, and Schleicher agreed.[29]

One reason for Schleicher's attitude was probably that, as Chancellor, with overt political responsibility for the Reich, he could no longer, in the last analysis, support irresponsible military demands. There was now no one else to take the onus of saying "no," and Ger-

[28] AA, Guidelines of the Reich government for the German disarmament delegation, and Neurath covering letter, 19 Jan 33, 7616/H188329-39. The guidelines continued to be official policy at least until Hitler's May 17 speech: see AA, Draft tel., [ca. 21 Mar 33], 7360/E536471; *ADAP*, C/I/1, No. 200.

[29] AA, Note by Neurath, 16 Jan 33, 4619/E197597; Marginal note by Neurath on note by Bülow, 19 Jan [33], 3154/D668413-14; Memo [by Nadolny], [Jan 33], 3154/D668415-19.

many could not afford to stand isolated, facing the hostility of the powers. But beyond this, his hands were tied by his overall need to reduce conflict and by the increased power of the other ministers, including Neurath. He had become, in effect, their nominee, and served at their pleasure, as well as Hindenburg's. If Schleicher did not agree with them, they could take their case to Hindenburg, who would now be more likely to agree with them than with him. After the five-power agreement was worked out in Geneva on December 10, Schleicher had accepted it even though it had not actually conceded practical equality, and in fact he had let Bülow send word of this acceptance to Geneva without consulting the cabinet. In doing this, however, he was really only agreeing with Neurath, the minister directly involved. If Schleicher had not been willing to accept the agreement, cabinet consultation would doubtless have been necessary—and perhaps this was another reason, along with his new national responsibility, why he did accept. Later, when he failed to insist on the Reichswehr's draft guidelines, and left the tactical handling of disarmament diplomacy in the hands of the Foreign Ministry, he was following the logic of the return to the conference, and also continuing to bow to Neurath's judgment.

3. SCHLEICHER'S FINANCIAL PROBLEMS

Another potential area of military-civilian controversy was that of military expenditure. The overt army (not army plus navy) budgets of the last Weimar years usually ran at about RM 500 million. Partly from extra fat in this sum, and to a lesser extent from the budgets of other ministries, a secret budget was assembled for the first (pre-1933) armament program, amounting to about RM 100 million per year.[30] A continuation at this level would not permit the execution of the Six-Week Program before 1938, and as we have seen, a more ambitious plan for material armament, the Billion Program, requiring RM 200 million each year, was already under discussion in late 1931 and early 1932. The prospects for an increase in the regular arms budget were fairly dim; Schwerin von Krosigk, the Finance Minister, advised Neurath in mid-November that it would be impossible to increase the army budget in the fiscal years 1933 through 1935, and that it was very un-

[30] Budget figures are analyzed in Castellan, pp. 58-66; see also Mueller-Hillebrand, I, 19; OKW, Wi/IF 5.499, TA 43/32, 15 Jan 32, 110/837917-20. The regular naval budgets between 1931 and 1933 were RM 186-191 million: International Military Tribunal, *Trial of the Major War Criminals before the International Military Tribunal, Nuremberg, 14 November 1945–1 October 1946*, 42 vols. (Nuremberg: Secretariat of the Tribunal, 1947-1949) (hereafter referred to as *Nuremberg Documents*), XXXV, 578.

likely there could be an increase in the three years after that.[31] But work-creation credits had been proposed as a device to finance public works and provide employment, and Schleicher had hoped to use some of this money to finance the additional costs of the Billion Program. He had fully endorsed the Billion Program shortly after he became Minister of Defense in June 1932, but the program was deferred until fiscal 1933.[32]

Papen, an economic and financial as well as a political conservative, failed to promote a significant governmental work-creation program. He held to the theory that the way to overcome unemployment is to assist private industry, and adopting a procedure favored by Hans Luther, the President of the Reichsbank, his principal means of doing this was to grant corporate taxpayers tax-credit certificates, good for the payment of taxes in future years. These certificates could be discounted and used as credit instruments. This and other measures favoring the private sector, particularly a decree permitting wage cuts of 12½ percent below contract levels, had increased Papen's unpopularity on the left. One Papen program provided for granting up to RM 700 million in tax-credit certificates as "new employment premiums" to employers who hired new hands, but it was soon obvious that most of this allocation would not be used, and the program failed to bring any significant improvement in employment figures. Hindenburg threatened to resign if something were not done to provide "work and bread," and Schleicher now made the creation of work the primary announced objective of his government, naming Günther Gereke, a leading advocate of large-scale, local governmental work-creation projects, to develop a major governmental program as Commissioner for Work Creation.[33] Such a program could also serve

[31] AA, Note by Neurath, 15 Nov 32, 7474/H185474-76.
[32] Dülffer, pp. 226-227. The deferral may have been partly due to the diplomatic situation, but in any case, funds were tight, and aside from certain overeager projections in the spring of 1932, the Billion Program had always been proposed for the five years beginning 1 April 1933.
[33] Petzina, "Hauptprobleme," pp. 20-27; Michael Schneider, *Das Arbeitsbeschaffungsprogramm des ADGB* (Bonn: Verlag Neue Gesellschaft, 1975), pp. 192-198; C[laude] W[illiam] Guillebaud, *The Economic Recovery of Germany: From 1933 to the Incorporation of Austria in March 1938* (London: Macmillan, 1939), pp. 32-34, 36-38. Krosigk (*Staatsbankrott*, pp. 140-144; *Memoiren*, p. 149) claims much of the credit for the "Papen Plan." The wage reduction proviso was revoked by Schleicher in December. Papen had launched a direct governmental work-creation program, but too small (RM 302 million) and too long-term to produce an effect. Papen also moved toward Gereke's ideas at the end of his chancellorship. These ideas reportedly stemmed from a Reichswehr Ministry expert on economic mobilization. Gereke was prominent in the unstable "cross-front" of intellectuals, trade unionists, and dissident Nazis, which Schleicher hoped would provide a new political base. Bredow expressed to Schleicher grave doubts on Gereke and his associates: Bredow Papers, ltrs. to Schleicher, 21, 25 Oct 32; Orientation notes, 23, 25 Nov 32.

Schleicher's actual primary objective, the development of the Reichswehr.

The key problem in work creation was that of finance. At this point in the depression, no German governmental body could raise new money by public long-term loans, bank loans, or tax revenues. And the trauma of the inflation of the 1920s barred an open resort to the printing press; specifically, the Reichsbank's statutes now forbade it to discount treasury bills. But the Reichsbank could discount private bills of exchange. Taking advantage of this, Gereke and the conventional financial experts worked out a clever scheme for "prefinancing" work creation. Certain public-credit institutions (notably the Deutsche Gesellschaft für öffentliche Arbeiten) were to grant loans running up to twenty-five years to the local, Land, and Reich governments for socially useful projects. In this connection, the private contractors for these projects were to draw "work-creation bills," up to the value of their contracts, on the credit institutions. These bills could be renewed up to five years. And, stretching the law somewhat, the Reichsbank agreed to regard them as private bills of exchange and to discount them, which made them discountable for all private banks, and virtually as good as cash. The Reich Treasury (qua guarantor) committed itself to redeem the bills later by depositing tax-credit certificates or "labor-treasury bills" with the credit institutions or the Reichsbank, in return taking over the long-term claim against the various governments, including the Reich government (qua project sponsor and debtor). Theoretically, the Treasury would later be repaid from budget surpluses of the governments or by long-term loans. Actually the debt was never paid off, only translated into new debt. This complicated arrangement made it possible to finance public works through a "noiseless" increase in debt.[34] The same procedure, greatly extended, was to finance a large share of Hitler's rearmament.

During Schleicher's chancellorship, however, the conservative guardians of German finance—Luther and the Ministers of Finance and Economics, Krosigk and Hermann Warmbold—held a strong position in finance matters, just as Neurath did in diplomatic questions. These men were wary of inflation, although they had been ready to pump sizable amounts into the private economy, and industrial pressure

---

[34] Petzina, "Hauptprobleme," pp. 23-24, 27-29; Guillebaud, pp. 34-36; René Erbe, *Die nationalsozialistische Wirtschaftspolitik 1933-1939 im Lichte der modernen Theorie* (Zurich: Polygraphischer Verlag, 1958), pp. 22-23, 42-45; Karl Schiller, *Arbeitsbeschaffung und Finanzordnung in Deutschland* (Berlin: Junker & Dünnhaupt, 1936), pp. 59-63. Discounted treasury bills had "covered" inflation from 1914 to 1923: Peter Czada, "Ursachen und Folgen der grossen Inflation," in Harald Winkel, ed., *Finanz- und wirtschaftspolitische Fragen der Zwischenkriegszeit* (Berlin: Duncker & Humblot, 1973), pp. 12-14.

groups supported their views, on grounds of economic principle, private interest, and fear of radical experiments. Gereke wanted to use some of the unused tax certificates allocated for new employment premiums to enlarge the immediate governmental work-creation program, but Luther, Krosigk, and Warmbold rejected this and insisted on limiting the size of the work-creation program to RM 500 million. This would hardly get the economy moving again, and it left little leeway for arms expenditure. Also, although Gereke was popular with the left, he alienated private industry, and he did not necessarily serve Reichswehr interests. In January 1932, staff officers were disappointed to hear from Gereke that he would only be able to spare RM 50 million for the armed forces. The Reichswehr planners had been hoping to get RM 100 million for the Billion Program from this source, and indeed they had also asked the Finance Ministry for another RM 95 million in regular budget money to pay the costs of personnel reorganization or Umbau for the army and navy. Krosigk replied that he could only make RM 50-60 million avaliable for Umbau, using work-creation funds—and in view of Gereke's position, this would eliminate any contribution from work-creation funds toward the Billion Program; there was also doubt as to whether work-creation funds could legally be used for Umbau. In short, to carry out both the Umbau and Billion Programs (and without counting naval construction, which fell outside the Billion or Six-Week Programs), the Reichswehr needed about RM 300 million for the coming year, while as of mid-January, about RM 100 million of this would foreseeably be available from the long-established secret budgets of the Reichswehr and other ministries (the funds earmarked for the old armament program and the Six-Week Program), with a prospect of only a further RM 50-60 million from work-creation funds.[35]

Major Osterkamp, a staff officer of the high command, briefed Schleicher on this problem on January 16, and Schleicher generally agreed that the Reichswehr's demands for funds should be upheld vis-à-vis Gereke and Krosigk. Schleicher also at first approved Osterkamp's suggestion that a cabinet resolution be drafted, authorizing Umbau and thereby the funds required for it. By January 20, however, Schleicher had decided that no such resolution was needed, as supposedly the

[35] Petzina, "Hauptprobleme," pp. 22-29; Michael Schneider, pp. 198-202; Gereke, pp. 213-217; BA-MA, RH 15/40, Notes for report to Schleicher, 16 Jan 33, and Wehramt 101/33, [19 Jan 33], both with marginal notes. Gereke's memoirs (p. 215) say he first aimed at RM 3 billion, and the 16 Jan 33 Reichswehr notes say he had projects requiring RM 2 billion. Gereke claims (pp. 157-159, 234-235) that he opposed Hitler's and Göring's proposals to use work-creation funds for armament, implying that there was no such plan before Hitler. But cf. his position at the 9 Feb and 17 Mar 33 meetings of the Work Creation Committee: Bundesarchiv, Neue Reichskanzlei documents, R 43 II/540.

292—Schleicher Reaches a Dead End

mid-October 1932 guidelines for Landesschutz made it a duty for all Reich and Land officials to support his measures for national defense. Nevertheless, it seems very doubtful that Schleicher could have forced Gereke, Krosigk, and Luther to provide funds. On the twenty-fifth, Osterkamp's office expected to obtain credits of RM 50 million from work-creation funds, and of RM 100 million, outside the work-creation framework, from the Reichsbank. But on January 28, the work-creation contribution appeared about to be diverted, and although the Reichswehr secured a delay on this, President Luther of the Reichsbank was to refuse categorically on February 1 (two days after Schleicher's replacement as chancellor) to provide additional funds. At the end of Schleicher's tenure in office, the financing of the Umbau and armament plans was very much up in the air.[36]

If there had ever been a chance of speeding up the Umbau of the armed forces and inaugurating a large-scale six-month militia program in April 1933, the decisions made on Disarmament Conference policy and the problems of the budget eliminated that possibility. Aside from these obstacles, the army needed time to train the cadre that would later train the militia. Documents indicate that, as of the beginning of February, the army and navy planned to inaugurate their full militia plans only in 1934, and that three-month service remained the army's objective until December 1933.[37] One must wonder if Schleicher's talk of establishing a 300,000-man conscript militia within a few

[36] BA-MA, RH 15/40, Marginal notes on notes for report to Schleicher, 16 Jan 33, and Wehramt 101/33, [19 Jan 33]; Petzina, "Hauptprobleme," pp. 29, 43; Geyer, "Zweite Rüstungsprogramm," p. 163 (notes 70-72). Krosigk's MS Diary says (5 Feb 33) that just before the new cabinet was sworn in on 30 January, with Krosigk remaining as Finance Minister, he asked Hitler for assurances that the budget would be balanced, that there would be no currency experiments, and that Papen's tax credit certificate system would be continued; Hitler apparently agreed, indicating only some reservations about the form of the tax certificate system. On 3 Feb 33, Blomberg, the new Defense Minister, would warn the generals: "Amount of what we wish to and can build up at first is modest. Need time and money! T[he] present government, namely also the outstanding Finance Minister, would be most faithful pathmaker for the buildup [of] t[he] Wehrmacht. Struggles over money would indeed also be necessary; they would, however, be carried out without bitterness" (Vogelsang, "Neue Dokumente," p. 424). Krosigk was later said to have refused funds for military modernization in February (DBFP, 2/V, No. 494 [encl.]).

[37] ML, PG 34176, A I 517/33, 2 Feb 33, reel 27/439; F VIII a 11, Minister-Amt 464/33, 6 Sep 33, reel 54/229-232; Chef der ML, B Nr. A Va 6240/33, 22 Dec 33, reel 32/321-322; WKK, WK VII/732, Chef HL 3500/33, 18 Feb 33, 27/570-574; WK VII/741, Artillerie Führer (Bayer.) VII Ia 65/33, 24 Oct 33, 28/184; WK VII/4082, Wehrkreiskommando VII 1133, 21 Dec 33, 75/1033. On certain limited applications of three-month militia-style training in 1933, see Chapter Seven, Section 5, incl. note 74. Regular recruits entering the army on 1 Apr 33 were supposed to be given three-month training in line organizations to prepare the old regulars for giving such training later to the militia (BA-MA, RH 15/49, TA 737/32, 7 Nov 32, Annex 6), and this was in fact done after 1 April (see Chapter Seven).

months was ever much more than a castle in the air; this would not have been out of character for him. But in comparing his bold talk on some occasions with his moderate policies on others, we see another reason why he was mistrusted, and not least in army circles.[38]

4. THE ROLE OF BLOMBERG AND REICHENAU IN HITLER'S APPOINTMENT

Schleicher's main purpose in replacing Papen had been to win over at least some of the Nazis to the government, and he was not able to do this. Instead, Papen turned the tables on Schleicher by entering into direct negotiations with Hitler on January 4, 1933. Now it was Papen who might be the agent for reconciling the Nazis with the government. Since Nazi electoral strength had slipped, Hitler's bargaining position was now weaker than before, and he was ready to forego some of the ministerial demands he had made earlier, although he still insisted on becoming Chancellor. Papen eventually gave way on this point, still believing that he and a group of aristocratic ministers would be able to dominate a government nominally headed by the Nazi leader. Schleicher's ministers concluded for their part that they would rather have a Hitler government than a Papen-Hugenberg government, a rumored alternative which would face violent opposition from the Nazis and most other Germans. There was still the problem that, although Hindenburg was ready to drop Schleicher, who now faced more Nazi opposition than Papen had faced in November, the President had difficulty in bringing himself to give Hitler the chancellorship. Apparently this obstacle was finally overcome by making the old man believe that Schleicher and General von Hammerstein were planning a putsch, and that therefore the appointment of a Hitler-Papen cabinet had to be made, and quickly.[39] Hindenburg drew reassurance from the fact that men he trusted would also be cabinet members: Papen as Vice-Chancellor, Neurath as Foreign Minister, and now also in particular, General Werner von Blomberg, hitherto chief military delegate at the Geneva Disarmament Conference and commander of the First Division

---

[38] On Schleicher's personality, see Brüning, *Memoiren*, pp. 130, 159, 346, 427, 452-453, 473, 489, 538-540, 595-596, 648-649—a biased, but well-informed source. Military distrust of Schleicher: Carsten, *Reichswehr and Politics*, pp. 298-304, 319-320; Vogelsang, *Schleicher*, pp. 48-50, 54-55; Conze, "Politischen Entscheidungen," p. 231.

[39] See Vogelsang, *Reichswehr*, pp. 351-360, 366-396. Vogelsang concludes (*Reichswehr*, pp. 394-395) that Hindenburg could not have believed in an army putsch, but he had angrily raised the question of obedience with Hammerstein only three days before (*Reichswehr*, p. 378). Krosigk believed that the threat of a putsch, real or presumed, brought the participants in the new cabinet together: Krosigk MS Diary, 5 Feb 33. See also Bracher, *Auflösung*, pp. 719-726; Bullock, pp. 248-249.

in Königsberg, as Reichswehr Minister. The German Nationals or DNVP, led by Alfred Hugenberg, were also to participate and provide conservative ballast. The new cabinet was hastily sworn into office on January 30, 1933.

Certainly Blomberg's readiness to head the Defense Ministry did much to make the new cabinet possible. Hindenburg absolutely opposed the idea of naming a Nazi as Reichswehr Minister, but Blomberg was acceptable to him. Blomberg's experience as military delegate in Geneva, along with a certain façade of gentlemanly self-assurance, probably made him appear all the more suitable to the President, who had gotten to know him as East Prussian commander. Blomberg was also acceptable to Hitler, who later said that the general had been recommended by his "friends in East Prussia." The "friends" made no mistake; Blomberg had a strain of idealism that made him perhaps more susceptible to Hitler than any other German general would have been. He had earlier been attracted to the ideas of Rudolf Steiner, the anthroposophist, and he had been impressed, during a visit to the Soviet Union, by the Soviet system for mobilizing and controlling public opinion. In 1945, Field Marshal von Rundstedt would say disparagingly that Blomberg "was always somewhat strange to us," i.e., to the officer corps, and that "nobody could really put up with him." But no officer seems to have questioned Blomberg's credentials in 1933, and indeed, as a former Chief of the Truppenamt, a former candidate for the post of Chief of the Army Command, and the commanding officer of the key First Division in East Prussia, he appeared to be highly qualified. Like Hindenburg, he enjoyed the advantage of physical stature, standing well over six feet. But he had recently been shaken by the loss of his first wife and by a serious riding accident.[40]

The background of Blomberg's appointment has been hard to establish. His unpublished memoirs would have it that, when he was summoned back to Berlin from Geneva on January 29, he had no inkling that "a completely new segment of life was to begin." Speaking to the division commanders on February 3, he also gave the impression that his appointment had only been decided on January 30.[41] A number of historians have concluded, however, that he must have taken

[40] On Blomberg's background and character: Sauer, "Mobilmachung," pp. 711-714; Müller, *Heer und Hitler*, pp. 49-52, 54-55n; Wheeler-Bennett, *Nemesis*, pp. 295-297. Blomberg's own account: Blomberg MS Memoirs, Vol. IV. Acquaintance with Hindenburg: Hans Otto Meissner and Harry Wilde, *Die Machtergreifung: Ein Bericht über die Technik des nationalsozialistischen Staatsstreichs* (Stuttgart: Cotta'sche Buchhandlung, 1958), p. 180. Hitler's "friends": Adolf Hitler, *Hitler's Secret Conversations, 1941-1944*, Norman Cameron and R. H. Stevens, trans. (New York: Farrar, Straus & Young, 1953), p. 404.

[41] See Blomberg MS Memoirs, Vol. IV; Vogelsang, "Neue Dokumente," p. 433.

part in political negotiations before then. They note that, according to the memoirs of Otto Meissner, State Secretary to the President, Blomberg talked secretly with the President in "the last week of January" and told him that the formation of a national coalition cabinet under Hitler would provide a welcome solution from the standpoint of the armed forces. Blomberg also said that he regarded as hopeless the idea, attributed to Schleicher, of establishing a military dictatorship and combating the SA and SS (Schutzstaffeln: at this time an élite storm-troop formation). According to Meissner, Blomberg told Hindenburg that in view of the sympathy for Hitler within the Reichswehr, the latter, "in a combat of German against German, would break down internally and be defeated externally." The historians believe that this discussion took place before January 30, and they also note Hitler's later recollection that Blomberg had been recommended to him by his East Prussian friends. Generally, they also hold that Blomberg's divisional chief of staff, Colonel von Reichenau, played a leading part in the negotiations.[42]

Contemporary sources now available tend to confirm that there were earlier negotiations. They show that Blomberg stopped in Berlin in mid-January, when he consulted with Neurath and Schleicher in connection with his return to Geneva, where the Bureau of the Disarmament Conference was to resume its meetings on January 23, and where plenary sessions would resume on February 2. More to the point, a little-known Italian document shows that Blomberg was considered "available" for a Hitler cabinet by January 23. On that date, Major Giuseppe Renzetti, who was Mussolini's political agent in Germany, reported to the Italian dictator that "the conspirators" (apparently Papen and the Nazi and DNVP leaders) planned to make Blomberg ("Beremberg," further identified as divisional commander in East Prussia), classified as "apolitical," the Reichswehr Minister in a cabinet of national concentration to be led by Hitler.[43] Either Blomberg

---

[42] Bracher, *Auflösung*, pp. 713-714, incl. 713n; Vogelsang, *Reichswehr*, p. 375 and note; Carsten, *Reichswehr and Politics*, p. 389; Eyck, II, 476-477; Otto Meissner, *Staatssekretär unter Ebert-Hindenburg-Hitler* (Hamburg: Hoffman & Campe, 1950), p. 266.

[43] AA, Memo by Neurath, 14 Jan 32, 3154/D668400-01; tel., Nadolny to Bülow, 28 Jan 32, 3154/D668443-44; Bredow Papers, Orientation notes, 12 Jan 33; National Archives, Microfilm T-586, Segretaria Particolare del Duce (folder 442R—"Hitler Adolf"), report of 23 Jan 33, reel 491/050269-71. Renzetti's list of prospective cabinet ministers would indicate that negotiations generally had advanced farther by January 23 than most historians have supposed; it included Papen as Vice-Chancellor, Hugenberg as Minister of Economics, Seldte as Minister of Labor, and Krosigk as Minister of Finance, while Nazis (unidentified) were to have the Reich Ministry of Interior, and both "Presidency" (now Reichskommissariat) and Ministry of Interior in Prussia.

had directly indicated a willingness to serve by January 23, or someone such as Reichenau had spoken for him.

This still leaves many questions unanswered, and much room for speculation. In particular, what led Blomberg and presumably Reichenau into negotiations for Blomberg's entry into a Hitler cabinet? Blomberg had no great personal regard for Schleicher, and indeed he believed that Schleicher had earlier prevented his own rise to the post of the Chief of the Army Command. Joachim von Stülpnagel, who had exactly the same grievance, and Werner von Fritsch, who described one of Schleicher's briefings as "all lies," are also said to have been potential candidates for the ministerial post; hostility to Schleicher was common enough.[44] But Blomberg himself was hardly the sort of man who seeks vengeance, least of all at this time of recent personal misfortune.

Aside from personal motives, did Blomberg seek to replace Schleicher because he was dissatisfied, as chief military delegate in Geneva, with German armament policy, in diplomacy and in practice? This would be particularly interesting from the perspective of this study, but a problem for this theory is that, aside from Blomberg's possibly pro forma accolades (both on February 3, 1933, and years later in his memoirs) for Schleicher's firm insistence on equality of rights, Blomberg's own acts in January and February give no indication of dissatisfaction with the armament policy of Schleicher's government. If anything, he would ask at first for less than Schleicher had sought.[45]

There is another explanation that may appear obvious: that Blomberg and Reichenau sympathized with Hitler and his movement. Blomberg would soon (as he later recorded) find "faith, veneration for a man [Hitler], and complete adherence to an idea [National Socialism]." As for Reichenau, Blomberg's thrusting and ambitious chief of staff, even at the time of Brüning's downfall, he was saying, according to

---

[44] On other candidates and their attitudes toward Schleicher: Sauer, "Mobilmachung," p. 711; Stülpnagel, *75 Jahre*, pp. 277-279; Robert J. O'Neill, *The German Army and the Nazi Party, 1933-1939* (London: Cassell, 1966), p. 28. Schleicher himself was ready to serve under Hitler: Bracher, *Auflösung*, p. 712; Vogelsang, *Reichswehr*, pp. 383-384.

[45] Blomberg's praise of Schleicher: Vogelsang, "Neue Dokumente," pp. 432-434; Blomberg MS Memoirs, Vol. IV. Note also the Blomberg comment later recalled by Ott (cited by Sauer, "Mobilmachung," p. 716n) that Schleicher's line was the "only possible." Heineman, "Neurath," p. 214, and Post, pp. 309-310n, report a dispute between Blomberg and Neurath over disarmament policy; there was (as described earlier) a quarrel between the two ministries, but I do not find much evidence of Blomberg's personal involvement at this time. To judge by Neurath's version of their discussions, Blomberg did not contest Neurath's judgment on disarmament matters in mid-January (AA, Memo by Neurath, 14 Jan 33, 3154/D668400-01), and accepted Neurath's views on 8 February (AA, Note by Bülow, 8 Feb 33, 4619/E197606-07). See also Chapter Seven, Section 5.

contemporary report, that "he could not control the Reichswehr and population of East Prussia if the Nazis were not admitted to the government."[46] Yet Blomberg and Reichenau were never committed, ideological National Socialists, such as Goebbels, Hess, Alfred Rosenberg, or Himmler were. For one thing, their professional training and tradition of being "apolitical" were too strong; for another, Blomberg was too open-minded and changeable, and Reichenau too perceptive and hard-headed, for such a faith.[47]

In the last analysis, three considerations probably explain Blomberg's and Reichenau's support of Hitler in January 1933. One consideration, perhaps the most important, was the particular concern of these two officers for the defense of East Prussia. The problem of defending the isolated province had recently arisen in the Ott war game, and in East Prussia the army depended heavily on the Nazis, both for Grenzschutz and for field-army mobilization.[48] The regular First Division itself was probably also, as Ott was to say in his postwar testimony, the least immune of army units to the spirit of Nazism. The second consideration was that a Hitler government might improve prospects for military development generally, especially by freeing the army from any domestic police responsibility.[49] The third consideration was that Hitler seemed to have a clearer conception of the con-

[46] Telford Taylor, *Sword and Swastika: Generals and Nazis in the Third Reich* (New York: Simon & Schuster, 1952), p. 80; Co[nze], "Zum Sturz Brünings," p. 284. Reichenau's statement was not quite as presumptuous as it sounds: Blomberg was at that time in Geneva, so that Reichenau was acting commander.

[47] See Telford Taylor, pp. 77-78, 152-154; Müller, *Heer und Hitler,* pp. 49-53, and Deutsch, *Hitler and His Generals,* pp. 8-13.

[48] Carsten argues (in *Reichswehr and Politics,* pp. 352, 381) that the Nazis did not preponderate in the East Prussian Grenzschutz, and cites Vogelsang, *Reichswehr,* pp. 157-158, and Blomberg's MS Memoirs. But Vogelsang's discussion relates to an earlier period (ca. 1930), and although Blomberg wrote that his main reliance was on the DNVP (perhaps meaning the Stahlhelm), he added that all elements were needed in the threatened province. With three field-army divisions and six Grenzschutz Verbände (see Castellan, *Réarmement clandestin,* pp. 400, 408-409) to be raised from the local population (2¼ million), this was doubtless true enough. Reichenau thought that the Nazis were crucial, as shown by his statement at the time of Brüning's fall (see above) and his communication to Hitler (see below). As a possible measure of Nazi strength in the province, they gained 47.1 percent of the East Prussian vote in the July 1932 election, and 39.7 percent in November (Bracher, *Auflösung,* p. 647).

[49] As soon as the new regime was established, Blomberg and Reichenau laid great stress on the nonresponsibility of the Reichswehr for domestic order: RK, Cabinet Protocol, 30 Jan 33, 3598/D791607; Sauer, "Mobilmachung," p. 729. In April 1933, Reichenau issued a directive stating the national government's intention that the duties for which the army had been destined at the time of the (Ott) war game would be handled in future by the police, reinforced if necessary by the national Verbände: ML, PG 34097, TA 135/33, 21 Apr 33, reel 31/934 (given wider dissemination on 31 May 33 [reel 31/936]).

nection between national defense and foreign policy, and of the need for a broad, popular, ideological mobilization for total war than Schleicher had.

In early December 1932, Hitler recorded his views about military, foreign, and domestic policy in a letter to Reichenau. This letter not only expresses Hitler's own ideas, but also suggests some of the likely preoccupations of Reichenau and Blomberg, and some of the considerations that motivated their collaboration in overthrowing Schleicher. Apparently under the impact of the Ott war game, Reichenau had sent a message to Hitler (through the division chaplain, Ludwig Müller) asking if the Nazi leader could not direct his organizations in East Prussia to take more account of the Polish threat. This request evidently showed serious concern over East Prussian defenses, while nothing in Hitler's reply suggests familiar relations between Reichenau and Hitler, or that Hitler considered Reichenau already won over. The request may have embarrassed Hitler, who may have been unsure of his control over East Prussian SA formations, but it also gave him a chance to explain and propagate his views. In his reply, dated December 4, 1932, he said that he would do what he could, but he went on to argue at length that the defense of East Prussia was really a national problem, involving Germany's overall foreign, domestic, and military policies. Drawing on ideas he had expressed in *Mein Kampf* and elsewhere, he maintained that the conservative-military policy of close relations with the Soviet Union was unwise, and that Italian and British ties were preferable. France (he continued) had deliberately surrounded Germany with countries that had territorial claims or quarrels with Germany, and the French had recently succeeded in tying Russia down in the Far East, and in bringing about a Soviet-Polish nonaggression pact. Under these conditions, any moment in which Germany repelled the sympathies of the world, for example by a revival of the Hohenzollern dynasty, would create favorable conditions for a Polish attack on East Prussia; the province would not have the strength to resist, and under existing political conditions the capacity of the rest of the Reich to provide military support was zero.

> In this connection I consider the theoretical forcing of German rearmament as the worst danger. It is conceivable that France is today no longer in a position to sabotage the theoretical concession of Germany's equality of rights. Since in this case practical, technical, and organizational rearmament is to follow on theoretical equality of rights, the succeeding time span will be the most dangerous period in German history. If ever there are grounds for a preventive attack, then [there are] in this case for a French attack on Germany.

Only such a military act creates the desired new conditions, and the same world that today extends to us its theoretical good wishes will take care not to try to correct the fait accompli with, say, armed force.[50]

Besides criticizing the demand for theoretical equality as likely to provoke a French invasion, Hitler also suggested that Schleicher's Umbau plans approached the problem from the wrong end. As he had argued before, he argued now that the first needs were to eliminate Marxism, forcibly establish a single Weltanschauung, and morally rearm the German people. As matters stood, more than half the German people were more or less pacifist, if not actively hostile to defense. In what can now be seen as a reference to army mobilization plans, he wrote that it was wrong to think of turning to nationalist Verbände (like the SA) for military reinforcements; as more or less untrained cannon fodder, they could have no effect on the battle, while their withdrawal to the front "means turning the homeland over to the red mob." The civil and military authorities regarded rearmament as a technical and organizational task, and had no comprehension of the need for a moral rearmament; Schleicher in particular lacked any understanding of this. To involve the army in moral rearmament made it appear partisan, and was futile anyway: "For neither the police nor the military have ever destroyed Weltanschauungen, and still much less created Weltanschauungen." And it was only after stamping out Marxism and establishing a single Weltanschauung that Germany should proceed next to carry out technical rearmament, after that to muster organizationally the power of the people for national defense, and only lastly to obtain legal recognition from the rest of the world for what had been done. Hitler's implication was that, as Nazi leaders had been urging, the army should give the SA a free hand to deal with the Marxists; the military itself should abstain from police activity.[51]

---

[50] Vogelsang, "Hitlers Brief an Reichenau," pp. 433-437. A slightly different version is available at the National Archives, Microfilm T-81, Records of the National Socialist German Labor Party, Adolf Hitler, Kanzlei, 7/11542-52. Gen. Adams stated in March 1933 that in case of war, field-army forces in the rest of Germany would have to leave East Prussia to its fate (Krupp Trial, Defense Exhibit 62, Adam memo). On Hitler's ideas expressed in his earlier writings, see Chapter Seven.

[51] Addressing a meeting of industrialists on 27 Jan 32, Hitler expressed similar ideas on the priority of internal organization and training, saying that Germany needed 8 million army reserves who were not ideologically corrupted: Max Domarus, ed., *Hitler, Reden und Proklamationen 1932-1945*, 2 vols. (each in two parts) (Munich: Suddeutscher Verlag, 1965), I/1, 68-90. See also his open letter to Papen in *Völkischer Beobachter*, 21 Oct 32. On the Nazi desire for a free hand: Vogelsang, *Reichswehr*, pp. 250-255, 330 and note.

We hardly think of Hitler as a sincere or candid man, but he had worked out a system of basic ideas, a picture of how peoples and powers interacted, and when he stated something which fitted his system, he was probably expressing his actual belief. Of course, his order of priority, putting the establishment of a single Weltanschauung ahead of rearmament and army expansion, served his political interest, and it would not keep him, when once in power, from moving rapidly toward rearmament. But the need for imposing a Weltanschauung was proven, in Hitler's system, by the alleged experience of the breakdown of the home front in 1918, the "stab in the back." There is every reason to think that he really believed in the need and the proof. Considering Hitler's jungle view of the world, and his belief in French malevolence toward Germany, along with his ignorance of actual French conditions, there is also reason to think that he did indeed believe at this time in the possibility of a French preventive attack; he would twice express this concern shortly after taking office, and it appears to have influenced his actions for some time thereafter.[52]

Blomberg presumably saw Hitler's letter of December 4 to Reichenau; Blomberg was with the First Division, as its commander, from August 1932 to January 1933, and his divisional chief of staff would surely have shown such a letter to him. The initial policies of Reichenau and Blomberg after January 30 suggest that they accepted Hitler's arguments quite literally. They would immediately seek to divorce the army from domestic political activity and responsibility—a move that also certainly corresponded with their own predilections and with Hindenburg's intentions—and they accepted without demur the idea that the SA should have a "free hand" in the streets. This acceptance seems to show that they fully agreed with the Nazi idea of purging the "antidefense" elements and forcibly imposing a single Weltanschauung. And Blomberg even seems to have agreed that material armament should follow next, although Krosigk rather than Hitler

---

[52] See Chapter Seven, Section 2. Joseph Goebbels told a small group of journalists in April 1940 that the domestic enemies of the Nazi party could easily have suppressed it in 1925, and he added: "It was the same in foreign policy. . . . In 1933, a French Premier should have had to say (and I would have said it if I had been French Premier), 'The man who wrote the book *Mein Kampf*, in which it says so and so, has become Chancellor of the Reich. The man cannot be tolerated in our proximity. Either he disappears, or we march.' That would have been thoroughly logical. They abstained from that. They let us alone, they let us pass through the danger zone unhindered, and we were able to sail past all the reefs, and when we were ready, well armed, better than they, they began the war": Andreas Hillgruber, *Deutschlands Rolle in der Vorgeschichte der beiden Weltkriege* (Göttingen: Vandenhoeck & Ruprecht, 1967), pp. 76-77.

may have influenced him here.⁵³ Neither Blomberg nor Reichenau was dominated by loyalty to a closed caste and, more than most officers, each of them—Blomberg through his Russian observations and Reichenau through his fertile imagination—could see in a broad popular movement a welcome basis for national mobilization. Like many other Germans—though not a majority—they were ready to try something new.

THIS STUDY has argued that the military objectives of rearmament and, particularly, of preparing for national mobilization, were a major consideration—possibly the largest consideration—in bringing about the replacement of Hermann Müller by Heinrich Brüning, and of Heinrich Brüning by Franz von Papen. It has been suggested that military concerns also counted heavily in fomenting impatience with the Braun-Severing government in Prussia, and in forcing Papen's resignation. But did military considerations have a comparable influence in bringing about the replacement of Kurt von Schleicher by the Hitler-Papen government? The question gains some urgency from the fact that a sharp debate has arisen over the role of economic interest groups, particularly heavy industry, in bringing Hitler to power.

Some writers have asserted that the leading heavy industrialists and their alleged tools, such as the "Keppler Circle," furnished large amounts of money to Hitler, exerted pressure on Hindenburg by sending him a petition-letter urging the appointment of Hitler, and brought about the crucial meeting between Hitler and Papen. With the skepticism of classical historical method, Henry Ashby Turner, Jr., has pointed to weaknesses in the arguments and documentation of this school of thought: he has shown that the Keppler Circle served Hitler rather than the major industrialists, that solid evidence of major financial contributions has not been produced, that the NSDAP did not depend on such contributions anyway, that few leading industrialists signed the letter to Hindenburg, and that Papen and Hitler sought each other out for their own reasons. The replies to Turner have usually criticized him for failing to see the socioeconomic forest for the trees. Some of his opponents clearly feel that one can only see the forest if one dons Marxist spectacles.⁵⁴

---

⁵³ See Vogelsang, "Neue Dokumente," pp. 432-435; also Bundesarchiv, Koblenz, Neue Reichskanzlei Records, R 43 II/540, Record of Work Creation Committee meeting, 9 Feb 33. On post-January attitudes generally, see Chapter Seven, Section 2 below.

⁵⁴ Writings blaming heavy industry include George W. F. Hallgarten, *Hitler, Reichswehr und Industrie* (Frankfurt: Europäische Verlagsanstalt, 1962); G.W.F. Hallgarten and Joachim Radkau, *Deutsche Industrie und Politik von Bismarck bis*

Yet the debate has also produced consensus on some points. Turner agrees that one should take account of the antirepublican, anti-Socialist, and antilabor bias of the industrialists.[55] Two of his principal opponents agree that most of industry preferred a Papen government to a Hitler government, at least until Schleicher came to power. And most important, they also concede that it was agrarian (and perhaps military) influence that did most to overcome Hindenburg's reluctance to appoint Hitler, although they still maintain that agrarian and industrial interest groups were closely linked.[56] Of course, from the kind of perspective they exemplify, a narrow concentration on political details only diverts attention from the broader theoretical or historical necessities.[57]

---

heute (Frankfurt: Europäische Verlagsanstalt, 1974); Eberhard Czichon, *Wer verhalf Hitler zur Macht?*, 2nd ed. (Cologne: Pahl-Rugenstein Verlag, 1971); Dirk Stegmann, "Zum Verhältnis von Grossindustrie und Nationalsozialismus 1930-1933," *Archiv für Sozialgeschichte*, 13 (1973), 399-482; Sohn-Rethel, *Ökonomie und Klassenstruktur*. Discussions of the literature more or less sympathetic with this view: Eike Hennig, "Industrie, Aufrüstung und Kriegsvorbereitung im deutschen Faschismus: Anmerkungen zum Stand der neuen Faschismusdiskussion," in Forstmeier and Volkmann, eds., *Wirtschaft und Rüstung*, pp. 385-415; and Hans-Erich Volkmann, "Politik, Wirtschaft und Aufrüstung unter dem Nationalsozialismus," in Funke, ed., *Hitler, Deutschland und die Mächte*, pp. 269-291. For the other side: Henry Ashby Turner, Jr., esp. "Big Business and the Rise of Hitler," *AHR*, 75 (1969-1970), 56-70; Turner, "The *Ruhrlade*, Secret Cabinet of Heavy Industry in the Weimar Republic," *CEH*, 3 (1970), 195-228; Turner, "Grossunternehmertum und Nationalsozialismus 1930-1933: Kritisches und Ergänzendes zu zwei neuen Forschungsbeiträgen," *HZ*, 221 (1975), 18-68; and Turner and Horst Metzerath, "Die Selbstfinanzierung der NSDAP 1930-1932," *GG*, 3 (1977), 59-92; also, Wilhelm Treue, "Der deutsche Unternehmer in der Weltwirtschaftskrise 1928 bis 1933," in Conze and Raupach, pp. 82-125.

[55] Turner, "Big Business," pp. 57, 70; Turner, "Grossunternehmertum," p. 68.

[56] Hallgarten in Hallgarten and Radkau, pp. 201, 209, 218; Stegmann, pp. 431-432, 439.

[57] See, e.g., Sohn-Rethel, *Ökonomie und Klassenstruktur*, pp. 94-95. But Sohn-Rethel dismisses (p. 78) the idea that all-powerful wire-pullers in the form of monopoly capitalists used the Nazis as mere agents: "There were in fact wire-pullers, but that they had the power to direct matters according to their purposes is again no more than a legend."

I would suggest that it was not through lobbies, "circles," and politicians in the pay of bankrupt heavy industry that the interests of business might affect the decisions of the presidial government and perhaps of Hindenberg himself; it was through top government and Reichsbank officials who were motivated not by bribes but by social paranoia, conventional economic wisdom, and fears for business confidence. Krosigk is the only relevant figure on whom we have much contemporary information, but Warmbold, Luther, and Papen himself, as well as Friedrich Wilhelm Dreyse and Friedrich Syrup, evidently shared many of Krosigk's attitudes. Krosigk consistently argued that the Nazis should be brought into the government for the sake of business confidence, but while sympathetic to rearmament, he could not envision any large new arms appropriations, and he (like Warmbold, also Luther) shared Gustav Krupp von Bohlen's opposition to "credit and currency experiments" for financing work creation. When Krosigk accepted a Hitler-Papen

If we take our cue from concrete circumstances, rather than from theory, then on this occasion it was Paul von Hindenburg who decided the course of events. And one fact stands out: Hindenburg had personally come first to dislike and distrust Schleicher, and then finally to fear him. The President turned to advisers who were not from the Bendlerstrasse. This meant, among other things, that for the first time since the formation of the Müller Grand Coalition, the choice would not be directly affected by high-level armament and mobilization planning. This kind of military consideration would have little or no influence at this juncture.

It does indeed seem likely that Hindenburg's sympathy for agrarian interests strongly influenced him to accept Hitler. The problems of Ostelbian agriculture were being agitated, with more attention to scandals than in May. Hindenburg himself was proud to be a landed aristocrat, and accessible to agrarian spokesmen. When Schleicher revived plans for foreclosure and settlement on Eastern estates, the leader of the agrarian Landbund (who had already advocated a Hitler chancellorship) denounced this as a "crime," and took his case to the President. Meissner reports that Hindenburg was later approached by the arch-Junker of them all, Elard von Oldenburg-Januschau, who made light of the President's concern over the "robust and violent manners" of the Nazis.[58]

Still, Hindenburg was an ex-Field Marshal as well as a landowner, and he reportedly listened to Blomberg as well as to Oldenburg-Januschau. The President doubtless shared the concern of the Königsberg command for the defense of East Prussian borders. Here the generalities of national mobilization became a specific problem, and when Blomberg advised that the army would be defeated internally and externally if it fought the Nazis, he spoke as the expert at the critical spot. Blomberg's counsel gave Hindenburg a higher, more disinterested rationale for accepting a Hitler chancellorship, while Blomberg's selection as Defense Minister also appeared to Hindenburg to provide a control on Hitler. Thus, while agrarian arguments touched Hinden-

---

government, it was not for the sake of imperialist expansion, much less an inflationary armament program, but because he thought this offered the least danger of disturbance. See Vogelsang, "Neue Dokumente," p. 434; AA, Note by Neurath, 15 Nov 32, 7474/H185474-76; Krosigk MS Diary, 27 Nov, 4 Dec 32, 29 Jan, 5 Feb 33 (for this, see note 36 above); Krosigk, *Staatsbankrott*, pp. 138, 140-143, 146-148, 150, 154, 163, 169; Michael Schneider, pp. 199-200. Specific evidence is lacking, but the willingness of Krosigk as well as Papen to serve under Hitler (probably as early as 23 January—see note 43) may have helped to overcome Hindenburg's hesitations.

[58] Vogelsang, *Reichswehr*, pp. 358-360, 368-369, 375; Stegmann, p. 435; *DBFP*, 2/IV, No. 229; Meissner, p. 265.

burg's pocketbook and pride, Blomberg provided salve for Hindenburg's conscience and sense of national responsibility, and this may have been equally important.

In any case, the priority given earlier to military plans had severely limited the possible choices. Largely because of that priority, parliamentary government had been curtailed, and the prorepublican forces weakened, while the Nazis had been declared worthy of sharing in government. A popular non-Nazi government had been effectively ruled out, and of the remaining possibilities—a reactionary Papen-Hugenberg "conflict" government, a "social" dictatorship under Schleicher, hoping for Nazi support, or a regime led by Hitler himself—only the last was practical politics. This situation was mainly the work of Schleicher, who had thought since 1931 that he could work with and use the Nazis, in the military interest. The General had accepted in August the idea of a Hitler chancellorship, and it had been Hindenburg who had prevented this. At the end of the year, Schleicher undertook to govern without Hitler, not because of any principled opposition to the Nazi leader, but because he now thought Hindenburg would never accept Hitler as Chancellor. Schleicher himself was willing in late January to serve under Hitler as Reichswehr Minister: he preferred to have Hitler, rather than Papen, succeed himself.[59] Perhaps Schleicher had begun to realize that only Hitler could dragoon the public, the financial experts, and his own following into supporting a massive program of training and rearmament.

[59] Vogelsang, *Reichswehr*, pp. 382-384, 388-389.

PART II

CHAPTER SEVEN

Hitler's Accommodation with the Military

1. HITLER'S IDEAS

As we turn to Hitler's accession to power, we face a broad problem, which German historians have warmly debated: did Hitler's rule represent a break with Germany's past, or did it fit in with the established patterns of German history? Before, during, and after World War II, some British and American writers found antecedents of Nazism in German history—in the German reformation, in the Thirty Years War, in Hegel and Herder and Nietzsche, in the Germany of Frederick the Great and Bismarck and William II.[1] After the war, some German historians, naturally enough, came to the defense of Germany's past, arguing that there were healthy as well as destructive currents, and that all of the destructive currents could be found also in the past histories of other nations. The most notable of these defenders was Gerhard Ritter, who had already begun a study distinguishing between German statesmanship and German militarism. He found the sharpest expression of this duality in the World War I conflict between Chancellor Theobald von Bethmann Hollweg and General Erich Ludendorff, while he dissociated the German soldiers of the 1920s and 1930s from Hitler, who to him represented the ultimate and fatal exaggeration of militarism.[2] This defense was somewhat shaken, however, by the work of Fritz Fischer, who exposed to the German public Bethmann's endorsement of annexationist aims.[3] Other historians, particularly Hans-Ulrich Wehler, have picked up the class-conscious, antiimperialist analysis pioneered by Eckart Kehr just before the Nazi takeover; this school has found an explanation for im-

---

[1] Particularly outspoken were Robert, Lord Vansittart, *Black Record: Germans Past and Present* (London: Hamish Hamilton, 1941); Rohan D'O. Butler, *The Roots of National Socialism* (New York: Dutton, 1942); William M. McGovern, *From Luther to Hitler: The History of Fascist-Nazi Political Philosophy* (Boston: Houghton Mifflin, 1941). But more influential have been such works as Telford Taylor's *Sword and Swastika*, Wheeler-Bennett's *Nemesis of Power*, and A.J.P. Taylor's *The Course of German History: A Survey of the Development of Germany since 1815* (New York: Coward-McCann, 1946).

[2] Aside from Ritter's four-volume opus, *Staatskunst und Kriegshandwerk*, see his article, "Das Problem des Militarismus in Deutschland," *HZ*, 177 (1954), 21-48.

[3] In his *Griff nach der Weltmacht*; his *Krieg der Illusionen: Die deutsche Politik von 1911 bis 1914* (Düsseldorf: Droste, 1969) makes charges of preventive war planning.

perial Germany's foreign and naval policy—and more or less implicitly, for the emergence of Nazism—in the social and economic conflicts within Germany.[4] Still others, however, have concluded from their closer study of Hitler's own policies that his actions cannot be explained in terms of domestic socioeconomic pressures. They point out that, although he may have left domestic problems to be fought out between his henchmen, he had an obsessive interest in foreign and race policy, and controlled these policies himself. They distinguish between "Wilhelminian" imperialism and Nazi radicalism, especially in the race issue, seeing this last as a new departure, a decisive break in continuity.[5]

In light of the policy conflicts of 1932 and 1933, I think there is much to be said for Ritter's basic contrast between the responsible civilian policy makers and the militarists. Without fully accepting his attempted exoneration of the former, we cannot deny the existence of a continuing conflict between these two groups. Ritter did base his work on a thorough study of available sources, military as well as civilian, and he did somewhat modify his own views in the light of experience and the evidence. Unfortunately, Ritter did not live to explore the new sources on the diplomatic and military policy of the 1920s and 1930s. Contrary to what he tended to assume, militarism —which he defined as "the one-sided determination of political decisions by military-technical considerations," as well as a prevalence of warlike traits[6]—did exist in the late Weimar era, and helped to bring about Hitler's rule. The position of Bülow vis-à-vis Schleicher is dis-

[4] On Kehr, see Introduction, note 1. Recent "Kehrite" writings include Helmut Böhme, *Deutschlands Weg zur Grossmacht: Studien zum Verhältnis von Wirtschaft und Staat während des Reichsgründungszeit 1848-1881* (Cologne: Kiepenheuer & Witsch, 1966); Hans-Ulrich Wehler, *Bismarck und der Imperialismus* (Cologne: Kiepenheuer & Witsch, 1969) and the same author's *Deutsche Kaiserreich*; Volker R. Berghahn, *Der Tirpitz-Plan: Genesis und Verfall einer innenpolitischen Krisenstrategie unter Wilhelm II* (Düsseldorf: Droste, 1971).

[5] Andreas Hillgruber, in *Kontinuität und Diskontinuität in der deutschen Aussenpolitik von Bismarck bis Hitler* (Düsseldorf: Droste, 1969) and his *Deutschlands Rolle in der Vorgeschichte der beiden Weltkriege*; Klaus Hildebrand, in "Hitlers Ort in der Geschichte des preussisch-deutschen Nationalstaates," and his *Deutsche Aussenpolitik 1933-1945: Kalkül oder Dogma*, 2nd ed. (Stuttgart: Kohlhammer, 1973). The distinctiveness of Hitler's foreign policy has been argued particularly by Hans-Adolf Jacobsen, *Nationalsozialistische Aussenpolitik 1933-1938* (Frankfurt: A. Metzner, 1968).

[6] Ritter, "Militarismus," pp. 23, 45-46; Ritter, *Staatskunst und Kriegshandwerk*, I, 13; II, 118-119. Eckart Kehr had quite a different definition, holding that militarism existed where a class of officers considered themselves apart from and above civilians, with their own code of honor, law, and convictions, and where a significant part of civil society willingly accepted this superiority, an attitude linked to the new alignment of forces in Prussian capitalist society after the rise of the industrial proletariat: "Zur Genesis des Königlich Preussischen Reserveoffiziers," in *Primat der Innenpolitik*, p. 54.

tinctly reminiscent of that of Bethmann Hollweg or of Richard von Kühlmann vis-à-vis Ludendorff, as described by Ritter. Ritter's basic thesis had more validity than he knew, and it applies precisely in the case in which he did not wish it to apply.

Thus there was not a single "Wilhelminian" tradition; there were two continuing and partially contradictory—as well as partially complementary—traditions, that of diplomatic state interest and that of military interest. Both diplomats and soldiers were certainly nationalist and expansionist. Undoubtedly, both served the interests of the dominant classes, particularly of the large landowners, but also of the industrialists. But both were also inspired by the ideal of national unification and expansion, and by the problem of defending a country surrounded by other military powers.[7] It should not be forgotten that both believed they were doing their duty to the nation.

If the two traditions continued up until 1933, then did Hitler break with both of them? It would seem hard to maintain that Hitler broke with militarism as such, however much he may have departed in tactics and style from the customs of the officer corps, or however much he may later have ignored military advice. Here, surely, was a man who believed in military power, and who often sacrificed other interests to it. Did he, then, break with the diplomatic tradition while upholding the military tradition? Not necessarily. In his letter to Reichenau, we have already seen him stressing the problems and dangers of the diplomatic situation, which he believed Schleicher overlooked. We shall see him pursue a policy of diplomatic restraint in his first months in power. Is it not possible that he retained, combined, and yet expanded on both traditions?[8] Hitler was certainly a unique personality, and no "typical German." But he acted within the context of German society, and therefore of German tradition. He was unusual in integrating varied elements into a system, but I believe that all these elements, including racialism, had a firm background in German life, and indeed in German public policy.

Racialism was a continuing phenomenon in all "advanced" nations, with the United States ranking among the worst. But well before

[7] This last point is made not only in Ritter, *Staatskunst und Kriegshandwerk*, I, 19-20, but also in Fritz Fischer's 1968 lecture, "A Comparison of German Aims in the Two World Wars," quoted in J.C.G. Röhl, ed., *From Bismarck to Hitler* (London: Longman, 1970), p. 146.

[8] Ritter would surely have denied this, since to him the diplomatic tradition was that of statesmanship, with a sense of responsibility for a lasting social order; see *Staatskunst und Kriegshandwerk*, I, 13-22. I am here regarding the diplomatic tradition as involving the conscious practice of the "art of the possible" in international affairs. Ritter also suggested that though one person might try to combine both approaches, the conflict would continue within him: *Staaskunst und Kriegshandwerk*, I, 15.

World War I, German and Austrian racialism was remarkable for its overtness and for the high-level acceptance it enjoyed. Prominent men like Richard Wagner, the composer, and Heinrich von Treitschke, the historian, publicly embraced anti-Semitic ideas. Anti-Semitism dominated the politics of Vienna during Hitler's formative years in that city. In Berlin, the court chaplain, Adolf Stöcker, had espoused anti-Semitism, if with less political success, and Bismarck was not above expressing and exploiting the prejudice, while Kaiser William II was notorious for his denunciations of the Yellow Peril. On one famous occasion, he told German troops departing to combat the Boxer Rebellion to emulate the Huns and give no quarter to Chinese prisoners. Sophisticated observers may have recognized that William was unstable and impulsive, but his example encouraged the utterance and acceptance of simplistic and violent ideas. Racialism was also latent in the belief in German cultural and sanitational superiority to Slavs and Latins. More seriously, racialism could combine fatally with the toughness and concern with prestige fostered by the German military. In German South-West Africa in 1904-1905, the acting governor, General Lothar von Trotha, backed by Count Alfred Schlieffen, the Chief of Staff in Berlin, deliberately used a military task force to destroy a rebellious tribe of 80,000, the Hereros, reducing them to a starving refugee remnant of 15,000. Trotha and Schlieffen were not, to be sure, so much guided by a theory of racial superiority—that was simply assumed—as by military theories of annihilating the enemy. But as with Hitler—and unlike the repression of natives elsewhere at that time—the doctrinaire Trotha-Schlieffen policy was tempered neither by economic considerations nor, in practice, by objections on diplomatic or humanitarian grounds, which did manifest themselves in Germany. Certainly there was a great difference between German great-power diplomacy before 1914 and Hitler's racially motivated foreign-policy program. But Hitler's policy had German precedents. And contrary to a recent assertion by Andreas Hillgruber, other Western colonial powers had nothing in their experience quite comparable to the high-level, deliberate, doctrinaire ruthlessness shown toward the Hereros.[9]

[9] On the general background of German anti-Semitism: Bracher, *Deutsche Diktatur*, pp. 35-48; Wehler, *Deutsche Kaiserreich*, pp. 97-98, 112-113. Wehler also points out (pp. 114-117) precedents for the "ruthless Germanization" of the East. On the Kaiser's "Hun" speech, and the dissemination of a bowdlerized form to the public: Bernd Sösemann, "Die sog. Hunnenrede Wilhelms II.: Textkritische und interpretatorische Bemerkungen zur Aussprache des Kaisers vom 27. Juli 1900 in Bremerhaven," *HZ*, 222 (1976), 342-358. On the Herero campaign, see Germany, Kriegsgeschichtliche Abteilung des Grossen Generalstabes, *Die Kämpfe der deutschen Truppen in Sudwestafrika*, 2 vols. (Berlin: E. S. Mittler & Sohn, 1906-1907), I, esp. pp.

Adolf Hitler had begun to form his political ideas in childhood, and (despite the impression given in *Mein Kampf*) he did not end this process until the mid-1920s. His father and his history teacher encouraged him to become a pan-German nationalist, looking to the Bismarckian Reich rather than to the Dual Monarchy of which they and he were subjects. On the other hand, he reacted against his early schooling by becoming anticlerical, and a poor showing in his French classes may have contributed to his later antipathy for that nation. Despite his admiration for Prussian tradition, he did not grow up a disciplined person or a respecter of rules. His schooling was never completed, and he did not undergo peacetime military training; one suspects that two peacetime years in the "school of the German nation" might have made a different man of him. He seems to have developed his virulent hatred of the Jews during his years of artistic failure and vagabond existence in Vienna, where he read a large quantity of anti-Semitic literature. Marxism, founded by a Jew and with a large number of Jewish leaders, was for him a related evil. The Marxists and the Jewish liberal intellectuals—from Karl Marx and Sigmund Freud to Viktor Adler, Arnold Schönberg, and Karl Kraus—were indeed threats to the Wagnerian-romantic world of Hitler. He also had his social-Darwinist side, manifested in a belief that those nations which did not fight and win were likely to starve. His dislike for the multinational Habsburg empire grew during the Vienna years, although it was probably not so much this as a desire to evade conscript service in the Austrian army that led him to Munich in May 1913. As a volunteer in the Geman army during the war, he anticipated the "stab-in-the-back" legend, believing as early as 1915 that "the invisible enemies of the German people were more dangerous than the largest enemy cannon." After the war he always insisted that an internal "reform"—i.e., the overcoming of Jewish, internationalist,

---

132, 153, 186, 207-208, 214; Horst Drechsler, *Sudwestafrika unter deutscher Kolonialherrschaft: Der Kampf der Herero und Nama gegen den deutschen Imperialismus (1884-1915)* ([E.] Berlin: Akademie Verlag, 1966); Helmut Bley, *South-West Africa under German Rule, 1894-1914*, Hugh Ridley, ed. and trans. (Evanston: Northwestern University Press, 1971), esp. pp. 150-169. Data on Herero losses are in Union of South Africa, Administrator's Office, Windhuk, S. W. Africa, *Report on the Natives of South-West Africa and Their Treatment by Germany, January 1918* (Cmd. 9146) (London: H.M.S.O., 1918), pp. 34-35, 67. This report was not of course disinterested, but the data themselves are mainly from German sources and were not contested in a 1919 German counterpublication (Drechsler, p. 252). Trotha had served in China and East Africa (*Die Kämpfe*, p. 131), and the Kaiser awarded him the *Pour le Mérite* for his work in South-West Africa (Bley, p. 165n). On the question of whether Hitler's policy constituted a "break" despite the Herero War, cf. Hillgruber, *Kontinuität*, pp. 23-24; Hildebrand, "Hitlers Ort," pp. 624-625.

and democratic elements—was a necessary precondition for alliances, as well as for Germany's redemption generally.[10]

In 1919 many other Germans resembled Hitler in being obsessed with the need to overthrow the peace treaty, but he recognized more clearly than other German rightists that Germany would need allies, particularly in her weakened postwar situation. France was ruled out as an ally; she was the implacable hereditary enemy, and the nation that brought black troops to Europe. Hitler believed Germany could earlier have had Russia and Italy for allies if she had only freed herself from her useless alliance with Austria-Hungary. Indeed the Bismarckian model still suggested a Russian alliance, an idea that appealed to the Baltic German emigrés Hitler knew in Munich, and he called for an Italian alliance as early as 1920, well before Benito Mussolini became prominent. A Germany associated with these powers could also seek an overseas empire to meet its need for resources and land for colonization. But with Russia, there was the awkward fact that the Bolsheviks—radical Marxists and, in Hitler's eyes, mostly Jews—had seized power; Hitler hoped that the Whites, the counter-revolutionaries, would defeat them, but by the end of 1920 it was the Whites themselves who were defeated, and the Japanese evacuation of Vladivostok in October 1922 ended the last foreign intervention. For Hitler, a Russian alliance then became an unacceptable solution.[11]

As Hitler gained more experience in politics, he became more realistic and more confident in his own judgment. By late 1922 he realized that the securing of alliances would involve an ostensible renunciation, or at least postponement, of some of the all-encompassing goals of the pan-Germans and of his fellow racists. In mid-November 1922, following Mussolini's "March on Rome," the Nazi leader stated publicly that it was necessary for Germany to renounce the German population in the Italian-ruled South Tyrol. In the same speech he said

[10] See Bracher, *Deutsche Diktatur*, pp. 60-72; Konrad Heiden, *Der Fuehrer: Hitler's Rise to Power* (Boston: Houghton Mifflin, 1944), pp. 44-82; Axel Kuhn, *Hitlers aussenpolitisches Programm: Entstehung und Entwicklung 1919-1939* (Stuttgart: Ernst Klett, 1970), pp. 11-12, 34-39, 40-41, 76-77, 104-109, 118-121; Adolf Hitler, *Hitlers Tischgespräche im Führerhauptquartier 1941-1942*, transcribed by Henry Picker, ed. Percy E. Schramm, 2nd ed. (Stuttgart: Seewald, 1965), pp. 46-52, 74-77; Alan Bullock, *Hitler: A Study in Tyranny*, rev. ed. (New York: Harper & Row, 1962), pp. 36-49.

[11] Adolf Hitler, *Hitlers zweites Buch*, Gerhard Weinberg, ed. (Stuttgart: Deutsche Verlags-Anstalt, 1961), p. 185; Kuhn, pp. 32-45, 45-59, 61-62, 67-69; Günter Schubert, *Anfänge nationalsozialistischer Aussenpolitik* (Cologne: Verlag Wissenschaft und Politik, 1963), pp. 33-34; Eberhard Jäckel, *Hitlers Weltanschauung* (Tübingen: Rainer Wunderlich, 1969), pp. 33-34. The anti-Semitism and anti-Communism of the Balts could also lead, even in 1921, to thoughts of a German-Soviet conflict; see Konrad Heiden, *Geschichte des Nationalsozialismus: Die Karriere einer Idee* (Berlin: Rowohlt, 1933), pp. 45-47.

that British as well as Italian agreement was a precondition for Anschluss with Austria, and by December 1, he reportedly favored friendship with Britain. This would, as Hitler realized, involve foreswearing the goals of overseas colonies and a large-scale navy.[12] Such renunciations were not popular with the elements to which Hitler was appealing for support, but he was a man with a program, not an opportunist.

British opposition to the French occupation of the Ruhr in January 1923 certainly encouraged Hitler's thoughts of an alliance with England, and since a British alliance entailed in Hitler's eyes hostility to Russia, his fading hopes of overseas colonies could be supplanted by the thought of seizing land for colonization from the Soviet Union. In 1924, in *Mein Kampf*, Hitler defended the British against their German denigrators and strongly advocated an alliance with them and the Italians. Although the first targets for such alliances were France and the Soviet Union, Hitler's *Second Book* (dictated in 1928 but withheld from publication) also envisioned a combination of states of "high national value," clearly including England, against the United States. By printing his alliance plans in *Mein Kampf* and defending them against party critics, Hitler in effect froze his foreign-policy program, which remained basically the same for the rest of his life.[13]

In opting for alliances with Italy and England, Hitler did not act simply from a rational calculation of the best way to oppose France. In the case of Italy, there was now the attraction of the Fascist movement led by Mussolini, whom Hitler regarded as a model. As for England, although it has sometimes been said that Hitler's attitude toward the British was a mixture of love and hate,[14] it seems to the present writer that this is insufficient, and that Hitler's irrational feeling was one of almost indomitable, if rather misconceived, admiration. At times he concealed this feeling for tactical reasons or to maintain his credibility as a nationalist leader, and he did not curb his dislike or scorn for individual Englishmen who disappointed his hopes. But

[12] Kuhn, pp. 72-74, 81-95, 99-104; Wolfgang Horn, "Ein unbekannter Aufsatz Hitlers aus dem Frühjahr 1924," *VfZ*, 16 (1968), 291.

[13] Adolf Hitler, *Mein Kampf*, 35th ed. (Munich: Franz Eher, 1933), pp. 154, 691-700, 705, 755-757; Hitler, *Zweites Buch*, pp. 128-130, 160, 164-218; Kuhn, pp. 87-91, 125-128. Recent writings by Dülffer (*Weimar, Hitler und die Marine*), Dietrich Aigner (*Das Ringen um England*), Josef Henke (*England in Hitlers politischen Kalkül 1935-1939* [Boppard: Harald Boldt, 1973]), and Andreas Hillgruber ("England in Hitlers aussenpolitischer Konzeption," *HZ*, 218 [1974], 65-84) indicate that although Hitler faced up to the possibility of war with the British in the late 1930s, he continued to hope for an ultimate alliance.

[14] Hitler, *Hitlers Tischgespräche*, p. 54; Schubert, p. 36; Fritz Hesse, *Hitler and the English*, F. A. Voigt, ed. and trans. (London: Allan Wingate, 1954), pp. 112-113.

the underlying Anglophilia was probably there from the time of the war, if not before. The war gave him respect for the British as soldiers and for David Lloyd George as a public tribune and master propagandist. Small actions sometimes reveal the emotional wellsprings behind elaborate policies: where shortly after the war Brüning revived his sagging spirits with Sunday morning walks past the monuments of Potsdam, Hitler in the same period reportedly paid an exorbitant price to obtain for himself an authentic British trench coat from an Englishman residing in Munich. When Hitler started on his postwar career as a Reichswehr-subsidized informer and patriotic orator, he devoted himself mainly to an assault on the Treaty of Versailles, and his speeches castigated England along with France as an oppressor of the German people. Even some of his most virulent attacks, however, revealed an admiration for British national feeling, the British maintenance of racial barriers in their colonies, and their supposed genius for using economic power for political ends and for winning over defeated opponents as allies. In his own plans to exploit alliances, Hitler probably believed himself to be following British practice. Hitler admired British rule in India, as he understood it, and in later years he watched the film *The Lives of a Bengal Lancer* several times. He was also impressed by T. E. Lawrence's *Seven Pillars of Wisdom*. His hope for an alliance with Britain persisted even after hostilities began between Germany and the United Kingdom, and a week before his death in 1945, he reportedly pointed to the necessity for Anglo-German friendship.[15]

Hitler's hopes for British friendship dovetailed with his racist ideas, both positively and negatively. He regarded England as a Germanic nation, if sometimes dangerously influenced by Jews, while the plan to resort to East European territory for colonization purposes, involving the expulsion or worse of the existing population, encouraged him to plan the destruction of the Jews and the Slavs. For the next few decades, at least, he would let the British rule the world beyond the

[15] On Hitler's admiration for the British or desire for English friendship, see *Mein Kampf*, pp. 80-81, 158-159, 168, 199, 533-534, 746, 753; *Zweites Buch*, pp. 107-108, 126, 164-166, 172-173, 183; Reginald H. Phelps, "Hitler als Parteiredner im Jahre 1920," *VfZ*, 11 (1963), 297; Jäckel, pp. 35-37; Schubert, pp. 35-36; Kuhn, p. 46; Aigner, *Ringen*, pp. 31-37; A.V.N. van Woerden, "Hitler Faces England," *Acta Historiae Neerlandica*, 3 (1968), 141-159. Persistence of this desire during the war is shown in *Hitlers Tischgespräche*, pp. 136, 145, 218, 244-245; further material in Hitler, *Secret Conversations*, pp. 42, 76, 243-244, 541, 555-557. Brüning in Potsdam: Brüning, *Memoiren*, p. 657. Trenchcoat story: Bernard Lansing, "Adolf Hitler: War Highlights the Character of Today's Most Important Man," *Life* (25 Sep 39), 47-48. Subsidization as spy and orator by Reichswehr: Heiden, *Der Feuhrer*, pp. 88-89, 451; Sefton Delmer, *Trail Sinister: An Autobiography*, 2 vols. (London: Secker & Warburg, 1961), I, 63-64. Bengal Lancers and Lawrence: *Hitlers Tischgespräche*, p. 54 (Schramm information).

seas while he inaugurated a German domination of the European continent. Unfortunately for him and for Europe, he had no comprehension, let alone appreciation, of the humanitarian and pacifist ideas that actually dominated public thought in the England of the 1920s and 1930s, and that in fact were serving German interests. He thought the British had won their empire by guile and were ruling it with the whip. Thus he was seeking a partnership with a Britain that existed only in his imagination. Although he occasionally made disparaging remarks, as on the dismal British showing at the 1936 Olympic Games, Hitler would be slow to comprehend the fading of the British will to rule and of British military strength. Even in late 1939 (as he admitted in 1942), he was ready to believe there were thirty-five to forty British divisions in France, when in reality there were only twelve to fifteen.[16]

In one respect, Hitler's view of British policy was not so unsound. Critics of Hitler's program within the Nazi party argued in the late twenties that, in accordance with the balance-of-power doctrine, Britain would side against Germany once the latter had become the dominant power on the continent. Hitler maintained, however, that Britain's past policy toward Prussia showed that the British would not oppose a strong power on the continent, as long as it only pursued continental goals. This was in fact a fairly accurate reading of the tendency of current British policy. Hitler linked this view, however, with a belief that Germany could best win British friendship by demonstrating her *Bündnisfähigkeit*, her fitness for an offensive alliance, by proving her resolution and increasing her military strength. This line of thought was quite foreign to British thinking.[17]

Whatever his misconceptions about the world, Hitler was a master propagandist and political showman, and a highly skilled political tactician. Indeed, he combined propaganda with political action, and in doing so he used both carrot and stick, enticement and terror. He

---

[16] Olympic Games: Jacobsen, *Nationalsozialistische Aussenpolitik*, p. 352 (information from Hans Thomsen). B.E.F.: *Secret Conversations*, p. 491. The official British history of the 1940 campaign (L. F. Ellis, *The War in France and Flanders: 1939-1940* [London: H.M.S.O., 1953], p. xv, Appendix I) states that there were fourteen B.E.F. divisions during the May 1940 campaign, of which thirteen were in France before May; at the end of April, British forces in France numbered 394,165 (p. 19).

[17] See Kuhn, pp. 125-128; *Mein Kampf*, pp. 691-720. Kuhn implies that Hitler blindly ignored sound criticism, not least because he could not give in to critics. But although the law of balance of power finally compelled England to oppose him, British policy did indeed for a long time disregard the balance-of-power doctrine. Paul Schroeder's article, "Munich and the British Tradition," pp. 225-228, 235, supports the argument that Britain earlier tolerated Prussian or Prusso-German predominance, at least in East Central Europe. On Hitler's *Bündnisfähigkeit* objective, see also Section 2.

was more flexible in tactics than in his basic ideas. After the army failed to support his 1923 putsch, he recognized that the army would be a standing danger to anyone who seized power illegally.[18] He gave up the idea of a coup d'état and adopted a principle which might be expressed as: "Don't fight city hall; *become* city hall"—make your own movement the center to which others come. Another tactical rule he learned from 1923 was never to let his own position be controlled by others. Still another practice he followed was to stress secret preparation and surprise, taking advantage of the idleness, preconceptions, or preoccupation elsewhere of his rivals. He used all these methods in gaining control of his own movement and building it up within Germany, and he also used these tactics later in his relations with other governments. Usually he showed skill in discerning the intentions of others. Where he erred, it was usually in assuming that others would act with the same ruthlessness as he would have shown had he been in their place.

Some retrospective observers have described Hitler as a racial bigot, a petty bourgeois spokesman for reaction, a professional failure, or a sexual misfit. These observations all have some truth, but it would be a mistake to dismiss Hitler on such grounds. It is easy to denigrate or dispose of an individual by placing him in a despised category; at one time, Hindenburg also wrote Hitler off as a Bohemian or Austrian corporal, a category in low esteem among Prussian aristocrats. At a minimum, Hitler had a real mastery of propaganda and tactics, backed by passionate beliefs and an iron will. If he had a secret, it was that he believed he had found the right system, and could provide the right answers. His ideas may seem gross and superficial to normal observers today, but German normality had been upset by a series of catastrophes. He offered faith and hope to a nation hit by the cumulative impact of a lost war, a change in governmental system, an inflation run amok, and an abysmal depression. He addressed a people that had lost its idols, and whose certainties had been called into question, and he gave at least an impression of positive objectives. He professed to stand for a unity of all Germans, as opposed to the sectional and class divisions that were rife, and he did in fact show a lack of status-consciousness unusual among Europeans of his generation. The slogan "The common interest before self-interest *(Gemeinnutz geht vor Eigennutz)*" had a genuine moral appeal. He appeared to have an answer to the problems of the depression, though it was not too clear what this was specifically, aside from ceasing all pay-

[18] On Hitler's recognition of the danger of the Reichswehr to putsch makers: *Hitlers Tischgespräche*, pp. 365-366.

ments to foreign creditors and putting the Jews out of business. He seemed to be able to bring Germany's problems into focus, and to be able to reach decisions, in contrast to leaders like Brüning, who appeared unable to cope with the overwhelming detail in which they immersed themselves. He offered symbolic satisfactions, drama, and movement.[19] And although he never won a national majority in a free election, he won enough followers to give him a strong claim to power in the eyes of conservative and military leaders.

Hitler had ideas on military policy, as well as on foreign and racial affairs. He admired Scharnhorst as the man who had led the Germans away from being a people of poets and thinkers. From Clausewitz, he drew the lesson that surrender was a lasting moral encumbrance. Before he came to power, however, Hitler's recorded thoughts in the realm of military technique centered particularly on one interest. This was in giving all the eligible manpower of the nation a thorough military training in peacetime. Hitler's strong views on this point sprang from personal experience. As a volunteer with a Bavarian regiment, he had participated in late October 1914 in the battles known generically in Germany as Langemarck, and referred to in British military history as the First Battle of Ypres. In this so-called "child-slaughter of Ypres (*Kindermord von Ypern*)," thousands of hastily drilled volunteers, young idealists who had not been trained in peacetime, advanced (in some cases cheering and singing) row on row and were mowed down by the rapid, aimed musketry of British regulars. Hitler mentioned this experience or its lesson at least four times in *Mein Kampf*, and it was still prominent in his mind when he wrote to Reichenau in December 1932, saying that the Verbände could only serve in the army as cannon fodder. He did not blame the British, much less the war, for the German losses at Ypres; he blamed the prewar Reichstag as having failed in peacetime to provide funds for training all the potential soldiers in the nation. Like the military leaders, he now worried because no annual class of Germans had received official military training since the war.[20]

[19] I am indebted to Robert Wolfe of the National Archives for some of the ideas in this paragraph.
[20] Scharnhorst and Clausewitz: *Secret Conversations*, p. 548; *Mein Kampf*, pp. 759, 761. On volunteers and peacetime training: *Mein Kampf*, pp. 180-181, 297-298, 368, 581-583, 604-605; Vogelsang, "Hitlers Brief an Reichenau," p. 456. In contrast to Hitler, Eckart Kehr attributed the failure to train all the nation to the unwillingness of the War Minister to dilute the exclusive officer corps: "Zur Soziologie der Reichswehr," *Primat der Innenpolitik*, p. 237. Hitler also mentioned Langemarck in early 1934: O'Neill, *German Army and the Nazi Party*, p. 41 (quoting from an unpublished record by Field Marshal von Weichs). For details on the Langemarck battle: James E. Edmonds et al., *History of the Great War: Military Operations*,

318—Hitler's Accommodation with the Military

Hitler did not ordinarily regard his storm troops, the SA, as a military training organization or as a potential substitute for the Reichswehr. Precisely because of his war experience, he held that at least two years of strict training were needed to make a man a proper soldier. He also believed that official authority for punishment was essential for military training; despite their military airs, the SA did not undergo real military discipline. For Hitler, the SA had primarily a propaganda value, both constructive and destructive. For those who sympathized with them, they were a heartening encouragement, while to those the Nazis opposed, they were a force of brute terror, of perverse sadism, a nightmare made flesh. Hitler's belief in the usefulness of political violence very likely stemmed from his early exposure to the Austrian street politics of Georg von Schönerer and Karl Lueger. His political interest contrasted with the military orientation of Ernst Röhm, a close friend and a leading early member of the party, who hoped to develop the SA into an armed force. Until 1923, Röhm had been an army officer and a key figure in the Bavarian command in guiding and sustaining illegal armament and paramilitary organizations for a possible expansion of the army; for him, the SA had originally been a Reichswehr reserve formation. Hitler's wish to confine the SA to a politico-propagandist role may have been influenced by fears of losing control over his movement; the ex-corporal and "drummer (*Trommler*)" might have had difficulty exercising effective authority over an armed militia, the upper ranks of which were largely filled by ex-officers. Hitler was ready, however, to profit from the interest of the army in the SA as a potential source of manpower. Since his wartime service, he had become acquainted with the army's concerns, particularly for developing private preparations for military training, and for maintaining the supreme authority of the German state.[21]

France and Belgium, 1914, 2 vols. (London: Macmillan, 1923-1925), II, 123-124, 154-159, 265-267, 277, 315-316; Germany, Kriegsgeschichtliche Forschungsanstalt des Heeres, *Der Weltkrieg*, V, 272-274, 323; the British official account by Edmonds et al. has much more detail, even on German units, than the German. See also Fridolin Solleder, *Vier Jahre Westfront: Geschichte des Regiments List, R.I.R. 16* (Munich: Verlag Max Schick, 1932), pp. 18-46; Hans Mend, *Adolf Hitler im Felde 1914-1918* (Munich: Jos. C. Hubers Verlag, 1931), pp. 19-20. Strictly speaking, Hitler's unit was not at Langemarck itself, which is NNE of Ypres, but ESE of Ypres at Messines and Wytschaete. In any case, he was part of the same general offensive in which the German command took a calculated risk in the use of poorly trained volunteers.

[21] Heiden, *Nationalsozialismus*, pp. 88, 196-197, 224-225, 270-273; Bracher, *Deutsche Diktatur*, pp. 95, 103-104, 116-117; Harold J. Gordon, Jr., *Hitler and the Beer Hall Putsch* (Princeton: Princeton University Press, 1972), pp. 61-64, 160-164. Gordon points out, however, esp. on pp. 84-86, that egalitarianism was strong in the Nazi party and the SA. On Schönerer and Lueger: *Mein Kampf*, pp. 105-133; Carl E.

Hitler seemed, at least at first glance, to be somewhat ambivalent on the role of material armament in warfare. He proclaimed in *Mein Kampf*: "The best arms are dead and useless material as long as the spirit is missing which is ready, willing, and determined to use them." Later in the same work, however, one of his principal arguments against a Russian alliance was the industrial backwardness of the Soviet Union. A tie with Russia would mean that, as in World War I, Germany would have to supply its allies as well as itself: "The universal motorization of the world, which in the next war will be overwhelmingly decisive in the struggle, could hardly be met by us [i.e., alone]." Considering Hitler's love of fast Mercedes cars, his *Popular Science* interest in technology, and his megalomania, expressed in his sketches of triumphal arches and domes, he would inevitably in practice be a strong advocate of material armament. It would be he who would ultimately propose the building of 100,000-ton battleships, and stake his hopes on "wonder weapons." Despite many clarifications, a legend persists that he did not interest himself in armaments when he first became chancellor, turning instead to the construction of autobahns. In actual fact, he emphasized material armament from the very start.[22]

Strengthening the national will did remain a fundamental aim for Hitler, however. In 1933, he regarded the actual preparation and leadership of the army as the province of professionals. His own self-assigned mission was to develop the German people's will to fight, and to him, the establishment of this will—by a totalitarian regime—was a precondition for a successful foreign and military policy. The lack of such a will, he believed, had led to the loss of World War I. He thought that the immediate cause of the breakdown of willpower in wartime lay in faulty propaganda and, over and beyond this, in the influence of liberal democracy, Marxism, and international Jewry. His solution lay in developing a more powerful propaganda and in overcoming these, as he thought, negative forces. Hitler was probably the most outspoken advocate of propaganda and coercion the world

---

Schorske, "Politics in a New Key: An Austrian Triptych," *JMH*, 39 (1967), 346-365. On some occasions, esp. in the early 1920s when Hitler had to take account of the army's wish to be able to mobilize reserves from the SA, he accepted the idea that the SA should be a defense organization: Bennecke, *Hitler und die SA*, pp. 30-31. Hitler's expressed views on the SA: *Mein Kampf*, pp. 603-608, 620.

[22] *Mein Kampf*, pp. 365, 748-749. Hitler's childish interest in technology emerges frequently in his *Tischgespräche* and *Secret Conversations*, and in Albert Speer's *Inside the Third Reich* (New York: Macmillan, 1970). On his architectural megalomania, see Speer's Chapter 5 and accompanying illustrations. On the battleships, Dülffer, pp. 383-386, 438. Schacht fostered the autobahn legend in his own defense. Hitler's actual interest in armament is discussed below.

has seen, although he can also be said to have only followed the logic of modern warfare to its ultimate conclusion.[23]

2. HITLER BIDS FOR MILITARY SUPPORT

Hitler took care to establish good relations with the army when he came to power. If not before the formation of the new cabinet, Hitler and Blomberg arrived soon afterwards at an understanding on the place of the Reichswehr within the new state. The two men conferred together after the swearing-in of the new cabinet on January 30, and at Hitler's first cabinet meeting later that day, they agreed that the army should not be used to suppress a general strike. On February 1, a Blomberg directive affirmed that the Reichswehr would remain above parties, that it would be broadened by training the people to bear arms, and that the armed services would be rendered capable of maintaining national seurity. By February 3, at the latest, it was evidently understood between Hitler and Blomberg, explicitly or tacitly, that Hitler would give priority generally to rearmament, would abstain from diluting the army with SA formations, and would uphold the authority of the army in national defense. In return, the army was to refrain from interference in internal politics. To many officers, even the last point doubtless appeared as a gain for the army, removing it from the political mire and eliminating an impediment to defense preparations. In April, Reichenau was to pronounce, with obvious satisfaction, that the plans for using the army for domestic duties, further developed after the Ott war game, had been overtaken by events; the "National Government" intended that such duties would be handled by the police, reinforced if necessary by the nationalist paramilitary organizations. Since Hitler had already won control of the police through the appointment of Nazis as Ministers of Interior in the Reich and Prussian governments, such a division of responsibilities meant that there was no force left which might prevent the SA from terrorizing the population.[24] Hitler believed later that

[23] See *Mein Kampf*, esp. Chapters 6, 7, and 10 of the "first volume," and Chapters 6, 7, 9, and 11 of the "second volume," also Vogelsang, "Hitler's Brief an Reichenau," pp. 456-457. On Hitler's views on military service and officers: Bennecke, *Reichswehr und "Röhm-Putsch*," pp. 7-11. Hitler's stress on propaganda: Bracher, *Deutsche Diktatur*, p. 105.

[24] Sauer, "Mobilmachung," pp. 708-712, 715-717; also ML, PG 34097, MA 135/33, 21 Apr 33, reel 31/934 (reissued generally as TA 3346/33, 31 May 33, reel 31/936). Aside from General Liebmann's account of the 3 Feb 33 meeting in Vogelsang, "Neue Dokumente," pp. 432-435, which Sauer refers to, the unpublished Liebmann papers include a copy of a Blomberg directive (Reichswehrminister 1549/33) of 14 Mar 33, which stated, with regard to "developments in recent weeks," that Hitler

he owed much to the "neutral" attitude of the army. In September 1933, he said: "On this day, we would particularly remember the part played by our army, for we all know well that if, in the days of our revolution, the army had not stood on our side, then we would not be standing here today." Speaking within his own entourage in 1942, he emphasized how important it had been for him to be appointed legitimately by the President, primarily because this put him in a position (through the loyalty of the Reichswehr to Hindenburg) to restrict the activities of the Wehrmacht "to its purely military tasks."[25]

Hitler first spoke to the assembled leaders of the army and navy at a party in honor of Neurath's sixtieth birthday, held at Hammerstein's house on February 3. This meeting gave Hitler an opportunity to try to clinch the direct support of the military leaders, while the latter had a chance to hear at first hand the voice and ideas of the new Chancellor. Some of Hitler's points were well suited to curry favor with his audience: he indicated he would end pacifist activity, he asserted that youth must be made fit and that the desire for military preparedness (*Wehrwille*) must be strengthened, he proposed to apply the death penalty to cases of treason (*Landes- und Volksverrat*), he said there had to be a return to conscription, and he promised there would be no amalgamation of the SA with the army. The task of the SA was to conduct the political battle within the country, while the army should be apolitical and nonpartisan; the task of the generals was to build up the army.[26]

But he was not just telling the generals what he thought they wanted to hear. The ideas he aired were consistent with *Mein Kampf* and with the major positions he had previously taken, including in his letter to Reichenau of December 4. The central theme of his talk was political power, and how it was to be recaptured for Germany. This, he said, was the aim of his policy. In his usual way, he gave priority to the elimination of dissident antinational activity on the domestic

---

at the 3 Feb 33 meeting had declared that the Wehrmacht "will be maintained unchanged in its significance and special position, and that he [Hitler] would oppose his unconquerable resistance to all other efforts, wherever they might come from."

[25] Bullock, p. 249; *Hitlers Tischgespräche*, pp. 365-366.

[26] Vogelsang, "Neue Dokumente," pp. 434-435. Evidently Hitler did not believe that Hammerstein had been planning a putsch against his chancellorship. In addition to the Liebmann record in "Neue Dokumente," there is another contemporary source on the meeting, the notes of Horst von Mellinthin, Institut für Zeitgeschichte, in Zeugenschrifttum 105, pp. 3-6. These notes are less coherent, but confirm Liebmann's record in its essentials. They also indicate that Hitler warned, as in his 4 Dec 32 letter to Reichenau, that too much overt emphasis on the idea of equality of rights might bring on a preventive attack; economic and military preparations should be secret.

scene. As General Liebmann (who was present) noted in telegraphic style, Hitler said: "Whoever [among the opposition] will not let himself be turned, must be bent. Uprooting of Marxism root and branch. . . . Strictest authoritarian state leadership, elimination of the cancerous sore of democracy." On foreign policy, Hitler promised to struggle against Versailles, to seek equality of rights in Geneva, and to find allies; he added, however, that equality of rights was useless unless the people had a desire for military preparedness. In economic matters, he favored a program of resettling the unemployed on the land, while indicating that he did not think this would be a substitute for an increase in Lebensraum. The foremost precondition for regaining political power was to build up the armed forces. The armed services were the most important and most "socialist" part of the state. Conscription had to return; however, the national leaders must first take care that those who would be called up would not already be infected by pacifism, Marxism, and Bolshevism, and also that they would not be exposed to these infections after their service was completed. Hitler also took this occasion to make the earliest recorded statement, after his entry into power, on his ultimate objectives. As Liebmann noted, Hitler said:

> How should political power be used when it has been won? Cannot yet be said. Perhaps fighting for and winning new opportunities for export, perhaps—and probably better—seizure of new Lebensraum in the East and its ruthless Germanization. Certain that only with political power and struggle can present economic situation be fundamentally changed. Everything that can happen now—resettlement—stopgap expedient.

It has been suggested that the military leaders should have risen and challenged Hitler's suggested objectives.[27] That would surely have been morally desirable, but there was no possibility that they would do so. Hitler's domestic program, as he described it, did not go much farther than most of the generals wanted to go. They wanted an end to opposition to national defense, and apparently he did not on this occasion attack the Jews, or threaten the use of shootings or concentration camps. The talk of conquering territory in the East and ruthlessly Germanizing it was, to be sure, something new coming from a German chancellor. But the Pan-German movement had long advocated Eastern annexations, and Ludendorff had pursued them during World War I. Even Brüning, in 1930, had told his cabinet that Germany (in replying to French proposals for European union) should indicate clearly, though in carefully chosen words, that a just and last-

[27] Sauer, "Mobilmachung," pp. 765-766.

ing European order must provide Germany with her "sufficient, natural Lebensraum."[28] Even Groener, as we have seen, was willing to initiate a war if favorable conditions could be arranged. And the violence of Hitler's tone was no shocking novelty to men who had served under William II. In any case, given the current strength of the German army, Hitler's talk of large-scale conquest may have seemed to many in Hitler's audience to be a kind of oral daydreaming.

Testimony as to the actual reaction of Hitler's listeners has varied. According to postwar recollections, some were shocked and alarmed at Hitler's proposals; some were disgusted by his market-square hawking of his ideas, and some thought he was merely trying to butter up his audience. One listener is said to have remarked, quoting (approximately) from Schiller's *Wallenstein*, "Talk was always bolder than the deed." Admiral Erich Raeder was impressed by statements Hitler made in honor of Hindenburg, and the admiral maintained after the war that the speech had had a very reassuring effect. With respect to all these reports, it must be said that a period of thirteen to twenty years can considerably cloud the recollection, particularly when in the interim there has been a drastic change in accepted standards and in the self-interest of the witnesses.[29] General Liebmann himself wrote one interesting comment at the time, however. A soldier who neither then nor later became a blind follower of Hitler, Liebmann noted:

> Gen. impression: At first unimpressive and insignificant. While talking, strong will and vivid imagination emerge, and one has the impression of a man who knows what he wants and who is resolved to turn his ideals into reality with the greatest energy. Whether along with this there are the *capacities* which are necessary to overcome the overwhelming difficulties facing each of his plans, one can only hope. Only the future can tell.[30]

We may suspect that Liebmann's reaction was actually shared at the time by many, even most of his colleagues. Though they would

---

[28] RK, Cabinet Protocol, 8 Jul 30 (8 p.m.), 3598/D784289. Apparently as a result of this discussion, the final German reply, of 11 Jul 30, stated that the existing economic and political system on the continent hindered a development "which would correspond with the natural life requirements [*Lebensbedingungen*] of the peoples" (AA, 3241/D702899-905).

[29] For discussions of the postwar testimony, see Sauer, "Mobilmachung," pp. 719-720, 735, 764; Gerhard Meinck, *Hitler und die deutsche Aufrüstung 1933-1937* (Wiesbaden: Franz Steiner, 1959), p. 18; Müller, *Heer und Hitler*, p. 42.

[30] For Liebmann comment: Liebmann MS Papers, p. 192 (typewritten). So far as I know, Sauer is the only writer who has made any allusion to Liebmann's comment, quoting the words "strong will and vivid imagination" (*starker Wille und idealer Schwung*). Vogelsang, "Neue Dokumente," which publishes Liebmann's report of Hitler's remarks (pp. 434-435), omits the comment.

forget this later, perhaps in some cases within a few months, it was natural at that juncture to feel some of the intoxication of the "national uprising *(nationale Erhebung)*," the belief, spread broadside by every means of propaganda, that a reform movement of young Nazis and mature Nationalists, together representing the true Germany, had finally broken through, replacing the alien and discredited Weimar democracy. Earlier on the same day on which Hitler spoke, Blomberg had expressed this "national uprising" mood by telling the generals that the new cabinet could be seen as an "expression of broad national desires & realization of that which many of the best [people] had striven for for years."[31] No doubt Liebmann (as he explained in 1953) had no idea what methods Hitler would use, at least in "ruthlessly Germanizing" eastern territories.[32] Even in 1933, however, it must have been evident that Hitler's program was likely to require a war.

Two things stand out about this meeting. First, Hitler indicated his intention to use the army as an active instrument of an expansionist foreign policy. As a dictator, he did not intend merely to provide bread and circuses for the people, or to find military sinecures for unemployed storm-troop leaders. Second, most of Hitler's audience either did not take Hitler's remarks seriously, or else actually sympathized with them. Army leaders had looked forward to the day when they might be ready to fight a successful foreign war, and they were relieved to have Hitler recognize this as their function and their exclusive responsibility.

Hitler struck one other note in his February 3 talk. After speaking of the importance of the military forces, he commented, much as he had to Reichenau:

> Most dangerous time is that of Wehrmacht expansion. Then it will be seen whether France has *statesmen*; if yes, she will not give us time, but will assail us (presumably with eastern satellites).

A few days later, Hitler returned to this idea yet again, giving it a new twist. At a meeting on February 9 of the Reich Committee on Work Creation, Hitler stressed the usefulness of a work-creation program for the concealment *(Tarnung)* of national defense measures:

> One must consider concealment to be especially valuable in the near future, because he was convinced that precisely the period between the theoretical recognition of Germany's equality of rights and the

---

[31] On the "national uprising," see Bracher, "Stufen der Machtergreifung," p. 46. Blomberg statement: Vogelsang, "Neue Dokumente," p. 432.

[32] Liebmann MS Papers, p. 326a (typed).

regaining of a certain state of armament would be the most difficult and most dangerous. The main difficulties of rearmament would only be overcome when Germany had rearmed to such a degree that she became fit for alliance in combination with another power [*für den Zusammenschluss mit einer anderen Macht bündnisfähig werde*], if necessary also against France.[33]

Thus Hitler expressed concern that France might wage a preventive war against Germany, while he also looked forward to the day when Germany would be able, through her military strength, to attract an ally against France. Only two nations might serve such a role: Italy and Britain.

3. CAUTION ON THE PART OF DIPLOMATS, SOLDIERS, AND CHANCELLOR

Hitler might deprecate Schleicher's public demand for arms, and try to conceal his own aggressive aims from public view, but the Nazi leader's arrival in power would doubtless make other governments more suspicious than before. German military policy faced several problems, all related to the danger that other governments might detect German rearmament and take preventive action. First, there was the matter of the position that should be taken toward foreign governments, their arms limitation proposals, and their other diplomatic moves. Second, there was the problem of financing and executing material rearmament, along with Umbau and the training of reservists. This was connected with, third, the question of the proper relation between the Reichswehr and the Nazi SA, now that Hitler had come to power. The army had planned to take major steps on April 1, and such steps had implications for diplomatic, budgetary, and training policy.

The French plan, which the Disarmament Conference would take up next, had approved—indeed demanded—a reduction in the term of service in the German army, and the introduction of conscription. The French also appeared to have accepted the idea of eventual parity or near-parity in personnel strength. But the French plan was not of course designed actually to satisfy Germany—that would have been a strange abdication. It was designed to undercut German claims and at the same time to weaken Germany. The plan purported to concede equality in a regime of security, but the equality, as the Germans would tirelessly point out, would not be genuine; France would retain

---

[33] BA, R 43 II/540, Record of Work Creation Committee meeting, 9 Feb 33. By contrast, Hitler told the cabinet on 8 Feb 33 that "the world, especially France, was entirely prepared for German rearmament and regarded it as a matter of course" (*ADAP*, C/I/1, No. 16).

her heavy arms (theoretically under League auspices) and would still have major additional forces available in nearby North Africa. The proposed security apparatus would probably never come into existence, due to British opposition, and if it did not, this failure would be used (the Germans thought) as an excuse for a French refusal to disarm or concede German rearmament. If by any chance the French plan did prevail, it would freeze Europe into the 1919 status quo.

The French desire to weaken Germany was expressed in the proposal for standardized, short-service, lightly armed militias, combined with small, heavily armed international contingents, all under close international supervision. The aim of these proposals was to slay that mythical dragon, the élite professional striking force of General von Seeckt, which the Germans supposedly intended to supplement with a separate defensive militia army. But quite unwittingly, the French organizational proposals also threatened to frustrate the real German plan to prepare for the mobilization of a single, integrated army, including twelve-year professionals, three-year regulars, and three-month militiamen. The standardized national militias of the French plan would have included a few professionals as instructors and specialists, but nothing like the proportion the Reichswehr wanted. According to the information given to Neurath by Schleicher in October 1932, the Reichswehr planned to have, after a five-year period, and with three-year service, four reservists with three years of training to every five with militia training only. These ideas seem to have been slightly scaled down in November, but the principle remained the same.[34]

The real nature of the Reichswehr plan was, however, a closely guarded secret. Since Germany had been asking for a militia, and since the French now proposed to concede one to her, it was difficult for outsiders to understand why she should not agree to the French proposals—unless she was trying to preserve the army of General von Seeckt. Even German diplomats had not been informed of the plan, excepting the one disclosure in October to Neurath, intended for his eyes only. This lack of explanation was one major cause of friction and misunderstanding between General Schönheinz, who succeeded

[34] AA, List of questions for Schleicher, [12 Oct 32], 3154/D672029-30; ltr., Schleicher to Neurath with atts., 15 Oct 32, 3154/D671232-42; Note by Bülow, [16 Oct 32], 3154/D671228-31; BA/MA, RH 15/49, TA 737/32 and Annex 6, 7 Nov 32. See Chapter Five, Sections 5 and 8, for a fuller discussion of these documents. After the MacDonald plan took up the French proposals for standardization, Schönheinz objected in a memo that the plan would reduce all armies to either eight-month men or long-service professionals, and that the proportion between these two categories would be the same for all continental states: "This will make it impossible for Germany to continue to maintain a cadre superior to those of other states" (ML, Shelf 5939, Memo by Schönheinz, 18 Mar 33, reel 18/83). See also *ADAP*, C/I/1, No. 97.

Blomberg as chief military delegate, and the civilian delegates. But Schönheinz also aggravated differences by taking a much more unbending position than was necessary for the protection of German plans. He was to maintain that he had separate instructions from Blomberg, barring any acceptance of the standardization of armies in any form. There is every reason to think that Schönheinz was working deliberately to bring about a German withdrawal from the conference.[35]

The wisdom of considering standardization soon became an issue, within the delegation and in Berlin. The conference reconvened in plenary session on February 2, taking up the French plan, and on February 17 the French presented a resolution calling for a vote on the principle of standardization. Nadolny, the diplomat at the head of the German delegation, concluded that flat opposition would tend to isolate Germany, while under the French plan (as he telephoned Berlin), "a framework seems to be offered for the extensive reorganization and numerical expansion of our own Wehrmacht. . . . Actually French fear of the Reichswehr offers us a good chance of asserting our essential military interests." Schönheinz did not admit this, however, and he and Nadolny came to Berlin soon after to confer with Reichswehr and Foreign Ministry chiefs.[36]

Blomberg, as chief military delegate, had hitherto strongly supported military claims; he might perhaps have been expected to oppose all negotiations on standardization, as Schönheinz did. But Blomberg did not do this. Even before discussing the question with the Foreign Ministry, Blomberg decided that Germany should "participate in the discussion of the standardization of army systems," while he added that before giving agreement in principle, it was desirable

---

[35] *ADAP*, C/I/1, No. 97; AA, tel., Schönheinz to Blomberg, 25 Apr 33, 3154/D668891-92. On Schönheinz generally, see also Rautenberg, *Rüstungspolitik*, p. 36. As suggested by a case described in Chapter Six, Section 2, Schönheinz may sometimes have thought he had gotten a commitment from his superiors when he had not. On Schönheinz's desire for a withdrawal, see his 1 Dec 32 memo, discussed with Schleicher on 3 Dec 32 and described in Chapter Five, Section 8, and his intransigence and apparent approval of leaving the conference in May: ML, Shelf 5939, ltr., [Freyberg] to Raeder, 13 May 33; note by Schönheinz, 13 May [33], reel 18/126, 129. By contrast, Admiral Freyberg thought the French plan could be bent enough to accommodate German plans, and that the immediate (1 April) steps could be claimed as in line with French ideas: ML, Shelf 5939, ltr., [Freyberg] to Raeder, 1 Mar 33, reel 18/29-31.

[36] ML, Shelf 5940, Delegation report No. 36, 22-26 Feb 33, reel 18/668-672; Shelf 5939, ltr., [Freyberg] to Raeder, 1 Mar 33, reel 18/28-33; *ADAP*, C/I/1, No. 23; AA, ltr., Frohwein to Kreutzwald, 14 Feb 33, 7360/E536216-17. AA, Handwritten notes on the Defense Ministry/Foreign Ministry meeting, 20 Feb 33, 3154/671283-90, also show the differing views. For the French statement of views (by Pierre Cot), and Nadolny's public response on 22 Feb 33, see *Records of Conference*, Series B, II, 279-284, 288-290.

to make reservations and demand first of all the clarification of other questions, such as the abolition of heavy offensive weapons. The delegation should try to avert a vote on the principle; if a vote came, the German delegation should abstain. Both Blomberg and the diplomats wanted to delay decisions on standardization, to bargain, and to divert attention to French reluctance to disarm. At a meeting between the Reichswehr and Foreign Ministry representatives on February 20, Schönheinz argued for an outright rejection of the French standardization resolution, and explained that the French proposals would eliminate too many professional soldiers, disrupting the army's plans to retain 75 percent of its regulars for three years and the rest for longer. This explanation should have enlightened some of the diplomatic participants, particularly Nadolny, although it is not certain that it did. The explanation did not, however, necessarily justify the rigid diplomatic tactics Schönheinz advocated, and although Blomberg seems to have been susceptible to pressure from his unbending subordinate, the Defense Minister agreed with the strategy of diplomatic delay. Nadolny returned to Geneva to argue that a decision on heavy weapons should precede one on standardization; this was a stiff and unhelpful position, but it did not constitute a final rejection of standardization. In giving his support to the diplomats, Blomberg was probably guided by what he knew of Hitler's views. In discussing a draft statement at the meeting with the Foreign Ministry representatives, Blomberg suggested that he present it to Hitler who might, he thought, make it "softer" (*Möglich dass er weicher macht*).[37]

After the March 5 Reichstag election, Hitler became more free to consider diplomatic and military affairs; these affairs also became more urgent. The London government had announced on March 3 that MacDonald and Simon would shortly go to Geneva, out of a concern "to enable the Conference to reach early decisions," and the British leaders obviously hoped to meet there with heads of other governments and their foreign ministers. The German Foreign Ministry seems to have feared that the British or other governments might sponsor a limited agreement, such as Norman Davis had proposed in December, which would make it harder for Germany to charge the others with failure to disarm, and which would not achieve any increase in German armament. Hitler refused absolutely to go to Ge-

[37] *ADAP*, C/I/1, No. 26; AA, Handwritten notes on meeting, 20 Feb 33, 3154/D671283, D671288-290; *SIA, 1933*, pp. 239-240; *Records of Conference*, Series B, II, 288-290. Unfortunately, the notes are highly abbreviated, in part illegible, and apparently out of order. From content and appearance of pages, the correct frame order is probably: D671283, -84, -89, -90, -85, -86, -87, -88. "Softer": D671288.

neva himself, and apparently there was no desire to send Neurath there at a time when German policy was not yet clear.[38] Werner von Rheinbaben and Karl Schwendemann, members of the disarmament delegation, were summoned to Berlin, ostensibly for coordinating press statements on British disarmament initiatives and directing disarmament propaganda, but probably also to provide first-hand information on the Geneva situation.[39]

In the next few days, three other developments further complicated the foreign situation. First, an incident occurred in which Nazi agitation in Danzig and a denunciation of a Danzig harbor police agreement by the German-dominated government of the free city led the Poles on March 6 to reinforce their garrison at Westerplatte, at the harbor entrance.[40] This launched a ten-day minor crisis, and encouraged rumors of Polish plans to wage a preventive war, rumors which continued through March and April and which apparently received credence from Hitler. Second, Benito Mussolini, the Italian dictator, secretly sent word to Berlin on March 14 that he was considering a proposal for a four-power pact, under which a great-power directorate of Italy, France, Germany, and Great Britain would encourage the revision of the peace treaties and recognize that Germany's equality of rights had to be effectively applied. The Italian proposal, which we will examine in the next chapter, was not really as revisionary and as favorable to effective equality as the Germans would have liked.[41] But for Hitler, with his admiration for Mussolini, it may have appeared at least as a portent of Italian—and indeed, perhaps also of British—support. Thirdly, the British were to unveil a disarmament plan on March 16 that conceded to Germany an army of 200,000 men,

[38] *ADAP*, C/I/1, Nos. 20, 46, 49 (and notes); AA, tel., Nadolny to Neurath, 1 Mar 33, 3154/D668607-09; Note by Bülow, 4 Mar 33, 7360/E536255-57.

[39] AA, Note by Bülow, 4 Mar 33, 7360/E536257; tel., Neurath to Nadolny, 7 Mar 33, 3154/D668645; tel., Bülow to Nadolny, 9 Mar 33, 3154/D668666. Schwendemann had just had an important talk with French delegates on standardization (ltr., Schwendemann to Koepke, 7 Mar 33, 7360/E536339-42), although his summons had apparently already been under discussion (ltr., Frohwein to [Koepke], 8 Mar 33, 7616/H188369). Personal tension between Neurath and Nadolny, and the difficulty of directing the latter, as well as the conspicuousness of calling home the head of the delegation (the explanation given to Nadolny) probably explain the summoning of Rheinbaben. Rheinbaben's memoirs (*Viermal Deutschland*, p. 272) say he was called because Nadolny's manner got on Hitler's nerves, and Neurath and Blomberg did not want to take the responsibility themselves of showing Hitler the advantages of working through Geneva.

[40] Gerhard Weinberg, *The Foreign Policy of Hitler's Germany* (Chicago: University of Chicago Press, 1970), pp. 57-63; *SIA, 1933*, p. 187.

[41] *ADAP*, C/I/1, Nos. 83 (and encl.), 84, 87, 88. On the German assessment of the proposal, see also Jens Petersen, *Hitler-Mussolini: Die Entstehung der Achse Berlin-Rom 1933-1936* (Tübingen: Max Niemeyer, 1973), pp. 150-151.

equivalent to the metropolitan contingent to be allowed to France. This again was far from meeting German demands,[42] but at a minimum it would have to be debated before Germany could plausibly claim that the conference had failed, and Hitler may already have seen in it a possible point of departure for direct talks with England.

Under these conditions, the need to avoid provocation and to gain time was obvious. Bülow completed a lengthy memorandum on March 13 which said that, although Germany's basic aim was the revision of Versailles, the danger of preventive war should never be underestimated, and that German armament should be remodeled "relatively slowly and as quietly as possible, in order to avoid intervention, preventive actions, and diplomatic convulsions." Bülow still thought that, for financial reasons, Germany would be unable to rearm extensively in the next five years, and indeed he suggested that for this reason it would be more effective to bring about the disarmament of others. Bülow's views apparently did not circulate until April 7, when Neurath repeated most of them to the cabinet as his own, leaving out those that reflected unfavorably on the new regime.[43] But when Hitler met on March 15 with Blomberg, Neurath, and the visiting Rheinbaben, the Chancellor concluded (according to Rheinbaben's contemporary account) that the delegation at Geneva should not sabotage the conference, that it was free to engage in substantive discussion on the French plan, exploiting its potentialities for Germany, and that Germany might accept certain temporary restrictions in materiel, although continuing to insist on equality of rights in principle. When Rheinbaben reported this to the delegation in Geneva, Schönheinz—who had evidently been fostering a totally different conception of Berlin policy —angrily telegraphed Blomberg asking for a clarification. Blomberg and Neurath took the position that the talk with Hitler had only been for purposes of information, and did not affect policy. But a Foreign Ministry official noted that, for the delegation, "there will no longer in future be an 'unacceptable' with respect to a change in preexisting Reichswehr organization and in connection with the question of materiel."[44]

[42] *ADAP*, C/I/1, Nos. 103, 238; ML, Shelf 5939, Memo by Schönheinz, 18 Mar 33, reel 18/80-85.

[43] Günter Wollstein, "Eine Denkschrift des Staatssekretärs Bernhard von Bülow vom März 1933: Wilhelminische Konzeption der Aussenpolitik zu Beginn der nationalsozialistischen Herrschaft," *MGM*, 13 (1973), 77-94; *ADAP*, C/I/1, No. 142. Neurath apparently modified Bülow's statements on the preferability of having others disarm and glossed over a final suggestion that provocative activities of paramilitary groups close to the government be banned.

[44] Rheinbaben gave two contemporary accounts of the meeting, one to a Berlin colleague on 15 March (AA, Note by Kreutzwald, 15 Mar 33, 7360/E536423) and one later in Geneva (*ADAP*, C/I/1, No. 94). After the war, Rheinbaben stated (in

Rheinbaben, in opposition to the Schönheinz viewpoint, had probably made the most of Hitler's readiness to negotiate, but there can be no doubt that the Chancellor did indeed want to proceed cautiously. On the same day (March 15), Hitler told the Mayor of Hamburg that under existing conditions, Germany had to conclude a truce with the European powers for at least six years, and he added that the saber-rattling of nationalist circles was wrong, as Germany lacked weapons. Hitler wanted of course (as Rheinbaben also reported) to increase Germany's military strength as rapidly as possible, so that she could emerge from her vulnerable position. But rearmament could also endanger Germany's foreign relations in the near term. On March 16, Blomberg ordered that "out of regard for the foreign policy situation," the expansion of the army's air force should be stretched out over a longer period. Later, after Hermann Göring succeeded on May 15 in having his new Air Ministry absorb the air force sections of the army, the former pilot and irrepressible political adventurer would put very different plans into effect; Göring cared little about foreign reactions. In mid-March, however, the watchword—undoubtedly stemming from Hitler—was caution.[45]

Military appraisals of the situation confirmed the need to gain time, and underlined the uncomfortable dilemma between the dangers of Germany's military weakness and the risks of doing anything about that weakness. In mid-March, the Chiefs of the Truppenamt and of the Army Command, Generals Adam and von Hammerstein, submitted memoranda at Blomberg's request on the overall military picture. Only Adam's report seems to have survived, but he testified after the war that Hammerstein's views had been similar to his. Adam's memorandum argued strongly that Germany was not prepared for war, either in terms of personnel or materiel. Current planning called for the mobilization of the "A-Army" (field army) of twenty-one divisions, but the difficulties of mobilization were very great:

Viermal Deutschland, pp. 273-275) that Hitler showed violent opposition to budgetary limitation and, implicitly, to the conference as a whole. On the subsequent repercussions: *ADAP*, C/I/1, Nos. 97, 106; AA, Notes by Kreutzwald, 20, 22 Mar 33, 7360/E536476, E536474-75. Schönheinz was particularly bothered by the implication in Rheinbaben's Geneva report that the long-service German forces were to be separate from the militia; this, he wrote, "is not entirely understandable according to the plan for the reorganization of the army."

[45] Carl Vincent Krogmann, *Es ging um Deutschlands Zukunft 1932-1939* (Leoni am Starnberger See: Druffel-Verlag, 1976), p. 54; National Archives, Von Rohden Collection, VR 4406-525, Reichswehrminister 380/33, 17 Mar 33, and R. M. f. d. Luftfahrt, L. A. 1305/33, 29 Jun 33 (this collection has now been microfilmed as T-971); Völker, *Militärische Luftfahrt 1920-1933*, pp. 200-206, 210-213. Blomberg's order, mentioned in a document of 17 March (above), may have come after MacDonald announced his disarmament plan (which made no provision for any German air force) on the afternoon of 16 March.

There is lacking any replacement organization, and consequently any possibility for planned, orderly inductions. Trained reserves are scarcely available. In the A-Army, the ratio of trained to untrained is 1:3; in many arms, it is still worse. One cannot believe that the training in the Wehrverbände, which provide the bulk of our war volunteers, is much more than play-soldiering.

"The A-Army, thus mobilized, is not ready for action; it requires training, which must be estimated at 4 weeks at a minimum.

The Grenzschutz, which only existed in the East, would only amount to unguided agglomerations of men unable to maneuver. Materially, Adam stated, the mobilized army would only have emergency equipment. Artillery batteries would only have two or three cannon, there was a serious shortage of light machine guns, and there were many other gaps. The army lacked reserves of weapons, and only had a two-week supply of ammunition.[46]

Adam's paper stated that, in view of the numerical and material superiority of France, Poland, and Czechoslovakia, "the possibilities of carrying on war even merely to gain time are limited." The Poles alone could be held off while ammunition lasted, but Czech and especially French intervention would make the situation much worse. The French (Adam wrote) would immediately occupy the Rhineland with mobile forces and seize the center of the German arms industry in the Ruhr; they could even move well beyond the Rhine without a heavy commitment of men. Adam ridiculed romantic ideas of guerrilla warfare or of a Sicilian vespers against the French. In one place, Adam made a general statement which has often been cited[47] to show German military impotence in 1933:

We cannot wage a war at this time. [*Wir können zur Zeit keinen Krieg führen.*] We must do everything to avoid it, even at the cost of diplomatic defeats. We must shun trumpet flourishes which unnecessarily irritate the enemy and intoxicate the [German] people.

[46] The Adam memorandum (dated only as "mid-March 1933") and accompanying postwar affidavit constitute Krupp Trial Defense Exhibit 62. Adam appears to have provided a notarized copy (not a photographic reproduction or a 1933 copy) of a document stated to be in his possession. According to Adam's postwar memoirs, Blomberg seemed shocked—indeed surprisingly so—at Adam's depiction of German military weakness, but he nevertheless at this time cancelled the execution of all the plans Schleicher had authorized for the fall of 1933: Institut für Zeitgeschichte, ED-109, Wilhelm Adam Erinnerungen, Vol. I, p. 184. This last recollection seems dubious, but the stretch-out in air force expansion, described just above, indicates some cutting back for diplomatic reasons.

[47] E.g., by Gerhard Meinck, in his *Aufrüstung*, p. 19; Esmonde Robertson, *Hitler's Pre-War Policy and Military Plans, 1933-1939* (London: Longmans, 1963), p. 10; Walter Bernhardt, *Die deutsche Aufrüstung 1934-1939* (Frankfurt: Bernard & Graefe, 1969), p. 35.

But although Adam stated that Germany could not wage a war at this time, his meaning seems to have been that Germany could not rationally choose to start a war, not that Germany could not fight. He went on to state that it was necessary to work tenaciously, patiently, and cautiously to strengthen Germany's military power and prepare the people for the fateful hour (*die schwere Stunde*). Even if Germany did everything to avoid war, he noted, she could not stop others from waging it preventively, if they wished to do so. And while he wrote that it was idle (*müssig*) to draw up operational plans or concentration plans for the eventuality of a preventive attack, this was because in that case Germany's initial actions would depend on what the enemy did. Historians have failed to note that he did expect to resist a preventive attack, and indeed went on to sketch out the kind of campaign he would try to conduct.[48] At first taking the defensive and harboring German strength, he would leave East Prussia (with the three field army divisions it would mobilize) to its fate, while trying to hold the line of the Oder and the mountains on the Czech border. Adam's strategy was to fend off attacks in the East with the Grenzschutz and six to eight divisions of the field army, two or three of them being held in reserve east of Berlin. The remaining ten to twelve divisions "make initially no deployment toward the West. They are held ready in the Reich." (Adam described eventual operations in the present tense.) If the line of the French advance could be discerned, detachments of these divisions would be led forward toward the enemy in order to upset him with pinpricks, increase his native caution, and thus gain time. If the French advanced farther, the German forces would spy out the chances for limited attacks (*Teilschlägen*) and exploit them. Meanwhile German defense power would be enhanced by

[48] One of the documents on the 1932 transport exercise (see Chapter Six, Section 1) sums up the Truppenamt's views on operational plans: "The military situation of Germany makes it impossible to draw up a fixed operational plan. How we operate will depend on what weak points the enemy offers": ML, PG 34095, TA 739/32, 17 Oct 32, reel 31/649.

Meinck, *Aufrüstung*, p. 19, misquotes the Adam report as saying that drawing up operational plans or concentration plans was "*unsinnig*" (nonsense, crazy), rather than "*müssig*" (idle). (Bernhardt, p. 35, says the same; the source he cites proves in turn to be based on Meinck, *Aufrüstung*, p. 19.) This stronger word *unsinnig*, and the omission by Meinck of Adam's actual sketch of a defensive campaign, make the passage Meinck reproduces appear as an elaboration of the idea that Germany was simply unable to fight. Meinck also says that Adam made no reference to the "Black Reichswehr" (meaning presumably unofficial reinforcements) and fails to mention Adam's statement that the Wehrverbände "provide the bulk of our war volunteers" (*die Masse unserer Kriegsfreiwilligen liefern*); arguing that the Wehrverbände were not to be mobilized (as closed units), Meinck passes over the intent to mobilize their individual members: *Aufrüstung*, pp. 5-6, 11. To be sure, the Wehrverbände members were not well trained, but they were the best resource available, and the army expected to use them in case of war.

raising second- and third-line formations. "Germany must become an armed camp." Although many have believed that French action in 1936 would have led the Germans to give up the reoccupation of the Rhineland without a struggle, we see here that even in 1933, and despite grave reservations about German strength, the German chief of staff proposed to resist a French invasion.

Adam was one of the German officers least sympathetic to Nazism, and one of the objectives of his memorandum clearly was to warn against provocative foreign policies and "hurrah patriotism," such as might lead to a conflict before Germany was ready. He wrote explicitly of the dilemma Germany was in, in that any visible effort to improve German strength in personnel and materiel increased tension and risked provoking an attack, while caution would not improve German armament. But despite risks, and his advice to tread quietly, he felt very strongly that German rearmament should be pushed forward. In this, his views corresponded closely with those of Hitler: Hitler did not want to delay German rearmament, but he did want to conceal it rather than to make an issue out of it. The need for caution arose of course largely from Germany's current military inferiority, and it was both galling and alarming that the Poles could consider—or appear to the Germans to consider—the possibility of preventive war. The sooner this state of affairs could be ended (so Hitler must have thought), the better. A record by Rheinbaben of the March 15 meeting indicates that Hitler agreed "to the thesis that the immediate German aim is the quick creation of a military power equal to any eventuality in the East and in the Baltic."[49]

4. Was Preventive War a Possibility?

Aside from apprehensions of a Polish preventive attack, springing from the Westerplatte affair, other incidents evoked similar fears with respect to France, and although some German observers doubted that danger existed, these apprehensions tended to increase, fed by the propensity of officers and officials to protect themselves by anticipating the worst contingencies.[50] Was there actually a possibility of a preventive attack? It seems worth a digression to consider this question.

[49] *ADAP*, C/I/1, No. 94. On Adam's outlook and background, see Telford Taylor, *Sword and Swastika*, pp. 202-205, 228; Müller, *Heer und Hitler*, pp. 45n, 335, 359. On Adam's unpublished memoirs, see Bibliography.

[50] Comment on preventive war danger (with + or − indicating balance of support or doubt in source as to the existence of danger): *ADAP*, C/I/1, Nos. 68 (−), 70 (+), 74 (−), 83 (−), 87 (+), 111 (−), 120 (+), 122 (+), 142 (+), 177 (+), 180 (+), 183 (−), 184 (+), 192 (+), 201 (+). Nos. 68, 83, 111, 120, and 122 suggest that Mussolini was trying to foster fears of preventive attack in order to induce the Germans to support his four-power pact; Ambassador Ulrich von Hassell in Rome was

The concern of German leaders over a possible attack seems ironic when we look at the actual situation. Neither the French nor the Poles appear to have seriously considered such an enterprise. In a discussion with the British military attaché in Paris in March 1933, General Weygand, regarded abroad as the personification of French militarism, observed that war was full of unexpected pitfalls and surprises, so that even if you thought, as the Germans (in Weygand's view) had thought in 1914, that you had all the trumps in your hand, you could never be sure of success.[51] French operational planning in this period was primarily defensive. If the need arose to coerce Germany, there were plans for "taking gages" (e.g., reoccupying the Saar and adjacent areas), and even for a "supporting attack" toward Mainz in case of a German offensive eastward. But such action, involving partial mobilization, could not be taken without public support, and there would have been little public support in France for a preventive attack.[52] None of the French political leaders of 1933 had any desire to undertake a punitive expedition in alliance with Marshal Józef Piłsudski, probably least of all Edouard Daladier, Premier from February to October. The natural pacifism of the Radical Socialist cabinets of 1932-34 was reinforced by their dependence on the support of the Socialist party, and instead of increasing the French army, the government was in the process of reducing it, for budgetary reasons and also to save some of the prospective recruits for the "lean years" to follow.[53]

---

skeptical. Neurath tended at first to minimize but later to stress the danger, as in No. 142, also RK, Cabinet Protocol, 25 Apr 33, 3598/D792415. Apparently because of the diplomatic situation, Blomberg barred the navy from sending ships on a training cruise to Spain: ML, PG 34168/1, Memo on discussion between Blomberg and Raeder, 28 Mar 33, reel 27/1003; PG 33994, Draft orders and note of 4 Apr 33, reel 28/37-41.

[51] FO 371, Memo, Heywood to Tyrrell, 15 Mar 33, C2627/245/18.

[52] *Rapport . . . sur les événements survenus en France*, I, 40-41, 47-50; Tournoux, pp. 248-250, 336-337; *DBFP*, 2/V, No. 154.

[53] Bankwitz, *Weygand*, pp. 86-105; Gamelin, II, 98-99, 103-105. On Daladier's views, see Chapter Nine below. Lord Tyrrell told his Belgian colleague in February 1933 that France would not move if Germany attacked Poland or occupied the demilitarized Rhineland; the French would only fight if Germany violated the French frontier: Ch. De Visscher and F. Vanlangenhove, eds., *Documents diplomatiques belges, 1920-1940: La Politique de sécurité extérieure*, 5 vols. (Brussels: Académie royale, 1964-1966), III, No. 12 (hereafter *DDB*). Tyrrell reported similar views to his own government in May: Cab 24/242, CP 151 (33), disp., 9 May 33; also *DBFP*, 2/V, No. 161. On the other hand, Tyrrell told Austen Chamberlain he had warned MacDonald that for the first time since the end of the war he could not assure the Prime Minister that the French would in no case precipitate a conflict, and the ambassador seemed to think preventive action possible in October: Austen Chamberlain Papers, ltr. to Ivy Chamberlain, 26 Apr 33, AC 6/1/985; FO 800/288, Simon Papers, ltr. from Tyrrell, 11 Oct 33. Boncour wrote later that preventive war would not have been supported by as many as ten votes in the Chamber: Paul-Boncour, II, 344.

On the Polish side, leaders in Warsaw seem to have cared little, in early 1933, about the state of French military planning, even for their own support. As with the French, Polish military planning was generally defensive.[54] Piłsudski had personal reasons for resenting French tutelage, as did his new Foreign Minister, Józef Beck. At the beginning of 1933, before Hitler became chancellor, Polish diplomacy pursued a course of increasing independence from France, especially scorning the French disarmament proposals, while it already envisioned a détente with Germany, which was seen as weakened by internal dissension.[55] Indeed, there were cogent reasons for Polish distrust of Paris. France was now governed by ministers who disliked the authoritarian Polish regime, who did not always refuse to participate in British- or Italian-sponsored "Great Power" meetings, and who might be disposed to make compromises on the subject of the Corridor. Certain financial-industrial circles in France were definitely ready for concessions at Polish expense. Moreover, ever since the mid-twenties, French military support for Poland had really depended on British backing, something unlikely to be forthcoming.[56] Perhaps out of the human tendency to find some ground for hope in the most hopeless situations, Piłsudski and Beck embraced and then clung to the belief that Hitler, as an ex-Austrian, was chiefly interested in achieving an Austro-German An-

[54] Harald von Riekhoff, *German-Polish Relations, 1918-1933* (Baltimore: Johns Hopkins Press, 1971), pp. 342-343.
[55] Piłsudski resented having to share credit for Polish military success in 1920 with Weygand; Beck had left the post of military attaché in Paris under a cloud in 1923. See Weygand, II, 101-107, 109-112, 151-152, 156, 166; Bankwitz, *Weygand*, pp. 22-23; Weinberg, *Foreign Policy of Hitler's Germany*, p. 58; Henry L. Roberts, "The Diplomacy of Colonel Beck," in Gordon A. Craig and Felix Gilbert, eds., *The Diplomats, 1919-1939* (Princeton: Princeton University Press, 1953), p. 580. According to Ian Colvin, *None So Blind: A British Diplomatic View of the Origins of World War II* (New York: Harcourt, Brace & World, 1965), p. 245, Beck's offense in Paris had been in removing a document from the desk of the Chief of the Deuxième Bureau during the latter's momentary absence. On cooling of Franco-Polish and warming of German-Polish relations: Castellan, *Réarmement clandestin*, pp. 474-475; *DDF*, 1/I, Nos. 87, 277 (and Annex); 1/II, Nos. 200, 266, 276; Czech docs., disp., Girsa to Foreign Ministry, 18 Feb 33, 2376/D497188-89; Marian Wojciechowski, *Die polnisch-deutschen Beziehungen 1933-1938* (Leiden: E. J. Brill, 1971), pp. 2-9; Józef Lipski, *Diplomat in Berlin, 1933-1939: Papers and Memoirs of Józef Lipski, Ambassador of Poland*, ed. Wacław Jędrezejewicz (New York: Columbia University Press, 1968), pp. 19, 51-52; Joseph Beck, *Dernier Rapport: Politique polonaise, 1926-1939* (Neuchâtel: Eds. de la Baconnière, [1951]), p. 24.
[56] Roberts, pp. 587-588, 593-594; Zygmunt J. Gasiorowski, "The German-Polish Nonaggression Pact of 1934," *Journal of Central European Affairs*, 15 (1955), 4; Lipski, *Papers and Memoirs of Józef Lipski*, pp. 6-7, 8-11, 13, 19; Jacques Bariéty and Charles Bloch, "Une tentative de réconciliation franco-allemande et son échec," *Revue d'histoire moderne et contemporaine*, 15 (1968), 454-465; Wojciechowski, pp. 2-5; Jacobson, *Locarno Diplomacy*, pp. 29-30; Schuker, pp. 389-390.

schluss, and that he wanted good relations with Poland, in contrast to the Prussian Junkers, who always stressed their grievance over the Polish Corridor.[57]

A story has circulated and has been debated since the 1930s, asserting that Piłsudski covertly approached the French in the spring (and also the fall) of 1933 to propose the joint waging of a preventive war against Germany, a proposal the French refused.[58] But a recent study by Marian Wojciechowski, making use of Polish archives and now available in German, concludes firmly that Piłsudski never intended to wage war but only to bring pressure on Hitler to reach an agreement.[59] Recent volumes of the French documents also tend to discredit the preventive-war story. In November 1933, Piłsudski, after having supposedly told his diplomats that he would ask the French military attaché "as one soldier to another" what France "really wants," proceeded to impress the attaché with his trust in Hitler, saying that he hoped the German leader would remain in power as long as possible, and in no way asking if France was interested in preventive action. It is quite possible that the secretive Piłsudski may have deliberately deceived some of his own officials by pretending to sound the French on the subject of preventive war, thus concealing his actual reliance on faith in Hitler's word. Of course, the rumors of preventive war were useful to Piłsudski's diplomatic policy in that they helped predispose Hitler to reach an accommodation. Meanwhile the Polish leaders began to excuse their *tour de valse* with Hitler by making constant complaints of French weakness. Although the French would almost certainly have rejected any Polish preventive-war proposals, the story that they actually did so seems to be part of

[57] See *DDF*, 1/III, Nos. 285, 429; 1/IV, No. 166; 1/V, Nos. 59 (and Annex), 62, 65, 288, 328, 337, 442; *Papers and Memoirs of Józef Lipski*, pp. 71-72, 95; Beck, *Dernier Rapport*, pp. 29, 101. The last reference indicates that, even after the defeat of 1939, Beck did not entirely give up his belief that Hitler himself had favored conciliation with Poland; see also Lipski's postwar statement in L. B. Namier, *Europe in Decay: A Study in Disintegration, 1936-1940* (London: Macmillan, 1950), pp. 282-283n.

[58] The story of the proposal to the French was aired privately as early as November 1933 by a Polish journalist (*DBFP*, 2/VI, No. 59) and seems to have been first publicized widely by André Simon [Otto Katz] (*J'Accuse! The Men Who Betrayed France* [New York: Dial, 1940], pp. 52-53). A variant in Sohn-Rethel, *Ökonomie und Klassenstruktur*, pp. 154-156, makes the French willing, but held back by the British. Notable arguments for the story are those of Hans Roos, "Die 'Präventivkriegspläne' Pilsudskis von 1933," *VfZ*, 3 (1955), 344-363, and Wacław Jędrezejewicz, "The Polish Plan for a 'Preventive War' against Germany in 1933," *Polish Review* (New York), 11 (1966), 62-91. Criticisms of the story are made in Roberts, pp. 612-614; Zygmunt J. Gasiorowski, "Did Pilsudski Attempt to Initiate a Preventive War in 1933?" *JMH*, 27 (1955), 135-151; Boris Čelovsky, "Pilsudskis Präventivkrieg gegen das nationalsozialistische Deutschland (Entstehung, Verbreitung und Widerlegung einer Legende)," *Welt als Geschichte*, 14 (1954), 53-70.

[59] See Wojciechowski, pp. 16-20, 64, 75.

338—Hitler's Accommodation with the Military

this same Polish tactic of using accusations to excuse and cover a change of policy in Warsaw that had begun before Hitler came to power.[60]

5. The Reichswehr Gets Its Money

As befitted a good staff officer, General Adam seems to have had in mind, in writing his memorandum for Blomberg, not only the political and military forces at hand, but also the coming budget for the fiscal year beginning on April 1. We have seen in Chapter One that he was preoccupied in the winter of 1931-32 with the need for more rapid material preparation. In his March 1933 memorandum, he wrote that the active defense measures he described presupposed an improvement in the ammunition and weapons situation: "This must be embarked on immediately, and indeed according to the precisely worked-out priority plan"—presumably a reference to the Billion Program.[61] It should be noted here that the actual figures on German armament show that the army was by no means barehanded. Some of the worst shortcomings were in new-model light machine guns and in antiaircraft artillery (which last was forbidden by treaty). The numbers of these weapons ran 9 and 14 percent respectively under the figure required for initial equipment for an army of twenty-one infantry and three cavalry divi-

[60] The *Papers and Memoirs of Józef Lipski* quote (p. 100) a 1942 copy of a November 1933 note by Jan Szembek, according to which, as described in my text, Piłsudski told his diplomatic advisers he would ask the French military attaché, "soldier to soldier," what France really wanted; further talks with the Germans would depend on the outcome. There are also vague, passive-voice references to preventive-war discussion on p. 75. But I find no direct affirmation in this volume from an informed source that Piłsudski actually did sound the French on preventive war. For the French attaché's report of the actual discussion: *DDF*, 1/V, No. 59 (Annex). Jędrezejewicz, "Polish Plan," pp. 85-86, reports that Piłsudski sent Ludwik Morstin, a writer and poet, to Paris in late October or early November to ask whether the French would respond to a German attack on Poland with a general mobilization and a massing of troops on the Franco-German border. Supposedly, the French government held two high-level meetings on this subject, one presided over by President Lebrun, and reached a decision to provide only staff, materiel, and propaganda support. But there is no reflection of this in published French documents, and even if true, the approach did not amount to a proposal for preventive war. Relevant French documents include *DDF*, 1/III, Nos. 24, 87, 91, 156, 238, 320; 1/IV, Nos. 59, 288. A number of the French documents were published earlier in Jules Laroche, *La Pologne de Pilsudski: Souvenirs d'une ambassade, 1926-1935* (Paris: Flammarion, 1953). On the date of origin of the shift in Polish policy, note esp. *DDF*, 1/II, No. 200.

[61] Dülffer reports (p. 242) that in mid-March the navy heard Blomberg was planning to increase the army budget by RM 400 million per year for five years, i.e., double the Billion Program. Such a plan may have been discussed, but as will be shown below, the Billion Program was put into effect as such, without doubling, for fiscal 1933.

sions, and for "a certain number" of Grenzschutz formations. Medium trench mortars—an easy weapon to produce—were 31 percent under strength. But Adam's picture of two- or three-cannon batteries was overdrawn; light artillery guns were up to strength, heavy guns were nearly so, and supplies of munitions were adequate for an estimated two weeks for heavy artillery, three weeks for trench mortars, four weeks for light artillery, and six weeks for small arms (rifles and machine guns). In fact, 610,637 rifles were available to meet fully the need for original equipment, apparently representing the number of those who would carry this arm in the field army and Grenzschutz, and 47,260 additional rifles nearly sufficed for six weeks' replacement requirements.[62] But the less real ground there was for complaint over shortages, the more reason, presumably, for fears of foreign detection and intervention.

As noted earlier, Schleicher and his officers had hoped to get RM 200 million, over and beyond the established budget total of recent years, in order to launch Umbau and the five-year Billion Program. Due largely to the limitation of the work-creation program to 500 million by conservative financial officials, they had failed to get that amount. At the outset of the new government, Gereke, still in charge of work creation, returned to his idea of expanding his program to meet both local and national needs by drawing on an unused tax credit certificate allocation for premiums for new employment; military finance officers strongly supported this idea. At the February 9 Work Creation Committee meeting, attended by Hitler, Krosigk, and Blomberg, Gereke said that another RM 300 million would be needed to meet the demands of Reich agencies, including RM 100 million for the Reichswehr and RM 30 million for "air transport" (i.e., the embryo Luftwaffe), and he suggested the termination of the employment premium. Krosigk, however, doubted that the Reichsbank (i.e., Luther) would agree to this termination, and he argued that it would weaken confidence, especially in small and medium industry, in the consistency of government policy. Krosigk also maintained that the demands of the Reich agencies cited by Gereke were much too high, and that the Reichswehr "could not actually in the current budget year spend more than RM 50 million [i.e., over the regular budget] for purposes of national defense." Contrary to what might have been expected, Blomberg then acquiesced, saying that he "could not actually spend more than RM 50 million in the budget year 1933," although he would need further sums later for a big, long-term arma-

[62] OKW, Wi/IF 5.383, Notes for report to cabinet, [March 1933], and att., 86/ ?10179-90.

ment program. Blomberg's acquiescence is the more surprising because Hitler, on the day before, had told the cabinet that, for the next four or five years, the guiding principle was, everything for the armed forces. Considering the statements and circumstances, it seems likely that Krosigk and Blomberg had discussed the matter earlier, and that Blomberg—despite the position of Reichswehr staff officers—had himself previously given Krosigk the figure of RM 50 million. It is possible that Krosigk and Blomberg were here referring to the remnant of the budget year 1932-33.[63]

Hitler, in any case, understood Blomberg to refer to fiscal 1933-34, and he was displeased. He stressed again the absolute priority of national defense, and said that billions were needed for German rearmament.

> Germany's future hung exclusively and alone on building up the armed forces again. He could only accept the limited character of the means now asked by the Reichswehr Ministry on the grounds that the tempo of rearmament could not in the coming year be accelerated more sharply. In any case, he took the view that in future, in case of a collision between the claims of the Wehrmacht and claims for other purposes, the interests of the armed forces under all conditions had to come first. The granting of the resources of the immediate [work-creation] program was also to be decided in this sense. He considered the combatting of unemployment by public expenditures to be the most suitable means of assistance. The 500 million program was the largest of its kind and especially suited to serve the interests of rearmament. It served best to make possible the concealment of work for the improvement of national defense.

Here Hitler added the statements already cited on the need for concealment during the dangerous period before Germany was rearmed and able to attract allies. But Hitler had not yet become an unchallenged dictator, and he agreed that the premium program should not be revoked before the March 5 election. Until he had safely passed that point, he had to take care not to offend business interest groups or the DNVP. More generally, he needed to allay fears of dangerous economic experimentation, fears that had been encouraged by some

---

[63] BA, R 43 II/540, Record of Work Creation Committee meeting, 9 Feb 33. Position of military finance officers: Geyer, "Zweite Rüstungsprogramm," pp. 134, 163 (note 72). Geyer does not refer to the basic record of the 9 Feb 33 meeting, or to Petzina's account of it ("Hauptprobleme," pp. 43-44), and his explanation of Blomberg's remark (that he was referring to Umbau only) does not stand up in the light of that record, or of the previous Reichswehr requirements for Umbau.

radical Nazi statements in the past.[64] And so the Reichswehr received only RM 50 million.

Once the election was past, Hitler moved rapidly to expand work creation and its use for armament. He told the cabinet on March 7 that a change in the Reichsbank leadership was "absolutely necessary"; a more flexible man was required. Luther was offered the embassy in Washington, and he resigned on March 16, hinting in a letter to Hindenburg that he could not accept Hitler's policies on currency, credit, public finance, and economics. As a matter of fact, on Gereke's motion, and with Hitler's strong support, the cabinet had decided on the fifteenth to end the employment premiums on April 1 in favor of the work-creation program. Krosigk went along with this, advising that this would make RM 100 million available, and on March 17 Gereke's committee awarded this additional sum to the Reichswehr. Luther was promptly replaced by Hjalmar Schacht, the previous Reichsbank President, who had long supported Hitler, and who had probably promised abundant credit. Soon afterward, Gereke, the advocate of large-scale local government projects, was also removed. He received no such tender consideration as Luther: Göring had him arrested on March 24 on charges of embezzlement, and although the Reich Supreme Court reversed his conviction in September 1933, he did not emerge from prison until two years later.[65]

[64] Krosigk stated at the 9 Feb 33 meet meeting that small and middle industry had claimed most of the RM 45 million in employment premiums so far applied for, and heavy industry had scarcely participated. Gereke said almost all Reichstag parties except the DNVP had criticized the premiums. At another work-creation meeting on 17 Mar 33, reference was made to opposition from "the Reichsverband der Deutschen Industrie and other interest groups" to ending the premiums (R 43 II/540). On the position of the RVDI and of "die Wirtschaft," see also Petzina, "Hauptprobleme," pp. 27, 41-42. Although heavy industrialists may not have had much stake in the employment premiums, they did have a stake in the tax credits they received under Papen's principal, RM 1.5 billion program for corporate taxpayers, and those might be the next to be abandoned.

[65] BA, R 43 II/540, Record of Work Creation Committee meeting, 17 Mar 33; *Secret Conversations*, pp. 349-350; RK, ltr., Luther to Hindenburg, 16 Mar 33, K1028/K267541-45; Cabinet Protocols, 7, [15], 24 Mar 33, 3598/D791922, D792047-48, D792114-15; *DBFP*, 2/IV, No. 267; Lutz Graf Schwerin von Krosigk, *Es geschah in Deutschland* (Tübingen: Rainer Wunderlich, 1951), p. 187; Gereke, pp. 240-256; Petzina, "Hauptprobleme," pp. 43-44; Schulz, "Anfänge des totalitären Massnahmenstaates," pp. 453, 659-660; Sauer, "Mobilmachung," pp. 786-787; Earl R. Beck, *Verdict on Schacht: A Study in the Problem of Political "Guilt,"* Florida State University Studies, No. 20 (Tallahassee: Florida State University, 1955), pp. 36-37. The embezzlement was supposedly from the Landesgemeindeverband Ost and the 1932 committee to reelect Hindenburg. Gereke was offered freedom from charges if he would forgo pay and pension rights, and the real objective was doubtless his speedy removal: see Michael Schneider, p. 207. He did not in fact oppose rearmament, but

342—Hitler's Accommodation with the Military

So far the Reichswehr had been granted RM 150 million in work-creation funds, of which the army share was around RM 110 million.[66] But now, probably in late March, the army was given to expect that RM 200 million would be available for the coming fiscal year, aside from the money allotted in the regular budget, and apparently aside from the RM 110 million already awarded. Staff officers now suggested that RM 226 million be requested, as a first annual installment on the Billion Program, and they outlined the total cost of that program. It would be possible, they said, to make the initial stage more rapid, but this would involve industrial training courses and changing factories from one shift to two or three shifts, and these activities would create

---

he preferred communal projects, and his presence in the work-creation system was an obstacle to Nazi propaganda exploitation of it.

Bracher, "Stufen der Machtergreifung," pp. 155-157, indicates that 15 Mar 33 witnessed Hitler's decisive establishment of power in the cabinet, at the very time when he was also, as we have seen, laying down his foreign policy of outward moderation. The main part of the record of the 15 March cabinet meeting is in *Nuremberg Documents*, XXXI, 402-409,. but the pages pertinent here were left in the Reich Chancellery files (RK, 3598/D792047-48).

[66] According to a 1944 lecture by a Flottenintendant Thiele (*Nuremberg Documents*, XXXV, 590-591, full text on National Archives Microfilm T-988, reel 39/B041067-68), the navy received additional amounts in February and March 1933 of 14 and 21.7 million marks respectively; Dülffer, p. 242, gives 28 million for March. These were evidently the naval shares of the amounts granted on 9 Feb and 17 Mar 33, representing the normal naval proportion of two sevenths of total Reichswehr expenditure. Thiele's figures would leave RM 114.3 million for the army; Dülffer's, RM 108 million.

Other work-creation programs (the First and Second Reinhardt Programs) followed later in 1933, but these apparently provided little finance for rearmament. Michael Wolffsohn, "Arbeitsbeschaffung und Rüstung im nationalsozialistischen Deutschland: 1933," *MGM*, 22 (1977), 10-14, argues that, despite the decisions of 9 Feb and 17 Mar 33, the pioneer work-creation program that has been discussed here (the Sofortprogramm) also did not in practice provide funds for rearmament. He cites a record of a discussion at the Wehramt of 19 May 33 (OKW, Wi/IF 5.370, 81/804849-50) that called for new work-creation proposals for a "Four-Year Plan"; the Sofortprogramm was said to represent the planning for 1933. Direct arms demands, such as procurement for the armed forces or the preparation of factories (for arms production) were not, according to this record, to be counted under "this program." The document is ambiguous, but "this program" appears to me to represent the projected four-year scheme (which, moreover, never materialized) and not (as Wolffsohn supposes) the already-launched Sofortprogramm. Wolffsohn also refers (p. 13) to a 1937 Finance Ministry memorandum on work-creation measures in the years 1932-1935. This memo, he relates, enumerated the Sofortprogramm expenditures, totalling RM 598.3 million; of this sum, listed Reich expenditures consisted of two items, RM 5.8 million for the Reich waterways administration and RM 190 million for "Other Reich measures." Although this RM 190 million corresponds exactly with the total grants made on 9 Feb and 17 Mar 33 for the Reichswehr and "air transport," Wolffsohn asks us to believe that these "Other Reich measures" did not cover significant military expenditure—largely on the grounds that, in the case of the First Reinhardt Program, a much smaller amount under this heading did not include such expenditure. He makes no reference to the Flottenintendant Thiele lecture, or to Dülffer.

too great a risk of disclosure at a time when the Disarmament Conference was reaching a critical stage.[67] Another RM 200 million was in fact authorized, probably in May; in May the navy was granted its corresponding further sum of RM 80 million, with assurances of the same amount for each of the following five years (i.e., a five-year total for the navy of RM 400 million). In July 1933, an Ordnance Office paper indicated that the details of the payment of the "200 000 (sic) Mill." were not yet cleared up. But a report from November 1933 stated that so far the army had received and committed funds for stockpiling from the first installment of RM 200 million for the acceleration of rearmament; this was expected to produce an overall backup supply for four weeks of combat, and with further funds—apparently not yet available—the goal of a six-week supply would be reached by the end of 1934.[68]

The additional RM 200 million meant that the army would get what it had hoped to get under Schleicher, with about RM 110 million to spare; the extra RM 110 million could now be spent on Umbau, since distinctions between regular budget and work-creation funds now no longer mattered. Blomberg sought and gained from the cabinet, on April 4, a general authority to proceed as he wished, with only Hitler's

[67] OKW, Wi/IF 5.383, Notes for report to minister, "ca. March 1933," notes for report to cabinet, and att., 86/810177-83, 810190. An exception to the restraint on extra shifts was made for artillery munition factories, one of which had already been working around the clock in 1932, while the others had at times previously worked more than one shift.

[68] *Nuremberg Documents*, XXXV, 591; BA-MA, RH 15/28, Wa A 898/33, 13 Jul 33; OKW, Wi/IF 5.383, Notes for report, 10 Nov 33, 86/810149. See also Geyer, "Zweite Rüstungsprogramm," p. 158 (Docs. 10 and 11). There are indications that some other supplementary amounts for defense were available: see *Nuremberg Documents*, XXIX, 3-5; OKW, Wi/IF 5.701, Reichswehrminister TA 616/33, 177/912272-73. Dülffer reports (p. 243) that the navy received, presumably for fiscal 1933 (1933-1934), 311.8 million marks, an increase of RM 124.4 million over fiscal 1932. Thiele reported the navy as receiving 301 million marks in 1933. Possibly these figures include the February and March supplements (actually in fiscal 1932; see note 66) as well as the May ones. OKW, Wi/VI 104, WH $\frac{58a\ 40\ 38\ g.K.\ Beih.\ 3}{1073/38\ g.K.}$, 4 Nov 38, 657/1857888, states that total 1933-1934 Wehrmacht expenditures (not including Air Ministry) were RM 914.8 million, i.e., 246 million more than the published budget: cf. *DBFP*, 2/VI, No. 374 (encl.).

The Reichswehr did not admit to affluence. In July, a Blomberg letter to Krosigk stated that the Reichswehr could not repay RM 400,000 due on the 1931 advance to Borsig (Tegel) because reasons of foreign and domestic policy had prevented a budget increase to cover Umbau when the 1933 budget was drawn up, and Blomberg now had to meet Umbau costs with a budget smaller than in 1932. Blomberg's letter added a veiled threat: "With the change which has since occurred in the domestic political situation, I assume that the limited, indirect relief which will be afforded the army budget [by nonrepayment] . . . will meet with your full agreement, and also that of the cabinet [i.e., Hitler]": OKW, Wi/IF 5.516, Reichswehrminister 701/33, 20 Jul 33, 116/845424.

approval, without being bound by fixed legal or budgetary provisions as to what funds could be expended for what. New laws on defense organization could come later, "as soon as the foreign policy situation is sufficiently clarified."[69] The Defense Minister still had to negotiate the annual defense budget total with the Reichsbank President and the Finance Minister, but with Schacht's willingness, now shared by Krosigk, to proceed with concealed inflation, there would be no serious limit to military spending. Schacht soon substituted the dummy "Metallurgical Research Company" (*Metallurgische Forschungsgesellschaft* or "Mefo") for the public-credit institutions of the work-creation scheme, and by this means conjured up billions for armament. Contrary to what has sometimes been believed, however, "Mefo" credits probably did not appear, at least on any significant scale, until 1934.[70]

[69] Cabinet material shows that Blomberg reported on Umbau and "Wehrmacht questions": RK, Cabinet Protocols, 29 Mar 33 (4:15 p.m.), 4 Apr 33 (4 p.m.), 3598/D792210, D792254; on April 4, the cabinet approved the establishment of a Council of Reich Defense (see below, including note 86). For draft cabinet resolution and draft Blomberg statement, see Geyer, "Zweite Rüstungsprogramm," pp. 156-157, from BA-MA, RH 15/40, and cf. (earlier) att. 2 from that file. The cabinet passed the resolution in a slightly less headstrong form, providing for review by Hitler: see Michiyoshi Oshima, ["Reichswehr und Rechnungshof: Die Bedeutung des Kabinettsbeschlusses vom 4. April 1933,"] *Mita Gakkai Zasshi*, 69 (1976), 317-318. Detailed sections in the earlier att. 2, omitted from the later versions (and by Geyer) provided for the shortening of enlistments after 1 Apr 33, the institution of a new replacement organization on 1 Oct 33, and an increase in inductions on 1 Apr 33, all of which measures were in fact carried out.

[70] On origin of Mefo, see Erbe, p. 46; Gerhard Kroll, *Von der Weltwirtschaftskrise zur Staatskonjunktur* (Berlin: Duncker & Humblot, 1958), p. 586; Beck, *Verdict on Schacht*, pp. 50-51; Hjalmar Schacht, *76 Jahre meines Lebens* (Bad Wörishofen: Kudler & Schiermeyer, 1953), pp. 384, 400-401; Schacht, *1933: Wie eine Demokratie stirbt* (Düsseldorf: Econ, 1968), pp. 81-82, 101; Lutz Graf Schwerin von Krosigk, "Wie wurde der zweite Weltkrieg finanziert?" in *Bilanz des Zweiten Weltkrieges: Erkenntnisse und Verpflichtungen für die Zukunft* (Oldenburg: Gerhard Stalling, 1953), pp. 315-316. 1938 documents, OKW, Wi/VI 104, WH $\frac{58a\ 40\ 38\ g.K.\ Beih.\ 3}{1073/38\ g.K.}$, 4 Nov 38, 657/1857888, and Wi/VI 104, Memo on "Endangering of Exchange, with Special Regard for Rearmament," 3 Nov 38, 657/1857889-900, show that (1) there was by later standards no major extension of credit in fiscal 1933; (2) Schacht undertook in 1934 to finance (not only by Mefo) about RM 35 billion in armament in a period of about eight years, on condition that this limit was maintained, that the money and capital markets were controlled, that the first arms credits should be retired after five years at the latest, and that price and wage levels be kept completely unchanged; (3) the figure foreseen for eight years was overstepped in four and a half years (i.e., between 1 Apr 34 and 30 Sep 38); and (4) the five years maximum extension of the first Mefo bills expired on 1 Jan 39. This last would indicate that they were issued no earlier than 1 Jan 34. Geyer holds that Mefo credits were used in 1933 ("Zweite Rüstungsprogramm," p. 163 [n. 81]; "Militär, Rüstung und Aussenpolitik," p. 249 [incl. n. 50], but does not cite specific contemporary sources. Admittedly, we do need an explanation of how the extra RM 200 million was raised in 1933.

As far as arms expenditure was concerned, 1933 witnessed mainly a fulfillment of those Reichswehr wishes that already existed in 1932.

April 1 marked not only the start of the new fiscal year, and now of the Billion Program, but also the launching of major new steps toward Umbau, or personnel reorganization. In view of the international situation, and especially the MacDonald proposals, this was no time for overt actions that would reveal German intentions. One of Blomberg's arguments to the cabinet for budgetary flexibility was the uncertainty as to the future course of the Disarmament Conference, and therefore as to the nature of future changes in the armed forces. Political decisions within Germany, respecting the SA, might also affect reorganization plans. And the shortage of trained officers and noncommissioned officers continued to impede any acceleration of the militia program. If there had been any thought during the winter of launching a large militia program at this point, it was not pursued.

Enlistments did definitely increase on April 1, in the way that had been planned in November 1932. Available figures vary widely, but from 2,500 to over 22,000 more men were inducted on that date than were needed to maintain a strength of 100,000, and 10,000 to 20,000 extra men also enlisted on October 1.[71] The recruits signed the same commitment to serve for twelve years as in the past, but they also signed a "supplementary obligation," recognizing that they could be discharged after shorter service. As it turned out, aside from those who

---

[71] To maintain an enlisted strength (excluding officers) of 96,000 with twelve-year service, the army had to induct 4,000 men every six months. As seen in Chapter Four, army leaders had decided in July 1932 on a 15 percent increase in army strength in April 1933. On the actual number inducted, the wide bracket given here reflects differing and incomplete information. At December 1933 army briefings on new Umbau plans, General Ludwig Beck stated that 22,000 men, "the larger part of the 1 Apr 33 inductees," would be discharged on 30 Sep 34, and this, considering the need to replace old soldiers, implies a total 1 Apr 33 induction of a still larger number: WKK, WK VII/4082, Discussions in Berlin, 21-22 Dec 33, I. Teil, 75/1065. Hans-Jürgen Rautenberg, who quotes this whole document in "Drei Dokumente zur Planung eines 300 000 Mann-Friedensheeres aus dem Dezember 1933," *MGM*, 22 (1977), 119-129, suggests (p. 135, n. 142) interpreting the phrase as meaning that of the 22,000 men to be discharged, the larger part were 1 Apr 33 inductees. Elsewhere in the same briefings, it was stated that there should have been 4,500 officers on hand on 1 Oct 33 to maintain a 4 percent proportion of officers, implying a total strength after the October inductions of 112,500 (III. Teil, 75/1073). Later, a Blomberg directive stated that the men inducted on 1 April and 1 October 1933 amounted to about one fourth of previous army strength (i.e., about 24,000) and constituted nearly 35 percent of some divisions: WKK, WK VII/2091, Der Reichswehrminister 404/34, 16 Apr 34, 49/657. See also Liebmann MS Papers, Remarks by Fritsch, 27 Feb 34, p. 220 (typed); pp. 96-101 (holographic). Meinck, *Aufrüstung*, p. 89, gives a total for 1 Apr 33 inductees of 6,500, and for 1 Oct 33 of 14,000-24,000, both inclusive of replacements. He bases this on correspondence with Burkhart Mueller-Hillebrand, who, however, gives no figure in his own published study.

became noncommissioned officers, they were kept on for one and a half to two and a half years.[72] Presumably their intended function was to provide cadre backing for the later training and eventual mobilization of militiamen. Their own enlistments began with three months of militia-style training. A regimental history reports that for the first time, new recruits were no longer trained in special training battalions, but instead—as Hammerstein's November orders had foreseen—in the ordinary line battalions, a policy which crowded the barracks; the end of July found them in the field, at Münster Lager. The number of officer candidates increased, and new stress was placed on training noncommissioned officers.[73] We should note that the increase in new long-term recruits gives a very incomplete picture of Reichswehr training activity after April 1, 1933. On July 1, the three-month training of reserves within the army, i.e., the militia training devised by Manstein and others, actually began within the first battalion of the Nineteenth Infantry in Munich, and probably also in the third battalions of the Third Infantry (Osterode, East Prussia) and Ninth Infantry (Spandau). These may have been the only units actually to launch training of this kind, but there was to be a great deal of other training activity.[74]

6. Emergency Preparations and the Role of the SA

Not everything could continue as planned in 1932. There were two conditions, affecting the Reichswehr, which developed within a few

---

[72] WKK, WK VII/2091, Der Reichswehrminister 404/34, 16 Apr 34, 49/657, and BA-MA, II H 645/2, Chef HL 535/35 (draft), — Mar 35, note that the 1933 recruits signed the "supplementary obligation" on entering the service. Rudolf Absolon, *Die Wehrmacht im Dritten Reich*, 3 vols. (Boppard: Harald Boldt, 1969–), I, 152-153, says the April recruits were not asked to sign the supplementary obligation until fall, but very likely the regulation he refers to was a registration of measures already secretly taken. On actual length of service, also Liebmann MS Papers, Remarks by Fritsch, 27 Feb 34, p. 220 (typed); OKH, H 24/6, Reichswehrminister 30 m 17 AHA Vers. Id. 676/34, 14 Jul 34, 248/6227567-69.

[73] Gerhard Lubs, *I. R. 5: Aus der Geschichte eines pommerschen Regiments 1920-1945* (Bochum: Gerhard Lubs, [1965]), pp. 65-66. Lubs says (p. 66) that the cadre of the former training battalion of the Fifth Infantry Regiment was largely sent to train the Grenzschutz; this only seems to have applied, however, to regiments near the Polish and Czech borders. Elsewhere, the training battalions were to be slowly transformed into line battalions, or else dissolved (see BA-MA, RH 15/8, TA 355/33, 21 Nov 33).

[74] WKK, WK VII/4082, WKK VII/7. Div. 562, 6 Sep 33, 75/1173; WKKdo VII 518, 29 Aug 33, 75/1174-1175; also WK VII/3422, II/J.R. 20, order of 29 May 33, 64/700; WK VII/2670, Wehrkreiskommando VII Ib/Ia 1006, 11 Jul 33, 55/641; Wehrkreiskommando VII 1532/Ib, 11 Jul 33, 55/643; and note OKW 845, Reichswehrminister

months of Hitler's coming to power. One, which has already been touched on, was the foreign reaction against Hitler, which made it seem, at least from Berlin, that preventive war was a serious possibility. Such anxieties seem to have increased in April. On April 7, Neurath gave the cabinet a sobering report (that written by Bülow on March 13), suggesting that either Poland or France might take preventive action. On the next day, Hitler (in Blomberg's presence) told Ambassador François-Poncet that Germany was threatened in the East, and he asked why Germany should not, like France, have the right to organize her eastern frontier. Alarming reports in the third week of April led to a cabinet discussion of the situation on the eastern border on April 25, and a high-level meeting on the twenty-seventh reportedly considered the steps to be taken in case of a Polish incursion, as well as precautions to prevent the SA from giving the Poles a pretext for action. The sense of being vulnerable to a Polish attack doubtless led to Hitler's first assurances to the Poles on May 2.[75] The international situation, particularly the MacDonald proposal, made it impossible to march out of the conference and openly rearm. Yet the same situation made covert rearmament more urgent than ever.

The second changing condition affecting the Reichswehr was the evolving status of the SA, the Wehrverband affiliated with the ruling Nazi party. Formerly the various Wehrverbände had been competing, financially insolvent formations that the army might hope ultimately to bend to its own purposes. When Blomberg first became Reichswehr Minister, he had still expected that the Reich Institute for Youth Training would continue much as before, and indeed he thought that responsibility for this organization should properly belong to Franz Seldte, the Stahlhelm leader, who was the new Minister of Labor. When the retired General Edwin von Stülpnagel, the original head of the Reich Institute, died in early March, he was succeeded as acting head by Colonel Georg von Neufville, a Stahlhelm functionary, while the titular leadership of the Institute passed to Seldte on April 4. But Ernst Röhm seems to have early decided that the Institute belonged in his organizational empire, and Seldte was no bulwark against Nazism; in 1932 he had discussed with Hitler the possibility of assigning ten Nazi Reichstag seats to the Stahlhelm, apparently in exchange for

---

301/33, 31 Aug 33, 793/5522465-66. Battalion locations: Georg Tessin, *Deutsche Verbände und Truppen 1918-1939* (Osnabrück: Biblio Verlag, 1974), pp. 189, 191, 194.

[75] *ADAP*, C/I/1, Nos. 142, 177, 180, 183, 184 (incl. note); *DDF*, 1/III, No. 105; Jacobsen, *Nationalsozialistische Aussenpolitik*, pp. 403, 768; Brammer Collection, Z Sg 101/26, Information report, 27 Apr 33. Cf. Weinberg, *Foreign Policy of Hitler's Germany*, pp. 62-63. On the Neurath report, see note 43.

votes. By mid-April 1933 he was proposing the turnover of the Stahlhelm to Hitler as a "gift," and in a few months, it and the other Wehrverbände were merged into the SA—or, in the case of the Reichsbanner, outlawed.[76]

Röhm had his own personal and organizational imperatives. This former Bavarian staff captain, who had been severely wounded during the war, regarded himself as a front-line—as opposed to a staff or parade-ground—soldier. Handicapped by his social and educational background, his interrupted career, and now more decisively by his overt homosexuality, Röhm could only have a future outside the Reichswehr. For him, if not for Hitler, the SA was a potential army, and would be a better one than the Reichswehr. Certainly something had to be done with the SA. Now that the last multiparty, if terror-ridden, election campaign was over, these bully-boys had lost their political raison d'être. In March they attacked Jews, Socialists, and in some areas even Stahlhelm members, and they began to be a problem for the regime, at home and abroad. But unemployment was still severe, and the SA could not be disbanded, especially since it regarded itself as the spearhead of the successful Nazi movement. One way to occupy the SA was to intensify its military training. The absorption of the functions of the Youth Institute would also help the organization financially, as so far the SA had no place in the governmental budget.[77]

Reichswehr leaders wanted to expand the training of the SA for military ends. To be sure, even those like Blomberg who were most enthusiastic about Hitler and the "national revolution" had no intention of sharing responsibility for national defense with Röhm and the SA. The Italian example made them fear that an armed party militia might be formed, even though Hitler had taken pains, in his February 3 remarks, to assure the officers this would not happen. Indeed, Röhm's ambition to have an independent military force ran directly counter to the Manstein plan to prepare Wehrverbände members as reserves by training them for three months within the army, and at some point these rival claims were likely to collide. But this problem did not have

---

[76] Thilo Vogelsang, "Der Chef des Ausbildungswesens (Chef AW)," in *Gutachten des Instituts für Zeitgeschichte*, II (1966), 147-149; National Archives, Microfilm T-580, Berlin Document Center Records (hereafter, BDC), Friedrich-Wilhelm Krüger Papers, Ordner 9a (1. Teil), Memo by Krüger, 10 Oct 35, reel 264 (no frame numbers); Sauer, "Mobilmachung," pp. 886-887 (incl. 886n); Bredow Papers, Orientation note, 2 Jul 32. According to *DBFP*, 2/V, Appendix, iii, military circles attributed the Neufville and Seldte appointments to President von Hindenburg. On the Reichsbanner's end: Rohe, pp. 461-471.

[77] On Röhm and the SA: Bracher, *Deutsche Diktatur*, pp. 95, 103-104, 116-117, 144, 206-207; Sauer, "Mobilmachung," pp. 813, 835-837, 850-855, 880-886, 942; Müller, *Heer und Hitler*, pp. 88-91.

to be faced as yet. In the immediate situation, the army, by directing the weapons training within the SA, could try to ensure that all weapons remained under its own control. And training the SA and (while they lasted) its rivals would also help to satisfy, unobtrusively, the urgent need for trained reserves in the event of foreign attack.[78] The continuing Disarmament Conference prevented an overt expansion of military training, even if the cadre had been ready for a large training program within the army. But instruction of the SA outside the army could now be expanded, following the procedures already used for Grenzschutz training. To Blomberg and Reichenau, with their East Prussian background, such a solution would have seemed particularly appropriate.[79]

Generally speaking, the training of SA men was of two kinds, although these varieties of training apparently often overlapped. One kind was Wehrsport training, of the type that had recently been sponsored by the Institute for Youth Training; here instruction was supposed to be largely by ex-officers or by instructors from the paramilitary groups themselves. This was intended to prepare young men for military training within the army, and it placed more emphasis on physical conditioning than on instruction in weapons. The other kind of training was actual military instruction in weapons and battle exercises, conducted outside the army by Reichswehr members detailed for this purpose. This sort of training, not necessarily for young men, had hitherto been given mainly to Grenzschutz participants, who had been drawn largely from the Stahlhelm and the SA. Outside the border areas it had only been given indoors, in great secrecy, but the army had

[78] Blomberg disseminated Hitler's assurance: Liebmann MS Papers, Der Reichswehrminister 1549/33, 14 Mar 33, p. 48 (copy is among handwritten notes). In regard to the need for all possible reserves, General Liebmann stated in March that in an emergency the army would also need the elements currently being persecuted for their political beliefs by the organizations of the right: Commanders' discussion, 15-17 Mar 33, pp. 195-196 (typed). On the other hand, according to postwar testimony by Ott, cited by Sauer, "Mobilmachung," p. 729, Reichenau warned officers in February against letting the persecuted find refuge with the army. But in fairness, it should be added that OKW, OKW 872, [Ministeramt], "Survey of Internal Situation," 22 May 33, 874/5621537-38, initialed by Reichenau, suggests at least some army concern for Jews who were front-line veterans or who were respected parts of the Jewish elements long settled in Germany. In 1932, Jewish leaders had sought to get an official rejection of "wild anti-Semitism" through Bredow, who replied with reassurances, but also with pressure on the leaders to get the "Jewish" and foreign press to end its criticism of the Reichswehr; finally, noting Jewish support of Wehrsport, he urged Schleicher to make a statement (which was never made): Bredow Papers, Orientation notes, 22 Jun, 13 Jul, 9 Dec 32. On the army interest in using the SA for defense, see also Müller, *Heer und Hitler*, pp. 91-93.

[79] Aside from superior Grenzschutz preparations, East Prussia was doubtless also the area best prepared for field-army mobilization, as through the listing of men eligible to serve. See AA, Extract from record of ministerial meeting, 26 Feb 27, 4565/E164074; Remarks by Fritsch, [prior to 21 Nov 27], K6/K000307.

planned a major expansion of such training in 1933.[80] This was the program most immediately important for expanding the Reichswehr to meet an enemy attack, and it seems to have often resembled the quick emergency training that had earlier been planned as a part of the mobilization process.

A new directive, dated April 10, amended an old order of December 1, 1932, on Grenzschutz and external training, and effective May 1, extended full external training on the Grenzschutz pattern to all parts of the Reich, excepting the demilitarized zone. The document stated that where the training was not for the Grenzschutz, it would be "for the field army," indicating a training of reserves for the mobilization of twenty-one divisions. Thus, while the trainees would come mainly from "the Verbände," this was not by any means a plan to establish an independent SA army. Aside from training detachments already existing or planned, each battalion-level unit of the regular army was to form a training troop consisting generally of an officer and three to six picked men or NCOs. The officers conducting the courses were to have complete authority over the trainees, not shared with any Verband leaders. The training courses were to take two to three weeks and, as concealment from foreign intelligence remained important, the courses were to be held as far away as possible from military installations, in the country. In some cases, however, supposedly as an exception, the training might take place at military posts; also, participants in the courses were to be clothed by the regular forces or from available supplies. Nevertheless, the training remained "external" even when on a regimental parade ground, in that the trainees were not members of the army. In Bavaria, and presumably elsewhere, a card was to be made out for each trainee, a record of his performance was to be attached, and the cards were to be sent to the Landesschutz offices—obviously for use in an eventual field-army mobilization. Aside from the training detachments, drawn from line organizations, the former training-battalion cadres of regiments near the border provided three-month training to the Grenzschutz. Some training troops were sent to train university students during the 1933 summer semester; generally these students would be required to show they had undergone preliminary training at one of the camps operated or recognized by the Youth Training Institute.[81] This training of students would help meet the urgent need for more officers.

[80] See BA-MA, RH 15/49, TA 232/33, 10 Apr 33; WKK, WK VII/2670, Wehrkreiskommando VII Ib 524, 27 Apr 33, 55/535-536.
[81] BA-MA, RH 15/49, TA 232/33, 10 Apr 33; WKK, WK VII/2670, Wehrkreiskdo VII Ib 524, 27 Apr 33, 55/535-553; WK VII/3422, Kommandantur Münsingen I/63a, 12 Aug 33, 64/746; Lubs, p. 66. Three-month training in Grenzschutz: BA-MA, RH 15/8, T2 1113/33, 14 Dec 33. On the idea of an independent SA army, see Appendix.

This system was not so much a concession to the military ambitions of Ernst Röhm as a sort of undercover harking-back to the formation of Jaeger detachments and Landwehr in 1813. It was also in the Reichswehr's established pattern of extending "existing Grenzschutz" to the rest of the country. It has been estimated that in 1933 nearly 60 percent of the young line officers were engaged in external training.[82] On June 1, Blomberg remarked to his divisional commanders that, as they knew, the army was very extensively engaged in such training, and he told them that under no circumstances would he accept the objection that the limit of what could be done in this respect had been reached, that no greater effort was possible. According to General Liebmann, Blomberg added: "In the next years Wehrmacht will have to give itself over completely to the task of creating the reserves which hitherto we were unable to create." He was not yet, however, able to say what system of recruitment and organization would be adopted—which sounds as though he may have been having some doubts about the eventual introduction of the militia system.[83]

The SA increased its sphere of activity after the beginning of April, and the new emphasis that army policy was giving to external training doubtless encouraged this development. On April 28, a law was issued providing the SA and SS with legal internal punishment powers, thus overcoming what Hitler had regarded as a major obstacle to their military usefulness. On the same day, an SA leader was named Reich Sport Commissioner. May 17 witnessed an agreement between Blomberg and the SA leaders that the SA might retain rifles and pistols (but not heavy weapons) as far as needed for their training and their "internal political duties," although in theory all military weapons still belonged to the Wehrmacht.[84] Meanwhile, Seldte and the Stahlhelm were losing ground. On June 21, Seldte agreed that the Young Stahlhelm (Jungstahlhelm), i.e., members of military age, would become part of the SA, and the male teen-age Stahlhelm youth group, the Scharnhorstbund, would be absorbed by the Hitler Youth (Hitler

---

[82] 1813: Dennis E. Showalter, "The Prussian *Landwehr* and Its Critics, 1813-1819," *CEH*, 4 (1971), 8-12; Shanahan, pp. 199-213. On "existing Grenzschutz," see Chapter Six, Section 1. The estimate on young line officers is by Rautenberg, in *Rüstungspolitik*, p. 225.

[83] Liebmann MS Papers, Remarks of Blomberg, 1 Jun 33, p. 200 (typed). On the priority given by the army to external training, see also BDC, Box 120, Ordner 309, Führertagung speech by Obersturmbannführer Blum, [ca. 16 Dec 38], reel 864. In late April, members of the American Consulate General staff driving within a fifty-mile radius of Berlin saw men engaged in drilling, target practice, and military exercises in practically every village or town (*FRUS, 1933*, I, 121). French diplomats had similar experiences (*DDF*, 1/III, No. 462).

[84] Sauer, "Mobilmachung," p. 886; Liebmann MS Papers, Remarks of Blomberg, 1 Jun 33, pp. 199-200 (typed).

Jugend). The leader of the latter, Baldur von Schirach, who was under Röhm's orders, had become Reich Youth Leader on June 17. Seldte retained control for the moment of the older Stahlhelm members, but Röhm now had paramilitary youth training under his banner, and he formed a new organization for *Ausbildungswesen* (literally: "training affairs") to replace the once nonpartisan and public Reich Institute for Youth Training; the Nazis could be relied on to devise a less cumbersome title for this activity. From July 1 to 3, a meeting of SA and SS leaders, attended by Hitler and also by Seldte and Reichenau, was held at Bad Reichenhall. At this meeting, Friedrich-Wilhelm Krüger, an SA leader, was named to be Chief of the Ausbildungswesen of the SA, and he received from Hitler the assignment of training 250,000 SA men and leaders within a year, so that they would be available to the Reichswehr "in an emergency." Krüger was also given authority over the Wehrsport training of students, and generally the SA assumed responsibility for premilitary and reserve activities outside the armed forces. In future, every man who joined the Reichswehr was to come from the SA; every man who left it was to become a member of the SA. Later in the year, army personnel were detailed to Ausbildungswesen "SA-Sport" camps, and the training there became more directly military than it had been in Reich Institute days.[85]

These developments may appear as part of the general *Gleichschaltung* ("alignment" with the party) process, which was everywhere eliminating potential opposition to the regime, but the army was very much interested in these matters from the point of view of the training and reserve problem. All that we know of Hitler's own ideas at this time suggests that he would have listened with respect to the opinions of the military on the subject of training. And there is evidence suggesting that the Bad Reichenhall decisions resulted from the pursuit of long-standing army plans.

A Reich Defense Council (Reichsverteidigungsrat), an objective of the Reichswehr since the days of General Heye, was established on April 4, and although the council, made up of ministers, accomplished

---

[85] Sauer, "Mobilmachung," pp. 886-888; Vogelsang, "Chef des Ausbildungswesens," pp. 149-150; *Nuremberg Documents*, XXIX, 1-6. According to a 1935 statement by Krüger (BDC, Krüger Papers, Ordner 9a [1. Teil], Memo by Krüger, 10 Oct 35, reel 264), the assignment was to give the 250,000 men training with weapons. Blomberg's official directive for army support of the program confirms this (and also that future recruits would all be drawn from the SA), but on 6 October, as will be discussed later, Hitler forbade weapons training in the SA-Sport camps, a change which doubtless disappointed Krüger and also Röhm: see WKK, WK VII/1453, Der Reichswehrminister TA 533/33, 27 Jul 33 (copy), 37/1118-1122; Wehrkreiskommando (Bayer.) VII 7. (Bayerische) Division Ib/Ia 1679, 5 Oct 33, 37/1113-1117. On assignment of army personnel to AW camps, see Lubs, p. 66; Meier-Welcker, "Briefwechsel," 88n, and the WKK documents just cited.

Hitler's Accommodation with the Military—353

little itself, the Reichswehr attempted with some success to use a working committee of this council to direct other ministries in preparations for national mobilization. At the third meeting of the working committee, on June 21, Colonel Wilhelm Keitel disclosed that in view of the newly created domestic political situation, the Reichswehr was promulgating a new edition of the "Guidelines for the Preparation of the Defense of the Reich." These guidelines were in fact only the most recent in that succession of arrangements on Landesschutz questions which went back to the Gessler-Severing agreement of 1923; we have seen Schleicher issuing new guidelines in October 1932. At the next working-committee meeting on July 12, Keitel again spoke of the latest guidelines, which had meanwhile been issued officially on July 1, and he said: "Foundation remains the same, they are only adapted to the domestic political conditions of the present time." The new guidelines now stated that it was the "national duty of all authorities *and Wehrverbände* [emphasis added] to take part in [defense] preparations." They referred to the Reich Defense Council, and added provisions for securing the western and southern borders and for training specialists for Göring's Air Ministry. Also, catching up with the directive of April 10, they authorized external training in all parts of the country (except the demilitarized zone) for rank and file as well as for officers, NCOs, and specialists. Thus the Reichswehr had adjusted its plans for conducting Landesschutz activities to exploit the opportunities afforded by a one-party state and to assign a role to the party's own Wehrverband, although—in the old above-party, official tradition—that organization was not specifically named.[86]

The previous history of the guidelines, involving controversy with Land authorities, makes it evident that, at least as far as Landesschutz was concerned, the army leadership must have welcomed the new do-

[86] Sauer, "Mobilmachung," pp. 793-794; Castellan, *Réarmement clandestin*, pp. 395, 400; OKW, Wi/IF 5.701, Reichswehrminister TA 582/33, 5 Jul 33 (Working Committee meeting of 21 Jun 33), 177/912288; Reichswehrminister TA 616/33, 18 Jul 33 (meeting of 12 Jul 33), 177/912270-71; Wi/IF 5.335, Chef HL, TA 851/33, 7 Oct 33, att. (Reichsminister der Finanz, G II RV, 2 Sep 33), 73/794778-79. Where the 14 Oct 32 changes required the sending out of a whole new text covering ten pages (73/794782-91), the 1 Jul 33 changes only involved two pages of amendments. Castellan comments (*Réarmement clandestin*, p. 395) that the detailed provisions of the guidelines "contained the seed of the whole war effort of the Reich"; the seed, however, had been sown as far back as the Gessler-Severing agreement, if not before that—as Castellan himself indicates (p. 397). Contrary to Sauer, "Mobilmachung," p. 797n, the "Hitler order" described on Castellan's p. 395 is the same as the guidelines mentioned on Castellan's p. 400, and Hitler's reported demand for 250,000 men to be trained for the Reichswehr was not contained in the formal guidelines. The guidelines do not show that the supposed Reichenau plan was approved, but rather that old defense-mobilization plans were adjusted to fit the new institutions of the Reich while retaining their essential character under Reichswehr conrol.

mestic situation. Certainly, army leaders were happy to take advantage of the unification of the Verbände to further their own plans for training. At the beginning of June, Blomberg told the divisional commanders that it would be a good thing if the Nazi movement would soon achieve its goal of totality by an elimination of the DNVP and the Center party; such an attitude implied that the Stahlhelm should also be eliminated. By June 19, the SA and Reichenau had agreed that the Reich Institute should be dissolved, and Reichenau was saying that the Reichswehr Ministry thought it important "henceforth to be able to work with the SA alone as *the* defense association, out of concern for a unified training." After the Bad Reichenhall meetings, army officers showed eagerness to get the Krüger programs launched, and in particular they urged the avoidance of further delay in winding up the affairs of the Reich Institute. In the light of this background, one is led to conclude that the Bad Reichenhall meeting amounted basically to a briefing of the party on, and its rallying to, the program of the army. This coordination under army leadership would have followed naturally from an edict signed by Hitler in May, stating that all the paramilitary organizations had to follow Blomberg's orders on Landesschutz matters.[87]

IN MOST areas of military concern, matters continued through 1933 in accordance with the old plans. The Billion Program was finally launched more or less on schedule, and the old problem of the guidelines for national mobilization of men and materiel had now been resolved in a way satisfactory to the military leadership. From the army's viewpoint, the expanded external training program introduced in April seems to have been simply a temporary adjustment in longstanding plans, a makeshift inspired by fear of attack, by the need to continue negotiating in Geneva, and by the availability of the SA. It was an emergency means of dealing with that problem of a lack of reserves which had so long preoccupied the military thinkers in the Bendlerstrasse. Blomberg, possibly influenced by Hitler, seems to have regarded the future organization of the army as something still open, and dependent on developments, both at home and abroad.[88] Through most of the year, however, the army still planned to introduce large-scale three-month militia training within its own ranks in 1934, a move

[87] OKW, OKW 872, [Ministeramt], "Survey of Internal Situation," 22 May 33, 874/5621537; Liebmann MS Papers, Remarks by Blomberg, 1 Jun 33, p. 199 (typed); Vogelsang, "Chef des Ausbildungswesens," p. 149; *Nuremberg Documents*, XXIX, 1-4.

[88] Aside from Liebmann MS Papers, Remarks of Blomberg, 1 Jun 33, p. 200 (typed), also *ADAP*, C/I/1, Nos. 26, 231.

that might provide a more controlled solution to the reserve problem, although it would be hard to reconcile with Röhm's ambitions.

Germany had become a dictatorship, however, and in such a state the dictator may decree sudden changes—not least of all, to keep his officials on the defensive. Hitler accepted for the time being the external training program, which accorded with his beliefs that Germany should avoid forcing the issue of equality of rights abroad, and should place priority on the mustering of forces at home. The program presumably also improved at least somewhat Germany's immediate capacity for meeting a preventive attack, which he apparently regarded as an imminent danger. But the veteran of Langemarck can hardly have regarded two to three weeks of external training, or for that matter, three months of training within the army, as adequate preparation for combat—and particularly not for the offensive combat he had in mind for the future. In 1914, he himself had had two and a half months in the army before boarding the troop train for Flanders, and he had stated in *Mein Kampf* that four to six months of training was only enough if those who had undergone it served side by side with old soldiers.[89] He doubtless continued to prefer a system of universal service, such as he had called for in his February 3 remarks to the generals, and there is no sign that he ever accepted Röhm's idea of a militia army. As for the scheme of three months' training in the army, it had lost much of its point, now that the French accepted national conscript service for Germany, and now—what was more important from Hitler's perspective—that the British (in the MacDonald plan) were adopting that proposal. For the time being, Hitler did not interfere, but we shall later see him change the German Umbau program, while bidding for an understanding with the British. Even the execution of his new plan may have been another way of trying to bring about the hoped-for British alliance, by showing that Germany was a worthwhile ally (bündnisfähig). In any case, it would also be a way of better preparing the Reichswehr for offensive tasks on the continent.

[89] Bullock, pp. 50-51; *Mein Kampf*, p. 605.

CHAPTER EIGHT

Britain Reconsiders Its Policy

1. NEW PROPOSALS AND A NEW FACE

The foreign policy of a nation might be compared to a loaded supertanker: once the ship of state is in motion in a given direction, the men on the bridge cannot suddenly bring it to a stop, or quickly turn it in a radically different direction, nor would they likely want to do these things. Changing course is particularly difficult in a democracy like Great Britain, as it requires a change in public opinion. Thus we should not be surprised if British policy showed no immediate reaction to Hitler's accession. It is more remarkable that there should have been, as we shall see, a change within a few months.

As 1933 began, British eyes focused less on Berlin than on the objective of a compromise agreement on armament. The Germans had been brought back to the Disarmament Conference, but when that gathering reconvened, the British would have to deal with the French proposals. In January, the British cabinet expressed its aversion to continental ties by rejecting any commitment to the French disarmament plan.[1] Whitehall officials recognized that plan to be a revival of the earlier Geneva Protocol, and an effort, if indirect, to gain new guarantees from Britain. Sir Maurice Hankey and some others regarded the French plan as dangerous. Under it, the League Council, to which Britain belonged, was to decide by simple majority on the use of international police contingents, including naval forces, against an aggressor, and Hankey thought that this feature and the existence of a League general staff would involve Britain in quarrels in which she had no interest; indeed, the council might even conceivably vote for sanctions against Britain herself. He argued that the British government should openly oppose the conclusion of rigid arrangements between continental powers for sanctions, even though Britain had not been asked to be a party to those arrangements. In Hankey's eyes, the main goal of disarmament seems to have been the reduction of French air power. Under the influence of this idea, Stanley Baldwin would say in June 1933 (in the ministerial committee on disarmament) that parity with France in the air was one of the two big questions in any disarmament agreement, the other being the prevention of German rearmament.[2]

[1] See Cab 23/75, meeting of 19 Jan 33, 1(33)3; meeting of 23 Jan 33, 2(33)1.
[2] FO 371, ltr., Vansittart to Simon, 24 Nov 32, W12577/1466/98; ltr., Sir M. Hankey to Vansittart, 28 Nov 32, W13336/1466/98; ltr., H. J. Creedy (War Of-

In the Foreign Office, however, some officials had more sympathy for French anxieties over security. They recognized that the French would not cooperate unless they obtained at least a hearing for their plan, and Vansittart endorsed the view of Allen Leeper and C. Howard Smith that Britain could hardly tell the continental governments that they could not reach an agreement for sanctions among themselves; she would thereby assume the onus of officiously blocking an agreement that was most unlikely to be reached in any case. In December 1932, Leeper and Alexander Cadogan concluded that if the British government wished to push on from the December 11 agreement, promote disarmament, and seize the initiative, it should draw up a comprehensive scheme that would combine French, American, and British ideas on arms reduction or limitation, and allow the continental states to reach an agreement among themselves while providing safeguards against British entanglement.[3]

Even these ideas were by no means in line with French plans for the "organization of peace," and Cadogan believed that it would be necessary to put pressure on France to disarm. He thought that nothing would be accomplished at the conference unless a small group of powers took the lead and reached decisions, as the five powers had done in December at Geneva. With this in mind, he drafted a "programme of work," which combined various proposals already made and gave the task of coordinating and refining them mainly to the fourteen-member Bureau, the steering committee of the conference. This smaller body would presumably function more efficiently than the General and Political Commissions, in which every delegation was represented, or the technical committees, where experts defended parochial interests. Cadogan further urged that, since Simon would presumably be unable to remain permanently in Geneva, some eminent figure should be sent there as Permanent Delegate; this personage would be able to conduct informal but high-level discussions at any time. Simon was indeed unwilling to spend more time in Geneva, and in fact wanted to unload some of that burden, but he was also reluctant to dispose of any of his authority. He supported Cadogan's "pro-

---

fice) to Vansittart, 2 Dec 32, and minute by A. Leeper, 12 Dec 32, W13356/1466/98; ltrs., Sir M. Hankey to Vansittart, 11 Jan 33, and to A. Leeper, 10 Jan 33, memo by Hankey, 9 Jan 33, minute by Leeper, 11 Jan 33, and ltr., Vansittart to Hankey, 18 Jan 33, W583/40/98; Cab 27/505, DC (M) 32, meeting of 19 Jan 33. Roskill, *Hankey*, III, 71, makes Vansittart and the Foreign Office appear blind at this time to the German danger, and plays down Hankey's primary concern with the French plan; cf., however, pp. 77-78 of the same volume.

[3] FO 371, Minutes by A. Leeper, 12 and 22 Dec 32, W13356/1466/98; Memo and minute by Leeper, 22 and 31 Dec 32, minute by Vansittart, 2 Jan 33, and ltr. by Cadogan, 29 Dec 32, W14230/1466/98; Minutes by Leeper, C. Howard Smith, and Vansittart, 11 Jan 33, and ltr., Vansittart to Sir M. Hankey, 18 Jan 33, W583/40/98; Cab 24/237, CP 3 (33), 12 Jan 33; CP 9 (33), 20 Jan 33.

gramme of work," but instead of some possibly obstreperous elder statesman, he sent his Parliamentary Under-Secretary, the relatively junior Anthony Eden, to be acting head of the British delegation. It seems safe to say that Simon did not realize where this appointment would lead.[4]

At this point thirty-five years old, Eden had served in the 1914-1918 war, gaining the rank of captain and winning the Military Cross. As a veteran, a graduate of Eton and (after the war) Oxford, and a handsome and articulate member of a prominent family in County Durham, he had had no difficulty becoming a Conservative candidate for Parliament, and after one defeat, he had gained a seat in 1923. He soon took part in debates in foreign affairs, and leaders of his party, including Baldwin and Sir Austen Chamberlain, seem to have early recognized in him a possible future Foreign Secretary. Eden's maiden speech had supported the 1923 program for parity in air power with France, and he continued to speak up for air defense, but as time went on he would develop an interest in disarmament and the League of Nations, and his once-bristly moustache would become more discreet and diplomatic.[5]

Young, energetic, and ambitious, Eden would soon acquire an influence on British foreign policy out of proportion to his experience and official status. This process was assisted by the characteristics of his seniors: Simon was intelligent and an able advocate, but suffered poor health through much of 1933 and seemed to have no firm foreign-policy objectives of his own; MacDonald was now aging and unsure of his position, anxious to achieve some success to wind up his career; Baldwin did not follow foreign affairs closely and was apparently very open to proposals from the young Eden, who frequently wrote him letters. Baldwin's strong sense of the emotions and ideals of the English people probably led him to see Eden as a Conservative who could satisfy and embody the aspirations of the public.

Certainly Eden genuinely loathed the idea of another war; in addition to his own experience in the front line, he had lost two brothers in action.[6] In domestic politics, he warmly supported the centrist policies of Baldwin and MacDonald, aimed at overcoming (some would say papering over) class divisions. Eden was no insular Englishman; he was a connoisseur of modern French painting, and fluent in French,

---

[4] FO 371, Memo by Cadogan, 29 Dec 32, W14239/1466/98; ltr., Cadogan to Leeper, 29 Dec 32, W14230/1466/98; Eden, p. 28.

[5] Eden, pp. 3-22. Austen Chamberlain, in a letter he wrote to his wife in 1933, described Eden as his protégé (Austen Chamberlain Papers, ltr. to Ivy Chamberlain, 4 May 33, AC6/1/970).

[6] Eden, pp. 4, 69.

though less skilled in German.⁷ This background, and a strong repugnance for loud talk and threats, may have influenced him in favor of France as opposed to the Germany of Hitler and Nadolny. Although he would be ready later to reach an arms agreement with Hitler and, unavoidably, to accept the German reoccupation of the Rhineland, his energy and idealism made him stand out against the dismal background of the 1930s.

In his first speech in a plenary conference session, Eden gave more comfort to Germany than to France, but this was really a statement of cabinet views. The conference recommenced on February 2 by debating the security side of the French plan. Various delegates stated their reservations, and Eden declared clearly (on February 3) that his government could undertake no new commitments, agreeing with Nadolny that there were already enough guarantees of security to permit a reduction in armaments. After the general airing of views on the French plan, Eden also had the task of expounding the "programme of work," which had already been circulated to the conference. But Paul-Boncour, who was now Foreign Minister as well as chief French delegate, privately protested to Eden that, under the "programme," the discussion of disarmament in the Bureau was likely to get ahead of the discussion of security in the General Commission. Eden did not strongly defend the "programme" procedure with Boncour, but instead indicated that adjustment would be possible. When the Bureau debated the matter, Boncour proposed that the General Commission should assume the responsibilities assigned by the "programme" to the Bureau, perhaps delegating those responsibilities to subcommittees. Eden agreed, and Boncour's suggestion, which was supported by the smaller states, was adopted.⁸ This defeated Cadogan's intention of using a smaller body to expedite decisions and hasten French disarmament.

Boncour also had his way, with Eden's support, in an important procedural controversy with the Germans, involving the diplomatic strategies of the French and German governments. In 1933, France, Germany, and Britain, too, each wanted to avoid being blamed in the altogether likely event that the conference failed. Neurath now stressed to Nadolny that every effort should be exerted to make France admit

---

⁷ Eden's memoirs and the British published documents indicate that he conversed fluently and frequently with French spokesmen. The memoirs state (p. 64) that his German was limited.

⁸ Cab 23/75, meetings of 19 and 23 Jan 33, 1(33)3, 2(33)1; *SIA, 1933*, pp. 232-233, 236-237; *DBFP*, 2/IV, No. 280; *DDF*, 1/II, No. 284. Eden seems to have understood that Boncour wanted the General Commission to turn over different aspects of disarmament to various subcommittees.

she would rather see the conference fail than agree to disarm. Accordingly, Nadolny sought to have the conference discuss first the question of (French) material disarmament, a matter on which the British and Americans would presumably side with him. This would compel the French either to make the first concessions, or else admit their unwillingness to do so. Boncour, on the other hand, wanted the conference to take up first the French proposals for standardizing (German) effectives on a system of short-term service. He had the thwarting of the supposed Seecktian plans much at heart, and he no doubt expected that a debate on effectives would provide occasions for pointing out dangerous aspects of the Reichswehr and of German police and paramilitary formations, thereby winning international backing to his side and isolating the Germans. In this case, it would be the Germans who would have only the alternative of uncompensated concessions. In accordance with French wishes, and over German and Italian objections, the conference agreed that it would first discuss security proposals and questions arising from the French plan, and leave the question of reducing materiel for later.[9]

Aside from the German preference for attacking the French on materiel, the French proposals for standardization were in themselves bound to evoke serious German opposition. Although the Reichswehr did not in fact plan a Seecktian army, officers like General Schönheinz feared that standardization might prevent Germany from having a superior cadre. Blomberg was not quite so rigid, but he thought that since the French placed so much stress on standardization, it should not be conceded until something was gained in return. Nadolny seems also to have shared Blomberg's views, and although he privately thought that the French proposals offered possibilities for Germany, he could strive fiercely for a hard bargain.[10]

2. THE BACKGROUND OF THE MACDONALD PLAN

Eden tried at first to be optimistic about the prospects in Geneva. When he made a quick trip to London to report in mid-February, he told the ministerial committee on disarmament that the critical

---

[9] *ADAP*, C/I/1, No. 20; *DDF*, 1/II, Nos. 188, 316; *SIA*, *1933*, pp. 237-238. Boncour's statements to Eden from February 6 on (*DBFP*, 2/IV, Nos. 280, 281, 283) suggest that he was preparing the way for a retreat to a minimum program. As of 8 March, a member of the American delegation had gathered that Massigli thought the Germans were likely to leave the conference, and that there was no point in trying to keep them (SDP, Ferdinand Mayer Geneva Diary, 8 Mar 33, 500.A15A4/1469½.

[10] On Schönheinz, see Chapter Seven, notes 34 and 35. On Blomberg's position: *ADAP*, C/I/1, No. 26. On Nadolny's: *ADAP*, C/I/1, No. 23.

stage would be reached during the next few weeks, with a Franco-German confrontation likely, but he also expressed the view that the French would accept significant disarmament if they could get approval for their continental security proposals, and he thought that in any case they would want some convention binding Germany. In actuality, Boncour seems now to have had little expectation that his continental security proposals would be accepted and put into effect; perhaps, now, he had no real wish to see a maximum plan applied. A week after Hitler's accession, Boncour argued to Eden that the situation in Germany had changed since December, and he referred to a report that Germany intended to enroll some 600,000 youths in her armed forces by April. He suggested that a protest was in order, and he asked how they could continue to discuss disarmament under such conditions. Boncour's thoughts turned more and more to proceeding, after an expected rejection of his proposals, to negotiations on a minimum program—i.e., not a program binding Germany, but one that was unacceptable to her and yet endorsed, the French would hope, by other governments.[11]

Eden, in spite of his initial hopes, soon came to share in the gloom which was spreading in Geneva; the conference seemed to be reaching a stalemate, and reports of violence and repression were pouring in from Germany. By February 22, he was beginning to believe that drastic action would be needed to prevent a breakdown of the conference. One reason for this conclusion was probably Nadolny's combative demand for debating material disarmament ahead of effectives, and the impression he gave of unbending opposition to standardization. Nadolny's attitude made it seem likely that Germany might withdraw if she continued to be outvoted in conference meetings. Eden soon had another reason for concern. Boncour told him plainly on February 24 that German opposition to the French plan might very soon compel France to fall back on her minimum proposals; this would end chances for material disarmament.[12]

Eden and Cadogan concluded that the only way to forestall a failure of the conference would be to present the assemblage with a completed draft convention, not only combining ideas suggested in various

[11] Cab 27/505, DC (M) 32, meeting of 17 Feb 33; *DBFP*, 2/IV, Nos. 280, 281, 283; *DDF*, 1/II, No. 358.

[12] *DBFP*, 2/IV, No. 283; *DDF*, 1/II, Nos. 292, 333, 356; *SIA, 1933*, pp. 238-240; Eden, pp. 30-32. As shown in the last chapter, the Germans did not utterly refuse to agree to standardization, as Boncour (believing in Seeckt's influence) seems to have concluded (*DBFP*, 2/IV, No. 299). An extract from a letter (quoted by Cowling in *Impact of Hitler*, p. 63) indicates that Eden warned Baldwin that if the conference reached its last gasp, Henderson would present a convention himself which would have awkward repercussions in domestic politics and be unacceptable to the British government.

proposals, as the December suggestions and the "programme of work" had done, but also inserting actual figures for effectives and weapon sizes. They thought this approach would overcome the obstacle that no nation would permit a limitation of its own armament until it knew what it would get in terms of constraints on the others; in the immediate situation, this meant simultaneously solving the questions of effectives and materiel. This method also offered a way around the delays introduced by committees of experts and the objections raised by minor powers. A revival of high-level five-power negotiation might have been another solution to these problems, but the French, supported by many smaller nations, were making plain their adamant opposition to any further five-power discussions. If the proposal failed, at least the British government could claim before the Anglo-American public to have made a fair proposal and a major effort. Together with diplomatic, legal, and military experts, Eden and Cadogan worked out a draft plan on the weekend of February 25 and 26.[13]

As finally published, the Eden-Cadogan draft plan, to be known as the MacDonald plan, contained the following salient points:

1. *Security*. In the event of a threatened breach of peace, the signatories of the convention would confer to determine who was responsible and what steps should be taken. Decisions would only be valid if concurred in by the American, British, French, German, Italian, Japanese, and Soviet governments, and by a majority of the other governments represented, excluding the parties to the dispute.
2. *Land Effectives*. Aside from officers and professional cadres, the land forces expected to serve in continental Europe would generally serve no longer than eight months. For service in continental Europe, Germany, France, Poland, and Italy would each be permitted a strength of 200,000, while the Soviet Union would be permitted 500,000 and Czechoslovakia 100,000. In addition, France would be allowed a strength of 200,000 (and Italy a strength of 50,000) for overseas service; men in these forces might serve longer.
3. *Land Armaments*. Mobile artillery over 155 mm would be abolished, and mobile guns from 105 mm to 155 mm could be retained but not replaced; the maximum weight for tanks would be 16 tons. Implicitly, Germany would be allowed tanks and large nonmobile guns.
4. *Naval Armaments*. Generally, provisions in this chapter were to run until December 31, 1936, in accordance with the London Naval Treaty of 1930, and the signatories of that treaty would remain

[13] Eden, pp. 30-32, 34-35; *DBFP*, 2/IV, No. 281, Appendix IV (MacDonald speech, 16 Mar 33).

subject to its limitations. As the French and Italian governments had not signed the London Treaty, specific limitations were proposed for them in cruisers, destroyers, and submarines, with Italy being permitted one additional battleship. By indirect language, the ban on German possession of submarines, created by the Treaty of Versailles, was maintained.

5. *Air Armaments.* Each of the major powers would be permitted 500 airplanes, excepting that the Versailles ban on German possession of military aircraft was maintained.
6. *Inspection.* Inspection or supervision would not be automatic, but would occur if a violation of the convention was suspected.
7. *Duration.* The convention would replace Part V of the Treaty of Versailles, and run for five years (except the naval provisions). A signatory could unilaterally suspend most of the restrictions if it went to war or if changed conditions menaced its national security.

Thus the plan contained only a vestige of the French proposals for mutual assistance and supervision, although regional agreements could also be reached. The plan proposed to limit French land and naval armaments, reduce the French army, and bring French air strength down to the British level. Germany would have gained a doubling of her authorized strength under arms and a reduction in the French preponderance in weapons. On the other hand, German effectives and arms would be restricted, Germany would have nothing comparable to the French overseas force close at hand in North Africa, and she was not to have qualitative equality in 155 mm mobile guns, in submarines, or in the air. Eden and Cadogan seem almost to have forgotten the German demand for all arms allowed others. And the provisions on effectives would, like the French plan, have interfered with German military planning for an integrated army of militia, three-year men, and professionals.[14] Yet, Eden and Cadogan were trying to strike a balance between the German and French positions, very much in the pattern of previous British policy. There was something of both pride and naiveté in this attempt to lay down, much as Hoover had, a master plan for other nations to marvel at and accept.

Eden and Cadogan took their draft to London, where Simon circulated it to the cabinet, along with a memorandum (largely drafted

[14] Text: League of Nations, Conference for the Reduction and Limitation of Armaments, *Conference Documents*, 3 vols. (Geneva: League of Nations publications, 1932-1936), II, 476-493. From the point of view of other governments, the provisions on air strength were particularly favorable to Britain, as the air forces of the dominions and of India were not included in the British total. For the German view of the plan, see *ADAP*, C/I/1, No. 103, and esp. ML, Shelf 5939, Note by Schönheinz, 18 Mar 33, reel 18/80-85.

by Vansittart), which stated that, once again, the conference was in crisis, and that if no convention were signed, Germany would rearm; since she could not afford expensive weapons like battleships, she would concentrate on submarines and aircraft, the arms most dangerous to England. The memorandum also used the usual argument that the British public would blame the government for any breakdown, and that an intense arms race would follow, with serious effects on Anglo-American relations and on "that economic and financial recovery which is essential not only to peace but to the very existence of civilisation." The paper warned that "if we cannot be friends with both France and Germany, we shall end by having to be friends with France alone, keeping in with the U.S.A. as far as we can." Simon gave Eden's draft a somewhat left-handed endorsement, saying that if a plan were offered and rejected, at least it would "stand as irrefutable proof" that the government had offered a means of rescue; he did not know if the proposed plan would save the conference, but he did not think it could be saved in any other way.[15]

The Eden-Cadogan plan made less of an impression on the cabinet at first than the warnings that the conference was in extremis and that German withdrawal and rearmament were imminent possibilities. In view of the reports coming from Berlin, such as on the Reichstag fire, the worst dangers seemed credible. Rather than relying on the disarmament plan, as Eden suggests in his memoirs, the thinking of the ministers, and especially of the Prime Minister, ran toward reviving the five-power discussions of the previous December. MacDonald privately had little faith in the conference, and at a meeting of the ministerial disarmament committee on March 2, he stated that if the situation was so serious as Eden said, then no mere production of a plan could save it:

> Something altogether heavier would be required. Some means ought to be found of getting the big nations at Geneva to realise that the matter was of cardinal importance, and that they must all rally round in a final effort to save the Conference.

After the leaders of the powers convened, he might produce a plan, but mainly so as to give the impression of having a concrete proposal. MacDonald indicated that if a big effort were not made and the conference not saved, "he might have to go to the Cabinet and warn them that they would have to be prepared for war in two years' time." This

---

[15] Eden, p. 32; Cab 24/239, CP 52 (33), 28 Feb 33. The memorandum was prepared "in consultation" with Vansittart. The concluding comment on the Eden draft was more likely Simon's.

was an appalling idea. The ministerial committee considered the situation so critical that, in violation of British political custom, it decided on March 3, without consulting the full cabinet, to announce that MacDonald and Simon were leaving as soon as possible for Geneva.[16]

It was less unusual that the British leaders should make such an announcement before taking soundings with other governments. Fortunately, although MacDonald intended to bring about five-power meetings, and the French, German, and Italian governments were later privately informed of this, a certain caution was observed, and the public announcement did not explicitly propose the attendance of leaders of other nations. Several ministers felt that the Germans might be unwilling to come, and Eden mentioned that it might be better not to speak specifically of five-power meetings, as the French had said they would not agree to such sessions. The doubts proved amply justified in the cases of both Germany and France. The Germans suspected that the British were considering an interim convention, which would fail to meet their demands and undermine their argument for equality. Also, it was not Hitler's style to attend multilateral meetings in Switzerland, even if he could have spared the time from his current task of consolidating his power. The French were determined not to become entangled in another five-power conclave, where they would be bereft of their small allies and easily isolated. René Massigli of the French delegation had had a premonition that the British were contemplating a proposal for five-power meetings, and after he had informed Boncour, the latter responded that such a plan should be avoided "at all costs." As soon as Simon told Ambassador de Fleuriau that he and MacDonald were going to Geneva, and before the Foreign Secretary could explain the British desire that other leaders come as well, Fleuriau gave him a message from Boncour expressing the hope that no five-power gathering was being proposed. The German response was slower to come, and for a few days Hitler may have considered sending Neurath, but while Boncour at least expressed interest in a private discussion with Simon, Neurath finally declined on March 7 even to come to the conference site at this juncture. As for Italy, by long-established custom, Mussolini, who was now Foreign Minister as well as Premier, never went outside the borders

---

[16] Cab 27/505, DC (M) 32, meetings of 2 and 3 Mar 33; Marquand, p. 752; MDP, 8/1, Pocket diary, 27 Dec [32]; Diary, 2, 5 Mar [33]. MacDonald informed Norman Davis: "I am going to Geneva to see if the Conference can be saved from immediate failure. The point of danger is that unless some agreement can be come to at once, Germany may draw out and start rearming immediately" (SDP, tel., Mellon to Dept., 9 Mar 33, 500.A15A4/1745). Aside from press coverage, reports like *DBFP*, 2/IV, Nos. 243-246 conveyed the first full impact of Nazi rule.

of his own country. An Italian alternative proposal would emerge a little later.¹⁷

Thus MacDonald and Simon had to set out for Geneva without assurance of meeting anyone of comparable rank to talk with. They had thought that at least the French Premier, Edouard Daladier, would come to Geneva a little later, but when they met with the French leaders while passing through Paris, it appeared that Daladier would probably not come. At this Paris encounter, Boncour said that it would be necessary to settle for a much more modest agreement than France had proposed, while Daladier claimed repeatedly that Germany, what with the SA and some police elements, now had more men under arms than France. The French ministers wanted to announce that the two governments would prevent rearmament as well as promote disarmament, and they showed in every way their readiness for that division into two camps which MacDonald wished to avoid. The French opinion was particularly aroused at this juncture, because an SA detachment had provocatively occupied some old army barracks in Kehl, in the demilitarized zone just across the Rhine from Strasbourg. Paris was also considering a protest before the League Council against the incorporation of SA members into the German police, and against the provision of arms to these men, both steps which could amount to an increase in German effectives, and the French evidently wanted British support for their proposed action. The British, however, had absolutely no interest in raising such an issue, which might break up the conference. For them, the objective was still to reach an agreed solution, to find some way of satisfying both the Germans and the French, saving hopes for international harmony.¹⁸

More discouragement was in store for MacDonald and Simon in Geneva. There they met with demands from Nadolny that France disarm, and the German delegate stated that "he believed he would be understood when he said that the German government would not

¹⁷ Cab 27/505, DC (M) 32, meeting of 3 Mar 33; *ADAP*, C/I/1, Nos. 46, 49 (and notes), 56, 67; AA, tel., Nadolny to Neurath, 1 Mar 33, 3154/D668607-09; Note by Bülow, 4 Mar 33, 7360/E536255-57; *DDF*, 1/II, Nos. 356, 358, 369; *DBFP*, 2/IV, Nos. 285, 287, 288, 289; 2/V, Nos. 38, 45; *FRUS, 1933*, I, 22-24.

¹⁸ *DBFP*, 2/IV, Nos. 286, 290, 291; *DDF*, 1/II, Nos. 390, 392, 399; Marquand, p. 753. The French waited several weeks for a British comment on their proposed action, and then finally lodged a protest with Neurath, who assured Poncet that the auxiliary police were temporary. On hearing of the intent to proceed with this protest, the British gave a belated reply which argued that the French case was weak, and that protest in either Berlin or Geneva might give the Germans a pretext to assert their immediate equality of rights. See *DDF*, 1/III, Nos. 79, 95, 130; *DBFP*, 2/V, Nos. 24, 25, 26; FO 371, Minutes by Wigram, Sargent, Howard Smith, Vansittart, Simon, 1-10 Apr 33, C3297/245/18. On MacDonald's objectives, see also *FRUS, 1933*, I, 37-38.

be satisfied with a limited result or a simple adjournment"—a hint that Germany might withdraw and openly rearm. MacDonald noted afterward that Nadolny "made me feel that he was out for freedom and wished to agree to nothing."[19] MacDonald's hopes for a five-power meeting had been disappointed, and his situation at the Disarmament Conference appeared highly awkward.

3. Mussolini Intervenes

MacDonald was to be extricated, however, and the idea of great-power negotiation was to receive a new exposition, thanks to Benito Mussolini. The Italian Premier shared MacDonald's desire for a direction of affairs by the great powers, although not under League of Nations auspices in Geneva. On October 23, 1932, the Fascist leader, or Duce, had made a speech in Turin arguing that the four great powers of Europe could assure the political tranquility of the continent,[20] and he now revived this idea. Such four-power direction clearly involved, in his eyes, a larger role for Italy and himself in international politics. Like Britain, Italy might profit by serving as a broker between France and Germany; indeed, although Britain had more standing as a power than Italy, the accession of Hitler, with his ideological affinities to Mussolini, might give the Duce a better claim to mediate. The Italian dictator did in fact have some chance of influencing Hitler, who regarded him as a sort of mentor, and who had already proposed making a personal visit to Rome, a project soon known to the French and British.[21]

In this period, Mussolini, a one-time socialist and journalist, drew up his calculations with a healthy awareness of Italian weaknesses and foreign dangers. At the same time, he was always ready to seize any opportunity for Italian or personal aggrandizement that might perchance be offered. With regard to Europe, his principal hope in 1932

---

[19] *DBFP*, 2/IV, Nos. 295, 297, 299; *ADAP*, C/I/1, No. 76; MDP, 8/1, Pocket diary, 11 Mar [33].

[20] *SIA, 1933*, pp. 207-208. In reporting the Turin speech (on 24 Oct 32 in *DDF*, 1/I, No. 271) the then French ambassador to Italy commented that in proposing a sort of European directorate, Mussolini "was no doubt inspired by M. MacDonald's project of a four-power conference; in defending [the idea of a directorate] M. Mussolini no doubt hopes to please London; he also thinks that in such a meeting England and Italy would find themselves in a position to arbitrate between the opposing theses of Germany and France, a position which would be as profitable to Italian interests as it would be satisfying for the Italian policy of prestige." Mussolini had spoken of a four-power agreement to the French chargé d'affaires in April 1931: Petersen, *Hitler-Mussolini*, p. 137.

[21] On Hitler's proposed Rome visit: Weinberg, *Foreign Policy of Hitler's Germany*, pp. 48-50; also *DDF*, 1/II, No. 332; *DBFP*, 2/V, No. 37.

and 1933 seems to have been for influence in Austria and Hungary, and for the disintegration of the conglomerate state of Yugoslavia, to the benefit of Austria, Hungary, and Italy herself. A major danger, however, was that Germany, through the agency of the Austrian Nazi party, might seize control in Austria and bring about a de facto annexation of that country. An enlarged Germany on the Brenner would be tempted to press claims to the German-speaking South Tyrol and might also seek an Adriatic port in the city of Trieste, formerly Austria's outlet to the sea. Mussolini had some specific ideas for diverting Hitler's interest elsewhere, for example to the Polish Corridor.[22] Overseas, Mussolini had already decided on infiltrating Ethiopia and apparently had authorized preparations for a military conquest that would upgrade Italy's empire of colonial leftovers, and avenge the defeat of Adowa in 1896. He saw, however, that a campaign in East Africa would be a risky undertaking unless Italy's rear was secure against a German move in Austria, and unless French and British acquiescence was first obtained.[23] Italian plans in both Europe and Africa would be threatened if France took sanctions against Germany, thereby presumably enhancing French predominance and reducing French willingness to consider concessions to Italy.[24] And an "outlawing" of Nazism would have implications for Fascism.

[22] See Weinberg, *Foreign Policy of Hitler's Germany*, pp. 48, 50-51.

[23] General Emilio De Bono had proposed armed action to Mussolini in early 1932 and had officially put the preparations in motion on 29 Nov 32, although action would depend on French and British approval: Giorgio Rochat, *Militari e politici nella preparazione della campagna d'etiopia: Studio e documenti 1932-1936* (Milan: Franco Angeli, 1971), pp. 26-27. This study indicates a firmer commitment in 1932 than is indicated in Emilio De Bono, *Anno XIIII: The Conquest of an Empire* (London: Cresset, 1937), pp. 3-13, or than is perhaps suggested by the 3 Jan 33 entry in the diary of Baron Pompeo Aloisi (p. 45), which reports Mussolini giving instructions to his new minister to Ethiopia to conceal Italian plans and Rome's policy of infiltration, and adds: "Without in any case sharing the optimism of military circles, Mussolini believed that a military operation by us in Ethiopia would succeed, on condition that we are completely free in Europe. He says that the military committee presided over by De Bono has already studied all the plans. . . ." Aloisi's statement itself would seem to indicate that Mussolini had seriously considered military action by the end of 1932, contrary to George W. Baer, *The Coming of the Italo-Ethiopian War* (Cambridge, Mass.: Harvard University Press, 1967), pp. 35-36, which gives September 1933 as the earliest date for such consideration. On the other hand, the fact that preparations for war were made in the 1920s in Eritrea and Somalia does not mean there was definite planning then for *offensive* action, as Manfred Funke seems to suggest in *Sanktionen und Kanonen: Hitler, Mussolini und die internationale Abessinienkonflikt* (Düsseldorf: Droste, 1970), p. 9; see Segretaria Particolare del Duce (folder 224/R—"S. E. Federzoni, On. Dott. Luigi"), att. to memo from Federzoni to Volpi, 4 Feb 28, reel 1134/070678, which appears to be the document cited by Funke.

[24] There are indications some Italian officials feared in early March that France, assisted by Yugoslavia, might launch a preventive attack on Italy: Aloisi, pp. 81, 105; *DDF*, 1/III, No. 136; *ADAP*, C/I/1, No. 122. But although Mussolini saw to it

Mussolini had not yet learned to admire Hitler's capacity for executing diplomatic coups, and although he had more sympathy for Germany than did his diplomatic advisers, he was repelled by the racism and primitivism of the Nazis.[25] Fascism sometimes employed violence and brutality at home, and it glorified war, but when off the balcony, Mussolini was a rational man of the world, not a self-taught prophet. It is striking how little there was in 1933 to attract Mussolini to Hitler's side. The outward affinity of Nazism and Fascism could, however, arouse exploitable fears in Britain and France that Mussolini would now become a close partner of Hitler's. Even before Hitler became chancellor, a French diplomat had noted that "in the course of the last five years, the demands of Italy have increased at the same rate as those difficulties which in her eyes should lead us to set a higher price on her friendship."[26]

France was indeed interested in Mussolini's friendship. It had been a tradition of French diplomacy, from the alliance with the Sultan Suleiman in 1536 to that with Tsar Nicholas II in 1893-1894, to seek favorable alliances and ententes regardless of ideology. Boncour, along with other French leaders, apparently decided in late 1932 or early 1933—before Hitler's accession—that the time had come to mend the poor relations existing between Paris and Rome, especially in view of the experience of the December 1932 meetings in Geneva, and British unwillingness to provide any additional guarantees to France. In Europe, France might foster Italian resistance to German influence among the small Central European states, while in Africa, the Italians had recently revealed to the French their interest in moving into Ethiopia, a project that might enable France to secure compensation, most likely an elimination of the privileged status of Italians in French Tunisia. In early 1932, Philippe Berthelot, then Secretary General at the Quai d'Orsay, and Pierre Laval, at that time Premier, had encouraged Italian hopes with respect to Ethiopia, although French officials recognized that international standards of behavior had changed since the pre-League of Nations days when European powers had bargained freely between themselves over the rights of third parties in Egypt, Morocco, Tripolitania, and Turkey. In January 1933, Henry de Jouvenel, a French senator, was sent to Rome as ambassador for a predetermined six-month term, and he was resolved to do what he

---

that "preventive war" dangers were brought to German attention (*ADAP*, C/I/1, Nos. 68, 83, 87, 111, 120, 122), this probably did not express genuine fear on his own part. Aloisi's record does not suggest that he or the Duce were greatly concerned, and they do not seem to have expressed concern to the British.

[25] See *DDF*, 1/IV, No. 113; *DBFP*, 2/V, Nos. 52, 144; Aloisi, p. 138.
[26] *DDF*, 1/II, No. 182.

could in that period to further a Franco-Italian entente.[27] In an interview with Jouvenel on March 2, Mussolini too expressed his interest not only in settling colonial and naval differences, but also in establishing an entente between the two countries. He suggested that this relationship should include Great Britain, and that it should oppose an *Anschluss* or integration of Austria into Germany, and try to solve the Polish Corridor, Danubian, and Balkan questions, as well as secure Italy's position in the Adriatic. He concluded this catalogue of Italian policy aims by saying, "Let us avoid above all letting the small nations instigate quarrels between the large ones who have sacrificed only too much for interests which do not always concern them directly."[28]

Only the recent publication of French diplomatic documents has fully revealed the French interest in 1933 in seeking an entente with Italy and also, in particular, a discernible Italian desire, even at this date, to clear the way diplomatically for a conquest of Ethiopia. This information on the respective interests of France and Italy in an understanding helps to clear up two hitherto puzzling aspects of the diplomacy of the spring of 1933. It does much to explain, first, why the French did not—after firmly opposing new British suggestions for four- or five-power negotiations—flatly reject the more drastic proposals that Mussolini now made for a four-power superintendence of treaty revision. It also helps to elucidate (though there are other explanations too) why Mussolini soon accepted French amendments to his proposals, even though these amendments seemed to deprive the proposals of most of their effect.[29]

Relations with England were also important to Mussolini. Aside from Rome's traditional pursuit of friendship with London, Britain (as Mussolini would point out to the Germans) could help to influence the French response to Italian proposals. MacDonald's transpar-

---

[27] Paul-Boncour, II, 335-341; *DDF*, 1/I, Nos. 119, 214, 317; 1/II, Nos. 182 (incl. Annexes I-III and V-VII), 236, 288, 339; Konrad H. Jarausch, *The Four Power Pact, 1933* (Madison: State Historical Society of Wisconsin, 1965), pp. 31-32. Legally, Jouvenel could not serve as ambassador longer than six months without losing his senatorial seat. He had had experience in calming a troubled situation as High Commissioner in Syria, perhaps also as the second husband of Colette. On Boncour's attitude, also Aloisi, pp. 91, 125.

[28] *DDF*, 1/II, No. 368. Wollstein, in *Weimarer Revisionismus zu Hitler*, p. 69, argues that Mussolini expressed his essential ideas to Jouvenel twelve days before he transmitted the text of his proposal to the German government; Wollstein concludes from non-French documents that Jouvenel must have accepted the idea of a four-power directorate, including Germany.

[29] Of the French documents cited in note 27 above, *DDF*, 1/II, No. 182 and annexes are especially important. Ulrich von Hassell, the German ambassador to Italy, had recognized "the danger of an Italo-French agreement without us and at our expense" (*ADAP*, C/I/1, No. 51 [incl. n. 19]).

ent bid for great-power negotiation on disarmament apparently encouraged the Italian dictator, and on March 4, the day after the announcement of the MacDonald-Simon plan to go to Geneva (and two days after his conversation with Jouvenel), Mussolini drew up a draft for a four-power treaty. He later sent an explanation to Berlin, which claimed that he had drafted his plan in response to an appeal from MacDonald for help in solving the disarmament impasse.[30]

The British believed that Mussolini could do much to restrain the new German government, but they were apprehensive over what might happen when Hitler visited Rome, and they received disturbing reports, stemming directly and indirectly from the Italian ambassador in London, Dino Grandi, that Mussolini was willing even to agree to an Anschluss.[31] On March 9, an editorial in the London *Times* suggested that a high-level Anglo-Italian meeting in Turin or Milan would be a possibility, and similar hints appeared in the Italian press. Overall, it must be suspected that the Italians were skillfully leading the British government on, but Vansittart may have been the immediate source of the *Times*'s suggestion. When Sir Ronald Graham, the British ambassador in Rome, suggested on March 11 that Simon and, if possible, MacDonald come to Italy, this suggestion was very warmly taken up by Vansittart, who telegraphed Simon in Geneva:

---

[30] *ADAP*, C/I/1, No. 83 and encl.; *DBFP*, 2/V, No. 45; Jarausch, *Four Power Pact*, pp. 34-35. Mussolini also reportedly told the Fascist Grand Council (presumably on 9-10 Mar 33; see *ADAP*, C/I/1, No. 68) that the British had proposed a four-power directorate to Italy, through Ambassador Grandi, which he had then taken up: see *DDF*, 1/II, No. 427. But according to Aloisi (p. 86), Grandi reported after an 8 Mar 33 meeting with Simon that the British had no plan except to do everything to save the conference (see also *ADAP*, C/I/1, No. 68); Simon seems in fact to have warned Grandi against the idea of forcible revision, while also saying that peaceful progress depended on close cooperation between the leading states (*DBFP*, 2/V, No. 38). Poles in the League Secretariat stated that MacDonald had discussed the four-power pact before leaving Geneva for Rome (SDP, tel., Gibson to Dept., 21 Mar 33, 740.0011 Four Power Pact/1; *FRUS, 1933*, I, 396), but at the most, MacDonald, encouraged by Aloisi, probably made only general statements in favor of great-power cooperation and leadership in Europe. Jarausch (*Four Power Pact*, pp. 41-43n) and Petersen (*Hitler-Mussolini*, p. 145) maintain, no doubt rightly, that MacDonald was not the actual author of the four-power pact. Yet there is also truth in the thesis of Arthur H. Furnia (*The Diplomacy of Appeasement: Anglo-French Relations and the Prelude to World War II, 1931-1938* [Washington, D.C.: University Press of Washington, 1960], pp. 65-66, 79-84), that a four-power agreement was already an objective of British policy—if we mean not an explicitly revisionist pact such as Mussolini drew up but a general understanding for the direction of European affairs, or a consultative pact, as sponsored by the British in July 1932.

[31] FO 371, tel., Vansittart to Simon, 13 Mar 33, C2287/2287/22; Memo by Sargent, 13 Mar 33, C2409/2287/22. That Grandi was the source of British fears that Italy would sanction Anschluss: *DBFP*, 2/V, No. 38; *DDF*, 1/II, Nos. 360, 421; *Papers and Memoirs of József Lipski*, pp. 60-62. Grandi claimed to strongly oppose a pro-Anschluss policy himself.

> In view of developments in Germany and their evident effect upon Italian policy I most earnestly trust that both you and the Prime Minister will find it possible to pay the suggested visit to Rome before Hitler goes there in April. Nothing less will probably suffice now. It seems to me of the highest ultimate importance to make the attempt and it should be made soon since we may well soon be faced with an attempt to make the Anschluss an accomplished fact (by a Nazi putsch in Vienna) possibly with *Italian approval.*

Aside from the Anschluss problem, Vansittart hoped the British ministers could help to improve the sad state of Italo-Yugoslav and Italo-French relations; he apparently did not recognize the significance of Jouvenal's appointment.[32]

At Geneva, as noted, Nadolny's demeanor had strengthened MacDonald's fear that the conference would fail, and that the Germans in turn would announce their intent to rearm. On the other hand, Boncour's insistence that Germany was already augmenting her forces, and his rigid assertions that France had already offered all she could, aroused MacDonald's anti-French feelings, and led him to reply to the French Foreign Minister with a heated denunciation of French policy on naval disarmament.[33] By contrast, Pompeo Aloisi, Mussolini's representative, ingratiated himself and his master with the British ministers. MacDonald asked Aloisi to help keep Germany in the conference, and when Aloisi suggested that the great powers should take up the political problems of disarmament and treaty revision, this seemed right in line with MacDonald's own ideas. Actually, Aloisi had for some time advocated a conference adjournment, and he now tried to forestall any major steps in Geneva, such as might interfere with great-power negotiations, hamper preparations for a four-power agreement under Mussolini's sponsorship, or steal some of the Duce's thunder. But although MacDonald on the whole agreed he should go to Italy, and he embraced Mussolini's invitation when it came, he also feared that a simple adjournment would bring German rearmament, and Eden convinced the Prime Minister that he could not afford, politically, to leave the conference without first launching the draft disarma-

---

[32] *Times* (London), 9 Mar 33; FO 371, tel., Graham to Simon, 11 Mar 33; tel., Vansittart to Simon, 13 Mar 33 (emphasis in source), C2287/2287/22; tel., Simon to Graham, 13 Mar 33, C2309/2287/22; tels., Graham to Simon and Vansittart to Simon, 14 Mar 33, C2362/2287/22. According to Aloisi (p. 84), Mussolini was planning on 8 March to have a meeting with MacDonald and Daladier in Turin to seal the four-power accord, after reaching agreement with the Germans.

[33] *DBFP*, 2/IV, Nos. 295, 297, 299; *DDF*, 1/II, Nos. 401, 404. The British record seems to have played down the force of Boncour's demands that they take a stand. On the other hand, the French mistook MacDonald's angry response to Boncour's pressure and inflexibility for a deliberate broaching of naval disarmament problems.

ment plan. MacDonald did this with a speech to the conference on March 16, after a hasty and inconclusive discussion with Daladier, who finally came to Geneva at urgent British request. The British seem to have desired Daladier's presence mainly to make it appear that there was no French resentment of MacDonald's launching a new disarmament proposal, or of his going to Rome, and the French wanted information on that proposal. Daladier was not at first shown the whole MacDonald plan (as it now began to be called), and he agreed to wish the Prime Minister a bon voyage to Italy. Much as the French disliked the British disarmament proposals when they studied them, they did not want an open rift, and Boncour hoped the Germans would dislike the proposals more.[34]

4. THE FOUR-POWER PACT AND THE UPSET OF BRITISH POLICY

A few hours before MacDonald's and Simon's arrival in Rome on March 18, the French and British ambassadors received copies of Mussolini's draft for a four-power pact; in Jouvenel's case, the dictator himself handed over the draft personally. Jouvenel and Graham were told that the German Ambassador, Ulrich von Hassell, had likewise just received a copy, which was nominally true but very misleading. The text had in fact been given to Neurath in Berlin four days before, with the message that Italy would only present the proposal to Britain and France if Germany concurred in principle. The German government had approved, while suggesting some changes. The draft did not become a joint Italo-German product, however, but remained unmodified, and in sending it to Berlin, Mussolini's main objective was apparently to make the Germans think he was really biased on their side; perhaps, too, Mussolini may have been glad to use the occasion of this prior notification to postpone Hitler's projected visit to Rome.[35]

Mussolini's draft called for four-power cooperation to maintain peace in the spirit of the Kellogg Pact and the pact against the use of force proposed by Simon. The first article was the most significant: it stated that the four signatories would "undertake to act in the

[34] Aloisi, pp. 73, 85-96; *DBFP*, 2/IV, Nos. 293, 297, 299-306, 310; *DDF*, 1/II, Nos. 322, 409, 411, 412, 418, 420; MDP, 8/1, Pocket diary, 12, 13, 14, 16 Mar [33]; Eden, p. 34; Marquand, pp. 753-754; SDP, tel., Gibson to Dept., 18 Mar 33, 500.A15A4 General Committee/218 (on French view of British plan). American influence supported Aloisi: *FRUS, 1933*, I, 35, 37-38, 40-41.

[35] *DDF*, 1/III, Nos. 2-4; *DBFP*, 2/V, No. 45; *ADAP*, C/I/1, Nos. 83, 84, 87, 88, 98. Jarausch also suggests (*Four Power Pact*, p. 47) that Mussolini wanted to protect himself from unwelcome surprise moves from Berlin while negotiating with Britain and France. On 27 April, German Foreign Ministry officials disclosed to the French counselor of embassy in Berlin the prior informing of Germany (*DDF*, 1/III, No. 199). François-Poncet had suspected this earlier (*DDF*, 1/III, No. 33).

sphere of European relations in such a way that this peace policy, if necessary, is adopted by others." This amounted to saying that four-power decisions would be imposed on smaller states in the interests of peace, as the great powers conceived those interests. Under Article II, the signatories were to reaffirm, "in accordance with the provisions of the Covenant of the League of Nations," the principle that peace treaties should be revised where conditions might lead to conflict; this principle, however, should only be applied within the framework of the League. Article III provided that "in case the Disarmament Conference should lead to partial results only, the equality of rights conceded to Germany must have effective application," while Germany would agree to implement this equality of rights only by stages agreed upon by the four powers; similar rules would apply to Austria, Hungary, and Bulgaria. The four powers would also adopt a common line of conduct in political and nonpolitical (i.e., economic) matters, and in extra-European and colonial questions. The pact was to be registered with the League Secretariat, and to run for ten years.[36]

On their first impromptu examination of the draft, after their arrival in the Italian capital, MacDonald and Simon found little to object to. Provisionally and tentatively, they suggested adding to Article II a prefatory bow in the direction of the sanctity of treaties, and to Article III a resolve to cooperate at the Disarmament Conference. They asked for the deletion of the reference to extra-European and colonial questions, and suggested an explicit pledge to find solutions to economic problems. These points were mostly hortatory, and hardly affected substance. The reaction of the British leaders was generally favorable: they already supported great-power leadership, revision, and (with some practical reservations) the equality of rights. MacDonald was suspicious of most foreign leaders, including Mussolini, but he came away from Rome on March 20 with a good impression of the proposal and of what he had been shown of Fascist Italy. On his way home, at a meeting with the French leaders in Paris, he spoke approvingly of the "new energy" appearing in Italy, Russia, and also Germany, in contrast to the "old-fashioned" France and Britain. In Paris, and again at a cabinet meeting in London, he commended a mot of Mussolini's to the effect that all treaties were holy, but none were eternal. He told his cabinet that, under Fascism, Italy had experienced "a regeneration which affected not only the whole administration and system of government but had resulted in a widespread spiritual de-

[36] *ADAP*, C/I/1, No. 83 (encl.). One addition was made between the time the pact was given to the Germans and the time it was given to the British and French: this specified that the pact should be submitted for ratification within three months (of signature) (cf. *DBFP*, 2/V, No. 44 [encl. A]; *DDF*, 1/III, No. 2 [Annex]).

velopment among the Italian people," and he also said, "The greatest service we could render to France at the moment was to get her to realise what was going on in Europe outside her own country."[37]

But Mussolini's proposal did not for long escape more drastic proposals for change. Official and more far-reaching British amendments were forwarded on March 31, eliminating the claim to impose four-power policy on other states, and substituting a four-power adoption of the MacDonald disarmament plan for Mussolini's Article III, a suggestion that MacDonald himself had failed to make. The French government was most suspicious of the proposal, and although Boncour did not want to reject it, in view of his desire to avoid isolation and to achieve an entente with Italy, he had to deal with violent protests from Poland and the Little Entente against the idea that a four-power directorate should carry out treaty revision, revision that was likely to affect them directly. The French felt compelled to press for major amendments striking out the claim to four-power authority, placing more stress on the League and the Covenant, and protecting their own and their allies' interests; they submitted these suggested changes on April 11. The Germans naturally opposed these British and French proposals, and wanted instead to strengthen the provisions for revision and equality of rights.[38] However, a flare-up of concern over Nazi infiltration in Austria, along with reports of militaristic rallies and anti-Semitic incidents, had made the atmosphere more unfavorable to concessions to Germany, and may have caused Mussolini

---

[37] *DBFP*, 2/V, Nos. 44 (incl. enclosures), 45, 46; Cab 23/75, meeting of 22 Mar 33, 20(33)10; MDP, 8/1, Pocket diary, 9, 14, 19, 21 Mar [33]. To Boncour, MacDonald appeared to have been "very struck by Fascism and rather beguiled by Mussolini" (Paul-Boncour, II, 340). Years later, at the height of the Ethiopian crisis, Grandi reported that MacDonald disparaged to him the anti-Italian direction which British policy had taken, saying he was opposing this in the cabinet and was still convinced that the four-power pact would have been "the right road": National Archives, Microfilm T-586, Segretaria Particolare del Capo del Governo (folder 205/R—Dino Grandi), ltr., Grandi to Mussolini, 18 Oct 35, reel 449/026782-85. But British cabinet records of the period (Cab 23/vol. 82) do not show that MacDonald was in fact very vocal in resisting the policy of opposition to Italy. Davis and the U.S. delegation in Geneva, although not required to take a position on the four-power pact, evidently sympathized with it: see *FRUS, 1933*, I, 83, 403-404; SDP, Memo of conversation between Davis and Herriot, 6 Apr 33, 550.S1 Washington/359; Ferdinand Mayer Geneva diary, 27, 28 Mar 33, 500.A15A4/1469½; Aloisi, pp. 116, 126. Like MacDonald, Davis was an advocate of great-power negotiation. Vansittart advised Simon that Mussolini's text "is to be welcomed as a practical form of approach to the realities of the European situation as it at present exists." Respecting Article I, Vansittart commented that pressure on the smaller powers "would necessarily result from agreement among the four Powers on any topic, but it is better not to say so": MDP, 1/521, tel., Vansittart to Simon, 21 Mar 33.

[38] *DBFP*, 2/V, Nos. 46, 49, 54, 56, 58, 61, 64; *DDF*, 1/III, Nos. 7, 12, 35, 38, 42, 43, 45, 47, 48, 53, 62, 81, 108, 111, 112; *ADAP*, C/I/1, Nos. 84, 88, 95, 108, 111, 115, 126, 128, 132, 133, 138; Nixon, ed., *Roosevelt and Foreign Affairs*, I, 121-122.

himself to become more cautious about revision and German rearmament.[39] As we have seen, the Duce was interested all along in reaching an understanding with France. When he first showed his Foreign Ministry officials his draft of a four-power pact, he stated his intention to follow it up with a lesser accord with France on "Tunisia, etc."; presumably this would have included arrangements on Italian influence in Ethiopia.[40]

Mussolini's four-power pact proposal was not really as far-reaching as it pretended to be. Article II provided that revision should be applied only within the framework of the League of Nations, and Article III stated that equality of rights should be implemented by stages and should result from successive agreements between the four powers—one of which, of course, was France. This article would have bound Germany not to leave the conference and rearm freely. Nevertheless, to the allies of France in Eastern Europe, these articles appeared as a direct threat, especially when coupled with the initial claim in Article I that the four great powers should impose decisions on the minor states. And it was not only Mussolini's draft that caused concern within the Polish, Czech, Rumanian, and Yugoslavian governments; it was also the evident sympathy of MacDonald for this approach, and the failure of France to be as hostile as might have been expected. Suspicions of the French doubtless encouraged the Poles to seek a direct discussion with the German government a few weeks later. In Geneva, the supporters of the League and the representatives of the Little Entente were in a turmoil. Eduard Beneš had been reluctant to believe that MacDonald would go as far as Mussolini, but after the British Prime Minister expounded the merits of the Mussolini plan in the House of Commons on March 23, Beneš told a French diplomat there was now no doubt that the British leader had purely and simply accepted the policy that Mussolini had followed for years.[41] A telegram from Simon, who was back in Geneva again, informed Vansittart: "You can have no idea of strength of suspicion here that outcome of Rome visit is a plot to dish the League of Nations." Beneš and Nicolae Titulescu, the Rumanian Foreign Minister, protested to Simon, and Titulescu went on from Geneva to Paris where, according to Jouvenel, he got into contact

---

[39] *DBFP*, 2/V, Nos. 39-41; *DDF*, 1/III, Nos. 19, 20; *ADAP*, C/I/1, Nos. 107, 112; Aloisi, pp. 96-97.

[40] Aloisi, p. 84. By the beginning of April, Jouvenel was discussing a Franco-Italian pact with Mussolini; concrete discussion at this time seems to have been on a disarmament agreement (*DDF*, 1/III, No. 112; see also Nos. 312 and 316 in the same volume).

[41] *DDF*, 1/III, No. 47; Czech docs., tel., Beneš (Geneva) to Prague for President and Prime Minister, 21 Mar 33, 1809/413896-97. On the Geneva atmosphere at this juncture: SDP, Ferdinand Mayer Geneva diary, 27 Mar 33, 500.A15A4/1469½.

with the French opposition and stirred it up.[42] These protests no doubt influenced the British and French proposals for amending the Mussolini draft, and in particular, encouraged both countries to oppose the Article I provision for compelling other states to accept four-power policies. This meant the elimination of the key feature of the pact, and its emasculation.

The British and French amendments responded even more to a domestic reaction in each country against the pact. An adverse reaction in France was quite predictable, irrespective of any propaganda from Titulescu. Whenever the Italians informed a political leader or a diplomat of their plan, the first response (except in the case of Jouvenel) was invariably that France would oppose the scheme. The French press was generally hostile, and even most ministers were opposed.[43]

More important, and less expected, was the opposition that developed in England. Here, loyalty to the League and dislike for the open exercise of power politics were reinforced, and indeed overshadowed, by a surge of disapproval for dictatorship, and in particular for the Nazi anti-Semitism and militarism manifested during the month of March, for example in the SA's picketing of Jewish shops, in discrimination against Albert Einstein and Bruno Walter, and in atavistic parades and rites at the Garrison Church in Potsdam. This disapproval emerged in the House of Commons debate following MacDonald's March 23 report on his trip, and it resounded more loudly in another debate on April 13. Clement Attlee, the leader of the opposition, opened the latter discussion, saying: "I think this House and this country ought to say that we will not countenance for a moment the yielding to Hitler and force what was denied to Stresemann and reason." Winston Churchill said:

> When I read of what is going on in Germany—I feel in complete agreement in this matter with hon. Gentlemen opposite—when I see the temper displayed there and read the speeches of the leading Ministers, I cannot help rejoicing that the Germans have not got the heavy cannon, the thousands of military aeroplanes and the tanks of various sizes for which they have been pressing in order that their status may be equal to that of other countries.

He indicated that he no longer supported the idea of an adjustment of the Polish Corridor. In 1933, the influence of Sir Austen Chamberlain was greater than that of Churchill, and Sir Austen stated:

---

[42] *DBFP*, 2/V, No. 53. On Titulescu, see also Georges Bonnet, *Le Quai d'Orsay sous trois républiques* (Paris: Fayard, 1961), p. 125.

[43] *DBFP*, 2/IV, No. 301; 2/V, No. 44 (encls. 4 and 5); *ADAP*, C/I/1, Nos. 78, 87, 102; *DDF*, 1/III, No. 3; Aloisi, pp. 92-93; Bonnefous, V, 159; Paul-Boncour, II, 340-350.

What is this new spirit of German Nationalism? The worst of the All-Prussian Imperialism, with an added savagery, a racial pride, an exclusiveness which cannot allow to any fellow-subject not of "pure Nordic birth" equality of rights and citizenship within the nation to which he belongs. Are you going to discuss revision with a Government like that? Are you going to discuss with such a Government the Polish Corridor? The Polish Corridor is inhabited by Poles; do you dare to put another Pole under the heel of such a Government?[44]

MacDonald and Simon could hardly resist the strong wave of parliamentary and public opinion, with which they basically sympathized. During the April 13 debate, MacDonald stated, in indirect language, that there was room for doubt as to whether the new German government could be trusted, and he said: "It is no use talking about cooperation for peace unless you have had some experience which justifies you in accepting the word of those with whom you are to cooperate." Simon drew the conclusion at the end of the debate that there was a "deep, general and I might say, universal feeling" in Britain regarding the treatment of Jews and other minorities in Germany. These statements received full attention abroad: the German chargé d'affaires, on urgent instruction, made an official protest, while Mussolini told Aloisi that the House of Commons debate had reversed the situation, and that it was necessary to act fast in winding up negotiations for the four-power pact. MacDonald had already told Leopold von Hoesch, the German Ambassador, that while there had still appeared to be considerable sympathy for Germany in England at the time that he and Simon had left for Geneva, they had encountered a changed atmosphere upon their return ten days later, "and had been compelled to recognize that their actions in Geneva and Rome did not find the favorable reception that they had expected."[45]

The new direction of British opinion, and the consequent reversal of British policy, doomed Mussolini's four-power pact plan. He had launched the plan immediately after MacDonald had shown interest in great-power discussions. Mussolini had apparently hoped that British pressure, together with the French interest in an entente with

[44] Eden, p. 36; Winston S. Churchill, *While England Slept: A Survey of World Affairs, 1932-1938* (New York: Putnam's, 1938), pp. 62-63; *SIA, 1933*, p. 168; Jarausch, *Four Power Pact*, pp. 74-82.

[45] *SIA, 1933*, pp. 169-170; *ADAP*, C/I/1, Nos. 152, 158; *DBFP*, 2/V, Nos. 34, 129; Aloisi, p. 109; Petersen, *Hitler-Mussolini*, p. 155. Despite MacDonald's talk, after his Rome visit, of "new energy" in Italy, Russia, and Germany, his diary shows a strong rejection of Nazism: MDP, 8/1, Diary, 28 Feb, 2, 9, 13 Mar, 12, 13 Apr, 30 Nov [33].

Italy, would overcome the natural objections of France and her allies. Now there would be no British pressure, except to hasten the signature of a nominal agreement, and despite bitter protests from Berlin, Mussolini had to accept the substance of the French amendments, if there was to be a pact at all. Dictators have to avoid public humiliation, and after all the publicity, a failure to reach any agreement would have been an open setback for Mussolini. A bowdlerized pact was at least something the controlled Italian press could still describe as a diplomatic triumph. As for the German government, it finally agreed to the unwanted changes, because its primary objective was secret rearmament, and because it was anxious, while this was going on, to avoid provoking other powers. Blomberg told the division commanders on June 1 that although nothing concrete would result from the Disarmament Conference or Mussolini's four-power pact, it was extremely important to have diplomatic calm in which to carry out the measures (of rearmament) the army considered necessary: "Much that will happen in the next period (e.g., the Mussolini pact itself) will happen for this reason." The pact was initialed on June 7, after very arduous negotiation, and signed on July 15, but it was never ratified.[46]

5. British Opinion and Hitler

The turn of British opinion against Hitler did credit to the humanitarian and libertarian traditions of the country. This moral reaction, shared by MacDonald himself, gave a sharp check to the idea that the European powers should dictate revision in an attempt to satisfy the German appetite. In 1932, MacDonald had used the procedure of the four- or five-power meeting to force changes against the French interest and in favor of the German; now, at least for the moment, the British Government abandoned this procedure.

Moral condemnation was not all-pervasive, however. When the Commons discussed a new Anglo-German trade treaty a few weeks after the April 13 debate, many members criticized the treaty, but on the grounds that, without adequate consultation with British businessmen, it lowered tariff barriers to imports from Germany; there was almost no reference to Nazi misdeeds. There were fluctuations in disapproval: for example, when anti-Semitic outrages subsided after one day (April

[46] Liebmann MS Papers, p. 200 (typed). Jarausch's *Four Power Pact*, the standard monograph on the Mussolini proposal, describes the later negotiations in detail; see also Petersen, *Hitler-Mussolini*; Lothar Krecker, "Die diplomatischen Verhandlungen über den Viererpakt vom 15. Juli 1933," *Welt als Geschichte*, 21 (1961), 227-237. A contemporary diplomatic survey by Sir Ronald Graham is given in *DBFP*, 2/V, No. 216.

1), many thought this signaled a victory of the moderates within the regime.[47] And even though many people condemned Nazi excesses, there were other influences on their actions too. Attlee, for example, would be consistently hostile to Nazi methods, but he and his party remained very reluctant to consider British rearmament. While some, like Churchill and Austen Chamberlain, questioned the wisdom of weakening France, no one publicly advocated a British commitment to that country, and the MacDonald plan remained a compromise between France and Germany. The practical dangers of the power situation in Europe were perhaps beginning to effect British attitudes, but the balance of power was a long way from regaining wide recognition as a principle, and the inclination to hope for a reconciliation remained very strong.

Hitler, in any case, was highly interested in power relationships. Given his interest in reaching an understanding with England, his faith in propaganda, and his whole belief in the efficacy of will power, he would not watch passively while British opinion and policy developed in an adverse way. He believed that relations with any foreign country were too serious a matter to leave to the diplomats. One of his techniques, for which there were American and British precedents, was to hold forth before the world public, appealing to other peoples over the heads of their governments, demonstrating that his own nation was behind him, and making his own position irrevocable; he was to apply this in May and again in October 1933. Another technique was to try, on a direct and unofficial basis, to argue his case through foreign journalists and others who might be influential due to their social status or political connections. With England, he had begun to apply this second approach almost as soon as he came to power.

On February 6, the new dictator met with Colonel P. T. Etherton, representing the *Daily Mail* and the *Sunday Express*, and stressed his desire for disarmament and his determination to fight Communism. Four days after the Reichstag fire, in early March, he assured a *Daily Express* reporter that there would be no St. Bartholomew's eve in Germany. In early May, Hitler gave an hour-long interview to Sir John Foster Fraser, of the *Daily Telegraph*, and told him, "Nobody in this country who went through the last war wants the same experience."

---

[47] Wendt, *Economic Appeasement*, pp. 85-87, 104-110. The British negotiated a number of trade treaties in this period, but Germany was the only nation (until 1938) to win substantial reductions in British duties; in return, Germany admitted a small quota of British coal: Arndt, *Economic Lessons*, p. 113. Much of the apparent fluctuation in opinion, stressed by Wendt, did not represent changes of mind, but the views of different people with different interests.

Hitler said that he wanted equality, but would prefer a reduction of "the Entente armies" to an increase in the German army. He added that Germany had given up the idea of expansion beyond the seas, and did not want to enter into competition with England in naval strength.[48]

Hitler did not rely entirely on political arguments in dealing with journalists and others. In a meeting of German leaders in July on the subject of disarmament propaganda, he said it was necessary to exploit "natural opportunities for exerting influence," and to spend considerable sums of money. "It would be advisable, moreover, to keep a card file in Berlin of all persons in public life abroad, having in mind especially all journalists known to be open to bribes." Goebbels said that he had already been working on the project for months.[49]

Hitler believed that the British ruling caste controlled British policy, and that this caste should be won over. Accordingly, aside from reaching journalists through interviews in Germany and clandestine approaches abroad, more prominent persons were sought out. Alfred Rosenberg, who had recently been named to organize a Foreign Political Office (Aussenpolitisches Amt) for the Nazi party, went to London on May 5 for a week's stay. During the course of his visit, he saw many leading persons, including the War Secretary, Lord Hailsham, and MacDonald's Principal Private Secretary, James Barlow, as well as Vansittart and Simon, and he also held a press conference. His argument that Nazism was a barrier against Communism may have made some impression in conservative circles, and the anti-Nazi demonstrations which occurred in the last days of his visit, including the removal of the swastika-adorned wreath he had left at the Cenotaph, the London World War memorial, actually provoked a certain amount of reaction in his favor. But Rosenberg, who had to rely on interpreters and who was an outspoken anti-Semite, was a far from ideal emissary, and the overall result was clearly negative.[50]

---

[48] Jacobsen, *Nationalsozialistische Aussenpolitik*, pp. 765, 766, 769; Klaus Hildebrand, *Vom Reich zum Weltreich* (Munich: Wilhelm Fink Verlag, 1969), pp. 455-456; Domarus, I/1, 201-202; *Daily Telegraph* (London), 5 May 33. Hitler apparently indicated some moderate hopes for colonial revision to Etherton, but as Hildebrand says, this was probably only because Etherton raised the question.

[49] *ADAP*, C/I/2, No. 359.

[50] Aigner, *Ringen*, pp. 77-78; Jacobsen, *Nationalsozialistische Aussenpolitik*, pp. 74-75, 767, 769; *Daily Mail* (London), 11-14 May 33; *Daily Telegraph* (London), 12, 13 May 33; *News Chronicle* (London), 11, 12, 15 May 33; *ADAP*, C/I/1, No. 237; *DBFP*, 2/V, Nos. 118, 126, 138. Aigner's assertion that Hitler appealed to British realism rather than ideology seems dubious. On the Cenotaph incidents, they were regarded by the conservative press as self-serving leftist propaganda and as a discourtesy to a visitor—while this press at the same time joined in condemning anti-

Despite Rosenberg's failure, German efforts would continue. A characteristic Nazi essay in direct diplomacy was the reception extended in June to a group of twenty-five British private-airplane owners and enthusiasts, including three MPs and two peers. They were entertained by Göring at the Berlin Aero Club, they laid a wreath on the German war memorial on Unter den Linden, and they were addressed by Hitler. As the London *Times* reported:

> Herr Hitler said that, as a German soldier, he had had a personal opportunity during the War in Flanders of admiring and respecting the achievements of British airmen. He was convinced that such of his British guests as had also been at the front had in the same way felt respect for the chivalry of their German opponents. Front-line soldiers would always and everywhere understand each other. The sincere feeling of this mutual respect was the soundest foundation for the political relations of the two great Germanic nations in the coming years.

Arrangements were also made for the visitors to inspect the site of the Reichstag fire because, according to the German press, they had a particular interest in seeing that site; the reporter of the *Times* commented drily, "The visitors themselves were not aware of this."[51]

Though the British revulsion against Nazism was not all-pervasive or consistent, most of the Nazi propaganda to England was a dismal failure. If the Nazis won any converts on the basis of brotherhood of the battlefield, or of Germanic racial affinities, these converts were unlikely to influence British policy. Of course, the Nazi activity served other purposes, such as domestic propaganda for the German public, and the self-advancement of Nazi paladins who wanted to impress their leader. There was a good deal of puffery and self-promotion in the work of men like Rosenberg and Joachim von Ribbentrop and their "offices," even though Ribbentrop eventually became Foreign Minister. The real effect of such activity on British policy was always restricted by the personal limitations of the would-be manipulators and also of the persons they sought to manipulate. The most successful German propaganda would be that which followed the pre-Hitler pattern and appealed to general principles such as the equality of rights or the

---

Semitism. Some incidents, such as a leaflet-dropping at the station at Rosenberg's departure, bore all the earmarks of organized agitprop. Rosenberg had tried to arrange a trip to London in 1932: FO 800/286, Simon Papers, ltr., Maj. Norman Bray to Hailsham, 18 Jan 32, and reply, 19 Jan 32.

[51] *Times* (London), 7, 10 Jun 33. Domarus, I/1, 280, describes the visitors as "air force officers" and a number were probably ex-World War I pilots, but wealth, not war service, seems to have been the group's common denominator.

self-determination of peoples.⁵² Hitler himself recognized the appeal of such arguments, as he would demonstrate in his speech of May 17.

6. CRISIS AT THE CONFERENCE

Although MacDonald had hardly had his full heart in the disarmament plan that bore his name, he had publicly launched it in Geneva, and it became a focus of diplomatic activity. After a few days of general discussion, the conference had decided on March 27 that the British draft should form the basis for future discussions, to resume on April 25 after an Easter recess.⁵³ And shortly after resumption, the conference ran into a snag.

This was in part the fault of the United States. When the conference resumed, its first business was supposed to be a consideration of Part I of the British proposals, under which all the signatories were to consult together in case of a threat of war. The American government, however, could not commit itself to such consultation, and it was also unable, in the chaotic onset of the "100 days" of emergency action to deal with the depression, to work out fully a different position of its own. In mid-April, the American delegation sent home a series of diffuse and ill-organized cables criticizing the MacDonald plan and suggesting that, in view of Japan's apparent unwillingness to accept limitation, disarmament provisions should for the most part only apply to continental Europe. But the delegation also suggested that the United States should commit itself by treaty not to trade with an aggressor, if it agreed on who the aggressor was. In other words, to enable other governments to apply effective economic sanctions, or in case of war, a blockade, the United States would not insist on its right to trade with a designated aggressor, if it agreed with the designation.⁵⁴ But the delegation's proposals had two political drawbacks: the restriction to Europe would be an obvious step backward from general disarmament, and a treaty commitment would require Senate ratification by the constitutional two-thirds majority.

Shortly after this, MacDonald, accompanied by Vansittart, arrived in Washington, having come primarily to discuss the World Economic Conference and the war-debt problem. At a meeting on the presidential yacht, the visitors found Roosevelt unimpressed by his delegation's

⁵² See Aigner, *Ringen*, pp. 85-88. Margarete Gärtner continued to get effective results with her old methods: AA, ltr., Gärtner to Aschmann, 8 Jan 34, and att. rpt., 5849/E428997-9026.
⁵³ *SIA, 1933*, pp. 258-260.
⁵⁴ *FRUS, 1933*, I, 89-99, 100-101, 105; *DBFP*, 2/V, No. 83; Moffat MS Diary, 17, 18 Apr 33. Hugh Wilson was the main author of the proposals: SDP, Ferdinand Mayer Geneva diary, 28 Apr 33, 500.A15A4/1469½.

views, and willing to come out strongly in favor of the British disarmament plan, except for its Part I. Instead of assuming a contractual obligation to consult with other powers under this part, which would entail a Senate ratification of consultation, Roosevelt proposed to issue a presidential declaration that the United States, if it independently agreed with the judgment of the other powers, would undertake unilaterally not to trade with an aggressor. But high State Department officers (Under Secretary William Phillips and Jay Pierrepont Moffat), who had not been present at the key talks between MacDonald and Roosevelt, had grave doubts about the British plan, and they delayed and blurred the President's instructions, while Norman Davis (now official chairman of the American delegation) was still wedded to the ideas of the delegation, and evidently also preferred to delay any American concession for tactical reasons.[55] There was the further circumstance that former Premier Herriot, who had likewise come to Washington for talks on war debts and the Economic Conference, was also currently discussing disarmament with Roosevelt. So Davis told the conference that his government had its policy on consultation "under careful advisement"; holding out the hope of a later clarification of American policy, he suggested that the conference move on to other parts of the MacDonald plan. This was agreed to.[56]

The next part of the MacDonald draft was the first section of Part II, dealing with effectives, and thus, before any momentum could be gained, and contrary to the original intention of Eden and Cadogan, the conference returned to the principal problem it had debated in February. Once again, the conference agenda placed Germany rather than France under-pressure to make the first concession. A crisis followed which was much sharper than that two months earlier, and it only ended when Hitler delivered a "peace speech" on May 17. This crisis has sometimes been interpreted as a first essay in Hitlerian power politics, and as a case in which resolute opposition, unfortunately not maintained later, brought Hitler to terms.[57] Actually, Hitler did not try to take an aggressive stance, and he did not significantly

---

[55] *DBFP*, 2/V, Nos. 83, 86-88, 94, 96; *FRUS, 1933*, I, 102-104, 106-108, 112-117; Moffat MS Diary, 20, 23-25, 28 Apr 33; Franklin D. Roosevelt Library, Hyde Park, N.Y., Franklin D. Roosevelt Papers, President's Secretary's File (PSF), File 142, ltr. and encl., Phillips to Roosevelt, 20 Apr 33; *Roosevelt and Foreign Affairs*, I, 68-69; MDP, 8/1, Pocket diary, Sunday, [23] Apr [33]; FO 800/288, Simon Papers, Note to Vansittart, [early May 1933], note from Vansittart, 5 May 33. At Roosevelt's suggestion, ex-Secretary Stimson was consulted and he advised against a treaty commitment: Moffat MS Diary, 19, 20 Apr 33; *Roosevelt and Foreign Affairs*, I, 56-57.

[56] *DBFP*, 2/V, Nos. 93, 95; *FRUS*, 1933, I, 108-112; *DDF*, 1/III, Nos. 179, 182, 189.

[57] See G. M. Gathorne-Hardy, *A Short History of International Affairs, 1920-1939*, 4th ed. (London: Oxford University Press, 1950), p. 352; Bracher, "Stufen der Machtergreifung," p. 241. On beliefs and claims at the time, see note 89 below.

retreat under pressure. But a belief at the time that Hitler had retreated may have encouraged an underestimation of his resolve.

It would be a great mistake to suppose that Hitler directed or followed every step taken by German diplomats. It was characteristic that Hitler himself received an initial visit from François-Poncet on April 8 with Blomberg present, but not Neurath, and that the Foreign Ministry was left to try to piece together what might have transpired. The German records show little instruction being sent to Nadolny from Berlin, and his own reports indicate he was operating almost independently, with little or no current guidance. Neurath and his top permanent officials seem to have been inclined to dodge responsibility, to avoid dealing with the prickly Nadolny, and even to give him enough rope to hang himself. This was an anxious period in the Wilhelmstrasse when many officials were busy protecting themselves and their positions, and when some took advantage of others' problems.[58]

Of course, Nadolny had the well-established line of German disarmament policy to go by, as reconfirmed in the January guidelines for the delegation. The cardinal point of these guidelines, and of the whole doctrine of equality of rights, was that Germany could accept no agreement which continued to place restrictions on her alone, rather than generally on all powers. Contrary to this principle, the MacDonald plan implied a continued ban on German possession of military aircraft, submarines, and heavy mobile artillery without barring these weapons to other countries, and it proposed restrictions on German military organization that would not apply to forces (including French colonial troops) outside continental Europe. Nadolny and Neurath did, even so, regard the British plan as a basis for discussion. The German delegation would of course seek to ensure that standardization did not reduce German professional cadres, and that police and

[58] Hitler and Blomberg apparently sounded Poncet on the possibility of direct Franco-German talks on armament; Poncet was probably willing, but he played down this bid in his report to the Quai. Hitler told Neurath only that the meeting was unfruitful: *DDF*, 1/III, No. 105; *ADAP*, C/I/1, No. 163 and n. 2; AA, Memo by Koepke, 15 Apr 33, 3154/D668880; Memo by Bülow, 24 Apr 33, 4602/E189182-84; François-Poncet, *Souvenirs*, pp. 142-144. Blomberg had earlier told high naval officers that negotiations between "leading statesmen" or in connection with the Mussolini proposals might offer better prospects for substantive arms demands than talks in Geneva: ML, PG 34176, Note on briefing of Blomberg by Groos, 24 Mar 33, reel 27/452; PG 34168/1, Memo on discussion between Blomberg and Raeder, 28 Mar 33, reel 27/1002. Reports from Rumbold and Poncet indicate that the Nazis were regarding the Foreign Ministry with suspicion and hostility. One official, an expert on Russia and a Jew, had his house broken into, tried to protect his wife, and was beaten; he resigned. Neurath's own son, a Foreign Ministry official, learned that SA men were tapping his telephone. See *DBFP*, 2/V, Nos. 14, 29, 30; *DDF*, 1/III, No. 178. On the bureaucratic Gleichschaltung generally, see Bracher, "Stufen der Machtergreifung," pp. 171-174, and Schulz, "Anfänge des totalitären Massnahmenstaates," pp. 490-515.

SA members did not become counted as military effectives; if the German army was to number 200,000, any inclusion of SA men in the total would reduce the number of other men permitted.[59]

Nadolny was in Germany during the Easter recess of the conference, and although he does not seem to have conferred with Neurath, let alone with Hitler, he did meet with Blomberg on April 21. In their discussion, Blomberg did not rule out the idea of a standardization of armies as a future prospect, but he reportedly refused to enter into any commitment applying to the period of the first convention, and he insisted on leaving open the form in which standardization should and could take place in the more distant future; nevertheless, Germany would "come part way" to meet the Anglo-French proposals during the first convention by shortening the term of enlistment in the Reichswehr. To seek to delay the German concession without absolutely refusing it was a position somewhat like that taken by Blomberg in February, although a five-year delay before standardization was a lot to ask. Blomberg doubtless had a real concern that a sudden standardization would leave Germany with neither sufficient professionals nor, for a long time, reserves; he may also have feared that early negotiation on standardization would uncover too soon the German plan for an integrated army, combining militiamen and three-year men with long-service cadre. Aside from this question of standardization, Nadolny and Blomberg also discussed numerical demands with respect to effectives and aircraft, whereby German equality of rights—evidently numerical equality—would be achieved by the end of the first convention. This amounted to a revival of the old Reichswehr claim to equality with France, and even to a demand for an early practical application of it.[60]

Nadolny seems to have been inclined, if anything, to go beyond Blomberg's position, which he may not have fully understood. On the

[59] Acceptance by Nadolny and Neurath: *ADAP*, C/I/1, No. 103; *Records of Conference*, Series B, II, 391-394 (Nadolny speech); *DBFP*, 2/V, No. 83 (note); *FRUS, 1933*, II, 218. Police and SA men were only likely to be counted in part, as part-time soldiers. See, e.g., *DBFP*, 2/V, No. 108; *ADAP*, C/I/1, No. 239.

[60] AA, tel., Schönheinz to Blomberg, attn. Obstfelder, 25 Apr 33, 3154/D668891-92. On Blomberg's aims, note the position he took on May 8, described in text below. Our information on the 21 April discussion with Blomberg comes from what Nadolny later told Schönheinz, as then reported by the latter, who asked for confirmation. Despite the 20 Feb and 15 Mar 33 discussions in Berlin, Schönheinz still considered himself under orders to reject standardization. The report may exaggerate Blomberg's demands, and his stance in May was, if so, perhaps less a change in policy than a correction of a mistaken impression. But Blomberg had told Raeder in March that an army of 200,000 (as under the MacDonald plan) was insufficient: ML, PG 34168/1, Memo on discussion between Blomberg and Raeder, 28 Mar 33, reel 27/1002.

same day—whether before or after the discussion with Blomberg is not clear—Nadolny spoke confidentially to a group of journalists, and told them that Germany hoped for and expected an agreement which, practically speaking, would authorize a German standing army of up to 600,000 men. He said that Germany was in fact already well on the way to achieving this strength, counting the Reichswehr, the paramilitary formations, the labor service, and the militia, which was in the developmental stage. (This figure of 600,000 may have corresponded to the equality in effectives discussed with Blomberg; French standing strength at this time, including colonial troops, was 560,000-570,000.) Nadolny also told the newsmen that he would reject proposals to reach equality (by disarming the other powers) either at a level of 300,000 men with international inspection, or at Germany's current (100,000-man) level. In all, he indicated Germany was determined to rearm on a large scale, preferably within the framework of an agreement, or perhaps while spinning out the negotiations, but if necessary after leaving the conference and perhaps also the League, even though this might lead to serious complications, especially with Poland. This implied a policy of risk not in keeping with Hitler's intentions in the spring of 1933.[61]

After returning to the conference, Nadolny, on April 27, handed in a series of amendments and, in particular, one that referred standardization to the Permanent Disarmament Commission, a body proposed under the MacDonald plan for resolving disputes and conducting investigations after the convention went into operation. Such a procedure would certainly have had the effect, as Blomberg had wished, of avoiding any commitment to standardize during the first convention. To other delegates, this amounted to placing standardization on the Greek calends. Nadolny's speech the next day heightened the negative impact; perhaps designed for home consumption, it stressed the fourteen years Germany had been waiting for others to disarm, and said

[61] Brammer Collection, Z Sg 101/26, Information report, 22 Apr 33. French standing strength: Gamelin, II, 12. The Nadolny remarks are discussed by Weinberg (*Foreign Policy of Hitler's Germany*, pp. 161-162), who accepts them as a statement of German policy. But no other document I have seen indicates that Nadolny had authority to reject numerical equality at the 100,000 or 300,000 level; at Geneva, Nadolny only reserved his position on the MacDonald figure of 200,000, and as will be discussed later, Hitler's proposals of October and secret orders of December were not for any 600,000-man standing army. Nadolny may have thought that he could further his chances of succeeding Neurath as Foreign Minister (see note 63 below) by making jingo remarks to a group of reporters. Nadolny's number of 600,000 was not too far from Germany's projected overall *mobilized* strength, which would be perhaps 100,000 more than the total number of rifles needed and on hand for first equipment of both the army and Grenzschutz: 610,637 (Chapter Seven, Section 5).

that the German delegation had not yet presented all the demands that arose from the principle of equality of rights, although Nadolny denied that Germany was opposed to standardization in principle.[62] Nadolny's ambition was to replace Neurath as Foreign Minister,[63] and he may have intended by his speech to establish himself as a champion of German equality. He recognized the danger of isolation, and he hoped to have an opportunity for some high-level bargaining within a restricted circle of powers. But he had an unusual insensitivity to the impression he was producing on others. He had a reputation among German diplomats as a very difficult person, and Ernst von Weizsäcker, whose nerves and stomach were ruined by association with him, wrote that to judge by his character, tact, and sense of responsibility, Nadolny would have done better not to have become a diplomat.[64]

From April 29 to May 8, and again on May 11, Nadolny urged that the debate move rapidly on through a first reading of the whole British draft; under this procedure, attention would shift from the effectives question to the question of material disarmament on which the French and many other powers were vulnerable. This would ensure that the French would bear the blame if the conference broke down.[65] But thanks to the circumstance that the standardization of effectives had become the first item on the agenda, France could press the demand for standardization before disclosing any of her own objections to the provisions of the British plan; and she might never have to disclose them if the conference broke down. She could stand together with the British (and also the Americans) on the standardization issue, and a German refusal to standardize would help bring about a common anti-

[62] *SIA, 1933*, p. 262; Loosli-Usteri, pp. 392-396; *DDF*, 1/III, No. 207; *DBFP*, 2/V, No. 100; *FRUS, 1933*, I, 122-123; AA, tels. (2), Nadolny to Ministry, 28 Apr 33, 3154/D668907-14. Text of speeches: *Records of Conference*, Series B, II, 421-429. The German Foreign Ministry was apparently not informed of the text of the amendments until after they had been submitted.

[63] Aloisi, pp. 90-91, 117-118.

[64] *ADAP*, C/I/1, Nos. 103, 200; AA, tel., Nadolny to Ministry, 25 Apr 33, 3154/D668890. After the heated debate of April 28, Nadolny approached Massigli and expressed his hopes for an entente (*DDF*, 1/III, No. 207). Nadolny's memoirs (*Mein Beitrag*, pp. 132-133) say he acted on instructions, apparently from Blomberg, after protesting that to seek a postponement of standardization was a mistake. Schönheinz's telegram to Blomberg of 25 Apr 33 (note 60) does not, however, suggest any objections from Nadolny. On Nadolny's character: D. C. Watt, "Hitler and Nadolny," *Contemporary Review*, 196 (1959), esp. 54; Leonidas E. Hill, *Die Weizsäcker Papiere 1933-1950* (Frankfurt/M: Propyläen Verlag, 1974), p. 66. Watt's article generally stresses Nadolny's courage in speaking frankly to Hitler.

[65] *DBFP*, 2/V, Nos. 103, 107, 113, 117, 135; AA, Note on phone call, 5 May 33, 7360/E536596; tel., Hoesch to Ministry, 5 May 33, 3154/D668979-80; Note on phone call, 8 May 33, 3154/D669011.

German front.⁶⁶ Under such conditions, French delegates could also take a stiffer position on other issues. When Nadolny charged that other governments had failed to carry out their obligation to disarm, René Massigli replied that France was ready at any time to take up the question of whether each side had fulfilled its engagements, a strong hint that the famous French dossier on German violations might be opened. In the Effectives Committee, which was meeting simultaneously, the French pressed the argument that elements of the German police were military and, in particular, that the German Wehrverbände were really an illicit militia.⁶⁷

As for the British, Eden, Cadogan and their aides regarded their plan as one that should be accepted or rejected pretty much as it stood. The objectives were, domestically, to offer a compromise which would convince the British public that their government was making a serious and fair proposal, and diplomatically, to get over the obstacle that neither side would make a concession first. Far from compelling Eden to bargain or to bring pressure on the French to bargain, Nadolny's aggressive tactics, which seemed to endanger the effort at compromise, put Eden's back up. Aside from his general sympathies, Eden was now driven to side with Massigli by his high personal stake in the MacDonald plan and by the consideration that if the conference did not adopt the effectives section of the plan, the one part borrowed from earlier French proposals, there would be little hope that France would accept other parts less palatable to her. The British delegate obtained cabinet instructions to seek a decision on the standardization question before moving on to a consideration of materiel and other matters. Hoesch in London tried to forestall such instructions by pointing out the danger of pushing Germany to a formal rejection on effectives—presumably a warning that Germany might again leave the conference. But Hoesch did not achieve his goal, and in Geneva on May 8, Eden maintained his demand on standardization despite suggestions by Arthur Henderson that materiel be discussed without any offer of amendments.⁶⁸

⁶⁶ See *DDF*, 1/III, Nos. 249, 254, 257, 260, 265; *DBFP*, 2/V, No. 119. On 8 May, Massigli reported to Paris his concern that if the conference moved on to even a general discussion of naval matters, it would be impossible to maintain Franco-British solidarity (*DDF*, 1/III, No. 254).

⁶⁷ *DDF*, 1/III, Nos. 116, 207; *Records of Conference*, Series B, II, 425-426; *SIA, 1933*, pp. 263, 265-266; AA, tel., Nadolny to Ministry, 11 May 33, 3154/D669048; tel., Nadolny to Ministry, 12 May 33, 3154/D669079.

⁶⁸ *DBFP*, 2/V, Nos. 99, 103, 104, 110, 111, 113, 117, 119; Cab 23/76, meeting of 5 May 33, 33(33)4; *FRUS, 1933*, I, 131-132, 589-590; *ADAP*, C/I/1, Nos. 209, 239; AA, Note on phone call, 5 May 33, 7360/E536596; tel., Hoesch to Ministry, 5 May 33, 3154/D668979-80. The German government also made approaches in Washing-

Thus Nadolny's demands in Geneva had led to a situation by May 8 in which Germany stood in virtual isolation. From the German perspective, it seemed that the other powers were attempting to force a new *Diktat* on Germany, or else to blame her for a failure of the conference. Now, as we have seen, Hitler had no desire for a test of wills at a time when Germany was still militarily weak. One striking measure of Hitler's frame of mind was his meeting with the Polish minister in Berlin on May 2, when, presumably to dispel the danger of a Polish preventive attack, he agreed to a communiqué stating that the German government intended "to keep its attitude and its conduct strictly within the limits of the existing treaties"—an exceptionally conciliatory statement by the standards prevailing in interwar German-Polish relations.[69] This step was probably farther than the Foreign Ministry wanted to go in the Polish question, but Neurath, advised by Bülow, had been urging the need to avoid provocation and isolation until Germany had rearmed, and Blomberg, although unhappy with the MacDonald proposals, was not the man to question Hitler's wisdom. There appears to have been alarm in Berlin over the crisis in Geneva, and probably over dissatisfaction with Nadolny's conduct of negotiations. In a May 5 cabinet discussion of an American tariff-truce proposal, Blomberg warned against doing anything "which could further aggravate the already critical position of Germany in foreign politics." Gerhard Koepke, a senior official at the Wilhelmstrasse, gave the Belgian minister to understand on May 6 that Neurath and Blomberg would shortly go to Geneva and meet there with MacDonald to work out a solution directly. This meeting did not take place, but Blomberg did issue a statement on the evening of the eighth, saying that Germany was prepared to negotiate on the standardization of armies, while refusing any dictated solution.[70]

If Blomberg did tell Nadolny in April that there should be no com-

---

ton, and Roosevelt finally told Ambassador Luther and Hjalmar Schacht that he sympathized with the German procedural argument "as presented," but the American instructions that resulted were somewhat equivocal, and Wilson and Davis did not carry them out, partly because Nadolny had been "so unreasonable and . . . his position so untenable": *FRUS, 1933*, I, 128, 131, 137; *ADAP*, C/I/1, Nos. 209, 239; AA, tel., Luther to Ministry, 8 May 33, 3154/D669019. Moffat thought that Roosevelt had only spoken as he had to Schacht on the procedural point to "gild the pill" of his blunt opposition to German rearmament (Moffat MS Diary, 8 May 33).

[69] *ADAP*, C/I/1, No. 201 and encl.; *Papers and Memoirs of József Lipski*, pp. 78-81.

[70] Neurath's, Bülow's, and Blomberg's views: *ADAP*, C/I/1, Nos. 142, 210; Wollstein, "Denkschrift," pp. 82-94; Liebmann MS Papers, Remarks of Blomberg, 1 Jun 33, p. 200 (typed). Neurath-Blomberg trip: *DDB*, III, No. 27; also *DBFP*, 2/V, Nos. 122, 134. Blomberg 8 May statement: Karl Schwendemann, ed., *Abrüstung und Sicherheit: Handbuch der Sicherheitsfrage und der Abrüstungskonferenz*, 2 vols. (Berlin: Weidmannsche Buchhandlung, 1933-1935), II, 383-385.

mitment to standardize during the first convention, he now dropped that position. His May 8 statement said that Germany (as he had also told Nadolny) was ready to shorten her period of service, but that a long transition period was essential for her, since she alone lacked trained reserves (i.e., to expand a 200,000-man MacDonald-plan army), and this lack had to be remedied gradually. He also called for more information on the disarmament plans of the other powers in organization and materiel: the more general disarmament there was, i.e., by others, and the more flexible the organizational provisions of the transition period were, the easier it would be for Germany to make concessions on standardization. This was rather vague language, but Blomberg's statements in less-public documents show his main concern was to be able to retain at least the existing number of professionals, so as to have a relatively high proportion of officers and long-service men to total strength. Blomberg was apparently trying to conceal the real nature of the German Umbau and rearmament plans, and to rule out convention stipulations that would interfere with the execution of those plans. Since negotiation on organizational arrangements might well have exposed actual German intentions, we must suspect that Blomberg would have preferred simply to see the conference collapse; a few days later, according to a note by Schönheinz, the Defense Minister said that the delegation should not shrink from the danger of such a collapse, even if Germany were unjustly blamed.[71] But at this time, Hitler could not accept the risk of being blamed for a collapse, with the possible sequel of preventive action.

Blomberg's May 8 statement may have been meant as a warning to Nadolny, but if so, the Ambassador did not get the message. He complained to Berlin because Blomberg had spoken without giving him a chance to use the concession in his negotiations, and he now proceeded to what is known in Germany as a *Flucht nach vorn*, or a jump from the frying pan into the fire. Without consulting Berlin, he proposed to Eden that they drop the debate over procedure and that he tell Eden privately what Germany was prepared to accept in effectives and other matters; Eden could later inform the other principal delegations. Eden and Massigli, who were collaborating closely, both welcomed this idea as an easy means of avoiding the public discussion of materiel: they feared that such a discussion would enable Nadolny to put his case in a better light, as compared with theirs. But Nadolny

---

[71] Schwendemann, II, 384-385; *ADAP*, C/I/1, Nos. 231, 238; ML, Shelf 5939, Note by Schönheinz, 13 May [33], reel 18/129. Blomberg's remarks at the 12 May cabinet meeting (*ADAP*, C/I/1, No. 226)—in which he suggested that Germany be present at Geneva but take no part in negotiations—may also have reflected a certain anxiety as to the course negotiations on standardization might take.

apparently had the idea that, by bargaining in private discussions, he could make an acceptance of standardization contingent on far-reaching concessions to Germany in materiel, and still not have to face public accusations that he was seeking rearmament. At his first private sessions with Eden and Cadogan on May 9, he stated that if Germany were to make the great sacrifice of reorganizing the Reichswehr, she would want "complete equality" by the end of the first convention, including *quantitative* equality in aircraft (as he had discussed with Blomberg in April), and a certain number of submarines—both categories in which the British plan made no allowance for Germany at all. If the other powers were to have 155 mm guns, Germany would have to have them too, and on effectives, she would have to have compensation for the French reserves. Eden and Cadogan were upset at Nadolny's demands, and they told him that Britain and the United States, as well as France, would refuse to agree to the rearmament of Germany.[72]

Nadolny's demands were in line with the principles of equality of rights, but previously no one had made it so clear that these principles meant rearmament. Members of Nadolny's own delegation had advised him not to go so far, and one member reported to Berlin the following morning that the situation had become very unpleasant. In subsequent talks, Nadolny retreated somewhat, reviving the old assertion that Germany would gladly remain disarmed if others disarmed to the German level; he also gave some explanations on effectives, and he said that Germany would be content in practice with less than quantitative equality, including in aircraft.[73] But meanwhile his initial demands (of May 9) were reverberating in various quarters, and the impression spread that the conference was on the verge of a breakdown, for which the Germans could and should be blamed. Hugh Wilson (the acting head of the American delegation) observed to Eden that "if [the] Conference must be broken by Germany, it was essential that the world should know at the right time that it was Germany who had broken it." A Foreign Office official minuted, "If the Conference is to break down, it is better that it should do so now—from

[72] AA, tels. (3), Nadolny to Ministry, 9 May 33, 3154/D669020-23, D669027; *DBFP*, 2/V, Nos. 120, 123, 124; *DDF*, 1/III, Nos. 249, 254, 257. Admiral Freyberg also deplored the lack of notice to the delegation on Blomberg's statement: ML, Shelf 5939, ltr., [Freyberg] to Raeder, 13 May 33, reel 18/125; Shelf 5940, Naval delegation report, 8-14 May 33, reel 18/712. According to Hugh Wilson's report (*FRUS, 1933*, I, 133-134), Eden told him that Nadolny had asked for "equality in quantity with the major powers." This is probably a mistaken simplification by Wilson; both Nadolny's and Eden's own reports only speak specifically of quantitative equality in aircraft.

[73] AA, Notes on phone calls, 10:30 a.m. and noon, 10 May 33, 3154/D669028-29; tel., Nadolny to Berlin, 10 May 33, 3154/D669030-32; *DBFP*, 2/V, No. 125.

our point of view." The British cabinet decided that Eden should disclose Germany's new demands to Henderson and perhaps others, and in transmitting this decision to Eden, Vansittart advised: "It is vital that if there is a break, the blame should be plainly put on the right shoulders, and a public exposure of Germany's intransigence would mobilize opinion against her and show her isolation." Eden and Massigli agreed that negotiations should not be broken off without placing the German position clearly on record so that Nadolny could not deny it; at the same time, they also agreed it would be a mistake to engender an immediate public debate, which might embitter the situation.[74]

It was arranged on the tenth that the delegates of the five powers would meet with Henderson on the evening of the eleventh to review the situation. Meanwhile, Eden told Nadolny, with French and American concurrence, that if all demands for material rearmament were dropped, the Western powers would be willing to consider German demands on effectives. But Nadolny, still hoping to force a hard bargain, seems to have taken Eden's readiness to discuss his demands on effectives as virtual agreement on that point, and early on the eleventh he phoned his Foreign Ministry asking urgently that immediate pressure be brought to bear in London and Washington so as to get British and American acceptance of token qualitative equality in materiel by 5:30 p.m. Bülow told him that this was impracticable, although Hoesch, in London, did what he could. This effort created an impression in London that the conference was close to a breakdown, while officials in Berlin seem to have become more dubious about Nadolny's judgment.[75]

An aggravating factor in the crisis was the circulation in Geneva on May 11 of an article which had appeared over Neurath's signature in the *Leipziger Illustrierte Zeitung*. This argued that since the other powers were obviously unwilling to disarm, Germany would have to have the kinds of arms retained by others, including aircraft and

[74] *DBFP*, 2/V, Nos. 124 (and note), 125, 131, 132; FO 371, Minute by R. Stevenson and telephone message, Vansittart to Eden, 10 May 33, W5094/40/98; Cab 23/76, meeting of 10 May 33, 34(33)1; *DDF*, 1/III, No. 260.

[75] *DBFP*, 2/V, Nos. 132, 133, 137; AA, Notes on phone calls, 11 May 33, 3154/D669046, D669051-52; 7360/E536663-64; tel., Nadolny to Ministry, 11 May 33, 3154/D669047-50; tel., Neurath to Nadolny, 13 May 33, 7360/E536688. French objections to high-level five-power conferences did not apply to this kind of delegates' meeting within the conference framework. Possibly Nadolny's judgment was affected by having, as an American observer put it, two Nazi gunmen at his elbow (SDP, Ferdinand Mayer Geneva diary, 15 May 33, 500.A15A4/1469½). It was during this crisis that Nadolny had to contend with the raising of a Nazi flag at the delegation headquarters at the instance of Reinhard Heydrich and Friedrich-Wilhelm Krüger (soon to be head of the Ausbildungswesen): see Nadolny, pp. 131-132; Paul Schmidt, *Statist*, pp. 262-263.

heavy artillery, and that similar steps were necessary with respect to effectives. Eden considered this article to be an authoritative statement that Germany intended to disregard the British draft and simply to help herself to more arms, notably aircraft and guns. Eden's nerves, as well as Nadolny's, had become somewhat frayed by this time. When the delegates of the five powers met with Henderson on the evening of May 11, Eden, Massigli, and Hugh Wilson all expressed strong opposition to any German rearmament. Only the Italian representative accepted the idea of granting token quantities of hitherto-forbidden arms to Germany. Nadolny refused to agree to a standardization subject to the rest of the convention proving satisfactory; he insisted that the other powers must accept the rest of his demands on reserves, overseas troops, and equality of rights in materiel. It proved impossible to reach agreement, and the problem was referred to the Bureau of the Conference.[76]

Nadolny had managed to make Germany vulnerable to denunciation on both effectives and materiel. Surveying the situation, Admiral Freyberg (the head of the naval contingent in the German delegation) noted sadly that most of his German delegation colleagues felt there was no value left in the claim that other states should disarm, and that hostile propaganda, assisted by Neurath's article, had convinced the world that the German disarmament demand was not genuine; even the pacifist women's organizations (he wrote) no longer echoed the German proposals. Massigli had now indeed begun to aim for a general debate, probably in a plenary session, in which the whole German position could be exposed to the public. Nadolny had some second thoughts after the meeting on the evening of the eleventh, and he thereupon recommended to Berlin that when the Bureau met, and if that body did not prove too hostile, he should make the concession he had just refused to make—agree to standardization subject to a general reservation that the rest of the convention be satisfactory. Neurath and Bülow approved this suggestion on the morning of the twelfth, but Hitler was to intervene, superseding this with another approach.[77]

[76] *SIA, 1933*, p. 265; *Times* (London), 12 May 33; *ADAP*, C/I/1, No. 239; *DBFP*, 2/V, Nos. 135-137, 140; AA, tel., Nadolny to Ministry, 12 May 33, 3154/D669072-73; *DDF*, 1/III, No. 265; *FRUS, 1933*, I, 136-138. When challenged at the five-power meeting about the Neurath article, Nadolny said that one should not attribute too much importance to something no doubt written by an ill-informed official of the Wilhelmstrasse press service; Massigli thought this reply quite inadequate, but it may well have been true. At all events, the article was probably written well before its first appearance.

[77] ML, Shelf 5939, ltr., [Freyberg] to Raeder, 13 May 33, reel 18/127; *DDF*, 1/III, No. 265; *ADAP*, C/I/1, No. 239; AA, tel., Nadolny to Ministry, 12 May 33, 3154/

7. HITLER PLANS HIS REPLY

Circumstances converged on May 11 and 12 to bring the crisis home to Hitler. He probably had not followed the detailed course of events in Geneva, but Neurath doubtless consulted with him now over Nadolny's new suggestion of accepting standardization subject only to a general reservation. Aside from this issue, a meeting of the Effectives Committee on the eleventh went disastrously for Germany. Despite— or very likely because of—the testimony of Reinhard Heydrich for the SS and of Friedrich-Wilhelm Krüger for the SA, General Schönheinz had been quite unable to convince the other experts that the SS, SA, and Stahlhelm should not be counted, at least in some proportion, as effectives, and all the committee members except the Hungarian representative either voted against Germany or abstained.[78] Aside from the threat that including SA men would reduce the permitted number of Reichswehr members, the vote showed how the rest of the world suspected the Wehrverbände and, for that matter, how little trust it placed in assertions that Germany was disarmed.

Another experience of May 11 was Hitler's initial reception of Sir Horace Rumbold, the British ambassador. Hitler did not lose sight of his objective of a British alliance, and during the conversation, he stressed that he was a long-time advocate of close collaboration between Britain and Germany, although (he alleged) his own followers were now attacking him for this. As Rosenberg was doing in London, Hitler argued that Germany was serving as a bulwark against Communism. The Führer also showed his interest in friendly relations with Britain by suggesting that a British loss of India might have disastrous effects on the rest of the world, and he promised to take prompt action in the case of a British subject who had been arrested and held without trial. (The man was released two days later.) But a heated argument developed over the Jewish question, leading Rumbold to infer that the Chancellor himself was the main source of official anti-Semitism. Rumbold said that although British sympathy for Germany and her problems had been increasing, this sympathy had been forfeited by the recent treatment of Jews and political opponents. Rumbold also read a watered-down version of a message from Simon, urging Hitler to be moderate for the sake of the Disarmament Conference and the coming World Economic Conference. The question of the military character of the Wehrverbände came

---

D669072-73; Memo on phone call (10:30 a.m.), 12 May 33, 3154/D669075. Freyberg did not blame Nadolny for the loss of credibility, but rather the influence of army leaders in Berlin, along with French tactics and propaganda.

[78] *ADAP*, C/I/1, No. 239; Paul Schmidt, *Statist*, pp. 261-262.

up, too, and Rumbold pointed out that the SA and Stahlhelm members he had seen at a recent rally were equivalent in number to eight army corps, and appeared to be excellent potential reserves. Like the Effectives Committee decision in Geneva, this must have brought home to Hitler the suspicion being aroused by these organizations.[79]

If Hitler tried to impress Rumbold with his desire for British friendship, he reacted differently to another event of this day. Lord Hailsham, the British War Secretary, stated on the evening of May 11 in the House of Lords that if Germany left the Disarmament Conference, his personal opinion was that "she would remain bound by the provisions of the Treaty of Versailles, and that any attempt to rearm would be a breach of that treaty and would bring into operation the sanctions which it provided." On May 12, Paul-Boncour contributed another statement, asserting that the Treaty of Versailles would remain in effect if no disarmament convention was concluded. Actually, Hailsham's off-the-cuff talk of sanctions was ill-founded: the Treaty of Versailles made no provision for applying sanctions in the event of German rearmament.[80] But to Hitler, the Hailsham and Boncour statements were threats, and according to his code, the best way to deal with a threat was to reply with another, more effective one.

Hitler's response would be to threaten withdrawal not only from the conference, but also from the League, a more serious and irreversible step. Hitler's menace could be made more convincing by stating it in a speech before the Reichstag, and by obtaining a unanimous Reichstag endorsement for it. At the same time, he could dispel

[79] *ADAP*, C/I/1, No. 223; *DBFP*, 2/V, No. 139. Rumbold had edited out the sentence most likely to offend in Simon's message: cf. *DBFP*, 2/V, No. 126 with *ADAP*, C/I/1, No. 223, note 5, noting also Rumbold's report (*DBFP*, 2/V, No. 139) that he had read to Hitler "a German version in general terms" of Simon's message. Rumbold had refused to meet Hitler while he was in the opposition; he did meet Hitler briefly at a dinner on 8 Feb 33: Gilbert, *Rumbold*, pp. 351, 368. One of Rumbold's subordinates stated later that the first time he had ever seen Rumbold in a temper was after his first interview (presumably that of 11 May) with Hitler (Gilbert, *Rumbold*, p. 383).

[80] On statements: *Times* (London), 12 May 33; *SIA, 1933*, pp. 265-266. Hailsham's declaration was his own improvisation; the Foreign Office had provided background for a less far-reaching statement (FO 371, Minute by Stevenson, 11 May 33, W5311/40/98). The Foreign Office was aware of the lack of sanction for sanctions: FO 371, Minute by Sargent, 16 May 33, C4339/245/18. The occasion for this minute was a telegram from Rumbold of 13 May reporting the grave impression made in Germany by Hailsham's statement, as well as a wide concern over isolation and foreign hostility, which could make Hitler's position very difficult. But Hoesch on 16 May called Berlin's attention to British doubts and the lack of basis for Hailsham's statement (*ADAP*, C/I/2, No. 242). As will be discussed in the next chapter, it was doubtful that any action could legally be taken against German arms violations. Hitler may not have heard of Boncour's statement when he met with his cabinet on the twelfth, as he referred to threats from "English quarters" (not "English and French"), although he also spoke of replying to the "ministerial speeches of the last few days" (*ADAP*, C/I/1, No. 226).

foreign hostility by avowing his peaceful intentions, by saying something to allay suspicion of the Wehrverbände, and by giving a conditional acceptance of the MacDonald plan. Such a procedure would also serve a domestic purpose: the last open dissenters, the remaining Social Democratic Reichstag deputies, would not refuse to endorse "voluntarily" a peace resolution—particularly if privately told that their colleagues in the concentration camps would otherwise be murdered. Thus, outside and inside Germany, a public picture could be presented of a unified nation, supporting its leader. Hitler referred to much of his planned reply in a cabinet meeting on the twelfth.[81]

At the meeting, Hitler expressed a different outlook than that which had governed German disarmament negotiation in the past—so different that several of his ministers did not understand him. For him, at this juncture, Germany's status with world opinion was more important than the arduous negotiation of provisions respecting standardization or the Wehrverbände. He was very interested in retaining a professional cadre for a future offensive army, though he was much less concerned to maintain Röhm's storm-troop formations. But apparently, he did not expect that convention provisions, if they were concluded, would seriously hinder a nation resolved to rearm. This was presumably the significance of his Bismarckian-sounding words to the ministers, "The disarmament question will not be settled at the conference table." He was not, as Hugenberg seems to have supposed, calling for an open proclamation of German rearmament. Rearmament, Hitler thought, could not be carried out "by normal means," that is, openly; it would be necessary "to exercise the greatest restraint in proceeding with rearmament." Nor did he intend, as Blomberg and Neurath seem at first to have thought, to withdraw from negotiations at Geneva.[82] His idea was to carry on rearmament covertly, while outwardly negotiating and reassuring the world. To this end, he planned to give a "convincing refutation" of foreign charges that the

[81] *ADAP*, C/I/1, No. 226. Bülow, in informing Nadolny of Hitler's proposed statement, described the prospective content as "non-disarmament of the others, question of the Wehrverbände" (AA, tel., Bülow to Neurath, 12 May 33, 7360/E536740). On domestic implications, see Bracher, "Stufen der Machtergreifung," pp. 197-198, and *Deutsche Diktatur*, p. 242.

[82] Dülffer says (pp. 262-263) that, in contrast to Hitler's more conciliatory position, Blomberg and Neurath pleaded at the 12 May cabinet meeting for a withdrawal from disarmament negotiations. As I read the record of the meeting, particularly in the light of Blomberg's and Neurath's characters, they were not pleading against Hitler; they thought, mistakenly, that they were agreeing with him. On my interpretation, Blomberg still had the wrong idea when he spoke of withdrawal to to Schönheinz (ML, Shelf 5939, Note by Schönheinz, 13 May [33], reel 18/129), who evidently seized on his words (and quite likely strengthened them) as justification vis-à-vis the rest of the delegation.

Wehrverbände were military organizations—as it would turn out, by offering inspection of them under a future convention. This was easy enough for him, as the Wehrverbände were no longer central to his plans, and as he did not believe that any convention would be effective anyway. The important task was to convince non-Germans, and especially the English, of the merits of the German case. He, the dreamer of alliances and conquests, must have writhed at the tactics of Nadolny and Schönheinz, who fought so stubbornly against standardization and the counting of police and storm troops that they made Germany appear the obstacle to disarmament, and thereby helped to isolate her.

8. ALARM OUTSIDE GERMANY

On May 12, before Hitler's plan to speak was known in Geneva, the immediate pressure at the conference receded somewhat. At the Bureau meeting, Henderson suggested again, and the delegates of the powers now agreed, that the General Commission should proceed to a general discussion of materiel without taking up amendments. Thus the other powers did not maintain their position on procedure. But the prospect of an eventual public confrontation remained. Papen gave new cause for concern by delivering a militarist speech on May 13.[83] After the announcement on that day that Hitler would address the Reichstag on the seventeenth, the proceedings at Geneva were in effect suspended, and the diplomatic world speculated anxiously about what Hitler would say. In a series of reports, Rumbold concluded, rightly, that the Reichstag session would be used to rally domestic opinion and give an outward impression of unity, and that the speech would be moderate. Mussolini apparently sent Hitler a private message urging moderation. Massigli feared that Hitler might follow the Italian example and give an unqualified German acceptance of the

---

[83] League of Nations, Conference for the Reduction and Limitation of Armaments, *Records of the Conference*, Series C, *Minutes of the Bureau*, vol. I, 171-172; *ADAP*, C/I/1, No. 239; C/I/2, No. 242; AA, tel., Nadolny to Ministry, 12 Mar 33, 3154/D669077-78; Note on phone call, 12 May 33, 7360/E536676; *SIA, 1933*, p. 266. Despite the reprieve provided by the 12 May Bureau meeting, Nadolny decided that Germany could only escape isolation and an eventual humiliation by following the Italian example and generally accepting the British plan (AA, tel., Nadolny to Ministry, 12 May 33, 3154/D669079-80; ML, Shelf 5939, ltr., [Freyberg] to Raeder, 13 May 33, reel 18/126-127; *ADAP*, C/I/2, No. 241). As noted in the text just below, Massigli got wind of this idea, and was concerned. But Neurath threw cold water on Nadolny's suggestions and now warned him against making concessions beyond what Blomberg had indicated in his speech (AA, tel., Neurath to Nadolny, 13 May 33, 7360/E536688).

MacDonald plan, provided that it was not amended; if so, then the British and Americans might request a similar acceptance from France. During a telephone conversation with Berlin, a member of the German delegation asked what the speech would be like, but at the Foreign Ministry itself, officials did not know.[84]

The most dramatic response, and perhaps the least informed, was the American. Norman Davis urged President Roosevelt to issue, before Hitler's speech, a declaration on disarmament, "so as to take the wind out of his [Hitler's] sails and lead him to take a more reasonable and conciliatory attitude." Davis feared that Hitler's speech would bring about a clash that would break up the Disarmament Conference. Now, as Roosevelt had already disclosed to MacDonald, and a little later also to Herriot, he had planned to declare that if other nations determined collectively that a certain nation was an aggressor, and if the United States agreed with the determination, it would refrain from any action, or from the protection of its citizens in any action, which would interfere with the collective effort against that aggressor. Roosevelt had also decided that the United States should accept continuous inspection. The American government had tended to hold back on announcing such policies, for the sake of leverage, until the French had committed themselves to substantial disarmament; like the Germans and the French, the Americans preferred to await prior concessions by others. Davis thought the time had now come for an announcement on these matters, and he sent a draft of a presidential statement to Washington.[85]

Within the last few days, however, Roosevelt had turned to another idea, i.e., to propose that all governments should declare they would refrain from sending armed forces across their borders, providing there was no increase in armaments anywhere. This proposal was discussed with Ambassador Hans Luther and Hjalmar Schacht (also in Washington for financial discussions), who both said diplomatically that the idea was "well worth consideration." As officials in the State Department understood it, Roosevelt's underlying intention was, on the one hand, to assure Germany that her current disarmed condition would not expose her to unprovoked attack, and on the other,

---

[84] *DBFP*, 2/V, Nos. 142, 144; FO 371, tels., Rumbold to F.O. 13, 16 May 33, C4339/245/18, C4405/245/18; *DDF*, 1/III, Nos. 272, 292; AA, Note on phone call, 13 May 33, 7360/E536696.

[85] *FRUS, 1933*, I, 102-104, 106-110, 140-142, 145-146, 150-151; *DBFP*, 2/V, No. 86; *DDF*, 1/III, Nos. 179, 182, 189, 197; SDP, tels. (2), Davis to Dept., 15 May 33, 500.A15A4 General Committee/376, 378; Moffat MS Diary, 25 Apr, 15 May 33. The proposal not to aid aggressors had leaked through Herriot and had been reported in the French press: Moffat MS Diary, 26 Apr 33; *DBFP*, 2/V, No. 161.

to prevent her from rearming and to indicate to other powers that Germany could not rearm with impunity—indeed, as the French ambassador heard, to intimate that if she did rearm they would then be free to take sanctions against her. Even before Davis made his suggestion, Roosevelt contemplated anticipating Hitler's speech with an announcement of this nonaggression-pact scheme. This was hastily worked up into a message sent on May 16 to the chiefs of state (not chiefs of government) of all nations participating in either the Disarmament or World Economic conferences.[86]

As issued, the message did not explicitly state the above-mentioned underlying ideas—which would have been political dynamite in the United States—and it gave the impression of being a minor supplement to the Kellogg-Briand pact. No other government had been consulted. In the crisis at hand, the most significant part of the proposal was a provision that, during the period of disarmament negotiations, no nation should increase its armaments beyond the limitations of (its) treaty obligations. It was not made clear, however, why Germany, presumably the nation Roosevelt had most in mind, should honor this aspect of his proposed agreement if she was not going to honor her older commitments anyway. To British and French leaders, the message was a sharp disappointment, as it made no reference to the proposal previously outlined to MacDonald and Herriot for withholding supplies from aggressors. One passage suggested that the MacDonald plan was a first step, but inadequate; this probably helped to evoke from MacDonald the marginal note: "The whole thing is depressing & shows the unsatisfactory nature of Ameri[can] Diplomacy. They cannot keep out of the limelight, they are always prone to do things on their own in the middle of negotiations." The Roosevelt message also contained an attack on "offensive weapons," including war planes, heavy mobile artillery, and tanks, a line of argument that was distinctly unpalatable in Paris.[87] On the whole, the message served as unintended grist for the German mill. Hitler used the Roosevelt proposal in his speech as a cue to stress Germany's lack of arms and particularly of those offensive arms Roosevelt had named, to state Germany's readiness to agree to a nonaggression pact, and generally, to speak of her ardent desire for peace. As for treaty limitations, he said, "Germany does not wish to take any other path than that recognized as justified by the Treaties themselves"—a reference to the

[86] Moffat MS Diary, 6, 13-15 May 33; AA, tel., Luther to Ministry, 8 May 33, 3154/D669018; *DDF*, 1/III, Nos. 283, 284; *FRUS, 1933*, I, 143-146.

[87] *FRUS, 1933*, I, 145-149; SDP, Memo of conversation with French ambassador, 17 May 33, 500.A15A4/1934; *DBFP*, 2/V, Nos. 150, 152; FO 371, tel., Lindsay to F.O., 18 May 33, with minutes by MacDonald and F.O. officials, 18-19 May 33, W5529/40/98; *DDB*, III, No. 33; *DDF*, 1/III, Nos. 283, 284, 287.

right to national security under Article 8, and probably also to the implied obligation of the Allies to disarm.[88]

After Hitler delivered his speech, American, British, and Italian officials claimed credit for their respective leaders for having induced the German Führer to follow the path of moderation.[89] But Hitler's short-term policy was already a peaceful one, and he had not intended to storm out of the Disarmament Conference at this time. The apprehension in Berlin in the days before the speech is shown by Neurath's anxiety on May 13 lest Nadolny, by leaving Geneva too abruptly for Berlin, might give the impression that Germany considered a rupture at Geneva inevitable, as well as by Blomberg's approval on May 16 of a proposal transmitted by the delegation (and later disclaimed by Schönheinz), under which Germany would accept a transition within eight years to an army of 200,000 serving an average of eight months.[90]

9. HITLER'S "PEACE SPEECH"

Hitler's speech did not simply extol peace, it wrapped the mantle of peace around German foreign policy.[91] Yet it contained hardly any new points. The basic theme was that oldest standby of Hitlerian rhetoric, the lesson he had preached as a Reichswehr "drummer" in 1919, i.e., the ruin and injustice arising from the Treaty of Versailles. As he had with British visitors, but at much greater length, Hitler said

[88] *ADAP*, C/I/2, No. 243; Domarus, I/1, 270-279.

[89] *FRUS, 1933*, I, 165; Moffat MS Diary, 17 May 33; *DBFP*, 2/V, Nos. 153, 160; FO 371, tel., Graham to F.O., 19 May 33, C4557/245/18; tel., Rumbold to F.O., 20 May 33, C4562/245/18. Heinrich Brüning also claimed credit for influencing the speech, and indeed Hitler conferred with him on 16 May: Brüning, *Memoiren*, pp. 669-670; *FRUS, 1933*, I, 150; SDP, disp., Gordon to Dept., 6 Jun 33, 862.00/3000; *DDF*, 1/III, No. 314. No doubt Hitler consulted Brüning to enlist his influence over the remaining Social Democratic and Center deputies and make the vote on the seventeenth unanimous, rather than because he really wanted Brüning's advice on the speech, which remained fundamentally true to the ideas he had outlined to the ministers on 12 May. Brüning, however, who had told Ambassador Sackett in February that he thought he could work with Hitler (SDP, disp., Sackett to Dept., 13 Feb 33, 862.00/2914—a passage omitted from the publication in *FRUS, 1933*, II, 188-190), still hoped he could exert a moderating influence.

[90] AA, tel., Neurath to Nadolny, 13 May 33, 7360/E536741; tel., Rheinbaben to Ministry, 15 May 33, and notes, 16 May 33, 7360/E536742-45. Under the delegation's proposal, police and the members of the Wehrverbände (whose nonmilitary character would supposedly be confirmed by inspection) would not be included in the 200,000, and aircraft would be permitted. An exact boundary would be drawn between military training and physical-fitness training. According to a youth leader from Western Germany, later an exile, a French invasion was feared; no lead came from Berlin, but the two western area commands (Wehrkreise V and VI) and the Western Command of the Prussian Landespolizei took different precautions on their own: H[ans] Ebeling, *The Caste: The Political Role of the German General Staff between 1918 and 1938* (London: New European Publishing, 1945), p. 41.

[91] Domarus, I/1, 270-279.

war was to be avoided, and like Rosenberg in London, Hitler maintained that his government was a barrier to Communist upheaval. Germany had disarmed, and claims to the contrary were false; as with Rumbold on May 11, Hitler denied that the SA and SS were trained military reserves. To make this denial convincing, he extended a new offer: if other states would accept mutual supervision of armaments, he would agree to subject these organizations, along with German arms generally, to international supervision.

Hitler refused to agree to an abolition of Germany's existing defense organization unless she were granted qualitative equality. He explained that the reorganization of the Reichswehr should take place step by step at the same rate as the disarmament of other states. He stated: "Germany agrees in principle to a transitional period of five years during which to build up her national security, in the expectation that at the end of this period she will really be put on a footing of equality with other states"; he also said Germany would forgo any weapons that others would forgo. Like Nadolny and Schönheinz, Hitler argued that French troops in North Africa and fully trained (French) reserves should be counted as effectives, but German police should not. Hitler welcomed Mussolini's four-power pact proposal and (as noted) Roosevelt's nonaggression-pact scheme. But Hitler also warned that any attempt to apply sanctions against Germany, as threatened by newspaper articles and "regrettable speeches," would lead to the invalidation, morally and in practice, of the peace treaties. If there was an attempt to dictate to Germany through a majority vote, the German people would draw the only possible conclusion for a constantly defamed nation, and withdraw from the League of Nations.[92]

In all, Hitler had offered perhaps two new concessions. First, there was the idea of extending a general international inspection, applying to all states, to the SA and SS, as well as to German armament. In the light of what we know now of German preparations, it is clear he did not expect there would be effective control, and he may already have planned to "demilitarize" the storm troops. Second, he would now regard the British plan as a possible basis for the solution of the

---

[92] Other occasions on which Hitler voiced many of the same themes include an interview with Louis P. Lochner, 21 Feb 33, and in particular his speech to the Reichstag on 21 Mar 33, in which he said of the British plan: "We recognize it as a sign of responsibility and good will that the British government has made the attempt through its disarmament proposal to bring the Conference at last to rapid decisions. The Reich government will support every effort which is directed at carrying out effective general disarmament and securing the long-overdue claim of Germany for disarmament." See Domarus, I/1, 212-213, 229-237.

disarmament question, instead of as a basis for negotiation.[93] The meaning of this last is not too obvious, but the practical purport seems to have been an acceptance conditional on qualitative equality, and an abandonment of obstructive tactics. Nadolny's basic positions remained unchanged, and by publicly lending his authority to these positions, Hitler made them irrevocable. Hitler had, however, said a great deal about wanting peace, reconciliation between nations, and general disarmament, and this gave a good impression and diverted attention from such substantive issues as Reichswehr organization. Hitler's speech tended to reassure the non-Nazi majority within Germany and allay apprehensions in England. Experts in the French, British, and American diplomatic services knew that Hitler had conceded little or nothing, but they had no direct access to the public ear, and they too were relieved. Some public figures, such as Arthur Henderson and Norman Davis, did not recognize the emptiness of Hitler's speech, or if they did, thought it expedient, for the sake of disarmament, to pretend that Hitler had made a major contribution. The speech at any rate did dispel fears of an immediate collapse of the conference, and so it ended the crisis which had begun on April 27.[94]

FROM HIS own point of view, Hitler may have somewhat overdone the conciliatory side of his speech, for in succeeding months his threat of withdrawal was not clearly perceived by other governments. Also, although his speech made a favorable impression in England, particularly by making further negotiations on the MacDonald plan possible, it did not clear the path for a special Anglo-German understanding. This could not develop in the Geneva setting, or while Eden and Vansittart influenced British policy. Yet Hitler's speech salvaged something of Germany's standing as the aggrieved victim of Versailles, and prevented a public exposure of German rearmament and a solidification of the Anglo-French cooperation temporarily practiced in Geneva. One problem for a regime like Hitler's is that out of fear, ambition, or ideological intoxication, every subordinate tends to become a firebrand. Only the dictator can afford to appear relatively moderate.

[93] A circular instruction, approved by Hitler, also said that the plan was accepted as "a basis of the convention to be concluded": *ADAP*, C/I/2, No. 251.
[94] *SIA, 1933*, pp. 271-272; Meinck, *Aufrüstung*, pp. 29-31; Wendt, *Economic Appeasement*, pp. 86-87; *DDF*, 1/III, No. 318. Expert opinion: *DDF*, 1/III, Nos. 299, 314; *DBFP*, 2/V, No. 155; FO 371, Minute by A. Leeper, 17 May 33, C4671/245/18; Minute by Vansittart, 19 May 33, C4447/245/18; *FRUS, 1933*, I, 151, 159-164; Moffat MS Diary, 17 May 33. On Davis's views, aside from his perhaps unavoidable public endorsement, see *FRUS, 1933*, I, 165. Poncet gave a penetrating analysis of the significance of the speech in *DDF*, 1/III, No. 314.

In considering what to say in his speech, Hitler had apparently taken account of Rumbold's warning of the turn in British opinion and of the ambassador's comment on the appearance the SA gave of being potential reserves. There was more to the change in British opinion that Hitler did not know. Brigadier A. C. Temperley, the principal British military delegate at the conference, had written a memorandum arguing that the time had come for a revision of the British attitude toward Germany. Temperley pointed out that Germany had been rearming for some time, and that Hitler was apparently accelerating the process. Storm troops had been converted into auxiliary police, and were to receive military training. The German labor service would also receive physical and "Defence Sport" training, and this had been openly described as preparation for military service under a conscription system. Göring was actively organizing an air force, and planes were being built. Temperley, who had been dealing directly with Schönheinz, noted that the German attitude in Geneva had stiffened: "The increasing insolence of the Germans has brought the discussion of effectives to a complete standstill." No moment could be worse chosen for the disarmament of France, the Little Entente, and Poland. Temperley suggested that Britain, France, and the United States should sternly warn Germany that there could be no (non-German) disarmament, equality of rights, or treaty revision without a complete reversal of current preparations and tendencies in Germany. A crisis would follow such a declaration, and military pressure might have to be maintained for some years at considerable financial cost, but strong combined action at this time would be effective, and preferable to allowing things to drift, and "the sacrifices of the last war to be in vain and the world to go down in economic ruin." The British government had paid little heed to intelligence of German rearmament, but in the anxious period preceding Hitler's speech, Temperley's warning did not pass unnoticed. At Vansittart's suggestion, the memorandum was circulated to the cabinet on May 16; it was accompanied by a paper signed by Simon which reviewed recent German claims and demands and ended by suggesting that Hitler's speech might call for a concerted warning from Britain, France, and the United States "as a final effort to restrain the Nazi Government from embarking on a policy of open defiance."[95]

[95] *ADAP*, C/I/1, No. 223; *DBFP*, 2/V, Nos. 108, 127, 139; Cab 24/241, CP 129 (33), 16 May 33; FO 371, Minutes by Stevenson, G. M[ounsey], Sargent, Vansittart, 11-15 May 33, W5158/40/98. According to Sargent's minute, a subcommittee of the Advisory Committee on Trade Questions was set up to consider the possible exertion of economic pressure, with directions from MacDonald to test its proposals by their applicability to Germany.

Recent German speeches and actions, and not least the behavior of Nadolny and Schönheinz, had caused a severe chill in Anglo-German relations. Without full knowledge, but with his often impressive political intuition, Hitler had acted to reduce the suspicion under which Germany—quite justifiably—was laboring. It remained to be seen whether British leaders would maintain their new vigilance toward Germany, or return to placing trust in the hope of reconciliation.

CHAPTER NINE

France Attempts to Form a New Alliance

1. DALADIER SEEKS PROVISION FOR SANCTIONS

The French Foreign Minister, Joseph Paul-Boncour, had begun by the end of February to doubt the feasibility of his "constructive plan." Even if the German military had been more ready to accept the provisions of the French plan, the French government itself now had a more profound distrust of German denials and assurances. For example, the French regarded the SA and Stahlhelm as illicit militia organizations, whatever Hitler might publicly claim. Moreover, while Britain and the United States now seeemed to be somewhat more wary of the Germans, nothing that representatives of the British and American governments had said since the announcement of the French plan gave any real encouragement to French ideas for international security organization, even if limited to the European continent. Since Poland was widely thought to be hand in glove with France, the refusal of even the Polish delegate to support the French plan seemed to observers to prove that that plan could never win much of a following.[1]

The outlook of Edouard Daladier, who had become Premier on January 31, cast another shadow on the Boncour-Herriot plan. Daladier was also Minister of War, a post he had already assumed in Boncour's own preceding six-week cabinet, which had followed Herriot's overthrow on December 14, 1932. (Boncour's cabinet had fallen, as would its two successors, over the problem of balancing the budget.) Daladier had been a lycée professor, like Herriot, and in fact had studied under Herriot before teaching in the same school. Daladier

[1] *DBFP*, 2/V, No. 283; *SIA, 1933*, pp. 231-234. The presence of Boncour at the Quai is in itself a reason for skepticism about the argument in Wollstein, *Weimarer Revisionismus zu Hitler*, pp. 36-50, that the French made an offer of negotiation which was rejected by Neurath. Unlike Wollstein, I believe that suggestions by François-Poncet of a Franco-German mutual-assistance pact were almost certainly his own (see Chapter Ten, note 12, below), and that Pierre Cot offered nothing new with his speech of 17 Feb 32. No doubt French disarmament negotiators sought on occasion to make their plan attractive to their German counterparts, and they had some success with the German delegation (see AA, ltr., Schwendemann to Koepke, 7 Mar 33, 7360/E536339-42), but the French plan remained basically the same until a new approach was adopted with the instructions of 2 May 1933.

became, however, a challenger rather than a follower of Herriot within the Radical Socialist party, leading the party's left wing. Short, with a bulging forehead and a strong Provençal accent, he had an air of bluntness and forcefulness that won for him the campaign sobriquet of "the bull of the Vaucluse," and in contrast to the emotional tendencies of Herriot and the doctrinaire rigidity of Boncour, Daladier presented himself as a hard-headed, no-nonsense realist. This kind of personality was more in keeping with conventional British and American models, and Norman Davis wrote to President Roosevelt of Daladier, "He is not so well-informed or educated as Herriot, but he is able, direct and level-headed. MacDonald, who has seen more of him than I have, thinks he is the most satisfactory Frenchman to deal with he has ever known and considers him as more practical than Herriot." Daladier's realism tended to take the form of renouncing complicated or burdensome goals or obligations: in May 1927, he drew a contrast between forces making for peace and forces preparing for war, and declared that the French army should have a "strictly defensive organization, subordinated to the principle of arbitration, the first provision of which will proclaim that France will not declare war on any people, but that she is resolved, strongly resolved, to maintain her frontiers, to defend her territory, to prevent war from being brought to it again." This made no allowance for French commitments in Eastern Europe. He was suspicious of conservative army leaders like Weygand, and regarded the military as a highly overrated profession. Today, the received opinion on Daladier is that his forcefulness was a façade, or that as his opponents maliciously put it, his horns were those of a snail, rather than of a bull. In any case, Daladier worked hard at keeping up an appearance of decisiveness.[2]

Daladier looked elsewhere for security than to oft-unrealized proposals for expanding the powers of the League. In line with his views, a subcommittee of the Foreign Affairs Committee of the Chamber urged in mid-February that France not insist too strictly on trying to improve the system of collective security, but rather seek effective ar-

---

[2] Sir Edward L. Spears, *Prelude to Dunkirk, July 1939-May 1940* (New York: A. A. Wyn, 1954), p. 42; Bonnefous, V, 142-143; Chastenet, VI, 54-55; Paul-Boncour, II, 378, 386; Soulié, pp. 23, 32, 279, 292, 299-300, 303, 305, 307, 309, 315, 318-319, 323-324, 329-330, 338-340, 355; Roosevelt Papers, PSF, 142, ltr., Davis to Roosevelt, 13 Apr 33; MDP, 8/1, Pocket diary, 9, 16 Mar [33]. Views on military organization and leaders: Bankwitz, *Weygand*, pp. 86-87. 1927 quotation: France, *Annales de la Chambre des Députés, 13me législature, Débats parlementaires, Session ordinaire de 1927*, II, 222 (the antecedent of the "which" is equally unclear in the French); I am indebted to Emmanuel Beau de Loménie's *Les Responsibilités des dynasties bourgeoises*, 5 vols. (Paris: Ed. Denoël, 1943-1973), IV, 358-359, for pointing out this quotation.

rangements for controlling armaments and supervising the execution of the disarmament convention. Daladier declared in a speech two weeks later that he considered the effective supervision of armaments to be the most important step toward a general reduction. When he and Boncour met with MacDonald and Simon on March 10, he linked French concessions on materiel directly to the obtaining of guarantees respecting (German) effectives, and on March 16, at Geneva, he emphasized to MacDonald that France attached the greatest importance to the supervision of armaments and to the maintenance of her superiority in materiel while the Reichswehr was changing to a militia basis. In an April speech before the Chamber, Daladier stressed the importance of supervision in disarmament; when discussing this speech with Norman Davis, he agreed (according to the American record) that treaties of mutual assistance were not as reliable as supervision, and contrasted his view on this with Herriot's.[3]

The Daladier viewpoint was spelled out in a set of instructions first approved by the French cabinet on May 2. These guidelines repeated that France was prepared to disarm within an organization of security for all nations. But even without a pact of mutual assistance, France was still willing within a first stage of disarmament to give up the new construction of heavy artillery (over 155 mm) and of tanks exceeding a modest total tonnage, provided the following measures were adopted by the conference and inserted in the convention:

1. Budgetary control and technical supervision over armaments.
2. The elimination, or failing that, the strict supervision of private manufacture of and trade in arms.
3. The reduction of continental armies to a standard type, and the elimination of paramilitary formations.
4. The establishment of guarantees that the convention would be carried out, so that any violation of it would entail sanctions. In this respect, it would not be enough to say that, in case of violation, each signatory nation would recover its own freedom to rearm, as in that case the violator state would gain an advantage of several months in its preparations for war.

The instructions epitomized the Daladier approach in expanding on the last point, the demand for a provision of sanctions:

With respect to these guarantees of execution, it may be difficult to take up again the question of the pact of mutual assistance, since it would be rejected. But the point is to get an answer to the fol-

[3] *SIA, 1933*, pp. 247-248; *DDF*, 1/II, Nos. 392, 418; *DBFP*, 2/IV, No. 310; SDP, Memo of discussion between Davis and Daladier, 21 Apr 33, 500.A15A4 General Committee/321.

lowing question: "If Hitler rearms in spite of the convention, what are you going to do?"

In short, it is less a matter of expounding and upholding general theses with respect to security than of raising concrete questions calling for precise answers.[4]

After specifying these conditions, the instructions noted that putting them into operation would inevitably take a certain period of time. France would promise at the outset that after satisfactory completion of this first stage, requiring perhaps four years, she would proceed during a second four-year stage either to place her surplus materiel at the disposition of the League, or to destroy it, "so as to achieve the equality of rights." Such a total duration of eight (or better, ten) years would be needed to judge the efficacy of the supervision; also, if the convention only lasted for five years (as the British plan proposed), the Germans might reclaim their freedom at the end of that period. Therefore (the directive noted) it was well to take advantage of current conditions in international politics, which were favorable to France, and secure a longer term for the convention. In effect, the concept of two terms, or of a trial period before French disarmament, posed a fifth major precondition.

This directive did not propose another "new plan," but its conditions would certainly amount to a major change in the MacDonald plan, and the French military services were likely to request further amendments. The demand for standardization, coupled with a ban on paramilitary forces, should by now be familiar. In the context of this instruction, we can see that standardization would not only thwart the Seecktian designs supposedly followed by the Reichswehr, but would also, like the other stipulations, facilitate supervision; standardization would leave no room for argument as to whether an unusual weapon or form of organization was a violation, or merely an acceptable alternative. The best-understood feature of the new French position was to be the idea of a trial period or of stages. The idea was not new. In early 1932, the British had proposed that Germany should remain "voluntarily" at her existing strength during a first convention. The French general staff and Boncour had considered two four-year stages in October 1932, and Sir John Simon had proposed stages in his House of Commons statement of November 10, 1932, with each subsequent stage "being justified and prepared for by the proved consequences of what has gone before." The 1932 Boncour-Herriot

---

[4] *DDF*, 1/III, No. 229. These general instructions had doubtless been germinating for some time, and can hardly have been a response to the (nontelegraphic) note of the same date from Massigli (*DDF*, 1/III, No. 228).

plan had also called for stages, a call repeated by French delegates in the February discussions.⁵ For that matter, Hitler, in his May 17 speech, spoke of proceeding by stages, meaning that a first stage in French disarmament should accompany the first stage in German reorganization. But in the new French directive of May 1933, full standardization of the Reichswehr was to precede any reduction in French materiel. If applied, the French scheme of stages, coupled with strict supervisory provisions, would constitute an obstacle to secret German rearmament, and would probably expose Germany's past violations.

In Daladier's eyes, however, stages and supervision were apparently less vital for restraining Germany than the provision of sanctions was. Precisely this part of the French conditions would prove hard for disarmament-minded British and American negotiators to recognize, and perhaps even harder for them to report to governments opposed to any serious commitment.⁶ Only the recently published French documents have shown fully how much stress the Paris government placed on the idea of taking positive steps against illegal rearmament. For lack of these documents, most historical writing has done less than justice to the French recognition of, and will to oppose, the rearmament of Germany.

Even so, Daladier's position was not exactly simple and straightforward. For one thing, he had already intimated to the Germans in March, through Fernand de Brinon, a pro-German journalist, a readiness for "direct and discreet discussions" with Neurath or Papen, apparently on the lines Papen had suggested to Herriot in July 1932. In later months, Daladier was to show further interest in the idea of a direct Franco-German understanding. Aside from this, his expressed willingness to disarm was not all that it appeared to be. To a large extent, Daladier's disarmament offer represented an attempt to gain diplomatic benefits from steps he intended to take anyway. In the first stage, he was prepared—no doubt for urgent budgetary reasons—to curtail new construction of heavy artillery and tanks. Indeed, he had decided to pare down the number of officers on active duty by 5,000 out of 30,000, and he planned to delay a proportion of the

⁵ *DDF*, 1/I, Nos. 260 (pp. 522-523), 331 (pp. 716, 718); *DBFP*, 2/IV, No. 170; *SIA, 1933*, p. 239. Simon may have gotten the idea from Neville Chamberlain: Neville Chamberlain Papers, ltr. to Simon, 28 Oct 32, NC 7/11/25/39.

⁶ On Daladier's strong interest in sanctions, see, aside from *DDF*, 1/III, No. 229, the French record of the 8 June meeting with British and American representatives (*DDF*, 1/III, No. 376 [pp. 681-682]), and compare this with the British record (*DBFP*, 2/V, Nos. 207, 208) and, more especially, Londonderry's and Eden's reports to the British cabinet (Cab 23/76, meeting of 9 Jun 33, 39[33]1) and American reports and records (*FRUS, 1933*, I, 190-192; Davis Papers, Memos of conversation, 8 Jun 33).

annual inductions, not only for the sake of the budget but also to spread the available manpower more evenly over the "lean years" to follow; later, the age of induction could also be progressively lowered. But although these measures would be readily supported by his parliament, a destruction of French arms or an extension of the term of military service would not.[7] Since Daladier seems to have been firmly convinced that Germany was already rearming, he presumably had little expectation, if his conditions were accepted, of actually having to destroy French arms stocks some four years hence.[8] After President Roosevelt's statement of May 16, Alexis Léger remarked separately to British and American diplomats that any French government which at this time gave up even one of the historic cannons at the Invalides would be hounded from office, while Daladier himself flatly told the Belgian ambassador that as long as he was in power, France would not destroy the arms which the treaties forbade to Germany.[9] Protestations of a French readiness to eventually disarm would, however, help convince the British and American governments that German rearmament was the fly in the ointment, and divert them from attacks on French armament.

The avowals that France no longer sought treaties of mutual assistance were also less than candid. Although he claimed to be following a new policy of directness and realism instead of the traditional rationale of international security, Daladier was quietly pursuing the old goals of isolating Germany and obtaining watertight guarantees from London and Washington. Now that the facts of German militarism were so evident, it no longer seemed so necessary to argue in terms of general principles. The first task was evidently to establish a community of view, i.e., to show the world that Britain and the United States agreed with the French interpretation of the situation. The French had long tended to think that a lack of support for France from London encouraged Germany to challenge the treaties, and that if only Britain would endorse French policy, the Germans would mend their ways. If the British and Americans accepted sanctions in principle, this would not only show a general approval of French conceptions, it would also in practice align the "Anglo-Saxons" with

[7] Intimation through Brinon: *ADAP*, C/I/1, No. 92; the reference in the document to "Neurath" having made the original suggestion was probably an error for "Papen." Daladier told Tyrrell, apparently early in April, that he would be ready to go to Berlin and Rome to talk with Hitler and Mussolini, were it not for Herriot's opposition (*FRUS, 1933*, I, 495). On curtailments: Bankwitz, *Weygand*, pp. 94-96; Gamelin, II, 95-99, 103-104.

[8] *DDF*, 1/II, Nos. 392, 418; 1/III, No. 376; SDP, Memo of discussion between Davis and Daladier, 21 Apr 33, 500.A15A4 General Committee/321.

[9] *DBFP*, 2/V, No 157; *FRUS, 1933*, I, 147-148; *DDB*, III, No. 33.

France and commit them to participate if necessary in one kind of coercion or another—the old goal of French policy.[10]

The French government undoubtedly wanted a disarmament convention. It did not want the convention for the sake of disarmament, however; the objective in French eyes was to make better provision for security. In the fall of 1932, Herriot and Boncour had thought that if Germany would not agree to a suitable document, France should still draw up a convention together with the remaining principal powers.[11] The French now revived this idea. Léger told Lord Tyrrell in May 1933 that this procedure "would at least settle things between other powers" and prevent an arms race. Germany would have the "opportunity of returning to the fold if at any time she realized the wisdom of doing so; in the contrary event [this procedure] would definitely fix on her shoulders the responsibility for failure to attain more substantial results." Beyond fixing responsibility for any shortcomings in the convention, the French evidently intended that the curtailed convention should draw its signatories together into a common front opposing Germany. No doubt they hoped it would also offer definite guarantees of assistance, as foreshadowed by Daladier's demand that sanctions be provided for.[12]

Thus Daladier wanted a dissolution of the Reichswehr and effective supervision, along with a trial period to make sure that the Reichswehr *was* dissolved and disarmed, and that supervision *was* effective. And at least as important, he wanted a commitment to apply sanctions in the event of German rearmament, which would serve much of the funtcion the Geneva Protocol had been designed to perform.

[10] Expressions of belief that British support for France would restrain Germany include the request (mentioned in Chapter Two, Section 4) in July 1931 that Britain not encourage German demands (*DBFP*, 2/III, No. 214); Léger's suggestion (see Chapter Four, Section 6) in August 1932 that Britain and France together say "Thus far and no farther" (*DBFP*, 2/IV, No. 46); and a statement by General Weygand to the British military attaché in March 1933 that "a word from England would put off war for 30, 40 or even 50 years" (FO 371, Memo, Heywood to Tyrrell, 15 Mar 33, C2627/245/18). On desire in May 1933 for British support in any coercive action: *DBFP*, 2/V, Nos. 151, 154.

[11] *DDF*, 1/I, Nos. 163, 180, 244; *DBFP*, 2/IV, Nos. 82, 86, 91.

[12] *DDF*, 1/II, No. 358; *DBFP*, 2/IV, Nos. 283, 299; 2/V, No. 157. Tyrrell suggested in the last document that the idea of a convention without Germany was mainly Léger's, but it was also very much Boncour's. Boncour explained the rationale more frankly when he described in his 1945 memoirs his desire in October 1933 to draft a convention without Germany: "Either Hitler would agree [to sign it] and it would remain only to ensure the effectiveness of the supervision, which would be a necessary element. Or he would refuse, and France would then not be alone in drawing conclusions from a situation whose seriousness could not decently be concealed" (Paul-Boncour, II, 385). See also *DBFP*, 2/V, Nos. 474, 493, 500; *DDF*, 1/I, Nos. 244 (p. 442), 250 (p. 478), 260 (p. 518); 1/II, No. 358; 1/IV, Nos. 334, 341, 343, 359, 367, 368, 386.

If Germany refused to agree to a disarmament convention, France might still seek some sort of agreement, nominally on disarmament, which in practice would function as a treaty of guarantee.

2. AMERICAN POLICY ATHWART THE FRENCH PATH

While the French were thinking of securing guarantees, Hitler's speech made some others think anew of disarming France. René Massigli now believed that the Germans had adopted the Italian tactic of giving an insincere acceptance to the MacDonald plan, forcing the French to take a position on it, with its disarmament provisions and its lack of guarantees. A Quai d'Orsay memorandum pointed out that France would shortly have to state whether, on the assumption of standardization, she would reduce her armaments, and particularly her heavy artillery. Pressure for disarmament would come particularly from American quarters.[13]

The hard-working, insecure, and suspicious Hoover had been replaced, and with him the realistic, patrician Stimson. Stimson's successor, Cordell Hull, was interested chiefly in the reduction of tariffs, and he had indicated to Stimson before taking office that he accepted Roosevelt's desire to act as his own Secretary of State. Roosevelt relished all the prerogatives of national leadership, including that of dealing with foreign powers. He had more understanding of European views than his predecessor, and he even told the French ambassador in early April, "France cannot disarm at present, and no one will ask it of her." He tried to meet French needs by advocating continuous and automatic inspection.[14] But he was now also deeply immersed in domestic problems, which had unquestioned priority, and he had neither the time nor, probably, the patience to study foreign problems thoroughly. His economic advisers pushed him toward nationalist policies, and his April 1933 decision to leave the gold standard, while probably compelled by a domestic revolt against "hard money," added to tariffs and war debts another potential cause of friction with foreign nations.

There was, then, no longer an attentive, fully informed overseer of American foreign policy. Jay Pierrepont Moffat, the permanent official responsible for Western European affairs, complained in his diary that his current departmental superiors knew nothing of disarmament,

---

[13] *SIA, 1933*, pp. 276-277, 283-284; *FRUS, 1933*, I, 170; *ADAP*, C/I/2, No. 257; *DDF*, 1/III, Nos. 299, 301.

[14] Arthur M. Schlesinger, Jr., *The Coming of the New Deal* (Boston: Houghton Mifflin, 1958), p. 204; *DDF*, 1/III, No. 86. On inspection: *Roosevelt and Foreign Affairs*, I, 374-376; *FRUS, 1933*, I, 106-110; *DDF*, 1/III, No. 179; *DBFP*, 2/V, No. 86.

as Stimson had. But there was an American in Europe who was very much interested in the subject. Norman Davis had failed to become Secretary of State, but Roosevelt had appointed him head of the disarmament delegation with the personal title of ambassador; Davis considered himself Roosevelt's roving adviser in Europe on this and other matters, even though Roosevelt did not always follow his advice. He visited Berlin as well as London and Paris, and to give him due credit, he out-argued Hitler, successfully countering the Führer's claims on the injustice of Versailles and on the Polish threat, while maintaining that the best way to secure revision was by quiet and gradual negotiation, with no threat of rearmament. Unfortunately, Davis seems to have thought he had convinced Hitler, and he concluded that Hitler was open to reason on foreign questions, wanted to avoid war, and favored disarmament.[15]

Davis still strongly believed high-level five-power meetings were the way to solve the disarmament problem, and in early May he reported enthusiastically that Dino Grandi had told him that Mussolini, Hitler, MacDonald, and Daladier should meet to settle disarmament before the Economic Conference, with the United States joining in. Davis had already looked with favor on Mussolini's four-power pact proposal, yet he seems to have paid no attention to reports that German rearmament had already begun. His draft statement for Roosevelt held that Allied disarmament was overdue, and it expressed sympathy for the German impatience and demand for equality, while warning the Germans not to rearm if "the armed powers" did disarm. Elsewhere Davis wrote, "Nothing could be more dangerous than for the French to feel that they have our unqualified support for any amendments they may wish to offer directed against Germany." As in the previous winter, Davis also had personal reasons for wanting an early diplomatic success. He had been offered a chance to act as arbiter for the unfortunate investors in Kreuger and Toll, the bankrupt match trust founded by Ivar Kreuger, and this would be a more remunerative position than that of ambassador, but he did not feel free to take the post until his disarmament work had been completed.[16]

[15] Offner, *American Appeasement*, pp. 21-22; Wollstein, *Weimarer Revisionismus zu Hitler*, pp. 91-93; Moffat MS Diary, 1, 6, 19, 20, 23 Mar, 15 May 33; *DDF*, 1/III, Nos. 97, 376; *FRUS, 1933*, I, 85-89, 142, also 276; *DBFP*, 2/V, Nos. 83, 208; SDP, Memo of conversation between Davis and Daladier, 21 Apr 33, 500.A15A4 General Committee/321; Memo of conversations between Davis and François-Poncet, 9-10 Apr 33, 550.S1 Washington/359.

[16] *FRUS, 1933*, I, 83, 117, 164-166, 403-404, 410; SDP, Memo of conversation between Phillips and Rosso, 7 Apr 33, 550.S1 Washington/1; Memo of conversation between Davis and Herriot, 6 Apr 33, 550.S1 Washington/359; tel., Davis to Dept., 15 May 33, 500.A15A4 General Committee/378; Roosevelt Papers, PSF, File 142, ltr., Davis to Roosevelt, 23 Apr 33.

After the Roosevelt May 16 proposal and the Hitler May 17 speech, Davis was anxious to "take advantage of the initiative given by the President's message and Germany's more conciliatory attitude to press for decisions here and now." With authorization from Washington, he publicly announced on May 22 the proposition, previously disclosed to MacDonald and Herriot, that if armaments were reduced, then if other states found that one state was an aggressor, and if the United States agreed with this finding, the United States would refrain from any action which might defeat a collective effort—meaning economic sanctions or a blockade—against the aggressor. At the same time, Davis revealed that his government sympathized with the French idea of "effective, automatic, and continuous supervision." But Davis also repeated Hugh Gibson's 1932 thesis that a nation could best ensure its security by disarming, and he called for an abolition of aggressive weapons, particularly on the part of "the armed powers of Europe," thus indicating that France should now make her contribution.[17]

The French, however, were by no means ready to destroy their armament, certainly not without guarantees of security. On May 23, Boncour gave the Conference an all-inclusive statement of the French position, combining the new demands for supervision, stages, and sanctions with his old theses that the organization of peace in Europe must precede disarmament, and that the League should take over existing "aggressive" arms. The French seemed to be no more favorably impressed by the American offer to refrain from defeating a collective effort than they had been by Roosevelt's message. To them, the offer as defined by Davis was not so much a promise of valuable support as an indication that the American commitment would be negative and strictly limited. Also, although the British had often said that American insistence on neutral rights blocked any broader British commitment, the new American position did not evoke any new offers from Sir John Simon. And for that matter, the American commitment might well not materialize at all. Contrary to Roosevelt's intention, Davis gave the French the impression that the American declaration would have to be ratified by the United States Senate, and the French had bitter memories of the Senate's failure in 1920 to ratify the Treaty of Versailles. As if to reinforce French doubts, the Senate Foreign Relations Committee on May 30 reported out an arms embargo resolution with an amendment, sponsored by Senator Hiram Johnson, providing that any ban on exports in a dispute should apply impartially to all parties in the dispute. Although no vote was taken

[17] *FRUS, 1933*, I, 150-151, 154-158.

on the Senate floor, this appeared to foreshadow a senatorial veto of the Davis offer of May 22, and Roosevelt failed to oppose the amendment. Of course, Roosevelt in turn may have been discouraged by the cool response his initiatives received from France and Britain, and the apparent stalemate on war debts made all three nations distrustful.[18]

Like Schleicher a few months before, Davis tried to sustain a fictitious reality with further displays of confidence. He pushed vigorously for a high-level meeting, claiming that Hitler and Mussolini were willing, and latterly also that Daladier had specifically agreed to come by June 2. Davis tended to build his dream castles on insubstantial foundations. For example, he reported to Washington that he and Simon had found Boncour "less disposed toward [the projected five-power] meeting" than Daladier, but also that they hoped to bring Boncour around. Actually (as Davis admitted to Eden) Boncour had flatly refused to attend a five-power meeting; moreover, Simon had not even taken part in the discussion, having previously left for London.[19] Eden himself did support Davis's proposals for high-level talks, largely because the MacDonald plan was sustaining awkward criticism in conference debates; Eden hoped the Davis proposals might serve to postpone a "second reading and its embarrassments" until the talks had "clearly shown whether a Convention can be realized or not." But in London, interest in such talks was slow in developing, and the British were inclined to combine them with the World Economic Conference, due to convene in London on June 12. The unhappy experiences of March and the domestic reaction in April against Hitler and the great-power promotion of revision probably made such talks less attractive than they would have been earlier. On June 1, the Disarmament Conference decided to postpone the second

[18] *Records of Conference*, Series B, II, 491-493; *SIA, 1933*, pp. 277, 283-284; *DBFP*, 2/V, Nos. 170, 179, 205; Moffat MS Diary, 26, 31 May, 2 Jun 33; *FRUS, 1933*, I, 166-168, 378; *DDF*, 1/III, Nos. 315, 343; Robert A. Divine, "Franklin D. Roosevelt and Collective Security," *Mississippi Valley Historical Review*, 48 (1961-1962), 42-59; the same author's *The Illusion of Neutrality* (Chicago: University of Chicago Press, 1962), pp. 42-56; and Offner, *American Appeasement*, pp. 35-37. Davis seemed to have difficulty in fully accepting Roosevelt's aversion to contractual ties requiring ratification: see *FRUS, 1933*, I, 103-104, 116, 125, 167, 168. The American delegation tried to tell the French that the Senate committee resolution did not apply if the United States agreed that a given state was an aggressor (i.e., if the United States abandoned its neutrality) (*DDF*, 1/III, No. 343), but Secretary Hull admitted at a press conference that the amendment nullified the administration's plan (Divine, *Illusion*, p. 54).

[19] Cab 23/76, meeting of 24 May 33, 36(33)5; *FRUS, 1933*, I, 165, 170; *DBFP*, 2/V, Nos. 170 (n. 1), 172, 174, 188, 191. At the 8 Jun 33 Anglo-Franco-American meeting in Paris, Boncour sharply denied a suggestion by Davis that he had been willing to have five-power talks later on with the Germans. Another case where Davis seems to have been more enthusiastic than accurate was in reporting his meeting with Herriot on 22 May 32; see Chapter Three above.

reading of the MacDonald plan and to adjourn for a month in deference to the Economic Conference; this put off the British embarrassment. While this postponement was under discussion, a plan took shape for an Anglo-French-American meeting in Paris. Whereas Davis hoped this would be preliminary to a five-power meeting, Boncour evidently wanted it only to win British and American agreement to French ideas. The meeting was finally set for June 8.[20]

When the French, British, and Americans met in Paris, the occasion gave Daladier a chance to explain his position, even though Davis and the British, Lord Londonderry (the Air Secretary) and Eden, preferred to talk about French disarmament. Daladier described current French ideas on guarantees, sanctions, and stages. In fact, he read out the essential points of the instructions of May 2.[21] The French recommendations that evoked most discussion were those for a two-stage convention and for supervision. In an effort, probably, to mitigate the shock to the British and Americans, Daladier proposed a first stage of "at least three years," instead of the four years prescribed in the May 2 instructions. The British and American representatives found the general concept of stages quite understandable and even acceptable. But they could not quite accept the essence of the French scheme of stages, the proposal that a period should go by in which Germany would be bound by the new convention while France remained fully armed. They suggested a different timing, under which a partial German reorganization would be the signal for some French disarmament. The British opposed any regular system of supervision, although the Americans accepted it, and neither the British nor American representatives cared for the idea of budgetary control. But despite the differences, Davis commented to Washington that this meeting marked the first time the French had given "any precisions" on the reduction of materiel, and in a cable a week later, he stated that the talks had made him "more hopeful than ever as to the ultimate success of the Disarmament Conference."[22]

The reports of Davis (to Washington) and of Londonderry and Eden (to the British cabinet) were far from complete. As the French record shows, and as the British written record of the meeting in large part confirms, Daladier expressed quite strongly the idea that something had to be done by the other signatories if the Germans violated

[20] *DBFP*, 2/V, Nos. 172, 174, 182, 187, 188, 191, 193, 202; Cab 23/76, meeting of 24 May 33, 36(33)5; *FRUS, 1933*, I, 182-183; MDP, 1/525, Notes by MacDonald and J.A.B., 13-15 Jun 33.
[21] Transcripts of morning and afternoon meetings: *DDF*, 1/III, No. 376; *DBFP*, 2/V, Nos. 207, 208; *FRUS, 1933*, I, 190-192; Davis Papers, Memos of conversation with the French and British (a.m. and p.m.), 8 Jun 33.
[22] Week-later cable: *FRUS, 1933*, I, 192.

the proposed convention. Aside from reading the abridged French instructions, he asked three different times what would happen if a violation was established, and he maintained it was not enough to say (as Davis did) that France would then be freed from restrictions on her side. Yet Davis did not even mention this problem in his summary report home, and the detailed American report only recorded the matter as being raised once, and gave the impression that Davis had succeeded in allaying Daladier's apprehensions. Similarly, Londonderry and Eden apparently did not bring this problem to the attention of the British cabinet, and Vansittart passed over it in his minute on the record of the Paris meeting. Perhaps the American and British representatives feared that the actual French demands would provoke a hostile reaction from their own governments; perhaps they did not take in what did not coincide with their own hopes and objectives.[23]

At the end of the Paris meeting, Davis expressed the hope that France, Britain, and the United States could reach agreement before the Bureau reconvened on June 27, "because the British and Americans should then get in contact with the Germans and Italians to win them over to the solutions envisaged"; he suggested a meeting in London (i.e., during the London Economic Conference), a suggestion Eden supported. But Daladier said he could not consult his cabinet before going to London, and that idea failed. Davis was unhappy that Henderson would be in London trying to promote discussions on the fringe of the Economic Conference; he himself hoped to do the same thing, and he evidently feared that Henderson would queer his pitch. But when Davis inquired, presumably because of public attacks in Washington on his past financial dealings, whether President Roosevelt would be in any way embarrassed if he went to London during the Economic Conference, Roosevelt reluctantly agreed that Davis might go if he was not there at the opening of the conference and did not mix with the American delegation. Davis decided a few days later to come home for a short time, and to stop only a day in London on the way.[24]

[23] Cab 23/76, meeting of 9 Jun 33, 39(33)1; FO 371, Minute by Vansittart, 10 Jun 33, W6744/40/98; cf. the other records in note 21. On 29 June, Davis told a press conference in Washington that, contrary to what was generally believed in the United States, France was now convinced she could not get guarantees of security from the United States or Great Britain (*DDF*, 1/III, No. 437). In September, according to a British record, Davis said that the question of sanctions had not been raised at the June meeting (*DBFP*, 2/V, No. 407); according to the French record, he said the question had not been pressed so strongly (as now, in September) (*DDF*, 1/IV, No. 261 [Annex II]).

[24] *DDF*, 1/III, No. 376; Moffat MS Diary, 9, 21 Jun 33; *FRUS*, *1933*, I, 192-194. MacDonald belatedly supported the idea of meetings in London: MDP, 1/525, Notes by MacDonald and J.A.B., 13-15 Jun 33. On Davis's wish to go to London, see also

In view of what was to happen at the Economic Conference, in particular the damage that American relations with European governments would suffer there, Davis might later have congratulated himself on having stayed away. The story of the anarchic American delegation and of Roosevelt's July 3 "bombshell message," rejecting currency stabilization, has often been told and need not be recounted here, being in every sense, except the largest, tangential to the present story. In that largest sense, Roosevelt's message, referring to "the specious fallacy of achieving a temporary and probably arbitrary stability" and to "old fetishes of so-called international bankers," showed that however international his instincts, he had decided to follow a nationalist, populist policy. Although the urgent human problems of the depression and the economic nationalism of other countries, including the earlier French and British exploitation of their devaluations, might justify Roosevelt's basic policy on stabilization, these considerations did not oblige him to display a cavalier attitude toward foreign sensitivities and to pander to domestic prejudice.[25]

Stabilization itself was not the most burning issue between the United States and the West European democracies; the really difficult problem was war debts. This ostensibly financial question was in fact highly political, an international controversy in which the different parties held utterly different views, embittering their whole relationship. The broad American public believed more than ever that the Allies, as President Coolidge was said to have put it, had "hired the money," and should pay it. On the other hand, the British and French thought the United States should write off the debts owed them as part of its very modest—in their eyes—share of the war's cost. The French resented the loss of reparations, precipitated by Hoover's intervention in 1931. The British considered it most urgent that the United States complete the work, begun at Lausanne, of ending the

Nixon, ed., *Roosevelt and Foreign Affairs*, I, 236. Davis was attacked over early business dealings in Cuba and on accepting favors from J. P. Morgan: Moffat MS Diary, 2, 9, 21, 24-26 Jun 33; Arthur M. Schlesinger, Jr., *The Crisis of the Old Order, 1919-1933* (Boston: Houghton, Mifflin, 1957), p. 467, and *Coming of the New Deal*, pp. 436-437.

[25] Text of message: *FRUS, 1933*, I, 673-674. Schlesinger writes (*Coming of the New Deal*, p. 192): "Though deeply pragmatic . . . , Roosevelt remained at bottom a Wilsonian internationalist." Robert A. Divine (*Roosevelt and World War II* [Baltimore: Johns Hopkins Press, 1969], pp. 6-7; "Roosevelt and Collective Security," pp. 43, 59) cites Charles A. Beard's *American Foreign Policy in the Making, 1932-1940* (New Haven: Yale University Press, 1946) in judging that Roosevelt was actually an isolationist in the 1930s. Perhaps closer to the truth (and to Beard's opinion) is the view of James MacGregor Burns in *Roosevelt: The Lion and the Fox* (New York: Harcourt, Brace & World, 1956), p. 262: "As a foreign policy maker, Roosevelt during his first term was more pussyfooting politician than political leader."

disruptive effects of intergovernmental payments. All three parties had grievances, and all would nurse them over the years—with Hitler the ultimate beneficiary. The problem may have been insoluble, but no American president was willing to stake his prestige on trying to solve it. Before his inauguration, Roosevelt had evaded Hoover's efforts to involve him in war-debt negotiations, even though, unlike Hoover, he privately admitted that wiping out the debts would be the best solution.[26] When MacDonald came to Washington in April, he came mainly because he hoped, even without the prior assurances desired by Neville Chamberlain and other ministers, that he could work out an agreement on the debts. He thought, apparently with reason, that he obtained the President's promise to seek from Congress the power to suspend the June 15 debt payment.[27] Yet weeks passed and Roosevelt took no steps in this direction; certainly he and Congress did have many pressing domestic matters to deal with. Then, on June 4, the British ambassador learned that, in view of the balky mood of Congress, Roosevelt would not seek authority to defer the British June 15 debt payment.[28] The British were to tender two more token payments, but there was now little hope that Roosevelt would actively seek a solution.

In other areas, too, Roosevelt showed unwillingness to risk a congressional revolt by pushing internationalist policies. As we have seen, he failed to oppose the Johnson amendment to the arms-embargo resolution. And despite the ambition of Secretary Hull to negotiate tariff reductions at the Economic Conference, Roosevelt decided in June not to ask Congress for authority for that purpose.[29] In the matter of

[26] On Hoover-Roosevelt negotiations: Schlesinger, *Crisis of the Old Order*, pp. 444-448; Ferrell, *Diplomacy in Great Depression*, pp. 232-240, 245-246. FDR's admission: *DBFP*, 2/V, No. 524. On Coolidge's statement: Howard H. Quint and Robert H. Ferrell, eds., *The Talkative President* (Amherst: University of Massachusetts Press, 1964), p. 176.

[27] *FRUS, 1933*, I, 474-476, 477-478, 480-482, 483-485; *DBFP*, 2/V, Nos. 539, 542, 546 (and n. 1); *Roosevelt and Foreign Affairs*, I, 44-48, 100, 153-155 (esp. notes 1, 3, and 5); MDP, 8/1, Pocket diary, Sunday, [23], [Tuesday, 25] Apr [33]; Davis Papers, Memo of conversation between Davis and MacDonald, 4 May 33. *FRUS, 1933*, I, 479, carries a message from Washington to Davis in London expressing satisfaction that a previous Davis discussion had eliminated debt discussions from (the agenda of the proposed) visit. Davis did not relay this part of the message to MacDonald (SDP, Memo of phone conversation between Davis and MacDonald, 2 Apr 33, 550.S1 Washington/359).

[28] *DBFP*, 2/V, Nos. 549, 551, 555, 559, 561, 563.

[29] *FRUS, 1933*, I, 922-924; Herbert Feis, *1933: Characters in Crisis* (Boston: Little, Brown, 1966), pp. 164-165, 174-175. The French and German ambassadors, André Lefebvre de Laboulaye and Hans Luther, both noted the shift to isolationism, blaming it partly on the lack of response to the Davis-Roosevelt proposals: *DDF*, 1/III, No. 347; AA, disp., Luther to Ministry, 23 Jun 33, 7360/E537040-45.

monetary stabilization, it was not merely that Congress was sympathetic to cheap money; Roosevelt himself was discovering that whatever might be the eventual merits of various recovery programs, devaluation—which recent legislation had equipped him to pursue—seemed a quick and efficacious way of alleviating the symptoms of depression. Stabilization assumed critical importance at the Economic Conference largely because, as Robert H. Ferrell has pointed out, war debts and tariffs had already been excluded from the conference agenda; when Roosevelt dismissed even the prospect of temporary stabilization, there was virtually no reason left for holding the conference.[30]

For the British in particular, who believed in the priority of finance and economics, and in the need to restore international trade, the "bombshell" epitomized American obstructiveness.[31] MacDonald had sought American goodwill over the years, in the hope and expectation that the United States would forgo war debts. At the opening of the Economic Conference, he made a last futile attempt, counter to the agreed limitation of the agenda, to bring the war debts under discussion. After Roosevelt's message, the policy of cooperation with America seemed nearly bankrupt, and MacDonald's morale and political authority were badly shaken. On the other hand, Roosevelt's sharp language evoked approval from an American public that had lost faith

---

[30] Ferrell, *Diplomacy in Great Depression*, p. 275.

[31] The ill-feeling and distrust that rankled in France and Britain after the "bombshell" can be found in many sources: *DDF*, 1/III, No. 470; Bonnet, *Quai d'Orsay*, p. 127; Feiling, *Chamberlain*, pp. 223-226, 253-254; Middlemas and Barnes, p. 729; Vansittart, *Mist Procession*, pp. 464-468; Cecil Papers, 817B/51108, ltr. to Noel-Baker, 13 Jul 33; Davis Papers, Memo of conversation with MacDonald, 18 Sep 33; Austen Chamberlain Papers, ltr. to Hilda Chamberlain, 3 Jul 33, AC 5/1/596; *Roosevelt and Foreign Affairs*, I, 339-341. Schlesinger (*Coming of the New Deal*, pp. 223-224) cites the approval of the Keynesians and of Churchill, but says the general European reaction was violent. The American refusal to stabilize should in itself have been no surprise to British ministers; a cabinet paper prepared by the advisory Committee on Economic Information under Sir Josiah Stamp (Cab 24/241, CP 131 [33]) had predicted on 16 May that if recovery did not proceed far in the next few months, and if the Economic Conference ended without agreement on monetary policy, the United States government would be likely to devalue much farther. Experience with the "bombshell" and also the war debts may explain in part why the British Treasury under Neville Chamberlain and Sir Warren Fisher led a movement for friendship with Japan at the expense of Anglo-American relations, and for meeting Germany with an air deterrent. See D. C. Watt, *Personalities and Policies*, Essay 4, and the same author's *Too Serious a Business* (Berkeley: University of California Press, 1975), pp. 55-56, 73, 98, 112-114. Fisher could be on occasion a venomous critic of American "Yahoodom": see Public Record Office, Papers of the Defence Requirements Sub-Committee of the Committee of Imperial Defence (Cab. 16, vols. 109-112), Vol. 109, DRC 12, encls. 1 and 2, 30 Jan 34. As Prime Minister, Chamberlain later showed considerable coolness toward the United States: see William L. Langer and S. Everett Gleason, *The Challenge to Isolation* (New York: Harper, 1952), pp. 26, 125; Feiling, *Chamberlain*, p. 325; Eden, pp. 548-568.

in financiers, and that was disgusted by the default or quasi-default of the European debtors.[32]

3. British Unease and Henderson's Mission

We have seen that the British delegation faced criticism at the Disarmament Conference in the last week of May and first days of June. Trouble arose particularly with the question of aerial bombardment: the British government, in the face of overwhelming sentiment at the conference, wished to maintain a right to bomb in outlying areas—i.e., a right to use bombing against rebellious tribesmen. Also, the Japanese resented the British proposal to continue until 1936 the London Naval Treaty arrangements that denied parity to them, and the British position appeared narrowly self-serving with respect to aircraft and tank strengths. The MacDonald plan remained in conflict at several points with the doctrine of equality of rights. Further, the government was clearly unwilling or unable to give France assurances that would lead the French to reduce their arms, and unlike the United States, Britain had refused to agree to permanent and automatic supervision.[33] Under these conditions, British disarmament policy was failing in its most serious objective, to reassure the British public that the government was doing all in its power to promote disarmament.

The unpopularity of British policy evoked separate reactions in the Foreign Office and among ministers. Since the British position was hard to defend, and yet the cabinet seemed unwilling to modify it, Allen Leeper in the Foreign Office thought the best solution might be to end the conference by exposing German rearmament. Picking up from the conclusion of the Temperley memorandum of early May, he suggested that "after Herr Nadolny has made one of his speeches in the best Hitlerian manner expounding Germany's peaceful intentions, the British and French representatives should stand up one after another and declare that, much as they appreciate the sentiments which have been expressed, it is impossible for disarmament discussions to proceed when to their knowledge Germany is actually rearming." This

---

[32] Schlesinger, *Coming of the New Deal*, pp. 213, 224-225; William E. Leuchtenburg, *Franklin D. Roosevelt and the New Deal, 1932-1940* (New York: Harper & Row, 1963), p. 202; Moffat MS Diary, 5 Jul 33. On MacDonald's hopes of debt cancellation or moratorium: MDP, 2/12, 13, ltrs. to Baldwin, 15 Jun 32, 31 Mar 33; 8/1, Diary, 20 Nov [32]. He had earlier hoped to use the Economic Conference to compel the U.S. to negotiate on war debts: MDP, 8/1, Pocket diary, 27 Dec [32]. On his reaction and situation after the "bombshell," esp. Davis Papers, Memo of conversation between Davis and MacDonald, 18 Sep 33; the relevant portion is omitted from the publication in *FRUS, 1933*, I, 214-217.

[33] *DBFP*, 2/V, Nos. 173, 176, 177, 179, 191, 194, 196; Cab 24/241, CP 150 (33), 30 May 33; *SIA, 1933*, pp. 277, 283, 285-288.

would be sensational, but preferable to a delay of weeks or months in which the conference would drift to "its certain death," and after which the British government would be blamed.[34]

Although the immediate problem of defending British policy was eased when the Disarmament Conference decided to adjourn during the Economic Conference, Vansittart thought well of Leeper's suggestion, and noted that it would be well to find out what material was available that could be publicly used in an accusation:

> I am not sure that it is very much, for a great deal of our material could not be used. I don't think the French have much more than we have. Our material shd. therefore be very carefully vetted in case the need for such tactics shd. be considered imminent.

Simon, in turn, minuted that Leeper's paper was "a most impressive piece of reasoning" and he agreed with Vansittart's comment. It was found, however, that the information was not in the Foreign Office, and in any case that department was not competent to say whether any given intelligence on this subject could be publicized. Queries were therefore drawn up for the Admiralty, War Office, and Air Ministry.[35]

Leeper had suggested that the British and French would propose to bring the German violations before the League Council, but a paper by Foreign Office legal experts, which appeared the day after Leeper's suggestions, cast doubt on the possibility of the Council's taking effective action. Article 213 of the Treaty of Versailles provided that Germany should facilitate any investigation which a majority of the League Council might consider necessary, but neither this nor any other article authorized action if such investigation uncovered violations. In many cases, for example if France were pressing for action against Germany, other agreements would require that the dispute be next submitted to an arbitral tribunal or to the Permanent Court of International Justice. If the resultant ruling was against Germany, France could then bring a proposal for action before the League Council. But under various treaty commitments not to use force, such as in the Treaty of Locarno, military sanctions could only be authorized by a unanimous vote of the League Council (the votes of the parties to

---

[34] *DBFP*, 2/V, Nos. 179, 200, 201 (encl.); on the latter document, a statement of the German point of view, Leeper minuted an explanation of the weakness of the British position on equality of rights (FO 371, 8 Jun 33, W6558/40/98). Leeper may have intended to lead the government to reconsider Simon's recent refusal to undertake further commitments by confronting ministers with the only plausible alternative.

[35] FO 371, Minutes by Vansittart, Simon, and R.M.A. Hankey, 30, 31 May, 8 Jun 33, W6536/40/98; *DBFP*, 2/V, No. 253.

the dispute not being counted). Prospects of obtaining such a vote were obviously dim. There would not be such a legal requirement of unanimity before measures of economic pressure could be taken, but economic measures would not, of course, be practical without broad international support.[36] Ironically, the network of international treaties and agreements, devised by sagacious men who thought they were going to bring international relations under the rule of law, tended in reality to bar action against flagrant treaty violations.

The French government was also aware of the bars to action, which it regarded as proof of the need for new guarantees against a clandestine German rearmament. No doubt Daladier's demand for specific sanctions against arms violations had been partly inspired by the lack of such provisions in the Treaty of Versailles.[37] With the British, the difficulty was more likely to serve as a rationale for inaction than as a spur for seeking new arrangements. But even if action was impractical, disclosures could still serve to justify a British (or French) refusal to disarm more drastically, and they would tend to educate British opinion and to isolate Germany. The threat of exposure was indeed a serious danger to the German government as long as it pretended that it was not rearming. A revelation of treaty breaches would damage the government's standing abroad, and even weaken it at home. Even in a dictatorship, power does not come solely from the barrel of a gun, but also from respect, including a regard for the competence and dexterity of the ruler, if not always for his integrity.

Like Leeper, the British ministers had anxieties about a continuation of Disarmament Conference sessions. They noted, as he had, the critical attitude at the conference toward British policy. A little later, they listened to accounts by Londonderry and Eden of the June 8 meeting with Daladier; evidently the French were seeking extensive changes in the MacDonald plan. Then, after the Economic Conference had met for a few days, it became obvious that Henderson was not succeeding in his efforts to negotiate in the corridors of that conclave.[38]

[36] *DBFP*, 2/V, No. 185.

[37] *DDF*, 1/III, No. 448 (encl.). French comment on Hailsham's 11 May speech indicates awareness of the problem: *DBFP*, 2/V, No. 161; FO 371, Minute by R. Stevenson, 17 May 33, C4339/245/18. French documents of September 1933 would stress the need to substitute for the Versailles system another that would "in no way contain the same gap in the matter of guarantees of execution" (*DDF*, 1/IV, Nos. 240, 242).

[38] Cab 23/76, meeting of 31 May 33, 38(33)3; meeting of 9 Jun 33, 39(33)1. On Henderson's experience at the Economic Conference, Wheeler-Bennett's description (*Pipe Dream*, p. 164) is justly famous. The French ambassador, Charles Corbin, reported that Henderson strolled through the corridors, accosting delegates, but "he met with no enthusiasm, and people seemed rather to avoid him" (*DDF*, 1/III, No. 412). One person who successfully avoided him was Daladier, with whom he was unable to get an appointment (Cab 24/242, CP 167 [33], 20 Jun 33).

Soon he would return to Geneva, probably in a disgruntled state, to resume the direction of the Disarmament Conference. Reviewing the situation on June 19, the ministerial committee on disarmament found it to be most unsatisfactory. The conference seemed to tend toward attacking the MacDonald plan piecemeal, and the French were making new demands while contributing nothing. Although the ministers do not seem to have known of the French bid for sanctions, the French call for permanent supervision annoyed them, and they could hardly believe the Americans supported it. Baldwin said that

> if it was possible to see agreement coming on what to his mind were the two big questions, firstly parity in the air with France, and secondly, no re-armament of Germany, then he thought it would be worth our while to consider any concessions which would make this achievement possible. But he could not help disliking intensely the way in which we were being chivvied from pillar to post at the moment and achieving no results of any value.

Unlike Leeper, the ministers did not consider using an exposure of German armament to end the conference. They thought that both Germany and France were being troublesome—and evidently that France was currently the most obstructive. The sense of the meeting was that Britain should stand pat on her offer and seek an extension in the conference adjournment while letting France and Germany hammer out their differences. MacDonald concluded, and the others agreed, that Henderson "should be told that the whole trouble arose over the fundamental differences between France and Germany, and that it was up to him to get agreement between these two." This negotiation should be done in private conversations, which Britain would be ready to help bring about.[39]

Eden accordingly spoke tactfully but energetically with Henderson. The latter, who admitted he had been unable to initiate talks in London, said that he was quite ready to ask the Bureau for a longer adjournment, if he could be assured of British and American support. Eden then talked with Allen Dulles; no serious objections to a further adjournment were raised by the United States.[40] When the French learned that Henderson had been urged to agree to a long adjournment, Boncour agreed that it was unwise to proceed with conference sessions until the "indispensable conversations" had taken place—

---

[39] Eden, pp. 42-43; Cab 27/505, DC (M) 32, meeting of 19 Jun 33; SDP, tel., Dulles to Dept., 22 Jun 33, 500.A15A4 General Committee/503. On the British interest in air parity with France, see also *FRUS, 1933*, I, 112.

[40] Cab 24/242, CP 167 (33), 20 Jun 33; SDP, tel., Dulles to Dept., 22 Jun 33, 500.A15A4 General Committee/503; *FRUS, 1933*, I, 197-199, 201-202.

which in his mind were certainly not Franco-German, but rather Franco-British. But at this stage, Boncour wanted to keep discussions under the aegis of the conference leadership—partly to show that conference activity was continuing, so that Germany would have no excuse for rearming, and partly to avoid the five-power talks that Davis so much desired.[41]

The conference convened briefly at the end of June and agreed (despite some German objections) to extend the adjournment. Meanwhile, Henderson received authority from the conference to conduct further private negotiations himself.[42] The procedure he chose was to undertake a tour of European capitals, beginning with Paris and including Rome, Berlin, Prague, and also Munich (in order to meet Hitler personally), before returning to Paris. Like the British ministers, he seems to have regarded France as the greatest obstacle, and at first he planned to wait in Paris until he obtained a substantial concession from the French before proceeding further. The French government, however, delayed his initial reception until July 11, and then gave him an aide mémoire on its position along the lines of the May 2 instructions, together with explanations which showed that France would not readily change its position; there was a marked absence of social and other courtesies in Paris.[43] Mussolini, though more sympathetic, was preoccupied with the signature of the four-power pact and with other visitors, his officials seemed uncooperative, and the hapless traveler suf-

[41] *DDF*, 1/III, Nos. 412, 416, 421, 466; *DBFP*, 2/V, Nos. 220, 221; SDP, tel., Dulles to Dept., 24 Jun 33, 500.A15A4 General Committee/506.

[42] *SIA, 1933*, pp. 292-293. Nadolny's opposition to further adjournment was somewhat pro forma. See *DBFP*, 2/V, No. 224; *DDF*, 1/III, No. 430; *FRUS, 1933*, I, 198-199, 201-202. Rheinbaben twice (27 and 28 Jun 33) told Ferdinand Mayer, a member of the American delegation, that adjournment would not endanger the future of the conference: *FRUS, 1933*, I, 199; SDP, Memo by Wilson, 29 Jun 33, 500.A15A4 Steering Committee/338. A clue to the explanation appeared when Rheinbaben on the twenty-eighth asked what Mayer thought the reaction would be in Britain and France if, during the adjournment, Germany acquired small numbers of the arms that would be permitted under the British draft convention. Mayer said the reaction would be highly adverse, and strongly advised against such a course.

[43] *DBFP*, 2/V, Nos. 226, 230, 232, 251; *DDF*, 1/III, No. 486; *FRUS, 1933*, I, 204-205; and SDP, Memo, Wilson to Dept., 25 Jul 33, 500.A15A4 General Committee/572 report on Henderson's trip. Eden's initial idea had been that Henderson establish himself in London, Paris, or Geneva and summon delegates to talk with him, whereas Massigli had at first thought that conversations might be held between delegates with Henderson and Beneš "presumably" present: SDP, tels., Dulles to Dept., 22, 24 Jun 33, 500.A15A4 General Committee/503, 506. Eden's memoirs state that he went with Henderson to Paris and took part in the talks between Henderson and the French on 11 Jul 33, but the contemporary records do not show this, and in fact American documents show he was in London that day (SDP, tels., Bingham to Dept., 11, 12 Jul 33, 500.A15A4 General Committee/541, 539). Henderson had expressed interest in having a British official spokesman accompany him on his journey, at least after Paris (*DBFP*, 2/V, Nos. 226, 230, 232), but this did not take place.

fered "quite a serious indisposition" while in Rome.[44] After this, Henderson was favorably impressed by the friendly reception and assurances of good will he encountered when he met with Neurath and Blomberg in Berlin. But the Germans clearly insisted on their right to all weapons that would be retained by others. Hitler, in Munich, rejected the French idea of a trial period, and he became irritated when Henderson stressed the ill-feeling existing abroad against Germany; the Führer said that England had never gone beyond theoretical and Platonic declarations of sympathy with his country. One byproduct of the trip was that Daladier, reportedly on Henderson's prompting, told the traveling emissary that he would be ready to meet alone with Hitler, and Henderson, while in Berlin, made a public suggestion that the two leaders should meet. Hitler, however, told Henderson that extensive preparation would be needed before any such meeting.[45]

Henderson naturally put the best possible face on his trip, but he had made no dent in the French position, and his transmission of it to Germany had in turn evoked Hitler's opposition. Although Boncour had earlier wanted to have the private discussions conducted under the auspices of the conference leadership, he seemed at his July 11 meeting with Henderson, and also when he met him again after the circuit on July 22, to regard Henderson's activity as uninformed and superfluous. Part of this was doubtless an incompatibility of viewpoint and personality. Also, by the time of the July 11 meeting, the danger of five-power talks had faded. Davis, the arch-advocate of such talks, had gone home, and on July 6 the government in Washington concluded that Davis would not be needed in Europe until September. Boncour quickly made sure the United States would not object to negotiations in Davis's absence, and then he moved forward with plans for talks with the British. Massigli told Henderson on July 11 that the French Foreign Ministry wished to hold direct private Anglo-French discussions, that is, without Henderson's participation; Henderson did not appear pleased at this idea.[46]

[44] *DBFP*, 2/V, No. 269; *DDF*, 1/IV, No. 14; *ADAP*, C/I/2, No. 367; Aloisi, pp. 138-139.

[45] *DBFP*, 2/V, Nos. 258, 269; *DDF*, 1/IV, Nos. 16, 17; *ADAP*, C/I/2, Nos. 370, 374. On 12 July, before leaving Paris, Henderson gave a member of the German embassy what was apparently Berlin's first authoritative information on the French proposal for a trial period; Hassell also received the information from Henderson in Rome: AA, tel., Köster to Ministry, 12 Jul 33, 3154/D669449-50; *ADAP*, C/I/2, No. 367. For earlier German reporting, see Chapter Ten, note 1.

[46] *DDF*, 1/III, Nos. 421, 466, 471, 486; 1/IV, No. 36; *Roosevelt and Foreign Affairs*, I, 281-282; SDP, tel., Straus (Paris) to Dept., 7 Jul 33, 500.A15A4 General Committee/531; tel., Phillips to Paris embassy, 8 Jul 33, 500.A15A4 General Committee/534; Memo by Moffat, 10 Jul 33, 500.A15A4/2171; Moffat MS Diary, 5, 6, 10 Jul 33; *DBFP*, 2/V, Nos. 220, 251, 269.

Boncour and Massigli probably intended that a first exchange of views between Paris and London should serve to win the British over to the French view of the German danger, and Boncour seems (as we shall see presently) to have had a certain lever already in mind for accomplishing this. Also, from a French perspective, specific changes were needed at many points in the British plan. But Boncour recognized that the British might want to delay until Davis returned, or out of regard for Henderson's feelings. Simon, Eden, and Vansittart evidently did feel some qualms about Henderson's sensitivities, or at least about his power to retaliate, and they had to bear in mind the ministerial committee's desire that Henderson try to resolve the differences between France and Germany. Consequently, no arrangement was made for further Anglo-French talks until after Henderson's tour, when the British ministerial committee again considered the situation.[47]

Eden advised the committee that a disarmament convention could only be obtained by "persuading the French to make such an offer as can be reasonably forced down the German throat with an Italian spoon." But the committee was unenthusiastic about exerting influence on France to disarm, as, with the existing German temper, this might be asking her to do something that was very inadvisable for her. Germany might well rearm in any case, and a convention would give Hitler the additional advantage of an arms reduction by the rest of the world; some members were inclined to feel there might be a better chance of preventing a war if there was no convention. The ministers had now received a hint of the French interest in providing for sanctions, mainly from the aide mémoire given to Henderson in Paris, and their reluctance to press France was clearly a consequence of the firm opposition they expressed to undertaking any new commitments. Both Baldwin and MacDonald were opposed to "bargaining negotiations" with France. But until there had been a meeting with the French, the possibility of an agreement on the MacDonald plan could not be ruled out. Therefore the ministers accepted the idea of a "frank talk" with the French in September. In the committee's view, the purpose of this meeting, to take place without Henderson, was to find out if there was a chance of getting agreement on the MacDonald plan without any new commitments; if not, tactics would have to be considered for making plain the responsibility for failure. A full cabinet meeting two days later reached much the same conclusions; Neville Chamberlain pointed

[47] *DDF*, 1/III, Nos. 443, 486; 1/IV, No. 36; SDP, tel., Bingham to Dept., 13 Jul 33, 500.A15A4 General Committee/545; Cab 24/242, CP 176 (33), 6 Jul 33; *DBFP*, 2/V, Nos. 251, 381. As Boncour and Massigli planned to take vacations in August, a failure to meet in July meant there could be no meeting before September.

out that to push the French would be to incur a responsibility toward them. On July 28, the Foreign Office proposed to the French government a meeting on September 18.[48]

4. Vansittart Challenges German Policy

Sir John Simon took no part in these last discussions: in the hope of ending the ill health that had plagued him for months, he had departed on July 15 for a six-week sea cruise.[49] But events did not stand still during the period of his absence, nor during the interval before the meeting with the French. Even before Simon's departure, Eden was practically in full charge of Disarmament Conference matters, while Vansittart was in effect (and after the departure, in fact) the acting chief of the Foreign Office. Eden's statements in the ministerial discussions show he was anxious, despite the hesitations of the older ministers, to try to work out some agreed position with the French that might have a chance of German acceptance. As he wanted very much to conclude a convention, and as he had more sympathy for the French viewpoint than most ministers (he was, for example, more ready than they to agree to continuous and automatic supervision), his discussions with the French would likely, despite all cabinet directives and reservations, lead to some modification of the British position pleasing to Paris.

Moreover, events in Central Europe, together with the leadership of Vansittart, were giving a particular anti-German impetus to British policy at this time. More than Eden, Vansittart hoped to exploit Italian influence in Berlin, or better, to arouse Mussolini against Berlin, forming a three-power front to restrain Germany. One promising basis for such a policy was the question of an Anschluss of Austria with Germany. In Austria, in contrast to Hitler's general foreign policy in this period, the Nazi party was aggressively trying to expand its influence and weaken the government of Engelbert Dollfuss. Austria was an area in which Italy, too, was vitally interested, and where Italians would naturally tend to oppose German influence. The Austrian situation became increasingly acute from May through July. When the Dollfuss government objected to a propaganda visit by a German Nazi leader, Hitler retaliated by imposing a prohibitive visa tax of RM 1000 on German travel to Austria, which played havoc with the Austrian tourist

---

[48] Cab 27/505, DC (M) 32, meeting of 25 Jul 33; Cab 23/76, meeting of 26 Jul 33, 48(33)3; *DBFP*, 2/V, No. 276; *DDF*, 1/IV, No. 50. The doubts of MacDonald and the ministers about reaching a convention and negotiating with the French are also shown in *FRUS, 1933*, I, 217, and *DDF*, 1/IV, No. 103.

[49] Eden, p. 38; *DDF*, 1/III, No. 354; *DBFP*, 2/V, No. 254n; *ADAP*, C/I/2, No. 391.

trade and created severe economic difficulties. Radio propaganda and aerial leaflet-dropping, both from German bases, were stepped up, an "Austrian Legion" was organized in Bavaria, and terrorist acts took place within Austria itself. After futile discussion of League action, on July 24 Dollfuss appealed to Britain, and shortly thereafter to Italy, to support him and make representations in Berlin. Vansittart had been following Austrian developments with increasing indignation, and he thought he saw an opportunity to enlist Mussolini's backing. He tried to organize a three-power protest, including France, but although the French were willing enough, he did not succeed with Mussolini, who instead undercut Vansittart's plan by passing the contents of the British proposal on to the Germans, along with a suggestion that the best means of obviating the British initiative would be for Germany quickly to give assurances on Austria to Italy.[50]

In the field of armament, Vansittart had obtained mixed results from his call for intelligence information to prove German violations publicly. The Admiralty had very little to report. The War Office had some data on secret armament and munitions production, and it suspected that the Reichswehr was now enlisting men for less than twelve years, but much of its information was old, and the Army Council drew the correct but short-sighted conclusion that the picture was more one of preparation for mobilization and expansion than of an actual expansion of peacetime forces. The Air Ministry, however, reported that the number of violations in air activity was "almost innumerable," and that the German army and navy had respectively 270 and 120 planes. The Ministry was aware (as was the War Office) of training and arms production in Russia.[51]

[50] On the Austrian crisis: Jürgen Gehl, *Austria, Germany, and the Anschluss, 1931-1938* (London: Oxford University Press, 1963), pp. 52-68; Dieter Ross, *Hitler und Dollfuss: Die deutsche Österreich-Politik 1933-1934* (Hamburg: Leibniz-Verlag, 1966), pp. 62-63; Robertson, *Hitler's Pre-War Policy*, pp. 13-16; *ADAP*, C/I/2, No. 383. Aloisi's published diary makes no reference to Vansittart's proposal. A French document (*DDF*, 1/IV, No. 46) shows that, contrary to what the British understood (*DBFP*, 2/V, No. 270), Dollfuss did not appeal to France for aid when he appealed to Britain and Italy. But Bülow had the "impression" by July 27 from "a very secret source" that Dollfuss had appealed to all three governments (*ADAP*, C/I/2, No. 376; see also No. 385 and *DBFP*, 2/V, No. 301). This supposed knowledge, four days before notification by Italy, suggests that he had the text of the British telegram either from decryption or an agent source, most likely the former in view of the short time involved. In another case, the British embassy suspected German foreknowledge of the protest made to Bülow on 29 Jul 33: *DBFP*, 2/V, Nos. 287, 327 (encl. 3). On Vansittart's strategy with Austria and Italy, see esp. *DBFP*, 2/V, No. 316. Vansittart stated his policy of enlisting Italy against Germany most explicitly in connection with the air-armament question; see below.

[51] See preceding section. The information from the Admiralty is contained in FO 371, files C5555/245/18, C3082/653/18, C4535/653/18; for the War Office, in C5997/245/18; for the Air Ministry, in C6021/245/18. A summary of some of the material reported is contained in *DBFP*, 2/V, No. 253, Annex I.

The Air Ministry paper also reported a recent conversation between the British air attaché in Berlin, Group Captain J. H. Herring, and a German Air Ministry official, a retired captain named Carl Bolle, in which Bolle frankly admitted that Germany was rearming in the air, specifically including high-powered single-seater aircraft. Soon afterward, Herring had lunch with Bolle and the State Secretary of the German Air Ministry, Erhard Milch; on this occasion, Milch contended that the German people were on the verge of committing a violent breach in the Treaty of Versailles. Milch's remarks seemed the more ominous because he had just publicly announced that two foreign airplanes had mysteriously appeared over Berlin, dropping leaflets, and he had argued that this showed the defenseless condition of Germany and the need for an immediate grant of equality of rights in the air and on land. In connection with this incident, Hermann Göring himself publicly indicated that he would do his utmost to build one or two police airplanes. British and American observers believed that the supposed raid was a hoax designed to justify the building of fighter airplanes.[52]

It is now clear that most of these remarks of German Air Ministry officials violated official German policy. The official line, stated in propaganda guidelines of June 23 and July 12, 1933, was to keep German rearmament strictly concealed and to await the reconvening of the Disarmament Conference in October, when Germany would be in a position to expect a decision on her demands. The Propaganda Ministry told the German press not to print Göring's statement about doing his utmost to construct one or two planes.[53] But Göring's new Air Ministry had little patience with the relatively cautious policy of Blomberg and the Foreign Ministry, and the alleged leaflet drop was just the kind of independent caper that Göring loved to indulge in. The Air Ministry's impatience was a function of its recent history and its ambitious rearmament plans. Although the German navy remained under the Reichswehr Ministry, Göring had managed in a few months, in the face of army opposition, to organize an independent air ministry

---

[52] *DBFP*, 2/V, Nos. 219 (incl. note), 222, 223, 231, 253; SDP, disp., Berlin embassy to Dept., 26 Jun 33, 862.00/3018. The latter document pointed out that the patent spuriousness of this "air raid" tended also to encourage doubts about the Reichstag fire's origins. According to Manfred Messerschmidt's *Die Wehrmacht im NS-Staat* (Hamburg: R. v. Decker's Verlag, 1969), p. 46, which names no source, Bolle had partly Jewish antecedents and resigned in 1934 when asked to present a certificate of Aryan origins. Homze, *Arming the Luftwaffe*, p. 61, reports that Bolle was a former fighter pilot and Cambridge scholar who had a Jewish wife and soon left the Air Ministry for private industry because of differences with Milch.

[53] RK, ltr., Reichsminister of Interior (Frick) to Land Ministers of Interior, 6 Jul 33, and att. of 23 Jun 33, 3650/D813084-87; ltr. by Blomberg, 15 Jul 33, and memo, "Fur die öffentliche Behandlung der Abrüstungsfrage," 12 Jul 33, 3650/D813131-37; Brammer Collection, Z Sg 101/1, Order for Hamburg, 24 Jun 33.

and to have it absorb the function of military aviation. On June 19, Milch obtained an agreement with the Reichswehr Ministry that a fleet of 600 planes should be completed by the fall of 1935 (the Air Ministry was already drafting plans for a 1,000-plane program), and on June 29, he set in motion a scheme to organize forty-five land and six naval squadrons to use the 600 planes when completed. This plan superseded a much more modest program for thirty-one squadrons by the fall of 1937, approved in April when the air force had still been under the Army Command.[54] Bolle and Milch may have thought that by being frank and telling Herring things he already knew or would soon find out anyway, they were preparing the way for future collaboration with British airmen. As we have seen, visiting British aviators and air enthusiasts had recently been entertained by Göring and addressed by Hitler with references to understanding between "front-line soldiers" and to "the two great Germanic nations."

The British Foreign Office had no idea that such indiscipline could prevail within some German quarters. The thought did occur to officials that it was "even possible" Milch might have spoken irresponsibly and without Göring's approval, but generally, Vansittart and his subordinates assumed that the utterances of Bolle, Milch, and Göring represented a calculated move on the part of the German government. They feared that the German government was testing whether there would be any British reaction, in which case British silence would be regarded as consent. They recommended an official démarche in Berlin, which would state that the British government presumed the Air Ministry statements did not imply that the German government intended to rearm unilaterally, manufacture military aircraft, or violate treaty obligations. A reassuring German reply would not be any real guarantee, but at least Britain would then have protected her position should she later, after a failure to obtain a disarmament convention, wish to oppose German rearmament.[55]

There were other reasons for taking this approach. Most other War Office and Air Ministry information on German rearmament could not be used without compromising secret sources; the step now proposed would only disturb Herring's relations with the German Air Ministry —although this was serious enough in British Air Ministry eyes. Also, to British leaders, and not least to Vansittart himself, the German air

---

[54] Völker, *Entwicklung der militärischen Luftfahrt 1920-1933*, pp. 200-206, 210-213. The order creating the Air Ministry had made it subordinate to "the Reich Defense Minister and commander of the whole armed forces," i.e., Blomberg, but as this was secret and as Göring had his own staff and budget, as well as a driving personality and access to Hitler, the proviso was virtually a dead letter.

[55] *DBFP*, 2/V, No. 253; FO 371, Foreign Office draft memo, 7 Jul 33, C6219/245/18; Minutes by Perowne, Wigram, Sargent, Vansittart, 11-12 Jul 33, C6099/245/18.

threat was much more a danger to Britain than preparations for land warfare would be. Finally, Vansittart hoped to use this démarche as an instrument of general European policy, enlisting Italy as well as France against Germany in a tripartite action:

> To my mind [he minuted], the essential thing is to get *Italy* in. Nothing else will be in any way fruitful: we shall run the risk of damage with no chance of profit. The profit wd. be in fact to detach Italy from Germany—since Italy also has pronounced against German rearmament. If that idea is to be abandoned I wd. rather abandon the whole thing.

The service departments gave grudging consent to the plan, providing the cabinet approved, and the cabinet considered the question on July 26. By that time, Vansittart (with MacDonald's approval) had already sent to Rome and Paris his proposal for a joint démarche in connection with Austria, and Mussolini, who had already shown some reluctance to join in concerted action against Germany, had not yet replied. It was hardly the moment to suggest another tripartite action. On the other hand, Göring had made a new assertion to Herring that Britain had sold police aircraft to Austria and hence should now sell them also to Germany; aside from requiring a specific denial and refusal, this seems to have strengthened the case for a diplomatic query on German air armament generally. So the cabinet decided that the planned query should be made, but that this should be an independent British action, the French and Italian governments only being notified.[56]

When the British démarche took place on July 29, State Secretary von Bülow (in charge during Neurath's absence) was angry, not least because the démarche referred explicitly to the unwelcome British strictures of September 19, 1932, against German unilateral rearmament. As in 1932, Bülow argued that such representations illustrated in themselves the need to establish German equality. Privately, he was probably also annoyed because Göring and Milch had failed to abide by the official policy of caution and concealment. But the State Secretary firmly asserted that Milch's public demand for an immediate grant of equality was no different from German official policy, that any idea that Milch and Göring were indicating imminent violations of treaty restrictions was a misunderstanding, and that no German Air Ministry

---

[56] FO 371, Minutes by Sargent and Vansittart, 10 Jul 33, C6219/245/18; *DBFP*, 2/V, Nos. 246, 249, 253, 255-257, 259, 261-263, 265 (and note), 270, 271 (and note), 274; Cab 23/76, meeting of 26 Jul 33, 48(33)4, 48(33)5. On the eve of his departure, Simon wrote a rather half-hearted endorsement for the action proposed by Vansittart: Cab 24/242, CP 184 (33), 14 Jul 33.

official could have made the statements attributed to Bolle. Bülow demanded a confrontation between Herring and Bolle; when this took place on August 9, in Milch's presence, Bolle flatly denied having made the reported remarks, and Milch gave a general denial that military aircraft were in fact being built. Following this performance, the British air attaché was deprived of access to the German Air Ministry, a restriction that was not to be lifted until April 1934, after the British had taken corresponding countermeasures against the German military and naval attachés in London.[57]

5. Paul-Boncour's Scheme for Expert Discussions

The main effect of the British démarche was not to discourage German rearmament, which continued as before, nor to protect the British capacity to bring charges in future, which never came near to being brought; it was rather to provide an opening for French political tactics. British diplomats explained to the French and Italian governments that their action was taken unilaterally because it was largely based on information given only to a British attaché, but they added that Britain would attach importance to "mutual consultation and joint action" if the German reply proved unsatisfactory. This suggestion probably resulted from Vansittart's desire to involve Italy, but the French were the ones who immediately welcomed the idea, offering full cooperation and further information on German air effectives. They had already drawn encouragement from Vansittart's proposal for a tripartite protest against German-based Nazi attacks on the Dollfuss government, and from British requests for support in blocking the export of aircraft and aircraft parts to Germany. On this last question, Massigli had suggested on July 26 that a joint protest against German air rearmament might be desirable, and two days later a French memorandum proposed to the British that the two governments prepare for their protest by jointly examining the available material on German air armament violations.[58]

From the perspective of Paris, Vansittart's active policy must have seemed to fit beautifully with a plan Boncour had been developing. At the time of the June 8 meeting, Londonderry and Eden had shown a desire to be more completely informed on German military policy, and this had given Boncour the idea of having a dossier drawn up on the rearmament of German land forces, containing only information

---

[57] *ADAP*, C/I/2, No. 380 (and n. 10); *DBFP*, 2/V, Nos. 277, 280, 281, 284, 287, 289, 298, 305, 326, 327 (and encls.). See also note 50. In calling for a confrontation, Bülow may have wanted to give the Air Ministry a lesson, as well as reply to the British.

[58] *DBFP*, 2/V, Nos. 268, 271, 275, 278, 282, 286; *DDF*, 1/IV, Nos. 37, 40, 41, 51.

that was incontrovertible and that could be passed confidentially to the British government. Indeed, the French General Staff prepared this select dossier from material which, for the most part and with certain precautions, could be made public. Boncour now directed his ambassador in London, Charles Corbin, to give Vansittart an outline of this dossier, to point out the seriousness of the violations described, especially in terms of effectives, and to ask if the British government did not think the time had come "to obtain precise answers from Berlin as to the goals of the military policy of the Reich." The dossier itself could not be sent to London ("for obvious reasons"), but would be available in Paris for British expert examination.[59]

Boncour argued that if explanations were not asked from Germany before the conference resumed, it would be necessary to seek them at that point, as it was useless to discuss general disarmament if those countries whose disarmament was supposed to make general disarmament possible were in fact rearming. He had also requested from the French Ministry of War a list of its documents suitable for exposing German rearmament in Geneva. If such action was under consideration, it was certainly prudent to try to secure British support, and if this was not forthcoming, Boncour wanted to be able to say later that he had offered the British a chance to compare French data with their own.[60] In any case, the mere threat to bring charges could serve as a means of pressure to induce British cooperation.

But Boncour seemingly also hoped to affect British policy more fundamentally. Within the Quai d'Orsay, understanding with Britian was explicitly considered to be the basis for all French policy.[61] By showing the French information to the British, Boncour no doubt hoped to persuade them that French concern was well founded, and the French position correct. In view of legal obstacles, it might not actually be feasible to demand an immediate international investigation of German armament, but apparently it did now seem possible, particularly after

[59] DDF, 1/IV, No. 65. The outline, which gave generalized source descriptions, indicated, inter alia, that the actual strength of the Reichswehr itself, owing to larger than usual enlistments in April, was 120,000; that there existed a mobilization plan and mobilization organs; that the mobilized army was to include a field army of twenty-one divisions, a "reinforcement" army (*Ergänzungsheer*), and a frontier protection force (*Grenzschutz*); that the German army had at least double the number of automatic weapons permitted; and that the army was supplemented by police (greatly increased by auxiliary police), paramilitary organizations (numbering over a million men), labor service (287,000 in December 1932), and a disguised air force. This information seems to have been generally accurate and well substantiated; unfortunately, the dossier, as Boncour's staff noted (*DDF*, 1/IV, No. 203), was weak on the side of materiel, and this was what would count most in British eyes.
[60] DDF, 1/III, No. 448; 1/IV, Nos. 65, 175, 202.
[61] See DDF, 1/IV, No. 20, in which it was a matter of not risking good relations with Britain by a secret agreement with the Soviet Union.

Vansittart's recent proposals for joint action, to convince the British of the need for supervising German arms.[62] If the British could be led to accept the necessity of a trial stage with effective supervision, that would be, from the French point of view, a definite step forward; just as "samples" were the thin end of the wedge of qualitative equality, the trial stage was the thin end of the wedge of investigation and control. But above all, the French hoped that the dossier would lead the British to "accept their responsibility" and agree to sanctions in the event of violations.[63]

Although it is difficult to document how far it played a role in French thinking, there may have been yet another argument for showing secret intelligence to the British: to *involve* the British could be at least as useful as to convince them. The very act of viewing such material in Paris would tend, regardless of the contents' validity, to align Britain with France, particularly if the fact of the viewing were presently leaked to the press. It would tend to prejudge the question of whether Britain should join in denouncing Germany. It would create a certain obligation on the British side toward France. It would bring the British into a close cooperation between experts, which might have some of the entangling ramifications that the pre-1914 staff talks had had.[64]

Boncour later protested that there was no immediate question of committing the British to take any action, and that France did not seek in any way to compel the British government to assume responsibilities on which only it could decide. Later still, he said that the French government had in no way intended to commit the British government in advance to a predetermined policy.[65] But obviously the

---

[62] On pertinence of the dossier to the need for supervision: *DDF*, 1/IV, No. 173. As for whether Boncour actually intended to bring charges, there remained the legal problems (*DDF*, 1/III, No. 448 [encl.]). Probably charges would have been brought if Britain had agreed, but it is also possible such action seemed more useful as a threat than for any good it might do.

[63] The close link between the dossier and sanctions emerges esp. in the French records of the 18 and 22 Sep 33 meetings: *DDF*, 1/IV, Nos. 260 (Annex), 261 (Annex). See also *DDF*, 1/IV, No. 241 and *DDB*, III, No. 47.

[64] A French Foreign Ministry internal memo of 22 Sep 33 indicated the French interest in staff talks, although recognizing that the British were unwilling to accept them under existing conditions; it suggested that Britain might declare she would agree to such talks as soon as it was confirmed that a violation of the (future) disarmament convention had occurred (*DDF*, 1/IV, No. 240); see also Section 7 and note 84 below. There is some evidence the British were aware of the risk of entanglement. Orme Sargent noted in November the danger of French leakage of any talks on German violations (FO 371, Minute by Sargent, 14 Nov 33, C9893/245/18), and in September, War Office and Air Ministry representatives argued that to enter into discussions would tend to prejudge the question of whether charges should be brought in Geneva (FO 371, Minute [by O'Malley], 16 Sep 33, ltr., Vansittart to Simon, 18 Sep 33, C8272/245/18).

[65] *DDF*, I/IV, No. 400; I/V, No. 66.

matter of commitment was at least on his mind. He believed sincerely and with increasing passion that German violations should be exposed, and the least that can be said is that he was trying to lead the British government to join in looking at the facts of German rearmament, so that that government would commit itself to helping keep Germany in check.

The dossier proposal got off to a slow start. When Corbin on August 4 gave Vansittart the outline description of the French dossier and proposed a joint investigation, the final results of the British démarche on the Bolle-Milch-Göring matter were not yet in, and Vansittart observed to Corbin that it would be necessary to await them. The British Under-Secretary suggested that the French should secure Italian collaboration, his favorite solution even though very recent experience in the Austrian matter showed that such collaboration would be hard to obtain. Then, a few days later, he told Corbin it would be difficult to conduct any joint investigation with most ministers and many Foreign Office officials absent, and that in any case no decision could be taken until the cabinet met at the end of August. Vansittart was privately aware that other departments of the British government would be very dubious about any joint investigation, and he must have realized (as Corbin did) that this and other considerations could lead to resistance in the cabinet.[66]

Nevertheless, Vansittart avowed to Corbin that he was personally not far from agreeing that it was time for joint discussions and for studying joint measures, and he also remarked that in any case the problem of German armament could be given priority in the discussions scheduled for September 18. This observation, and also some statements by Lord Tyrrell to Léger, encouraged Boncour to think that the questions of German violations and what should be done about them could become the main topic in the September Anglo-French conversations; if so, then it might be well to have talks between experts before these conversations.[67]

While Tyrrell was giving encouragement to Léger and Boncour, he was doing the same for Eden. He produced a memorandum which

[66] *DBFP*, 2/V, Nos. 289, 298, 316, 326; *DDF*, 1/IV, No. 103. Respecting Vansittart's advocacy of collaboration with Italy, he had been informed by 3 August that Italy had "already" made a separate approach in Berlin (*DBFP*, 2/V, Nos. 288, 296), instead of the joint approach he had suggested on 25 July (*DBFP*, 2/V, No. 271); he did not learn until 8 August, however, that the Italians had divulged his proposal to the Germans (*DBFP*, 2/V, Nos. 312, 316). Even after that, he continued to advocate working with Italy: see below. The Air Ministry—bitter over the exposure of Herring—and also the War Office were already reacting against the French proposal of July 28 for a joint examination of German armament violations: Public Record Office, Air Ministry Records, AIR 2/1354, Note by Squadron Leader A. Boyle, 4 Aug 33; FO 371, Memo [by Perowne], 18 Sep 33, C8272/245/18.

[67] *DDF*, 1/IV, Nos. 103, 125.

made no reference to the French dossier or to the French demand for sanctions, and which held that France was genuinely anxious, for idealistic and financial reasons, to disarm. Tyrrell maintained that the essential French desiderata were automatic and permanent supervision (for Germany, not necessarily for England) and a probationary period. "If these are accepted and proved," the Ambassador concluded, "the French Government will be able to face public opinion and proceed to reduction to the German level." Eden also heard privately from Massigli that France might go farther in disarmament than had been indicated to Henderson. He, Eden, circulated the Tyrrell memorandum to the cabinet, along with a covering note arguing that it was essential before the conference resumed to encourage the French to offer disarmament in return for supervision and a trial stage. He suggested that if the Germans refused such an offer, they would put themselves in the wrong with Italy, the United States, and "ourselves."[68] Although he wrote as though a breakdown in negotiations was likely, Eden himself probably hoped that the Germans would come around when confronted by Anglo-French agreement on such a position.

Eden wanted to use the September 18 discussions to draw the French into making promises of eventual disarmament, rather than demands for security or for an investigation of German armament, and at first he opposed the French suggestion of preliminary expert discussions on the dossier. But to judge by the French record, Eden nevertheless finally accepted the possibility of prior expert discussions, and according to another French document, Foreign Office officials were talking on August 25 of a meeting of experts before mid-September, a meeting of British and French ministers around September 15, and then a Franco-British-American meeting on September 18. Ten days later, however, the French chargé in London, Roger Cambon, had to report that there had been no further move toward a meeting of experts. He concluded, with some accuracy, that the British feared a public disclosure of German rearmament would provoke strife with Germany, terminate the conference, show the incapacity of the League to act, and confront British opinion with responsibilities it was not prepared to assume.[69]

[68] Cab 24/243, CP 205 (33), 21 Aug 33; Eden, p. 43. Tyrrell did mention that the French still wanted a continental security pact, but said they had given up hope that Britain would accept and act on a definition of aggression. Although Eden's covering paper circulated to the cabinet displayed what is called "hard-headed realism" as to the prospects for a convention, a French document (*DDF*, 1/IV, No. 68) and Eden's memoirs seem to indicate that he was really more optimistic.

[69] *DBFP*, 2/V, Nos. 348, 359; *DDF*, 1/IV, Nos. 130, 137, 173 (incl. note). There are considerable discrepancies between Eden's and Cambon's accounts of their two meetings, illustrating what might be called the iron law of official record-making: no maker of a record ever shows himself losing an argument.

6. ANGLO-FRENCH PROBLEMS AND AN ITALIAN SOLUTION

As Simon and other ministers returned from their vacations, the coolness of British officials toward a meeting of experts doubtless increased, but Vansittart did not weaken in his efforts to organize an anti-German front. In a cabinet paper of August 21, he was still advocating a common policy on disarmament between Britain, France, and Italy, saying: "At moments such as these our only chance of preventing a dictator situated like Hitler from committing himself irrevocably to a policy of foreign adventure and provocation is to pull him up sharp before his plans have crystallised, and this sharp pulling up can be best and most painlessly done by joint action of the Great Powers." On August 28, the day before Simon's return, Vansittart raised the question of whether some action might not have to be taken through the League, principally on the Austrian question, but also on that of German illegal rearmament. League action, he indicated, would probably mean economic sanctions, and for the immediate future he preferred to encourage Italian opposition to Germany, and Franco-Italian cooperation.[70] When the cabinet convened on September 5, it concluded that if any action had to be taken on the Austrian problem—which by now seemed less urgent—it should not be through the League, but through Italy; Neville Chamberlain said that economic sanctions were unlikely to be successful. Thus the ministers were not well disposed toward the idea of an appeal to the League. With regard to disarmament, Simon at the same meeting—no doubt relying on briefing material from Eden—painted a rosy picture, saying there was a new factor in that the French government now clearly wanted to appear eager for the success of the Disarmament Conference, and might make a reasonable proposal. The ministers agreed that Eden should report back after September 18 on whatever proposals the French might make, without himself going beyond existing cabinet decisions. On the question of an Anglo-French meeting of experts, Simon seems to have understood (again, probably from Eden's record) that this had already been given up in favor of a frank discussion of policy between governmental representatives.[71]

Simon's impressions of the French outlook did not jibe well with the indications now coming in. Boncour was telling Ronald Campbell, the British chargé in Paris, that it would soon be necessary to refer both the Austrian question and German arms infractions to Geneva, i.e., to the League, and on September 1 and again on September 5, *Le Temps* suggested a preliminary investigation of German arma-

[70] Cab 24/243, CP 204 (33), 21 Aug 33; *DBFP*, 2/V, No. 371.
[71] Cab 23/77, meeting of 5 Sep 33, 50(33)2; *DBFP*, 2/V, Nos. 381, 387.

ment. Campbell talked with Daladier, but although the Premier was believed to be much more flexible than Boncour, he did not rule out the possibility of bringing charges of arms infractions. By September 12, Campbell had concluded that the French cabinet was indeed on the verge of demanding an investigation of the existing state of German armament. As for the dossier, Cambon on Boncour's instruction vigorously took the matter up again with Vansittart on September 13, and left a supplement to the original outline at the Foreign Office on September 15. Vansittart got the impression that the French might be about to take their case to Geneva or otherwise publicize it, and he now considered this very undesirable at a time when disarmament negotiations were being resumed. He concluded that expert discussions might be a way to "keep the French in play" and spin matters out until the danger of independent French action was past. But he was unable to get the agreement of the War Office and of the Air Ministry before the September 18 meeting.[72]

Before going to Paris for this Anglo-French meeting, Eden had expected the French would stress the subject of German violations.[73] This expectation was fully borne out on the eighteenth, and (as appears more clearly in the French than in the British record) he was definitely subjected to pressure. According to the French record, prolonged argument took place, and indications were wrung from the British negotiators that Britain would discuss the dossier and join with France in bringing eventual accusations, before Boncour, Massigli, and (later) Daladier agreed that public disclosure might wait until after the conference resumed, when the Germans would un-

[72] *DBFP*, 2/V, Nos. 379, 382, 386; SDP, Ferdinand Mayer Geneva diary, 7 Sep 33, 500.A15A4/1469½; *DDF*, 1/IV, Nos. 202, 209; FO 371, Note on cover of French supplement of 12 Sep 33, C8225/245/18. On Daladier's supposed flexibility as compared with Boncour, see Cab 27/505, DC (M) 32, meeting of 25 Jul 33; Cab 24/243, CP 205 (33), 21 Aug 33. The principal French proposal for joint expert discussion, that concerning land armament transmitted on 4 Aug 33, did not originally propose any contribution of British intelligence information (see *DDF*, 1/IV, No. 65). Thus, strictly speaking, the argument most used in London against discussions—that British sources might be compromised—did not apply. When the French finally succeeded at the end of December 1933 in getting a very limited commentary on illegal armament from the British government, they replied bitterly that the British had misunderstood; they had never asked if British information confirmed their own, but rather if the British would join in examining the French material (*DDF*, 1/V, No. 178). This "never" was not correct, however; after their 4 Aug 33 proposal, the French, from Boncour on down, had frequently spoken of a comparison of French and British information: *DDF*, 1/IV, Nos. 125, 137, 202, 221, 242, 400. Also, the earliest proposal, for a joint study of information on air violations, seemed to imply contributions on both sdes: *DDF*, 1/IV, No. 51.

[73] Davis Papers, Memo of conversation with Eden and Cadogan, 14 Sep 33; see also *FRUS, 1933*, I, 217.

doubtedly either claim to have disarmed, or protest against the idea of stages. Anyway, the exposure of German armament was not an end in itself for the French. Instead, to the French, disclosure of the existing state of German armament would serve principally to satisfy French opinion and to convince the British of the need for positive action if the disarmament convention were violated. From the French point of view, a convention that would legalize German rearmament would be worse than having no new convention at all, but the best solution would be a convention that would commit the British government to cooperate in restraining any violations. When Daladier joined the conversations, he placed great emphasis on sanctions, citing French beliefs that the Germans were violating Locarno as well as Versailles by military activity in the Rhineland. On this more important question of sanctions, Eden said he would consult his government.[74] He must have known very well what the answer would be.

As far as French disarmament was concerned, Daladier generally maintained the May 2 position, given to Henderson on July 11. There could now be no mistaking the French insistence on provision for sanctions, and Britain would have to state her position on this.[75] This French demand might be somewhat hard to explain to the cabinet in London, as the Foreign Office had tended to conceal from ministerial eyes the full extent of French desires. Eden would have had a difficult report to make in London if he had not received personal assurances from Daladier that preventive war was ruled out,[76] and if it had not been for new information on Franco-Italian negotiations.

[74] 18 Sep 33 meeting: *DBFP*, 2/V, No. 399 (Annex); *DDF*, 1/IV, Nos. 227, 260 (Annexes). There appears to be a misprint in the first sentence of the first numbered paragraph (following "1⁰. Manquements au traité.") in *DDF*, 1/IV, No. 227. On the French point of view, also *DDF*, 1/IV, Nos. 240, 241; *DDB*, III, No. 47. The British report of the meeting is less complete than the French and may have been edited to avoid alarming the cabinet, while the French participants presumably wanted *their* record to show that they really pressed for the French desiderata. The French threat of making an exposure was also a lever to get British and American acceptance of the trial period; in talking with Davis the next day, Boncour directly linked the trial period with a demand for an investigation, saying that if France did not get agreement to the first, she would have to ask for the second: *FRUS, 1933*, I, 219; cf. *DDF*, 1/IV, No. 229 (II).

[75] Davis reported (*FRUS, 1933*, I, 231) that at one of the following meetings on 22 Sep 33 Simon said he had not come prepared to discuss the question of action in case the convention were violated. There would seem to be some misunderstanding on someone's part here; Simon had come to Paris prepared to reject sanctions and did so in the earlier meeting the same day, which Davis did not attend.

[76] Daladier's assurance on preventive war (*DBFP*, 2/V, No. 399) was welcome not only in the light of recent suggestions of French initiatives at Geneva but also in view of a private statement by Pierre Comert (chief of the Quai's press bureau), also on 18 Sep 33, that if there were no convention, France, having to choose be-

Mussolini had proposed to the French a disarmament scheme that agreed with the French proposals on stages and on permanent and automatic supervision. The scheme left existing French arms intact for four years; Mussolini also later seemed willing to recognize the principle of sanctions. The catch was that Mussolini also proposed to give the Germans an allotment of "defensive" arms, probably light tanks and fighter aircraft. This would have gone far to defeat any effort at supervision: once prototypes were permitted, troops could be trained with them, further specimens could be hidden, and preparations could be made for full-scale, wartime production. Mussolini argued persuasively for his scheme, saying to the new French ambassador, the Comte de Chambrun:

> Germany is arming. If she continues to arm, what will you do? A preventive war? An occupation of territory? An action in Geneva which would lead Germany to leave the Council and thus be a fatal blow to the League of Nations? The solution I propose, while fulfilling the promises foreseen toward Germany, leaves you all your superiority.

Aloisi, claiming to speak on the Duce's authority, also told Chambrun that if the worst came to the worst, "we would be on your side."[77]

The Italians obviously cared more about sponsoring a successful compromise than about reducing armaments or restraining German rearmament.[78] The French government could hardly have agreed to the Italian scheme. Critics in France would quickly have charged that the government had been duped, and British and American opinion might have reacted badly to an apparent failure to disarm. Interest-

---

tween attacking and being attacked, would "march"; FO 371, disp. (with encl.), Harvey to F.O., 19 Sep 33, and minutes by Vansittart and Perowne, 20, 22 Sep 33, C8343/245/18.

[77] *DDF*, 1/IV, No. 177, 224. French reaction: *DDF*, 1/IV, No. 160, 187, 203, 213 (Annex); No. 203 esp. describes the problems with samples. Aloisi, p. 144, mentions his talk with the ambassador, the Comte de Chambrun, but does not record the assurance. The previous negotiations are covered in *DDF*, 1/IV, Nos. 113, 160, 167. The meaning of "defensive" arms, implicit in the first discussions, was made clearer in succeeding talks (*DBFP*, 2/V, No. 395; *DDF*, 1/IV, No. 179, 224).

[78] Their prime objective, apparently, was to make a proposal that all four signatories of the four-power pact would accept, thereby giving more significance to the pact. Mussolini also had the realistic idea that the French government simply did not want to disarm. See Aloisi, pp. 139, 143; *DDF*, 1/IV, Nos. 113, 178, 234, 274; *DBFP*, 2/V, No. 395; *ADAP*, C/I/2, Nos. 382, 396. According to Aloisi's record, the initiative for the Italian proposal came from him and his colleagues rather than from Mussolini, and his own special efforts to secure French agreement (*DDF*, 1/IV, Nos. 177, 178) seem to show he was bending every effort (and fact) to assure a success.

ingly enough, officials at the Quai d'Orsay did not refuse to consider a concession of new arms to Germany, but they thought that France could gain more by offering such a concession directly to Germany, rather than through Mussolini.[79] Still, both the Italians and the French hoped the difference might be bridged over, and the French probably thought the negotiations useful for evoking British interest in, and perhaps agreement to, stages, supervision, and sanctions. Eden had only learned of the Mussolini proposals the day before his meeting with the French and only saw the diplomatically worded French reply at that meeting itself; he concluded enthusiastically that the Italian proposals created a "new situation."[80]

7. THE BRITISH REJECT SANCTIONS

The news of the Franco-Italian negotiations was very welcome in London, where Vansittart in particular had been strongly urging common Franco-Italian action, and where the cabinet preferred cooperation with Italy to working with the League. When the British cabinet considered the result of Eden's talks on September 20, Simon stressed that the French wanted a convention, and that they were nearer to an accord with Italy than they had been for a long time. There was an urgent need to pursue talks in Paris, and Simon could do this on his way to the September League Assembly meetings. Following his usual pattern of argument, Simon suggested that Britain now faced a choice between failure to reach a convention, followed by German rearmament, and agreement to a two-stage convention, with supervision on the lines of the Franco-Italian proposals, but perhaps less strict for Britain and the United States. He thought that to gain German acceptance, it might be necessary to concede specimens of forbidden weapons, at least in the second period; if others would not reduce, Germany had to be allowed some increase. The ministers showed strong resistance to supervision, and above all to the acceptance of any new commitments. The cabinet finally agreed that Simon could tell the French ministers in Paris that Britain would accept stages, and even close supervision if that was the only remaining obstacle to agreement. But if Daladier raised the question of what would happen in the event of a breach in the convention, Simon should reply that a serious breach verified by the Permanent Disarmament Commission would simply release the other signatories from restric-

[79] *DDF*, 1/IV, No. 167.
[80] Aloisi, pp. 144-145; *DDF*, 1/IV, Nos. 179, 190, 203, 213 (Annex), 224, 260 (Annex I); *DBFP*, 2/V, Nos. 395, 397, 399 (Annex).

tions. He would have to make it clear that Britain could accept no new commitments.[81]

When Simon, Eden, and also Stanley Baldwin (on his way back to England from Aix-les-Bains) met with Daladier and Boncour on September 22, Simon clearly stated the British cabinet's position on stages, supervision, and also sanctions. Baldwin suggested that Germany would find it difficult to accept a convention that contained sanctions. If, however, she did rearm during the first period, then a new situation would arise which Britain would have to consider very seriously. Daladier replied firmly that Germany was already rearming and that France would have to insist on guarantees. At Simon's request, Daladier explained what he thought these should be: it was not enough that each signatory should regain its freedom to rearm, as then the violator would have a head start; there should also be a collective guarantee that economic and diplomatic relations would be broken off with the violating nation, and if the violations did not cease, the other states should all declare themselves ready to compel compliance, if need be by force. Boncour argued that the current French position was really a great concession, as compared with the French proposals for mutual assistance advanced in November 1932, but Eden aptly pointed out that while the earlier French plan of concentric circles had left Britain in an outside position, "the present proposals reduced everything to one single circle." The difference over sanctions was now open and unresolved, although the French cherished a hope that the British might reconsider.[82]

Although the French maintained that Germany was already rearming, and thought it important to establish the existing level of German armament, they do not seem to have wished to take sanctions against the violations Germany had already committed. Daladier, and also Boncour, said they did not want to wage a preventive war.[83] They did hold that if Germany freely agreed to the new convention, then there should be no such legal inhibitions on enforcement as existed

[81] Cab 23/77, meeting of 20 Sep 33, 51(33)2. Simon described some of the cabinet's conclusions to Ambassador Corbin on 21 Sep 33 (*DBFP*, 2/V, No. 403; *DDF*, 1/IV, No. 233). Corbin reported to Paris that Simon was not opposed to a concession of defensive arms to Germany (in which stage was not stated), and that Simon had expressed with some spirit the British disagreement with the French concept of sanctions. Simon made no response when Corbin again raised the question of the joint examination of the dossier.

[82] *DBFP*, 2/V, No. 406; *DDF*, 1/IV, No. 261 (Annex I). Neither record indicates that Daladier called clearly for a general declaration of readiness to use force before a specific situation arose requiring its use. At this meeting, Boncour also raised again the matter of the examination of the dossier, and Simon promised an early reply.

[83] *DBFP*, 2/V, No. 399; *FRUS, 1933*, I, 219; *DDF*, 1/IV, Nos. 229, 246; AA, tels., Nadolny to Ministry, 10 Oct 33, 3154/D670032-35, D670040.

under current treaties. A Quai d'Orsay memorandum suggested the possibility of getting a firmer British commitment under the Treaty of Locarno, including agreement that staff talks should begin in the event of a duly confirmed, serious violation of the arms convention.[84] No doubt the French hoped that the British, American, and other support of their position would impress the Germans and deter them from both rearmament and eventual aggression, even if no convention was concluded.

It was support against eventual aggression that was most important. The French were quite willing to agree to a later German rearmament if they could gain guarantees of support by such a (perhaps nominal) concession. After Simon had indicated that he did not completely oppose the German possession of samples, Daladier and the British agreed on September 22, largely on Daladier's initiative, that weapons retained by other powers might be granted to Germany in the second period. Norman Davis had returned to Europe, and on the same day, according to French and British records, he also stated that German acquisition of all generally permitted weapons in the second stage was logical.[85] On the twenty-third, the Quai d'Orsay sent word to Rome stating that France would offer new arms to Germany in the second stage as a "very great concession" to the Italian point of view. This concession could only be made, however, if the convention contained very clear-cut guarantees against violations. If only Italy would agree on this last point, "the key to everything," then it might be possible to get British and American agreement as well, as in the case of supervision.[86]

THE FRENCH hopes for guarantees were vain. Mussolini continued to think that some new types of arms would have to be conceded in the first period, and his representatives declined to commit themselves on sanctions beyond what would be accepted by all nations.[87] Officials in London sent a proposal to the cabinet for a pronunciamento by the signatories to the eventual convention that, in case of a violation,

---

[84] *DDF*, 1/IV, Nos. 240, 242. Massigli tentatively raised with Cadogan the idea of reinforcing Locarno but met with a rebuff: *DBFP*, 2/V, No. 413.

[85] *DDF*, 1/IV, Nos. 233, 261 (Annexes I and II); *DBFP*, 2/V, Nos. 403, 406, 407. The American record (*FRUS, 1933*, I, 228-229) reports the proposed second-stage concession, but does not indicate that Davis agreed with the idea.

[86] *DDF*, 1/IV, No. 241.

[87] *DBFP*, 2/V, Nos. 409, 412, 424, 444; *DDF*, 1/IV, Nos. 237 (Annex), 243, 246, 254, 261, 265, 274; Aloisi, pp. 146-148. Mussolini did finally, on 10 October, give Chambrun some further hope of support on "guarantees of execution," while suggesting that the term "sanctions" be avoided (*DDF*, 1/IV, No. 291). He also proposed a compromise on samples at the last moment: *ADAP*, C/I/2, No. 494 (and encl.); *DBFP*, 2/V, No. 450.

they would consult together "for the purpose of restoring the situation" and to keep the convention provisions in operation. But the idea was stillborn, and meanwhile, when the French made new appeals, Cadogan and Eden firmly repeated that Britain could not undertake any new hard and fast commitment.[88] Although Norman Davis suggested there might be some provision for consultation in case of violation, he maintained that the United States could give no undertaking beyond what he had indicated on May 22, and he did not admit that that statement of position had become somewhat flyblown.[89] Davis indeed had the same outlook on disarmament that he had had in the spring of 1932. In London he urged Simon and MacDonald to press the French to disarm, and in Paris he showed Daladier a letter from President Roosevelt, threatening an American return to isolation if the other powers did not proceed with disarmament. He evidently still supposed that the German government wanted disarmament; he could not understand why Italy should propose a plan which failed to provide disarmament, "as normally one would not expect Mussolini to propose a disarmament plan unless he had some reason to believe that the Germans would accept."[90]

[88] Cab 24/243, CP 228 (33), 3 Oct 33, and Annex; *DBFP*, 2/V, Nos. 413, 420, 426; *DDF*, 1/IV, No. 272. See also note 84. The proposed declaration was part of a compromise formula, and was probably inspired by French pleas to Eden and Vansittart for at least a window-dressing declaration providing something to show the French public (see *DBFP*, 2/V, No. 420; MDP, 1/504, Note from Vansittart, 29 Sep 33). Since Eden soon after stressed that the French should not hope for any new British commitments (*DBFP*, 2/V, No. 426; *DDF*, 1/IV, No. 272), the declaration would not apparently have had binding force; see also Simon's paper in CP 228 (33), which said that German violations would meet with "the moral condemnation of the world." In any case, the cabinet (in the wake of a new statement clarifying the German position) decided not to propose the compromise (*DBFP*, 2/V, No. 440; Eden, p. 45).

[89] *FRUS, 1933*, I, 220, 224, 229-230.

[90] *FRUS, 1933*, I, 212-224, 226; *DBFP*, 2/V, No. 381; Davis Papers, ltr., Roosevelt to Davis to show Daladier, [ca. 30 Aug 33]. The editors of *FRUS, 1933*, I, published letters for MacDonald (pp. 209-211) but stated that a letter to show Daladier had not been found (p. 208n). The letter in the Davis Papers, cited above, presumably the letter in question, was less diplomatic than those for MacDonald. It said that the United States did not ask France to give up its "present military predominance" without the establishment of constant and thorough inspection, but if the Germans, under such an inspection, gave proof that they did not intend to seek revenge, "then the French must accept alone the responsibility of a refusal to reduce military strength. . . . We have helped in the past and we remain ready to help to the extent that we have made clear. But if various of the other powers decide that they are unable to carry the movement for disarmament to fruition, I urge that they should in all frankness tell us so in order that we may go our own way." The letter ended by saying it would be a "tragic error" if any power allowed "differences of opinion on policy in economic or financial matters" to influence their disarmament policy. The Davis Papers show that Davis made suggestions for the various Roosevelt letters. Allen Dulles apparently also thought the Germans were in favor of pre-

French diplomacy did seem to have made some positive gains. Britain, Italy, and the United States agreed sufficiently with the French arguments to accept a trial stage and a measure of supervision. Also, Britain and America generally agreed with France in refusing samples until the second stage.[91] From a French point of view, these results, though less than had been hoped, still provided an apparent chance to form a united front against the Reich. The support of a trial period by the other powers might lead to a German withdrawal from the conference, and to what would be nominally a disarmament convention and in fact (since Germany would refuse to sign it) an anti-German alliance.[92]

But diplomatic and political stratagems can hardly bring about basic policy changes within a democratic government. As shown by Herriot's experience with the Anglo-French consultative agreement in 1932, it was useless for Boncour to try to maneuver British spokesmen into a position for which British opinion was not prepared, and which even the spokesmen themselves did not genuinely and wholeheartedly accept. Like the declarations of MacDonald at Lausanne in July 1932, the intimations by Eden and his colleagues at Paris on September 18 —that Britain would discuss the dossier and join in an eventual accusation—were forced and fundamentally untenable. Similarly, although Eden tried vis-à-vis the cabinet to play up the supposed French interest in disarmament, and to play down French demands for guarantees, the reality of the French position was bound to emerge eventually. And Vansittart's attempts to develop an effective anti-German policy foundered partly on Mussolini's opportunism, but also on Vansittart's own lack of support at home. He made valiant efforts to overcome indifference, but his management of British policy in the summer of 1933 seemed erratic and ineffective to a veteran (and strongly anti-German) British diplomat, while one of his memoranda at the end of August made MacDonald fear that the Permanent Under-Secretary might "break down under the strain or show nerves

---

venting an increase in armies; his cryptic exposition of his ideas on this baffled Bülow, but the latter sent Dulles the comment that Germany opposed rigid standardization of length of service and size of armies: *ADAP*, C/I/2, Nos. 433, 434.

[91] British and American diplomats did entertain the idea of granting some minor samples in the first stage, but this did not become policy. See Cab 23/77, meeting of 20 Sep 33, 51(33)2; *DBFP*, 2/V, Nos. 428, 432; also AA, Memo by Frohwein, 27 Sep 33, 3154/D669912; Memo by Neurath, 29 Sep 33, 3154/D669927-29.

[92] The French continued to press for sanctions, but the importance to them of a united front is evidenced in a number of documents: *DDF*, 1/IV, Nos. 259, 261, 287, 300, 301; *DBFP*, 2/V, No. 437. On the idea of concluding a convention without Germany, aside from note 12 above, see SDP, Ferdinand Mayer Geneva diary, 26 Oct 33, 500.A15A4/1469½; *DDF*, 1/IV, No. 337; *DBFP*, 2/V, No. 465; *FRUS, 1933*, I, 270.

in dealing with the [Austrian] matter." Vansittart's very zeal laid the seeds for later skepticism among the ministers, who came to believe that Foreign Office views simply represented his (to them) very opinionated outlook. The British nation had been shocked by Hitler's regime, and it opposed German rearmament, but it was not ready to give up the idea of a disarmament convention, to join in an alliance against Germany, and to take steps to improve its own military potential.[93]

[93] On Vansittart's stability: Sir Walford Selby, *Diplomatic Twilight, 1930-1940* (London: John Murray, 1953), pp. 13-17; MDP, 2/13, ltrs., to Vansittart, 25 Aug 33; from Vansittart, 26 Aug 33; to Simon, 26 Aug 33. Vansittart's loss of influence with ministers after 1934 is stressed by Keene, "Foreign Office and the Making of British Policy," pp. 117, 369-383. A German study, Oswald Hauser, *England und das Dritte Reich* (Stuttgart: Seewald, 1972-) argues (esp. I, 35-38) that the British accepted the French policy on samples and stages out of a concern to maintain an indispensable alliance with France, which was endangered by their refusal of sanctions and which they preferred to the saving of the conference. But to refuse to agree to a provision for sanctions was in itself a refusal of an alliance with France. If the British accepted other parts of the French position, this was in large part to induce the French to agree to a compromise convention, so as to save the conference: see Cab 23/77, meetings of 5, 20 Sep 33, 50(33)2, 51(33)2; Cab 24/243, CP 228 (33), 3 Oct 33. And the British did not recognize at this point that a refusal of an early qualitative equality to Germany would bring about a German withdrawal.

CHAPTER TEN

Hitler's Hand Is Forced, and He Frees It

1. GERMAN HOPES AND CONCERNS

If British ministers and officials had difficulty in fixing their eyes on the ugly reality of German rearmament, German observers found it hard to fathom British intentions. Despite Hitler's desire for an Anglo-German understanding, his warlike ideology obviously hobbled his comprehension of Britain's peace-seeking tendencies. In particular, neither he nor even his more traveled officials adequately sensed the British resistance to the idea that the first result of a disarmament convention should be Germany's acquisition of new and more destructive weapons. They also suffered from a plain shortage of factual information on British official policy. And the most mysterious and unsettling part of British policy, from a German perspective, was that which concerned Britain's relationship with France.[1]

German information on British and French policy was especially imperfect in the summer of 1933. To begin with, there was the thinness of official contact. Private conversations, such as flourished between Eden or Cadogan and Massigli, seem to have scarcely taken place between British and German or French and German diplomats. And no ministerial discussions were proposed or carried out with Germany such as those that occurred on June 8, September 18, or September 22 in Paris. For that matter, neither Britain nor France officially supplied the German government with any account of the June 8 or later discussions. Berlin did receive at least three separate reports on what had happened on June 8, but these reports emphasized Anglo-French differences over the French unwillingness to disarm and the British unwillingness to undertake guarantees. In a paper of July 5, prepared by German delegates with Reichswehr participation, it was observed that, for the time being, the British, Americans, and Italians appeared to reject the French proposal for a trial period.[2]

---

[1] Robertson, in *Hitler's Pre-War Policy*, p. 24, writes, "German statesmen dreaded nothing more than those occasions when the British and French were closeted together in negotiations on German affairs behind their back . . ."; see also his p. 78.

[2] *ADAP*, C/I/2, No. 297; AA, tel., Frohwein to German delegation in London, 19 Jun 33, 3154/D669406; tel., London to Ministry, 23 Jun 33, 3154/D669411-12; Foreign Ministry memo and cover ltr., 5 Jul 33, 7360/E537029-34. See also *DDF*,

450—Hitler's Hand is Forced, and He Frees It

German leaders were left free to suppose that France was isolated with her plan.

The July 5 paper, entitled "The Status of the Disarmament Question," suggested that the French would probably cling to their proposal of stages, despite the rejection by other powers, to avoid disarming themselves; conference circles now thought that the French preferred a measure of de facto German rearmament to any reduction on their own part. The document argued that if Henderson's efforts failed, as was likely, Germany should not proclaim the failure of the conference and assert her freedom to rearm. Aside from the diplomatic objections to such a course, there would probably be new proposals on the basis of the four-power pact, or from Britain or the United States, so that negotiations would continue and some kind of nonspecific agreement would be reached. Germany might in that case seek legalization of the arms measures she had already secretly taken or begun. Propaganda should be pushed abroad, not merely demanding the disarmament of others, which the French could readily answer (with charges of German illegal rearmament), but broadening the base and putting more stress on the (supposed) war-guilt lie and on French efforts at hegemony.

These ideas were by and large also those of Hitler—for example, to avoid a break over disarmament, to proceed with secret rearmament, to emphasize propaganda, and to try to take the offensive by raising new issues. Hitler appears to have approved of the paper, since he authorized the use of an attachment to it that supplied propaganda guidance to further the same general policy. This propaganda guideline concluded with a warning against giving any pretext for a charge that Germany wished to rearm. On the other hand, it also pointed out that since a suitable draft was to be prepared for a second reading by October 16, when the conference was to resume, Germany would have to expect the nations to decide then on her thesis of "Disarmament and Equal Rights." The guideline also called for more stress on Germany's need for security; military circles considered this last point especially important.[3]

The need to reach foreign audiences was underscored when German leaders met on July 12, partly to consider these guidelines. It was

---

1/IV, No. 182. The League of Nations Section (Group Army) immediately rejected any form of stages that would deny qualitative equality: OKW, OKW 845. TA 282/33 and att., 21 Jun 33, 793/5522432-33.

[3] AA, att. to Foreign Ministry memo, 5 Jun 33, 7360/E537035-39; circulated in OKH, H24/6, Reichswehrminister 209/33, 15 Jul 33, and att. memo, 12 Jul 33, 248/6227760-67. On military views, see Blomberg's comments in *ADAP*, C/I/2, No. 359, also Dülffer, p. 265, quoting from a naval directive of 12 Jul 33.

at this meeting that Hitler brought up the necessity of developing an apparatus which could influence key people abroad, establishing a card index on them, and especially on all bribable journalists. General von Blomberg pointed to the danger that there might be a call in the League Council for an investigation of Germany, and he suggested that it was important to influence Council members in good time. The French were in fact giving new indications that they might seek to expose German violations of Part V.[4]

German officials followed the recommended policy when they showed Henderson every courtesy during his stay in Berlin on July 17-18. At the same time, they made it clear that Germany would insist on qualitative equality—on acquiring any hitherto-forbidden weapons retained by other nations—a point which Hitler also emphasized to Henderson at their meeting in Munich on July 21. Henderson implied to German officials that the French would make some slight reductions during the first period in aircraft, guns, and tanks, and he said he thought that this period might be limited to three years; the Germans for their part provided him with a statement, accommodating in tone, of the German point of view, and this statement was moderated somewhat further in discussion with him. Even Hitler's irritation at Henderson's reports of German unpopularity showed at least an assumption that England should naturally support Germany's equality of rights, in practice as well as in principle.[5]

From the German point of view, events in the succeeding weeks were discouraging, especially the foreign protests on the questions of Austria and aerial rearmament. Privately, Neurath and Bülow did not at first take the British diplomatic moves too tragically, as they attributed these moves to the absence of Simon and MacDonald, combined with hyperactivity on the part of Vansittart; Bülow gave the Italian ambassador the interesting, indeed Gothic, explanation that Vansittart was known to be strongly under the influence of the Austrian minister in London, Baron Franckenstein. Bülow and Neurath found it necessary, however, to urge a moderation of pro-Anschluss activity, and Hitler in fact ordered this in the political sphere, al-

---

[4] *ADAP*, C/I/2, No. 359. On bribery, see Chapter Eight, Section 5. Massigli had recently shown a new tendency to take the offensive in Geneva: *DDF*, 1/III, No. 430; 1/IV, No. 260 (Annex I [p. 447]); see also *ADAP*, C/I/2, No. 360.

[5] *ADAP*, C/I/2, Nos. 370 (and encl.), 374. During the summer of 1933, the German government usually used the term "defensive weapons" to refer to arms that other governments intended to retain, and that Germany now also claimed; the tongue-in-cheek theory was that if the other nations were retaining the arms under the disarmament convention, those arms must of course be defensive. Although the term "defensive weapons" is less clumsy than "hitherto-forbidden weapons," I do not use it here because it was a polemical device. Also, the term was not always understood, e.g., by Henderson (*ADAP*, C/I/2, No. 374 and note 2).

though he kept the exorbitant-visa charge.[6] Reports indicated that a press campaign against German rearmament had commenced in mid-July in Paris, that something similar had begun in London at the end of the month, and that the press in both Paris and London was being informed on the British proposals for démarches. It appeared from Berlin that this campaign might foreshadow a French demand for an investigation by the League, and German diplomats were instructed to say that Germany would find such an investigation, or a one-sided trial period, unacceptable.[7] On August 22, Bülow wrote to Neurath that it was not easy to draft a foreign-policy section for Hitler's Nuremberg rally speech, "since after all the foreign-policy situation does not look very favorable"; there would probably be difficulties at the League Assembly session in September and at the Disarmament Conference in October. Bülow observed that something was surely brewing in France, that the British were making all kinds of protests, and he feared Mussolini would become an arbitrator rather than a conciliator.[8]

German concerns became more specific as the conference resumption approached. A report from Paris, dated September 11, stated that all information on French intentions indicated they would try to win the British and American governments over to the idea of establishing at an early date what arms were ready and in preparation in Germany. Also in early September, reports came from Paris that the French and British, in preparation for the coming conference sessions, were trying to get evidence from the Soviet Union on past German armament activity in Russia. The German embassy in Moscow was directed to discuss this matter with the Soviet government, and the Commissar for Foreign Affairs, Maxim Litvinov, gave a formal assurance that his government had made no such disclosures; he added that they would not do so in future, out of regard for their own prestige, no matter how Franco-Soviet relations developed. But Britain

[6] *ADAP*, C/I/2, Nos. 376, 381, 385, 390-392, 398.

[7] OKW, OKW 872, TA 279/33, 27 Jul 33, 874/5621631; AA, Wolff's Telegraphic Bureau reports, 3 Aug 33, 3154/D669510-11; tel., Washington embassy to Ministry, 6 Aug 33, 3154/D669512-13; ML, Shelf 5939, Memo by Schwendemann (Foreign Ministry), 21 Jul 33, reel 18/229-230; Shelf 5930, Foreign Ministry circular dispatch, 22 Aug 33, reel 12/671-673. An informant close to the American embassy in Berlin reported to German military intelligence (the Abwehr) that British and French intelligence services had for months been collecting material for a dossier on the militarization of Germany, which the Poles or Czechs would present to the second or third session of the conference in October: ML, Shelf 5996, Abw Va 3186/33, 2 Aug 33, reel 18/1345-1346.

[8] AA, ltr., Bülow to Neurath, 22 Aug 33, 4619/E197721-23. The foreign-policy section of the Nuremberg speech draft was eventually used instead by Neurath on 15 Sep 33: Heineman, "Neurath," p. 307; text in Schwendemann, *Abrüstung und Sicherheit*, II, 440-450.

and France already had considerable information, as the Germans themselves established. Over the summer, the Abwehr found, in interrogating its double agents, that they had received assignments from Western intelligence showing extensive knowledge of German training and rearmament activity. Schönheinz's League of Nations Section (Group Army) made a tabulation of the arms violations alleged in the British and other foreign press in recent weeks, and sent it to the Reichswehr Ordnance Office on September 18 with a query as to whether the charges were well founded and, in cases where they were, as to how the violations could be explained; the reply, on October 7, was far from reassuring.[9]

2. SOUNDINGS WITH BRITAIN AND FRANCE

German officials were aware, however, that there was more than one point of view in Britain, and indeed also in France, and they did not have an impression of solid opposition. Ambassador von Hoesch sent home a very lengthy memorandum on August 16, in which he frankly described Germany's recent loss of sympathy in many formerly friendly quarters, but also pointed to a persistence (or in some cases increase) of regard for Germany in military, court, and other aristocratic circles, and in a segment of the conservative press. Hoesch noted the extreme sensitivity of the British government to its public opinion, and he recommended moderate policies and discreet propaganda as the most likely ways of bringing the British around.[10] After Hoesch went on leave, Prince Otto von Bismarck, a grandson of the Iron Chancellor, became chargé d'affaires, and he had two talks with Eden, one before and one after Eden's September 18 discussion with the French. Bismarck's reports gave the impression that British policy was still undecided, and that the September 18 and 22 meetings with the French were for the purpose of gathering information. Eden denied that the British and Americans had accepted French views on supervision and a trial period; while he said that Britain shared French objections to German demands for prototypes in the first period, he added that there would need to be an adjustment of the several points

---

[9] AA, tel., Koester to Ministry, 11 Sep 33, 3154/D669776; *ADAP*, C/I/2, No. 459 (and note); OKW, Wi/IF 5.3683, Abw 40/9.33, 2 Sep 33, 437/1300549. Abwehr information from double agents: ML, F VIII a 11, Abw 237/33, [12] Jun 33, reel 54/192-195; Abw 311/33, 2 Aug 33, reel 54/206-210; Abw 354/33, 17 Aug 33, reel 54/212-214; Abw 414/33, 25 Sep 33, reel 54/225-228. In the same file, a directive signed by Raeder (Chef ML, B Nr. A II m 3762/33, 22 Sep 33, reel 54/223-224) stated that recent incidents showed the need for tighter security. Query to Ordnance Office: OKW, Wi/IF 5.3683, VGH 583/33, and atts., 18 Sep 33, 437/1300787-811. The reply is described in Section 7 below.

[10] *ADAP*, C/I/2, No. 406.

of view if an agreement was to be reached. The German chargé heard a different description of the situation from Edgar Lewis Granville, Simon's parliamentary private secretary, who said that powerful forces were at work, including Simon himself, for establishing a common front of states in opposition to Germany, and that Daladier proposed to bring up the dossier if the Germans claimed at Geneva to be disarmed. But Granville added that there was strong resistance within the cabinet to supervision, and he thought the Germans could isolate the French with skillful tactics.[11]

Whether from a wish to forestall French preventive action, from a desire to compromise the French government in order to separate it from the British, or from a hope of actually reaching an understanding with Paris, Hitler, encouraged by Blomberg, became very interested in discussions with France. The French ambassador, André François-Poncet, continuously sought a rapprochement: he was one possible channel. Beginning before Hitler's accession, the idea of some direct Franco-German understanding came up in a number of conversations between Poncet and Hitler, Blomberg, Neurath, and Bülow. The Quai d'Orsay did not show interest, however, and the Wilhelmstrasse seems to have concluded that there was little prospect of achieving any results through the Ambassador.[12] But although Neurath

---

[11] Bismarck's reports: AA, tels., 15, 21 Sep 33, 3154/D669780-82, D669837-39. After returning from Paris, Eden told Bismarck on September 21 that he had been pleasantly surprised by the reasonable attitude of the French. According to Eden's record of the second conversation (*DBFP*, 2/V, No. 402), Bismarck was not unduly perturbed at the idea of a long first period, but thought that prototypes (i.e., samples) were the crux of the problem. Bismarck said he was confident that Hitler himself was "eminently reasonable" on disarmament. A British informant of the Abwehr's, allegedly in close touch with at least one British cabinet member, reported that under cover of pro-French press coverage, the British together with Davis were pressing the French to make substantial concessions on disarmament, and to accept the shortest conceivable trial period, if possible one year: NA-VGM, VGM 35, Abw 3752/33/Va, 26 Sep 33, 7792/E564683.

[12] Where François-Poncet's reports to the Quai showed that a bilateral agreement was discussed, he usually indicated it was the Germans who had proposed it, while he himself had (supposedly) only suggested, if anything, an acceptance of Boncour's November 1932 proposal for a multilateral agreement: see *DDF*, 1/II, Nos. 205, 260; 1/III, Nos. 105, 145, 191, 481. The German documents, on the other hand, state that it was Poncet who suggested a bilateral agreement, perhaps between soldiers: AA, Memo by Bülow, 18 Jan 33, 7360/E536148-49; *ADAP*, C/I/1, Nos. 9, 163, 165, 190; C/I/2, No. 360. Poncet had come to Berlin in 1931 hoping to bring about a direct Franco-German understanding on the basis of the interlocking interests of French and German heavy industry (François-Poncet, p. 23; Bennett, *Diplomacy*, pp. 94-98; Franklin L. Ford, "Three Observers in Berlin," in Craig and Gilbert, eds., *The Diplomats, 1919-1939*, p. 462), and in the summer of 1932 he had favored direct negotiations with Schleicher and Neurath (*DDF*, 1/I, Nos. 77, 125 (Annex). Thus it is credible he might have been the advocate of a bilateral agreement in 1933. But if so, he must have realized that his current chief, Paul-Boncour, was opposed to such a policy; this realization might explain a failure to report

and other officials disliked Poncet, Hitler did remember the Ambassador years later with respect.[13] In mid-September, Poncet insisted on seeing Hitler to protest the dedication at the Nuremberg rally of an SA standard (actually of the Kehl Standarte of the SA) bearing the name "Strassburg." Hitler assured Poncet that this standard had been withdrawn, and that he had no intention of reclaiming Alsace-Lorraine; for the first time, he also explicitly stated that he would honor the Treaty of Locarno. He made flattering statements about his esteem for the national spirit, intellectual activity, and soldierly valor of France, and he even managed to say he could imagine no more beautiful monument for himself than to have it said later that he had reconciled Germany and France.[14]

The main hope for some actual dialogue with the French government lay, however, in Daladier. Poncet himself, who went home on leave after his interview with Hitler, saw Daladier on September 19, and reported on the discussion he had had; he later told Tyrrell that he had urged his government to talk with the Germans and try to educate them, rather than refuse to have anything to do with them. Poncet had already suggested to the Germans that some French parliamentary personage might serve as an intermediary between Neurath and Daladier or Boncour, and in Paris, Ambassador Roland Köster had gotten in touch with Joseph Caillaux, a pre-World War I premier and a senior Radical Socialist, who had long been associated with a policy of reconciliation with Germany—to the point where he had been arrested in January 1918 and tried in 1920. Caillaux agreed to approach Daladier, but Köster heard no more and soon concluded that this approach would have to be written off.[15]

Other intermediaries were somewhat more successful. As noted, Daladier had indicated in March, through Fernand de Brinon, a willingness to meet with Neurath or Papen, and in July he had told Henderson that he would be willing to confer with Hitler. Blomberg reacted favorably to this last feeler, even though Hitler seemed dubious at first. Meanwhile, Brinon continued active, and he and Joachim von Ribbentrop worked to arrange a Hitler-Daladier meeting. With

---

having raised the subject himself. Perhaps Poncet meant to suggest an eventual bilateral agreement as a possible alternative for later, and the Germans took him to be speaking more concretely and immediately than he intended. In any case, Neurath early concluded that Poncet was merely speaking for himself, rather than on instruction: AA, tel., Neurath to Nadolny, 10 Feb 33, 7360/E563202.

[13] See *ADAP*, C/I/1, Nos. 163, 165; C/I/2, No. 430 (and note 1); *Secret Conversations*, pp. 225-226.

[14] *ADAP*, C/I/2, No. 430; *DDF*, 1/IV, No. 215. Neurath's record does not report the Locarno assurance.

[15] *FRUS, 1933*, I, 221; *DBFP*, 2/V, No. 416; AA, Memo by Koepke, 20 Sep 33, 5669/H014244; ltr., [Koester] to Neurath, 5 Sep 33, 3154/D669758-60.

Daladier's knowledge, Brinon went to Germany and saw Hitler at Berchtesgaden on September 9, with Blomberg present. Hitler said that Germany had no desire for war or for territorial acquisitions from France, although the Saar, which was German, should be transferred to Germany; further, if there was a Franco-German entente, Germany would not ask for any reduction in French military power. To this, Blomberg added that Germany would not ask for any of the heavy artillery capable of destroying French fortifications. Brinon gave Daladier a report on this meeting, and the session presumably encouraged Hitler to make his effusive declaration to Poncet on September 15.[16]

Brinon returned to Munich and met again with Blomberg on September 27. Blomberg now spelled out in more detail the military arrangements he wanted to embody in a Franco-German agreement. He said that although Germany had to insist at Geneva on the possession of all arms retained by others because she had always advanced a claim to equality of rights, she could accept less than equality in a direct Franco-German agreement. She would not ask for the heaviest artillery (over 210 mm). Germany would need observation and pursuit aircraft in any case, and bombers if others retained them. On effectives, Germany would accept a figure "around" 200,000, and a term of military service three months shorter than the French, while asking to be able to enlist some regulars "for several years, more or less copying the system in effect in the French colonial army"; here we see Blomberg pursuing the old objective of combining short-service men with a strong contingent of cadre and specialists. Rather than speaking of samples, which he said was fundamentally a dishonest approach, Blomberg claimed that he preferred to be frank about the maximum sizes and numbers Germany wanted, which would not be high. He gave Brinon a message from Hitler to the effect that although Mussolini was proposing four-power talks, perhaps in Stresa, Hitler preferred a direct Franco-

[16] Brinon described his activity at his postwar trial for collaboration, at which a record of the 9 Sep 33 conversation was also produced: Maurice Garçon, ed., *Les Procès de collaboration: Fernand de Brinon, Joseph Darnand, Jean Luchaire, compte rendu sténographique* (Paris: Ed. Albin Michel, 1948), pp. 53-55, 78-80, 200-202. At the trial, Daladier denied having met personally with Ribbentrop, or having intended to conclude any other agreement with Germany than the general one France was seeking at the conference (*Les Procès*, pp. 203-206). Daladier may in fact have thought Germany could be induced to join the general accord; see *DDF*, 1/IV, No. 259 (antepenultimate sentence). He did not deny that Brinon had sounded Hitler out, however, and Blomberg's reported views are partly confirmed by *DBFP*, 2/VI, No. 67 and AA, Reichswehrminister 384/33, 25 Sep 33, 7360/E537467. The Czech minister in Paris reported on 25 Nov 33 that Daladier, during his premiership, had almost decided to negotiate directly with Hitler: Czech docs., 2376/D496980. On Blomberg's and Hitler's reactions to Henderson's message: *DBFP*, 2/V, No. 258; *DDF*, 1/IV, No. 17; *ADAP*, C/I/2, No. 374.

German agreement. One gets the impression that Blomberg had more interest in the talks with the French than Hitler. The General did not believe an agreement could be reached in Geneva and probably did not believe that any government would ever willingly agree to disarm. Frustrated by the delays of the conference, he was trying to activate old plans for an agreement "between soldiers"—or for driving a wedge between the French and other nations.[17]

After the September 9 meeting, according to Brinon, Daladier became apprehensive about meeting with Hitler himself, being concerned about the reaction in parliament and on the part of Paul-Boncour. Such concern would seem indeed to have been well founded. But contrary to Brinon's belief, Daladier and Boncour did discuss the possibility of a direct understanding with Germany, and although the Germans presumably did not know it, the Quai did consider (as we have seen) the possibility of conceding specimens of forbidden arms in direct negotiation with Germany. Still, Boncour feared that a Franco-German understanding would destroy the common front he was trying to uphold, i.e., with the British and Americans, and this remained his position—and that of the French government—despite an overture from Neurath and a fervent approach by Goebbels, both in Geneva, in late September. Even in explaining Germany's withdrawal from the conference in October, Hitler made a speech which seemed to show that he still hoped for a bilateral discussion with Daladier.[18] But probably the

[17] Garçon, *Les Procès*, p. 54. A more complete record of the 27 Sep 33 conversation, apparently written at the time, is contained in a booklet published anonymously by friends of Brinon's, *La Vérité sur Fernand de Brinon* (Paris: n.p., 1947), pp. 27-30. Germany had in fact received proposals from Mussolini for a four-power meeting at Stresa: AA, tel., Bülow to Neurath, 23 Sep 33, 7360/E537465-66. At one of the meetings with Brinon, Blomberg apparently proposed an exchange of French and German officers: *Vérité sur Brinon*, p. 32; *DBFP*, 2/VI, No. 67. In the latter document, Blomberg was reported as saying that he had given his views to Weygand; he may have regarded Brinon as an emissary of Weygand's, as well as of Daladier's. Léger told the Belgian chargé in Paris that the military, if left to themselves, would easily reach an understanding: *DDB*, III, No. 47. As explained in Chapter One, a shorter service period than the French one would have enabled Germany to train more reserves than France with a lower, or the same, nominal average daily strength.

[18] Garçon, *Les Procès*, pp. 55, 201-202; *DDF*, 1/IV, Nos. 167, 259; *ADAP*, C/I/2, No. 466. According to Léger, solicitations from German milieux were arriving at the Quai daily: *DDB*, III, No. 47. At some point in Brinon's negotiations, a plan was drawn up for a possible sequence of steps toward an eventual Daladier-Hitler meeting and a communiqué; the communiqué would have included a mutual recognition of the right to national defense—which meant, no doubt, a German withdrawal of demands for French disarmament, and a French acceptance of the German right to rearm: *Vérité sur Brinon*, pp. 31-32. It was planned that the first step would be certain conciliatory remarks in public by Daladier, and in a speech on 8 Oct 33, Daladier actually made some of the projected statements (along with much that was less conciliatory): he called Germany a great nation and indirectly disclaimed any wish to interfere in German domestic affairs (text of Daladier's

main result of these soundings was to assure Hitler that Daladier preferred negotiation to confrontation, so that France, under Daladier's leadership, was unlikely to launch actual preventive action.

3. NEURATH'S TALKS IN GENEVA

Generally, then, the international situation in September appeared from a Berlin perspective to be disquieting, but in both the British and French governments, certain elements seemed ready to go some way to accommodate German desires. Aside from the picture Brinon doubtless painted of a receptive Daladier, the Germans believed that Hitler had made a favorable impression on the French cabinet through his statements to Poncet.[19] While in Paris, the latter (who personally believed in the concession of samples to Germany) indicated to Köster that, in his view, the French public was not completely unwilling to concede prototypes to Germany.[20] Neurath expected to attend the League Assembly sessions at the end of the month, and that would provide opportunities for further soundings and maneuvers. Joseph Goebbels was to accompany Neurath to Geneva, apparently to try out his own methods in an international milieu, and perhaps to give Hitler an opinion on the feasibility of pursuing Nazi goals in such surroundings.[21]

When Neurath saw Simon and Eden in Geneva on September 23, differences in outlook emerged clearly, but negotiation appeared possible. Neurath said he was opposed to one-sided supervision, but he accepted the idea of stages, including that substantial disarmament (i.e., by France) would take place only in the second stage. The chief point at issue was samples. The German Foreign Minister said that his

---

foreign policy remarks as wired to Berlin: AA, tel., Forster [Paris] to Ministry, 8 Oct 33, 7360/E537680-83). This seems to have been interpreted in Berlin as the first prearranged signal (see *DDF*, 1/IV, No. 301), and in his speech on the German withdrawal from the conference and League on 14 Oct 33, Hitler gave the second signal by stressing his wish to avoid bloodshed. Poncet suspected that the aim was to compromise France (*DDF*, 1/IV, Nos. 314, 328), but Propaganda Ministry directives indicate a genuine interest in talks. A propaganda instruction of 17 Oct 33 told the German press not to seize on Daladier's first hostile reaction to the withdrawal, but await a later foreign policy declaration; after Daladier's fall on 24 Oct 33, German journalists were advised on the twenty-fifth that if he should be taken into a new cabinet, they should not embarrass him with too much German praise: Brammer Collection, Z Sg 101/1, 101/2. See also *ADAP*, C/II/1, No. 27.

[19] *ADAP*, C/I/2, No. 474; *DBFP*, 2/V, No. 411. Poncet had also sent word to Berlin that Daladier and the French cabinet had been much impressed, and his chargé d'affaires passed this on to the Wilhelmstrasse (AA, Memo by Koepke, 20 Sep 33, 5669/H014244).

[20] SDP, Memo of conversation between Davis and Poncet, 9-10 Apr 33, 550.S1 Washington/359; AA, tel., Köster to Ministry, 26 Sep 33, 3154/D669909.

[21] *ADAP*, C/I/2, Nos. 426, 430.

country wanted samples in the first stage of whatever types of arms were retained by other countries, whereas Simon, according to his own account, expressed firm opposition to any German rearmament in the first stage, particularly considering the change in British opinion in recent months. The British suggested, however, that perhaps the first period could be shortened and the second extended, so that Germany would still get samples before too long. Simon understood that Neurath was only seeking "a limited number of samples," and he now asked what quantities of hitherto-forbidden weapons the Germans actually wanted. Here an important misunderstanding occurred: Neurath understood Simon to ask how many of these weapons Germany desired in the first stage, but Simon thought that Neurath agreed to find out the number of weapons Germany desired on the assumption that she would wait for them until the second stage. Simon spoke in terms of amendments to the MacDonald plan, and Neurath gained the impression that no Anglo-French agreement had been reached. Neurath surmised, however, that Britain and France were both solidly opposed to the German procurement of additional weapons, and he concluded his report to Berlin by writing, "the negotiations here will presumably collapse over this point."[22]

Simon and Neurath each had talks with Davis and the Italian Under Secretary, Fulvio Suvich, in the next few days, and what the Germans learned indirectly confirmed that the British opposed any rearmament in the first period. Davis himself seemed definitely ranged against this, and even the Italians tried to talk Neurath out of it. Neurath did not meet directly with Boncour until nearly the end of his stay, but meanwhile other sources doubtless gave him a good idea of Boncour's firm opposition.[23]

When Neurath did meet with Boncour in Geneva on September 28, he said that since he saw little prospect of concluding a multilateral convention, he thought it necessary to reach a direct understanding

---

[22] *ADAP*, C/I/2, No. 447; *DBFP*, 2/V, No. 411. Other evidence confirms that Neurath accepted the idea that major disarmament by the "highly armed powers" would only take place in the second period: *ADAP*, C/I/2, No. 445; Davis Papers, Memo of conversation between Davis and Neurath, 25 Sep 33. According to Neurath's report, *he* said that a first stage lasting four years was not acceptable (a claim which seems to be confirmed by *DBFP*, 2/V, No. 419); Neurath did not record at this time that the British thought it possible to shorten the first period. A solution through such a shortening was discussed by Simon with the Italians (*DBFP*, 2/V, Nos. 409, 412; Aloisi, p. 146); they suggested instead a subdivision of the first stage into two parts. With regard to the conclusion of Neurath's report to Berlin, he had made a similar prediction during the negotiations in December 1932 (see Chapter Five, Section 9).

[23] *DBFP*, 2/V, Nos. 409, 412, 421; *ADAP*, C/I/2, Nos. 454, 469; *FRUS, 1933*, I, 232-234; *DDF*, Nos. 246, 254; Davis Papers, Memo of conversation with Neurath, 25 Sep 33; Aloisi, p. 147.

with France. He recalled Hitler's statement to Poncet on Alsace-Lorraine, and suggested that Boncour consider the possibility of reaching an understanding between high military figures on the two sides. On specific questions of disarmament, Neurath indicated that he attached no importance to the French promise of disarmament in the second period, and he maintained that Germany could not remain deprived during the first period of arms retained by others. Probably no one could have been less likely to accept Neurath's proposals and demands than Paul-Boncour, the champion of the League and of the multilateral approach, the republican distrustful of the military in politics, and the would-be organizer of Germany's isolation. He replied that he would inform Daladier of the proposals, but he also said (as he informed Daladier) that the Disarmament Conference was not just a Franco-German affair, so that it would be difficult to extract the question from that forum, whereas if—as the latest French proposals led him to hope—there could be success there, that would create an atmosphere of detente conducive to German-French negotiations. He rejected any rearmament during the first period, noting that this was also the British and American position. And then, according to Neurath, Boncour brought up the French dossier on German violations and "embarked on tirades" about France having continually yielded to German demands. Neurath answered in the same vein, and he concluded his report on the meeting with the observation, "Further talks with M. Paul-Boncour appear to me useless at this time."[24]

4. HITLER SUPPORTS FOREIGN MINISTRY VIEWS

Neurath's comment on Boncour, or his prediction that negotiations would collapse, should not be taken as showing a definite plan to leave the conference. Actually, Neurath and his officials were working to keep Germany in the conference.[25] They faced in truth a very awkward problem: how could the longstanding German insistence on qualitative equality—now made more urgent by the rearmament under way—be reconciled with the insistence of other nations that Germany remain disarmed during a trial period? The diplomats hoped, as before, to keep conference negotiations alive by giving other nations the impression that German desires were really modest. But the

---

[24] *ADAP*, C/I/2, No. 466; *DDF*, 1/IV, No. 257, 259. According to Aloisi (p. 147), Neurath received a telephone call from Hitler during the day; this may have influenced the German Foreign Minister to propose bilateral negotiations to Boncour.

[25] At a cabinet meeting of 12 Sep 33, Neurath outlined arguments for remaining in the League, and said that the time for leaving it would probably come only after "the complete collapse of the Disarmament Conference and after the final settlement of the Saar question": *ADAP*, C/I/2, No. 426. See further below.

Reichswehr Ministry, also as before, tended to insist on quantitative equality, regardless of the effect on the conference.

The difference reemerged after the two ministries discussed how to answer Simon's question on the actual quantities of hitherto-forbidden arms Germany desired. Believing the question related to the first stage, Blomberg and Bülow agreed that no concrete figures could be given until the MacDonald plan had been worked out in greater detail. But after the discussion, Bülow informed Neurath that the British could be told the Reichswehr wanted the same relative strength in weapons as the French (for example, the same number of 155 mm guns for each division, or for each 1,000 men). As Bülow noted, German arms under this rule would not approach the French in numbers, because even after the reorganization to 200,000 men, the German army would have fewer units or effectives than the French. Bülow probably misconstrued Blomberg's position in his effort to send Neurath something that might be negotiable. Blomberg and Schönheinz declared Bülow's proposition to be a new and unwelcome suggestion, and they prepared a substitute formulation which seemed to ask for a number of arms in the first period equal in absolute terms to the number the French would have at the end of the second period.[26]

Thus the old issue of numerical equality with France arose again. Indeed, as Neurath pointed out in a telegram from Geneva, the Reichswehr formula implied that Germany was even planning to go on to exceed the French level in materiel at the end of the second stage. Neurath argued that Blomberg's formula was politically unfulfillable and would be regarded as a German attempt to break up the conference. Most of the German diplomats involved in the disarmament question—Albert Frohwein, Friedrich Gaus, Werner von Rheinbaben, Ernst von Weizsäcker, and Karl Schwendemann—thought that the military position should be opposed, and Schwendemann wrote that to ask for quantitative equality, when the German standing in the equality-of-rights question was weaker than in December 1932, was "naturally an impossible thing." Schönheinz, on the other hand, claimed that Blomberg had presented the question to Hitler, and that Hitler had ruled in favor of the Reichswehr formula. This assertion seems dubious: when Blomberg on September 30 sent Bülow a written

[26] AA, tel., Bülow to Neurath, 25 Sep 33, 7360/E537463-66; Reichswehrminister 384/33, 25 Sep 33, 7360/E537467; Memo by Schwendemann, 26 Sep 33, 7360/E537469-71; Reichswehrminister 404/33, 30 Sep 33, 7360/E537559; ltr., Schwendemann to Frohwein, 28 Sep 33, 7360/E537476-77. When Nadolny (at this time in Berlin) questioned Schönheinz, the general confirmed that quantitative rather than qualitative equality was asked for. In May, Bülow had been close to resigning over the trend of German policy; see Peter Krüger and Erich J. C. Hahn, "Der Loyalitätskonflikt des Staatssekretärs Bernhard Wilhelm von Bülow im Frühjahr 1933," *VfZ*, 20 (1972), 397-400.

clarification of his stand, he made no reference to Hitler, and waived any claim to exceed the French level at the end of the second stage. Blomberg did assert, however, that even full equality with France would not satisfy German security needs, and he maintained that new claims were justified by the longer duration—six to eight years—of the "new plan."²⁷ By using this title, Blomberg implied that a whole new proposal had been advanced by the other side—a circumstance which would release Germany from Hitler's May 17 acceptance of the MacDonald plan as a basis for solving the disarmament question. For his part, Blomberg seems to have been seeking a good pretext for German withdrawal from the conference.

Blomberg's attitude was very much at odds with the diplomatic view; Bülow believed that the eight-year term had not yet been decided, and he continued to hold *his* view that Germany should be ready to negotiate on quantities. But the difference did not have to be explicitly resolved, since Bülow and Blomberg did agree that it would be a mistake to give Simon what they thought he had asked for, specific figures on German demands in the first period. When Albert Frohwein in the Foreign Ministry finally worked out a position on October 4 for communication to the British and also the Italian governments, it neither gave specific numbers, nor said anything about the relation of German to French numerical strength in arms. Instead, on the question of numbers, it stated that Germany could only give information on the armament of the new short-service army after the provisions of the convention on materiel were known. Yet, in deference to Reichswehr wishes, and in particular to a concern apparently expressed by Hitler to Neurath, the reply did indicate that Germany would want far more than samples or a mere doubling of Versailles equipment.²⁸

---

²⁷ AA, tel., Neurath to Ministry, 27 Sep 33, 7360/E537499; Memo by Schwendemann, 26 Sep 33, 7360/E537469-71; ltr., Schwendemann to Frohwein, 28 Sep 33, 7360/E537476-78; draft ltr., Bülow to Blomberg, 28 Sep 33, 7360/E537500-01; Reichswehrminister 404/33, 30 Sep 33, 7360/E537588-59. That Blomberg had Hitler's full support is also rendered doubtful by what Hitler reportedly told Neurath on 30 Sep 33; Blomberg may, however, have successfully made the point with Hitler that a limitation to an equality in armament per unit or per 1,000 men would not provide for Germany's future reserves: see text below. Schwendemann's 28 Sep 33 letter to Frohwein indicates that Schönheinz had been present at the original meeting between Blomberg and Bülow, and implies that Schönheinz instigated the revival of the claim to quantitative equality.

²⁸ AA, draft ltr., Bülow to Blomberg, 28 Sep 33, 7360/E537500-01; Reichswehrminister 404/33, 30 Sep 33, 7360/E537558-59; *ADAP*, C/I/2, No. 480 (and notes). Bülow was later to amend the position paper with an additional statement that Germany was ready to discuss quantities: see Section 6 below and note 43. Christine Fraser, *Der Austritt Deutschlands aus dem Völkerbund, seine Vorgeschichte und seine Nachwirkungen* (Doctoral diss., University of Bonn, 1969), pp. 189-190, and

On Friday, September 29, before Blomberg sent Bülow his final clarification, Neurath had taken an overnight train from Geneva to Berlin, so as to report to Hitler before the Führer left for a visit to Hindenburg's estate in East Prussia.[29] Quite likely, Neurath had decided that it would be well to get his own point of view before Hitler, alongside Blomberg's. Neurath briefed Hitler late on Saturday afternoon, and afterwards recorded in short form the results:

> The Chancellor was entirely in agreement with the course adopted by me at the negotiations. On the alternatives of concluding a disarmament convention, or prolonging or breaking off the negotiations, the Chancellor expressed himself to the effect that it would be desirable in any case to conclude a disarmament convention even if all our wishes were not fulfilled by it.
>
> Concerning the question of equality of rights with regard to materiel, it would be wrong, in the opinion of the Chancellor, to ask for more than we are able for technical, financial, and political reasons actually to procure in the next few years. The Chancellor stated that he had always held this view and also wished to see it upheld in the further course of negotiations.

Thus Hitler supported the Wilhelmstrasse in its desire to remain at the conference.[30] In taking this position, he may have been influenced by Goebbels, who, returning to Berlin by air on September 29, apparently saw Hitler before Neurath did.[31]

We could wish for a fuller account of this discussion between Neu-

---

Wollstein, *Weimarer Revisionismus zu Hitler*, p. 191, assume that this 4 October position was a result of the Hitler-Blomberg-Bülow meeting of 4 Oct 33; but see note 41 below.

[29] *DBFP*, 2/V, No. 422 (encl.); *DDF*, 1/IV, No. 257; *ADAP*, C/I/2, No. 472 (that Neurath left by rail late on the twenty-ninth). It was arranged that Bülow and Koepke would meet Neurath at the Anhalter Bahnhof at 3:29 p.m., 30 September, for a quick discussion before Bülow's own departure for Hameln and Neurath's briefing of Hitler (AA, tels., Bülow to Neurath and vice versa, 29 Sep 33, 3154/D669933-35).

[30] *ADAP*, C/I/2, No. 475 (RM 1373). On one copy of this document, a marginal note in Bülow's hand alongside the first sentence in the second paragraph says: "RK [Reich Chancellor] said [this] to me (and Blomberg) on Oct. 4" (4619/E197771). A note for the Wolff's news agency from Bülow and Völckers, dated 2 Oct 33, reported that Neurath had briefed Hitler on Saturday on the Geneva situation, and that Hitler "has approved the position of the Foreign Minister in every respect" (AA, 3154/D669952).

[31] Goebbels's arrival at Tempelhof was witnessed by the American Consul General, George Messersmith, who later believed that Goebbels, out of anger and resentment over a talk he had had with Simon in Geneva, had urged Hitler to leave the conference; Messersmith thought the decision to withdraw was made on the evening of the same day, before Neurath's return: *FRUS, 1933*, I, 303. But documents now disprove this. As to Goebbels's position, a Propaganda Ministry press guidance of 2 Oct

rath and Hitler; the Foreign Minister, who generally spent as much time as possible hunting in southern Germany, seems to have been anxious to get away from his office again himself, and he was absent from Berlin for the following week.[32] But in briefing Hitler, Neurath probably followed background papers prepared for his use. One paper, by Neurath himself, noted that Britain, France, Italy, and the United States were all agreed in wanting a two-stage convention, in accepting a 200,000-man Reichswehr, and on automatic and periodic inspection. Except for France, the other powers would agree to stages of three years' duration. Neurath also wrote that Britain and Italy were opposed to French proposals for investigating the existing state of armament, and that Britain and France were at odds over sanctions. While Italy professed to support the German claim for samples, France and—so far—Britain both opposed granting Germany new arms in the first period; the British would perhaps relent on certain arms such as antiaircraft guns, but would absolutely oppose a grant of aircraft. Another background paper reported signs that the British were divided as to whether they should meet Germany part way on the question of arms in the first period.[33] The suggestions of some variance between the British (or some of them) and the French would have intrigued Hitler; aside from his interest in a separate meeting with Daladier, he had just discussed with Colonel T.C.R. Moore, a visiting Conservative M. P., the possibility of restoring friendly relations between Britain and Germany.[34]

But there is also evidence that Hitler took a further position, one less in keeping with Foreign Ministry objectives. Before leaving his office on Saturday afternoon, Neurath left a second note directing that

---

33 followed faithfully the restraint ordered in the 12 Jul 33 propaganda guidelines, and also pursued the then-current official line that Germany had formulated no counterdemands, as these would depend on the degree of disarmament by others (Brammer Collection, Z Sg 101/1).

[32] Neurath had informed Bülow on Friday that he was thinking of leaving Berlin Sunday night (AA, tel., Neurath to Bülow, 29 Sep 33, 3154/D669933). He did go to Balderschwang, which is in the Alps of the Allgäu (AA, Marginal note by Bülow on tel. to Washington embassy, 5 Oct 33, 3154/D669977).

[33] AA, Memos by Neurath, 29 Sep 33, and Frohwein, 27 Sep 33, 3154/D669927-29, D669911-14. The second paper, by Frohwein, concluded that the questions of Germany's right to the arms permitted to others and of the numbers she should have of them were the only matters directly involving Germany on which agreement was not in sight. Neurath may also have used the argument in a 29 Sep 33 paper by Schwendemann (7360/E537514-16) that Germany could take advantage of conference delays to improve quietly her actual arms position, while the hands of the others were in practice tied as long as the conference lasted.

[34] Jacobsen, *Nationalsozialistische Aussenpolitik*, pp. 334, 776. According to the German record, Moore told Hitler that he and his friends in Parliament were ready to use their influence for an improvement in Anglo-German relations: AA, Note by Thomsen, 29 Sep 33, 5740/H031119.

Weizsäcker (the German minister in Berne) should approach Simon and Aloisi in Geneva and inform them orally that Germany had to insist on equality of rights in materiel for the first period. Germany was ready to discuss quantities, but a mere doubling of Versailles strengths in men and materiel would not suffice, and would even perpetuate the discrimination of the Treaty of Versailles. This note suggests that Hitler had confirmed to Neurath the requirement for weapons permitted to others in the first period, and also that he stressed the point—to which Neurath had evidently given little previous attention—that Germany would need the ability to equip not merely an army of 200,000, but the much larger eventual army resulting from a mobilization of reserves.[35] This last question was to get more attention henceforth and Hitler himself would emphasize, to the Italian ambassador on October 12 and to the cabinet on October 13, the need to be able to equip reserves; apparently he took an active interest in this matter. Actually, as we have seen, German arms already far surpassed the quantities needed for 200,000 men, while a surplus in already authorized types of arms would not be very easy to detect.[36] But the stress in the reply to Simon on the need for unrestricted quantities of these arms was to make a serious impression on the British Foreign Secretary.

Hitler was not at this juncture planning and preparing for a break with Geneva. He hoped there would be division between the French and the other powers, especially the British, and he concurred in the reluctance of the diplomats to make a public issue out of German arms demands. He did not want to arouse other governments unnecessarily, as would be done by abruptly breaking off negotiations, and he had accustomed himself to the idea of a convention on the general lines of the MacDonald plan, perhaps as a stepping-stone to an Anglo-German entente. If a window-dressing agreement were reached decreeing short stages for material disarmament, and allowing Germany new types of arms from the start, he could adapt himself to that, even though it purported to provide for a regular inspection of all nations. A level of inspection acceptable to all could hardly be very stringent, and it would have been hard to detect quantities. In the near future, Germany's greatest need in material rearmament would be to have enough specimens of new types of arms to work out the problems of production and to train future reserves in their use.

But a basic contradiction existed between longstanding German pol-

---

[35] AA, Memo by Neurath (RM 1374), 30 Sep 33, 3154/D669947.
[36] *ADAP*, C/I/2, Nos. 480, 486, 494, 499; AA, Memo by Bülow, 5 Oct 33, 3154/D669984; tel., Neurath to London, Paris, Rome, and Washington embassies, 13 Oct 33, 7360/E537797-98. On actual German armament, see Chapter Seven, Section 5.

466—Hitler's Hand is Forced, and He Frees It

icy and the desire of other nations that Germany remain disarmed. Neither Hitler nor any other German nationalist could accept an agreement which combined a regular and automatic inspection system with a complete four-year ban on new weapons for the Reich. Politically, these conditions, and perhaps also a limitation of authorized weapons to the numbers needed for 200,000 men, would have been an obvious step backward from what Schleicher had apparently won in December 1932, discrediting the new regime in the eyes of the German nationalist public and particularly of the army. Practically, it would have meant that only Germany would be inspected (not only a discrimination against her but also a bar to any retaliation by her against others' overzealous inspection), and an inspection where new types of weapons were totally banned would in fact expose past violations. The Reichswehr could not train short-service troops in the use of new arms and keep the existence of these arms concealed. And if Hitler had to reject an agreement on these grounds, it would be desirable, from his point of view, to conceal why he was rejecting it.

5. An "Ems Dispatch" Arrives in Berlin

Blomberg was presumably disappointed by Hitler's approval of Neurath's position. But the Reichswehr Minister soon received an effective counterweapon in the form of a telegram from London. Bismarck, the chargé d'affaires there, reported on the night of October 3 that according to various sources, including an unnamed informant especially close to Simon, the British were revising their draft convention, or indeed drawing up a "new draft convention," in collaboration with President Roosevelt. The French were not participating in this. This new draft (Bismarck reported) would be shown to Baldwin and MacDonald the next day and brought to Geneva by Simon on October 7; it would not be made public until then to "prevent a weakening of its effect by prior criticism on the part of the other powers." The British would not grant the sanctions the French wanted, but they might invoke a provision of the League Covenant against violations, and they would reject the German demand for samples of weapons. The British reportedly believed that "Germany was weaker than she had ever been in the field of foreign policy, since she was not only disarmed but also isolated," and "this realization would necessarily make it much easier for us to do without samples of weapons during the first four years."[37]

Now although some details of Bismarck's report, such as the sup-

[37] *ADAP*, C/I/2, No. 478.

posed collaboration with Roosevelt, seem to have been mistaken, the Foreign Office had in fact drawn up a new proposal, and a memo on this proposal seems to have been the principal basis of the Bismarck report. Ironically, Simon and Eden thought of the new proposal as a compromise between the conflicting positions of France and Germany. This "compromise," in the form of amendments to the MacDonald plan, would certainly have restrained Germany from acquiring new types of weapons in the first period, although it would have permitted Germany in the second period to acquire all hitherto-forbidden weapons retained by other signatories. This included aircraft, which the British had previously sought to deny to the Germans altogether. To try to satisfy France, the draft (as mentioned in Chapter Nine) provided for consultation in case of violation, with the avowed object of exchanging views "for the purpose of restoring the situation and maintaining in operation the provisions of the Convention." The compromise—which would hardly have been regarded as such in Berlin—was thus largely between the French and British points of view, although the generalized promise to exchange views would probably not have satisfied Paris either. Simon and Eden discussed the proposal with Baldwin and MacDonald on October 4, but Simon did not plan to submit it to the full cabinet until October 9.[38]

Bismarck's handling of his information calls to mind his grandfather's editing of the Ems dispatch in 1870, although the later Bismarck probably did not fully realize what he was doing, and the influence of his editing would be exerted not on foreign and domestic public opinion but on the leadership in Berlin itself. He could hardly have sent a message more provocative to Hitler. Although Simon's refusal to concede samples was serious enough for Germany, and would certainly rule out agreement if confirmed by the British cabinet, Bismarck might have pointed out that this confirmation was as yet by no means certain. Beyond this, the assertion that the British were producing a new draft proposal was bound to come as a blow to Hitler, who had accepted

[38] Cab 24/243, CP 228 (33), 3 Oct 33 (incl. Annex and notes of discussion, 4 Oct 33); Eden, p. 45; *DBFP*, 2/V, No. 440 (note). The Annex to CP 228 (33), the proposal itself, is not entirely clear on the concession of weapons to Germany in the second stage, but the 4 Oct 33 discussion between MacDonald, Baldwin, Simon, Eden, and Hankey clearly indicated that this was the Foreign Office intention, with Simon implying that this had been a hard-won concession wrested from the French. Bismarck's unnamed source had probably seen Simon's 3 Oct 33 covering memo. The reference to Roosevelt was apparently a distortion of statements in the memo that America was absorbed in domestic problems but could be expected to approve of British policy; certain other elements in Bismarck's report, including on Italy and on the course Britain would take if the proposal failed, resemble elements in the Simon covering memo.

the MacDonald plan as a possible basis for a solution. The Führer's hope for British friendship and support were disappointed; some might say he had been duped. Bismarck's allegations that the British believed Germany to be weak and isolated, and easily compelled to wait for four years—views it is doubtful Simon or Eden actually expressed—must have further wounded Hitler's ego. For Blomberg, on the other hand, the telegram confirmed the view he had already stated, that what was under discussion was a "new plan."[39]

Hitler and Blomberg discussed the telegram together on October 4, and Blomberg probably argued that the introduction of such a new plan would justify a German departure from the conference. As Neurath was absent, Hitler summoned Bülow to consult with them. By the time Bülow joined them, Hitler and Blomberg had agreed that Germany should not negotiate at all on a new plan which was known to be unacceptable in the last analysis. The other powers might accept it, and try to force Germany to accept. And, Hitler added, a breakdown of the conference on the grounds of a German rejection of supervision or of German rearmament demands "must be absolutely avoided." To prevent this development, Germany should now return to the "original question," as Blomberg put it: she should demand by ultimatum the disarmament of others, and declare that she would leave the conference and also the League if they did not disarm, if they denied Germany equality of rights, or if they brought up an unacceptable draft for debate. Hitler also spoke of making another major speech before the Reichstag, as on May 17, "in order to appeal to world opinion."[40] Obviously, the Führer was much concerned with how his government would appear before the world public.

[39] On the charge that the British thought Germany isolated and easily compelled to wait, it only seems to have been true in the sense that Simon thought disinterested world opinion would not sympathize with German unwillingness to wait (*DBFP*, 2/V, No. 419). Simon and Eden had no genuine understanding of German impatience, and imagined they could influence Berlin by getting the French to tell the Germans what reductions France would make in the same period. Massigli promised Eden that he or Boncour would see Neurath before the latter left Geneva; Massigli actually saw only Weizsäcker, who spoke to Neurath and Frohwein at the railroad station and sent a dispatch on the French position three days later, which arrived in Berlin on 6 October: *DBFP*, 2/V, Nos. 420, 422; *ADAP*, C/I/2, No. 472. After receiving a preliminary account by Frohwein, Bülow did express a rather favorable first reaction to British and French diplomats on 5 October, but concessions in the second stage were a far cry from qualitative equalization in the first stage: *DBFP*, 2/V, No. 427; AA, Memo by Frohwein, 4 Oct 33, marginal note by Bülow, 5 Oct 33, and tel., Schwendemann to Frohwein, 5 Oct 33, 7360/E537594-95; Memo by Bülow, 5 Oct 33, 3154/D669984-85; *DDF*, 1/IV, No. 276.

[40] *ADAP*, C/I/2, No. 479. Bülow was summoned from a cabinet meeting which had begun at 4:30 p.m., and the summons probably came only at its end, as he contributed an amendment on the last (fifth) point on the agenda (RK, Cabinet

Bülow for his part called attention to the statement of position drafted by Albert Frohwein, which was now finally on its way to Weizsäcker for transmission to Simon and the Italians, in line with Neurath's second note of the afternoon of September 30. It is important to realize that this statement did not, as some historians have supposed, result from this October 4 Hitler-Blomberg-Bülow meeting itself; thus the statement did not reflect a Hitlerian decision to force a break. Instead, this statement antedated the October 4 meeting, and it constituted the overdue answer which Neurath had promised to provide to British and Italian questions about Germany's actual demands. This statement called for a total duration of five years for the convention as proposed in "the English plan," rejected the idea of a trial period, and repeated (as noted earlier) that the type and extent of German armament could only be settled after the disarmament of others was settled. It said that Germany would forgo all arms which others agreed to destroy within the term of the convention, and, together with a supplementary message of clarification, it emphasized that if other nations were permitted to retain particular types or limited numbers of any weapons, Germany must also be permitted to have a quantity of those weapons. Likewise, weapons available without limit to other nations must also be available without any restriction to her; limiting Germany to double her Versailles armament, while leaving other nations unrestricted, would preserve discrimination and also fail to meet German security requirements. This reply was a new, sharper, and more indigestible statement of the significance of equality of rights, but the basic points had all been present implicitly or explicitly in earlier German declarations, including Hitler's May 17 speech, the statement of the German viewpoint given to Henderson in July, and a speech by Neurath of September 15. At the end of the message, the recipients were at least warned not to give the impression that these were counterproposals in the nature of an ultimatum, "since we must attach importance to the negotiations being continued." Significantly, this concluding section was omitted from the copy sent by messenger from the Foreign Ministry to Blomberg on October 5.[41]

Protocol, 4 Oct 33, 3598/D793926-35). The beginning of the second paragraph of Bülow's record (No. 479) is mistranslated in the English language edition of the published *Documents on German Foreign Policy*, and instead of "agreed" should read "had agreed," thus indicating previous discussion between Hitler and Blomberg.

[41] *ADAP*, C/I/2, Nos. 479, 480 (and notes). The statement of position, which had evidently been in preparation several days, was not actually dispatched to Berne, Rome, and London until 5:40 p.m. (AA, 7360/E537566), but Bülow can hardly have seen it or changed it following his talk with Hitler and Blomberg, which in turn apparently followed the 4:30 p.m. cabinet meeting (see note 40). As *ADAP*, C/I/2,

By mentioning this paper to Hitler and Blomberg, Bülow apparently meant to point out that the Foreign Ministry had already defined the demand for equality of rights in such a clear way that, if the British were taking the position reported by Bismarck, a break-off of negotiations might follow very soon; the State Secretary said that the break might come on Saturday or Monday, October 7 or 9. If anything, the instruction to Weizsäcker was rather too clear if there was to be an open confrontation, and Bülow seems to have thought that a more moderate-appearing stance would be better for public consumption. Hitler himself observed (as he had to Neurath) that it was a mistake to ask for more arms than Germany could actually produce in the next few years. Bülow now suggested that as a first step, Germany should retreat to the MacDonald plan, as she interpreted it and without any change for the worse, and Hitler and Blomberg accepted this. After the meeting, Bülow sent word to Neurath in the Allgäu Alps: "A new, much worsened British disarmament plan is threatened. RK [Chancellor] and RWehrM[inister] are therefore considering necessity of a complete break with Geneva." But although Hitler was now seriously weighing, and preparing for, a break, he had not yet committed himself. Hitler and Blomberg were anxious to find out more about the new Simon plan—and Hitler seems to have still hoped there was some mistake.[42]

No. 479 implies, Bülow no doubt believed, when he talked with Hitler and Blomberg, that the statement had already been sent. Internal evidence (an assumption that Simon would be in Geneva) indicates the instruction was drafted without consideration of Bismarck's 3 Oct 33 report; see also marginal note on Bismarck's report (*ADAP*, C/I/2, No. 478, note 4). Copy sent to Blomberg: AA, 7360/E537567-69; this copy did include the added statement on readiness to discuss quantities (see text below and note 43). On Neurath's 15 Sep 33 speech, see note 8.

[42] *ADAP*, C/I/2, No. 479; AA, Handwritten note by Bülow, 4 Oct 33, 3154/ D669974. That Neurath was in the Allgäu Alps: see note 32. For Hitler's observation, see note 30. Bülow also suggested to Hitler and Blomberg a démarche in the form of an ultimatum to the main delegation leaders before the withdrawal. Bülow's first memo on the talk with Hitler and Blomberg seemed to assume that a break would follow, but by the time he communicated with Neurath, he was saying that it was under consideration. Aside from the first memo, there seem to be no grounds for the view of Fraser (pp. 188-189), Wollstein (*Weimarer Revisionismus zu Hitler*, p. 191), and Geyer ("Militär, Rüstung und Aussenpolitik," p. 250) that the decision to withdraw was made final on 4 October; probably their view arises from a belief that the 4 Oct 33 instructions to Weizsäcker resulted from the Hitler-Blomberg-Bülow meeting of that day (see notes 28 and 41). Weinberg, *Foreign Policy of Hitler's Germany*, p. 165, Jacobsen, *Nationalsozialistische Aussenpolitik*, p. 399, and Kimmich, *Germany and the League of Nations*, p. 187, make the final decision later. Fraser is probably correct, however, in suggesting that debates on German refugees in the League in early October increased Hitler's hostility to the League and to Geneva generally.

6. THE BRITISH TAKE POSITION

Because Simon had delayed his return from London to Geneva (actually, he was not to return to the conference site until October 11), Weizsäcker could not present the October 4 statement of position to him, and the task of doing this therefore fell to Bismarck in London. Bülow, after talking with Hitler and Blomberg, had tried to ease matters slightly by making the clarification of the statement add that Germany was ready to discuss quantities.[43] And as we have seen, the German statement did not disclose the Reichswehr's actual desire for quantitative equality with France. Still, the statement was bound to come as a shock to the British Foreign Secretary, and the bearer of the message did little to soften the impact. The younger master of Schönhausen and Friedrichsruh repeated the statement to Simon on the morning of October 6, and the clarification that Germany wanted her share of any weapons which were limited by number struck the Foreign Secretary as "a very serious novelty." Simon also took a serious view of the demand for unrestricted quantities of the weapons not restricted for others—apparently all the more because he did not realize, and Bismarck did not explain, that this was for the reserves which would be created by a short-service system. Bismarck chose aircraft, the weapon most feared by the British, as an example of what Germany wanted. Simon said that Neurath had frequently mentioned samples to him, and the Foreign Secretary hinted that samples of antiaircraft guns might be conceded, but where, he asked, was the reference to samples in the present communication? Bismarck readily admitted that he could no more find reference to samples in it than could Simon. It also apparently did not help matters when Bismarck (acting on a suggestion from Schwendemann in Berlin) evoked Simon's own statement of November 17, 1932, which had conceded that Germany might have tanks.[44]

---

[43] The draft of the supplement to para. 3b of the instructions (*ADAP*, C/I/2, No. 480, note 3), sent out at 8:45 p.m., shows that the last clause was added in Bülow's own hand (7360/E537570).

[44] *DBFP*, 2/V, No. 434; AA, tel., Bismarck to Ministry, 6 Oct 33, 3154/D670001-03; Note by Schwendemann, 5 Oct 33, 7360/E537597; *DDF*, 1/IV, No. 279. Weizsäcker, though more tactful, made a very serious impression on Aloisi and Davis with his explanations in Geneva: Aloisi, pp. 150-151; *FRUS, 1933*, I, 238-240; AA, tels., Weizsäcker to Ministry, 5, 6 Oct 33, 3154/D669986, D669997-670000. Hassell's exposition likewise impressed Fulvio Suvich: AA, Hassell to Ministry, 5 Oct 33, 3154/D669988. See also *DBFP*, 2/V, Nos. 433, 435, 436. Simon evidently thought that antiaircraft artillery was only for defensive use against aircraft, but ground-force veterans of World War II will well remember the German 88s, which could be used against either aircraft or troops. On Simon's ignorance of the need for more arms for reserves, see note 50.

There are signs that Bülow had doubts about Bismarck's competence. An urgent

But if Simon was gravely disappointed by the German position, this was not Bismarck's fault. Basically, what Bismarck did, in the cases of both Hitler and Simon, was to break through crusts of illusion, crusts that were bound to crack soon anyway. Whether from excessive tactfulness and feigned acquiescence on Neurath's side, or wishful thinking and self-deception on his own, Simon had gotten the impression at Geneva that the German government would accept a probationary stage (though thinking four years to be too long), and might be willing to wait for samples until the second stage. Simon thought that shortening the probation period would be a way of reaching a compromise, and in asking for the number of samples desired, he presumably hoped that modest demands might offer some chance for a negotiated solution.[45] He and Eden evidently believed that Germany's just desires should be satisfied by a grant of qualitative equality in the second period. They had little sense of the way the trial period impinged on German sensitivities. They also had trouble imagining that the Germans could dare to ask for substantial rearmament. They seem to have had no awareness that Germany had begun preparations for mobilizing a large, integrated wartime army, that she had long been planning to provide modern arms for reserve training and for mobilization, and that the production of these arms could not be entirely concealed. Yet these were considerations which, from the German government's point of view, made the combination of a delay in permission for new arms and an effective inspection impossible to accept. Bismarck's démarche removed Simon's immediate delusions on the German view of stages and samples, but it did not of course rectify all of the deeper misunderstandings.

Simon's reaction went beyond disappointment, into a deep if rather short-lived resentment. He had understood Neurath to be asking only for samples, but now the Germans evidently wanted large quantities of arms: thus he described the German demands as "vastly enlarged."

---

message was sent on 5 Oct 33, telling Bismarck to use the instruction orally only, and in no case to leave a copy (with Simon) (7360/E537598); there seems to have been no need to send such a warning to Weizsäcker. Bülow also made a point of telling the British ambassador on 5 October that Simon would get an answer to his questions in Geneva, and that although Bismarck was presenting the instructions in London, he was naturally not so well versed in disarmament matters as Weizsäcker, who had been involved with the subject for years. See also *ADAP*, C/I/2, No. 480 (n. 5). On 7 October, after Bismarck's report had indicated the possibility of negotiations in London, he was told to tell the Foreign Office that he was not in a position or authorized to debate and interpret the content of his instructions: Weizsäcker and Nadolny were in Geneva for that purpose (3154/670008). It seems quite possible that, in Bülow's eyes, Bismarck's 3 Oct 33 report had clumsily disrupted the Wilhelmstrasse's efforts to keep Germany in the conference.

[45] *DBFP*, 2/V, Nos. 409 (encl.), 411 (encl.), 412 (encl.), 419, 422, 436, 438, 443.

And the German statement of position seems to have embarrassed him personally in several ways. First, Bismarck had pointed out that the British had changed their policy from what he, Simon, had enunciated the previous November. Although Simon retorted that there had then been security plans as well, this did not apply as well to the November 1932 statement as it did to the December 1932 agreement, and it would hardly deflect public criticism in England. Also, the MacDonald plan had indeed been modified, as Simon would admit a month later. Further, the German position as it now appeared destroyed the prospects for the "compromise" Simon and Eden had just expounded to MacDonald and Baldwin, and in the process tended to show that that compromise amounted essentially to a modification of the French position, without serious consideration of German views or even much attempt to ascertain them. In November, Baldwin and MacDonald would each stress the need to get into touch with Germany, implying that this effort had recently been neglected.[46] But in early October, Simon indignantly committed himself to the policy of "forcing Germany into the open," of exposing her as seeking rearmament.[47]

In the immediate situation at any rate, the leading ministers staunchly supported Simon's position. In a speech before a conference of the Conservative and Unionist party in Birmingham on October 6, Baldwin spoke of the necessity of arriving at a disarmament convention in order to restore political security and bring about economic confidence; a failure of the conference would mean an arms race and no tax cuts for a generation. And he stated that anyone who violated the convention would have no friend in the civilized world, adding, in a transparent reference to Germany, that "the same is true of any nation which deliberately prevents such an agreement being reached by putting forward demands which might be acceptable after a time, but which would not be acceptable today to the other co-signatories." Baldwin also specifically reaffirmed Britain's resolve to honor her commitments under the Treaty of Locarno.[48]

---

[46] Later positions: Cab 27/505, DC (M) 32, meeting of 7 Nov 33; see also *ADAP*, C/II/1, Nos. 59, 76.

[47] The idea of "forcing Germany into the open" had already been accepted in the May crisis (Cab 23/76, meeting of 10 May 33, 34[33]1), and it became a frequently voiced objective in this October crisis (Cab 23/77, meeting of 9 Oct 33, 52[33]1; *DBFP*, 2/V, Nos. 440, 447; FO 371, Minutes by A. Leeper, Sargent, Vansittart, 11-12 Oct 33, W11472/40/98).

[48] Text of speech: *Times* (London), 7 Oct 33. At the meeting between Baldwin, MacDonald, Eden, and Simon on 4 Oct 33, there was some discussion of Baldwin's prospective speech. Baldwin thought his Conservative audience cared too little about disarmament, and he planned to say that a failure to get a convention would mean an arms race and an increase in government expenditure; he was thinking of asking, in a burst of brutal frankness, what Europe would do if Germany rearmed and whether Britain would expect Germany to observe the Treaty of Ver-

The British soon formalized their position and communicated it to the Germans. Acting on Simon's report on his interview with Bismarck, the British cabinet agreed on the ninth that the Germans should be compelled to disclose their rearmament plans, either by point-blank questions later in Geneva or through a prior warning to Hoesch that the British position would be openly declared in Geneva. One senses the ministers looked forward gratefully to the prospect that the conference might be concluded without any blame falling to England.[49] On October 10, Simon told Ambassador von Hoesch (who had just returned to his post) that the cabinet believed that Bismarck's communication widened the breach between the British and German positions. Germany had now refused the concept of stages and had asked for substantial rearmament instead of mere samples, and neither of these positions appeared acceptable to the British government. Simon—who believed Nadolny would soon make a public statement—proceeded to indicate that he might himself reply with a "frank and full" pronouncement at the conference. He ended his remarks by reading to Hoesch a special appeal from MacDonald, asking the German government to remember past British intercession on the German behalf, and beseeching Berlin not to refuse to cooperate in what were now the only methods and ways by which a convention might be reached.[50] The indication that the cabinet rejected the German position, the warning that Simon might make a statement in Geneva, and the solemnity of the final appeal all clearly implied that a die had been cast, and that a result might soon be read—very loudly—off its top face.

---

sailles if the conference broke down. (Apparently he assumed that Britain would not.) Simon thought this would have a bad effect on the French, who already suspected Britain of being lukewarm because of her refusal to discuss the dossier. Baldwin then said he also wanted to recall that Britain would never go back on her signature (i.e., of Locarno). See Cab 24/243, CP 228 (33), notes of conversation, 4 Oct 33. MacDonald wanted to make British views plain to the Germans and to remind them of past British support, pointing to the possibility of its withdrawal (MDP, 8/1, Diary, 8, 9 Oct [33]).

[49] The 9 Oct 33 British cabinet discussion is recorded in Cab 23/77, 52(33)1. As Eden (in Geneva) gathered (*DBFP*, 2/V, No. 440), the cabinet dropped all idea of using the "compromise." Eden himself may now have welcomed the prospect of a breakdown as a means of evading the apparently unresolvable differences between Britain and France on sanctions: see *DDF*, 1/IV, No. 272.

[50] *DBFP*, 2/V, No. 443; *ADAP*, C/I/2, No. 486. During the conversation, Hoesch explained that a militia army would produce reserves who would also need arms, and that if the reserves were to be of any use, there would have to be rifles for them as well as for the 200,000 of the peacetime force; Simon's record implies he had never heard before of this explanation for the wish for more than a doubling of arms, and even suggests he may not have known that reserves were to be created under the MacDonald plan. That Simon believed the Germans would make a public declaration is also apparent in *DDF*, 1/IV, No. 279.

7. HITLER UNDER COMPULSION

We are apt to think of Hitler as always the moving force in the European politics of his time. But there is little evidence that he had intended, before receiving Bismarck's report, to withdraw from the Disarmament Conference. After receiving it, he waited for confirmation. On October 6, he approved guidelines for the German delegation that treated British policy as still uncertain. If the British submitted a new plan containing the basic French theses, then the delegation should refuse to negotiate on this and demand a return to the MacDonald plan, or to the original conference objectives of disarmament and equal national security, "intimating" that if this were not done, Germany would withdraw from the conference and the League. Hitler reserved for himself the decision on whether to carry out this threat. On the other hand (the guidelines continued), if the British presented new demands merely in the form of amendments to the MacDonald plan, the delegation should deal with these on their merits, in the light of the existing German position.[51] Probably Hitler would still have preferred negotiations on the old MacDonald basis, under which secret rearmament could have quietly continued. Indeed, even if the report of the new British plan was correct, there was much to be said for staying in the conference and seeking to break down the opposing front, perhaps forcing a decision between the new plan and the MacDonald plan. This was the view that German diplomats tended to take.[52]

But Hitler was afraid that conference debates might expose Germany as seeking rearmament. When he learned of Bismarck's report on the fourth, he emphasized that a conference breakdown over a German rejection of supervision or German rearmament demands "must be absolutely avoided." Again, the October 6 guidelines stated that a situation had to be avoided in which Germany would reject a series of points and possibly withdraw from the conference because her demands for "equal rights to armaments" were refused. In explaining his actions to his cabinet on October 13, he would stress the need to pre-

[51] New directive: *ADAP*, C/I/2, No. 484. A note on an original draft indicates that this document was drawn up by General Schönheinz (AA, 7360/E537604). (A crossed-out heading on the same copy, referring to a talk with Hitler on 25 August, presumably represents only reuse of an old piece of paper.) Other drafts by Nadolny and Schwendemann urged a continuation of negotiations (7360/E537586-90, E537602-03, E537605-06).

[52] Weizsäcker had made a suggestion along these lines: see AA, tel., Weizsäcker to Bülow, 6 Oct 33, 3154/D669995-96. Marginal notes by Koepke indicate that he held similar views as late as 10 October: AA, Delegation memo, 4 Oct 33, 7360/E537586-90. According to Paul Schmidt, German diplomats did not think a break was necessary: *Statist*, pp. 282-283. See also Jacobsen, *Nationalsozialistische Aussenpolitik*, p. 399.

vent other nations from exploiting the situation propagandistically against Germany.[53] The recently stated German demands for arms, while quite in keeping with the principles of equality of rights and the right to national security, did indeed amount to a proposal for rearmament. If a debate on these demands took place in a general session of the conference, a series of denunciations of Germany would probably be made. No doubt many other nations would side against Germany in Geneva, as had happened in recent League debates on the Jewish-refugee question.[54]

Further, not only were the German proposals vulnerable to attack; other governments might also reveal that Germany was already rearming. Blomberg had pointed out in July that there might be a call in the League Council for investigating Germany. Hitler demonstrated his own awareness of foreign suspicion and his fear of exposure, particularly at this juncture, by now ordering, on October 5, that the SA-Sport (or Ausbildungswesen) schools should drop the planned training in weapons from their courses, which were due to begin in about a week. This step presumably reduced the risks, but the abnormal inductions just carried out by the Reichswehr on October 1 doubtless provided other chances for disclosures. Moreover, Hitler probably learned from Blomberg about the report the Ordnance Office completed on October 7, on the charges of German rearmament appearing in the foreign press. The report showed that there was in fact a foundation for many of the foreign press charges, including that firms were producing unauthorized munitions, that tanks were being built on British patterns, that poison gas had been produced and had been stockpiled in Russia, and that German antitank guns now had real barrels, not wooden ones. The Ordnance Office described the disclosure of German tank production as "very unwelcome," noting that about 150 tanks were being produced.[55]

Reports from Nadolny, now back in Geneva, also pointed up the danger that Germany might soon be placed in the dock. When he

[53] *ADAP*, C/I/2, Nos. 479, 484, 499.
[54] Recent League debates: Fraser, pp. 181-187.
[55] *ADAP*, C/I/2, No. 359; WKK, WK VII/1453, Wehrkreiskommando (Bayer.) VII 7. (Bayerische) Division Ib/Ia 1679, 5 Oct 33, 37/1117; OKW, Wi/IF 5.5, Wa A 1279/33, 7 Oct 33, incl. marginalia and att., 2/713988-90; the query is in OKW, Wi/IF 5.3683, VGH 583/33, and atts., 18 Sep 33, 437/1300787-811. On this subject, see Robertson, *Hitler's Pre-War Policy*, pp. 21-22. About one third of the total of press stories on material rearmament submitted for checking had information appearing on this list of confirmed accusations; some stories had several confirmed accusations in them. Two other attachments to the Ordnance Office's reply are missing; they may have listed further confirmed accusations, or accusations that could not be checked, but as matters stand we cannot tell whether the other press stories were checked and found to be true or untrue. Blomberg could also have referred to the Abwehr material in notes 7 and 9.

spoke with Boncour and Massigli on October 9, Boncour told him that the existing level of German armament had to be established in order to have a point of departure. Nadolny objected that they were there to conclude a new convention, not to check on the execution of the Treaty of Versailles, and Massigli retorted that according to French information, Germany had already rearmed more than the convention would ever permit, and was also violating the Treaty of Locarno (i.e., the demilitarized zone). Nadolny claimed that this was slander. On the next day, Boncour again told Nadolny that the current situation must be established and that the French dossier must be published, although Nadolny managed to divert the discussion to other subjects.[56]

Aloisi had noted on October 7 that Boncour "was in a very good humor in spite of the German reply," i.e., that delivered by Weizsäcker in Geneva and Bismarck in London. Indeed, the good humor may have been *due to* the German reply, which promised to make it easier to organize an anti-German front. Henderson had suggested five-power discussions under his own direction, but Boncour (as the Germans also heard) made it plain that France would not enter any multilateral discussions. By French account, Boncour told Henderson that this would imply a readiness to make concessions; he also refused to specify more clearly how France would disarm in the second period. He considered that Germany now knew the French position, that she had already refused to accept it, and that it was now in order for the other nations to exert pressure upon her. To Eden, Boncour said that the German reply delivered by Bismarck and Weizsäcker would have been different if Italy had stood firm against German rearmament in the first period, as had Britain, the United States, and France.[57] Boncour plainly wanted to force the Germans to choose between retreat and exposure. The Germans did not know all that Boncour was saying to other governments, but his strategy was evident enough.

Along with the reporting from Geneva, there now came Hoesch's account of Simon's October 10 statements, which arrived in Berlin on the morning of the eleventh.[58] Now Hitler could scarcely have any illusions about the British viewpoint. And the British support of the French position made the diplomatic situation very hazardous for Germany, as she would stand in isolation. Aside from the risks of a public exposure of German rearmament, Hitler and Blomberg feared that informal negotiations might bring a series of piecemeal concessions, leading to an unsatisfactory final result, as had happened with the

[56] AA, tels., Nadolny to Ministry, 10 Oct 33, 3154/D670032-35, D670040.

[57] Aloisi, p. 152; AA, tel., Nadolny to Ministry, 10 Oct 33, 3154/D670032-35; *DDF*, 1/IV, No. 287; *DBFP*, 2/V, No. 437.

[58] *ADAP*, C/I/2, No. 486.

four-power pact.[59] Hitler was on the defensive, and he would have to take the initiative if he was to get off of it.

Hitler must have mulled the situation over on the eleventh, and by evening he had apparently reached some tentative decisions. At 8 p.m. on October 11, Neurath sent a telegram to Nadolny advising that Germany would withdraw from the conference if, as Baldwin's speech and Simon's statements to Hoesch suggested, the British proposals for amending the MacDonald plan were incompatible with German equality. This warning, for Nadolny personally, also indicated that any statement on withdrawal would be made by Hitler in Berlin. This message to Nadolny seems to indicate that by this time Hitler had virtually decided on withdrawal, although he still kept open the possibility of changing his mind, if the situation should change.[60]

That same evening, Simon, now in Geneva, repeated to Nadolny what he had told Hoesch, that the British cabinet had made its decision, and Simon added that this was dictated by public feeling against Germany. Nadolny, according to his own report, "pointed out in a very serious way the great responsibility which England took upon herself if she persisted in this attitude," thus "intimating" (as the October 6 instruction had directed) that Germany might leave the conference. But Simon held to his position. The next morning, October 12, presumably after studying the report of this conversation, Hitler ordered Nadolny to come to Berlin for consultation. Apparently Hitler and Neurath considered that Simon had been warned, and that, unless he now backed down, the occasion for withdrawal was almost at hand.[61] Probably Hitler's main reason for recalling Nadolny was to avoid getting drawn into negotiations and unsatisfactory compromises,

---

[59] See *ADAP*, C/I/2, Nos. 479, 484, 499.

[60] *ADAP*, C/I/2, No. 489. The original document shows editorial changes by Neurath himself, something he only troubled to make in the case of crucial messages. On 11 October, the German Foreign Ministry prepared a message that would have directed Hoesch to try to approach MacDonald personally and ask if Britain was no longer ready to stand by the 11 Dec 32 agreement. But that evening, shortly after its transmission to London and Geneva, Neurath sent word that the recipients should hold the message without deciphering it. The next day, recipients were directed to destroy the cipher text. Presumably Neurath had decided that the message was inconsistent with Hitler's latest thinking. See AA, tels., Neurath to London and Geneva (3 tels. each), 11, 12 Oct 33, 7360/E537754-56, E537758-60.

[61] *ADAP*, C/I/2, Nos. 493, 495 (note 1). Alluding apparently to the 11 Oct 33 discussion, Bülow on 14 Oct 33 told Newton of the British embassy that Nadolny had made clear to Simon the consequences of the British position, so that the German withdrawal should have come as no surprise: AA, Note by Bülow, 14 Oct 33, 3154/D670093. (According to the embassy, Bülow said Nadolny had "hinted" at such a development [*DBFP*, 2/V, No. 454].) The editors of the *DBFP* report (2/V, No. 447, note 4) that they found no record of the 11 Oct 33 Simon-Nadolny talk in the Foreign Office archives.

though Hitler may also have had some interest in hearing Nadolny's account of the Geneva situation so as better to plan his own moves.

Nadolny saw some signs of weakening and division in the opposition alignment, and before and after his return to Berlin on Friday the thirteenth, he tried unsuccessfully to dissuade Hitler from withdrawal. He does seem to have convinced the Chancellor that it would be safe and wise to await a statement by Simon, now scheduled to be made in the Bureau on Saturday morning, October 14. It became clear by the thirteenth that Simon would only sum up the results of the Geneva discussions, rather than replying directly to the German demands, or presenting an Anglo-French-American declaration, and Nadolny suggested that Simon might retract his "discriminatory" position.[62]

Indeed, Hitler could hardly justify a withdrawal action until there had been a public report on the current position of the other powers. But once such a report was given, Hitler had an interest in announcing his withdrawal as rapidly as possible and from outside the conference hall, so as to forestall a session that might turn into a public trial of Germany. As he told his ministers, he wanted to capture world attention in a different manner than before; thus he would try to get away from the issue of rearmament. He would use the occasion to dissolve the Reichstag and the Land parliaments, conduct new Reichstag elections, and hold a plebiscite on his foreign policy. These steps would show the support of the German public for his "peace policy" —and in the process not so incidentally dispose of non-Nazi membership in the Reichstag and eliminate the parliaments of the Länder.[63]

Hitler would also take the initiative by leaving the League as well as the Disarmament Conference. Aside from Hitler's dislike of the whole League of Nations machinery, and his (and Schleicher's) earlier warnings that Germany might leave the League, a departure from the League would show that Germany would not again, as in 1932, resume negotiation in Geneva under the cover of League activity. Germany would also, in effect, be serving notice that she would not cooperate with a League Council investigation or abide by an arbitral

---

[62] Nadolny, pp. 139-140; AA, tel., Neurath to London, Paris, Rome, and Washington embassies, 13 Oct 33, 7360/E537797-98; Memo by Frohwein, 19 Oct 33, 7360/E537899. The decision to await Simon's speech was confirmed at the cabinet meeting on Friday evening, 13 Oct 33: *ADAP*, C/I/2, No. 499; *DDB*, III, No. 61; *DDF*, 1/IV, No. 328; *DGFP*, 2/V, No. 469; ML, Shelf 5940, Diary, 10-13 Oct 33, reel 18/778-779. Indications of opposition weakening and moderation of Simon's views: *ADAP*, C/I/2, No. 495; AA, tel., Nadolny to Ministry, 12 Oct 33, 3154/D670066; Memos by Völckers, 13 Oct 33, 3154/D670070-72; *Deutsche Allgemeine Zeitung* (Berlin), 14 Oct 33.

[63] *ADAP*, C/I/2, No. 499. On the domestic significance of the plebiscite, see Bracher, "Stufen der Machtergreifung," pp. 348-368.

decision. In this connection, and because Britain and the United States did not support sanctions, as well as because of Daladier's attitude, Hitler could now at least feel fairly sure that France would not take the ultimate step of preventive action, in which she would find herself quite alone. Certainly, withdrawal from Geneva did not need to imply an end to negotiations—an end to the spinning out of talks while rearmament quietly continued. Instead, withdrawal from the conference could promote the transfer of negotiations to the bilateral format he much preferred, and a break with the League would help to ensure such a transfer. He intended in any case to use a recent (October 8) speech by Daladier as a springboard for further proposals for direct negotiation with France.[64]

8. DAVIS LEADS SIMON TO MODERATE HIS POSITION

After Bismarck had presented the German statement of position on October 6, Simon had decided to "force the Germans into the open," so as to expose their intention to rearm, and he had gotten cabinet approval for this policy. One problem with this course was that the Germans were not so obliging, or so foolish, as to proclaim their position publicly in the terms of their October 4 instruction to Weizsäcker and Bismarck.[65] Nevertheless, Simon himself could have publicly exposed the significance of the German insistence on immediate qualitative equality, showing that it meant an immediate and significant increase, quantitative as well as qualitative, in armament. His experience as an attorney often appeared of little use in diplomacy, but that experience would presumably have served well in cross-examining the Germans before the Conference Bureau, or in making an arraignment of them there. Since the British cabinet wanted any breakdown of the conference to be blamed on Germany, not Britain, and since Simon would need to justify the modification of the MacDonald plan before the British public, he had ample reason to make such an exposure. When he arrived in Geneva, he planned to force the Germans to come out into the open, and for a time he intended to reply to the German position in the Bureau.[66] Yet he changed his

[64] On plans for bilateral negotiations, see esp. *DDF*, 1/IV, Nos. 301, 312; also *DDB*, III, No. 61.

[65] As noted in note 44, Bülow had in fact sent Bismarck very explicit directions only to transmit the instruction orally, and in no case to give the British a copy (AA, tel., Schwendemann to Bismarck, 5 Oct 33, 7360/E537598).

[66] An exposure could probably have been made without specifically citing Bismarck's démarche, although the circumstances might have justified such a breach of courtesy. A timely leak to the press might have helped; there was in fact a belated disclosure in the *Echo de Paris* of 20 Oct (*Times* [London], 20 Oct 33). On Simon's initial positions, see notes 70 and 71 below.

mind, and never made a public denunciation of the German proposal to rearm. Norman Davis seems to have been the chief agent in persuading Simon to change his course.

Davis, as before, was anxious to reach some kind of disarmament settlement. He agreed with Boncour that Britain, France, and the United States, and perhaps Italy should agree on a set of proposals to present to the Germans. In contrast to Boncour, however, Davis wanted to offer a proposal the Germans would freely and willingly accept—perhaps in five-power talks. Hugh Wilson, in the American delegation, devised a plan that would have avoided overt references to stages, instead tying German access to new arms to a gradual, regulated reorganization of German effectives. This part of the plan actually resembled the idea that Bülow had advanced, and Blomberg had vetoed, at the end of September: the Germans would have had the right to the same weapons for each 1,000 men as other armies. At the same time, the destruction of the very largest French guns (over, say, 364 mm) would begin in two or three years.[67] Davis telephoned President Roosevelt in Washington, vaguely outlining this scheme, and Roosevelt—who had allowed a minute and a half for a preparatory briefing—gave his assent, but warned Davis (as advised by the briefing) against "breaking up the solidarity" with France, England, and Italy. To the British, Davis meanwhile expressed confidence that the Germans could be induced to enter an agreement satisfactory to Britain and France, although he was afraid right-wing elements in Berlin were seeking a breakdown of the conference in the hope that this would bring Hitler down.[68]

The Wilson-Davis proposal was indeed a possible solution if the object was to find a way of satisfying Germany and reaching a nominal disarmament settlement. Noel Field, then in the State Department's Division of Western European Affairs, pointed out that the plan appeared to give the Germans substantial new concessions, and in particular, complete qualitative equality. Rheinbaben later observed that he saw no conflict between this program and the German

[67] Davis Papers, Memos of conversations between Davis and Paul-Boncour, 27 Sep, 3 Oct 33; *FRUS, 1933*, I, 245-246, 248-249; *DBFP*, 2/V, Nos. 432, 436, 441.

[68] *FRUS, 1933*, I, 245-247; Hooker, *Moffat Papers*, pp. 103-104; *DBFP*, 2/V, Nos. 436, 445; *DDF*, 1/IV, No. 304; Davis Papers, ltr., Davis to Viscount Astor, 10 Oct 33. While Davis had illusions about Hitler's desire for disarmament, Blomberg doubtless hoped for an end to the conference, and Neurath now probably did too. Bismarck (AA, tel., Bismarck to Ministry, 6 Oct 33, 3154/D670001-03), Hoesch (*ADAP*, C/I/2, No. 486), and Nadolny (in transparent form: Nadolny, p. 139) were reporting that Neurath had seemed ready in Geneva in September to accept stages and samples, and he now seems to have felt compelled to show himself an ultranationalist: see *DBFP*, 2/V, No. 489; *DDB*, III, No. 61; also the tone of *ADAP*, C/II/1, Nos. 8, 23, 54.

position.⁶⁹ Boncour's objective, however, and at first also Simon's, was to expose the Germans as rearming.

After Simon arrived in Geneva on the morning of the eleventh, he told Davis of the British cabinet's decision, and he seemed to Davis to "evidence considerable feeling" over the German statement of position. That afternoon, Simon suggested to Davis and Boncour that a draft resolution, providing for a trial period with no qualitative rearmament, be shown to Nadolny. If the Germans refused this, Simon believed (as Davis recorded) "that in that event the issue should be forced and a resolution clearly stating their opposition to German rearmament should be presented to the General Commission on Monday and brought to a prompt vote. This would let the Germans see how they stood." Davis argued, however, against trying to impose a solution.⁷⁰ The next day, the twelfth, Simon told Davis he was thinking of making a speech of reply (at the Bureau meeting on the fourteenth) to the German statement of position; he would give notice of the British government's opposition to the German claim to rearm, and demand that this issue be decided before the conference proceeded. Davis reported to Washington that he talked Simon out of speaking on the German statement specifically, or making a direct issue out of it; in return, Davis loyally supported Simon when they talked with Nadolny just before the latter's departure for Berlin. Nevertheless, Davis still feared on the thirteenth that Simon was going to use his Bureau speech to reply to the German statement, and the American delegate devoted himself to inducing Simon to make a conciliatory statement, leaving the path open for negotiation. In this, he was strongly supported by the Italian delegation. During the day, Davis also for the first time gave the Germans, in the person of Rheinbaben (the acting head of the delegation), some idea of the Wilson proposal; Rheinbaben listened receptively, thus encouraging American hopes, but it does not appear that he reported the proposal to Berlin at the time.⁷¹

Simon was nervous as the hour came near for his public statement at the Bureau meeting on the morning of the fourteenth. At his request, Eden showed a draft of the statement, which Davis had helped

⁶⁹ SDP, Memo by N. H. F[ield], 11 Oct 33, 500.A15A4 General Committee/641; AA, Memo by Rheinbaben, 21 Oct 33, 7360/E537938-41.

⁷⁰ *FRUS, 1931*, I, 249-251; Davis Papers, Memo relating to draft resolution, 11 Oct 33; *DDF*, 1/IV, No. 300; MDP, 1/504, ltr. from Simon, 11 Oct 33.

⁷¹ *FRUS, 1931*, I, 256; *DBFP*, 2/V, No. 447; *DDF*, 1/IV, No. 304; AA, Note by Schwendemann, 13 Oct 33, 7360/E537791-92; Memo by Frohwein, 19 Oct 33, 7360/E537898; Memo by Rheinbaben, 21 Oct 33, 7360/E537938; SDP, Ferdinand Mayer Geneva diary, 15 Oct 33, 500.A15A4/1469½.

to modify the night before, to the French and Italian delegations, and Cadogan gave a copy to Frohwein of the German delegation. The meeting was delayed until Simon and Davis secured the agreement in principle of the Italian delegation; some amendments were made at Italian request, and Cadogan gave these, too, to the Germans. Boncour also wanted some changes, but it was only after Simon arrived at the meeting, forty-five minutes late, that Davis persuaded him to accommodate the French Foreign Minister. Eden recorded that Simon "then made P.-B. purple with rage by going to sit ostentatiously on opposite side of room to whisper to German (God knows what about!) and holding up business for this." The comment speaks volumes for Paul-Boncour's desire to see Germany isolated, and also for Eden's own views at this point. The unnamed "German" was Rheinbaten: as Rheinbaben recorded a few days later, Simon expressed to him the hope that the German government would recognize the "spirit of conciliation" in his text, and contribute to the continuation of negotiations. But Rheinbaben considered the substance of the statement still unacceptable; in particular, in proposing a numerical increase in (already authorized) Reichswehr arms in the first period, proportionate to the increase in effectives (to 200,000), the text implicitly closed the door to any immediate change in German qualitative armament. Rheinbaben now urged Simon to strike out the word "numerical." Simon answered that he could not do this at this stage because of understandings with other delegations, but that this and other German demands could be discussed later.[72]

These preliminaries suggest that Simon had become concerned over the possible repercussions of his speech, and such a concern also seems apparent from the speech itself. Clearly, much of the tone of Simon's statement can be attributed to the influence of American and Italian diplomats, and especially of Davis. Simon's own background and sympathies made him susceptible to Davis's arguments against trying to force "a new Treaty of Versailles." And it began to seem that the Germans might not simply back down. The British in Geneva had heard rumors that Germany might withdraw from the conference before the Bureau met, and Frohwein told Cadogan of his impression that

> people in Berlin consider the situation extremely serious, and under no circumstances want to expose themselves to the situation of be-

---

[72] *FRUS, 1933*, I, 262, 266; Eden, p. 46; *DDF*, 1/IV, No. 305; AA, Notes by Völckers on phone calls, 14 Oct 33, 3154/D670080-81; Memo by Frohwein, 19 Oct 33, 7360/E537897-904; Memo by Rheinbaben, 21 Oct 33, 7360/E537936-41.

ing confronted on Monday, in the General Commission, with a united front of the great powers on the basis of the proposals, completely unacceptable to us, which the British had made in talks up to now with the Reich Minister [Neurath; perhaps Hoesch was meant] and the Ambassador [Nadolny].

Cadogan asked what Germany would do to evade a common front; Frohwein's personal view was that a prior withdrawal from the conference could be expected.[73] As the time for the speech drew near, Simon's own anger no doubt had decreased while his awareness that he might be blamed at home had risen. One ominous sign was an article by David Lloyd George, published prominently in the *Daily Mail* of October 13, which argued that France was unilaterally scrapping the Treaty of Versailles through her refusal to disarm and that a Polish-Czech-French occupation of Germany and the overthrow of Hitler would only lead to Communism.

In his speech, Simon made only the most general kind of reference to the "present unsettled state of Europe," drawing on a recent statement by Henderson, and he also played down the division of the period of the convention into two stages, emphasizing instead the change in overall length from five to eight years; "so far as I recall," he said, "no serious objection to this extension has been raised." Although he indicated that four years had been mentioned as a possible length for the first stage, he did not commit himself to this, and he said that others had suggested a shorter term. He did state clearly the principle of no rearmament, while adding (despite Rheinbaben's warning) that the Reichswehr should nevertheless have "a proportional numerical increase" in its armament as it changed into a more numerous short-service army. Massigli commented to the Quai that Simon had framed his exposé so as to win over the largest possible number of delegations, and that "the circumstances explain the timidity of expression in certain passages." Davis, the Marquis di Soragna (for Italy), and Boncour added their own statements generally supporting Simon's position, although Boncour laid more stress on the division into stages, as did also Maurice Bourquin (Belgium), Beneš, and Nicolas Politis (Greece).[74]

[73] Davis on no "new Treaty of Versailles": *DDF*, 1/IV, No. 304. Possible German withdrawal: *DBFP*, 2/V, No. 447; AA, Memo by Frohwein, 19 Oct 33, 7360/E537898-99. Frohwein's memo betrays some apprehension that he might be thought to have disclosed too much, and in writing of his prediction of withdrawal, he used the word "previous" (i.e., prior to the Monday General Commission meeting) and then crossed it out.

[74] Text: *FRUS, 1933*, I, 260-263. On the other delegates: *DBFP*, 2/V, No. 455; *FRUS, 1933*, I, 263-264, 266; *DDF*, 1/IV, No. 305.

9. Hitler's Diplomatic Triumph

The teletyped English text of Simon's speech arrived in Berlin at 11 a.m. At Geneva, Rheinbaben merely said, when his turn came to speak, that he would report Simon's remarks to his government, which would examine them in the light of the German demands for disarmament of the heavily armed powers, and for giving immediate practical effect to the principle of equality while leaving quantities open to negotiation. In Berlin, Hitler told a noon cabinet meeting that Simon would presumably strike a calmer note, but he added that the basic position remained the same; he therefore proposed to carry out the measures already planned. By this time, Hitler was doubtless set on withdrawal, particularly for domestic reasons, and it seems unlikely that knowledge of Hugh Wilson's scheme, or a dropping of the word "numerical," as suggested by Rheinbaben, would have changed Hitler's mind. Simon's shift of emphasis from stages to the overall length of the convention did not improve matters in German eyes, and Neurath—perhaps embarrassed personally by Simon's implication that he had acquiesced in the eight-year term—was to interpret the statement publicly as postponing disarmament for eight years.[75]

Goebbels announced the withdrawal to the Berlin press at 1 p.m., newspapers in Geneva were out with the news by 2 p.m., and Neurath communicated it to Henderson by a telegram received in Geneva by 3 p.m., just after the Bureau meeting. The announcement surprised and disoriented diplomats and newspapermen.[76] The counselor of the British embassy in Berlin telephoned Bülow at his apartment and asked excitedly if he or the ambassador could come over and get some information on the German withdrawal. The German embassies in London and Paris noted that although the newspapers the next day, Sunday, all reported the sensational news, they showed no evidence of governmental guidance. No immediate official response appeared from either capital, not only because of the weekend but also no doubt because both foreign ministers were in Geneva. The disarray in Paris was probably also aggravated by the fact that Hitler, in a radio broadcast on the evening of the fourteenth, made an unexpected and dis-

---

[75] AA, Note by Völckers, 14 Oct 33, 3154/D670080; tel., Delegation to Ministry, 14 Oct 33, 3154/D670082-88; *ADAP*, C/I/2, No. 499; *DDF*, 1/IV, No. 318. The Simon statement may also have been telephoned; this had been planned. I take Hitler's remarks to the cabinet, along with the timing, to indicate that he had been told of the general tenor of the advance copy of the Simon speech, and did not think it worth waiting to have a confirmed copy of the speech as actually delivered.

[76] *SIA, 1933*, pp. 305-306; *DDF*, 1/IV, No. 307; *DDB*, III, No. 61; *DBFP*, 2/V, Nos. 455, 458; Kimmich, *Germany and the League of Nations*, p. 189.

concertingly friendly reference to Daladier and to France.⁷⁷ Hitler did not, in this case, carefully choose a weekend for his move. The German announcement occurred on Saturday because Simon made his statement that day. But Western disorientation on this occasion must have encouraged Hitler to plan later "Saturday surprises" to take advantage of the weekend holiday custom.

Certainly the delegates in Geneva showed confusion. The British, American, French, and Italian representatives began to confer, and they continued to do so into the early hours of Sunday morning; then they gathered again before noon on Sunday and carried on until late that night. Simon, sounding rattled, telephoned MacDonald and asked to be summoned "urgently" home, but MacDonald refused to agree. Boncour and Massigli urged that the next step was to proceed immediately to draw up a convention without Germany, on the basis of Simon's statement. But Soragna indicated that his government now doubted there was any use proceeding on that basis; he said that Italy could envisage other procedures, meaning a meeting of four or five powers. Davis, too, apparently expressed doubts about the approach the French suggested; he still favored five-power talks, while he feared that the French were moving toward coercing Germany. Simon at first still hoped that the conference, in plenary session, might endorse his position, and show that the world was opposed to German rearmament, but the British representatives were skeptical about proceeding to write a treaty, and they favored adjournment. On Monday, October 16, the General Commission did in fact decide on a ten-day adjournment, which ultimately was extended until May 29, 1934, when the conference began its last, futile plenary sessions. Before adjourning, the General Commission voted approval of a letter from Henderson to Neurath, in which Henderson politely asserted that the conference had been on the verge of agreement, and concluded that he could not accept as valid the German arguments for withdrawal.⁷⁸ This was to be the only official censure Germany would receive.

Norman Davis, in his effort somehow to arrange a compromise and reach an agreement, had already privately dissuaded Simon from accusing the Germans of demanding rearmament. As a result, the spokes-

---

⁷⁷ AA, Note by Bülow, 14 Oct 33, 3154/D670093; tel., Forster (Paris) to Ministry, 15 Oct 33, 3154/D670102; tel., Hoesch to Ministry, 16 Oct 33, 3154/D670117; *SIA, 1933*, pp. 307-308. In Washington, Secretary Hull did provide an official reaction on 14 October, but the administration soon backed down from his mildly critical stance: Hooker, *Moffat Papers*, pp. 105-108.

⁷⁸ Roosevelt Papers, PSF, File 142, ltr., R. H. Leigh to Standley, 17 Oct 33; *FRUS, 1933*, I, 267-269, 270-272; *DBFP*, 2/V, No. 458; Czech docs., tel., Beneš (Geneva) to Foreign Ministry, 16 Oct 33, 2376/D497202-03; MDP, 8/1, Diary, 15 Oct [33]; Boncour, II, 385-386; AA, Phone report from Krauel (Geneva), 16 Oct 33, 3154/D670112; *SIA, 1933*, pp. 311-312; *SIA, 1935*, Vol. I, 33-58.

man of Great Britain, the pivotal nation in European politics, did not make the case against the Germans, and without a lead, the conference floundered. This was not the end of American influence. As noted, the Italian and American delegates continued to urge moderation toward Germany in the post-withdrawal discussions. Then, a few hours after Henderson's message to Neurath, the United States opened the first public breach in the Franco-British-American front. Now, however, the objective was no longer to perpetuate disarmament negotiation, but rather to avert American exposure to the stormy weather that appeared to be coming in Europe. Following charges in the Hearst press that the State Department was plunging the United States "into the ominous European imbroglio," Roosevelt informed Davis that the United States should not get involved in great-power conferences, or in meetings concerned with imposing a treaty on Germany. A statement was sent from Washington for Davis to issue publicly in Geneva, supposedly on his own initiative, denying any American involvement in "political" or "European" (as opposed to "disarmament") negotiations, and stating that the United States was "in no way politically aligned with any European powers"; this was given to the press Monday night. It ended any remote chance of a joint condemnation.[79]

Not having shown that the Germans were claiming a right to rearm, let alone that they were already rearming, Simon had failed to explain to his own countrymen why the MacDonald plan should be modified, or in particular, why French disarmament should be delayed. And alongside the older moral deprecation of Nazism, there was a sudden fear that too much of the crusading spirit might lead to calamity. The British government did not and could not immediately turn away from the position Simon had taken on the fourteenth, but Hoesch reported that the British press, even after it began to reflect official guidance on Monday, showed no demand for action. Some voices called with great effect for sympathy for the German position. Lloyd George repeatedly expressed this point of view in this period, and on the night of the fourteenth Vernon Bartlett, a commentator for the British Broadcasting Corporation, told his listeners that Ger-

[79] AA, Phone report from Krauel (Geneva), 16 Oct 33, 3154/D670112; Hooker, *Moffat Papers*, pp. 106-108; *FRUS, 1933*, I, 273-278. Reports to Berlin on the initial Hull statement (see note 77), the American press reaction, and official disavowal of American involvement: AA, tels., Luther to Ministry (4 tels.), 14, 15, 16 Oct 33, 3154/D670096-98, D670103, D670120-21. In the transatlantic phone conversation over issuing the statement on American nonalignment, Hull observed that the German state of mind made it impossible to deal with them in the ordinary way, and Davis replied: "That is perfectly true, but my information is that Hitler is the best one of the lot and this election is probably going to get rid of some of the worst part of his group. He certainly wants to make peace with France" (Davis Papers, Memo of phone conversation, 16 Oct 33; cf. *FRUS, 1933*, I, 276).

many "would fight if necessary with fists and pitchforks—in defence of what she sincerely believes to be honour and justice." Bartlett called for fair play for Germany and said, "The great majority of my listeners want peace and are fed up with politics. . . . It is worth swallowing any amount of pride if peace is at stake. Civilisation is more valuable than prestige."[80]

Hitler had shown some hesitation in deciding on withdrawal, and he had mentioned the danger of sanctions to his cabinet on October 13, saying that dealing with this threat was only a matter of keeping cool and remaining true to principles. His remarks convey a forced, zero-hour bravado, and when, in his Saturday night broadcast, he made a public bid for Franco-German friendship and direct negotiation, he probably did this mainly as an extra precaution against a French reaction. Although the odds were favorable, the withdrawal was a gamble, the first serious gamble Hitler had taken in foreign affairs. But by the afternoon of Tuesday, October 17, it was clear that the gamble had succeeded. Then, with the confidence of hindsight, Hitler jubilantly told his ministers that the political situation had developed as was to be anticipated:

> Threatening steps against Germany had neither taken place nor were they to be expected. Already in the note of reply sent to us by the President of the Disarmament Conference, the internal conflicts between the leading powers in the Disarmament Conference were evident. Germany could now let events take their course. No step by Germany was necessary. Germany was finding herself in the pleasant situation of being able to watch how the conflicts between the other powers turned out.

Along with this jubilation, the ministers did also authorize Blomberg to take some precautions against the disclosure of secrets, and to prepare plans against the danger of military sanctions.[81] But Hitler had won a diplomatic victory. The feeble tone of Henderson's note

---

[80] AA, Hoesch to Ministry, 16 Oct 33, 3154/D670117-19; Cab 24/244, CP 252 (33), 1 Nov 33. The latter document, quoting from Bartlett's broadcast, was a paper from Simon suggesting that such "terrible power" as Bartlett had wielded should be under direct government control. See also *DDF*, 1/IV, No. 32. Bartlett's broadcast ended his career as a regular BBC broadcaster, although 90 percent of the letters he received supported his views: Martin Ceadel, "Interpreting East Fulham," in Chris Cook and John Ramsden, eds., *By-Elections in British Politics* (London: Macmillan, 1973), p. 128. In 1938, he won a by-election as an "anti-Munich" candidate.

[81] *ADAP*, C/II/1, No. 9; cf. No. 39. Blomberg's 25 Oct 33 directive, prescribing resistance to attack (*Nuremberg Documents*, XXXIV, 487-490) may thus be a result of this meeting. As shown in Chapter Seven, Section 3, an intention to resist attack already existed.

did indeed give evidence of disunity among the Western powers. Hitler had received other indications of disunity too, and some of them probably counted for more in his own mind: the American disclaimer of any alignment, the Italian interest in a discussion between the powers, and in particular, the position taken by Lloyd George. Hitler had long regarded Lloyd George as a skillful politician and a pioneer propagandist, and he probably considered him to be a "true" leader of British opinion, as opposed to Simon, who in this period was often (mistakenly) described in the German press as Jewish.[82]

Like the British Colonel Nicholson in Pierre Boule's *The Bridge on the River Kwai*,[83] Hitler had outfaced his opponents and established his ascendancy. He knew that he would soon gain a further psychological advantage by winning overwhelming support in a November 12 plebiscite on his foreign policy. The approval of 93.4 percent of the voters was to be obtained in part by coercion and falsification, but the announced vote was to give the impression that most of the German public supported a withdrawal from Geneva.[84] Already, by

[82] On the Italian position, also *ADAP*, C/I/2, No. 502; *SIA, 1933*, pp. 309-310; Mussolini's later irritation (see, e.g., *ADAP*, C/II/1, No. 18) was not yet apparent. Hitler's impression of American nonalignment may have been strengthened during his initial reception, earlier on 17 Oct 33, of William E. Dodd, who agreed with him that there was "evident injustice" in the French attitude: William E. Dodd, Jr., and Martha Dodd, eds., *Ambassador Dodd's Diary, 1933-1938* (New York: Harcourt, Brace, 1941), p. 50. The Propaganda Ministry warned journalists on 17 October that Davis's statement had not been confirmed: Brammer Collection, Z Sg 101/1, Directives of 17 Oct 33. But this warning was probably issued in ignorance of Luther's reports of 16 October, which arrived on the morning of the seventeenth (3154/D670120-21); also cf. *FRUS, 1933*, I, 285. On view of Lloyd George, see *Mein Kampf*, pp. 533-534; *Secret Conversations*, pp. 212, 470-471, 534, 550. The report of Lloyd George's 1936 visit with Hitler (Gilbert, *Roots of Appeasement*, Appendix 2) is very revealing. On allegations that Simon was Jewish: *FRUS, 1933*, I, 304; *DDB*, III, No. 61. In late November, in a letter which also expressed interest in a meeting with Hitler, Simon told Phipps that he was not Jewish, and hinted that the German Propaganda Ministry might be informed; Phipps advised Simon against seeking a meeting too soon after the attacks on him, but did get the Propaganda Ministry to instruct the German press on Simon's non-Jewish origins: FO 800/288, Simon Papers, ltr. to Phipps, 27 Nov 33, and reply, 7 Dec 33; Brammer Collection, Z Sg 101/2, Instruction No. 119, 15 Dec 33.

[83] (New York: Vanguard, 1954).

[84] The percentage given here was reported by the American embassy (*FRUS, 1933*, II, 265), and represents a contemporary non-German perception. Bracher, in "Stufen der Machtergreifung," pp. 348-368, analyzes the plebiscite and the means of coercion and falsification employed. As he considers abstention oppositional, he compares "Ja" votes with the total of those entitled to vote, rather than the total of actual voters, and for this, using data from the final report of the Ministry of Interior, now in the Hauptarchiv, Berlin-Dahlem, he arrives at a national percentage of 87.8 percent. The corresponding American embassy figure was higher: 89.95 percent. Therefore the percentage in my text, of "Ja" votes to total actual votes, is no doubt also a little higher than the final Ministry of Interior data would show.

October 17, the date of Hitler's jubilant cabinet statements, the successful diplomatic coup bore out the correctness of his judgment, as compared with the hesitations of Foreign Ministry officials.[85] The success also strengthened Hitler's hand for dealing with Röhm and the SA, and on this same day, the Führer—probably with reference to his recent cancellation of weapons training in SA-Sport courses—flatly told the SA leaders that the military training of their following brought a risk of disclosures to the enemy. Röhm reacted with a threat of barring his followers from training within the army, that is, from pilot courses for the projected militia program. But now, with Hitler's prestige involved, Reichenau could retort that Röhm's projected action would make the training of the SA in special courses pointless and the training in the Grenzschutz impossible[86]—implying that the SA would lose its army instructors and indeed any role in national defense.

For that matter, Hitler's success also gave him new stature and independence vis-à-vis the Reichswehr leadership itself. Although Blomberg had apparently encouraged Hitler to withdraw, Hitler made the move in his own way, claiming to reject force, to seek disarmament, and to be ready for negotiations. Some Reichswehr officers would apparently have preferred an open assertion of Germany's right and intention to rearm. Ostensibly to avoid an impression of limitless rearmament—and actually, in all likelihood, to ensure an end to clandestinity and diplomatic hobbles—General Schönheinz had suggested on October 12 an announcement that Germany would rearm to a limited extent, along with a promise to destroy all arms that others might later agree to ban. In May 1934, in deploring proposals for accelerating the clandestine expansion of the army, General Ludwig Beck, then Chief of the Truppenamt, would write: "The failure on October 14 simultaneously to lay the card of rearmament on the table [i.e., alongside that of withdrawal] is least of all, in my opinion, to be made good in this way." The remark suggests that Beck and other officers thought Germany should have frankly and openly claimed its right to rearm in October 1933.[87] But they had to do things Hitler's way now.

[85] On 23 Oct 33—well after the withdrawal—a Foreign Ministry paper did assess French action as very unlikely: NA-VGM, F VII a 11, [Unsigned], 7792/E565652-56.
[86] On this, see *Nuremberg Documents*, XXIX, 10-12; Wollstein, *Weimarer Revisionismus zu Hitler*, pp. 203-206; Sauer, "Mobilmachung," p. 932.
[87] AA, tel., Schönheinz to Obstfelder, 12 Oct 33, 3154/D670062-65; Bundesarchiv-Militärarchiv, Ludwig Beck Papers, N28/1, Memo by Beck, 20 May 34. The Beck sentence is omitted from a quotation in Wolfgang Foerster, *Ein General kämpft gegen den Krieg: Aus nachgelassenen Papieren des Generalstabschefs Ludwig Beck* (Munich: Münchner Dom Verlag, 1949), p. 22, also in Gert Buchheit, *Ludwig Beck: Ein preussischer General* (Munich: List Verlag, 1964), p. 44.

10. Hitler's New Armament Proposals

Hitler soon gave a diplomatic expression to his new self-confidence and independence; he was also impelled to take new steps to keep this independence from becoming isolation. His October 14 bid for French friendship led to nothing: Daladier stated in reply on October 17 that France was "deaf to no appeal, but blind to no act." If Hitler had ever seriously hoped to negotiate with France, and not merely to impede a French reaction, he now deferred that hope.[88] Instead, having extricated Germany from the toils of Geneva, he would seek to destroy Geneva as a focus for the other powers. He would bring the pressure of other powers to bear on France. He would seek to bring England, in particular, over to his side. And he would gain more time for secret rearmament. In return for not asking others to disarm, he now asked for more than he would have received under the MacDonald plan.

Hitler gave his new demands their first diplomatic airing in a meeting with the new American Ambassador, William E. Dodd, on October 17. This was probably an impromptu trial exposure of his thoughts, and in any case Dodd, whose German was rusty and who lacked experience in disarmament, diplomacy, and the ranting of the German dictator, barely took anything in. When Dodd checked back with Neurath three days later to find out what Hitler had said, he was told that Hitler had asked for an army of 200,000 men in five years, and 300,000 in eight.[89]

Then, on October 24, Hitler spoke for the first time with the new British Ambassador, Sir Eric Phipps. The Führer argued that "the highly armed states"—he named France, Poland, and Czechoslovakia —had shown their unwillingness to disarm, leaving Germany only the

---

[88] On Hitler's overall position: Geyer, "Militär, Rüstung und Aussenpolitik," p.251. An article in the 15 Oct 33 *Sunday Chronicle* (Manchester), purportedly by Hitler, argued that Britain and Italy were Germany's natural allies, while France was seeking European hegemony. Apparently fabricated from extracts from *Mein Kampf*, the article did correspond with Hitler's long-term views. Léger, at the Quai d'Orsay, spoke of the article to a German diplomat in Paris, who inquired at Berlin and was informed that a *démenti* had been issued. He was told on 21 October to inform Léger of this, but significantly, he was also told not to seize the opportunity to pursue with Léger the possibility of Franco-German negotiations: AA, tel., Forster (Paris) to Ministry, 19 Oct 33, 7360/E537929-31; *Démenti*, 17 Oct 33, 7360/E537932; tel., Ministry to Paris embassy, 21 Oct 33, 7360/E537934.

[89] *FRUS, 1933*, II, 396-397; SDP, tel., Dodd to Dept., 20 Oct 33, 500.A15A4/2260; *Dodd's Diary*, pp. 48-51, 63. The Dodd-Hitler meeting preceded Hitler's meeting with his ministers. Neurath was present, but seems to have left no record (see *ADAP*, C/II/1, p. 12). Neurath also told Dodd on 20 October that Germany would renounce "offensive weapons," including large planes and tanks over six tons. If Hitler (or Neurath later) mentioned a linked demand for a ceiling on the "heavily armed states," Dodd did not report that point. Hitler later advised Mussolini that he had informed Dodd: *ADAP*, C/II/1, No. 40.

alternative of securing enough armament to defend herself. He proposed eight-year agreements under which the "highly armed states" would not increase their armament, while Germany would accept specific limits: an army of 300,000 with one-year service, no artillery over 15 cm or tanks over 6 tons, and no bombers. All parties would agree not to use gas or aerial bombing against civilians. Hitler also spoke of the economic danger to Germany of Russian competition, and he said that he wanted "certain possibilities of expansion in Eastern Europe," although he denied any wish to alter the Polish Corridor by force.[90] Thus Hitler hinted at his plan for eastern conquest, for which he hoped to get British acquiescence.

The experts of the Reichswehr and the Foreign Ministry had not approved these proposals. For the army, a combination of a limit to 300,000 with one-year (instead of three-month) service would mean a restriction in the possibilities for rapidly training reserves.[91] But Hitler believed in thorough training, and he wanted to propose an army size which Britain, Italy, and the United States might accept, thus isolating France. On October 26, he sent Göring to Rome, armed with a personal letter to Mussolini and also with a list, provided by Blomberg, largely repeating what had been said to Phipps. In a speech in Berlin on November 6, Neurath stated that the German government was making "an honest and trustful offer" to the other great powers. Blomberg himself also talked with Phipps on November 20, and tried to ascertain the British reaction to Hitler's proposition.[92] Hitler only got around to discussing the proposal with François-Poncet on November 24, when he also suggested that Franco-German rela-

[90] *ADAP*, C/II/1, No. 23; *DBFP*, 2/V, Nos. 485, 489.

[91] When informed of what had happened, Schönheinz expressed surprise; such ideas had been discussed in his ministry, and Blomberg might have talked them over with Hitler, but he, Schönheinz, had not known of anything definite. Hitler's proposals did in fact resemble somewhat those which Blomberg had given to Brinon in September, but they were contrary to the longstanding demand that Germany should have everything allowed to the other powers, and in particular, they fell short of Schönheinz's goal of quantitative equality. Three years later, lecturers at the War College attributed the 300,000-man-army plan to Hitler's political requirements, and described the plan as a "political limitation," with which it had been necessary to reconcile the military need for a peace army so organized and composed that an adequate war army could be developed within reasonable time. The problem lay presumably in the number of reserves that could be trained with one-year service and a 300,000-man ceiling. In the Foreign Ministry, Bülow advised selected German embassies that Hitler's remarks had been a spontaneous reply to a question and did not represent a finished German proposal; Bülow's main concern seems to have been that the British might circulate the proposals as evidence of a German intent to rearm. See *ADAP*, C/II/1, Nos. 26, 29; Liebmann MS Papers, Army Area V briefing, 15 Jan 34, p. 210 (typed); OKH, H1/324, Briefings by Majors Ochsner and Bühle, Oct-Nov 36, 395/6397904, 6397920-22.

[92] *ADAP*, C/II/2, Nos. 40, 45, 50; *SIA, 1933*, pp. 312-313; cf. Kuhn, p. 147.

tions would be improved if France would cede the Saar area to Germany without awaiting the plebiscite scheduled for 1935.[93]

Hitler's offers of negotiation and Neurath's November 6 speech struck British leaders as a very welcome development. The East Fulham by-election on October 25 has sometimes been thought to have forced the government's hand by showing an overwhelming public insistence on further efforts for disarmament, but two days before it, the cabinet had endorsed Foreign Office recommendations that Britain should still seek to limit and reduce armaments, and oppose the French policy of continuing the conference without Germany, keep in contact with Germany to forestall German rearmament, and encourage talks between the powers.[94] Individuals like Eden and Allen Leeper, who had recently advocated "forcing the Germans into the open," now changed their tune. When Hitler's proposals to Phipps arrived at the Foreign Office, these proposals did not win acceptance as such, but Foreign Office officials quickly drafted a new "compromise plan," which would have conceded to Hitler most of his apparently modest demands for "defensive weapons," while trying to induce him to forgo his increase in effectives by carrying out reductions in French armament. Eden's and Leeper's Francophilia was not as strong as their commitment to disarmament, which would eventually lead, in January 1934, to yet another British proposal.[95]

In November, however, this "compromise plan" had a cold recep-

[93] In answer to questions from the French ambassador, Neurath may also have outlined the terms of the armament proposal to François-Poncet on 9 Nov 33; Poncet's reports indicate that this occurred (*DDF*, 1/IV, No. 413; *DBFP*, 2/V, No. 29), while Neurath's record implies it did not (*ADAP*, C/II/1, No. 54). See also *ADAP*, C/II/1, Nos. 61, 62, 86, 100, and *DDF*, 1/V, Nos. 52, 61. By ceding the Saar, France would of course have lost a valuable lever on Germany, and let down the local opponents of reunification.

[94] Cab 24/243, CP 240 (33), 20 Oct 33; Cab 23/77, meeting of 23 Oct 33, 54(33)1. The view that the East Fulham election was pivotal has been stated most explicitly by Middlemas and Barnes, pp. 744-747. Cf. Richard Heller, "East Fulham Revisited," *JCH*, 6 (1971), no. 3, 172-196; C. T. Stannage, "The East Fulham By-Election," *Historical Journal*, 14 (1971), 165-200; Ceadel, pp. 118-139. The Australian spokesman, Stanley Bruce, probably helped to inspire the cabinet decision: MDP, 1/533, Note for MacDonald from Hankey, 21 Oct 33 (copies to Baldwin and Simon).

[95] On pre-14 Oct 33 positions: Eden, pp. 45-47; FO 371, Minutes by Leeper, 11 Oct 33, W11462/40/98, W11472/40/98. Reaction to Hitler proposal: FO 371, Minutes by Stevenson, Leeper, M[ounsey], 25 Oct 33, W12113/40/98. That the Hitler proposal largely led to the Foreign Office's compromise "Draft Plan" (Cab 24/244, CP 255 [33], 1 Nov 33) was stated in a 9 Nov 33 minute by Leeper (FO 371, W12762/40/98). On Eden's support of the new plan: FO 371, Minute by Eden, 15 Nov 33, C9893/245/18. At his suggestion of 9 Nov 33, a short version was drawn up: FO 371, Minute by Leeper, 5 Dec 33, C10605/245/18. There is a clear continuity between the "Draft Plan" and the British memorandum on disarmament of 25 Jan 34; see also *DBFP*, 2/VI, Nos. 54, 55, 57, 105, 157, 206, Appendix I; FO 371, Minute by Eden, 12 Dec 33, C10740/245/18.

tion from the cabinet. Isolationist as well as pro-disarmament currents were affecting public opinion, and Lord Beaverbrook's papers advocated a denunciation of the Treaty of Locarno. The cabinet rejected this, which would (in MacDonald's words) have been "a bombshell on the mind of Europe." But senior ministers like MacDonald, Neville Chamberlain, and Baldwin were unhappy with the conference, and thought that Britain, and also France, should get into direct touch with Germany; the first two, at least, tended to criticize Simon. MacDonald worried lest Europe be divided. Chamberlain found the main German objections reasonable, paraphrasing them as follows:

> You have insisted on our undergoing a 4 years period of penance and humiliation during which we are to remain unarmed while everybody else remains armed and when it is over the French General Staff will find some excuse for saying that they are released from their obligations and we shall be no nearer disarmament than before.

Simon himself remarked at a ministerial disarmament committee meeting that since the Germans had never rejected the original March 1933 MacDonald plan, but only the recently proposed modifications, it was wrong to say that it was useless to go on trying to negotiate. Increasingly, the British government took the position that it had never abandoned the original plan, and that Simon, on October 14, had only reported on the talks he had had, and on the sort of amendments which "certain powers" might move.[96]

Boncour and his aides did not fully divine the trend of British policy, but unsurprisingly, they did not like what they perceived of it, or of American or Italian policy either. They resisted the idea that they should negotiate directly with Berlin on disarmament, or enter into great-power talks, and they continued to think that the proper course was to complete a convention without Germany. They were especially unhappy over the continuing refusal of the British government to discuss the French dossier. Daladier's cabinet fell on October 24, and in the two short-lived succeeding cabinets of Albert Sarraut and Camille Chautemps, Boncour, carried on as Foreign Minister.

[96] Cab 27/505, DC (M) 32, meeting of 7 Nov 33; Cab 23/77, meetings of 2, 6, 15 (2) Nov 33, 59(33)3, 5, 60(33)1, 2, 62(33)2, 63(33)1; *DBFP*, 2/V, No. 487; MDP, 8/1, Diary, 18, 19, 20, 22, 23, 26 Oct, 5, 10, 17, 22, 30 Nov, [8?], 11, 17 Dec [33]; Neville Chamberlain Papers, ltr. to Hilda Chamberlain, 18 Nov 33, 18/1/851. MacDonald and Baldwin talked privately with the German ambassador, and MacDonald intimated that he disagreed with recent Foreign Office policy; he also suggested a Hitler visit to London, an idea Neurath characterized as "absurd." See *ADAP*, C/II/1, Nos. 57, 59 (and note), 76; RK, tel., Hoesch to Foreign Ministry, 15 Nov 33, 3650/D813171-74; *FRUS, 1933*, I, 272-273, 290-291; Marquand, p. 756.

On October 31, he began to sound the Soviet Union as to the possibility of closer ties, and he stepped up his efforts in this direction in late November, following a disappointing outcome to talks with Britain and Italy in Geneva.[97]

It is not certain how far Boncour would have gone against Germany if he had had full British and American support. Léger told the Czech minister in Paris on October 21 that a preventive war would be "madness," and that he was seeking a common front of the former allied great powers. Some other statements, by Boncour and Weygand, implied a willingness to go to the League Council, and even to invade Germany, at least if British support could be secured.[98] But as already indicated, invasion or acquiescence were not the only possible alternatives; an exposure and condemnation of German armament by the remaining League powers and the United States would have rallied opinion in those countries, and might have embarrassed Hitler, casting doubt on his competence. Also as Hitler well realized, the French still held a valuable hostage in the shape of the Saar. And some non-military collective action might have been possible: in December, the French would suggest economic and financial sanctions for the event of violations of an arms agreement,[99] and such procedures, while certainly difficult, as events in 1935 would show, might have succeeded against Germany if Western leaders had had the courage to support them.

[97] William E. Scott, *Alliance Against Hitler* (Durham, N. C.: Duke University Press, 1962), pp. 132-134; *DBFP*, 2/VI, Nos. 5, 14, 23, 27, 40, 45, 53, 54, 64, 92; *DDF*, 1/IV, Nos. 316, 323, 327, 338, 355, 360, 367, 368, 375, 377, 378, 379, 386, 393, 395, 397, 400, 402, 405, 408, 409, 412, 419, 422; 1/V, Nos. 4, 5, 18, 20, 23, 24, 28, 32, 33, 38, 64, 66, 67, 68, 69, 71, 73, 76, 79, 84, 88; *FRUS, 1933*, I, 306-319; M. Andreyeva and L. Vidyasova, eds., "The Struggle of the U.S.S.R. for Collective Security in Europe during 1933-1935," *International Affairs* (Moscow), 9 (1963), no. 6, 108-109.

[98] Czech docs., tel., Osusky to Foreign Ministry, 21 Oct 33, 2376/D496970; *DBFP*, 2/V, No. 508; 2/VI, Nos. 3, 53, 64; *DDF*, 1/IV, No. 368; 1/V, No. 38; *FRUS, 1933*, I, 312. Léger, in talking with Osusky, said that without British support France would either have to follow an independent policy or reach a direct agreement with Germany. As earlier, the French seem to have deliberately hinted they might make a composition outside the League with Germany, but informed observers considered this unlikely: see, e.g., *DBFP*, 2/VI, No. 3; *FRUS, 1933*, I, 280. The American tendency to pull back from Europe doubtless gained force from reports that French preventive action was fairly likely; see *FRUS, 1933*, I, 271, 279-281, 289, 301. The American military attaché in Paris reported, however, that French general-staff policy was not to occupy a foot of foreign territory unless France were attacked: SDP, tel., Marriner to Dept., 19 Oct 33, 500.A15A4/2290.

[99] See *DDF*, 1/IV, No. 324; 1/V, Nos. 85, 109; also *DBFP*, 2/VI, No. 172 (and note). Corbin had suggested at the end of September that provision for economic and financial measures would satisfy the French demand for sanctions: MDP, 1/504, Note by Vansittart for MacDonald, 29 Sep 33. Léon Blum seemed to think the powers might agree on economic sanctions: FO 800/288, Simon Papers, ltr. from Tyrrell, 11 Oct 33.

496—Hitler's Hand is Forced, and He Frees It

As matters stood in November 1933, however, the disunity of the West was all too apparent, and this encouraged great expectations in Berlin. Blomberg happily pointed out to his fellow ministers that the British press was showing sympathy for revision, and he suggested that each ministry should be preparing its demands; the Reichswehr's demand was simple, the removal of Part V. Encouraged also, no doubt, by reports from Joachim von Ribbentrop of his discussions with British leaders, particularly with Baldwin and MacDonald, Hitler's hopes for an Anglo-German understanding seem to have risen. Presumably with his sanction, Admiral Raeder told the British naval attaché on November 29 that neither the navy nor the politicians of Germany wished to build a fleet in competition with England. Raeder also expressed the idea, which he believed to be shared by the "political leadership," that, in view of the parity between the British and American navies, a squadron of German battleships, accompanied by a good Anglo-German understanding, might signify "a political plus for England." On December 5, Hitler himself spoke again with Phipps and, aside from answering some questions on his armament proposal, he stated that Germany must "be in a position to throw her weight into the scales at some future time, and, in this connection, might not Great Britain herself be glad of other alternatives to her present friendships?" He also hinted at an Anglo-German alliance by saying that the standstill he proposed for the "heavily armed" powers only applied to France, Poland, and Czechoslovakia, and that as far as Britain was concerned, "he would even welcome considerable additions to the British fleet and air force." Phipps could say nothing as yet on his government's views on Hitler's armament proposals, much less on the bids for an alliance.[100]

11. Hitler Proceeds with His Own Plan

Hitler did not confine himself to arms proposals and bids for a British alliance; he now also secretly put the German side of his arms proposal into actual effect. At some date between November 30 and December 9, he approved the abandonment of previous Umbau plans, largely inherited from the Schleicher era, and ordered instead the actual development within four years of a twenty-one-division, 300,-000-man peacetime army based on one-year service. This entailed

[100] RK, Cabinet Protocol, 1 Dec 33 (4:15 p.m.), 3598/D794391-92; Middlemas and Barnes, pp. 748-749; Marquand, p. 756; Dülffer, pp. 267-269; DBFP, 2/VI, Nos. 97, 99; ADAP, C/II/1, No. 99. Blomberg had also assured Phipps that Germany would never again build a fleet against England: DBFP, 2/V, No. 67.

sweeping changes in the peacetime forces and wartime mobilization plans. As far as land forces were concerned, the Reichswehr now dropped the idea—still not generally implemented—of three-month, so-called militia service in the ranks. The army also turned away from the concept of a uniform trebling in the event of mobilization, and instead proposed to mobilize two different grades of troops, thirty-two or thirty-three first-class field divisions and thirty "occupation" or Landwehr divisions: thus there was some reversion to the ideas of General von Seeckt. There would also be cavalry, armor, and other formations. For the time being, the Grenzschutz would be retained, although its future was unclear. Much more certain was the eventual introduction of compulsory service, which the Organization Section of the Truppenamt proposed to initiate on October 1, 1934. General Beck, however, wanted to draw on willing recruits so long as possible; those who didn't want to be soldiers would be placed in the Labor Service, where they would become potential second-class reserves.[101]

The diplomatic situation in early December probably encouraged Hitler's decision to implement his 300,000-man proposal. There was little danger. Poland and Germany had jointly issued, on November 15, a declaration renouncing the use of force. This, the British willingness to negotiate, and Poncet's personal readiness for talks probably convinced Hitler that preventive action could now be entirely ruled out.[102] With his proposals already known to other governments, their implementation, if detected, would show that he was not to be

[101] BA-MA, RH 15/8, TA, T2 1113/33,* 14 Dec 33; Chef HL, TA 381/33,* 18 Dec 33; WKK, WK VII/4082, Discussions in Berlin,* 21-22 Dec 33, 75/1062-1083; ML, PG 34108, Chef ML, B Nr A Va 6020/33, 14 Dec 33, reel 32/290-292; B Nr A Va 6240/33, 22 Dec 33, reel 32/321-326; Liebmann MS Papers, Army Area V briefing, 15 Jan 34, pp. 209-211 (typed). Among the above documents, those listed with asterisks (*) are now available in published form in Rautenberg, "Drei Dokumente," pp. 115-127. Speaking of the accumulation of reserves under the new plan, Liebmann said: "Faster with 3 month service, but better this way." Material in Chef HL, TA 381/33 and B Nr A Va 6240/33, also BA-MA, RH 15/8, TA 355/33, 21 Nov 33, indicates the termini post and ante quem for the decision to proceed; see also Rautenberg, "Drei Dokumente," p. 131 (n. 93). Landwehr, like field, divisions would take cadre from the peacetime army, but with a much higher proportion of reservists to cadre. The navy retained the three-month plan for personnel for coastal and land air defense. Respecting Hitler's responsibility, Rautenberg, in "Drei Dokumente," p. 107, points out that no proof has been found in the sources to show that Hitler gave Blomberg an express order for a 300,000-man army, but Rautenberg does not here take account of *ADAP*, C/II/1, No. 26, or of the 1936 briefings by Majors Ochsner and Bühle (OKH, H1/324, 395/6397904, 6397920-22). See note 91. Hitler's involvement doubtless explains the suddenness of the shift and the lack of intra-departmental discussion, which Rautenberg now (p. 113) finds baffling. Cf. Rautenberg's *Rüstungspolitik*, p. 306n.

[102] On the Polish-German declaration: Wollstein, *Weimarer Revisionismus zu Hitler*, pp. 271-277. Re Poncet, see Rautenberg, *Rüstungspolitik*, p. 306.

played with. If the execution of the new plan became publicly known, Hitler could still claim that his aims remained modest, and that he had had to go ahead to protect German security.

And there was little to gain by waiting. The British did not respond to Hitler's overtures for an alliance, and on December 8, Phipps passed on the view of his government that 300,000 was too high a figure. But Hitler probably took the British rebuff as a consequence of German weakness, and inferred that the British would only respect him and regard him as a worthwhile ally (bündnisfähig) after German armed power had been substantially increased.[103] He probably concluded, too, that although France would not act against German rearmament, there was no chance of her agreeing to it, at least unless Daladier became Premier again. The Paris government had lately seemed to encourage anti-German press revelations, and on December 5, Boncour sent Poncet a message expressing flat opposition to German rearmament and to any bargaining for a cession of the Saar to Germany; Hitler may well have had a decryption of Boncour's message, from either French or British telegrams, when he was deciding to proceed.[104] In view of the French attitude, Hitler probably now thought it a waste of time to await an agreement before going ahead with the 300,000-man plan. No diplomatic advantages could be gained by pursuing an outdated militia program, designed to maintain a pretense of continued German disarmament.

The new army plan probably served military and domestic purposes even more than diplomatic ones. Tired of waiting for an arms convention, Blomberg and Beck wanted to move forward, out of what Beck called—with some exaggeration—"the state of complete defenselessness." This would mean ending the diffusion of army cadres into external training, and concentrating instead on giving effective training within the army itself, so as to accumulate reliable reserves as rapidly as possible.[105] The ad hoc emergency training programs with

---

[103] See Kuhn, pp. 148-149, 165-166. Blomberg told the generals on 2 Feb 34 that Hitler's aim was to secure peace for a number of years for reconstruction and rearmament, not with the idea of then attacking someone, but of enabling the Reich "to intervene more actively in grand policy": Liebmann MS Papers, p. 212 (typed).

[104] See Liebmann MS Papers, Army Area V briefing, 15 Jan 34, p. 208 (typed); Blomberg briefing, 2 Feb 34, p. 212 (typed); Weinberg, *Foreign Policy of Hitler's Germany*, pp. 171-172. In explaining the new program to area commanders in December, Blomberg began by saying that it was unlikely a disarmament convention would be achieved; if not, Germany must proceed independently: WKK, WK VII/4082, Discussions in Berlin, 21-22 Dec 33, 75/1062. Possible intercepts: *DDF*, 1/V, No. 81; *DBFP*, 2/VI, No. 104. On German reading of French and British telegraphic traffic, see Chapter One, note 102, and Chapter Nine, note 50.

[105] ML, PG 34108, B Nr A Va 6240/33, 22 Dec 33, reel 32/321, rather implies that Blomberg proposed the new plan to Hitler. On general views of Blomberg and Beck: WKK, WK VII/4082, Discussions in Berlin, 21-22 Dec 33, 75/1062, 1064-1065.

the SA had proved unsatisfactory, and Röhm's ambition to build a new SA army was intolerable. Historians, while noting earlier portents, have usually considered the end of February as the date when Hitler decisively limited the role of the SA in national defense.[106] But he and Blomberg took the critical step in reducing the SA's role —which, moreover, was never as important as sometimes supposed— when they decided, in early December, to proceed with the 300,000-man army. To officers from the area commands, Blomberg explained frankly that two problems remained: the regulation of the Grenzschutz and the efforts of the SA to establish an armed force of its own. Blomberg added that Hitler agreed with him that, aside from premilitary training, everything, including leadership, mobilization preparations, and military training, should be in the hands of the Reichswehr: "Leadership in wartime can only rest with the Wehrmacht." The new plan put these principles into effect by carrying out the measures Reichenau had hinted at in October: it virtually ended the assignment of Reichswehr officers and training detachments to the SA, thus freezing the SA out of serious military training. Wehrsport would continue, but it was to be deemphasized: all army assignments to SA Wehrsport camps would end on March 31, 1934. The SA would be removed from army training grounds. Those cadres which had been sent to give three-month training within the Grenzschutz would be recalled, effective April 1. Generally, the new plan, with its one-year service, was designed, as a directive put it, to build up the new peacetime army "on a sound basis"; "only the army should remain as sole bearer of arms and highest means of power of the state authority."[107]

Army leaders had hoped, from 1930 on, to make use of the "good elements" in the SA. They had expected the SA and the other Wehrverbände to give young Germans a premilitary preparation for three-month or emergency training, to furnish reserves for mobilization in case of need, and generally to encourage a "national" spirit. It is hardly too much to say that the Weimar Republic was sacrificed for these hopes. But the results were disappointing, and the intended tools

---

[106] On disillusionment with the SA: Müller, *Heer und Hitler*, pp. 91-96. Stress is placed on the 27-28 Feb 34 events in Sauer, "Mobilmachung," pp. 941-945; O'Neill, *German Army and Nazi Party*, pp. 35-43; Müller, *Heer und Hitler*, pp. 94-100.

[107] WKK, WK VII/4082, Discussions in Berlin, 21-22 Dec 33, 75/1062-1063, 1068-1070, 1082-1083; BA-MA, RH 15/8, TA, T2 1113/33, 14 Dec 33. Instructors were to remain at the SA's Leaders' School until October. The recall of Grenzschutz training battalion cadres was decided after the issuance of TA, T2 1113/33. Army leaders had already hoped to reduce sharply army involvement in SA-Sport after March 1934, and to this end they had stressed the training of SA leaders, including in short courses in the army: WKK, WK VII/1453, Wehrkreiskommando (Bayer.) VII 7. (Bayerische) Division Ib/Ia 1679, 5 Oct 33, 37/1116.

had wills of their own. The leaders of the Reichswehr had realized for some time now that the SA might become a rival, and they would literally go to any lengths to prevent this. Perhaps the Reichswehr exaggerated the danger of an SA army, which Hitler never supported. But the military really did have serious problems in working with the SA.

One of the main reasons for working through the Wehrverbände had been that they would supposedly supply their own instructors. But at least initially, it had proven necessary to detail regulars for training purposes; apparently most SA leaders were not qualified even to give Wehrsport instruction. Nor were they fit to serve as military commanders in the Grenzschutz. Yet they wanted to retain their posts. Bavarian army area documents from the fall of 1933 show SA leaders asserting their claim in case of mobilization to remain in charge of their formations, which were now, in that event, to be taken as closed units into the Grenzschutz; at a minimum, they wanted the leadership function to be divided between themselves and military leaders. But the army ruled out such a division in December, only conceding that lower-level posts might be designated for qualified SA leaders.[108] Since Hitler and the generals wanted an effective military force, they could not let commands be assigned on the basis of position and service in a party formation.

Probably even more serious was the vulnerability of the Manstein three-month training plan to Röhm's obstruction. In the fall of 1933, military leaders had still planned to conduct three-month training within the army on a large scale, gaining recruits for such training from "the Wehrverbände"—i.e., from the SA. As noted, after Röhm had threatened to withhold men from training in the army, Reichenau retorted that this would mean the end of the special courses and Grenzschutz training. Available evidence does not show whether in fact an execution of Röhm's threat preceded the now-known execution of Reichenau's. But army leaders must have considered the danger. As long as there was no conscription, and as long as Röhm retained an autonomous position as head of the amalgamated Wehrverbände, he could block the flow of men into a three-month training program within the army. From his point of view, moreover, he had almost every reason to do this: the army's so-called militia scheme, along with

[108] Earlier hopes that the SA would supply its own instructors: OKW, Wi/IF 5.498, T2 III A 1207/31, att., 25 Nov 31, 110/837784-85. On Grenzschutz problem: WKK, WK VII/4082, Discussions in Berlin, 21-22 Dec 33, 75/1062, 1070; WK VII/182, Wehrkreiskdo. VII Ib 2288, 14 Dec 33, 10/1266-1268; Kommandantur Regensburg, Record of discussion, 14 Nov 33, 10/1271-1273; ltr., Kraus and Baumgärtl to "Sehr geehrter Oberst," 9 Nov 33, 10/1274-1275; the last document reported the angry reaction of SA leaders to the removal of an unqualified cohort.

its mobilization plan, would make the SA a mere recruiting organization and frustrate any attempt to set up a genuine, independent SA militia under SA leaders. But in army eyes, Röhm's threat to withhold men doubtless appeared as confirmation that he was trying to build up an armed force of his own. Three-month training would have provided the fastest accumulation of reserves, but it was now even more important to eliminate, finally, dependence on the SA, and to establish beyond question the exclusive claim of the army to defend the Reich.[109]

Hitler also had his own position to defend. Röhm's ambitions and his autonomous power seemed to threaten Hitler's supremacy, and it may be no coincidence that Hitler made his decision at about the time Röhm gave an address (on December 7) to foreign diplomats and press representatives, attacking private interests and urging a "soldier's socialism." The SA was also an embarrassment to Hitler's diplomacy: SA training activity could not be kept secret, and he constantly had to argue to skeptical diplomats, from Rumbold in May to Phipps now, that the SA lacked military significance. This must have been particularly annoying to Hitler because he himself had never regarded the SA as soldiers. His own firm conviction, based on personal experience, was that troops needed thorough military training. He was to refer specifically to his Langemarck experience when he declared to army and SA leaders, on 28 February, that a militia was not a suitable instrument for his plans.[110] This experience and these long-term plans may have been decisive all along in motivating his 300,000-man, one-year-service proposals.

In short, the army's earlier plans for raising reserves—especially the Manstein plans for three-month training preceded by Wehrsport and for trebling the army by mobilizing Wehrverbände members—were scrapped at the end of 1933. One reason for this shift seems to have been Hitler's wish to make a proposal on arms related to the Mac-

[109] Fall plans for three-month training within the army: WKK, WK VII/741, Artillerie Führer (Bayer.) VII Ia 65/33, 24 Oct 33, 28/184; WK VII/4082, WKK VII 1133, 21 Dec 33, 75/1033; ML, F VIII a 11, Minister-Amt 464/33, 6 Sep 33, reel 54/229-232 (described in Appendix). Röhm and Reichenau threats: *Nuremberg Documents*, XXIX, 10-11. The expansion of the army and accumulation of reserves by one year service was soon accelerated, in May 1934; see OKH, H1/324, Lecture by Major Ochsner, 20 Nov 36, 395/6397906.

[110] See Sauer, "Mobilmachung," pp. 884, 943; a fuller record of Hitler's statement is in O'Neill, *German Army and Nazi Party*, pp. 40-41. Hitler seems to have had for some time the idea that the first stage in armament should be a 300,000-man army. He had charged the Papen-Schleicher government with demanding such a force in his open letter of 21 Oct 32, published in the *Völkischer Beobachter*, although at that time Schleicher seems to have neither made, nor considered making, such a demand; the army's intended peacetime strength was then smaller, and its projected mobilized strength was larger.

502—Hitler's Hand is Forced, and He Frees It

Donald plan; he apparently hoped that this proposal would win British acceptance and lead to a British alliance. But Hitler was under no compulsion to make German military programs actually conform to his diplomatic offer. There were other reasons for changing plans. One was that the SA had proven unable to fulfill its expected military functions. Another was that the old plans were exposed to disruption by Röhm, especially since they were incompatible with his own ambitions. Beyond this, longer training within the army was inherently more desirable, in Hitler's view and doubtless also in that of many officers. And although longer training meant a slower buildup of reserves, the threat of an early preventive attack had almost entirely evaporated by December 1933. Instead of worrying about an early defensive war, Hitler could now think more about a later, offensive war.

In the June 30, 1934, purge, Schleicher would be murdered as well as Röhm. Assuming that Hitler ordered Schleicher's murder, he probably did so because he feared the general's capacity for intrigue and devisive maneuvers, and because he suspected that Schleicher had conspired with Röhm against him. These suspicions were probably unfounded.[111] But in past history, there was a link between Schleicher and the SA, which the general had believed would provide a reservoir of "defense-minded" youth. Schleicher and Röhm had each helped to bring Hitler to power. Now, like Schiller's Moor, they had done their historical duty, and they could go.

12. THE ROAD TOWARD WAR

Perhaps ironically, just when Hitler was starting to bid for a British alliance, his arms proposals were awakening British leaders to the danger of German rearmament; even though they would continue to seek agreements, they would be unable in future to forget the German threat. Phipps's first report of Hitler's proposals, on October 24, was inaccurate and incomplete, and it seems to have attracted little attention outside the Foreign Office. MacDonald appeared to Hoesch to be unfamiliar with the proposals when they conversed on November 15.[112] But after Neurath's November 6 speech, and especially after Blomberg raised the proposals again with Phipps, on November 20, the British began to study them intensively. On November 23, the ministerial committee on disarmament had a long discussion of the proposed 300,000-man German army. It was pointed out that these proposals in fact meant rearmament, and that (ignoring the cadre

[111] See O'Neill, *German Army and Nazi Party*, pp. 48-49; Vogelsang, *Schleicher*, pp. 101-105; but cf. Sauer, "Mobilmachung," pp. 921-923.

[112] RK, tel., Hoesch to Ministry, 15 Nov 33, 3650/D813178. Hoesch did not bring up the proposals himself, thinking that would be bad tactics.

element) a German army of that size could mean, with short service, six million trained men in twenty years. This thought, and the general idea that Germany was rearming, seem to have come as new realizations to the ministers. They asked the military experts for an opinion, and the latter soon reported that Hitler's scheme would give Germany a peace establishment of twenty-five to thirty infantry divisions, and enable her to mobilize fifty to sixty divisions in wartime— an estimate rather close to the actual new German plan. The experts also described what they believed to be the current state of German armament, and although this estimate was well below the German reality, it was well above the limits of Part V. An appendix to the report called attention to the (legendary) Prussian use of Krümper in the period from 1807 to 1813. But among the findings of the experts, what most impressed the ministers was not the prospective number of infantry divisions, but a calculation that the Germans already had 234 military aircraft and would have at least 400 by the end of 1934. In the interwar period, the threat of aerial attack inspired much the same terror as is now conveyed by the idea of a nuclear holocaust.[113]

This fear of air attack cut more than one way. It meant, on the one hand, that Hitler's dream of an Anglo-German understanding could never succeed, at least as long as he claimed a major air force. His talk of racial brotherhood fell on deaf ears, or repelled, and his offers to forgo naval rivalry no longer touched the heart of British concerns, while the threat of German air power withered any chances of friendship. On the other hand, fear of air attack worked to weaken British land forces and to enhance British isolationism vis-à-vis France.

By the fall of 1933, Neville Chamberlain considered Britain strong enough financially to permit some strengthening of defenses, and in 1934, British ministers debated what should be done. They concerned themselves mainly with the threat of an air attack on England herself, rather than with the growth of the German army. Chamberlain and Sir Warren Fisher favored a stronger defensive air force, but to avoid too heavy a burden on the Treasury, they propounded two ideas: first, that Britain should avoid a naval race by reaching an understanding with Japan, and second, that Britain should renounce any preparations for sending a land expeditionary force to the continent. As

[113] Cab 23/77, meeting of 22 Nov 33, 64(33)1; Cab 27/505, meetings of 23, 30 Nov 33; Cab 24/245, CP 291 (33) (DRC 5, 29 Nov 33). The British chiefs of staff had completed an annual review of defense policy just before the German withdrawal, emphasizing Far Eastern more than European problems; see Chapter Two, Section 10. This report, as well as the German action, led to the establishment of the Defense Requirements Sub-Committee of the CID, or DRC, which submitted the report described in the text. See Gibbs, *Rearmament Policy*, pp. 85-87, 93; Shay, pp. 28-30. For an illustration of the contemporary horror of aerial attack, see Toynbee's remarks in *SIA, 1932*, pp. 189-192.

Chamberlain saw it, trench warfare after 1914 had only led to a stalemate, and to a reliance on "financial, economic, and psychological attacks." It would be up to the French, presumably, to fight the bloody land battles. The ministers did not explicitly accept the Treasury theses—regard for naval tradition, for the United States, for dominion interests in the Far East, and for Belgium were too strong—and Chamberlain backed down in theory. But in practice, Treasury views prevailed through an assignment of priorities, and in the summer of 1934 the army's request for funds to meet deficiencies was cut in half. When an emergency defense loan was suggested, Chamberlain pointed to the political difficulty of forgoing another reduction in the income tax, and he gave a clear statement of his philosophy that no minister ventured to refute:

> It was necessary to cut our coat according to the cloth. He regretted that the suggestion of a defence loan had been put forward, as he regarded that as the broad road which led to destruction. No doubt it would be the easiest method of finding the money since it put upon succeeding generations the onus of paying it. He hoped we had not yet come to that stage and would be prepared to pay our own debts in our own generation.

Chamberlain also noted the lack of public support for an expeditionary force.[114]

There was a gulf between British conditions and German, between Chamberlain's outlook and that of Adolf Hitler. After 1933, the German army, in effect, wrote its own budget as large as it wanted, while Hitler, who gambled on paying all costs by conquest, was to remark (as he recalled in 1942): "No state has ever gone bankrupt for economic reasons—but only as the result of losing a war!"[115] Due to Chamberlain's policy, Britain could still have provided only two divisions in the spring of 1936, and this was a major consideration in the Rhineland crisis of that year. In late 1937, the British government

[114] Neville Chamberlain Papers, ltr. to Hilda Chamberlain, 21 Oct 33, NC 18/1/847; Cab 2/6 (1), 261st meeting, 9 Nov 33; Cab 24/247, CP 64 (34) (DRC 14, 28 Feb 34); Cab 24/248, CP 80 (34), 16 Mar 34; Cab 27/507, meetings of 3, 4, 8, 10, 15, 17 May, 11, 21, 25, 26 Jun, 2, 12, 17, 24 Jul 34; Cab 16/109, DRC 12 (encl. 1, 29 Jan 34); DRC 16, 12 Feb 34; DRC 19, 17 Feb 34. See also Gibbs, *Rearmament Policy*, pp. 93-99, 102-127; Watt, *Personalities and Policies*, Essays 4 and 5; Watt, *Too Serious a Business*, pp. 97-99; Middlemas and Barnes, pp. 761-781; Shay, pp. 28-44. It should be added that in early expert discussions (Cab 16/109, meetings of 14 Nov 33, 23 Jan 34) Fisher actually urged the army to increase its demands. Chamberlain may have been influenced against the experts by their rebuff of his proposal of a limited liability guarantee (Cab 27/506, meetings of 26, 28 Mar, 24 Apr, 1 May 34; Macleod, *Neville Chamberlain*, p. 166).

[115] Geyer, "Zweite Rüstungsprogramm," pp. 131-132, 134-135, doc. 8; *Secret Conversations*, p. 516.

was to decide against any continental role for the British army. Despite a belated reversal of policy, the adoption of conscription in April 1939, and a decision to form fifty-five divisions in September of that year, the neglect of the British army would contribute to the Allied defeat in the spring of 1940, and the expulsion of British forces from the continent. After the defeat and expulsion there would follow in turn Hitler's decision to attack Russia, the eventual Russian counter-attack, the land invasion by American, British, and Free French forces, and the present division of Europe.[116]

[116] On the later evolution, see esp. Gibbs, *Rearmament Policy*; Shay; Collier, pp. 25-73; Webster and Frankland, pp. 60-106; Dennis, Chapters 5 and 6; Correlli Barnett, pp. 494-511, 581-593; and F. Coghlan, "Armaments, Economic Policy, and Appeasement," pp. 205-216.

CONCLUSION

IN EARLY 1932, at the beginning of the Disarmament Conference, the impression was widespread in Britain and America that Germany had learned from its defeat in 1918, that the country had a moderate, responsible government, and that in military affairs, the Germans now wanted only the disarmament of others, for the sake of their own security. Germans did not want war, and they had no significant military power; thus there was no need to think of trying to balance German military power. If anything, it was French power that needed to be reduced.

The only element of truth in this general impression was that most ordinary Germans did not want another war. Far from accepting defeat, the German élite—and a sizable number of other citizens too—believed that, having vanquished Russia, they had nearly won the First World War, and that somehow, by trickery and betrayal, victory had been snatched from their grasp.[1] Military influence in the government was now strong, and growing. Also, what counted militarily was not the peacetime strength of the regular army, but the size of the force Germany could mobilize. Although Germany had suffered humiliating peace conditions, her population and industrial power were not much reduced, and in view of the temporary eclipse of Russian power, her potential military strength, relative to rival powers, was greater than before.[2] Furthermore, Germany's mere numbers did not provide an adequate idea of her potential, since, as German strategists had learned long before, organization, training, and above all initiative provide advantages. German accomplishments due to these advantages doubtless helped to foster myths of racial superiority—although after World War II, the Israelis would show that German soldiers had no monopoly on the ability to strike hard and efficiently.

The first major conclusion of this study is that German civilian and especially military leaders sought from early on to restore the nation's military strength as a means of revising the peace settlement. While the need to defend the country served as a justification for rearmament, German leaders also hoped to use military strength "as a precondition for [a] stronger foreign policy" and for "national liberation." They thought that a strong military force would strengthen Germany's diplomatic hand, and beyond this, some of them intended, under favorable circumstances, to use war again as an instrument of

[1] See Hildebrand, *Deutsche Aussenpolitik,* p. 16; also Bennett, *Diplomacy,* p. 1.
[2] See Taylor, *Origins,* Chapter 2; Gerhard L. Weinberg, "The Defeat of Germany in 1918 and the European Balance of Power," *CEH,* 2 (1969), 248-260.

policy. The presupposition of general disarmament was that Germany was now disarmed, and would remain disarmed. But German generals sought to retain the capacity for a large-scale mobilization, and they worked to take advantage of each successive removal of control—the end of the Ruhr occupation, the termination of the Inter-Allied Military Control Commission, the withdrawal of all foreign occupation in 1930—to improve their mobilization capability. They hoped likewise to take advantage of the Disarmament Conference.

Mobilization for modern war requires not only a mustering of vast material resources, but also, in the face of the horrors of battle, a willing supply of trained manpower, backed by a willing nation. This was the greatest problem confronting German military planners. At first there was a certain reserve supply of trained men, left over from the war, but the men were getting too old to serve, and any system for mustering them required civilian cooperation. The Reichswehr needed to train new reserves, and at the same time it also needed to conceal military preparations from other governments. In this situation, the unwillingness of the Socialist-dominated Prussian government to cooperate in Landesschutz preparations posed a major difficulty. The sudden development of the Nazi movement in 1930 encouraged the military to turn, more than ever, to the Wehrverbände for eventual reserves; the Verbände provided a means of carrying on concealed preparations, and an alternative to official civilian support. But the Socialists continued to oppose secret defense preparations, and they and other moderates sought to disband the SA, thus interfering with army plans.

Another major conclusion is that *on military grounds*, army spokesmen intervened in the German political structure to remove Socialists and moderates from power and to bring the Nazis into the government. Historians have given much attention to the rightist ideologies prevalent among German officers, but even more important, the Reichswehr had adopted plans which depended on the Wehrverbände, and which soon led to dependence on the SA. By the end of 1932, army policies had destroyed constitutional government and made it impossible to govern the country by force against Nazi opposition, especially since the Nazis were essential for defense against Polish attack.

The military leaders did not intend, before 1933, that Hitler should have exclusive power. But they wanted to use his movement, and they welcomed his ability to provide nationalist inspiration, and to overcome "antidefense" elements. Soon after his accession they also found him able to clear away the financial obstacles to rearmament, and many of them began to admire his capacity as a leader. He at first respected the professional expertise of the officers, and he consistently supported their objective of preparing and directing an integrated defense force, as against the claims of Röhm and the SA to form a new revolutionary army. Hitler wanted an effective military striking force, as did the Reichswehr leaders, and neither he nor they contemplated a separate militia. Military policy in the first ten months of Hitler's rule followed mainly the course already charted under Schleicher. When the policy changed, it took the direction of eliminating dependence on the SA by adopting one-year service entirely within the army. When once the SA leadership was purged, the army would cease to have domestic political concerns.

While Hitler continued to pursue the broad military objective of rearmament, which had helped to bring him to power, he also had the political sense to follow the policy of German diplomats, that of concealing German armament plans as long as possible, rather than making threats of rearmament, as Schleicher had done. This did not prevent Hitler from being strongly revisionist, as were the diplomats. Taking account of both military and diplomatic policy, a further conclusion is that there was continuity in German history, not only between Wilhelminian Germany and the Nazi era, but also specifically at the critical junction of the end of the Weimar era with Hitler's new regime. A further evidence of this continuity is that serious military and diplomatic affairs continued to be conducted by the professionals of the old regime; the SA and the "offices" of Rosenberg and Ribbentrop functioned more as welfare institutions than as real instruments of policy. There are many admirable things in German tradition, including German military tradition, but Hitler's ideas, including his racism, did not represent a radical break with the German past. Most of Hitler's ideas were anticipated by the Pan-German movement, which also persisted in Hugenberg's leadership of the DNVP. One major exception, his belief in brotherhood with the English, had a background in German liberal tradition.

Governments try to keep their military preparations secret, and Germany had an additional reason for secrecy, in that her preparations violated the Treaty of Versailles. If known, her activities would demolish her case for the disarmament of others, and perhaps lead to preventive measures. The need for secrecy, particularly during disarmament negotiations, tended to impede the rearmament which, once it was well advanced, would make concealment less necessary. One is led to ask, then, how far Western intelligence organs were able to detect German violations, and we have found that, although Western intelligence reporting showed some gaps and distortions, it did provide considerable data. But a real problem lay in the difficulty in fathoming German long-range intentions, something particularly difficult when those intentions ran counter to both the expectations and the hopes of the observers.

Of the three major Western powers, France was the closest to Germany, the most fearful of Germany, and the readiest to recognize indications of German preparation for another war. Despite this readiness, the war-weary French had modified their army in such a way as almost to rule out a preventive action, and as historians have often shown, they were disposed to take shelter behind the Maginot line. Still, and this is another major conclusion, French leaders did strive in both 1932 and 1933 to organize a united front in opposition to Germany. In their efforts, Herriot, Boncour, and Daladier tried particularly to commit the British, in 1932 by exploiting a tactical offer of consultation from MacDonald, and in 1933 by seeking a British commitment to sanctions, attempting to use private export discussions as a lever. Though reluctant to take military action, the French did have other cards they could have played with Anglo-American support: in particular, they could have made a public exposure of German violations. The fear of such an exposure seems to have precipitated Hitler's decision to leave the Disarmament Conference and the League.

Britain was now in the key position to decide whether the reviving power of Germany should be counterbalanced. But the British did not want to believe that German military power was reviving. Out of horror of war and fear of financial ruin, they pushed aside the evidence, evaded any firm continental commitment, and stripped their own defense capability to the bone. The British hoped to stay out of continental conflicts, and yet they attempted to manage continental affairs. The conclusion here is that the British government tended to appease German demands whenever they became pressing, not for Germany's sake, but to avoid new expenses and to maintain hope at home of a new world without war. Fashionable condemnation of the immorality of Versailles influenced the government. Up until April 1933, the London government frequently sought to gain concessions for Germany by fostering international discussions in a four- or five-power forum, which was a means of isolating France and overcoming her objections. After that—and despite Hitler's personal hopes for British friendship —Nazi excesses caused the British parliament and public to become more dubious about German intentions. But, as had happened after the first German departure from the conference in 1932, the German withdrawal from Geneva in October 1933 once again made British leaders anxious to get into touch with Germany. They had no idea of an alliance with Germany, and paid no attention to Hitler's bids for such an alliance, but they still hoped for disarmament—by the French.

The United States was far more remote from Germany, and in Washington there was no systematic official review of German intentions. Some individuals, such as Stimson and General MacArthur,[3] were concerned about German developments, but most American spokesmen looked at Europe as merely a stage from which political backing at home might be won—as by proposing a new disarmament scheme; or lost—as through entangling the United States in a dangerous alignment with European powers. Their insensitivity to practical power considerations is illustrated by the argument of Norman Davis (in his May 15, 1933, draft of a statement for Roosevelt): "It does not contribute to peace and stability in Europe to keep the largest and most populous of the nations [i.e., Germany] in a permanent condition of inequality."[4] Many of the American interventions had little serious effect, but the final major conclusion of this study is that the United States bears more responsibility for smoothing the path for Hitler than has been generally recognized. After Roosevelt yielded to isolationist pressures in June and early July, the United States in October 1933 torpedoed French hopes of exposing and isolating Hitler: Davis dissuaded Simon from arraigning Germany, and Roosevelt suddenly insisted, following press attacks, that the United States was taking no sides in European questions. Americans who blame MacDonald, Baldwin, and Simon for seeking an accommodation with Hitler thereafter should remember the role the United States government played at the moment when Simon was primed to condemn German rearmament.

More than anything else, it was British and American indifference to, or even revulsion from, the principle of the balance of power that permitted German rearmament. Of course, men like Davis were guided by their conviction that disarmament was an overriding objective, a panacea which would banish the danger of war. Looked at one way, the argument for disarmament is a corollary of the law of the balance of power: if an increase in the power of one nation leads the other nations to attempt to restore balance by alliances and by increasing their own power, then an acceptance by all of an equally low level of power should eliminate competition for it. The essence of the matter is to establish an agreed balance of force. The Washington and London naval treaties established such a relation to the general satisfaction of the civil governments in those two capitals, if not of the British and American navies, or of the Japanese. This was possible because neither government intended—or seriously suspected the other of intending—

[3] On MacArthur, see Chapter Three, note 25.
[4] SDP, tel., Davis to Dept., 15 May 33, 500.A15A4 General Committee/378.

to change the status quo, or to resort to force against the other. Unfortunately, no such state of mind existed in Berlin and Paris. German leaders considered the existing settlement to be unnatural, dictated by (temporarily) superior Allied force; in their view, the French had again resorted to naked force in the Ruhr occupation. French leaders, receiving intelligence of German violations and news reports of nationalist manifestations, believed—rightly—that Germany was preparing to take her revenge and resume her drive for the domination of Europe. The advocates of disarmament tended to take German demands for equality at face value, and they overlooked the basic incompatibility between the dominant German view of the way the world should be and the views of other nations. Winston Churchill recommended that the removal of grievances should precede disarmament,[5] but this too was impractical, since a removal of the grievances of the German right would create unbearable grievances in France, Poland, and much of the rest of Europe. To be blunt, efforts to give equitable treatment are misplaced when those who demand such treatment would deny it to others.

This does not mean that one should not try to comprehend their views. Aside from such themes as the German desire for a restoration of armed power, the French anxiety for guarantees, or the British wish to maintain a hope of harmony, a leitmotiv of 1932 and 1933 was the failure of the leaders of any one country to comprehend the points of view of the other nations. We have seen striking instances of misunderstanding, such as a German misconception that Stimson and MacDonald, at Bessinge, had fully accepted German demands for equality of rights; the French misapprehension that the Germans intended to rely on an élite professional army; American illusions as to the willingness of governments to disarm; and a mistaken British assumption that Germany would be able and willing to accept a convention which would effectively block rearmament. Perhaps most striking of all was Hitler's idea that there were bonds of brotherhood between Germany and England. One gets the constant impression of a dialogue of the deaf, as the French phrase puts it—of people living in different worlds.

Probably one could never expect a Hitler to understand the psychology of another nation—although sometimes the Führer did show insight into the way foreign leaders might react in given circumstances. But Britons and Americans might well have understood more about German traditions and institutions, and in particular about the formative role of military force in German history, the deference given to

---

[5] Churchill, *While England Slept*, pp. 31-32 (speech of 23 Nov. 32). Five months later, he stated that he had changed his mind respecting the Polish Corridor (p. 62, speech of 13 Apr 33).

German officers, and the means German staff planners had used in the past to build successful armies. One learns about the traditions and institutions of other countries by studying their cultures, social structures, and histories. Within nations, governmental and economic institutions form subsocieties; these have persisting viewpoints which arise from their functions, and in some cases, as in that of the German army, they have great influence in the general society. Those who strive for disarmament need to know the institutions, ideals, and fears that shape the policies of each of the various powers, and that lead them to build up armaments. The proper assessment of intelligence material also calls for a full knowledge of foreign traditions, institutions, and cultures. Indeed, what is known as statesmanship cannot exist unless a cultural and historical knowledge of others is widespread within the political public. It really is not surprising that there should be continuity in German history: because of the passing on of language, culture, and institutions from older to younger, continuity is the rule in all societies, even through apparent revolution. But the patterns in other societies must be understood if nations are to exist together.

APPENDIX

WOLFGANG SAUER, on pp. 718 (and note), 797 (and note), 885-886, and 888 of his "Die Mobilmachung der Gewalt" (in Bracher-Sauer-Schulz, *Die Nationalsozialistische Machtergreifung*, 1960), advances the thesis that General von Reichenau intended to develop the professional Reichswehr on Seecktian lines, backed up by an SA militia. Sauer's study is, deservedly, a prominent landmark in the historiography of the German army, and the thesis just mentioned has embedded itself in the literature. This thesis is, however, invalidated by documents now available.

Sauer's main source is the postwar recollection of Franz von Gaertner (Institut für Zeitgeschichte, Zeugenschrifttum 44). Gaertner says he wanted to rejoin the army, and that Reichenau urged him instead, in November 1933, to take a post with Krüger's Ausbildungswesen (AW) organization for training SA men and leaders, and also to report to a Reichswehr intelligence officer on SA attitudes. Reichenau and Blomberg probably did play with various ideas, and the term "militia" was ambiguous. But if Reichenau described the existing AW training to Gaertner as the preparation of a self-contained militia, with its own leaders, he was prettifying the AW work, which was primarily premilitary training of a Wehrsport nature, largely conducted by retread officers (like Gaertner). And if Blomberg ever talked to Ernst Röhm in this same vein, as Reichenau is said to have told Gaertner, he was deceiving Röhm and sowing the seeds of future trouble.

Actually, Blomberg's basic directive for the AW or SA-Sport program (WKK, WK VII/1453, Der Reichswehrminister TA 533/33, 27 Jul 33, 37/1118-1122) was entitled "Guidelines for Premilitary Training," and its whole content shows that the program was to prepare German youth for short-term training in the army. As for Reichenau's plans, his office issued a paper on 6 September 1933 (ML, F VIII a 11, Minister-Amt 464/33, reel 54/229-232) describing a new draft law to give a legal basis for leave from work "for participation in the premilitary SA training as well as in short-term military training in the Wehrmacht (three-month recruit training, exercises for training subordinate leaders and leaders)." Both SA and army training were to be described publicly as "Courses in Physical Training," and for extra cover of the army's program, there was to be no overt indication that two different kinds of training, SA and army, were involved. The directive indicated several times that the SA (i.e., the AW) was to have responsibility for premilitary or "sport" training. See also WKK,

WK VII/741, Artillerieführer (Bayer.) VII, Ia 65/33, 24 Oct 33, 28/184, and WK VII/4082, WKK VII 1133, 21 Dec 33, 75/1033, which show that the planning for three-month training within the army continued through the fall.

The idea of a separate SA force has some limited validity for the Grenzschutz. By the fall of 1933, most of the Grenzschutz reservists were SA men, and SA organizations were being reshuffled to correspond with the Grenzschutz structure. Here there was talk of mobilizing closed SA or SS formations as Grenzschutz formations. But a discussion between Bavarian army, SA, and SS leaders in November showed that there was no thought of such a procedure for the field army: "The situation is completely different for the mobilization of the field army [A-Heeres]. The [army] organizations will be formed as a skeleton by the division of the companies, etc., of the Reich army. The Reich army posts and names the leaders and supplementary leaders, SA and SS constitute the flesh, provide the needed supplementary [men]. Whole units cannot be taken over [i.e., from the SA and SS], but each [man] only as an individual. . . . The SS leaders agreed that available SS units outside the border area would lead their people to the [Landesschutz] Meldestellen Passau, Regensburg, and Ingolstadt": WKK, WK VII/182, Kommandantur Regensburg Ia 386, 20 Nov 33, 10/1260.

A mid-September speech by Blomberg to officers, after singing Hitler's praises, stated that the Wehrverbände were a welcome phenomenon, "whose human material needs, with an eye to the future, development and cultivation by us," but he also stressed that their members were not soldiers, and that they were there for the internal political struggle. The army, by contrast, had the tasks in peacetime of training German youth as soldiers and in wartime of defending German borders, and it also remained the ultima ratio of the Chancellor at home; on these points, Blomberg stressed that he and Hitler were in full agreement: WKK, WK VII/3272/2, Der Reichswehrminister 460, 22 Sep 33, 63/152-153.

In several ways, it is historically significant that army planning envisioned the mobilization of an integrated army, rather than of a Seecktian army accompanied by a separate militia. The decision to prepare an integrated field army meant, first, that the planners sought to mobilize a large army for serious warfare, a force which could ultimately meet and defeat the French and Polish armies; this implied the overthrow of the Versailles settlement. The planners did not see the Reichswehr primarily as a closed corporation, or as a guard for defending an existing class order or a restored monarchy against domestic opponents. Nor did they suppose that an élite force could defeat enemies who were less qualified, but ten times more numerous; in fact, they probably underestimated the possibilities for a Blitzkrieg.

Second, the mobilization plans adopted provided that the army would train young men as its own reserves, and this meant active intervention in domestic society. As the army could not openly assume the role of "the school of the nation," it was led to expunge Socialist opposition to secret defense preparations, and also to form closer ties with the Wehrverbände. Indeed, mobilization of the field army would depend on the Verbände for personnel. But the Verbände would remain, as Blomberg put it in his speech, sources of human "material," not potential armies in themselves.

Third, the fact that the basic plan was inaugurated in the early days of Brüning's chancellorship and retained for nearly a year under Hitler shows that military plans did not take inspiration from Nazi political success. The plan was based on *military* considerations, and these showed continuity through changes in political regime—certainly, the military plan contributed much to producing the political changes. When the plan was finally dropped in December 1933, this was done partly for foreign-policy reasons, and probably partly out of a wish to reduce Röhm's political potential, but military objectives—the desire to make training more effective, to free mobilization from dependence on the SA, and to speed the development of an effective striking force—appear to have had at least as much importance.

BIBLIOGRAPHY

I. UNPUBLISHED OFFICIAL DOCUMENTS.

A. *Germany.*

1. REICH CHANCERY (REICHSKANZLEI) RECORDS (RK).

I have used the following film serials (or files) from Microfilm T-120 at the National Archives, Washington, D.C. The T-120 microfilming project was a joint Anglo-American-French enterprise in connection with the publication of the *Documents on German Foreign Policy,* and copies of all T-120 material are available at the Public Record Office in England. I have dispensed with listing container or reel numbers; the reels corresponding to the serials may be found in George O. Kent, ed., *A Catalogue of Files and Microfilms of the German Foreign Ministry Archives, 1920-1945,* 4 vols. (Stanford, Calif.: Hoover Institution, Stanford University, 1962-1972). The original Reichskanzlei documents are now available at the Bundesarchiv, Koblenz (BA).

T-120 Serial	Alte Reichskanzlei File
3598	Kabinettsprotokolle (1932-38).
3617	Auswärtige Angelegenheiten: Auswärtige Politik. Allgemeines.
3642	Abrüstungskonferenz Genf 1932: Deutsche Gleichberechtigungsfrage.
3650	Auswärtige Angel.: Allgemeine Abrüstung.
9242	Auswärtige Angel.: Baseler Bericht, Reparationskonferenz, Konferenz Lausanne: Vorbereitungen und Ergebnis der Konferenz.
K951	Reichswehr, Volkswehr und Wehrpflicht.
K953	Landesverteidigung.
K1028	Akten betr. Reichsbankpräsident.

In addition, I have used portions of a Neue Reichskanzlei file on Arbeitsbeschaffung; this was not microfilmed, but it is held at the Bundesarchiv as R 43 II/540.

2. AIR MINISTRY (REICHSLUFTFAHRTSMINISTERIUM) RECORDS.

Von Rohden Collection—At National Archives, now as Microfilm T-971. Originals returned to West Germany.

3. DEFENSE MINISTRY (REICHSWEHRMINISTERIUM) RECORDS.

I gathered Defense Ministry and army material from Microfilms T-77 (Oberkommando der Wehrmacht—OKW), T-78 (Oberkommando des Heeres—OKH), and T-79 (Wehrkreis Kommando—WKK) records at the National Archives, and from the RH 15 and II H files of the Bundesarchiv-Militärarchiv (BA-MA) in Freiburg im Breisgau. (The BA-MA collection includes material formerly held by the Militärgeschichtliches Forschungsamt/Dokumentenzentrale Freiburg.) I gathered naval material from Microfilm 6342, Marineleitung Records, Project II (ML), University of Michigan Library, Ann Arbor, Mich., and also from serial 7792 (reels 3233-3234) (NA-VGM) of Microfilm T-120 at the National Archives. The filming of the T-77, T-78, and T-79 material (and of many other captured German records) has been sponsored by the American Historical Association. The Michigan naval films resulted from a joint project of the Universities of Michigan and Cambridge (England), undertaken through the enterprise of Prof. Howard M. Ehrmann and of F. H. Hinsley. The originals of all this microfilmed military material are now at the Bundesarchiv-Militärarchiv.

T-77 Serial	OKW File		
2	Wi/IF 5.5	436	Wi/IF 5.3682
5	Wi/IF 5.23	437	Wi/IF 5.3683
18	Wi/IF 5.126	657	Wi/VI 104
19	Wi/IF 5.132	793	OKW 845
73	Wi/IF 5.335	870	OKW 2139
86	Wi/IF 5.383	870	OKW 2140
92	Wi/IF 5.406	874	OKW 872
98	Wi/IF 5.422	T-78 Serial	OKH File
109-110	Wi/IF 5.498	146	H15/103
110	Wi/IF 5.499	247	H24/3
110	Wi/IF 5.501	248	H24/6
111	Wi/IF 5.502	324	H27/10
113	Wi/IF 5.509	395	H1/324
116	Wi/IF 5.515	395	H1/326
116	Wi/IF 5.516	395	H1/378
116	Wi/IF 5.517	409	H1/663
177	Wi/IF 5.701		
327	Wi/IF 5.1983	T-79 Serial	WKK File
347	Wi/IF 5.2234	6	WK IV/21
399	Wi/IF 5.3049	10	WK VII/182

T-79 Serial	WKK File	U of M film	Admiralty
27	WK VII/732	6342 Reel (ML)	File
28	WK VII/741	5	Shelf 5993
37	WK VII/1453	11	Shelf 5916
49	WK VII/2091	12	Shelf 5930
55	WK VII/2670	17-18	Shelf 5939
63	WK VII/3272/2	18	Shelf 5940
64	WK VII/3422	18	Shelf 5946
75	WK VII/4082	26	Levetzow Nachlass
		27	PG 34176
BA-MA Files		27	PG 34168/1
RH 15/8		28	PG 33994
RH 15/28		30	PG 34072
RH 15/40		31	PG 34095
RH 15/49		31	PG 34097
II H 226		32	PG 34108
II H 227		33	PG 34162/3
II H 228		33	PG 34016
II H 272		35	PG 49031
II H 285		48	Levetzow Nachlass
II H 296		54	(F VIII a 11)
II H 645/2		T-120 Serial 7792 (NA-VGM)
		(VGM 35, 36; F VII a 11)

4. FOREIGN MINISTRY (AUSWÄRTIGES AMT) RECORDS (AA).

The following film serials (or files) were used from Microfilm T-120 at the National Archives; see also Section I, A, 1 above. The original Auswärtiges Amt documents are held at the Politisches Archiv of the West German Foreign Ministry at Bonn.

T-120 Serial	Auswärtiges Amt File
1809	Politische Abteilung II, Pol. Verschluss Geheim: Tschechoslowakische Dokumente. [See Weinberg, *Foreign Policy of Hitler's Germany*, p. 366.]
2368	Büro des Reichsministers: England, Sudafrikanische Union.
2376	[Same as Serial 1809]
2406	Büro des Reichsministers: Frankreich.
3154	Büro des Reichsministers: Sicherheitskomitee—Abrüstung—Gleichberechtigungsfrage.
3243	Büro des Reichsministers: Reparation.

4565	Büro des Staatssekretärs: Landesverteidigung.
4602	Büro des Staatssekretärs: Aufzeichnungen St. S. von Bülow über Diplomatenbesuche.
4604	Büro des Staatssekretärs: Abrüstung.
4617	Büro des Staatssekretärs: Reise des Herrn Staatssekretärs von Bülow nach London (Donaukonferenz), Paris, Genf in April 1932.
4618	Büro des Staatssekretärs: Reparationsfrage.
4619	Büro des Staatssekretärs: Schriftwechsel mit dem Herrn Reichsminister sowie Aufzeichnungen des Herrn Reichsministers.
5669	Abteilung II: Frankreich, Politische Beziehungen zu Deutschland.
5740	Abteilung III: England, Politische Beziehungen zu Deutschland.
5849	Presse Abteilung: Die Presse in England.
7334	Stresemann Nachlass.
7360	Abteilung II F Abrüstung: Allgemeine Abrüstungskonferenz 1932.
7474	Abteilung II F Abrüstung: Verhandlungen über die Gleichberechtigungsfrage.
7616	Abteilung II F Abrüstung: Deutsche Delegation zur Abrüstungskonferenz 1932.
9095	Abteilung II F Abrüstung: Vorbereitung der allgemeinen Abrüstungskonferenz.
9284	Abteilung II F Abrüstung: Kommissionen— Allgemeines; Generalkommission und Büro der Konferenz.
K6	Geheimakten: Militärpolitik.
K936	Abteilung II: Frankreich, Politische Beziehungen zu Deutschland.
K957	Abteilung II F Abrüstung: Abrüstungsdelegationen, Beziehungen Deutschlands zu Frankreich.

5. NATIONAL SOCIALIST GERMAN WORKERS' PARTY (NATIONALSOZIALISTISCHE DEUTSCHE ARBEITERPARTEI) RECORDS.

Nazi records held at the Berlin Document Center, West Berlin, are available (in part) on National Archives Microfilm T-580. Other Nazi records have been filmed on National Archives Microfilm T-81.

T-580 Reel	File
264	Krüger Nachlass Ordner 9a (1. Teil).
864	Box 120, Ordner 309.

T-81 Serial	File
7	Adolf Hitler, Kanzlei.

B. *Great Britain.*

1. CABINET OFFICE RECORDS (CAB).

The following record series used are held at the Public Record Office, London. In the case of Cab 24 records (Cabinet Papers), and a few Cab 2 files (CID), copies are available on Microfilm X608 at the University of Michigan Library.

Record Series		Volumes
Cab 2	Minutes of the Committee of Imperial Defence (CID)	6
Cab 16	Defence Requirements Sub-Committee (DRC) of the CID	109
Cab 23	Cabinet Meeting Conclusions	69-73, 75-77, 79, 82
Cab 24	Cabinet Papers	225, 227-230, 233, 234, 237, 239, 241-245, 247
Cab 27	Ministerial Committee on Disarmament Conference, 1932-1935 (DC [M] 32)	505

2. AIR MINISTRY RECORDS.

Air Ministry Records are held at the Public Record Office. I used Bundle 1354 of record series Air 2.

3. FOREIGN OFFICE RECORDS.

Foreign Office papers are held at the Public Record Office. I have dispensed here with the five-digit volume numbers for the FO 371 series; finding aids at the PRO indicate the volumes in which the various files may be found.

FO 371 Files for 1932
—/29/62
—/211/18
—/10/98
—/1466/98
—/5810/18
—/5920/62
—/130/98
—/22/98

FO 371 Files for 1933
—/245/18
—/40/98
—/2287/22
—/653/18

FO 800
Vols. 286-288 (Private Office Papers of Sir John Simon)

522—Bibliography

C. *Italy.*

The National Archives holds microfilms of various captured Italian records, including records of the Duce's Secretariat (Segretaria Particolare del Duce) and of the Secretariat of the Chief of Government (Segretaria Particolare del Capo del Governo), which are on T-586. I have used the following:

T-586 Reel	Folder
449	205R
491	442R
1134	224R

D. *United States*

1. PRESIDENTIAL PAPERS.

The following boxes were used from the Roosevelt Papers at the Franklin D. Roosevelt Library, Hyde Park, N.Y.

President's Secretary's File	Boxes 51, 142
President's Personal File	Box 140
Official File	Boxes 198, 404

2. DEPARTMENT OF STATE PAPERS (SDP).

Now at the National Archives. Files used:
500.A15A4/—
500.A15A4 Steering Committee/—
500.A15A4 General Committee/—
550.S1 Washington/—
751.62/—
763.72119 Military Clauses/—
862.00/—

3. WAR DEPARTMENT RECORDS.

At the National Archives. I used U.S. Army military attaché reports from Berlin and Paris, 1931-1934.

E. *Other.*

1. INTERNATIONAL MILITARY TRIBUNAL RECORDS.

National Archives Microfilm T-988 records the full text of the trial documents. I used reel 39 (for Doc. 855-D).

2. NUREMBERG TRIAL 10 (KRUPP TRIAL) RECORDS.

At the National Archives. I used Defense Exhibit 62, Affidavit and document by Wilhelm Adam, 5 Mar 48.

II. Unpublished Private Papers.

Wilhelm Adam Memoirs (Erinnerungen): Institut für Zeitgeschichte, Munich, ED-109. (Not available during my last visit to Munich, but Herr Weiss of the Institut has kindly sent me a copy of a short extract.)

Werner von Blomberg Memoirs (Erinnerungen) (EAP 21-a-14/30c): on National Archives Microfilm T-84, serial 202.

Brammer Collection: Bundesarchiv, Koblenz, Z Sg 101. The Brammer and Traub collections include private reports and letters by journalists, and Propaganda Ministry instructions.

Ferdinand von Bredow Papers: Institut für Zeitgeschichte, ED-86. (These are photocopies; the originals are at the Bundesarchiv-Militärarchiv.)

Heinrich Brüning: Witness's statement (Zeugenschrifttum), Institut für Zeitgeschichte, ZS-20.

Lord Cecil of Chelwood Papers: British Library, London.

Austen Chamberlain Papers: University of Birmingham Library, Birmingham, England.

Neville Chamberlain Papers: University of Birmingham Library.

Norman H. Davis Papers: Library of Congress, Washington, D.C.

Franz von Gaertner: Witness's statement (Zeugenschrifttum), Institut für Zeitgeschichte, ZS-44.

Wilhelm Groener Papers (Nachlass): on National Archives Microfilm M-137.

Friedrich-Wilhelm Krüger Papers (Nachlass): see Section I, A, 5 above.

Magnus von Levetzow Papers (Nachlass): see Section I, A, 3 above.

Curt Liebmann Notes and Diary: Institut für Zeitgeschichte, ED-1.

James Ramsay MacDonald Papers and Diary: Public Record Office, London, PRO 30/69.

Horst von Mellenthin: Witness's statement (Zeugenschrifttum), Institut für Zeitgeschichte, ZS-105.

Jay Pierrepont Moffat Diary: Harvard University Library, Cambridge, Mass.

Hermann Pünder Papers (Nachlass): Bundesarchiv, Koblenz.

Sir Horace Rumbold Papers: private possession (see preface).

Herbert Louis Samuel, Viscount Samuel Papers: House of Lords Record Office, London.

Hans Schäffer Diary (Tagebuch): Institut für Zeitgeschichte, ED-93. (Transcribed from shorthand by Schäffer and Ernst Göhle.)

Kurt von Schleicher Papers (Nachlass): Bundesarchiv-Militärarchiv, Freiburg im Breisgau.

Lutz Graf Schwerin von Krosigk Diary (Tagebuch): in part in Rathmannsdorfer Hauschronik, microfilmed on National Archives film T-84, reel 426.
Sir John Simon Private Office Papers: see Section I, B, 3 above.
Henry L. Stimson Diary: Yale University Library, New Haven, Conn.
Gustav Stresemann Papers (Nachlass): see Section I, A, 4 above.
Joachim von Stülpnagel Papers (Nachlass): Bundesarchiv-Militärarchiv, Freiburg.
Traub Collection: Bundesarchiv, Koblenz. (See under Brammer above.)

III. PUBLISHED DOCUMENTS.

A. Belgium.

De Visscher, Ch[arles], and F[ernand] Vanlangenhove, eds. *Documents diplomatiques belges, 1920-1940: La Politique de sécurité extérieure.* 5 vols. Brussels: Académie royale, 1964-1966.

B. France.

France, Assemblée nationale, Commission chargée d'enquêter sur les événements survenus en France de 1933 à 1944. *Rapport fait au nom de la commission chargée d'enquêter sur les événements survenus en France de 1933 à 1945.* 2 vols. Paris: Presses universitaires de France, [1952].
France, Ministère des affaires étrangères. *Documents diplomatiques français (1932-1939).* Paris: Imprimerie nationale, 1963-.

C. Germany.

Domarus, Max, ed. *Hitler, Reden und Proklamationen 1932-1945, Kommentiert von einem deutschen Zeitgenossen.* 2 vols. (each in two parts). Munich: Suddeutscher Verlag, 1965.
Rothfels, Hans, et al., eds. *Akten zur deutschen Auswärtigen Politik, 1918-1945.* Göttingen: Vandenhoeck & Ruprecht, 1966-.
Schüddekopf, Otto-Ernst, ed. *Das Heer und die Republik: Quellen zur Politik der Reichswehrführung 1918 bis 1933.* Hanover: Norddeutsche Verlagsanstalt O. Goedel, 1955.
Schwendemann, Karl, ed. *Abrüstung und Sicherheit: Handbuch der Sicherheitsfrage und der Abrüstungskonferenz.* 2 vols. Berlin: Weidmannische Buchhandlung, 1933-1935.
Sontag, Raymond J., John W. Wheeler-Bennett, Maurice Baumont et al., eds. *Documents on German Foreign Policy, 1918-1945.* Washington, D.C.: Government Printing Office/London: Her Majesty's Stationery Office, 1949-.

The English-language edition of the German diplomatic documents.

Vernekohl, Wilhelm, and Rudolf Morsey, eds. *Heinrich Brüning: Reden und Aufsätze eines deutschen Staatsmanns*. Münster: Verlag Regensburg, 1968.

Vogt, Martin, ed. *Das Kabinett Müller II, 28. Juni 1928 bis 27. März 1930*. (In Karl Dietrich Erdmann and Wolfgang Mommsen, eds., *Akten der Reichskanzlei: Weimarer Republik*.) Boppard: Harald Boldt, 1970.

D. *Great Britain*.

Woodward, E. L., Rohan Butler, J.P.T. Bury, W. N. Medlicott, Douglas Dakin, M. E. Lambert, eds. *Documents on British Foreign Policy, 1919-1939*. London: Her Majesty's Stationery Office, 1946–.

E. *Union of Soviet Socialist Republics*.

Andreyeva, M., and L. Vidyasova, eds. "The Struggle of the U.S.S.R. for Collective Security in Europe during 1933-1935," *International Affairs* (Moscow), 9 (1963), no. 6, 107-116; no. 7, 116-123; no. 8, 132-139, no. 10, 112-120.

F. *United States*.

United States, Department of State. *Foreign Relations of the United States*. Washington, D.C.: Government Printing Office, 1861–.

Nixon, Edgar B., ed. *Franklin D. Roosevelt and Foreign Affairs*. 3 vols. Cambridge, Mass.: Harvard University Press, Belknap Press, 1969.

G. *International*.

League of Nations, Conference for the Reduction and Limitation of armaments. *Records of the Conference*. Series A: *Verbatim Records of the Plenary Meetings*. (Only Vol. I appeared.) Geneva: League of Nations publications, 1932.

———. *Records of the Conference*. Series B: *Minutes of the General Commission*. 3 vols. Geneva: League of Nations publications, 1932-1936.

———. *Records of the Conference*. Series C: *Minutes of the Bureau*. 2 vols. Geneva: League of Nations publications, 1935-1936.

———. *Conference Documents*. 3 vols. Geneva: League of Nations publications, 1932-1936.

International Military Tribunal. *Trial of the Major War Criminals before the International Military Tribunal, Nuremberg, 14 November 1945–1 October 1946*. 42 vols. Nuremberg: Secretariat of the Tribunal, 1947-1949.

Wheeler-Bennett, John W., et al., eds. *Documents on International Affairs*. Annual. London: Oxford University Press, 1928-1973.

H. Individual and unofficial.

Co[nze], W[erner], ed. "Zum Sturze Brünings," *VfZ*, 1 (1953), 261-288.
Craig, Gordon A., ed. "Quellen zur neusten Geschichte: Briefe Schleichers an Groener." *Welt als Geschichte*, 11 (1951), 122-133.
Garçon, Maurice, ed. *Les Procès de collaboration: Fernand de Brinon, Joseph Darnand, Jean Luchaire, compte rendu sténographique*. Paris: Editions Albin Michel, 1948.
Meier-Welcker, Hans, ed. "Aus dem Briefwechsel zweier junger Offiziere des Reichsheeres 1930-1938." *MGM*, 14 (1973), 57-100.
Phelps, Reginald H., ed. "Aus den Groener Dokumenten." *DR*, 76 (1950), 530-541, 616-625, 735-744, 830-840, 915-922, 1013-1022; 77 (1951), 19-31.
La Vérité sur Fernand Brinon. Paris: n. p., 1947.
Vogelsang, Thilo, ed. "Hitlers Brief an Reichenau vom 4. Dezember 1932." *VfZ*, 7 (1959), 429-437.
———, ed. "Neue Dokumente zur Geschichte des Reichswehr 1930-1933." *VfZ*, 2 (1954), 397-436.
Wollstein, Günter, ed. "Eine Denkschrift des Staatssekretärs Bernhard von Bülow vom März 1933: Wilhelminische Konzeption der Aussenpolitik zu Beginn der nationalsozialistischen Herrschaft." *MGM*, 13 (1973), 77-94.

IV. Selected Published Memoirs, Diaries, and Personal Accounts.

Aloisi, Baron Pompeo. *Journal (25 juillet 1932–14 juin 1936)*. Maurice Vaussard, trans. Mario Toscano, ed. Paris: Plon, 1957.
  Pending the publication of Italian diplomatic documents in this period, the major source on Italian policy.
Braun, Magnus, Freiherr von. *Weg durch vier Zeitepochen*. Limburg a.d. Lahn: C. A. Starke, 1965.
  One of the more honest memoirs.
Braun, Otto. *Von Weimar zu Hitler*. 2nd ed. New York: Europa Verlag, 1940.
  Braun's recollections are confirmed by the documents.
Brüning, Heinrich. "Ein Brief." *DR*, 8 (1947), 1-22.
  Now almost entirely supplanted by Brüning's memoirs.
———. *Memoiren 1918-1934*. Stuttgart: Deutsche Verlags-Anstalt, 1970.
  These memoirs are highly informative, but they do not always square with the documents. They show that Brüning was a subtle

conservative and an intense nationalist. They do not explain the "great constructive policy"—probably rearmament—which the military told Brüning they wanted to introduce in 1930. See my review in *CEH* 4 (1971), 180-187.

Eden, Anthony, Earl of Avon. *The Eden Memoirs: Facing the Dictators.* London: Cassell, 1962.

In writing these memoirs, Eden and his assistant, David Dilks, had access to official papers.

Edge, Walter E. *A Jerseyman's Journal: Fifty Years of American Business and Politics.* Princeton: Princeton University Press, 1948.

François-Poncet, André. *Souvenirs d'une ambassade à Berlin, septembre 1931-octobre 1938.* Paris: Flammarion, 1946.

Written with style, but not all-revealing.

Gärtner, Margarete. *Botschafterin des guten Willens: Aussenpolitische Arbeit 1914-1955.* Bonn: Athenäum, 1955.

Gamelin, Maurice. *Servir.* 3 vols. Paris: Plon, 1946-1947.

Gereke, Günther. *Ich war königlich-preussischer Landrat.* Berlin (E.): Union Verlag, 1970.

Gereke claims to have opposed the use of work-creation funds for rearmament.

Grzesinski, Albert. *Inside Germany.* New York: Dutton, 1939.

A good source on Reichswehr-SPD relations.

Herriot, Edouard. *Jadis.* 2 vols. Paris: Flammarion, 1948-1952.

Contains a few useful fragments.

Hitler, Adolf. *Hitler's Secret Conversations, 1941-1944.* Norman Cameron and R. H. Stevens, trans. Introduction by H. R. Trevor-Roper. New York: Farrar, Strauss & Young, 1953.

As Norman Rich has noted in his *Hitler's War Aims*, the strictures on this source by Percy Schramm, editor of the Picker *Tischgespräche*, appear unjustified. This volume provides a record of much of Hitler's "table talk" not represented in the *Tischgespräche*.

———. *Hitlers Tischgespräche im Führerhauptquartier 1941-1942.* Transcribed by Henry Picker. Percy E. Schramm, ed. 2nd ed. Stuttgart: Seewald, 1965.

The most scholarly version of a collection first published by Gerhard Ritter in 1951. A 1976 edition, edited by Picker himself, mixes the same basic text with Picker's recollections.

———. *Hitlers zweites Buch: Ein Dokument aus dem Jahr 1928.* Gerhard L. Weinberg, ed. Stuttgart: Deutsche Verlags-Anstalt, 1961.

———. *Mein Kampf.* 35th ed. Munich: Franz Eher, 1933.

"35th ed." here doubtless means the thirty-fifth printing. But *Mein Kampf* was altered slightly over the years; see Hermann

Hammer, "Die deutschen Ausgaben von Hitlers 'Mein Kampf,' " *VfZ*, 4 (1956), 161-178.

Liddell Hart, Basil H. *The Memoirs of Captain Liddell Hart*. 2 vols. London: Cassell, 1965.

    The record of Britain's foremost military journalist, the champion of the "indirect approach" and the "British way in warfare."

Lipski, Józef. *Diplomat in Berlin, 1933-1939: Papers and Memoirs of Józef Lipski, Ambassador of Poland*. Wacław Jędrezejewicz, ed. New York: Columbia University Press, 1968.

    Based on papers held at the Józef Piłsudski Institute of America.

Manstein, Erich von. *Aus einem Soldatenleben 1887-1939*. Bonn: Athenäum-Verlag, 1958.

    Revealing. Manstein's pride in his contribution to arms planning overcame his discretion.

Meissner, Otto. *Staatssekretär unter Ebert-Hindenburg-Hitler*. Hamburg: Hoffmann & Campe, 1950.

    Often demonstrably inaccurate; must be used with caution.

Moffat, Jay Pierrepont. *The Moffat Papers: Selections from the Diplomatic Papers of Jay Pierrepont Moffat, 1919-1943*. Nancy Harvison Hooker, ed. Cambridge, Mass.: Harvard University Press, 1956.

    Provides a picture of the thinking and attitudes in the old State Department.

Morgan, John H. *Assize of Arms: The Disarmament of Germany and Her Rearmament (1919-1939)*. New York: Oxford University Press, 1946.

    See Chapter Two, note 18, above.

Müller, Vincenz. *Ich fand das wahre Vaterland*. Klaus Mammach, ed. Berlin (E.): Deutscher Militärverlag, 1963.

    Memoirs of one of Schleicher's onetime associates, later commander of the East German army.

Nadolny, Rudolf. *Mein Beitrag*. Wiesbaden: Limes Verlag, 1955.

    Nadolny's prickly personality and tendency to self-glorification show through. His courage was genuine, but one must doubt that some of the reported conversations took place.

Papen, Franz von. *Der Wahrheit eine Gasse*. Munich: Paul List, 1952.

    Papen's principal apologia.

―――. *Vom Scheitern einer Demokratie 1930-1933*. Mainz: v. Hase & Koehler, 1968.

Paul-Boncour, Joseph. *Entre Deux Guerres: Souvenirs sur la troisième république*. 3 vols. Paris: Plon, 1945-1946.

    Written mainly during World War II, and apparently without substantial records at hand; more useful for atmosphere than fact.

Rheinbaben, Werner Freiherr von. *Kaiser, Kanzler, Präsidenten.* Mainz: v. Hase & Koehler, 1968.

———. *Viermal Deutschland: Aus dem Erleben eines Seemanns, Diplomaten, Politikers 1895-1954.* Berlin: Argon Verlag, 1954.
Rheinbaben's recollections appear to have been influenced by time and subsequent events.

Schacht, Hjalmar. *1933: Wie eine Demokratie stirbt.* Düsseldorf: Econ, 1968.

———. *76 Jahre meines Lebens.* Bad Wörishofen: Kindler & Schiermeyer, 1953.
The longest and most detailed of Schacht's several apologias, but still not very informative on the Mefo system.

Schmidt, Paul. *Statist auf diplomatischer Bühne: Erlebnisse des Chefdolmetschers im Auswärtigen Amt mit den Staatsmännern Europas.* Frankfurt: Athenäum, 1964.
Useful recollections by Germany's ace interpreter.

Schwerin von Krosigk, Lutz Graf. *Es geschah in Deutschland: Menschenbilder unseres Jahrhunderts.* Tübingen: Rainer Wunderlich, 1951.
More vignettes than solid source material.

———. *Memoiren.* Stuttgart: Seewald Verlag, 1977.

———. *Staatsbankrott: Die Geschichte der Finanzpolitik des Deutschen Reiches von 1920 bis 1945, geschrieben vom letzten Reichsfinanzminister.* Göttingen: Musterschmidt, 1974.
The most detailed of Krosigk's three memoirs.

———. "Wie wurde der zweite Weltkrieg finanziert?" In *Bilanz des zweiten Weltkrieges: Erkenntnisse und Verpflichtungen für die Zukunft.* [Essays by various writers; no editor named.] Oldenburg: Gerhard Stalling, 1953.

Seeckt, Hans von. *Gedanken eines Soldaten.* Berlin: Verlag für Kulturpolitik, 1929.
Seeckt's writings were mistakenly assumed to express Reichswehr policy.

Severing, Carl. *Mein Lebensweg.* 2 vols. Cologne: Greven Verlag, 1950.
Less frank about Reichswehr affairs than Otto Braun's or Albert Grzesinski's memoirs.

Simon, John Allsebrook, Viscount. *Retrospect: Memoirs of the Rt. Hon. the Viscount Simon.* London: Hutchinson, [1952].
Virtually useless. For information on Simon, see Edward B. Segel's dissertation.

Stülpnagel, Joachim von. *75 Jahre meines Lebens.* Oberandorf/Obb.: Joachim von Stülpnagel, 1960.

A useful source, but unfortunately published privately and hard to find. Available at the Bundesarchiv-Militärarchiv, Freiburg.

Treviranus, Gottfried Reinhold. *Das Ende von Weimar: Heinrich Brüning und seine Zeit.* Düsseldorf: Econ, 1968.

Largely a defense of Brüning, by a loyal friend.

Vansittart, [Robert], Lord. *The Mist Procession.* London: Hutchinson, 1958.

Vansittart used an opaque style, partly no doubt to protect official secrecy and partly because it came naturally to him, and this reduces the book's usefulness. But the observations are usually telling and true.

Weizsäcker, Ernst von. *Erinnerungen.* Munich: Paul List, 1950.

Provides little on 1932-1933.

———. *Die Weizsäcker Papiere 1933-1950.* Leonidas E. Hill, ed. Frankfurt: Propyläen Verlag (Verlag Ullstein), 1974.

Weygand, [Maxime]. *Mémoires.* 3 vols. Paris: Flammarion, 1950-1957.

Useful mainly where it quotes from documentary records.

Wheeler-Bennett, John W. *Knaves, Fools, and Heroes: In Europe Between the Wars.* London: Macmillan, 1974.

Recollections, particularly of Wheeler-Bennett's experiences in Germany.

V. Selected Secondary Discussions.

Absolon, Rudolf. *Die Wehrmacht im Dritten Reich.* 3 vols. Boppard: Harald Boldt, 1969–.

Aigner, Dietrich. *Das Ringen um England: Das deutsch-britische Verhältnis: Die öffentliche Meinung 1933-1939: Tragödie zweier Völker.* Munich: Bechtle Verlag, 1969.

Along with some questionable judgments, reflects extensive research and gives a good statement of Hitler's views on Britain.

Aldcroft, Derek H. *The Inter-War Economy: Britain, 1919-1939.* New York: Columbia University Press, 1970.

Arndt, Fritz. "Vorbereitungen der Reichswehr für den militärischen Ausnahmezustand." *Zeitschrift für Militärgeschichte,* 4 (1965), 195-203.

Bankwitz, Philip C. F. *Maxime Weygand and Civil-Military Relations in Modern France.* Cambridge, Mass.: Harvard University Press, 1967.

Bariéty, Jacques, and Charles Bloch. "Une tentative de réconcilation franco-allemande et son échec (1932-33)." *Revue d'histoire moderne et contemporaine,* 15 (1968), 435-465.

Barnes, John. See under Middlemas, Keith.

Barnett, Correlli. *The Collapse of British Power.* New York: Morrow, 1972.
A biting critique of British leadership and "the Establishment" in the twentieth century. Barnett suggests that Britain should have pursued a tough policy of national egotism, and upheld the balance of power.

Beck, Earl R. *Verdict on Schacht: A Study in the Problem of Political "Guilt."* Florida State University Studies, No. 20. Tallahassee: Florida State University, 1955.

Becker, Josef. "Zur Politik der Wehrmachtsabteilung in der Regierungskrise 1926/27: Zwei Dokumente aus dem Nachlass Schleicher." *VfZ,* 14 (1966), 68-78.

Bennecke, Heinrich. *Hitler und die SA.* Munich: Günter Olzog, 1962.
Bennecke was a member of the SA.

———. *Die Reichswehr und der "Röhm-Putsch."* Munich: Günter Olzog, 1964.

Bennett, Edward W. *Germany and the Diplomacy of the Financial Crisis, 1931.* Cambridge, Mass.: Harvard University Press, 1962.

Berghahn, Volker R. *Der Stahlhelm: Bund der Frontsoldaten 1918-1935.* Düsseldorf: Droste, 1966.

Berndorff, Hans Rudolf. *General zwischen Ost und West: Aus den Geheimnissen der deutschen Republik.* Hamburg: Hoffmann & Campe, 1951.

Bernhardt, Walter. *Die deutsche Aufrüstung 1934-1939: Militärische und politische Konzeptionen und ihre Einschätzung durch die Alliierten.* Frankfurt: Bernard & Graefe, 1969.
Apologetic and superficial.

Bloch, Charles. *Hitler und die europäischen Mächte 1933/1934: Kontinuität oder Bruch.* Hamburg: Europäische Verlagsanstalt, 1966.

———. See also under Bariéty, Jacques.

Boelcke, Willi A. "Probleme der Finanzierung von Militärausgaben." In Friedrich Forstmeier and Hans Erich Volkmann, eds., *Wirtschaft und Rüstung am Vorabend des zweiten Weltkrieges.* Düsseldorf: Droste, 1975.

Bonnefous, Georges and Edouard. *Histoire politique de la troisième république.* 7 vols. Paris: Presses universitaires de France, 1956-1967.

Bowley, A. L. *Some Economic Consequences of the Great War.* London: Thornton Butterworth, [1930].

Bracher, Karl Dietrich. "Das Anfangstadium der Hitlerschen Aussenpolitik." *VfZ,* 5 (1957), 63-76.

———. *Die Auflösung der Weimarer Republik: Eine Studie zum*

Problem des Machtverfalls in der Demokratie. 4th ed. Villingen: Ring-Verlag, 1964.

Still the top-ranking study on the end of the Weimar Republic. The chapter on the Reichswehr is by Wolfgang Sauer.

———. "Die deutsche Armee zwischen Republik und Diktatur." In K. D. Bracher, *Deutschland zwischen Demokratie und Diktatur: Beiträge zur neueren Politik und Geschichte.* Bern: Scherz, 1964.

———. *Die deutsche Diktatur: Entstehung, Struktur, Folgen des Nationalsozialismus.* Cologne: Kiepenheuer & Witsch, 1969.

A brilliant survey, much more readable than Bracher's earlier, detailed volumes. A full translation in English is available as *The German Dictatorship: The Origins, Structure, and Effects of National Socialism,* Jean Steinberg, trans. (New York: Praeger, 1970).

———. "Die Stufen der Machtergreifung." In K. D. Bracher, Wolfgang Sauer, Gerhard Schulz, *Die nationalsozialistische Machtergreifung: Studien zur Errichtung des totalen Herrschaftssystems in Deutschland 1993/34.* 2nd ed. Cologne: Westdeutscher Verlag, 1962.

Brecht, Arnold. *Federalism and Regionalism in Germany: The Division of Prussia.* New York: Russell & Russell, 1971.

Breitman, Richard. "On German Social Democracy and General Schleicher." *CEH,* 9 (1976), 352-378.

Bright, Charles C. "Britain's Search for Security, 1930-1936: The Diplomacy of Naval Disarmament and Imperial Defense." Ph.D. dissertation, Yale University, 1970.

Bucher, Peter. *Der Reichswehrprozess: Der Hochverrat der Ulmer Reichswehroffiziere 1929-1930.* Boppard: Harald Boldt, 1967.

Cairns, John C. "A Nation of Shopkeepers in Search of a Suitable France." *AHR,* 79 (1974), 710-743.

Provides a vivid picture of British distrust of the French.

Caro, Kurt, and Walter Oehme. *Schleichers Aufstieg: Ein Beitrag zur Geschichte der Gegenrevolution.* Berlin: Rowohlt, 1933.

A good specimen of "investigative journalism."

Carroll, Berenice A. *Design for Total War: Arms and Economics in the Third Reich.* The Hague: Mouton, 1968.

Based largely on the account of Georg Thomas and the records he preserved.

———. "Germany Disarmed and Rearming, 1925-1935." *Journal of Peace Research,* 3 (1966), 114-124.

Carsten, Francis L. *The Reichswehr and Politics, 1918-1933.* Oxford: Oxford University Press, 1966.

A good survey from a moderately left viewpoint. There are some discrepancies between the English edition and the German (*Reichswehr und Politik 1918-1933* [Cologne: Kiepenheuer & Witsch, 1964]).

Caspar, Gustav Adolf. *Die sozialdemokratische Partei und das deutsche Wehrproblem in den Jahren der Weimarer Republik*. Beiheft 11 of the *Wehrwissenschaftliche Rundschau*. Berlin: Mittler, 1959.

Castellan, Georges. *Le Réarmement clandestin du Reich, 1930-1935: Vu par le 2ᵉ Bureau de l'État-Major Français*. Paris: Plon, 1954.
An extremely valuable source of material on French intelligence reporting.

———. "Von Schleicher, Von Papen et l'avènement d'Hitler." *Cahiers d'histoire de la guerre*, no. 1 (January 1949), 15-39.

Čelovsky, Boris. "Pilsudskis Präventivkrieg gegen das nationalsozialistische Deutschland (Entstehung, Verbreitung und Widerlegung einer Legende)." *Welt als Geschichte*, 14 (1954), 53-70.

Challener, Richard D. *The French Theory of the Nation in Arms, 1866-1939*. New York: Russell & Russell, 1965.
A valuable guide to French thinking on military affairs.

Chastenet, Jacques. *Histoire de la troisième république*. 7 vols. Paris: Hachette, 1952-1963.

Churchill, Winston S. *The World Crisis*. 4 vols. 2nd ed. New York: Charles Scribner's Sons, 1955.

Coghlan, F[rancis] A. "Armaments, Economic Policy, and Appeasement: Background to British Foreign Policy, 1931-1937." *History*, 57 (1972), 205-216.

Colvin, Ian. *None So Blind: A British Diplomatic View of the Origins of World War II*. New York: Harcourt, Brace & World, 1965.
Colvin used papers in Lady Vansittart's possession. Begins in 1933.

Conze, Werner. "Die politischen Entscheidungen in Deutschland 1929-1933." In W. Conze and Hans Raupach, eds., *Die Staats- und Wirtschaftskrise des Deutschen Reichs 1929/33*. Stuttgart: Ernst Klett, 1967.

Cowling, Maurice. *The Impact of Hitler: British Politics and British Policy, 1933-1940*. Cambridge: Cambridge University Press, 1975.
Insular and sometimes obscure, but with a wealth of material.

Craig, Gordon A. *The Politics of the Prussian Army, 1640-1945*. New York: Oxford University Press, 1956.
Balanced, well written, masterly. Considering the time span covered, amazingly concrete in the treatment of particular periods.

Czichon, Eberhard. *Wer Verhalf Hitler zur Macht? Zum Anteil der deutschen Industrie an der Zerstörung der Weimarer Republik*. 2nd ed. Cologne: Pahl-Rugenstein Verlag, 1971.
An East German analysis, but published only in the West.

D'Amoja, Fulvio. *Declino e prima crisi dell'Europa di Versailles: Studio sulla diplomazia italiana ed europea (1931-1933).* Milan: Dott. A. Guiffrè Editore, 1967.

Deak, Istvan. *Weimar Germany's Left-Wing Intellectuals: A Political History of the "Weltbühne" and Its Circle.* Berkeley: University of California Press, 1968.

Deist, Wilhelm. "Brüning, Herriot und die Abrüstungsgespräche von Bessinge, 1932." *VfZ,* 5 (1957), 265-272.

———. "Internationale und nationale Aspekte der Abrüstungsfrage 1924-1932." In Helmuth Rössler, ed., *Locarno und die Weltpolitik 1924-1932.* Göttingen: Musterschmidt, 1969.

———. "Schleicher und die deutsche Abrüstungspolitik im Juni/Juli 1932." *VfZ,* 7 (1959), 163-176.

Demeter, Karl. *Das deutsche Offizierskorps in Gesellschaft und Staat 1650-1945.* 4th ed. Frankfurt: Bernard & Graefe, 1965.

Dennis, Peter. *Decision by Default: Peacetime Conscription and British Defence, 1919-1939.* London: Routledge & Kegan Paul, 1972.

Divine, Robert A. "Franklin D. Roosevelt and Collective Security, 1933." *Mississippi Valley Historical Review,* 48 (1961-1962), 42-59. See Chapter Nine, note 25, above.

———. *The Illusion of Neutrality.* Chicago: University of Chicago Press, 1962.

———. *Roosevelt and World War II.* Baltimore: Johns Hopkins Press, 1969.

Dorpalen, Andreas. *Hindenburg and the Weimar Republic.* Princeton: Princeton University Press, 1964.

Dülffer, Jost. *Weimar, Hitler und die Marine: Reichspolitik und Flottenbau 1920-1939.* Düsseldorf: Droste, 1973.
A thoroughly documented study of naval and political policy, showing continuities between imperial and Nazi Germany, but also arguing that Hitler ultimately pursued radically different aims.

Duroselle, Jean-Baptiste. *From Wilson to Roosevelt: Foreign Policy of the United States, 1913-1945.* Nancy Lyman Roelker, trans. Cambridge, Mass.: Harvard University Press, 1963.

Ehni, Hans-Peter. *Bollwerk Preussen? Preussen-Regierung, Reich-Länder Problem und Sozialdemokratie 1928-1932.* Bonn-Bad Godesberg: Verlag Neue Gesellschaft, 1975.

Erbe, René. *Die nationalsozialistische Wirtschaftspolitik 1933-1939 im Lichte der modernen Theorie.* Zurich: Polygraphischer Verlag, 1958.
Provides a full explanation of the work-creation bills and their effects.

Erfurth, Waldemar. *Die Geschichte des deutschen Generalstabes von 1918 bis 1945*. Göttingen: Musterschmidt, 1957.
Gives a few indications on military plans, but stresses the personal and political.

Eyck, Erich. *A History of the Weimar Republic*. Harlan P. Hanson and Robert G. L. Waite, trans. 2 vols. Cambridge, Mass.: Harvard University Press, 1962.

Feiling, Keith. *The Life of Neville Chamberlain*. Rev. ed. London: Macmillan, 1970.

Feis, Herbert. *1933: Characters in Crisis*. Boston: Little, Brown, 1966.

Feldman, Gerald D. "The Social and Economic Policies of German Big Business, 1918-1929." *AHR*, 75 (1969-1970), 47-55.

Ferrell, Robert H. *American Diplomacy in the Great Depression: Hoover-Stimson Foreign Policy, 1929-1933*. New Haven: Yale University Press, 1957.
Conveys a clear picture of Hoover's and Stimson's contrasting personalities.

Foerster, Wolfgang. *Ein General kämpft gegen den Krieg: Aus nachgelassenen Papieren des Generalstabschefs Ludwig Beck*. Munich: Münchner Dom Verlag, 1949.
Hagiographical. See Chapter Ten, note 87, above.

Ford, Franklin L. "Three Observers in Berlin." In Gordon A. Craig and Felix Gilbert, eds., *The Diplomats, 1919-1939*. Princeton: Princeton University Press, 1953.

Fox, John P. "Britain and the Inter-Allied Military Commission of Control, 1925-1926." *JCH*, 4 (1969), no. 2, 143-164.

Frankenstein, Robert. "A propos des aspects financiers du réarmement français (1935-1939)." *Revue d'histoire de la deuxième guerre mondiale*, [26] (1976), no. 102, 1-20.

Fraser, Christine. *Der Austritt Deutschlands aus dem Völkerbund, seine Vorgeschichte und seine Nachwirkungen*. Doctoral dissertation, University of Bonn, 1969.

Furnia, Arthur H. *The Diplomacy of Appeasement: Anglo-French Relations and the Prelude to World War II, 1931-1938*. Washington, D.C.: University Press of Washington, 1960.
Not as far off the mark as some critics have supposed.

Gaertner, Franz von. *Die Reichswehr in der Weimarer Republik: Erlebte Geschichte*. Darmstadt: Fundus-Verlag, [1969].
Mostly devoted to the first years.

Gasiorowski, Zygmunt J. "Did Pilsudski Attempt to Initiate a Preventive War in 1933?" *JMH*, 27 (1955), 131-151.
Gasiorowski says no.

Gasiorowski, Zygmunt J. "The German-Polish Nonaggression Pact of 1934." *Journal of Central European Affairs*, 15 (1955), 3-29.
Gatzke, Hans W. "Russo-German Military Collaboration During the Weimar Republic." *AHR*, 63 (1957-1958), 565-597.
———. *Stresemann and the Rearmament of Germany*. Baltimore: Johns Hopkins Press, 1954.
Exposes Stresemann's revisionism and concealment of German rearmament.
Gehl, Jürgen. *Austria, Germany, and the Anschluss, 1931-1938*. London: Oxford University Press, 1963.
George, Margaret. *The Warped Vision: British Foreign Policy, 1933-1939*. Pittsburgh: University of Pittsburgh Press, 1965.
Geyer, Michael. "Militär, Rüstung und Aussenpolitik: Aspekte militärischer Revisionspolitik in der Zwischenkriegszeit." In Manfred Funke, ed., *Hitler, Deutschland und die Mächte: Studien zur Aussenpolitik des Dritten Reiches*. Düsseldorf: Droste, 1976.
Many insights, but some of Geyer's theses need qualification or seasoning.
———. "Die Wehrmacht der deutschen Republik ist die Reichswehr: Bemerkungen zur neueren Literatur." *MGM*, 14 (1973), 152-199.
A review article which also makes some penetrating observations on Reichswehr development.
———. "Das Zweite Rüstungsprogramm (1930-1934)." *MGM*, 16 (1975), 125-172.
The first eleven pages constitute an important discussion of material armament and its financing; the rest is documents and footnotes.
———. Note: Geyer's draft dissertation, variously titled, has been seen and used by several German writers, but an accepted version was not available from the University of Freiburg in time for use in this study.
Gibbs, N. H. *Rearmament Policy*. (In Sir James Butler, ed., *History of the Second World War: Grand Strategy*, 6 vols., as Vol. I.) London: Her Majesty's Stationery Office, 1976.
Mainly post-November 1933, but surveys the stringent reductions of 1919-1922 and the operation of the ten-year rule.
Gilbert, Martin. *The Roots of Appeasement*. New York: New American Library, 1966.
Gilbert's second, and more sympathetic, thoughts, following *The Appeasers*.
———. *Sir Horace Rumbold: Portrait of a Diplomat, 1869-1941*. London: Heinemann, 1973.

———. *Winston S. Churchill*. 5 vols. Boston: Houghton, Mifflin, 1966–.
   The first two volumes were completed by Randolph S. Churchill; Volume V (*Prophet of Truth, 1922-1939*, 1977) is the most pertinent to this study. Companion volumes supply documents.
Gilbert, Martin, and Richard Gott. *The Appeasers*. London: Weidenfeld & Nicolson, 1963.
   A sharp critique of the immorality of appeasement.
Gilmore, Kenneth Harold. "Nazi Military Policies and Nazi-Reichswehr Relations, 1923-1933." Ph.D. dissertation, University of Florida, 1975.
Gilpin, Robert Peyton. "Christian Names vs. Führerpolitik: Anglo-German Relations, 1933-1935." Ph.D. dissertation, Duke University, 1972.
Glees, Anthony. "Albert Grzesinski and the Politics of Prussia, 1926-1930." *English Historical Review*, 89 (1974), 814-834.
Gordon, Harold J., Jr. *Hitler and the Beer Hall Putsch*. Princeton: Princeton University Press, 1972.
   Probably the definitive study.
———. *The Reichswehr and the German Republic, 1919-1926*. Princeton: Princeton University Press, 1957.
   Gordon pioneered in using the Seeckt papers. Most German comment has found him too uncritical on Seeckt.
———. "Reichswehr und Politik in der Weimarer Republik." *Politische Studien*, 22 (1971), 34-45.
   Acquits the Reichswehr of dishonorable, insubordinate, or illegal activity in politics. Ignores Reichswehr efforts to rearm, and political maneuvers to that end.
Gott, Richard. See under Gilbert, Martin.
Gunsberg, Jeffrey Albert. " 'Vaincre ou Mourir': The French High Command and the Defeat of France, 1919–May 1940." Ph.D. dissertation, Duke University, 1974.
Hahn, Erich J. C. See under Krüger, Peter.
Hallgarten, George W. F. *Hitler, Reichswehr und Industrie: Zur Geschichte der Jahre 1918-1933*. Frankfurt: Europäische Verlagsanstalt, 1962.
Hallgarten, G.W.F., and Joachim Radkau. *Deutsche Industrie und Politik von Bismarck bis heute*. Frankfurt: Europäische Verlagsanstalt, 1974.
   Hallgarten's contribution (which runs up to 1933) is polemical and, to me, unconvincing.
Hammerstein, Kunrat Freiherr von. "Schleicher, Hammerstein und die

Machtübernahme." *Frankfurter Hefte*, 11 (1956), 11-18, 117-128, 163-176.

Hauser, Oswald. *England und das Dritte Reich: Eine dokumentierte Geschichte der englisch-deutschen Beziehungen von 1933 bis 1939 auf Grund unveröffentlichter Akten aus dem britischen Staatsarchiv*. Stuttgart: Seewald, 1972-.

Hauser argues that Britain needed a French alliance; he slights Anglo-French frictions.

Heineman, John L. "Constantin von Neurath and German Policy at the London Economic Conference of 1933: Backgrounds to the Resignation of Alfred Hugenberg." *JMH*, 41 (1969), 160-188.

———. "Constantin von Neurath as Foreign Minister, 1932-1935: A Study of a Conservative Civil Servant and Germany's Foreign Policy." Ph.D. dissertation, Cornell University, 1965.

Helbich, Wolfgang J. "Between Stresemann and Hitler: The Foreign Policy of the Brüning Government." *World Politics*, 12 (1959), 24-44.

———. *Die Reparationen in der Ära Brüning*. Berlin: Colloquium, 1962.

Higgins, Trumbull. *Winston Churchill and the Second Front, 1940-1943*. New York: Oxford University Press, 1957.

Higgins uncovered the caution behind Churchill's "bulldog" reputation, and traced back the roots of that caution.

Higham, Robin. *Armed Forces in Peacetime*. London: Foulis, 1962.

———. *The Military Intellectuals in Britain: 1918-1939*. New Brunswick, N.J.: Rutgers University Press, 1966.

Hildebrand, Klaus. " 'British Interests' and 'Pax Britannica': Grundfragen englischer Aussenpolitik im 19. und 20. Jahrhundert." *HZ*, 220 (1975), 623-639.

———. *Deutsche Aussenpolitik 1933-1945: Kalkül oder Dogma?* 2nd ed. Stuttgart: W. Kohlhammer, 1973.

Finds continuity with the past in Hitler's early foreign policy, while adding that Hitler's dogmatic racism later brought a break. Available in English as *Foreign Policy of the Third Reich*, Anthony Fothergill, trans. (Berkeley: University of California Press, 1973).

———. "Hitlers Ort in der Geschichte des preussisch-deutschen Nationalstaates." *HZ*, 217 (1973), 584-632.

An extremely useful review of historical discussion on 19th and 20th century German history.

———. *Vom Reich zum Weltreich: Hitler, NSDAP und koloniale Frage 1919-1945*. Munich: Wilhelm Fink Verlag, 1969.

Shows that Hitler viewed colonies as either a hindrance to, or a lever for obtaining, a British alliance.

Hillgruber, Andreas. *Deutschlands Rolle in der Vorgeschichte der beiden Weltkriege*. Göttingen: Vandenhoeck & Ruprecht, 1967.

———. "England in Hitlers aussenpolitischer Konzeption." *HZ*, 218 (1974), 65-84.

———. *Kontinuität und Diskontinuität in der deutschen Aussenpolitik von Bismarck bis Hitler*. Düsseldorf: Droste, 1969.

A thought-provoking essay, based on recent studies.

———. "Militarismus am Ende der Weimarer Republik." In A. Hillgruber, *Grossmachtpolitik und Militarismus im 20. Jahrhundert: 3 Beiträge zum Kontinuitätsproblem*. Düsseldorf: Droste, 1974.

Uses an unpublished study of Michael Geyer's.

Homze, Edward L. *Arming the Luftwaffe: The Reich Air Ministry and the German Aircraft Industry, 1919-1939*. Lincoln: University of Nebraska Press, 1976.

Hoop, Jean Marie d'. "Frankreichs Reaktion auf Hitlers Aussenpolitik 1933-39." *Geschichte in Wissenschaft und Unterricht*, 15 (1964), 211-223.

Howard, Michael. *The British Way in Warfare: A Reappraisal*. London: Jonathan Cape, 1975.

The 1974 Neale lecture; an excellent critique of Liddell Hart and the "blue water" school.

———. *The Continental Commitment: The Dilemma of British Defence Policy in the Era of the Two World Wars*. London: Temple Smith, 1972.

A valuable survey of the British resistance to continental involvements.

Hughes, Judith M. *To the Maginot Line: The Politics of French Military Preparations in the 1920's*. Cambridge, Mass.: Harvard University Press, 1971.

Shows how preventive action was rendered impractical.

Immanuel, Friedrich. *Die deutsche Miliz der Zukunft: Eine Frage von entscheidender Bedeutung für das deutsche Volk*. Berlin: Mittler, 1933.

The view expounded here was not unlike the Manstein plan, only with six-month instead of three-month service.

Irving, David. *The Rise and Fall of the Luftwaffe: The Life of Luftwaffe Marshal Erhard Milch*. London: Weidenfeld & Nicolson, 1973.

Jablon, Howard. "The State Department and Collective Security, 1933-34." *Historian*, 33 (1970-1971), 248-263.

Jäckel, Eberhard. *Hitlers Weltanschauung: Entwurf einer Herrschaft.* Tübingen: Rainer Wunderlich, 1969.
   Brings out strongly the link between Hitler's anti-Semitism and his foreign-policy objective of gaining Lebensraum on the continent.
Jacobsen, Hans-Adolf. *Nationalsozialistische Aussenpolitik 1933-1938.* Frankfurt: A. Metzner Verlag, 1968.
   Dedicated to the thesis that Nazi foreign policy and diplomatic machinery differed sharply from their predecessors. Jacobsen seems to exaggerate the importance of the new developments and institutions.
Jacobsen, Otto. *Erich Marcks: Soldat und Gelehrter.* Göttingen: Musterschmidt, 1971.
   Uses Marcks's papers.
Jacobson, Jon. *Locarno Diplomacy: Germany and the West, 1925-1929.* Princeton: Princeton University Press, 1972.
Jarausch, Konrad H. *The Four Power Pact, 1933.* Madison: State Historical Society of Wisconsin, 1965.
   The most thorough monograph on Mussolini's proposal.
Jędrezejewicz, Waclaw, "The Polish Plan for a 'Preventive War' against Germany in 1933." *Polish Review* (New York), 11 (1966), 62-91. Argues that there was such a plan.
Jordan, W. M. *Great Britain, France, and the German Problem, 1918-1939: A Study of Anglo-French Relations in the Making and Maintenance of the Versailles Settlement.* London: Oxford University Press, 1943.
Keene, Thomas H. "The Foreign Office and the Making of British Foreign Policy, 1929-1935." Ph.D. dissertation, Emory University, 1974.
Kehr, Eckart. *Der Primat der Innenpolitik: Gesammelte Aufsätze zur preussisch-deutschen Sozialgeschichte im 19. und 20. Jahrhundert.* Hans-Ulrich Wehler, ed. Berlin: Walter de Gruyter, 1965.
   Includes two perspicacious articles, "Zur Genesis der Königlich Preussischen Reserveoffiziers" and "Zur Soziologie der Reichswehr"; the latter, however, is contradicted at some points by evidence now available.
Keynes, John Maynard. *The Economic Consequences of the Peace.* New York: Harcourt Brace, 1920.
   Probably the most influential book in English in shaping opinion on the Treaty of Versailles.
Kimmich, Christoph M. *Germany and the League of Nations.* Chicago: University of Chicago Press, 1976.
   Includes two chapters on the Disarmament Conference.

Kindleberger, Charles P. *The World in Depression, 1929-1939.* Berkeley: University of California Press, 1973.

Kitchen, Martin. *A Military History of Germany from the Eighteenth Century to the Present Day.* London: Weidenfeld & Nicolson, 1975.
Reflects a very critical view of the German military.

Klein, Burton H. *Germany's Economic Preparations for War.* Cambridge, Mass.: Harvard University Press, 1959.
An influential book, used by A.J.P. Taylor in answering his critics. But Klein's data on the prewar period have been questioned. See, for other analyses, Berenice Carroll, *Design for Total War*; Willi A. Boelcke, "Probleme der Finanzierung"; and Arthur Schweitzer, "Die wirtschaftliche Wiederaufrüstung Deutschlands von 1934-1936," *Zeitschrift für die gesamte Staatswissenschaft*, 114 (1958), 594-637.

Köhler, Henning. "Arbeitsbeschaffung, Siedlung und Reparationen in der Schlussphase der Regierung Brüning." *VfZ*, 17 (1969), 276-307.

———. "Sozialpolitik von Brüning bis Schleicher." *VfZ*, 21 (1973), 146-150.

Krecker, Lothar. "Die diplomatischen Verhandlungen über den Viererpakt vom 15. Juli 1933." *Welt als Geschichte*, 21 (1961), 226-237.
A good short treatment.

Krüger, Peter, and Erich J. C. Hahn. "Der Loyalitätskonflikt des Staatssekretärs Bernhard Wilhelm von Bülow im Frühjahr 1933." *VfZ*, 20 (1972), 376-410.

Kuhn, Axel. *Hitlers aussenpolitisches Programm: Entstehung und Entwicklung 1919-1939.* Stuttgart: Ernst Klett, 1970.
An able study, stressing Hitler's interest in a British alliance, while noting his growing conviction (in the later 1930s) of British weakness.

Lafore, Lawrence. *The End of Glory: An Interpretation of the Origins of World War II.* Philadelphia: Lippincott, 1970.

Landes, David. *The Unbound Prometheus: Technological Change and Industrial Development in Western Europe from 1750 to the Present.* Cambridge: Cambridge University Press, 1959.

Larmour, Peter J. *The French Radical Party in the 1930's.* Stanford, Calif.: Stanford University Press, 1964.

L'Huillier, Fernand. *Dialogues franco-allemandes, 1925-1933.* Strasbourg: Diffusions Ophrys, 1971.

Liddell Hart, Basil H. *The German Generals Talk.* New York: Morrow, 1948. Published in England as *The Other Side of the Hill.*

Link, Werner. *Die amerikanische Stabilisierungspolitik in Deutschland 1921-1932.* Düsseldorf: Droste, 1970.

An interesting and important study; the "stabilization" here is political as well as financial and economic.

Loosli-Usteri, Carl. *Geschichte der Konferenz für die Herabsetzung und die Begrenzung der Rüstungen 1932-1934: Ein politischer Weltspiegel.* Zurich: Polygraphischer Verlag, 1940.

An early study, sympathetic to the German case.

Lubs, Gerhard. *I. R. 5: Aus der Geschichte eines pommerschen Regiments 1920-1945.* Bochum: Gerhard Lubs, [1965].

Most histories of German military units only cover the war years. This privately published volume gives a view of the Weimar years, Umbau, and the later zenith and collapse of the Wehrmacht, all as seen from below.

Ludwig, Karl-Heinz. "Strukturmerkmale nationalsozialistischer Aufrüstung bis 1935." In Friedrich Forstmeier and Hans-Erich Volkmann, eds., *Wirtschaft und Rüstung am Vorabend des Zweiten Weltkrieges.* Düsseldorf: Droste, 1975.

Macleod, Iain. *Neville Chamberlain.* New York: Atheneum, 1962.

Maier, Charles S. *Recasting Bourgeois Europe: Stabilization in France, Germany, and Italy in the Decade after World War I.* Princeton: Princeton University Press, 1975.

Malanowski, Wolfgang. "Die deutsche Politik der militärischen Gleichberechtigung von Brüning bis Hitler." *Wehrwissenschaftliche Rundschau,* 5 (1955), 351-364.

Marcks, [Erich]. "Das Reichsheer von 1919-1935." In Karl Linnebach, ed., *Deutsche Heeresgeschichte.* Hamburg: Hanseatische Verlagsanstalt, 1935.

Marcon, Helmut. *Arbeitsbeschaffungspolitik der Regierungen Papen und Schleicher: Grundsteinlegung für die Beschäftigungspolitik im Dritten Reich.* Bern: Herbert Lang, 1974.

Says very little on the rearmament function of work creation.

Marquand, David. *Ramsay MacDonald.* London: Jonathan Cape, 1977.

Very well balanced, fills a longstanding need. Uses the MacDonald papers.

Matuschka, Edgar Graf von. See under Wohlfeil, Rainer.

Medlicott, W. N. *British Foreign Policy Since Versailles, 1919-1963.* 2nd ed. London: Methuen, 1968.

Meier-Welcker, Hans. *Seeckt.* Frankfurt: Bernard & Graefe, 1967.

Meinck, Gerhard. *Hitler und die deutsche Aufrüstung 1933-1937.* Wiesbaden: Franz Steiner, 1959.

Presents German rearmament as the work of Hitler, while minimizing earlier planning; this view is undercut by material now available. Beyond this, see Chapter Seven, note 48, above.

Messerschmidt, Manfred. *Die Wehrmacht im NS-Staat: Zeit der Indoktrination*. Hamburg: R. v. Decker's Verlag, 1969.
Not a history of the Wehrmacht as such, but a discussion of its—and the NSDAP's—methods of indoctrination and leadership.

Metzerath, Horst, and Henry A. Turner, Jr. "Die Selbstfinanzierung der NSDAP 1930-1932." *GG*, 3 (1977), 59-92.

Middlemas, Keith. *Diplomacy of Illusion: The British Government and Germany, 1937-1939*. London: Weidenfeld & Nicolson, 1972.

Middlemas, Keith, and John Barnes. *Baldwin: A Biography*. London: Weidenfeld & Nicolson, 1969.

Milward, Alan S. *The Economic Effects of the Two World Wars on Britain*. London: Macmillan, 1970.

Minart, Jacques. *Le Drame du désarmement français (ses aspects politiques et techniques): La Revanche allemande (1918-1939)*. Paris: Le Nef de Paris, 1959.

Mueller-Hillebrand, Burkhart. *Das Heer 1933-1945: Entwicklung des organisatorischen Aufbaues*. 3 vols. Darmstadt: Mittler, 1954-1969.
Not very informative on the pre-Hitler period or the developments of 1933.

Müller, Klaus-Jürgen. *Das Heer und Hitler: Armee und nationalsozialistisches Regime 1933-1940*. Stuttgart: Deutsche Verlags-Anstalt, 1969.
Carefully researched; unearths new material on the relation of the Reichswehr to Röhm.

Newman, Bernard. *The Sosnowski Affair: Inquest on a Spy*. London: Werner Laurie, 1954.

Nickolaus, Günter. *Die Milizfrage in Deutschland von 1848 bis 1933*. Berlin: Junker & Dünnhaupt, 1933.
Reviews published material now difficult to obtain.

Northedge, F. S. *The Troubled Giant: Britain among the Great Powers, 1916-1939*. London: London School of Economics and G. Bell, 1966.

Oehme, Walter. See under Caro, Kurt.

Oertzen, Karl Ludwig von. *Rüstung und Abrüstung: Eine Umschau über das Heer- und Kriegswesen aller Länder*. Berlin: Mittler, 1929.

Offner, Arnold A. *American Appeasement: United States Foreign Policy and Germany, 1933-1938*. Cambridge, Mass.: Harvard University Press, Belknap Press, 1969.
Shows that American diplomats at first mistook the character of Nazi rule.

O'Neill, Robert J. *The German Army and the Nazi Party, 1933-1939*. London: Cassell, 1966.

Emphasizes military personalities in its discussion of army-party relations. The work is sharply criticized in Klaus-Jürgen Müller's *Heer und Hitler*.

Osgood, Robert E. *Ideals and Self-Interest in America's Foreign Relations: The Great Transformation of the Twentieth Century*. Chicago: University of Chicago Press, 1953.
An important and instructive book, influenced by Hans Morgenthau.

Oshima, Michiyoshi. ["Reichswehr und Rechnungshof: Die Bedeutung des Kabinettsbeschlusses vom 4. April 1933."] *Mita Gakkai Zasshi*, 69 (1976), 298-318.
In Japanese. The cabinet decision is quoted in German on pp. 317-318, and a German-language summary of the article is included in the same issue of the journal. I am grateful to Herr Verlande of the Bundesarchiv for calling this reference to my attention.

Parker, R.A.C. *Europe 1919-45*. New York: Delacorte, 1970.

Petersen, Jens. *Hitler-Mussolini: Die Entstehung der Achse Berlin-Rom 1933-1936*. Tübingen: Max Niemeyer, 1973.

Petzina, Dieter. "Hauptprobleme der deutschen Wirtschaftspolitik 1932/33." *VfZ*, 15 (1967), 18-55.
A clear, concise, and balanced discussion.

Phelps, Reginald H. "Hitler als Parteiredner im Jahre 1920." *VfZ*, 11 (1963), 274-330.

Post, Gaines, Jr. *The Civil-Military Fabric of Weimar Foreign Policy*. Princeton: Princeton University Press, 1973.
Uncovers the revisionism common to diplomats and soldiers, and in particular, the willingness of the military to wage another war. I do think Post tends to overstate the amount of cooperation between the Reichswehr and Foreign Ministry.

Rabenau, Friedrich von. *Seeckt: Aus seinem Leben*. Leipzig: Von Hase and Koehler, 1940.
Supplanted as a biography by Meier-Welcker's study, this is now of interest primarily as a document on the accommodation of conservative officers to the Nazi regime.

Radkau, Joachim. See under Hallgarten, G.W.F.

Rautenberg, Hans-Jürgen. *Deutsche Rüstungspolitik vom Beginn der Genfer Abrüstungskonferenz bis zur Wiedereinführung der allgemeinen Wehrpflicht 1932-1935*. Doctoral dissertation, University of Bonn, 1973.
A very useful study.

———. "Drei Dokumente zur Planung eines 300 000 Mann-Friedensheeres aus dem Dezember 1933." *MGM*, 22 (1977), 103-139.
The first eleven pages of text discuss the December 1933 decision.

Riekhoff, Harald von. *German-Polish Relations, 1918-1933.* Baltimore: Johns Hopkins Press, 1971.

Ritter, Gerhard. "Das Problem des Militarismus in Deutschland." *HZ*, 177 (1954), 21-48.
For a fuller exposition of Ritter's views, see his *Staatskunst und Kriegshandwerk*. In spite of Ritter's apologetic tendencies and his efforts to impose conformity on the German historical profession, his thesis—of a contrast between civilian and military thinking—has much to be said for it. Not that the civilians were entirely moderate and peaceloving, but the soldiers were so much less so.

Roberts, Henry L. "The Diplomacy of Colonel Beck." In Gordon A. Craig and Felix Gilbert, eds., *The Diplomats, 1919-1939.* Princeton: Princeton University Press, 1953.

Robertson, Esmonde. *Hitler's Pre-War Policy and Military Plans, 1933-1939.* London: Longmans, 1963.
Robertson made early use of captured documents, preparing what was at first an official paper. Unfortunately, he could not fully footnote his study at the time it was published, and this diminishes its usefulness.

Rochat, Giorgio. *Militari e politici nella preparazione della campagna d'etiopia: Studio e documenti 1932-1936.* Milan: Franco Angeli, 1971.
An important study, presenting new material.

Rohe, Karl. *Das Reichsbanner Schwarz Rot Gold: Ein Beitrag zur Geschichte und Struktur der politischen Kampfverbände zur Zeit der Weimarer Republik.* Düsseldorf: Droste, 1966.

Roos, Hans. "Die 'Präventivkriegspläne' Pilsudskis von 1933." *VfZ*, 3 (1955), 344-363.

Roskill, Stephen. *Hankey: Man of Secrets.* 3 vols. New York: Walker, 1970-1974.
Provides extensive material from the Hankey papers.

———. *Naval Policy between the Wars.* 2 vols. (Vol. I: *The Period of Anglo-American Antagonism, 1919-1929*; Vol. II: *The Period of Reluctant Rearmament, 1930-1939*) (Vol. I:) New York: Walker, 1968. (Vol. II:) Annapolis: Naval Institute Press, 1976.

Rowse, A. L. *All Souls and Appeasement: A Contribution to Contemporary History.* London: Macmillan, 1961.
Personal and censorious.

Salewski, Michael. *Entwaffnung und Militärkontrolle in Deutschland 1919-1927.* Munich: R. Oldenbourg, 1966.
Another full-scale study on this subject, using unpublished British and French documents, would be desirable.

———. "Zur deutschen Sicherheitspolitik in der Spätzeit der Weimarer Republik." *VfZ,* 22 (1974), 121-147.

Sauer, Wolfgang. "Die Mobilmachung der Gewalt." In K. D. Bracher, W. Sauer, and G. Schulz, *Die nationalsozialistische Machtergreifung,* 2nd ed. Cologne: Westdeutscher Verlag, 1962.
Sauer's principal contribution to Reichswehr history. Contains many valuable insights, but sources now available invalidate it at some points. See Appendix.

———. "Die Reichswehr." In Karl Dietrich Bracher, *Die Auflösung der Weimarer Republik: Eine Studie zum Problem des Machtverfalls in der Demokratie.* 4th ed. Villingen: Ring Verlag, 1964.

Schiller, Karl. *Arbeitsbeschaffung und Finanzordnung in Deutschland.* Berlin: Junker & Dünnhaupt, 1936.
Provides the contemporary rationale for the mechanism for financing work creation. By the later West German Finance Minister.

Schlesinger, Arthur M., Jr. *The Age of Roosevelt.* 3 vols. completed. (Vol. I: *The Crisis of the Old Order, 1919-1933*; Vol. II: *The Coming of the New Deal*) Boston: Houghton, Mifflin, 1957-.

Schlueter, Rolf R. *Probleme der deutschen Friedensbewegung in der Weimarer Republik.* Doctoral dissertation, University of Bonn, 1974.

Schneider, Michael. *Das Arbeitsbeschaffungsprogramm des ADGB: Zur gewerkschaftlichen Politik in der Endphase der Weimarer Republik.* Bonn: Verlag Neue Gesellschaft, 1975.

Schramm, Percy Ernst. *Hitler: The Man and the Military Leader.* Donald W. Detwiler, ed. and trans. Chicago: Quadrangle, 1971.
A translation of the long and penetrating discussion of Hitler's personality that Schramm provided with his edition of the *Tischgespräche.*

Schröder, Hans-Jürgen. *Deutschland und die Vereinigten Staaten 1933-1939: Wirtschaft und Politik in der Entwicklung des deutschamerikanischen Gegensatzes.* Wiesbaden: Franz Steiner, 1970.
Brings out the initial Hitlerian policy of placating the United States.

Schubert, Günter. *Anfänge nationalsozialistischer Aussenpolitik.* Cologne: Verlag Wissenschaft und Politik, 1963.
Traces the early development of Hitler's views.

Schuker, Stephen A. *The End of French Predominance in Europe: The*

Financial Crisis of 1924 and the Adoption of the Dawes Plan. Chapel Hill: University of North Carolina Press, 1976.

This in-depth study of international political and financial relations in the 1920s provides valuable background for the 1930s. Knowledgeable and thorough. This may be the best study of interaction between nations since William L. Langer's *Diplomacy of Imperialism* appeared in 1935.

Schulz, Gerhard. "Die Anfänge des totalitären Massnahmenstaates." In K. D. Bracher, W. Sauer, and G. Schulz, *Die nationalsozialistische Machtergreifung*, 2nd ed. Cologne: Westdeutscher Verlag, 1962.

Scott, William E. *Alliance Against Hitler: The Origins of the Franco-Soviet Pact.* Durham, N.C.: Duke University Press, 1962.

Segel, Edward B. "Sir John Simon and British Foreign Policy: The Diplomacy of Disarmament in the Early 1930's." Ph.D. dissertation, University of California, Berkeley, 1969.

An excellent dissertation, drawing extensively on the private Simon papers.

Shay, Robert P., Jr. *British Rearmament in the Thirties: Politics and Profits.* Princeton: Princeton University Press, 1977.

Thoroughly explores the files of the British Treasury.

Skidelsky, Robert. *Politicians and the Slump: The Labour Government of 1929-1931.* London: Macmillan, 1967.

Sohn-Rethel, Alfred. *Ökonomie und Klassenstruktur des deutschen Faschismus: Aufzeichnungen und Analysen.* Johannes Agnoli, Bernhard Blanke, Niels Kadritzke, eds. Frankfurt: Suhrkamp, 1973.

An interesting collection of papers by a covert Marxist who worked for a lobbying organization serving heavy industry. Although purporting to date from the 1930s, internal evidence shows that some of the material has at least been revised since 1945.

Solleder, Fridolin. *Vier Jahre Westfront: Geschichte des Regiments List, R. I. R. 16.* Munich: Verlag Max Schick, 1932.

Soltikow, Michael Alexander Graf. *Rittmeister Sosnowski.* Hamburg: Verlag der Sternbücher, 1954.

Soulié, Michel. *La Vie politique d'Edouard Herriot.* Paris: Librairie Armand Colin, 1962.

Uses Herriot's papers.

Stegmann, Dirk. "Zum Verhältnis von Grossindustrie und Nationalsozialismus 1930-1933: Ein Beitrag zur Geschichte der sog. Machtergreifung." *Archiv für Sozialgeschichte*, 13 (1973), 399-482.

See the reply of Henry Ashby Turner, Jr., "Grossunternehmertum und Nationalsozialismus 1930-1933: Kritisches und Ergänzendes zu zwei neuen Forschungsbeiträgen." *HZ*, 221 (1975), 18-68.

Stürmer, Michael. *Koalition und Opposition in der Weimarer Republik 1924-1928*. Düsseldorf: Droste, 1967.

Sywottek, Jutta. *Mobilmachung für den totalen Krieg: Die propagandistische Vorbereitung der deutschen Bevölkerung auf den Zweiten Weltkrieg*. Opladen: Westdeutscher Verlag, 1976.

Taylor, A.J.P. *English History, 1914-1945*. Oxford: Oxford University Press, 1965.

———. *The Origins of the Second World War*. New York: Atheneum, 1961.

See Chapter Two, esp. note 1, above.

Tessin, Georg. *Formationsgeschichte der Wehrmacht 1933-1939: Stäbe und Truppenteile des Heeres und der Luftwaffe*. Boppard: Harald Boldt, 1959.

Thomas, Georg. *Geschichte der deutschen Wehr- und Rüstungswirtschaft (1918-1943/45)*. Wolfgang Birkenfeld, ed. Boppard: Harald Boldt, 1966.

Tournoux, Paul-Emile. *Haut Commandement: Gouvernement et défense des frontières du Nord et de l'Est, 1919-1939*. Paris: Nouvelles Editions latines, 1960.

An important source of information on French military plans.

Toynbee, Arnold J., et al. *Survey of International Affairs*. Annual. London: Oxford University Press, 1925-1971.

Treue, Wilhelm. "Der deutsche Unternehmer in der Weltwirtschaftskrise 1928 bis 1933." In Werner Conze and Hans Raupach, eds., *Die Staats- und Wirtschaftskrise des Deutschen Reichs 1929/33*. Stuttgart: Ernst Klett, 1967.

Trumpp, Thomas. *Franz von Papen, der preussisch-deutsche Dualismus und die NSDAP in Preussen: Ein Beitrag zur Vorgeschichte des 20. Juli 1932*. Doctoral dissertation, University of Tübingen, [1963?].

Turner, Henry Ashby, Jr. "Big Business and the Rise of Hitler." *AHR*, 75 (1969-1970), 56-70.

———. "Hitlers Einstellung zu Wirtschaft und Gesellschaft vor 1933." *GG*, 2 (1976), 89-117.

———. See also under Metzerath, Horst, and Stegmann, Dirk.

Völker, Karl-Heinz. *Die deutsche Luftwaffe 1933-1939*. Stuttgart: Deutsche Verlags-Anstalt, 1967.

———. *Die Entwicklung der militärischen Luftfahrt in Deutschland 1920-1933*. (In the same volume as Wiegand Schmidt-Richbert, *Die Generalstäbe in Deutschland 1871-1945*.) Stuttgart: Deutsche Verlags-Anstalt, 1962.

A standard history of Germany's illegal military aviation in the Weimar period.

Vogelsang, Thilo. "Der Chef des Ausbildungswesens (Chef AW)." In *Gutachten des Instituts für Zeitgeschichte*, 2 (1966), 146-156.
———. *Kurt von Schleicher: Ein General als Politiker*. Göttingen: Musterschmidt, 1965.
Brief, but the best available biography of Schleicher.
———. "Papen und das aussenpolitische Erbe Brünings: Die Lausanner Konferenz 1932." In Carsten Peter Claussen, ed., *Neue Perspektiven aus Wirtschaft und Recht: Festschrift für Hans Schäffer zum 80. Geburtstag am 11. April 1966*. Berlin: Duncker & Humblot, 1966.
———. *Reichswehr, Staat und NSDAP: Beiträge zur deutschen Geschichte 1930-1932*. Stuttgart: Deutsche Verlags-Anstalt, 1962.
Vogelsang's major work contains a wealth of material, including in the annexes. No one can work on the Reichswehr in these years without being in Vogelsang's debt: for this study, for his other articles and documentary publications, and for his helpful guidance at the Institut für Zeitgeschichte. But for lack of some documents now available, this study does not get to the bottom of the relation between the elements named in its title; it also understates, in particular, the friction between the SPD and the Reichswehr.
———. "Zur Politik Schleichers gegenüber der NSDAP 1932." *VfZ*, 6 (1958), 86-117.
———. See also under Section III, H.
Waite, Robert G. L. *Vanguard of Nazism: The Free Corps Movement in Postwar Germany, 1918-1923*. Cambridge, Mass.: Harvard University Press, 1952.
Waites, Neville. "The Depression Years." In Neville Waites, ed., *Troubled Neighbours: Franco-British Relations in the Twentieth Century*. London: Weidenfeld & Nicolson, 1971.
A good survey of Anglo-French relations, but I think it understates the friction between the two governments.
Warner, Geoffrey. *Pierre Laval and the Eclipse of France*. London: Eyre & Spottiswoode, 1968.
Watt, D[onald] C. "Hitler and Nadolny." *Contemporary Review*, 196 (1959), 53-56.
———. *Personalities and Policies: Studies in the Formulation of British Foreign Policy in the Twentieth Century*. London: Longmans, 1965.
A collection of essays. Essays 2, 4, 5, 6, 7, 8, and 12 are pertinent to, and were highly useful for, the present study.
———. *Too Serious a Business: European Armed Forces and the Approach of the Second World War*. Berkeley: University of California Press, 1975.

Very welcome in its emphasis on the political role of the military. But the overall thesis, that military leaders did not want war, now needs qualification for the German army.

Weinberg, Gerhard L. "The Defeat of Germany in 1918 and the European Balance of Power." *CEH*, 2 (1969), 248-260.
Shows Weinberg's imagination and grasp.

———. *The Foreign Policy of Hitler's Germany: Diplomatic Revolution in Europe, 1933-36.* Chicago: University of Chicago Press, 1970.
Thorough, based on an unsurpassed knowledge of the German sources. But Weinberg tends to exaggerate Hitler's calculation and control of the situation, and to treat other governments as passive bystanders to German diplomacy. Also, we are given little idea of, or understanding for, the widespread non-Nazi acceptance of revisionism, both inside and outside Germany.

———. "Schachts Besuch in den USA in Jahre 1933." *VfZ*, 11 (1963), 166-180.

Wendt, Bernd Jürgen. *Economic Appeasement: Handel und Finanz in der britischen Deutschland-Politik 1933-1939.* Düsseldorf: Bertelsmann Universitätsverlag, 1971.
Brings out new information on the influence of the interest in trade on British policy.

Wheeler Bennett, John W. *The Nemesis of Power: The German Army in Politics, 1918-1945.* 2nd ed. New York: St. Martin's, 1964.
Highly readable; emphasizes the personality of Schleicher, blaming his personal ambition for the fall of Weimar. Wheeler-Bennett was well acquainted with Brüning.

———. *The Pipe Dream of Peace: The Story of the Collapse of Disarmament.* New York: W. Morrow, 1935.
An early account of the Disarmament Conference, now a document itself of British hopes for disarmament.

Wight, Martin. "The Balance of Power." In Herbert Butterfield and M. Wight, eds., *Diplomatic Investigations: Essays in the Theory of International Politics.* Cambridge, Mass.: Harvard University Press, 1968.

Williams, William Appleman. *The Tragedy of American Diplomacy.* New York: Dell, 1962.

Woerden, A.V.N. van. "Hitler Faces England: Theories, Images, and Policies." *Acta Historiae Neerlandica*, 3 (1968), 141-159.

Wohlfeil, Rainer, and Edgar Graf von Matuschka. *Reichswehr und Republik (1918-1933).* (Vol. VI of Hans Meier-Welcker and Wolfgang von Groote, eds., *Handbuch zur deutschen Militärgeschichte 1648-1939.*) Frankfurt: Bernard & Graefe, 1970.

A good handbook on the Reichswehr, with an excellent critical bibliography. An illustrated version, by Wohlfeil and Hans Dollinger, is entitled *Die Deutsche Reichswehr—Bilder—Dokumente—Texte zur Geschichte des Hunderttausend-Mann-Heeres 1919-1933* (1972).

Wojciechowski, Marian. *Die polnisch-deutschen Beziehungen 1933-1938*. Leiden: E. J. Brill, 1971.

A recent study, based on Polish (and other) archival materials.

Wolfers, Arnold. *Britain and France between Two Wars: Conflicting Strategies of Peace from Versailles to World War II*. New York: W. W. Norton, 1966.

Still important and helpful; shows what valuable research can be done without using secret documents.

Wollstein, Günter. *Vom Weimarer Revisionismus zu Hitler: Das Deutsche Reich und die Grossmächte in der Anfangsphase der nationalsozialistischen Herrschaft in Deutschland*. Bonn: Edition Ludwig Voggenreiter im Verlag Wissenschaftliches Archiv, 1973. Wollstein advances some debatable theses, particularly on a supposed French bid for an agreement; see Chapter Nine, note 1, above.

Youngson, A. J. *The British Economy, 1920-1957*. Cambridge, Mass.: Harvard University Press, 1960.

VI. NEWSPAPERS (1932-1933).

Daily Mail (London)
Daily Telegraph (London)
Financial News (London)
Manchester Guardian
News Chronicle (London)
The Times (London)
Figaro (Paris)
Le Temps (Paris)
New York Times
Deutsche Allgemeine Zeitung (Berlin)
Frankfurter Zeitung
Völkischer Beobachter (Munich)
Vossische Zeitung (Berlin)

VII. GERMAN MILITARY JOURNALS (1932-1933).

(These contain many useful short articles, some of which are specifically named in my footnotes.)
Deutsche Wehr
Militär-Wochenblatt
Wissen und Wehr

Index

Abwehr (German military intelligence), 69, 261, 453
Adam, General Wilhelm, 45, 61, 184-185, 261, 299n, 331-334, 338-339
Adler, Viktor, 311
Akademisches Wissenschaftliche Arbeitsamt (AWA), 64
Allen of Hurtwood, Lord (Clifford Allen), 91
Aloisi, Pompeo Barone, 372, 378, 442
Alsace-Lorraine, 196, 455, 460
Altona "Bloody Sunday," 274
Alvensleben, Werner von, 34n
Anschluss, 313, 336, 370-372, 429-430. *See also* Austria
appeasement, 4; Western, 78; British, 101, 105-106, 217-218, 272, 509
Armour, Norman, 200
arms infractions, exposure of: German fear of, 18, 79-80, 257, 273, 324-325, 339, 350, 450-453, 466, 475-476, 508, 509; regarded as treason (*Landesverrat*), 18, 24, 26, 28, 31, 283; French attempts at, in press, 79, 84-85; French threaten, 81, 147, 389, 439-441, 451, 477; potential use of, 100-101, 424, 495, 509; risk minimized, 257-258; Hitler and, 321, 475-476; British consider, 404, 422-423, 430, 432-433
Asquith, Herbert H. (Earl of Oxford and Asquith), 115n
Attlee, Clement, 377, 380
Auden, W. H., 107
Australia, 115, 119
Austria, 20, 119, 313, 368, 370, 374, 429-430, 433, 439; projected customs union with Germany, 54n, 65, 96-97, 100. *See also Anschluss*
Austria-Hungary, 15, 310-312

balance of power, 7, 102-105, 113, 139-140, 509-511
Baldwin, Stanley, 229, 356, 361n, 425, 467, 496; outlook and characteristics, 124, 230, 358; talks of radical disarmament, 157, 161, 165; and talks with French, 428, 444; Birmingham speech (6 Oct 33), 473, 478; position after German withdrawal from conference, 494, 510
Bank of England, 97, 106

Baratier, Gen. Paul, 88n
Barlow, James, 381
Barnett, Correlli, 118
Bartlett, Vernon, 487-488
Baum, Vicki, 107
Bavaria, 71
Bavarian People's party (BVP), 279
Beaumarchais, Maurice Delarue Caron de, 367n
Beaverbrook (Lord) press, 494
Beck, Józef, 336
Beck, Gen. Ludwig, 22, 345n, 490, 497, 498
Belgium, 179, 205-206, 210, 212, 213, 249
Beneš, Eduard, 202n, 245, 246, 376, 484
Berber, Fritz, 57
Berg, Benita von, 82n
Bergsträsser, Arnold, 57
Berlin transport strike, 274, 278
Berliner Tageblatt, 233
Bernhardt, Sarah, 158, 269
Bernstorff, Albrecht Graf von, 192, 218
Berthelot, Philippe, 369
Beschnitt, Captain, 39n
Bessinge meeting (26 Apr 32), 59, 147-156, 159, 167-168, 172, 178, 203-205, 223
Bethmann Hollweg, Theobald von, 307, 309
Bibesco, Princess Marthe, 216n
Bismarck-Schönhausen, Otto Fürst von (1815-1898), 7, 171, 202, 275, 307, 310, 312, 467
Bismarck-Schönhausen, Otto Fürst von (1897-1975), 453-454, 466-468, 470-475, 477, 480
Blomberg, Gen. (later Field Marshal) Werner von, 84, 141n, 177n, 191n, 198n, 292n, 324, 329n, 338, 385, 388n, 397, 398n, 432n, 451, 476, 488, 513; belief in war on two fronts, 43; views on disarmament tactics, 55-56, 59; at Disarmament Conference, 174, 181, 182, 188, 189, 296, 327; wants negotiations in Berlin, 196; appointed Defense Minister, 293-296, 303-304; character, 294; reasons for supporting Hitler, 296-298, 300-301; restraint on arms programs, 296, 331, 339-340; and Hitler, 320, 328; and 1933 disarmament negotiations, 327-328, 330, 360,

Blomberg, Gen. (*cont.*)
379, 386-387, 390-391, 401, 427, 461-463, 466, 468-470, 477, 481; new budgetary powers, 343-345; and Wehrverbände, 347-349, 351, 354, 514-515; and Umbau plans, 351, 354; statement of 8 May 33, 390-391; and direct negotiations with France, 454-457; and withdrawal from conference, 462, 468, 490; and 300,000-man army, 492, 498-499, 502

Blum, Leon, 495n

Börsenzeitung (Berlin), 271

Bolle, Capt. Carl, 431-432

Bonar Law, 119n

Boncour, *see* Paul-Boncour, Joseph

Boncour plan, *see* France, new disarmament plan

Bonin, Col. Erich von, 25n, 31

Bonn, Moritz, 57

Borsig (Tegel), 13

Boule, Pierre, 489

Bourquin, Maurice, 484

Boxer Rebellion, 310

Boxheim incident, 70

Braun, Magnus Freiherr von, 169n

Braun, Otto, 28, 30, 31, 34, 186-187, 276, 282

Bredow, Ferdinand von, 283n, 285n, 289n

Brest-Litovsk, Treaty of, 139

Briand, Aristide, 96

Brinon, Fernand de, 410, 455-457

British air force (Royal Air Force), 108, 116, 252-253, 503

British army: Imperial General Staff, 38n, 82, 108, 128; controversies over organization and strategy for, 113-116, 117; outclassed by continental armies, 113; decline after World War I, 116, 509; mobilization strength of, 126-127; opposition to disarmament to German level, 252-253; further weakening after 1933, 503-505

British Broadcasting Corp., 487-488

British intelligence: and German military developments, 80-83, 89-90, 238-239, 404, 423, 430-432, 452-453, 503, 508-509; and German infractions, 89-91

British navy (Royal Navy), 110, 116, 125, 127, 139-140, 503, 510

Brüning, Heinrich, 47, 64, 171, 175, 301, 314, 317, 515; becomes Chancellor, 34-35; and Reichswehr plans, 40, 63, 66, 76, 151-152; reparations and disarmament strategy of, 48-52, 57-58, 75; and Hitler, 49, 401n; in domestic affairs, 50-51, 57, 62, 73; and SPD, 50-51; in Stimson's eyes, 50n, 193; and Bessinge meeting, 59, 147-155, 159, 178, 204, 241; and work-creation program, 61-62; reorganizes cabinet, 65-66; dismissal of, 74, 90, 125, 159-162, 164, 170; in Geneva, 135-136, 145, 147-150, 204, 238; and Herriot, 159-161; and final reparations payment, 177; and Lebensraum, 322-323

Bülow, Bernhard Fürst von (1849-1929), 55

Bülow, Bernhard Wilhelm von (1885-1936), 41n, 69, 135, 160, 196, 208n, 258n, 261, 267, 397n, 430n, 454, 480n, 485, 492n; background and outlook, 54-55; and approaches to French, 56, 172n, 173, 176-177, 179, 188n, 200-201, 204; and Nadolny, 57, 181-182, 393; and Bessinge meeting, 152-154, 205; talk with Poncet (23 Aug 32), 155, 200-201, 204, 231n; and Lausanne, 173, 175-177, 179; vs. Schleicher, 173, 175-176, 181-183, 191, 284, 308-309; and equality of rights negotiations (Sep-Nov 32), 221, 228, 256, 257, 259; and British statement of 19 Sep 32, 227, 433; and proposed four-power meeting (Oct 32), 234-235, 242, 244; and Reichswehr's detailed program (Oct 32), 235-237; and Wheeler-Bennett, 241-242; and British "no force" proposal, 255-256; and equality-security formula, 264-266; and five-power agreement, 269-270; advises unprovocative policy (1933), 330, 347, 390; and disarmament negotiation (Feb-Jun 33), 393, 394; and British démarches (Jul 33), 433-434, 451-452; and German conference policy (Sep-Oct 33), 461-462, 468-470, 472n, 481

Bülow, Field Marshal Karl von (1846-1921), 55

Bündnisfähigkeit, idea of, 47, 315, 325, 355, 498

Bulgaria, 374

Buschenhagen, Captain, 68n

Cadogan, Alexander (later Sir Alexander), 357, 361-362, 363, 384, 389, 392, 445n, 446, 449, 483-484

Caillaux, Joseph, 455

Cambon, Roger, 438, 440

Campbell, Ronald (later Sir Ronald), 200, 220, 238, 439-440
Canada, 110-111, 115, 119
Carillon lunch (22 May 32), 158-160
Carr, E. H., 223
Carsten, Francis L., 25
Catholic Center party, see Center party (Z)
Cecil of Chelwood, Lord (Robert Cecil), 102, 119-120, 226
Center party (Z), 12, 23, 34, 279, 354
Chamberlain, Sir Austen, 358, 377-378, 380
Chamberlain, Neville, 105n, 112n, 124, 125, 164n, 421n, 494; character of, 126; and arms expenditure, 126-128, 503-504; and war debt to U.S., 266, 269, 420; and talks with French, 428-429; and economic sanctions, 439; influence on post-1933 strategy, 503-505
Chambrun, Charles, Comte de, 442, 445n
Chapouilly, Col. Edouard, 196, 201n
Chastenet, Jacques, 99
Chautemps, Camille, 494
China, 138
Church, Maj. Archibald, 162n
Churchill, Winston, 113-114, 117, 122, 377, 380, 421n, 511
Claudel, Paul, 166
Clausewitz, Carl von, 7, 317
Comert, Pierre, 441n
Communist party of Germany (KDP), 274, 275, 277-279
Conservative and Unionist party (U.K.), 119, 120, 229-230, 473
Coolidge, Calvin, 419
Corbin, Charles, 424n, 435, 437, 444n, 495n
Cot, Pierre, 406n
Council of Foreign Relations, 206
Court of International Justice, 100, 222, 423
Curtius, Julius, 52, 54, 56n, 65, 68n, 96
Czechoslovakia, 19, 93-94, 119, 175, 201, 332, 376, 491, 496

Daily Express (London), 380
Daily Mail (London), 380, 484
Daily Telegraph (London), 380-381
Daladier, Edouard, 446; and German arms infractions, 85, 366, 411, 440, 444, 454; distrust of French military, 85n, 407-408, 410-411; and preventive war, 335, 441, 444; meets with MacDonald and Simon, 366, 373, 408; personality and views, 406-407; and guarantees of security, 407-408, 410-413, 417-418, 424, 441, 443-444, 509; and negotiations with Germany, 410-411, 427, 455-456, 460, 480, 491, 498; and great-power meetings, 414, 416; and later qualitative equality, 445; cabinet falls, 494
Danubian Confederation proposal, 144, 150
Davidson, J.C.C., 229
Davis, Norman, 112n, 375n, 390, 403, 407, 408, 446, 459, 489n; aims and methods, 142, 250-251, 414, 416; visit to Washington (Mar-Apr 32), 143, 145-146; and Bessinge meeting, 153; visits London and Paris with Gibson, 157-158; lunches with Herriot, 158-159; and great-power meetings, 160-161, 414, 416-417, 426, 427, 481, 486; and Hoover disarmament plan, 166; disarmament plan of, 250, 260, 262-264, 328; and new French plan, 250; at five-power meetings (Dec 32), 262-264, 266, 268; belief Germans seeking disarmament, 264, 414, 446, 481; and U.S. consultative declaration, 384, 399-400, 415-416, 446, 510; and disarmament of France, 414, 417, 446; and Paris meeting (8 Jun 33), 417-419; and Economic Conference, 418-419; accepts later qualitative equality, 445; urges compromise (Oct 33), 481-484, 486, 510; denies U.S. involved in Europe, 487
Dawson, Geoffrey, 102
DDP, 12, 23
De Bono, Gen. Emilio, 368n
Democratic party (U.S.), 250
Deutsche Allgemeine Zeitung, 283
Deutsche Demokratische Partei (DDP), 12, 23
Deutsche Friedensgesellschaft, 12n
Deutsche Gesellschaft für öffentliche Arbeiten, 290
Deutsche Volkspartei (DVP), 12
Deutscher Volksportverein, 64-65
Deutschnationale Volkspartei, see German National People's party
Deuxième Bureau, see French intelligence
Dietrich, Hermann, 71
Dietrich, Marlene, 107
Disarmament Commission, Preparatory, 131, 132, 137

556—Index

Disarmament Conference (1932-1933), 7, 36, 41, 51-59, 91, 95-99, 108-109, Chapter Three *passim*, 173, 174, 194, 196, 199, 206, 208, 209, 225, 232, 234, 241, 252, 253, 254, 256-257, 259, 261, 263-268, 270-272, 284, 286-287, 292, 328, 330, 345, Chapter Eight *passim*, 415, 429, 447, 457, 460-463, 468-470, 473-475, 479-486, 488-490, 493-494, 506, 507, 508; as new peace conference, 131-134; initial stagnation of, 135, 143; French-"Anglo-Saxon" opposition at, 143-146; Bureau of, 149, 198, 200, 221-224, 229, 243, 295, 359, 394, 479, 480, 482-485; French-British-American talks at (Jun 32), 163-165; negotiation of adjournment resolution, 166-167, 180-181, 183-184, 188, 190-193; takes up new French plan, 325, 327, 359-360; takes up MacDonald plan, 383, 384; criticism of British at, 416-417, 422-424; adjourns for summer (1933), 425-426; expected reconvening (16 Oct 33), 431, 450, 452
DNVP, *see* German National People's party
Dodd, William E., 489n, 491
Dollfuss, Engelbert, 429-430
Dreyse, Friedrich Wilhelm, 302n
Duisberg, Carl, 84n
Dulles, Allen W., 85, 267, 446n
DVP, 12

East Fulham by-election, 493
Economic Conference, London (1933), 204, 383-384, 395, 400, 414, 416, 418-419, 420, 423, 424
economic sanctions, 100, 247, 383, 400, 415, 424, 439, 444, 495
Eden, Anthony (later Earl of Avon), 251, 358-359, 426n, 449, 458, 474n, 477, 482-483; prefers France to Germany, 358-359, 389, 403, 429; and disarmament negotiation (Feb-Jun 33), 359-362, 365, 389, 391-394, 416-417, 425; and meetings with French, 417-418, 424, 428, 429, 434, 437-441, 443-444, 447, 453-454; rejects new commitments, 446; and draft "compromise" (Oct 33), 467-468, 473; position after German withdrawal, 493. *See also* MacDonald disarmament plan
Edge, Walter, 142, 144, 150
Einstein, Albert, 377

Ems dispatch (1870), 467
"equality and security" agreement, 267-272
equality of rights, *see* Germany, strategy for disarmament question
Etherton, Col. P. T., 380
Ethiopia, 368-370, 376

Falkenhayn, Benita von, 82n
Ferrell, Robert H., 421
Field, Noel, 481
Fischer, Fritz, 307
Fisher, Sir Warren, 107-108, 128n, 421n, 503
five-power ("equality in a system of security") agreement (11 Dec 32), 267-272
five-power conference (6-11 Dec 32), 262-268
Fleuriau, Aimé Joseph de, 200, 201n, 204, 238
Foch, Marshal Ferdinand, 87, 88
Foster Fraser, Sir John, 380-381
France: perception of German aims, 7, 78-85, 92, 201, 285, 361, 406, 509, 511; defeat in 1870, 15; Germans expect opposition from, 20; reaction to Brüning statement on reparations, 58, 69, 98-99; leaders' belief in Seeckt influence, 87-88, 248, 326, 360; financial problems, 92, 159, 166; alliances sought, 92-96, 99-100; view of preventive war, 92-93, 335, 337; Anglo-American guarantees to (1919), 94, 137; disarmament strategy, 95-96, 132-133, 202, 243, 325-326, 408-413; efforts to commit British, 95-96, 99-100, 211-212, 216, 222-223, 247, 249, 366, 430, 434-437, 440, 441, 445, 447; maximum and minimum disarmament plans, 97-99, 245-247; Supreme Council of National Defense (Conseil supérieur de la Défense nationale), 98, 248; Tardieu plan, 98-99, 109-110, 135, 143, 145, 147, 151, 154, 164, 168, 250; objections to four- or five-power conclaves, 99-100, 150-151, 159-160, 180, 259, 271, 362, 365; war debt to U.S., 121, 122, 138-139, 263, 266, 269, 384, 416, 419-420; tactics at Disarmament Conference, 146-147, 163, 168, 359-360, 388-389, 393, 394; on Hoover disarmament plan, 166; British undertaking to consult with, 179-180, 199-200, 204, 209-216, 243, 267; and consultative declaration (13 Jul 32),

180, 204, 209, 213-215; equality linked with security, 199, 259, 263-264; alliance sought through a disarmament convention, 243, 249, 361, 411-412, 441, 447, 486, 494-495; new disarmament plan (Herriot-Boncour plan), 245-249, 256, 264, 285-296, 325-328, 355, 356, 363, 406, 409-410; Supreme War Council (Conseil supérieur de la Guerre), 249; on direct negotiation with Germany, 258, 443; on entente with Italy, 369-370, 378-379; and four-power pact proposal, 375-377, 379; demand for sanctions, 408-410, 412, 424, 428, 436, 441, 443; instructions of 2 May 33, 408-410, 417, 426, 441; demand for stages, 409, 412, 436; demand for trial period, 443, 449; proposes discussion of dossier, 434-438, 440-441, 444n; position after German withdrawal, 486, 491, 494-495
Franckenstein, George von, 451
François-Poncet, André, 88n, 147, 149, 172, 207, 242, 347, 385, 403n, 406n, 492, 498; talk with Bülow (23 Aug 32), 155, 200-201, 204; on negotiation with Germany, 194-195, 197, 454-455, 497; and German résumé, 197, 202, 221; and concept of "samples," 231; meeting with Hitler (15 Sep 33), 454-455, 460
Frankfurter Zeitung, 283
French air force, 108, 356, 425
French army: war strength, 19, 332; German army's demand for parity with, 53-56, 175-176, 386, 461; general staff (Etat-major général), 82, 239, 409, 435, 494; loss of strength, 92-93, 96, 335, 410-411, 509; plans for coercing Germany, 92, 335; North African forces, 326, 363, 385, 402; German view of plans of, 332; standing strength, 387
French intelligence: aware of German reserve problem, 38, 83; knowledge of German military developments, 80-85, 452-453, 508-509; quality of, 84-85
French navy, 175-176
Freud, Sigmund, 311
Freyberg-Eisenberg-Allmendingen, Adm. Albrecht von, 174n, 394
Fritsch, Col. (later Gen.) Werner Freiherr von, 26-27, 296
Frohwein, Albert, 55, 461, 462, 468n, 483-484

Gaertner, Franz von, 513
Gärtner, Margarete, 107, 242
Gamelin, Gen. Maurice, 85, 88, 98, 246n
Gaulle, Maj. (later Gen.) Charles de, 87
Gaus, Friedrich, 461
Gayl, Wilhelm Freiherr von, 187, 275, 281
Geländesport-Verbände-Arbeitsgemeinschaft (GVA), 64-65
General Confederation of German Trade Unions (ADGB—Allgemeiner Deutsche Gewerkschaftsbund; Socialist-linked trade unions), 282
Geneva Protocol (1924), 95, 109, 250, 356
George V, King of England, 108n
Gereke, Günther, 289-292, 339-341
Germain-Martin, Louis, 210, 266
German air force (Air Ministry), 331, 430-434, 503
German army: Truppenamt (Troops Office), 16-17, 19, 20, 60, 171, 497; Ordnance Office, 37, 60, 62, 453, 476; League of Nations Section (Group Army), 56n, 261, 453;
political outlook of officers, 4, 12, 23, 277-280; general plans and preparations, 6-7, 13-15, 26, 76-77, 174-176, 472, 506-507; arms procurement, 12-15, 77; officer caste, 13; doctrines, 14, 20-21, 70, 75-76, 506; reserve and mobilization system, 15, 17-19, 22-23, 26-28, 36, 42-43, 75, 77, 151-152, 201, 235-237, 276, 281, 297, 299, 326, 332, 350, 472, 496-497, 499-502, 507, 514-515; ties with Soviet Union, 24, 81; hopes for waging war, 26, 46-47, 75, 324, 506-507, 514; decline in trained reservists, 27, 36, 38, 76-77, 235, 317-318, 351, 354, 507; and guerrilla warfare, 29, 332; impatience with Prussia, 32-35, 71-72, 301; indoctrination plans, 32, 65, 66, 283-284; first and second armament plans, 36, 37, 472; preparedness, 37, 43, 47, 52, 331-333, 338-339; external training, 39, 276, 349-351, 354-355; "militia" training, 40-42, 72, 77, 153, 174, 185, 200-201, 235-237, 276, 284-285, 292, 345, 346, 348-349, 354, 363, 386, 497, 500-502, 513-515; peacetime cadre expansion, 40, 42, 152, 174, 185, 201, 237, 285, 292, 326, 346, 360, 363, 386; and Manstein mobilization plan, 42-43; ultra-short-term training, 44, 65, 276, 332, 350; armament costs,

German army: (cont.)
52, 59-62, 288-292; demand for parity with France, 53-56, 175-176, 386, 392, 461; on disarmament negotiation tactics, 55-56, 181, 183, 190, 196, 257, 260-262, 327-328, 386, 490; Billion Program (*Milliarde-Program*), 60-62, 185, 288-289, 338-339, 342-345, 354, 391; Six-Week Program, 60-62, 288, 291; enlistment of Nazis, 67-70; "Hidden German Goal" paper, 181-182, 184-185; *Umbau* (new peacetime army) plan, 184-185, 195-196, 256-257, 291, 339, 343, 345-346, 355, 391, 496-497; compulsory service anticipated, 198, 257, 284-285, 292, 322, 255, 404, 497; detailed program (Oct 32), 235-237, 257; uses of qualitative equality, 235, 465; and new French plan, 285-286; 300,000-man plan, 496-499; arms infractions, *see* arms infractions, exposure of; general staff, *see* German army, Truppenamt. *See also* Abwehr, Grenzschutz, *Landesschutz, Wehrsport*

German Democratic party (DDP), 12, 23

German intelligence, 69, 261, 453

German National People's party (DNVP), 12, 34, 275, 283, 294, 324, 340, 354, 508

German navy, 29, 79, 175, 236-237, 257, 343, 431, 496

German People's party (DVP), 12

Germany: continuity in history, 4, 6, 307-310, 508, 512, 515; influence of military on domestic politics, 7, 11-12, 18, 25, 29-30, 33-35, 50-51, 71-75, 169-171, 187, 280-281, 294-298, 301-304, 506, 507, 515; military pressure on foreign policy, 7, 26-27, 52-57, 59, 76, 173-176, 180-185, 187-191, 193-194, 196-199, 203, 206-207, 256, 265, 386, 391, 455-457, 461-462, 468, 490, 506; influence of industrial interest, 12-23, 282, 301, 309, 339, 340; pacifism in, 12, 18, 321, 322; revisionism in, 20-22, 75-78, 506; presidial government considered, 25, 34, 74; diplomatic views on army plans, 27, 237; effects of depression in, 48, 61-62, 123-124; election of 14 Sep 30, 50, 52, 64; schedule for revision, 52, 195; strategy for Disarmament Conference, 52-59, 132, 146-147, 431, 450; demand for qualitative equality, 53, 197, 231-232, 235, 392-394, 413, 436, 451, 456, 458-460, 465, 466, 469, 471, 485; diplomatic views on disarmament tactics, 54-55, 167, 173-174, 188-191, 196-197, 359-360, 460-462; financial restraints on, 55, 129, 173, 185-186, 237, 330, 364; spying on own officials, 68, 261, 385; presidential election of 10 Apr 32, 71, 73; views on British and French policy, 69-70, 146, 154-156, 160-161, 171-173, 198, 202, 207, 227-228, 256, 258, 261, 265-266, 298, 300, 315, 324-325, 381, 449-453; *Land* elections of 24 Apr 32, 72, 73; influence of Ostelbian landowners, 74, 90, 162, 282, 302-304, 309; propaganda with British, 107, 241-242, 380-383, 453; elections of 31 Jul 32, 170, 195, 273; disarmament tactics for Lausanne Conference, 172-177; and military talks with France, 172, 176-177, 198; reparations strategy for Lausanne Conference, 177; proposal for negotiation on equality during adjournment, 178, 179, 183-184, 188, 190-191, 196-199, 204; and 13 Jul 32 consultative declaration, 179-180, 192, 199-200, 217, 219-220; cabinet decisions on disarmament tactics (Jul 32), 184; instructions for disarmament delegation, 188, 285-287, 385; adjournment declaration (22 Jul 32), 191, 256; résumé of disarmament position (Aug 32), 197-198, 202, 203; isolation threatens (Aug 32), 202-203, 206-207; warnings of leaving League, 207, 270-271, 387, 396, 402, 475; and British statement of 19 Sep 32, 227-228; and British "no force" proposal, 255-256; election of 6 Nov 32, 274; militarism in, 307-309; racialism in, 309-310; election of 5 Mar 33, 328, 340; and four-power-pact proposal, 373, 375-376, 379; demand for quantitative equality, 392; paper on status of disarmament (5 Jul 33), 449-450; statement of position (4 Oct 33), 462, 469-473, 477, 480, 482; demand for arms for reserves, 465, 469

Gessler, Otto, 22, 34n

Gessler-Severing agreement, 39n

Gibson, Hugh, 142-143, 153, 162-163, 193; Apr 32 statement, 143-146, 415; visits London and Paris with Davis, 157-158; letter to Brüning, 160-161; and Hoover disarmament plan, 166-167

Gilbert, Martin, 105-106
Gladstone, William Ewart, 120
Goebbels, Joseph, 297, 300n, 457, 463, 485
Göring, Hermann, 274, 291n, 331, 341, 353, 382, 404, 431-433, 492
Gott, Richard, 105
Graham, Sir Ronald, 371, 373
Grandi, Dino, 371, 375n, 414
Granville, Edgar Lewis, 454
Graves, Robert, 118
Gravina, Manfredi Conte, 69
Great Britain: perception of German aims, 7, 78-83, 87-91, 116, 130, 132, 237-242, 404-405, 502-503, 509; German hope for support from, 20; and 1931 financial crisis, 48-49, 96-97, 108-109, 123-125; power of cabinet, 90, 101-102; on Disarmament Conference, 91, 108, 133-134, 192, 219-226, 235, 240-241, 251-254, 356, 357, 364-365, 438, 448; on Versailles Treaty, 91, 119, 509; resistance to continental commitments, 94, 111, 117, 119, 136, 148, 208, 240, 247, 250, 326, 356, 380, 406, 428, 438, 443-444, 446, 448, 509; importance of U.S. friendship to, 95, 110-112; influence of British dominions on, 95, 110-111, 119; support for four- or five-power conclaves, 100, 150-151, 180, 212, 217, 219-220, 509; authority of Treasury, 102, 121, 126-129, 504; public pressure for disarmament, 102, 108, 133, 135, 163, 192, 219, 224, 229-230, 234, 240-241, 252, 422, 448; and balance-of-power principle, 103-105, 129-130, 209, 315, 380; sympathy for Germany, 106-107, 240; antipathy for France, 107-110, 127-128; war debt to U.S., 110, 112, 121, 124, 135, 138-139, 212, 263, 266, 269, 383-384, 416, 419-421; revulsion against war, 116-119; revisionism, 119, 217-218; economic and financial problems, 120-126, 509; concern for economic and financial problems, 120-126, 509; concern for economic recovery, 125-126, 128, 204, 219, 222, 225-226, 364, 421; on German financial weakness, 129; fear of air attack, 129, 433, 503; on Nazism, 130, 377-378, 509; and qualitative disarmament, 134, 136, 143, 145, 252-255; on Hoover disarmament plan, 165-166; undertaking to consult with France, 178-180, 199-200, 204, 209-216, 243, 267; and consultative declaration (13 Jul 32), 179-180, 204, 209, 213-215, 217-220; in "war-in-sight" crisis (1875), 202; statement of 19 Sep 32, 206-207, 209, 226-230, 243, 254, 433; wish to avoid blame for a conference failure, 225, 252, 272, 359, 362, 364, 392-393, 423, 474, 480; proposes four-power meeting (Oct 32), 234; and new French plan, 247, 253-254; equality and "no force" proposal, 254-256; "programme of work," 357-359, 362; and four-power-pact proposal, 374, 376-377, 379; query on German air rearmament (29 Jul 33), 432-434; draft "compromise" proposal (Oct 33), 466-467, 473; position after German withdrawal, 486-488, 493-494; second draft "compromise" plan (Nov 33), 493-494; strategic planning after 1933, 503-505. *See also* MacDonald disarmament plan
Greece, 119
Grenzschutz, 17, 21, 27, 28, 33, 75, 77, 275-276, 281, 332, 349-351, 497, 499; Gessler-Severing agreement on (1923), 22-23; and Wehrverbände, 26, 32, 45, 63, 349-350; in West, 27; and SA, 74-75, 277, 297, 500, 514
Grey, Sir Edward, 117n
Groener, Gen. Wilhelm, 21-22, 28, 32n, 35, 42n, 50, 51, 63-64, 66, 68, 71, 90, 170; and war-of-liberation idea, 22; anxious to implement new guidelines, 30, 36; envisions waging war, 40n, 46-47, 323; informs Brüning of Reichswehr plans, 40, 63, 151-152; on Disarmament Conference goals and strategy, 53-54, 56; and ban of SA, 71-73; resigns as Defense Minister, 73-74
Grzesinski, Albert, 23-24, 30-33, 38, 187

Haig, Field Marshal Sir Douglas (later Earl Haig of Bemersyde), 114
Hailsham, Lord (Douglas McGarel Hogg), 90n, 252, 381, 396
Hammerstein-Equord, Gen. Kurt Freiherr von, 44n, 52, 61-62, 65, 81-82, 83 177n, 280n, 293, 321, 331; fears Polish-Czech attack, 69-70; opposes arming Wehrverbände, 71; and ban of SA, 72; and Umbau plans, 184-185, 256-257, 284, 346
Hankey, Sir Maurice, 108n, 133, 205n, 219, 233-234, 239, 240n, 252, 356, 467n
Hankey, Robert, 239

Harzburg front, 275
Hasse, Gen. Otto, 25n, 86
Hassel, Ulrich von, 373, 427n
Havas (French news agency), 202-203
Hearst (William Randolph) press, 487
Heimatdienst, 203
Hellpach, Willy, 57
Henderson, Arthur, 272, 393-394, 398, 403, 450, 455, 469, 477, 484, 486, 488; appointed chairman of Disarmament Conference, 131; relations with British government, 219, 361n, 424-425, 428; at Economic Conference, 418, 424-425; trip to European capitals (Jul 33), 425-428, 451
Herero War, 310
Herring, Grp. Capt. J. H., 431-434
Herriot, Edouard, 157, 161, 233, 408, 410, 412; and French intelligence reporting, 85; with Davis and Wilson, 158-159, 416n; personality, 158, 199, 407; and Brüning, 159-161; and MacDonald, 159-160, 216, 222; and German claim to equality of rights, 160, 163, 167, 168, 179, 180, 184, 190, 191, 195, 199-203, 222-223, 243-245, 251, 259; and Hoover disarmament plan, 166; and bilateral negotiations with Germany, 176, 258; at Lausanne, 176-180, 182, 184, 209-213; and British undertaking to consult with France, 178-180, 199-200, 209-217, 243, 447, 509; fear of German rearmament, 196, 248n, 249; links equality of rights with security, 199, 259, 263-264; indiscreet with press, 203, 399n; and consultative declaration of 13 Jul 32, 213, 214n, 215-217, 243; consults British, 222-223; and proposed great-power meeting (Sep-Nov 32), 223-224, 235, 242-244, 259; reception of British statement of 19 Sep 32, 228; Gramat speech, 233; and new French plan, 243-244, 247-249; trip to London, 244-245, 251; and five-power conference (Dec 32), 262-263, 269; and war debt to U.S., 266, 269, 384; trip to Washington, 384, 399, 415
Herriot plan, *see* France, new disarmament plan (Herriot-Boncour plan)
Hess, Rudolf, 297
Heydrich, Reinhard, 393n, 395
Heye, Gen. Wilhelm, 26, 36n, 37, 38, 42, 46, 352
Hillgruber, Andreas, 310
Himmler, Heinrich, 297

Hindenburg, Oskar von Beneckendorff und von, 282
Hindenburg, Field Marshal Paul von Beneckendorff und von, 11, 24, 32, 34, 57, 79, 289, 300, 316, 323, 341, 348n; entourage, 25, 35, 51, 74; and Brüning, 35, 51, 73-74, 159-162; reelection, 71, 73; and Papen, 169, 274, 280-282, 293; and Neurath, 189-190; refuses to appoint Hitler (Aug 32), 274, 275, 304; and Schleicher, 281, 282-283, 293, 303; appoints Hitler, 293, 321; and Blomberg, 294-295, 303-304
Hitler, Adolf, 14, 47, 77, 87, 104, 105, 126, 291n, 343, 352, 356, 359, 361, 377, 409, 414, 416, 420, 484, 510; responsibility for World War II, 4, 6, 78; and Brüning, 49, 401n; forbids followers to serve republic, 63-64; policy of legality, 65; and Reichswehr, 66, 298-299, 316, 318, 320-325, 352, 354, 507-508, 514, 515; and Schleicher, 68, 275, 280, 298-299, 309; presidential candidacy, 71; and England, 104n, 298, 313-315, 325, 355, 380-382, 395-396, 432, 449, 464, 467-468, 491-492, 496, 498, 501-503, 508-511; and Papen-Schleicher government, 170, 196, 207; and chancellorship, 273-275, 282, 293, 302-304; and Papen, 293, 301; and Blomberg, 294-295, 320; system of ideas, 297-300, 310-322, 354-355, 380, 449; and France, 298-300, 311, 312, 324-325, 427, 454-456, 460, 491-493, 498; and Italy, 298, 312-313, 325; and preventive war, 298-299, 324-325, 329, 334, 337, 347, 355, 390; letter to Reichenau (4 Dec 32), 298-300, 309, 317, 321, 324; and Soviet Union, 298, 313; place in German history, 307-310; racism, 310-312, 314, 319, 322, 395, 508; personality and background, 311, 313-317, 319; war experience, 311, 314, 317, 355, 501; and Austria, 313, 336, 368, 429-430, 451-452; and Mussolini, 313, 329, 367, 369, 371, 373, 398-399; as tactician, 316-317, 380, 381, 383, 396, 398, 450; and Röhm, 318, 501-502; assurances to Reichswehr on SA, 320-321, 348, 354, 499; 3 Feb 33 remarks to Reichswehr leaders, 321-325, 355; and covert rearmament, 324-325, 331, 334, 339-341, 397-398, 450, 475, 491; and work-creation program, 324, 339-341; re-

Index—561

fusal to go to Geneva, 328-329, 365; initial diplomatic caution, 328-331, 384-385, 387, 390, 391, 401; and Poland, 336-338, 347, 390; and 300,000-man army, 355, 491-493, 496-499, 501-502; "peace speech" (17 May 33), 383-385, 394-398, 400-405, 410, 413, 415, 469; methods of leadership, 385, 395; warning of leaving League, 396-397, 402; meets Henderson, 426-427, 451; and German conference policy (Sep-Oct 33), 461-470, 472, 475, 477-478; and withdrawal from Geneva, 470, 475-476, 478-479, 485-486, 488-490, 504-505
Hitler Youth (Hitler Jugend), 276, 351-352
Höltermann, Karl, 276
Hoesch, Leopold von, 150, 190, 195, 255, 378, 389, 393, 453, 474, 477, 478, 487, 502
Hoetzsch, Otto, 57
Holborn, Hajo, 57
Hoover, Herbert, 112n, 139, 140, 156, 172, 413; supporter of Wilson, 137; disarmament plan of, 140-143, 164-167, 191, 247, 363; antipathy for France, 141, 208; scant military knowledge, 141; sympathy with Germany, 141; and war debts, 142, 165, 420; statement on equality of rights (20 Sep 32), 208, 228
Hoover moratorium on intergovernmental debts (1931), 48-49, 111, 125, 138, 165, 269, 419
Houghton, Alanson B., 97
Howard Smith, C., 357
Hugenberg, Alfred, 48, 49, 169, 293-294, 295n, 304, 397, 508
Hull, Cordell, 413, 416n, 420, 486n, 487n
Hungary, 368, 374

Il Resto del Carlino (Bologna), 203
India, 240
Inter-Allied Military Control Commission, 24, 37n, 79, 83, 89n, 153, 507
International Chamber of Commerce, 140-141
International Court of Justice, 100, 222, 423
Irish Free State, 218
Isherwood, Christopher, 107
Israel, 506
Italian navy, 175

Italy, 134, 179, 211n, 213, 215, 258, 433, 438, 449, 464, 481; as Locarno guarantor, 94; war debt to U.S., 121, 212; and equality of rights, 160, 172, 205; and great-power negotiation, 223, 224, 234, 251, 365-367; at five-power meetings (Dec 32), 263; Hitler's interest in, 298, 312-313, 325, 329, 367, 371-372; and four-power pact, 329, 367, 373-379, 414, 426, 450; and Mussolini's diplomacy, 367-373; and Austria, 368, 370-372, 429-430; and compromise arms plan, 442-443, 446; and discussions with Neurath (Sep 33), 459, 469; delegates urge compromise (Oct 1933), 482, 483, 486, 489. *See also* MacDonald disarmament plan

Jäckh, Ernst, 57
Japan, 90, 110, 127, 133, 138, 208, 240, 422, 510
Johnson, Hiram, 415
Jones, Thomas, 229
Jordan, W. M., 103
Jouvenel, Henry de, 369-370, 372, 373, 376

Kästner, Erich, 107
Kaltenborn, H. V. (Hans von), 196
Kapp putsch (1920), 278
Kehl incident, 366
Kehr, Eckart, 4, 307, 308n, 317n
Keitel, Col. (later Field Marshal) Wilhelm, 353
Kellogg-Briand pact, 137, 247, 400
Keppler circle, 301
Keynes, John Maynard (later Lord Keynes), 119, 121, 421n
Knesebeck, Karl Friedrich von der, 16n
Koepke, Gerhard, 27, 390, 463n, 475n
Köster, Roland, 258, 455
KPD, 274, 275, 277-279
Kraus, Herbert, 57
Kraus, Karl, 311
Kreuger, Ivar, 414
Krosigk, Lutz Graf Schwerin von, 237n, 280, 288, 290-292, 293n, 295n, 302n, 339-341, 344
Kruckenberg, Dr. Franz Friedrich, 107
Krüger, Friedrich-Wilhelm, 352, 354, 393n, 395, 513
Krümper system, 15, 41, 201, 503
Krupp, Friedrich, A.G., 107

Krupp von Bohlen, Gustav, 302n
Kühlmann, Richard von, 309

Laboulaye, André Lefebvre de, 177n, 420n
Laforest de Divonne, Commandant Louis de, 196
land settlement, plans for, on German eastern lands, 303
Landesschutz, 18, 33, 63, 67, 75, 77, 350, 354; Gessler-Severing agreement (1923), 22-23, 353; briefings on, 26-27; new guidelines proposed and adopted, 28-29, 30; guidelines resisted by Prussian government, 30-34, 507; and Wehrverbände, 32, 45, 63, 67; implementation of guidelines lags, 35-36; guidelines of Oct 32, 275-276, 353; guidelines of Jul 33, 353
Langemarck, 317, 355, 501
Langer, William L., 5
Lausanne Conference, 157, 159, 163, 186, 191, 192, 209-212, 221, 419; postponed, 57-58; British give priority to, 125, 164; equality of rights at, 173-180; course of, 176-179; question of final payment at, 177-178; settlement at, 180, 222; "Gentlemen's Agreement" at, 212
Laval, Pierre, 96, 139, 165, 369
Law, Andrew Bonar, 119n
Lawrence, T. E., 314
League of Nations, 27, 53, 95, 100, 131, 151, 248, 356, 358, 374-376, 443, 479, 509; failure to protect Germany, 19; Council, 52, 100, 131, 208, 222, 256, 258-259, 356, 366, 423-424, 451, 478, 495; Covenant of, 53, 94, 96, 105, 119, 126; receives report on suspected German violations of Part V, 79-80; France and, 95, 97-100, 132-133, 246-247, 326, 407, 409, 415; Britain and, 119-120; U.S. and, 120, 137; Assembly, 194, 198, 230, 443, 452, 458; talk of appeal to, 201, 222, 366, 423, 438-440, 451, 452, 478, 495; German talk of withdrawal from, 207, 270, 387, 396, 402, 475; debate on Jewish refugees, 470n, 476
League of Nations Union, 102, 120
Lebrun, Albert, 338n
Leeper, Allen, 134, 221n, 255, 357, 424; urges disarmament to German level, 220, 222, 251-252; favors a great-power meeting, 222, 251; proposes exposure of German arms violations, 422-423, 425; position after German withdrawal, 493
Léger, Alexis Saint-Léger (Saint-John Perse), 200, 220, 411, 412, 437, 457n, 491n, 495
Leipziger Illustrierte Zeitung, 393-394
Lewis, Sinclair, 107
Lewis, Wyndham, 107
Liberal party (U.K.), 230
Liddell Hart, Capt. B. H., 117
Liebmann, Gen. Curt, 22, 322-324, 349n, 351
Lithuania, 19
Little Entente, 93-94, 99, 249, 376
Lloyd George, David, 87, 94, 114, 117, 119, 212, 314, 484, 487, 489
Locarno Treaty, 56, 93-95, 105, 126, 131, 247, 423, 441, 455, 473, 477, 494
Londonderry, Lord (Charles Stewart Henry Vane-Tempest-Stewart), 153, 417-418, 424, 434
Louis XIV, King of France, 92
Ludendorff, Gen. Erich, 115, 307, 309, 322
Lueger, Karl, 318
Luther, Hans, 289, 290, 292, 302n, 339-341, 390n, 399, 420n

MacArthur, Gen. Douglas, 144n, 171n, 510
MacDonald, James Ramsay, 101, 160, 163, 174, 205, 221n, 222, 238, 244, 250, 259, 260, 358-359, 404n, 407, 433, 446-448, 467, 473, 474, 496, 502; on France, 109, 148, 214, 262-263, 372; and U.S., 112n, 124n, 400; and treaty revision, 119, 218; aims and character, 124-125, 358; and British war debt to U.S., 124, 164n, 266, 269, 383-384, 420, 421; and Bessinge meeting, 149-156, 159, 178, 204, 205, 511; and great-power discussion, 150-151, 216-217, 223, 243, 244, 364, 365, 367, 370-372, 378, 379, 414; eye operation, 156-157; at Lausanne, 176-179, 209-212, 214; suspicious of Franco-German negotiation, 176, 178, 210; and undertaking to consult with France, 178-180, 199-200, 204, 209-217, 243, 447; and consultative declaration of 13 Jul 32, 213-214, 217, 219; and Henderson, 219, 425; on Germany, 226, 239, 378, 379; political situation of, 229-230; and Herriot visit to London (Oct 33), 244-245;

cool to further British disarmament, 253; seeks German return to Disarmament Conference, 253, 268; at five-power conference (Dec 32), 262-264, 266-269; trip to Geneva and Rome (Mar 33), 328, 365-367, 372-375, 377, 378, 408; and MacDonald plan, 373, 383; and four-power-pact proposal, 374-377; trip to Washington (1933), 383-384, 399, 415; and talks with French (Summer 33), 428; position after German withdrawal from conference, 473, 486, 494, 510

MacDonald disarmament plan: and Great Britain, 329-330, 347, 355, 373, 380, 385, 387, 397, 399, 400, 402-403, 409, 413, 416-417, 422, 424, 428, 459, 462, 465, 467, 469, 470, 473, 475, 478, 480, 487, 491, 494, 501-502; and Anthony Eden, 361-364, 372, 384, 389; and Italy, 362-363, 398, 413; and J. R. MacDonald, 373, 383; and U.S., 383-384; at Disarmament Conference, 383-384

Maginot, André, 98
Maginot line, 93
Manchester Guardian, 79, 219, 229, 250
Manchurian question, 69, 208, 247, 255, 256, 258
Mann, Thomas, 107
Manstein, Maj. (later Field Marshal) Erich von, 40, 42, 43, 346, 348
Marcks, Maj. (later Gen.) Erich, 35n, 172n, 195-196, 198, 284
Marshall-Cornwall, Col. J. H. (later Gen. Sir John), 80-81, 83, 84, 90, 108
Martin, Germain, 210, 266
Marx, Karl, 311
Marx, Salomon, 172n
Massigli, René, 94, 177, 191, 251n, 258, 268, 272, 360, 365, 389, 391, 393-394, 398-399, 413, 427-428, 434, 440, 445n, 449, 451n, 468n, 477, 484, 486
Mayer, Ferdinand, 426n
Medlicott, W. N., 103
"Mefo," 344
Meissner, Otto, 74, 280, 295, 303. *See also* Hindenburg, Paul von, entourage
Mellon, Andrew, 157n
Memel, 19
Messersmith, George, 463n
Metallurgical Research Company (Metallurgische Forschungsgesellschaft: "Mefo"), 344
Middlemas, Keith, 103
Milch, Erhard, 431-432

Milne, Field Marshal Sir George, 80, 87n, 91, 105, 110, 129, 239-240, 252
Mittelberger, Col. (later Gen.) Hilmar Ritter von, 25n
Moffat, Jay Pierrepont, 250, 384, 390n, 413
Moltke, Helmuth von, 171
Moore, Col. T.C.R., 464
Morgan, Brig. Gen. J. H., 89n
Morstin, Ludwik, 338n
Müller (or Müller-Franken), Hermann, 29, 30, 33-35, 37-38, 301
Müller, Pastor Ludwig, 298
Mussolini, Benito, 312, 313, 365, 416, 426, 430, 433, 452, 456, 491n, 492; and Germany, 205, 367, 369, 373, 398, 430; four-power-pact proposal, 329, 367, 371, 373-379, 402, 414, 426; personality and aims, 367-369; and England, 370-371; and France, 370, 376, 442-443; and arms compromise, 442-443, 445, 446

Nadolny, Rudolf, 141n, 145, 153, 155n, 160-161, 177, 192, 287, 327, 328, 329n, 359, 402, 405, 422, 461, 474, 475n; personality, 57, 182, 385, 388; and "Hidden German Goal" guidelines, 181-182; and adjournment negotiation (1932), 184, 188-190; and adjournment declaration, 191, 193, 218; and disarmament negotiation (Feb-Jun 33), 359-361, 366, 372, 385-394, 398, 401, 403; and final Geneva negotiations, 476-479, 482; recalled to Berlin, 478-479, 482
Napoleon I, Emperor of the French, 15, 92
National Socialist German Workers' party (NSDAP), 12, 29, 50, 67-70, 77, 79, 169, 170, 273-282, 324, 347, 352, 354, 507
National-Sozialistischer Deutsche Studentenbund, 64
Nationalsozialistische Deutsche Arbeiterpartei (Nazi party), *see* National Socialist German Workers' party
Natzmer, Renate von, 82n
naval treaties, 56, 112, 133, 137, 166, 250, 362-363, 422, 510-511
Nazi party, *see* National Socialist German Workers' party
Netherlands, the, 81
Neufville, Col. Georg von, 347
Neurath, Constantin Freiherr von, 192n, 196, 205, 242, 262, 280, 287-288,

Neurath, Constantin von (cont.) 290, 293, 296n, 321, 329, 330, 365, 366n, 373, 385, 387n, 388, 397, 410, 433, 454, 455, 468-471, 486-487, 491, 492, 502; becomes Foreign Minister, 172-173; and Reichswehr programs, 175n, 182, 236-237, 326; at Lausanne, 177, 184; and adjournment negotiation, 184, 188-191; character and position, 189-190; shows diplomatic caution, 189, 283n, 347, 390; and German résumé, 197, 202, 203; and equality-of-rights negotiations (Sep-Nov 32), 221, 228, 231-233, 256-261; denies Umbau begun, 238; rejects meeting in Geneva, 244; and British "no force" proposal, 255-256; at five-power conference (Dec 32), 262-267; and five-power agreement, 269-270, 273; and disarmament negotiation (Feb-Jun 33), 359-360, 385, 390, 394-395, 401; *Leipziger Illustrierte Zeitung* article by, 393-394; and disarmament negotiation (Jul-Oct 33), 427, 451-452, 455, 457-461, 463-466, 469, 471, 472, 481n; and withdrawal from conference, 478, 485

New Statesman, 229
New York Times, 195
News Chronicle (London), 225, 229
Norman, Montagu, 106

Oldenburg-Januschau, Elard von, 303
O'Malley, Owen, 221, 222, 224, 225, 227, 230, 232-235, 238
Osgood, Robert E., 139
Ossietzky, Carl von, 79
Osterkamp, Maj. Herbert, 291
Ott, Maj. Eugen, 277-278, 280-281, 297
Ott war game, 277, 280-281, 297, 298, 320
Ottawa Conference, on British Commonwealth trade, 221

pan-German movement, 312, 322, 508
Papen, Franz von, 76n, 162, 182, 194, 196, 258, 283, 293, 301-302, 398, 410, 455; reasons for selection of, 169, 172; and Nazi party, 169n, 170, 293; at Lausanne, 176-181, 183, 198; and adjournment negotiation, 183, 184; economic recovery plans of, 186, 289, 341n; and Hitler, 196, 293, 301; and equality-of-rights issue (Sep-Nov 32), 233, 241-242; resigns, 262, 281, 301; trusted by Hindenburg, 274, 282, 293; unpopularity, 274, 280, 289, 293; ideas of conflict and "reform," 275, 280, 281, 304

Paris, Treaty of (1808), 15
Part V, *see* Versailles Treaty, military provisions of
Paul-Boncour, Joseph, 166, 168, 191, 285n, 335n, 396, 412, 454n; and negotiation with Germans, 163, 455, 457, 459-460, 477; at Lausanne, 179, 184; and "samples," 230; character, 245, 258, 407; and new French disarmament plan, 245-249, 361, 406, 409-410; at five-power meetings (Dec 32), 262-263, 266, 268; and disarmament negotiations (Feb-Jun 33), 359-361, 366, 372, 408, 415; and German arms infractions, 361, 434-440, 460, 477, 482; opposes five-power talks, 365, 416, 426, 477; and Italy, 369; and preventive war, 375n, 444, 495; cabinet of, 406; and program of 2 May 33 instructions, 415, 444, 484; seeks to commit British, 417, 425-427, 434-437, 440-441, 447, 509; and anti-German front, 457, 460, 477, 483, 486, 494-495, 509; and Soviet Union, 495; rejects Hitler's proposals (Dec 1933), 498
Permanent Court of International Justice, 100, 222, 423
Pétain, Marshal Henri-Philippe, 88
Phillips, William, 384
Phipps, Sir Eric, 105n, 203n, 489n, 491, 496, 498, 502
Piłsudski, Marshal Józef, 335-337
Pirmasens rally, 196
Planck, Erwin, 241
Poincaré, Raymond, 227
Poland, 93-94, 217, 249, 258, 387, 491, 496, 497; German fear of attack by, 19, 22, 27, 28, 69, 76, 129, 279, 298, 303, 329, 347, 390; and France, 336-337, 406; supposed preventive war proposals, 337; and four-power-pact proposal, 376
Polish army, 175; war strength of, 19, 332; general staff, 239
Polish Corridor, 11, 20, 21, 119, 139, 217, 218, 239, 254, 256, 279, 336-337, 368, 370, 492
Polish intelligence, 82
Politis, Nicolas, 484
Polonization, 11, 21
Poncet, *see* François-Poncet, André
Porter, Katherine Anne, 107
Potempa affair, 274

preventive action, question of, 7-8
preventive action short of war, feasibility of, 100-101, 424, 509
preventive war: possibility of, 92-93, 298-299, 324-325, 329, 330, 334-337, 339, 347, 390, 508; renounced, 441, 444; fading danger of, 480, 502
Prittwitz und Gaffron, Friedrich Wilhelm von, 193
Prussia: SPD-led government in, 12, 22-24, 28, 32-35, 71-72, 169, 186-187; 19th century defeat and revival, 15; police follow illicit army activity, 23, 29; and Landesschutz guidelines, 26-32, 276, 353, 507; dualism of Reich and, 48; elections of 24 Apr 32, 72, 73, 186; overthrow of Braun-Severing government (20 Jul 32), 187, 273, 301

Rabenau, Gen. Friedrich von, 25, 86-87
Radical Socialist party (of France: *Parti radical et radical-socialiste*), 157
Raeder, Adm. Erich, 174n, 198, 257, 323, 496
Red Front Fighters League, 32
Reich Defense Council (Reichsverteidigungsrat), 67, 352-353
Reich Institute for Youth Training (Reichskuratorium für Jugendertüchtigung), 186, 276, 347-350, 352, 354; antecedents, 45, 64, 66, 67, 170
Reichenau, Gen. (later Field Marshal) Walther von, 278, 295, 320, 352; supposed militia plans of, 40n, 513; reasons for supporting Hitler, 296-298, 300-301; and Wehrverbände, 320, 349, 354; threatens end of SA training programs, 490, 499, 500. See also Hitler, Adolf, letter to Reichenau (4 Dec 32)
Reichsbank, 290, 292
Reichsbanner, 31, 34, 70, 73, 276, 282
Reichsverband der Deutschen Industrie (RVDI), 84, 341n
Reichswehr, see German army; German navy
revisionism, historical, 4-5, 78
Rheinbaben, Werner Freiherr von, 329-331, 334, 426n, 461, 481-485
Rheinmetall, 13
Rhineland: demilitarization of, 16, 20, 56; foreign occupation of, 11, 20, 36; infringement on demilitarization of, 33, 195-196, 366, 477

Ribbentrop, Joachim von, 382, 455, 496, 508
Ritter, Gerhard, 307-309
Roberts of Kandahar, Lord, 113
Röhm, Ernst, 65, 318, 347-348, 351, 352, 354, 355, 397, 490, 499-502, 508, 513, 515
Roon, Gen. Albrecht Graf von, 15, 41, 77, 113, 201
Roosevelt, Franklin D., 112n, 137-138, 140, 390n, 407, 446, 466-467, 481; and proposed consultative declaration, 383-384, 399-400, 415; and nonaggression-pact proposal, 399-400, 402, 411, 415; and Davis, 400, 414, 418; on foreign affairs, 413; and Economic Conference, 418-420; drops international policies, 419-421, 510; and war debts, 420; disclaims involvement in Europe, 487, 510
Rosenberg, Alfred, 297, 381, 382, 395, 402, 508
Rowse, A. L., 105
Ruhr, Franco-Belgian occupation of, 19, 93, 158, 227, 507, 511
Rumania, 93-94, 119, 376
Rumbold, Sir Horace, 58n, 89, 205, 207n, 221, 227, 238, 255, 395-396, 398, 402, 404
Rundstedt, Gen. (later Field Marshal) Gerd von, 187, 294
Russia, Czarist, 93, 202, 312, 506
Russia, Soviet, see Soviet Union

SA (Nazi Sturmabteilungen—storm troops), 32, 70, 275-277, 295, 318, 320-321, 325, 345, 347-349, 352, 354, 366, 386, 395, 402, 404, 406, 455, 490, 499-502 507-508; and Landesschutz, 31; as source of manpower, 63-64, 71, 276, 299, 514-515; receives material and financial support, 64-65; and Wehrsport, 64-65, 68-69, 75, 170, 186, 276, 499-500; banning of, 71-75; and Grenzschutz, 74-75, 277; ban removed, 170, 186, 187; as auxiliary police, 277, 280, 299, 300, 366; and Hitler, 318, 476, 490, 499-501; kept separate from army, 320, 321; and other Verbände, 348-352, 354; Ausbildungswesen (SA-Sport), 352, 354, 476, 490, 499-500, 513; its militia status, 513-515
SA-Sport, see SA, Ausbildungswesen
Saar, 20, 456, 493, 495, 498
Sackett, Frederic, 160-162

Samuel, Sir Herbert, 199, 220n, 225
Sargent, Orme, 218, 238, 436n
Sarraut, Albert, 494
Sassoon, Siegfried, 118
Sauer, Wolfgang, 513
Schacht, Hjalmar, 14, 341, 344, 390n, 399
Scharnhorst, Gerhard von, 15, 16, 77, 113, 152, 201, 317
Scharnhorstbund, 351
Scheidemann, Philipp, 24, 25, 30n, 34
Schiller, Friedrich von, 323, 502
Schleicher, Gen. Kurt von, 19n, 24, 31n, 39n, 40n, 41n, 43n, 50, 52, 63, 71, 135, 149, 199, 225, 227-228, 238n, 242, 258, 261, 295, 298, 302, 308, 416, 501n, 508; and Prussian government, 23, 30, 34-36, 71-73, 186-187, 273; general aims and character, 25, 68-69, 171, 198-199, 202, 283; and the SPD, 25, 29, 33, 34, 51, 282; on training, 26-27, 38; and Groener, 28, 50, 51, 66, 71-74; and Brüning, 34-35, 72-75, 162; anxious to implement new guidelines, 33-36; and disarmament strategy, 54, 58-59; and proposed talks with the French, 56, 172, 198; demands recognition of equality of rights, 59, 167, 180-184, 188-191, 256-258, 265-266, 296; and work-creation program, 62, 288-292, 339; hope of integrating the Nazis, 64, 65, 68-69, 74-75, 170, 273-276, 280, 282-283, 293; and Röhm, 65, 502; and Hitler, 68, 196, 274, 275, 280, 299, 304, 309; and ban of SA, 71-75; conceptions of foreign attitudes, 155, 171-172, 174, 194, 198, 265-266; becomes Reichswehr Minister, 169; aims in forming Papen government, 169-171; seeks domestic support, 170-171, 189, 194, 207; inflexible on planning schedules, 171, 198, 257; insists on Reichswehr program, 173-176, 181-182, 184-185; and threat of rearming, 188-190, 193-195, 203, 206-207; yields to Neurath, 190, 287-288; broadcasts of, 193-194, 198, 218, 284; and German résumé, 197-198, 202; and Reichswehr's detailed plans (Oct 32), 236-237, 326; becomes Chancellor, 262, 281; and proposed great-power meeting (Dec 32), 262; and five-power agreement, 270, 271, 273, 466; warns of leaving League, 270-271, 479; opposes Papen, 280-281; unpopularity of, 282-283, 293, 296; limits on power, 282-283, 288, 292-293; talks of faster Umbau, 284, 292-293; dismissal, 293, 301-303; assessment of role, 304
Schlesinger, Moritz, 385n
Schlieffen, Gen. Alfred Graf von, 310
Schmitt, Carl, 278n
Schönberg, Arnold, 311
Schönerer, Georg von, 318
Schönheinz, Gen. Kurt, 42n, 141n, 248n, 285, 287, 330, 397n, 401, 404, 405, 453, 461, 475n, 490, 492n; views on disarmament tactics, 55-56; briefs Schleicher (3 Dec 32), 261-262; opposes standardization, 326-328, 360, 386n, 391, 398, 401
Schwab, Dipl. Ing., 65n
Schwendemann, Karl, 55, 329, 461, 471, 475n
Schwerin von Krosigk, Lutz Graf, *see* Krosigk, Lutz Graf Schwerin von
security, *see* France, disarmament strategy
Seeckt, Gen. Hans von, 25, 36; doctrine of, 24, 44, 86-87, 91, 285, 497, 513, 514; Western view of influence, 85-88, 239, 240, 248, 326, 409, 511
Seldte, Franz, 295n, 347-348, 351-352
settlement on German eastern lands, plans for, 74, 162, 282
Severing, Carl, 22-23, 29-32, 34; removed by Papen government, 186-187
Sèvres, Treaty of, 119
Simon, Sir John, 157, 160, 163, 172, 240, 262, 266, 363-364, 381, 395, 404, 416, 423 428, 429, 433n, 441n, 446, 454, 463n, 469; and continental commitments, 105, 148, 415; background and character, 135-136, 358, 489; initial disarmament policy, 136, 143, 145, 153, 156, 205; and Hoover disarmament plan, 165, 167; and undertaking to consult with France, 179-180, 211-213, 215-216; and adjournment resolution, 192, 219; and consultative declaration (13 Jul 32), 192, 213, 215, 217, 219, 234; temporary disinterest in German equality of rights, 192, 208; views on German initiative (Aug 32), 200, 204-205, 227; seeks German return to Disarmament Conference, 221, 224-226, 230-234, 252-255, 258-259, 268; and great-power discussion, 223-224, 243, 255, 259; faces domestic criticism, 225, 228-230; and "samples," 230-232, 248, 459, 471, 472; scant military knowledge, 238, 474n; and Herriot visit, 245; equality

and "no force" proposal, 254-255, 258, 259, 261, 263, 267, 270-272, 409, 473; trip to Geneva and Rome (Mar 33), 328, 365-367, 371-374, 376, 378, 408; and Eden, 357-358; criticizes Nazi regime, 378; meetings with French (Sep 33), 439, 443-445; meeting with Neurath (23 Sep 33), 458-459, 461; reaction to German demands (Oct 33), 465, 471-474, 477, 480, 482; and draft "compromise" (Oct 33), 467-468, 473; position after German withdrawal, 473, 486-487, 494, 510; Bureau statement (14 Oct 33), 479, 482-486, 494; decides not to denounce Germans, 480-482, 486-487, 510
Singapore base, 127, 129
Snowden, Philip (later Lord Snowden), 125
Social Democratic party of Germany (SPD), 12, 22-23, 25, 27, 29, 31-34, 45n, 50, 70, 76-77, 170, 276, 279, 282, 348, 397, 507, 515
Socialist party (of France: *Section française de l'internationale ouvrière*—SFIO), 335
Soragna, Marchese di (Antonio Meli Lupi di Soragna Terasconi), 484, 486
Sosnowski, Capt. Jerzy, 82
South Africa, 111, 119
South Tyrol, 312, 368
Soviet Union, 115, 134, 158, 191, 239-240, 294, 495, 506; German hopes for support from, 20; German army ties with, 24, 452, 476; as deterrent to Polish attack on Germany, 69, 298
Sozialdemokratische Partei Deutschlands (SPD), *see* Social Democratic party of Germany
Spectator, 229
Spender, Stephen, 107
SS (Nazi Schutzstaffeln—élite storm troops), 295, 351, 352, 395, 402, 514
Stahlhelm, 31-34, 45-46, 63, 65, 79, 276, 281, 347-352, 354, 395, 406
"Statistical Society" (secret industrial mobilization organization), 37
Stein, Karl vom, 15
Steiner, Rudolf, 294
Stimson, Henry L., 109, 112n, 119, 131, 143-144, 150, 154n, 157, 162, 231n, 250, 384n, 413, 414; esteem for Brüning, 50n, 193; and policy of strength, 133, 138, 167; and the Far East, 138, 143, 146, 148, 208, 247; visit to Europe (Apr 32), 143, 145-146, 148, 156; and Bessinge meeting, 149, 153, 155, 156, 204-206, 511; on the Germans, 193, 206, 228, 510; and Hoover statement on equality of rights (20 Sep 32), 208
Stöcker, Adolf, 310
Strasser, Gregor, 282
Stresemann, Gustav, 15, 21n, 27, 377; letters of, 158, 160
Stülpnagel, Gen. Edwin von, 186, 347
Stülpnagel, Gen. Joachim von, 19, 34n, 38, 43n, 86, 271n, 275n, 296
Sunday Express (London), 380
Suvich, Fulvio, 459
Swanson, Claude, 142
Sweden, 81
Swiss army, as model, 40, 41, 114n, 134, 153
Syrup, Friedrich, 302n

Tägliche Rundschau (Berlin), 283
Tardieu, André, 96, 100, 133, 135-136, 264n; pushes his disarmament plan, 98-99, 135; and U.S. disarmament policy, 138, 144, 146, 148; in Geneva, 144, 145, 147, 149, 204, 238; and Bessinge meeting, 147-151, 154; electoral defeat, 156, 157, 164
Tardieu plan, *see* France, Tardieu plan
Taylor, A.J.P., 78, 94, 103, 104
Technische Nothilfe, 277-279, 281
Temperley, Brig. A. C., memorandum by, 404, 422
Le Temps (Paris), 271, 439
ten-year rule, 122, 128
Thomas, Gen. Georg, 14, 84n
Thomas, J. H., 225n
Thompson, Dorothy, 107
Thyssen, Fritz, 13n
Tilsit, Treaty of (1807), 15
Times (London), 219, 229, 240, 241, 371, 382
Titulescu, Nicolae, 376
Tournès, Gen. René, 82-83
Toynbee, Arnold, 125, 131
Treitschke, Heinrich von, 310
Treviranus, Gottfried, 34, 50, 118n, 161
Trieste, 368
Trocadéro rally (Nov 31), 97
Trotha, Gen. Lothar von, 310
Tunisia, 369, 376
Turner, Henry Ashby, Jr., 301-302
Tyrol, South, 119
Tyrrell, Lord (William Tyrrell), 205n, 212, 213, 215, 216, 226, 228, 242-243, 260, 335n, 412, 437, 455

Ulm lieutenants' case, 21-22
United Press, 196
United States: view of Germany, 7, 107, 132, 154, 208; German hope for support from, 20, 48-49; fear for German private debts, 48-49, 138; Allied war debts to, 49, 138, 142, 165, 212, 217, 416, 419-421; sympathy for England, 112; and the depression, 123-124, 137, 419; interest in success of Disarmament Conference, 133, 143; public opinion and disarmament, 135, 137-138, 163, 510; in and out of world affairs, 137-140, 510; and balance-of-power principle, 139-140, 208, 510; rejection of consultative pacts, 139, 148; resistance to European commitments, 139, 140, 247, 250, 406, 446, 510; Gibson disarmament proposals (Apr 32), 143-145; and Tardieu plan, 154; and new French plan, 247, 250; Davis interim convention proposal, 250, 260, 263-264; racialism in, 309; and MacDonald plan, 383-384; and proposed consultative declaration, 383-384, 399, 400, 415; and Economic Conference, 418-422; Wilson-Davis disarmament proposal (Oct 33), 481-482; involvement in Europe disclaimed, 487, 489
United States Army, 115
United States intelligence, military attaché reporting from Berlin, 83-85
United States Navy, 110, 112, 138, 141, 166, 510
Upper Silesia, 20, 119, 239, 256

Vansittart, Sir Robert (later Lord Vansittart), 134, 135, 233, 234n, 237n, 251, 255, 357, 364, 375n, 376, 381, 428, 446n; on balance of power, 103-104; opposes French hegemony, 109, 128; opposes Germany, 129, 393, 403, 404, 447-448; favors pressure on France, 232-233, 245; and Wheeler-Bennett, 241; seeks Italian support, 371-372, 429-430, 433, 434, 437; visit to Washington (1933), 383-384; and exposure of German arms infractions, 404, 423, 430, 432-435, 439; and meetings with French, 418, 435, 437, 440; and anti-German front, 429-430, 433-437, 439, 447-448, 451
Versailles Treaty, 91, 218; territorial provisions of, 11, 119; military provisions of (Part V), 7, 11, 13, 14, 16, 27, 53, 55, 56, 74, 79-83, 87, 91, 96, 132, 134, 142, 146, 147, 153-155, 163, 190, 197, 204, 222, 226-227, 230-232, 237, 243, 245, 251, 253, 254, 270, 287, 363, 451, 465, 477; Allied moral obligations under, 52-53, 108, 131, 182, 484; and the United States, 137; lack of sanctions against rearmament, 396, 423-424; provision for investigation (art. 213), 423
Vickers Armstrong, Ltd., 107
Vilna, 19, 27
Vogelsang, Thilo, 30
Vogt, General, 65n
Voigt, Frederick, 79n, 207n, 219n
Vollard-Bockelberg, Gen. Alfred von, 62
Vorwärts (Berlin), 79, 186

Wagner, Richard, 310
Walter, Bruno, 377
"war-in-sight" crisis (1875), 202
War of Liberation, as model, 15, 19-22
Warmbold, Hermann, 290, 302n
Wehler, Hans-Ulrich, 307
Wehrsport, 39, 42, 44-46, 62-69, 71-72, 152, 170, 186, 276, 499-501. See also SA, Ausbildungswesen (SA-Sport)
Wehrverbände (private paramilitary organizations), 32, 63-67, 170, 281, 347-350, 353, 389, 395-398, 499-501, 507, 514-515; army ties with, 18, 24, 26, 27, 31, 45-46, 71, 77; and mobilization, 18, 45-46, 276, 299, 332, 333n, 348-350; army promises not to use, 22-23, 29. See also SA, SS, Stahlhelm, Reichsbanner, Grenzschutz, *Wehrsport*
Weizsäcker, Ernst von, 55, 461, 468n, 469-471, 475n, 477, 489
Weltbühne, 79
Westerplatte incident, 329
Weygand, Gen. Maxime, 88, 98, 166, 201, 248-249, 335, 412n, 495
Wheeler-Bennett, John W. (later Sir John), 139, 241-242
Wigram, Ralph, 180n, 211, 219, 251, 253n
William II, King of Prussia and German emperor, 307, 310, 323
Wilson, Hugh, 158, 230, 383n, 390n, 392, 481, 482
Wilson, Woodrow, 94, 112, 120, 137, 139
Wirth, Joseph, 63, 65-66
Wojciechowski, Marian, 337
Wolfe, Thomas, 107
Wolfers, Arnold, 57, 104

Wolff, Theodor, 233
work-creation program: use of to finance German rearmament, 61-62, 186, 288-289, 291-292, 324, 339-343; established, 289-291; scheme for funding, 290, 344
World Economic Conference, *see* Economic Conference, London (1933)
World War I: responsibility for, 3; German war aims in, 115
World War II: responsibility for, 3-4, 6-8

Young Stahlhelm (Jungstahlhelm), 351
Yugoslavia, 93-94, 119, 368, 372, 376

Library of Congress Cataloging in Publication Data

Bennett, Edward W
   German rearmament and the West, 1932-1933.

   Bibliography: p.
   Includes index.
   1. Germany—Politics and government—1933-1945.
2. Germany—Politics and government—1918-1933.   3. Germany
—Defenses—History.   4. World politics—1933-1945.
I. Title.
DD256.5.B375   355.03'3043   78-70227
ISBN 0-691-05269-7